Free Student Aid.

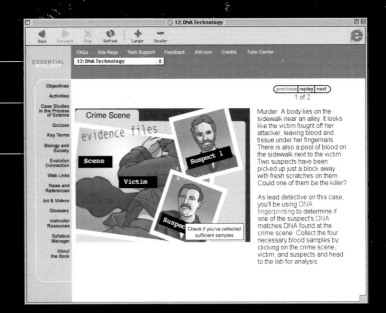

Log in.

Explore.

Succeed.

To help you succeed in introductory biology, every new copy of *Essential Biology*, comes with great media resources on an interactive CD-ROM and website. See the back of this insert for a description of the media resources designed to help you learn biology.

Minimum system requirements:

WINDOWS
266 MHz processor or greater
Windows 98, NT4, 2000, or XP
64 MB Minimum RAM
1024 x 768 screen resolution
Browser: Internet Explorer 5.x and 6.x; Netscape 6.2
Plug-Ins: Shockwave Player 8, Flash Player 5, QuickTime 4
Internet Connection: 56k modem minimum for website
NOTE: Use of Netscape 6.0 is not recommended due to a known compatibility issue between Netscape 6.0 and the Flash and Shockwave plug-ins.

MACINTOSH
266 MHz Minimum CPU
OS 8.6, 9.x, X
64 MB Minimum RAM
1024 x 768 screen resolution
Browser: Netscape 6.2; (Internet Explorer 5.x only on the web)
Plug-Ins: Shockwave Player 8, Flash Player 5, QuickTime 4
Internet Connection: 56k modem minimum for website
NOTE: Use of Netscape 6.0 is not recommended due to a known compatibility issue between Netscape 6.0 and the Flash and Shockwave plug-ins.

Got technical questions?

For technical support, please visit www.aw.com/techsupport and complete the appropriate online form. Technical support is available Monday-Friday, 9 a.m. to 6 p.m. Eastern Time (US and Canada).

Here's your personal ticket to success:

How to log in to www.essentialbiology.com

1. Go to www.essentialbiology.com
2. Click the book cover for your textbook.
3. Click **Register**.
4. Using a coin, scratch off the silver coating below to reveal your access code. Do not use a knife or other sharp object, which can damage the code.
5. Enter your pre-assigned access code exactly as it appears below.
6. Complete the online registration form to create your own personal user Login Name and Password.
7. Once your personal Login Name and Password are confirmed by email, go back to www.essentialbiology.com, type in your new Login Name and Password, and click **Log In**.

Your Access Code is:

If there is no silver coating covering the access code above, the code may no longer be valid. In that case, you need to either:
• purchase a new website subscription and CD-ROM (ISBN 0-8053-7504-X) at your campus bookstore.
• purchase access online using a major credit card. Go to www.essentialbiology.com, click the cover of your textbook and click Buy Now.

Important: Please read the Subscription and End-User License Agreement located on the "Log In" screen before using the *Essential Biology* website or CD-ROM. By using the website or CD-ROM, you indicate that you have read, understood, and accepted the terms of this agreement

FEATURES OF THE WEBSITE AND STUDENT CD-ROM FOR *ESSENTIAL BIOLOGY*

Activities

Explore approximately 200 activities, including animations, interactive review exercises, and videos. Test your knowledge with Activities Quizzes that can be e-mailed to instructors.

Thinking as a Scientist

Perform 56 virtual investigations that develop scientific thinking skills such as data collection, analysis, and communication of results. Answer questions in a Lab Report that can be e-mailed. With a subscription to Biology Labs On-Line, link to 12 more extensive virtual labs and complete assignments that can be e-mailed.

Quizzes

Assess understanding with over 2,000 multiple-choice questions. Each chapter includes a Pre-Test to diagnose current knowledge (10 questions per chapter), an Activities Quiz with graphics that tests understanding of the Activities in the chapter (20 questions per chapter on average), and a comprehensive Chapter Quiz (50 questions per chapter on average). Each quiz has hints, immediate feedback, grading, and e-mailable results.

Flashcards

Use electronic flashcards to test your knowledge of the key terms and definitions for each chapter or multiple chapters. Includes audio pronunciations for selected terms.

Terms & Definitions

Study the list of key terms for each chapter, looking up definitions and hearing selected audio pronunciations. Word roots are also provided to improve vocabulary skills.

Glossary

Look up definitions and hear selected audio pronunciations.

Additional Features of the Website

Web Links

Each chapter includes links to relevant websites with descriptions of the sites.

News and References

Access news links on recent developments in biology, organized by chapter. Consult an archive of relevant biology news articles and lists of further readings for each chapter.

Art and Videos

View art from the textbook and 85 videos. The art is provided both with and without labels. The art can be printed out to take to class for note-taking and the version without labels can be used as a self-quiz.

Biology and Society

Explore how the themes in each chapter are relevant to today's society by visiting a selected website and answering several questions. Answers are e-mailable.

Evolution Connection

Discover how evolution affects the themes in each chapter by visiting a selected website and answering several questions. Answers are e-mailable.

How to Start the Student CD-ROM

WINDOWS:

1. Insert the Student CD-ROM into the CD-ROM drive. If you have autorun turned on, the installer should launch automatically. If it does not, follow these steps:
   a) Navigate through My Computer to your default CD-ROM drive.
   b) Double-click on the CD-ROM icon.
   c) Double-click on the Windows_Installer.exe icon and follow the instructions in the dialog boxes.
2. Click the "Check Browser" button to launch your default browser and see if it meets the minimum system requirements.
3. If you need to install a new browser, click the "Install Browser" button and select a browser to install.
4. After you have a browser that meets the minimum requirements, click on the "Install Plug-ins" button and launch each of the installers for the required plug-ins.
5. Once all plug-ins have been installed, return to the Main Screen and click the "Launch" button to begin using the CD-ROM.

Once you have a valid browser and have installed all the required plug-ins you can skip the above steps and click on the "Launch" button directly to begin using the CD-ROM. Or, if you have the valid browser and plug-ins installed and you don't have autorun installed, you can click on essential_biology.html to skip the installation process.

MACINTOSH:

1. Insert the Student CD-ROM into the CD-ROM drive. Double-click the Macintosh_Installer.
2. Click the "Check Browser" button to launch your default browser and see if it meets the minimum system requirements.
3. If you need to install a new browser, click the "Install Browser" button and select a browser to install.
4. After you have a browser that meets the minimum requirements, click on the "Install Plug-ins" button and launch each of the installers for the required plug-ins.
5. Once all plug-ins have been installed, return to the Main Screen and click the "Launch" button to begin using the CD-ROM.

Once you have a valid browser and have installed all the required plug-ins you can skip these steps and click on the "Launch" button to begin using the CD-ROM. Or, if you have the valid browser and plug-ins installed, you can click on essential_biology.html to skip the installation process.

ESSENTIAL BIOLOGY

SECOND EDITION

Neil A. Campbell
University of California, Riverside

Jane B. Reece
Palo Alto, California

Eric J. Simon
New England College

PEARSON

Benjamin Cummings

San Francisco ▪ Boston ▪ New York
Cape Town ▪ Hong Kong ▪ London ▪ Madrid ▪ Mexico City
Montreal ▪ Munich ▪ Paris ▪ Singapore ▪ Sydney ▪ Tokyo ▪ Toronto

Executive Editor: Beth Wilbur
Acquisitions Editor: Chalon Bridges
Developmental Manager: Pat Burner
Editorial Project Manager: Ginnie Simione Jutson
Developmental Editor: Evelyn Dahlgren
Senior Producer, Art and Media: Russell Chun
Photo Image Editor: Donna Kalal
Publishing Assistant: Nora Lally-Graves
Producer: Aaron Gass
Managing Editor, Production: Erin Gregg
Senior Marketing Manager: Josh Frost

Production and Composition: Dovetail Publishing Services
Manufacturing Supervisor: Pam Augspurger
Copyeditor: Janet Greenblatt
Proofreaders: Pete Shanks, Bobbi Watkinson
Indexer: Charlotte Shane
Text Designer: Gary Hespenheide
Cover Designer: Yvo Riezebos
Illustrations: Precision Graphics
Cover Printer: Phoenix Color
Printer: VonHoffmann Press

On the cover: Photograph of a scarce copper butterfly (*Heodes virgaureae*). © Getty Images/Pam Hermansen

Credits continue in Appendix C

Library of Congress Cataloging -in-Publication Data

Campbell, Neil A.,
   Essential biology / Neil A. Campbell, Jane B. Reece, Eric J. Simon.—2nd ed.
     p. cm.
   Includes index.
   ISBN 0-8053-7495-7
   1. Biology. I. Reece, Jane B., II. Simon, Eric J. (Eric Jeffrey) III. Title.

QH308.2.C343 2004
570—dc21   2003057284

Benjamin Cummings
1301 Sansome Street
San Francisco, CA 94111
www.aw.com/bc

3 4 5 6 7 8 9 10—VHP—08 07 06 05 04

Neil A. Campbell has taught general biology for 30 years, and with Dr. Reece, has coauthored *Biology*, Sixth Edition, the most widely used text for biology majors. His enthusiasm for sharing the fun of science with students stems from his own undergraduate experience. He began at Long Beach State College as a history major, but switched to zoology after general education requirements "forced" him to take a science course. Following a B.S. from Long Beach, he earned an M.A. in Zoology from UCLA and a Ph.D. in Plant Biology from the University of California, Riverside. He has published numerous articles on how certain desert plants thrive in salty soil and how a sensitive plant (*Mimosa*) and other legumes move their leaves. His diverse teaching experiences include courses for non-biology majors at Cornell University, Pomona College, and San Bernardino Valley College, where he received the first Outstanding Professor Award in 1986. Dr. Campbell is currently a visiting scholar in the Department of Botany and Plant Sciences at UC Riverside, which recognized him as the university's Distinguished Alumnus for 2001. In addition to *Biology*, Sixth Edition, he is coauthor of *Biology: Concepts and Connections*, Fourth Edition.

Eric J. Simon is an Assistant Professor of Biology at New England College in Henniker, New Hampshire. He teaches introductory biology to both biology majors and non-biology majors, as well as upper-level biology courses in genetics, microbiology, and molecular biology. Dr. Simon received a B.A. in Biology and Computer Science and an M.A. in Biology from Wesleyan University, and a Ph.D. in Biochemistry at Harvard University. Currently, he is working toward an M.S.Ed. in Educational Psychology. Dr. Simon's diverse classroom experience includes teaching both biology majors and non-biology majors at numerous institutions, including St. John's University (Minnesota), Minneapolis Community and Technical College—where he earned an Outstanding Teacher Award—and Fordham College at Lincoln Center in New York City. Dr. Simon's research focuses on innovative ways for using technology to improve teaching and learning in the science classroom, particularly among non-biology major students.

Jane B. Reece has worked in biology publishing since 1978, when she joined the editorial staff of Benjamin Cummings. Her education includes an A.B. in Biology from Harvard University, an M.S. in Microbiology from Rutgers University, and a Ph.D. in Bacteriology from the University of California, Berkeley. At UC Berkeley and later as a post-doctoral fellow in genetics at Stanford University, her research focused on genetic recombination in bacteria. Dr. Reece taught biology at Middlesex County College (New Jersey) and Queensborough Community College (New York). During her 12 years as an editor at Benjamin Cummings, she played major roles in a number of successful textbooks. Subsequently, she was a coauthor of *The World of the Cell*, Third Edition, with W. M. Becker and M. F. Poenie. Dr. Reece also coauthored *Biology*, Sixth Edition, and *Biology: Concepts and Connections*, Fourth Edition.

ESSENTIAL
BIOLOGY

Neil Campbell, Jane Reece, and Eric Simon bring biology to life in *Essential Biology,* **Second Edition,** a brief non-majors text that focuses on four core biological topics: cells, genetics, evolution, and ecology. This manageable text combines clear writing with real-world applications and powerful media. The following walk-through of Chapter 12, DNA Technology, highlights key features of this innovative text.

CHAPTER 12

DNA Technology

Intriguing facts at the beginning of every chapter pique your interest.

Biology and Society: Hunting for Genes   217

Recombinant DNA Technology   217

From Humulin to Genetically Modified Foods

Recombinant DNA Techniques

DNA Fingerprinting and Forensic Science   224

Murder, Paternity, and Ancient DNA

DNA Fingerprinting Techniques

Genomics   229

The Human Genome Project

Tracking the Anthrax Killer

Genome-Mapping Techniques

Human Gene Therapy   234

Treating Severe Combined Immunodeficiency

Safety and Ethical Issues   235

The Controversy Over Genetically Modified Foods

Ethical Questions Raised by DNA Technology

Evolution Connection: Genomes Hold Clues to Evolution   237

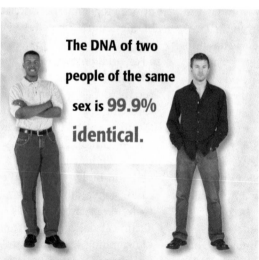

The DNA of two people of the same sex is **99.9%** identical.

The first use of **DNA fingerprinting** in a murder case proved one man innocent and another guilty.

Animals, plants, and even **bacteria** can be genetically modified to produce **human proteins.**

Genetically modified strains account for half of the U.S. corn crop.

216

Human applications reveal the world of biology

Biology and Society chapter-opening essays focus on topics of human interest and show you how biology affects your life and the world around you.

New! To reinforce the connection between biology and society, you can access the **Biology and Society web link** and essay questions from the Essential Biology Website (www.essentialbiology.com)

Biology and Society

Hunting for Genes

DNA technology, a set of methods for studying and manipulating genetic material, has brought about some of the most remarkable scientific advances in recent years: Corn has been genetically modified to produce its own insecticide; DNA fingerprints have been used to solve crimes and study the origins of ancient peoples; and significant advances have been made toward curing fatal genetic diseases. Perhaps the most exciting use of DNA technology in basic biological research began in 1990 with the launch of the Human Genome Project. The goal of the Human Genome Project is to determine the nucleotide sequence of all DNA in the human genome and to identify the location and function of every gene. A rough draft of the human genome sequence has already been completed. This ambitious project is revealing the genetic basis of what it means to be human.

Biology and Society on the Web Learn more about the Human Genome Project.

tial benefits of ha... ...te map of the hu...

Farm Animals and "Pharm" Animals While transgenic plants are used today as commercial products, transgenic whole animals are currently only in the testing phase. **Figure 12.7** shows a herd of transgenic sheep that carry a gene for a human blood protein. The human protein can be harvested from the sheep's milk and is being tested as a treatment for cystic fibrosis. Because transgenic animals are difficult to produce, researchers may produce a single transgenic animal and then clone it. The resulting herd of genetically identical transgenic animals, all carrying a recombinant human gene, could then serve as a grazing pharmaceutical factory—"pharm" animals.

While transgenic animals are currently used to produce potentially useful proteins, none are yet found in our food supply. It is possible that DNA technology will eventually replace traditional animal breeding—for instance, to make a pig with leaner meat or a cow that will mature in less time. Someday soon, scientists might, for example, identify a gene that causes the development of larger muscles (which make up most of the meat we eat) in one variety of cattle and transfer it to other cattle or even to sheep.

Recombinant DNA technology serves many roles today and will certainly play an even larger part in our future. In the next section, you'll learn more about the methods that scientists use to create and manipulate recombinant DNA.

Recombinant DNA Techniques

Figure 12.7 Genetically modified sheep.
These transgenic sheep car... they secrete in their milk. Thi... to lung dama...

Essential Biology **integrates human applications** into the narrative throughout the book. Here, the text discusses the genetic engineering of animals for pharmaceutical uses.

Evolution Connection

Genomes Hold Clues to Evolution

The DNA sequences determined to date confirm the evolutionary connections between even distantly related organisms and the relevance of research on simpler organisms to understanding human biology. Yeast, for example, has a number of genes close enough to the human versions that they can substitute for them in a human cell. In fact, researchers can sometimes work out what a human disease gene does by studying its counterpart in yeast. Many genes of disparate organisms are turning out to be astonishingly similar, to the point that one researcher has joked that he now views fruit flies as "little people with wings." On a grander scale, comparisons of the completed genome sequences of bacteria, archaea, and eukaryotes strongly support the theory that these are the three fundamental domains of life—a topic we discuss further in the next unit, "Evolution and Diversity."

Evolution Connection on the Web Investigate relationships among different life forms by examining genes.

As the capstone of each chapter, the **Evolution Connection** helps you connect every field of biology to *Essential Biology*'s overarching theme of evolution.

The website also has an **Evolution Connection web link** and essay questions. Responses to essay questions can be printed or e-mailed to instructors.

Text, art, and photographs work together to teach core concepts. Numbered steps in the text and art guide you through complex processes. The accompanying media activity allows you to see the process in motion.

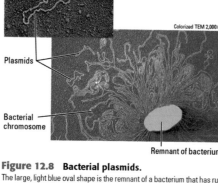

Plasmids

Bacterial chromosome

Colorized TEM 2,000×

Remnant of bacterium

Figure 12.8   Bacterial plasmids.
The large, light blue oval shape is the remnant of a bacterium that has ruptured and released all of its DNA. Most of the DNA is the bacterial chromosome, which extends in loops from the cell. Two plasmids are also present. The inset shows an enlarged view of a single plasmid.

Making recombinant DNA in large enough quantities to be useful requires several steps. Consider a typical genetic engineering challenge: A molecular biologist at a pharmaceutical company has identified a human gene *V* that codes for a valuable product—a hypothetical substance called protein V that kills certain human viruses. The biologist wants to set up a system for manufacturing the protein on a large scale. **Figure 12.9** illustrates a way to accomplish this using recombinant DNA techniques.

❶ First, the biologist isolates two kinds of DNA: many copies of a bacterial plasmid (to serve as a vector) and human DNA containing many genes, including gene *V*, the gene of interest. ❷ The researcher cuts both the plasmids and the human DNA. The human DNA is cut into many fragments, one of which carries gene *V*. The figure shows the processing of just three human DNA fragments and three plasmids, but actually millions of plasmids and human DNA fragments (most of which do not contain gene *V*) are treated simultaneously. ❸ Next, the human DNA fragments are mixed with the cut plasmids. The plasmid and human DNA join together, resulting in recombinant DNA plasmids, some of which contain gene *V*. ❹ The recombinant plasmids are then mixed with bacteria. Under the right conditions, the bacteria take up the recombinant plasmids. ❺ Each bacterium, with its recombinant plasmid, is allowed to reproduce. This step is the actual **gene cloning**, the production of multiple copies of the gene. As the bacterium forms a clone (a group of identical cells descended from a single ancestral cell), any genes carried by the recombinant plasmid are also cloned (copied). ❻ The molecular biologist finds those bacterial clones that contain gene *V*. ❼ The transgenic bacteria with gene *V* can then be grown in large tanks, producing protein V in marketable quantities.

Activity 12C on the Web & CD
View an animation that will help you further understand gene cloning.

Media References guide you through a wealth of media activities. Activities Quizzes on the website and CD-ROM can be assigned to encourage students to use the Activities.

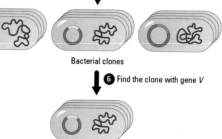

Bacterial cell   ❶ Isolate DNA from two sources   **Human cell**

Plasmid   ❷ Cut both DNAs   **DNA**

Gene V   **Other genes**   DNA fragments

❸ Mix the DNAs and join them together

Gene V

Recombinant DNA plasmids

❹ Bacteria take up recombinant plasmids

Recombinant bacteria

❺ Clone the bacteria

Bacterial clones

❻ Find the clone with gene *V*

❼ Grow bacteria and isolate protein V

Protein V

Figure 12.9   An overview of recombinant DNA techniques.

CHAPTER 12   DNA Technology   **221**

New! End-of-Chapter Visual Summaries use art to visually reinforce important concepts. The chapter summaries also refer you to media Activities and Case Studies in the Process of Science.

Chapter Review

Summary of Key Concepts

For study help, go to the Essential Biology Web site (www.essentialbiology.com) or CD-ROM to explore the Activities and Case Studies in the Process of Science.

Recombinant DNA Technology

• Recombinant DNA technology is a set of laboratory procedures for combining DNA from different sources—even different species—into a single DNA molecule.

Activity 12A *Applications of DNA Technology*

• **From Humulin to Genetically Modified Foods** Recombinant DNA techniques have been used to create non-human cells that produce human proteins, genetically modified (GM) food crops, and transgenic farm animals.

Activity 12B *DNA Technology and Golden Rice*

• **Recombinant DNA Techniques** Review the steps of recombinant DNA technology in the following diagram.

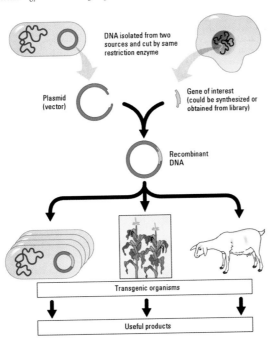

Case Study in the Process of Science *How Are Plasmids Introduced Into Bacterial Cells?*
Activity 12C *Cloning a Gene in Bacteria*
Activity 12D *Restriction Enzymes*

238 UNIT TWO Genetics

DNA Fingerprinting and Forensic Science

• DNA fingerprinting is used to determine whether two DNA samples come from the same individual.

Activity 12E *DNA Fingerprinting*

• **Murder, Paternity, and Ancient DNA** DNA fingerprinting can be used to establish innocence or guilt of a criminal suspect, identify victims, determine paternity, and contribute to basic research.

• **DNA Fingerprinting Techniques** Restriction fragment length polymorphism (RFLP) analysis compares DNA fragments using restriction enzymes and gel electorophesis.

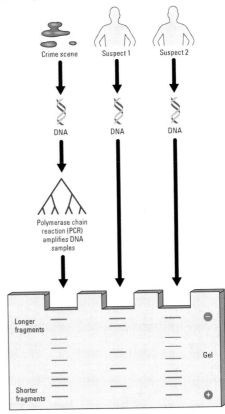

Activity 12F *Gel Electrophoresis of DNA*
Activity 12G *Analyzing DNA Fragments Using Gel Electrophoresis*
Case Study in the Process of Science *How Can Gel Electrophoresis Be Used to Analyze DNA?*

Questions that encourage self-assessment and help you think like a scientist

Case Studies in the Process of Science on the website and CD-ROM ask you to make observations, analyze data, and draw conclusions. Lab Report questions can be answered electronically.

Self-Quiz questions provide a variety of formats for your self-assessment at the end of each chapter. The CD-ROM and website also include approximately 70 multiple-choice quiz questions for each chapter.

CheckPoint questions within the chapter help you to assess your understanding of key concepts.

DNA fingerprints from an actual murder case are shown on a gel in **Figure 12.19**. The restriction fragments from the victim's DNA clearly match the fragments from the blood on the defendant's clothes. Furthermore, the markers from the defendant's DNA are clearly different. Thus, electrophoresis allows us to see similarities as well as differences between mixtures of restriction fragments, reflecting similarities as well as differences between the nucleotide sequences from two DNA samples.

Case Study in the Process of Science on the Web & CD
Conduct virtual gel electrophoresis.

CHECKPOINT

1. Why is only the slightest trace of DNA at a crime scene often sufficient for forensic analysis?

2. You use a restriction enzyme to cut a DNA molecule. The base sequence of this DNA is known, and the molecule has a total of three restriction sites clustered close together near one end. When you separate the restriction fragments by electrophoresis, how do you expect the bands to be distributed in the electrophoresis lane?

3. Put these three techniques in the order that would allow you to create a DNA fingerprint from a minuscule crime scene sample: gel electrophoresis, PCR, treatment with restriction enzymes.

Answers: 1. Because PCR can be used to produce enough molecules for analysis **2.** Three bands near the positive pole at the bottom of the gel (small fragments) and one band near the negative pole at the top of the gel (large fragment) **3.** PCR, treatment with restriction enzymes, gel electrophoresis

Self-Quiz

1. Suppose you wish to create a large batch of the protein lactase using recombinant DNA. Place the following steps in the order you would have to perform them.

    a. Find the clone with the gene for lactase.

    b. Insert the plasmids into bacteria and grow the bacteria into clones.

    c. Isolate the gene for lactase.

    d. Create recombinant plasmids, including one that carries the gene for lactase.

2. Why is an artificial gene made using reverse transcriptase often shorter than the natural form of the gene?

3. A carrier that moves DNA from one cell to another, such as a plasmid, is called a _____.

4. In making recombinant DNA, what is the benefit of using a restriction enzyme that cuts DNA in a staggered fashion?

5. A paleontologist has recovered a bit of organic material from the 400-year-old preserved skin of an extinct dodo. She would like to compare DNA

The Process of Science questions encourage you to think like a scientist by analyzing data, posing hypotheses and designing experiments. **Case Studies in the Process of Science** on the CD-ROM and website also reinforce these skills.

The Process of Science

1. A biochemist hopes to find a gene in human liver cells that codes for an important blood-clotting protein. She knows that the nucleotide sequence of a small part of the gene is CTGGACTGACA. Briefly explain how to obtain the desired gene.

2. Some scientists have joked that once the Human Genome Project is complete, "we can all go home" because there will be nothing left for genetic researchers to discover. Do you agree? Why or why not?

Case Study in the Process of Science on the Web & CD *Learn how plasmids are introduced into bacterial cells.*

Biology and Society

1. In the not-too-distant future, gene therapy may be an option for the treatment and cure of many inherited disorders. What do you think are the most serious ethical issues that must be dealt with before human gene therapy is used on a large scale? Why do you think these issues are important?

2. Today, it is fairly easy to make transgenic plants and animals. What are some important safety and ethical issues raised by this use of recombinant DNA technology? What are some of the possible dangers of introducing genetically engineered organisms into the environment? What are some reasons for and against leaving decisions in these areas to scientists? Who do you think should make these decisions?

Biology and Society on the Web *Learn more about the Human Genome Project.*

Biology and Society questions ask you to apply biological concepts to environmental problems, public policies, health concerns, and other social issues. The **Biology and Society web link** includes essay questions that encourage you to explore current issues and write about them. Responses can be printed or e-mailed to instructors.

We are privileged to help instructors share the story of life with students during this golden age of biology. It is an era of breathtaking progress in our understanding of life and a time when biology weaves into the fabric of our society as never before. Modern biology is remodeling medicine, agriculture, forensics, conservation science, anthropology, psychology, sociology, and even philosophy, including ethics. This is the best time ever to take a biology course!

This is also the most challenging time to teach and learn biology. The same discovery explosion that makes biology so much fun also threatens to suffocate students under an avalanche of information. With each of its many subfields bustling in research activity, biology grows larger every year, while the academic semester stays the same size. Something has to give.

In this era of ever-expanding biology, many instructors are opting to cover fewer main topics rather than compromise depth in the most important areas. We created *Essential Biology* to support this trend. Yes, it is a shorter biology text than most, but we did not achieve this brevity by trying to fit all of biology into less space. Instead, we focused on just four core topics: cells, genes, evolution, and ecology. In the context of these four main topics, students will encounter diverse organisms and their evolutionary adaptations. However, we have not included separate units on the anatomy and physiology of plants, animals, and other organisms. This enabled us to keep *Essential Biology* manageable in size without being superficial in developing the concepts that are most fundamental to understanding life. We take the "less is more" mantra in education today to mean fewer topics, not more diluted explanations.

The book's title, *Essential Biology*, partly reflects this look at life that is relatively brief, selective, and integrated. But the title has a second meaning. It announces an emphasis on concepts and applications that are essential for students to make biologically informed decisions throughout their lives—to evaluate various health and environmental issues, for example. From ethical and safety concerns surrounding genomics to debates about global warming, students will find biology in the news every day. Biology is more essential than ever in a general education, and *Essential Biology* spotlights this central place of biology in modern culture.

The success of the first edition of *Essential Biology* validated the book's counter-encyclopedic, culturally-connected approach. And with the help of reviewers and many other instructors and students who contributed suggestions, we found ways to make this second edition work even better. Here are just a few of the improvements you'll find in *Essential Biology*, Second Edition:

- *"Biology and Society" Sections:* Each chapter now begins with a human-interest story that engages students in the chapter topic.

- *Visual Summaries:* We have replaced the traditional all-text chapter summary with a review that includes diagrams to help students synthesize relationships among the key concepts. Student focus groups were particularly helpful in helping us refine our idea for this new kind of chapter summary.

- *Campbell Image Presentation Library:* The breakthrough package of *Essential Biology* supplements is now even more robust with the addition of an extensive archive of digital content for lecture presentations. This image library gives instructors easy access to over 1,000 photos, all text art with and without labels, selected figures layered for step-by-step presentation, all text tables, 110 animations, and 85 video clips.

- *New Coauthor Eric Simon:* Bringing his award-winning gift for teaching non-majors, Eric played an especially important role in strengthening *Essential Biology*'s connections to the social and personal issues that concern students.

We see our responsibility as science educators to be especially important in communicating with students who are not biology majors, because their attitudes about science and scientists are likely to be shaped by a single, required science course—this course. Long after students have forgotten most of the specific content of their college courses, they will be left with general impressions that will influence their interests, opinions, values, and actions. We hope this textbook will help students fold biological perspectives into their personal worldviews. Please let us know how we are doing and how we can improve the next edition of *Essential Biology*.

Neil Campbell
Dept. of Botany and
 Plant Sciences
University of California
Riverside, CA 92521

Eric J. Simon
Dept. of Biology
New England College
Henniker, NH 03242

Jane Reece
Benjamin Cummings
1301 Sansome St.
San Francisco, CA 94111

ACKNOWLEDGMENTS

As authors, we feel incredibly privileged to have worked with the very best publishing professionals and biology colleagues. Though the responsibility for any shortcomings of this book lies solely with us, its merits reflect the contributions of many associates who helped us with the textbook, supplements, and the integral media activities that complement the lessons.

First, we'd like to thank the *Essential Biology* team at Benjamin Cummings. Beth Wilbur oversaw the project as Executive Editor. Beth's talents as a consensus builder and team leader helped to move the book along at several critical junctures. Acquisitions Editor Chalon Bridges brought unending enthusiasm and positive energy to the book; her bright nature inspired us, especially during the most difficult stages of writing. Of course, publishing excellence flows from the top, and we are grateful to Linda Davis, President of Benjamin Cummings, for the example she provides to her entire company. Further guidance and inspiration was provided by Frank Ruggirello, Vice President and Editorial Director, and Kay Ueno, Director of Development. Benjamin Cummings authors are very fortunate to have such supportive leadership.

Developmental Editor Evelyn Dahlgren helped to craft every chapter of this book. Her handiwork can be found on every page and the final book is much the better for it. Editorial Project Manager Ginnie Simione Jutson guided the many facets of the project with just the right combination of firmness and compassion. Developmental Manager Pat Burner applied her broad knowledge of biology to help ensure the clarity and accuracy of our words.

If you find yourself admiring a wonderful piece of art in this book, you can thank the gifted Russell Chun, Senior Producer, Art and Media. Photo Image Editor Donna Kalal used her keen eye and discerning taste to provide the wealth of photographs used in the book and the supplements. Freelance developmental editors Betsy Dilernia and Robin Fox maintained their high standards despite the challenging schedule. Publishing Assistants Trinh Bui, Krystina Sibley, and the always cheerful Nora Lally-Graves (who also contributed significantly to the supplements) smoothed out the day-to-day operations of the whole team. The dedication to perfection shown by the entire editorial team inspired the authors to do our very best.

After our words and drawings were committed to paper, the production team transformed them into the book you hold in your hands. Erin Gregg, Managing Editor, Production, rose to the challenge of our very tight schedule with aplomb. For the production and composition of the book, we thank Jonathan Peck and Joan Keyes of Dovetail Publishing Services, whose professionalism and commitment to the quality of the finished product eased our authorial burdens tremendously. The authors owe much to the copyeditor, Janet Greenblatt, and proofreaders, Pete Shanks and Roberta Watkinson, for polishing our words. We thank Charlotte Shane for compiling the index. If you like the look of the book, it is thanks to text designer Gary Hespenheide and cover designer Yvo Riezebos. We thank Precision Graphics for creating the illustrations. In the final stages of production, the talents of manufacturing buyer Pam Augspurger shone.

More and more, the success of a textbook depends upon the quality of its supplements. Luckily, the *Essential Biology* supplements team is as committed to the core goals of accuracy and readability as the authors and editorial team. The multitalented Aaron Gass produced the website and student CD that accompany every copy of *Essential Biology*. The stylish and informative website is also the product of web developers Steve Wright and Marie-Beth Millares. Corinne Benson expertly coordinated the production of the supplements, a difficult task given their number and variety. We owe particular gratitude to two supplement authors: Ed Zalisko of Blackburn College wrote the Student Study Guide and Instructor's Guide, while Gene Fenster of Longview Community College wrote most of the questions found in the Test Bank. Additionally, Brian Shmaefsky contributed many valuable media teaching tips to the Instructor's Guide. As teachers, we know the value of well crafted PowerPoint lectures, so we thank Chris Romero and John Hammett for writing and editing those that accompany this book. Karl Miyajima ably oversaw the production of all of the visual supplements.

As educators and writers, we rarely think about marketing. But for what we try to do as authors, "market" translates as "the students and instructors we are trying to serve." Senior Marketing Manager Josh Frost and Marketing Development Manager Susan Winslow helped us achieve our authorial goals by keeping us constantly focused on the needs of students and instructors. For their amazing efforts in marketing, we also thank Director of Marketing Stacy Treco, Creative Director Lillian Carr, Marketing Specialist David Good, and Marketing Coordinator Ben Russo. We also thank the Benjamin Cummings field staff for representing *Essential Biology* on campuses. These representatives are our lifeline to the greater educational community, telling us what you like (and don't like) about this book and media. Their enthusiasm for sharing our words buoyed us during the writing process. In particular, we'd like to thank Science Project Specialists Scott Davidson, Kathryn Speers, and Melissa Young, and Field Market Specialist Jeff Howard for their dedication to presenting *Essential Biology* to instructors all over the country.

Numerous colleagues in the biology community also contributed to *Essential Biology*. In particular, Jennifer Warner of University of North Carolina at Charlotte wrote many of the critical thinking questions found at the end of the textbook chapters and in the Instructor's Guide. Amanda Marsh Simon provided nearly superhuman support to the project, as an incomparable research specialist and in other ways too numerous to mention. Furthermore, at the end of these acknowledgments you'll find a list of the many instructors who provided valuable information about their courses, reviewed chapters, and/or conducted class tests of *Essential Biology* with their students. We thank them for their efforts and support.

Most of all, we thank our families, friends, and colleagues who continue to tolerate our obsession with doing our best for science education.

Neil A. Campbell
Jane B. Reece
Eric J. Simon

Reviewers

Tammy Adair, *Baylor University*
Felix O. Akojie, *Paducah Community College*
William Sylvester Allred, Jr., *Northern Arizona University*
Estrella Z. Ang, *University of Pittsburgh*
David Arieti, *Oakton Community College*
C. Warren Arnold, *Allan Hancock Community College*
Mohammad Ashraf, *Olive-Harvey College*
Bert Atsma, *Union County College*
Yael Avissar, *Rhode Island College*
Barbara J. Backley, *Elgin Community College*
Gail F. Baker, *LaGuardia Community College*
Kristel K. Bakker, *Dakota State University*
Linda Barham, *Meridian Community College*
Charlotte Barker, *Angelina College*
S. Rose Bast, *Mount Mary College*
Sam Beattie, *California State University, Chico*
Rudi Berkelhamer, *University of California, Irvine*
Penny Bernstein, *Kent State University, Stark Campus*
Karyn Bledsoe, *Western Oregon University*
Donna H. Bivans, *East Carolina University*
Andrea Bixler, *Clarke College*
Brian Black, *Bay de Noc Community College*
Allan Blake, *Seton Hall University*
Sonal Blumenthal, *University of Texas at Austin*
Lisa Boggs, *Southwestern Oklahoma State University*
Dennis Bogyo, *Valdosta State University*
Virginia M. Borden, *University of Minnesota, Duluth*
James Botsford, *New Mexico State University*
Cynthia Bottrell, *Scott Community College*
Richard Bounds, *Mount Olive College*
Cynthia Boyd, *Hawkeye Community College*
Robert Boyd, *Auburn University*
B. J. Boyer, *Suffolk County Community College*
Mimi Bres, *Prince George's Community College*
Patricia Brewer, *University of Texas at San Antonio*

Jerald S. Bricker, *Cameron University*
George M. Brooks, *Ohio University, Zanesville*
Janie Sue Brooks, *Brevard College*
Evert Brown, *Casper College*
Mary H. Brown, *Lansing Community College*
Richard D. Brown, *Brunswick Community College*
Joseph C. Bundy, *University of North Carolina at Greensboro*
Carol T. Burton, *Bellevue Community College*
Rebecca Burton, *Alverno College*
Warren R. Buss, *University of Northern Colorado*
Miguel Cervantes-Cervantes, *Lehman College, City University of New York*
Bane Cheek, *Polk Community College*
Thomas F. Chubb, *Villanova University*
Pamela Cole, *Shelton State Community College*
William H. Coleman, *University of Hartford*
James Conkey, *Truckee Meadows Community College*
Karen A. Conzelman, *Glendale Community College*
Ann Coopersmith, *Maui Community College*
James T. Costa, *Western Carolina University*
Pradeep M. Dass, *Appalachian State University*
Paul Decelles, *Johnson County Community College*
Galen DeHay, *Tri County Technical College*
Cynthia L. Delaney, *University of South Alabama*
Jean DeSaix, *University of North Carolina at Chapel Hill*
Elizabeth Desy, *Southwest State University*
Edward Devine, *Moraine Valley Community College*
Dwight Dimaculangan, *Winthrop University*
Deborah Dodson, *Vincennes Community College*
Diane Doidge, *Grand View College*
Don Dorfman, *Monmouth University*
Terese Dudek, *Kishawaukee College*
Shannon Dullea, *North Dakota State College of Science*
David A. Eakin, *Eastern Kentucky University*
Brian Earle, *Cedar Valley College*
Dennis G. Emery, *Iowa State University*
Virginia Erickson, *Highline Community College*
Marirose T. Ethington, *Genesee Community College*
Paul R. Evans, *Brigham Young University*
Zenephia E. Evans, *Purdue University*
Jean Everett, *College of Charleston*
Dianne M. Fair, *Florida Community College at Jacksonville*
Joseph Faryniarz, *Naugatuck Valley Community College*
Phillip Fawley, *Westminster College*
Jennifer Floyd, *Leeward Community College*
Dennis M. Forsythe, *The Citadel*
Carl F. Friese, *University of Dayton*
Suzanne S. Frucht, *Northwest Missouri State University*
Edward G. Gabriel, *Lycoming College*

Anne M. Galbraith, *University of Wisconsin, La Crosse*
Kathleen Gallucci, *Elon University*
Gregory R. Garman, *Centralia College*
Gail Gasparich, *Towson University*
Kathy Gifford, *Butler County Community College*
Sharon L. Gilman, *Coastal Carolina University*
Mac Given, *Neumann College*
Patricia Glas, *The Citadel*
Ralph C. Goff, *Mansfield University*
Marian R. Goldsmith, *University of Rhode Island*
Andrew Goliszek, *North Carolina Agricultural and Technical State University*
Curt Gravis, *Western State College of Colorado*
Larry Gray, *Utah Valley State College*
Tom Green, *West Valley College*
Robert S. Greene, *Niagara University*
Ken Griffin, *Tarrant County Junior College*
Denise Guerin, *Santa Fe Community College*
Paul Gurn, *Naugatuck Valley Community College*
Peggy J. Guthrie, *University of Central Oklahoma*
Henry H. Hagedorn, *University of Arizona*
Blanche C. Haning, *North Carolina State University*
Laszlo Hanzely, *Northern Illinois University*
Sherry Harrel, *Eastern Kentucky University*
Frankie Harriss, *Independence Community College*
Lysa Marie Hartley, *Methodist College*
Janet Haynes, *Long Island University*
Michael Held, *St. Peter's College*
Michael Henry, *Contra Costa College*
Linda Hensel, *Mercer University*
Jana Henson, *Georgetown College*
James Hewlett, *Finger Lakes Community College*
Phyllis C. Hirsch, *East Los Angeles College*
A. Scott Holaday, *Texas Tech University*
R. Dwain Horrocks, *Brigham Young University*
Howard L. Hosick, *Washington State University*
Carl Huether, *University of Cincinnati*
Celene Jackson, *Western Michigan University*
John Jahoda, *Bridgewater State College*
Richard J. Jensen, *Saint Mary's College*
Tari Johnson, *Normandale Community College*
Tia Johnson, *Mitchell Community College*
Greg Jones, *Santa Fe Community College*
John Jorstad, *Kirkwood Community College*
Tracy L. Kahn, *University of California, Riverside*
Robert Kalbach, *Finger Lakes Community College*
Mary K. Kananen, *Pennsylvania State University, Altoona*
Thomas C. Kane, *University of Cincinnati*
Arnold J. Karpoff, *University of Louisville*
John M. Kasmer, *Northeastern Illinois University*
Valentine Kefeli, *Slippery Rock University*
Cheryl Kerfeld, *University of California, Los Angeles*
Henrik Kibak, *California State University, Monterey Bay*
Kerry Kilburn, *Old Dominion University*
Peter Kish, *Oklahoma School of Science and Mathematics*

Robert Kitchin, *University of Wyoming*
Richard Koblin, *Oakland Community College*
H. Roberta Koepfer, *Queens College*
Michael E. Kovach, *Baldwin-Wallace College*
Jocelyn E. Krebs, *University of Alaska, Anchorage*
Nuran Kumbaraci, *Stevens Institute of Technology*
Roya Lahijani, *Palomar College*
Vic Landrum, *Washburn University*
Gary Kwiecinski, *The University of Scranton*
Lynn Larsen, *Portland Community College*
Siu-Lam Lee, *University of Massachusetts, Lowell*
Thomas P. Lehman, *Morgan Community College*
William Leonard, *Central Alabama Community College*
Shawn Lester, *Montgomery College*
Leslie Lichtenstein, *Massasoit Community College*
Barbara Liedl, *Central College*
Harvey Liftin, *Broward Community College*
David Loring, *Johnson County Community College*
Lewis M. Lutton, *Mercyhurst College*
Maria P. MacWilliams, *Seton Hall University*
Michael Howard Marcovitz, *Midland Lutheran College*
Angela M. Mason, *Beaufort County Community College*
Roy B. Mason, *Mt. San Jacinto College*
John Mathwig, *College of Lake County*
Cyndi Maurstad, *Texas A&M University, Corpus Christi*
Bonnie McCormick, *University of the Incarnate Word*
Tonya McKinley, *Concord College*
Mary Anne McMurray, *Henderson Community College*
David Mirman, *Mt. San Antonio College*
Nancy Garnett Morris, *Volunteer State Community College*
Angela C. Morrow, *University of Northern Colorado*
Patricia S. Muir, *Oregon State University*
Jon R. Nickles, *University of Alaska, Anchorage*
Jane Noble-Harvey, *University of Delaware*
Jeanette C. Oliver, *Flathead Valley Community College*
David O'Neill, *Community College of Baltimore County*
Lois H. Peck, *University of the Sciences, Philadelphia*
Kathleen E. Pelkki, *Saginaw Valley State University*
Rhoda E. Perozzi, *Virginia Commonwealth University*
John S. Peters, *College of Charleston*
Pamela Petrequin, *Mount Mary College*
Bill Pietraface, *State University of New York, Oneonta*
Rosamond V. Potter, *University of Chicago*
Karen S. Powell, *Western Kentucky University*
Elena Pravosudova, *Sierra College*
Hallie Ray, *Rappahannock Community College*
Jill Raymond, *Rock Valley College*
Dorothy Read, *University of Massachusetts, Dartmouth*
Philip Ricker, *South Plains College*
Todd Rimkus, *Marymount University*
Lynn Rivers, *Henry Ford Community College*
Jennifer Roberts, *Lewis University*
Laurel Roberts, *University of Pittsburgh*
Maxine Losoff Rusche, *Northern Arizona University*
Michael L. Rutledge, *Middle Tennessee State University*

Sarmad Saman, *Quinsigamond Community College*
Leba Sarkis, *Aims Community College*
Walter Saviuk, *Daytona Beach Community College*
Neil Schanker, *College of the Siskiyous*
Robert Schoch, *Boston University*
John Richard Schrock, *Emporia State University*
Julie Schroer, *Bismark State College*
Karen Schuster, *Florida Community College at Jacksonville*
Brian W. Schwartz, *Columbus State University*
Sandra Seidel, *Elon University*
Wayne Seifert, *Brookhaven College*
Patty Shields, *George Mason University*
Cara Shillington, *Eastern Michigan University*
Cahleen Shrier, *Azusa Pacific University*
Jed Shumsky, *Drexel University*
Michele Shuster, *New Mexico State University*
Greg Sievert, *Emporia State University*
Jeffrey Simmons, *West Virginia Wesleyan College*
Frederick D. Singer, *Radford University*
Anu Singh-Cundy, *Western Washington University*
Margaret W. Smith, *Butler University*
Thomas Smith, *Armstrong Atlantic State University*
Deena K. Spielman, *Rock Valley College*
Minou D. Spradley, *San Diego City College*
Eric Stavney, *Highline Community College*
Andrew Storfer, *University of Washington*
Mark T. Sugalski, *New England College*
Marshall D. Sundberg, *Emporia State University*
Adelaide Svoboda, *Nazareth College*
Sharon Thoma, *Edgewood College*
Kenneth Thomas, *Hillsborough Community College*
Sumesh Thomas, *Baltimore City Community College*
Betty Thompson, *Baptist University*
John Tjepkema, *University of Maine at Orono*
Bruce L. Tomlinson, *State University of New York, Fredonia*
Leslie R. Towill, *Arizona State University*
Bert Tribbey, *California State University, Fresno*
Robert Turner, *Western Oregon University*
William A. Velhagen, Jr., *Longwood College*
Jonathan Visick, *North Central College*
Michael Vitale, *Daytona Beach Community College*
Stephen M. Wagener, *Western Connecticut State University*
James A. Wallis, *St. Petersburg Community College*
Jennifer Warner, *University of North Carolina at Charlotte*
Harold Webster, *Pennsylvania State University, DuBois*
Ted Weinheimer, *California State University, Bakersfield*
Lisa A. Werner, *Pima Community College*
Joanne Westin, *Case Western Reserve University*
Wayne Whaley, *Utah Valley State College*
Joseph D. White, *Baylor University*
Quinton White, *Jacksonville University*

Leslie Y. Whiteman, *Virginia Union University*
Rick Wiedenmann, *New Mexico State University at Carlsbad*
Judy A. Williams, *Southeastern Oklahoma State University*
Dwina Willis, *Freed Hardeman University*
David Wilson, *University of Miami*
Mala S. Wingerd, *San Diego State University*
E. William Wischusen, *Louisiana State University*
Darla J. Wise, *Concord College*
Bonnie Wood, *University of Maine at Presque Isle*
Mark L. Wygoda, *McNeese State University*
Samuel J. Zeakes, *Radford University*
Uko Zylstra, *Calvin College*

Class-Test Instructors

Marilyn Abbott, *Lindenwood College*
Sarah Barlow, *Middle Tennessee State University*
Suchi Bhardwaj, *Winthrop University*
Donna H. Bivans, *East Carolina University*
Steve Browder, *Franklin College*
Reggie Cobb, *Nash Community College*
Pat Cox, *University of Tennessee, Knoxville*
Ade Ejire, *Johnston Community College*
Carl Estrella, *Merced College*
Dianne M. Fair, *Florida Community College at Jacksonville*
Lynn Fireston, *Ricks College*
Patricia Glas, *The Citadel*
Consetta Helmick, *University of Idaho*
Richard Hilton, *Towson University*
Howard L. Hosick, *Washington State University*
Greg Jones, *Santa Fe Community College*
John Kelly, *Northeastern University*
Joyce Kille-Marino, *College of Charleston*
Katrina McCrae, *Abraham Baldwin Agricultural College*
Ed Mercurio, *Hartnell College*
Jon R. Nickles, *University of Alaska, Anchorage*
Jennifer Penrod, *Lincoln University*
Laurel Roberts, *University of Pittsburgh*
April Rottman, *Rock Valley College*
Mike Runyan, *Lander University*
Travis Ryan, *Furman University*
Walter Saviuk, *Daytona Beach Community College*
Sandra Slivka, *Miramar College*
Robert Stamatis, *Daytona Beach Community College*
Paula Thompson, *Florida Community College*
Joy Trauth, *Arkansas State University*
Virginia Vandergon, *California State University, Northridge*
Lisa Volk, *Fayetteville Technical Community College*
Dave Webb, *St. Clair County Community College*
Wayne Whaley, *Utah Valley State College*
Michael Womack, *Macon State College*

Supplements for the instructor

New! **Campbell Image Presentation Library with Visual Guide** (0-8053-7498-1)

The new Campbell Image Presentation Library is a chapter-by-chapter visual archive that includes more than 1,000 photos from the text plus additional sources, all text art with and without labels, selected figures layered for step-by-step presentation, all text tables, 110 animations, and 85 video clips. Thumbnail-sized images in the printed Visual Guide provide easy viewing of all resources in the Campbell Image Presentation Library.

More than 1,000 photos

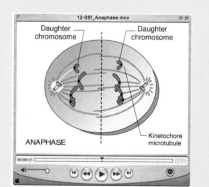

110 animations

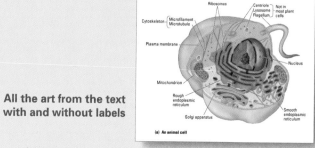

Selected figures layered for step-by-step presentation

85 video clips

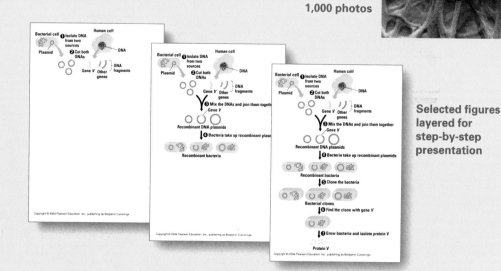

All the art from the text with and without labels

Visual guide to all resources

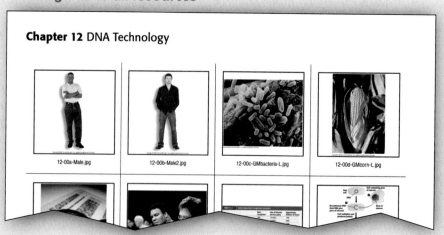

Chapter 12 DNA Technology

12-00a-Male.jpg 12-00b-Male2.jpg 12-00c-GMbacteria-L.jpg 12-00d-GMcorn-L.jpg

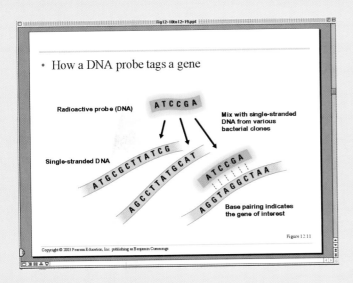

New! PowerPoint® Lectures (0-8053-7474-4)

Chris Romero, *Front Range Community College, Larimer Campus*

Prepared PowerPoint® Lectures integrate the art, photos, tables, and lecture outlines for each chapter. Art labels can be edited in PowerPoint®, and selected figures are layered for step-by-step presentation. These PowerPoint® Lectures can be used as is, edited, or customized for your course with your own images and text. Videos and animations can be added from the Campbell Image Presentation Library.

PowerPoint® lectures include art, photos, tables, and lecture outlines

New! Essential Biology Website Instructor Resources   www.essentialbiology.com

The Instructor Resources section of the *Essential Biology* Website provides one convenient location for adopters to download materials they need to teach their course: the Campbell Image Presentation Library, the PowerPoint® Lectures, the Instructor's Guide to Text and Media (in Word), the Test Bank (in Word), and additional critical thinking questions (in Word). The Instructor Resources section also includes worksheets (in Word) for the Case Studies in the Process of Science media investigations, which can be printed and given to students as assignments; suggested answers for the Case Studies worksheets; links to additional photo resources; and the Forum for Great Teaching Ideas, where professors can share ideas with their colleagues.

Instructor's Guide to Text and Media (0-8053-7475-2)

Edward J. Zalisko, *Blackburn College*

This comprehensive guide provides chapter-by-chapter references to all the media resources available to instructors and students plus a list of Transparency Acetates. The guide also includes objectives, additional critical thinking questions, lecture outlines, teaching tips, media tips, key terms, and word roots. A separate chapter offers suggestions for effective uses of technology in teaching introductory biology. The Instructor's Guide is available in print and in Word.

Transparency Acetates (0-8053-7484-1)

Over 650 full-color acetates include all illustrations and tables from the text, many of which incorporate photographs. New to this edition are selected figures illustrating key concepts broken down into layers for step-by-step lecture presentation.

Printed Test Bank (0-8053-7482-5)
Computerized Test Bank (0-8053-7477-9)

Eugene J. Fenster, *Longview Community College*

Thoroughly revised and updated, the test bank now includes more questions that emphasize critical thinking and an optional section with questions that test students on the Web/CD Activities. The Test Bank is available in print, on a cross-platform CD-ROM, and in the instructor section of CourseCompass™, Blackboard, and WebCT.

Annotated Instructor's Edition for Laboratory Investigations for Biology, Second Edition

(0-8053-6792-6)

Jean Dickey, *Clemson University*

The instructor's version of the lab manual includes the complete student version plus margin notes with instructor overviews, time requirements, helpful hints, and suggestions for extending or supplementing labs; answers to questions in the Student Edition; and suggestions for adapting the labs to a two-hour period.

Preparation Guide for Laboratory Investigations for Biology, Second Edition (0-8053-6771-3)

New! Course Management Systems

The content from the Essential Biology Website and Computerized Test Bank is available in these popular course management systems: CourseCompass™, Blackboard, and WebCT. Visit http://cms.aw.com for more information.

Supplements for the student

New! Essential Biology CD-ROM and Website

www.essentialbiology.com

The CD-ROM and Website included with the book contain:
- **200 Activities**, including animations, interactive review exercises, and videos. Activity Quizzes test students on the Activities.
- **56 Case Studies in the Process of Science** involve students in the process of science by asking them to make observations, analyze data, and draw conclusions. Lab Report questions can be answered electronically or on printable worksheets.
- **Over 2,000 quiz questions.** Each chapter includes a Pre-Test, an Activities Quiz, and a comprehensive Chapter Quiz. Each quiz has hints, immediate feedback, grading, and e-mailable results.
- **Glossary** with pronunciations
- **Flashcards** test knowledge of key terms.
- **Word Roots**
- **Key Terms**
- In addition, the Website provides access to **Biology and Society Web links** with questions, **Evolution Connection Web links** with questions, all the **art** from the book (with and without labels), **85 Videos, Web Links, News Links, News Archives, Further Readings,** and the **Syllabus Manager.**

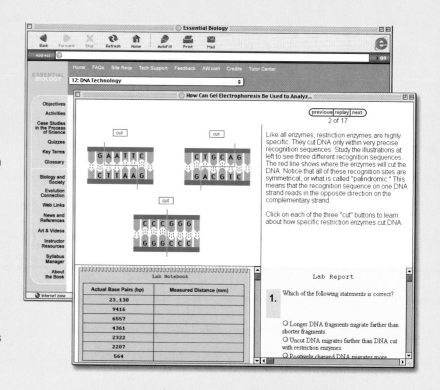

Addison Wesley Tutor Center

www.aw.com/tutorcenter

This center provides one-to-one tutoring in four different ways—phone, fax, email, and the Internet—during evening hours and on weekends. Qualified college instructors are available to answer questions and provide instruction regarding self-quizzes and other content found in *Essential Biology*.

Student Study Guide (0-8053-7479-5)

Edward J. Zalisko, *Blackburn College*

Students can master key concepts and earn a better grade with the thought-provoking exercises found in this study guide. Study advice, tables, quizzes, and crossword puzzles help students test their understanding of biology. The Student Study Guide also includes references to student media activities on the *Essential Biology* CD-ROM and Website.

Flexible options for the lab

Laboratory Investigations for Biology, Second Edition

(0-8053-6789-6)

Jean Dickey, *Clemson University*

An investigative approach actively involves students in the process of scientific discovery by allowing them to make observations, devise techniques, and draw conclusions. Twenty carefully chosen laboratory topics encourage students to use their critical thinking skills and the scientific method to solve problems.

New! Symbiosis Lab Authoring Kit—Customized Lab Manuals (0-321-10049-2)

Instructors can build a customized lab manual, choosing the labs they want, importing artwork from our graphics library, and even adding their own notes, syllabi, or other material. For more information visit http://www.pearsoncustom.com/database/symbiosis.html.

Biology Labs On-Line

www.biologylabsonline.com

These 12 virtual lab exercises enable students to perform potentially dangerous, lengthy, or expensive experiments in a safe electronic environment. The labs are available for purchase individually or in a 12-pack with the printed Student Lab Manual.

DETAILED CONTENTS

1 Introduction: Biology Today 1

The Scope of Biology 2

Life at Its Many Levels 2
Life in Its Diverse Forms 6

Evolution: Biology's Unifying Theme 8

The Darwinian View of Life 9
Natural Selection 10

The Process of Science 13

Discovery Science 13
Hypothesis-Driven Science 14
A Case Study in the Process of Science 15
The Culture of Science 16
Theories in Science 16
Science, Technology, and Society 18

Unit One CELLS

2 Essential Chemistry for Biology 20

Biology and Society: Fluoride in the Water 21

Tracing Life Down to the Chemical Level 21

Some Basic Chemistry 22

Matter: Elements and Compounds 22
Atoms 23
Chemical Bonding and Molecules 25
Chemical Reactions 27

Water and Life 28

The Structure of Water 28
Water's Life-Supporting Properties 29
Acids, Bases, and pH 21

Evolution Connection: Earth Before Life 32

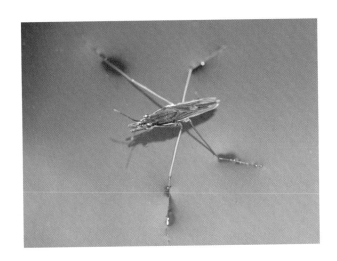

3   The Molecules of Life   35

Biology and Society: Got Lactose?   36

Organic Molecules   36

Carbon Chemistry   36

Giant Molecules from Smaller Building Blocks   38

Biological Molecules   39

Carbohydrates   39

Lipids   42

Proteins   44

Nucleic Acids   47

Evolution Connection: DNA and Proteins as Evolutionary Tape Measures   49

4   A Tour of the Cell   52

Biology and Society: Drugs That Target Cells   53

The Microscopic World of Cells   53

Microscopes as Windows to Cells   54

The Two Major Categories of Cells   55

A Panoramic View of Eukaryotic Cells   57

Membrane Structure and Function   58

A Fluid Mosaic of Lipids and Proteins   58

Selective Permeability   58

The Nucleus and Ribosomes: Genetic Control of the Cell   59

Structure and Function of the Nucleus   60

Ribosomes   60

How DNA Controls the Cell   60

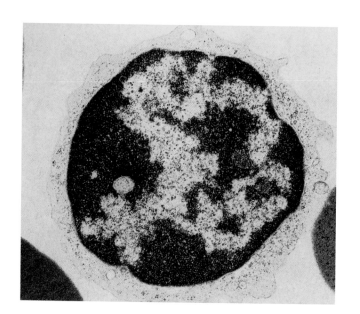

The Endomembrane System: Manufacturing and Distributing Cellular Products   61

   The Endoplasmic Reticulum   61

   The Golgi Apparatus   62

   Lysosomes   62

   Vacuoles   63

Chloroplasts and Mitochondria: Energy Conversion   64

   Chloroplasts   64

   Mitochondria   65

The Cytoskeleton: Cell Shape and Movement   65

   Maintaining Cell Shape   65

   Cilia and Flagella   66

Cell Surfaces: Protection, Support, and Cell-Cell Interactions   67

   Plant Cell Walls and Cell Junctions   68

   Animal Cell Surfaces and Cell Junctions   68

Evolution Connection: The Origin of Membranes   69

5   The Working Cell   72

Biology and Society: Stonewashing Without the Stones   73

Some Basic Energy Concepts   73

   Conservation of Energy   73

   Entropy   74

   Chemical Energy   74

   Food Calories   75

ATP and Cellular Work   76

   The Structure of ATP   76

   Phosphate Transfer   77

   The ATP Cycle   77

Enzymes 78

Activation Energy 78

Induced Fit 78

Enzyme Inhibitors 78

Membrane Transport 80

Passive Transport: Diffusion Across Membranes 80

Osmosis and Water Balance in Cells 81

Active Transport: The Pumping of Molecules Across Membranes 82

Exocytosis and Endocytosis: Traffic of Large Molecules 82

The Role of Membranes in Cell Signaling 83

Evolution Connection: Evolving Enzymes 84

6 Cellular Respiration: Harvesting Chemical Energy 87

Biology and Society: Feeling the "Burn" 88

Energy Flow and Chemical Cycling in the Biosphere 88

Producers and Consumers 89

Chemical Cycling Between Photosynthesis and Cellular Respiration 89

Cellular Respiration: Aerobic Harvest of Food Energy 90

The Relationship Between Cellular Respiration and Breathing 90

The Overall Equation for Cellular Respiration 91

The Role of Oxygen in Cellular Respiration 91

The Metabolic Pathway of Cellular Respiration 93

Fermentation: Anaerobic Harvest of Food Energy 98

Fermentation in Human Muscle Cells 98

Fermentation in Microorganisms 99

Evolution Connection: Life on an Anaerobic Earth 99

Contents

7 Photosynthesis: Converting Light Energy to Chemical Energy 102

Biology and Society: Plant Power 103

The Basics of Photosynthesis 103

Chloroplasts: Sites of Photosynthesis 103

The Overall Equation for Photosynthesis 105

A Photosynthesis Road Map 105

The Light Reactions: Converting Solar Energy to Chemical Energy 106

The Nature of Sunlight 106

Chloroplast Pigments 107

How Photosystems Harvest Light Energy 107

How the Light Reactions Generate ATP and NADPH 109

The Calvin Cycle: Making Sugar from Carbon Dioxide 111

Water-Saving Adaptations of C_4 and CAM Plants 112

The Environmental Impact of Photosynthesis 113

How Photosynthesis Moderates the Greenhouse Effect 112

Evolution Connection: The Oxygen Revolution 114

Unit Two GENETICS

8 The Cellular Basis of Reproduction and Inheritance 118

Biology and Society: A $50,000 Egg! 119

What Cell Reproduction Accomplishes 119

Passing On the Genes from Cell to Cell 120

The Reproduction of Organisms 120

The Cell Cycle and Mitosis 121

Eukaryotic Chromosomes 121

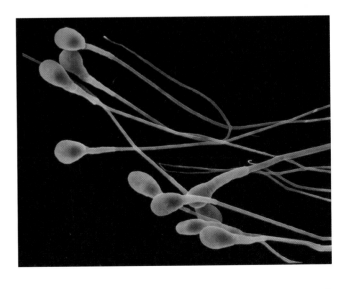

The Cell Cycle 122

Mitosis and Cytokinesis 124

Cancer Cells: Growing Out of Control 126

Meiosis, the Basis of Sexual Reproduction 128

Homologous Chromosomes 128

Gametes and the Life Cycle of a Sexual Organism 129

The Process of Meiosis 130

Review: Comparing Mitosis and Meiosis 133

The Origins of Genetic Variation 133

When Meiosis Goes Awry 134

Evolution Connection: New Species from Errors in Cell Division 137

9 Patterns of Inheritance 141

Biology and Society: Testing Your Baby 142

Heritable Variation and Patterns of Inheritance 142

In an Abbey Garden 143

Mendel's Principle of Segregation 144

Mendel's Principle of Independent Assortment 147

Using a Testcross to Determine an Unknown Genotype 149

The Rules of Probability 149

Family Pedigrees 150

Human Disorders Controlled by a Single Gene 151

Beyond Mendel 154

Incomplete Dominance in Plants and People 155

Multiple Alleles and Blood Type 156

Pleiotropy and Sickle-Cell Disease 157

Polygenic Inheritance 158

The Role of Environment 158

The Chromosomal Basis of Inheritance 159

Gene Linkage 160

Genetic Recombination: Crossing Over and Linkage Maps 162

Sex Chromosomes and Sex-Linked Genes 163

Sex Determination in Humans and Fruit Flies 163

Sex-Linked Genes 163

Sex-Linked Disorders in Humans 165

**Evolution Connection:
The Telltale Y Chromosome 166**

10 Molecular Biology of the Gene 170

Biology and Society: Sabotaging HIV 171

The Structure and Replication of DNA 171

DNA and RNA: Polymers of Nucleotides 172

Watson and Crick's Discovery of the Double Helix 173

DNA Replication 175

The Flow of Genetic Information from DNA to RNA to Protein 176

*How an Organism's DNA Genotype Produces
Its Phenotype 177*

*From Nucleotide Sequence to Amino Acid
Sequence: An Overview 178*

The Genetic Code 179

Transcription: From DNA to RNA 180

The Processing of Eukaryotic RNA 181

Translation: The Players 182

Translation: The Process 183

Review: DNA → RNA → Protein 184

Mutations 186

Viruses: Genes in Packages 188

Bacteriophages 188

Plant Viruses 189

Animal Viruses 190

HIV, the AIDS Virus 190

Evolution Connection: Emerging Viruses 192

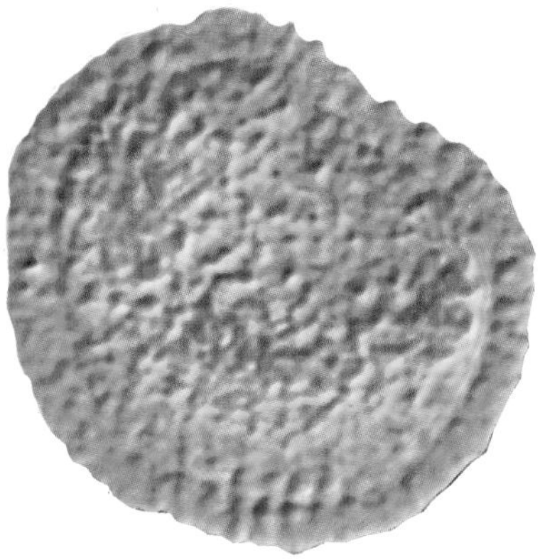

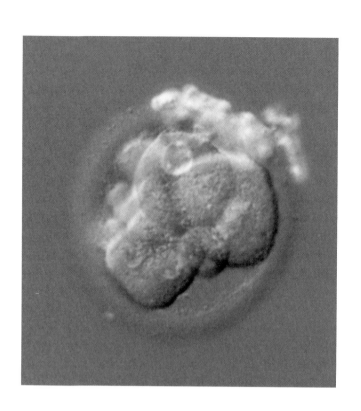

11 Gene Regulation 196

**Biology and Society:
Baby's First Bank Account 197**

**From Egg to Organism: How and Why Genes Are
Regulated 197**

Patterns of Gene Expression in Differentiated Cells 198

DNA Microarrays: Visualizing Gene Expression 198

The Genetic Potential of Cells 200

Reproductive Cloning of Animals 200

Therapeutic Cloning and Stem Cells 201

The Regulation of Gene Expression 203

Gene Regulation in Bacteria 203

Gene Regulation in the Nucleus of Eukaryotic Cells 204

Regulation in the Cytoplasm 206

Cell Signaling 207

The Genetic Basis of Cancer 208

Genes That Cause Cancer 208

The Effects of Lifestyle on Cancer Risk 211

Evolution Connection: Homeotic Genes 212

12 DNA Technology 216

Biology and Society: Hunting for Genes 217

Recombinant DNA Technology 217

From Humulin to Genetically Modified Foods 218

Recombinant DNA Techniques 220

DNA Fingerprinting and Forensic Science 224

Murder, Paternity, and Ancient DNA 225

DNA Fingerprinting Techniques 226

Genomics 229

The Human Genome Project 230

Tracking the Anthrax Killer　231

Genome-Mapping Techniques　232

Human Gene Therapy　234

Treating Severe Combined Immunodeficiency　2234

Safety and Ethical Issues　235

The Controversy Over Genetically Modified Foods　235

Ethical Questions Raised by DNA Technology　236

Evolution Connection: Genomes Hold Clues to Evolution　237

Unit Three EVOLUTION AND DIVERSITY

13 How Populations Evolve　242

Biology and Society: Persistent Pests　243

Charles Darwin and *The Origin of Species*　244

Darwin's Cultural and Scientific Context　245

Descent with Modification　248

Evidence of Evolution　249

The Fossil Record　250

Biogeography　251

Comparative Anatomy　251

Comparative Embryology　252

Molecular Biology　253

Natural Selection and Adaptive Evolution　254

Darwin's Theory of Natural Selection　254

Natural Selection in Action　255

The Modern Synthesis: Darwinism Meets Genetics　256

Populations as the Units of Evolution　256

Genetic Variation in Populations　257

Analyzing Gene Pools   258

Population Genetics and Health Science   259

Microevolution as Change in a Gene Pool   260

Mechanisms of Microevolution   260

Genetic Drift   261

Gene Flow   263

Mutations   263

Natural Selection: A Closer Look   263

Evolution Connection: Population Genetics of the Sickle-Cell Allele   266

14 How Biological Diversity Evolves   270

Biology and Society: The Impact of Asteroids   271

Macroevolution and the Diversity of Life   271

The Origin of Species   272

What Is a Species?   272

Reproductive Barriers Between Species   274

Mechanisms of Speciation   275

What Is the Tempo of Speciation?   278

The Evolution of Biological Novelty   280

Adaptation of Old Structures for New Functions   280

"Evo-Devo": Development and Evolutionary Novelty   281

Earth History and Macroevolution   282

Geologic Time and the Fossil Record   282

Continental Drift and Macroevolution   285

Mass Extinctions and Explosive Diversifications of Life   285

Classifying the Diversity of Life   287

Some Basics of Taxonomy   287

Classification and Phylogeny   288

Arranging Life into Kingdoms: A Work in Progress   290

Evolution Connection: Just a Theory?   292

15 The Evolution of Microbial Life 296

Biology and Society: Bioterrorism 297

Major Episodes in the History of Life 297

The Origin of Life 300

Resolving the Biogenesis Paradox 300

A Four-Stage Hypothesis for the Origin of Life 300

From Chemical Evolution to Darwinian Evolution 302

Prokaryotes 303

They're Everywhere! 303

The Two Main Branches of Prokaryotic Evolution: Bacteria and Archaea 304

The Structure, Function, and Reproduction of Prokaryotes 304

The Nutritional Diversity of Prokaryotes 307

The Ecological Impact of Prokaryotes 307

Protists 311

The Origin of Eukaryotic Cells 311

The Diversity of Protists 312

Evolution Connection: The Origin of Multicellular Life 317

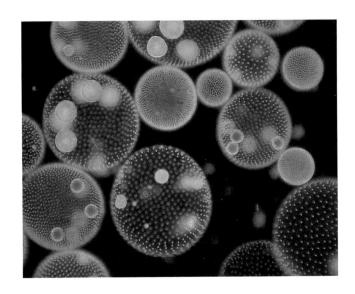

16 Plants, Fungi, and the Move onto Land 320

Biology and Society: The Balancing Act of Forest Conservation 321

Colonizing Land 321

Terrestrial Adaptations of Plants 321

The Origin of Plants from Green Algae 323

Plant Diversity 324

Highlights of Plant Evolution 324

Bryophytes 325

Ferns   *327*

Gymnosperms   *328*

Angiosperms   *330*

Plant Diversity as a Nonrenewable Resource   *332*

Fungi   334

Characteristics of Fungi   *336*

The Ecological Impact of Fungi   *337*

Evolution Connection: Mutual Symbiosis   340

17 The Evolution of Animals   343

Biology and Society: Invasion of the Killer Toads   344

The Origins of Animal Diversity   344

What Is an Animal?   *344*

Early Animals and the Cambrian Explosion   *346*

Animal Phylogeny   *347*

Major Invertebrate Phyla   349

Sponges   *350*

Cnidarians   *350*

Flatworms   *351*

Roundworms   *351*

Mollusks   *353*

Annelids   *354*

Arthropods   *356*

Echinoderms   *359*

The Vertebrate Genealogy   360

Characteristics of Chordates   *361*

Fishes   *363*

Amphibians   *364*

Reptiles   *365*

Birds   *366*

Mammals   *367*

The Human Ancestry   368

   The Evolution of Primates   368

   The Emergence of Humankind   370

Evolution Connection: Earth's New Crisis   376

Unit Four ECOLOGY

18 The Ecology of Organisms and Populations   380

Biology and Society: The Human Population Explosion   381

An Overview of Ecology   381

   Ecology as Scientific Study   382

   A Hierarchy of Interactions   383

   Ecology and Environmentalism   383

   Abiotic Factors of the Biosphere   384

The Evolutionary Adaptations of Organisms   387

   Physiological Responses   387

   Anatomical Responses   388

   Behavioral Responses   388

What Is Population Ecology?   388

   Population Density   389

   Patterns of Dispersion   390

   Population Growth Models   390

   Regulation of Population Growth   392

   Human Population Growth   395

Life Histories and Their Evolution   398

   Life Tables and Survivorship Curves   399

   Life History Traits as Evolutionary Adaptations   400

Evolution Connection: Testing a Darwinian Hypothesis   401

19 Communities and Ecosystems   406

Biology and Society: Reefs: Coral and Artificial   407

Key Properties of Communities   407

Diversity   408

Prevalent Form of Vegetation   408

Stability   408

Trophic Structure   409

Interspecific Interactions in Communities   409

Competition Between Species   409

Predation   410

Symbiotic Relationships   414

The Complexity of Community Networks   415

Disturbance of Communities   415

Ecological Succession   416

A Dynamic View of Community Structure   417

An Overview of Ecosystem Dynamics   417

Trophic Levels and Food Chains   419

Food Webs   419

Energy Flow in Ecosystems   421

Productivity and the Energy Budgets of Ecosystems   421

Energy Pyramids   422

Ecosystem Energetics and Human Nutrition   423

Chemical Cycling in Ecosystems   424

The General Scheme of Chemical Cycling   424

Examples of Biogeochemical Cycles   425

Biomes   428

How Climate Affects Biome Distribution   428

Terrestrial Biomes   428

Freshwater Biomes   434

Marine Biomes   435

Evolution Connection: Coevolution in Biological Communities   438

20 Human Impact on the Environment   442

Biology and Society: Aquarium Menaces   443

Human Impact on Biological Communities   443

Human Disturbance of Communities   444

Introduced Species   444

Human Impact on Ecosystems   446

Impact on Chemical Cycles   446

Deforestation and Chemical Cycles: A Case Study   447

The Release of Toxic Chemicals to Ecosystems   448

Human Impact on the Atmosphere and Climate   449

The Biodiversity Crisis   452

The Three Levels of Biodiversity   452

The Loss of Species   452

The Three Main Causes of the Biodiversity Crisis   454

Why Biodiversity Matters   455

Conservation Biology   456

Biodiversity "Hot Spots"   456

Conservation at the Population and Species Levels   457

Conservation at the Ecosystem Level   459

The Goal of Sustainable Development   461

Evolution Connection: Biophilia and an Environmental Ethic   462

Appendix A   Metric Conversion Table

Appendix B   Answers to Self-Quiz Questions

Appendix C   Credits

Glossary

Index

CHAPTER 1

Introduction: Biology Today

The Scope of Biology   2

Life at Its Many Levels

Life in Its Diverse Forms

**Evolution:
Biology's Unifying Theme   8**

The Darwinian View of Life

Natural Selection

The Process of Science   13

Discovery Science

Hypothesis-Driven Science

A Case Study in the Process of Science

The Culture of Science

Theories in Science

Science, Technology, and Society

Biologists have identified about **1.7 million species** of living organisms.

Scientists have determined the **complete DNA sequences** of humans, puffer fish, mosquitoes, and rice.

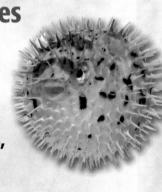

All organisms share a common chemical language for their genetic material, DNA.

Amoebas, molds, trees, and people are all made from similar cells.

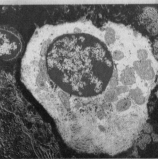

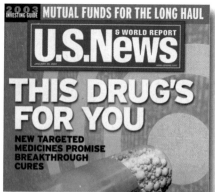

We are living in a golden age of biology. The largest and best-equipped community of scientists in history is beginning to solve biological puzzles that once seemed unsolvable. We are moving ever closer to understanding how a single cell becomes a plant or animal; how plants trap solar energy and store that energy in food; how organisms network in biological communities such as forests and coral reefs; and how the great diversity of life on Earth evolved from the first microbes. Exploring life has never been more exhilarating. Welcome to the big adventure of the twenty-first century!

Modern biology is as important as it is inspiring, with exciting breakthroughs changing our very culture. Genetics and cell biology are revolutionizing medicine and agriculture. Molecular biology is providing new tools for anthropology, helping us trace the origin and dispersal of early humans. Ecology is helping us evaluate environmental issues, such as the causes and consequences of global warming. Neuroscience and evolutionary biology are reshaping psychology and sociology. Biology has even entered the legal system, with terms such as *DNA fingerprinting* now part of our vocabulary. These are just a few examples of how biology is woven into the fabric of society as never before. It is no wonder that biology is daily news, as illustrated in **Figure 1.1**.

So, whatever your reasons for taking this course, even if only to meet your college's science requirement, you'll soon discover that this is the best time ever to study biology. To help you get started, this first chapter of *Essential Biology* defines biology, surveys its scope, introduces evolution as the theme that unifies all of biology, sets the study of life in the broader context of science as a process of inquiry, and illustrates the connections between biology and society.

> **Biology and Society on the Web**
> Find out how "wonky holes" affect coral reefs.

Figure 1.1 A small sample of biology in the news.

The Scope of Biology

Biology is the scientific study of life. It's a huge subject that gets bigger every year. We can think of biology's enormous scope as having two major dimensions. First, life is structured on a size scale ranging from the molecular to the global. The second dimension of biology's scope stretches across the enormous diversity of life on Earth, now and throughout life's history.

Life at Its Many Levels

In *Essential Biology*, we will probe life all the way down to the submicroscopic scale of molecules such as DNA, the chemical responsible for inheritance. At the other extreme of biological size and complexity, our exploration will take us all the way to the global scale of the entire biosphere, which consists of all the environments on Earth that support life—soil, oceans, lakes and other bodies of water, and the inner atmosphere. **Figure 1.2** takes us into this world of life through a series of views that zooms in a thousand times with each step. We start by approaching the biosphere from space. A thousand times closer, we can recognize Manhattan, with its Central Park. Another thousand-power jump puts us in a wooded area of the park. The next scale change takes us up close to an organism (living thing), an eastern gray squirrel. The squirrel is eating an acorn, which was produced by an oak tree, another organism. There are many other organisms living on, in, and around the squirrel. Our next size change vaults us into the squirrel to see that its body consists of microscopic units called cells. In each cell, you can see a nucleus, the part of the cell that contains all the genes that this squirrel inherited from its parents. And our last thousand-power change zooms us into the nucleus, where we can behold the molecular architecture of the DNA that makes up the genes. From the interactions within the biosphere to the molecular machinery within cells, biologists are investigating life at its many levels. Let's take a closer look here at just two biological levels near opposite ends of the size scale: ecosystems and cells.

Activity 1A on the Web & CD Play a card game to quiz yourself on the hierarchy of life.

Ecosystems   Life does not exist in a vacuum. Each organism interacts continuously with its environment, which includes other organisms as well as nonliving components of the environment. The roots of a tree, for example, absorb water and minerals from the soil. Leaves take in a gas called carbon dioxide from the air. Chlorophyll, the green pigment of the leaves, absorbs sunlight, which drives the plant's production of sugar from carbon dioxide and water. This food production is called photosynthesis. The tree also releases oxygen to the air, and its roots help form soil by breaking up rocks. Both organism and environment are affected by the interactions between them. The tree also interacts with other living things, including microorganisms (microscopic organisms) in the soil that are associated with the plant's roots and animals that eat its leaves and fruit. We are, of course, among those animals.

Ecology is the branch of biology that investigates these relationships between organisms and their environments. Ecologists use the term *ecosystem* for the biological level consisting of all the organisms and nonliving factors affecting life in a particular area, such as a wooded glen or grassland.

The dynamics of any ecosystem depend on two main processes. One is the cycling of nutrients. For example, minerals that plants take up from the soil will eventually be recycled to the soil by microorganisms that decompose leaf litter and other organic refuse. The second major process in an ecosystem is the flow of energy from sunlight to producers and

A view of Earth from space

Approaching Central Park (the red rectangle in the middle of this photo)

A Central Park woodland

An eastern gray squirrel

Cell   Nucleus within cell

Cells in intestine

DNA

Figure 1.2 Zooming in on life.   Each scale change in this series takes us a thousand times closer. Biologists explore life at levels ranging from the biosphere to the molecules that make up cells.

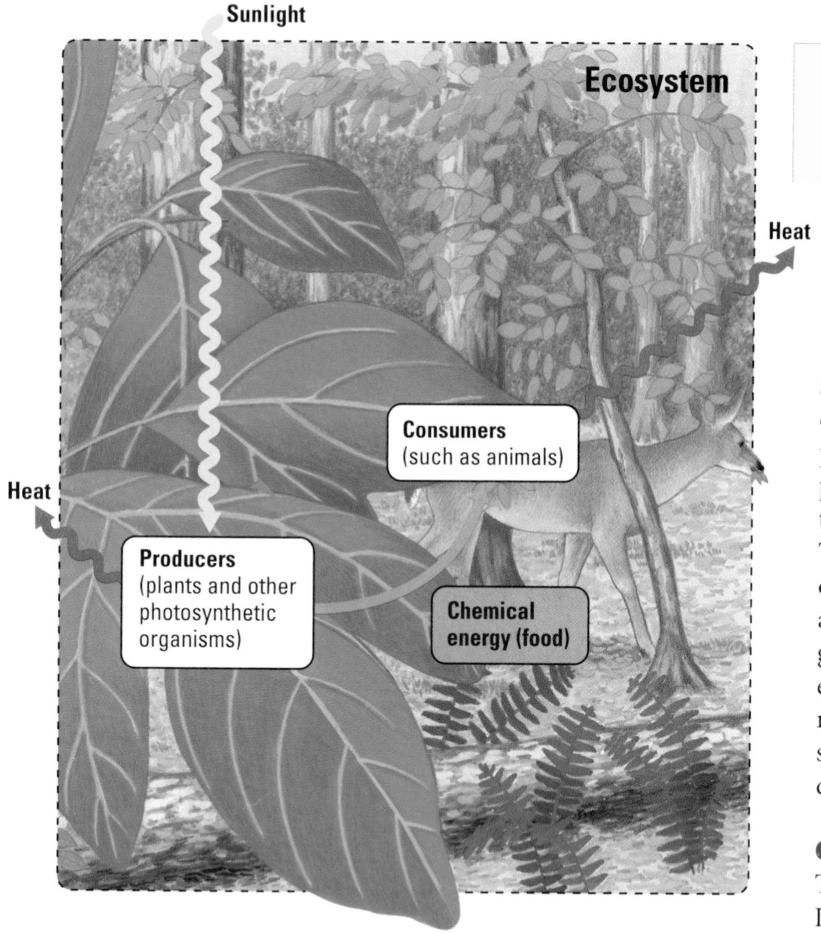

Figure 1.3 Energy flow in an ecosystem.

Sunlight

Ecosystem

Heat

Heat

Consumers (such as animals)

Producers (plants and other photosynthetic organisms)

Chemical energy (food)

Living is work, and work requires that organisms obtain and use energy. Most ecosystems are solar powered. The energy that enters an ecosystem as sunlight exits as heat, which all organisms dissipate to their surroundings whenever they perform work.

Activity 1B on the Web & CD
Explore energy flow and nutrient recycling.

then on to consumers. (**Figure 1.3** depicts this solar powering of ecosystems.) Producers are photosynthetic organisms, such as plants; consumers are the organisms, such as animals, that feed on plants, either directly (by eating plants) or indirectly (by eating animals that eat plants). Thus, energy flows one way (from sunlight to producers to consumers), whereas nutrients are recycled.

The biosphere is enriched by a great variety of ecosystems. A tropical rain forest in South America is an ecosystem. Very different from tropical forests are the ecosystems of deserts, such as those of the southwestern United States. A coral reef, such as the Great Barrier Reef off the eastern coast of Australia, is an ecosystem, and so is any small pond that may exist on your campus or in your city. Even a woodland patch in New York's Central Park qualifies as an ecosystem, small and artificial as it is by forest standards and disrupted as it is by human visitors. The fact is, humans are organisms that now have some presence, often disruptive, in all ecosystems. And the collective clout of 6 billion humans and their machines has an impact on the entire biosphere, which is the global ecosystem, the sum of dynamic processes in all ecosystems. For example, our fuel-burning, forest-chopping actions are changing the atmosphere and the planet's climate in ways that we do not yet fully understand, though we already know that this global vandalism jeopardizes the diversity of life on Earth.

Cells and Their DNA   Let's downsize now from ecosystems to cells. The cell has a special place in the hierarchy of biological organization. It is the lowest level of structure that can perform *all* activities required for life, including the capacity to reproduce.

All organisms are composed of cells. They occur singly as a great variety of unicellular organisms, mostly microscopic. And cells are also the subunits that make up the tissues and organs of plants, animals, and other multicellular organisms. In either case, the cell is the organism's basic unit of structure and function. The ability of cells to divide to form new cells is the basis for all reproduction and for the growth and repair of multicellular organisms, including humans.

We can distinguish two major kinds of cells: prokaryotic and eukaryotic. The prokaryotic cell is much simpler and usually much smaller than the eukaryotic cell. The cells of the microorganisms we commonly call bacteria are prokaryotic. All other forms of life, including plants, animals, and fungi, are composed of eukaryotic cells. In contrast to the prokaryotic cell, the eukaryotic cell is subdivided by internal membranes into many different functional compartments, or organelles. For example, the nucleus, the largest organelle in most eukaryotic cells, houses the DNA, the heritable material that directs the cell's many activities. Prokaryotic cells also have DNA, but it is not packaged within a nucleus (**Figure 1.4**).

Though very different in structural complexity, prokaryotic and eukaryotic cells have much in common at the molecular level. Most importantly, all cells use DNA as the chemical material of genes, the units of inheritance that transmit information from parents to offspring. Of course, bacteria and humans inherit different genes, but that information is encoded in a chemical

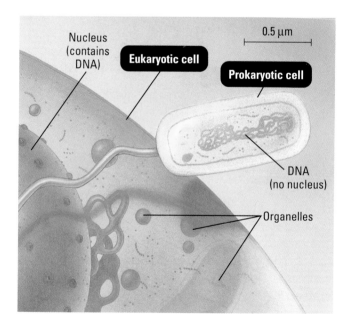

Nucleus (contains DNA)

Eukaryotic cell

0.5 µm

Prokaryotic cell

DNA (no nucleus)

Organelles

Figure 1.4 Two main kinds of cells: prokaryotic and eukaryotic.

The scale bar represents 0.5 micrometer (µm). There are 1,000 µm in a millimeter (mm). (You can review metric measurements in Appendix A.)

Activity 1C on the Web & CD
Watch a 3D animation of a rotating DNA molecule.

language common to all organisms. This language of life has an alphabet of just four letters, DNA's four molecular building blocks, their chemical names abbreviated A, G, C, and T (**Figure 1.5**).

An average-sized gene may be hundreds or thousands of chemical letters long. That gene's meaning to the cell is written in its specific sequence of these letters, just as the message of a sentence is encoded in its arrangement of letters selected from the 26 letters of the English alphabet. One gene may be translated as "Build a blue pigment in a bacterial cell." Another gene may mean "Make human insulin in this cell." Insulin, a hormone secreted into the bloodstream by cells of an organ called the pancreas, helps regulate your body's use of sugar as a fuel. In people who have certain forms of the disease diabetes, insulin is in short supply. The loss of normal regulation of blood sugar can be debilitating or even deadly. Many people with diabetes now bring their sugar levels back into balance by injecting themselves with insulin produced by genetically engineered bacteria. The engineering involves transplanting a gene for insulin production from a human cell into a bacterial cell. Each time the bacterial cell reproduces, it copies its new insulin gene, which is inherited by offspring along with the bacterial genes. Soon there is a massive colony of insulin-producing bacterial cells. This was one of the earliest products of a new biotechnology that has transformed the pharmaceutical industry and extended millions of lives (**Figure 1.6**). And it is only possible because biological information is written in the universal chemical language of DNA.

The entire "book" of genetic instructions that an organism inherits is called its genome. The nucleus of each human cell packs a genome that is about 3.2 billion chemical letters long. In 2001, scientists tabulating the sequence of these letters published a "rough draft" of the human genome. The press and world leaders acclaimed this international achievement as the greatest scientific triumph ever. But unlike past cultural zeniths, such as the landing of Apollo astronauts on the moon, the sequencing of the human genome is more a commencement than a climax. As the quest continues, biologists will learn the functions of thousands of genes and how their activities are coordinated in the development of an organism. Additionally, the genomes of other organisms (such as *E. coli* bacteria, fruit flies, and mice) have been fully sequenced, allowing scientists to compare the genomes of different species. This emerging field of genomics—a branch of biology that studies whole genomes—is a striking example of human curiosity about life at its many levels.

Figure 1.5 The language of DNA.
These simple shapes and letters symbolize the four kinds of chemical building blocks that are chained together in DNA. A gene is a segment of DNA composed of hundreds or thousands of these building blocks, of which we see only a short stretch here. Each gene encodes information in its specific sequence of the four chemical letters that are universal among all life on Earth.

Figure 1.6 DNA technology in the drug industry.
This employee of a biotechnology company is monitoring a tank of microorganisms that are genetically engineered to produce an ingredient of hepatitis B vaccine.

Life in Its Diverse Forms

Diversity is a hallmark of life. The eastern gray squirrel shown in Figure 1.2 is just one of about 1.7 million species that biologists have identified and named (the scientific name for the squirrel is *Sciurus carolinensis*). The diversity of known life includes over 280,000 plants, almost 50,000 vertebrates (animals with a backbone), and more than 750,000 insects (about half of all known forms of life). Biologists add thousands of newly identified species to the list each year. Estimates of the total diversity range from 5 million to over 30 million species. Whatever the actual number, the vast variety of life widens biology's scope.

Grouping Species: The Basic Concept   Biological diversity can be something to relish and preserve, but it can also be a bit overwhelming (**Figure 1.7**). Confronted with complexity, humans are inclined to categorize diverse items into a smaller number of groups. Grouping species that are similar is natural for us. We may speak of "squirrels" and "butterflies," though we recognize that each group actually includes many different species. We may even sort groups into broader categories, such as rodents (which include squirrels) and insects (which include butterflies). Taxonomy, the branch of biology that names and classifies species, formalizes this hierarchical ordering according to a scheme you will learn about in Chapter 14. Here we consider only kingdoms and domains, the broadest units of classification.

The Three Domains of Life   Until recently, most biologists divided the diversity of life into five main groups, or kingdoms. (The most familiar two are the plant and animal kingdoms.) But new methods, such as comparisons of DNA among organisms, have led to an ongoing reassessment of the number and boundaries of kingdoms. Various classification schemes are now based on six, eight, or more kingdoms. But as the debate continues on the kingdom level, there is broader consensus that the kingdoms of life can now be assigned to three even higher levels of classification called domains.

The three domains are named Bacteria, Archaea, and Eukarya (**Figure 1.8**). The first two domains, Bacteria and Archaea, identify two very different groups of organisms that have prokaryotic cells. In the five-kingdom system, these prokaryotes were combined in a single kingdom. But newer evidence suggests that the organisms known as archaea are at least as closely related to eukaryotes as they are to bacteria.

All the eukaryotes (organisms with eukaryotic cells) are grouped into at least four kingdoms (depending on the classification scheme) in the domain Eukarya. Kingdom Protista consists of eukaryotic organisms that are generally single-celled—for example, the microscopic protozoa, such as the amoebas. Many biologists extend the boundaries of the kingdom Protista to include certain multicellular forms, such as seaweeds, that seem to be closely related to the unicellular protists. (Other biologists split the protists into multiple kingdoms.) The remaining three kingdoms—Plantae, Fungi, and Animalia—consist of multicellular eukaryotes. These three kingdoms are distinguished partly by their modes of nutrition. Plants produce their own sugars and other foods by photosynthesis. Fungi are mostly decomposers that absorb nutrients by breaking down dead organisms and organic wastes, such as leaf litter

Figure 1.7 A small sample of biological diversity. Shown here are just some of the tens of thousands of species in the butterfly and moth collection at the National Museum of Natural History in Washington, DC. As diverse as the species are, they are all variations on a common anatomical theme. One of biology's major goals is to explain how such diversity arises while also accounting for characteristics common to different species.

Activity 1D on the Web & CD
Learn more about how organisms are classified.

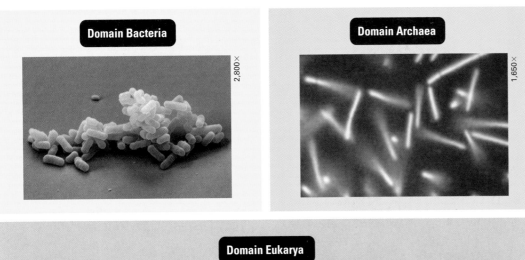

Domain Bacteria 2,800×

Domain Archaea 1,650×

Domain Eukarya 200×

Kingdom Protista

Kingdom Plantae

Kingdom Fungi

Kingdom Animalia

Figure 1.8
The three domains of life.

and animal feces. Animals obtain food by ingestion, which is the eating and digesting of other organisms. It is, of course, the kingdom to which we belong.

Unity in the Diversity of Life   If life is so diverse, how can biology have any unifying themes? What, for instance, can a tree, a mushroom, and a human possibly have in common? As it turns out, a great deal! Underlying the diversity of life is a striking unity, especially at the lower levels of biological organization. We have already seen one example: the universal genetic language of DNA. That fundamental fact connects all kingdoms of life, even uniting prokaryotes such as bacteria with eukaryotes such as humans. And among eukaryotes, unity is evident in many of the

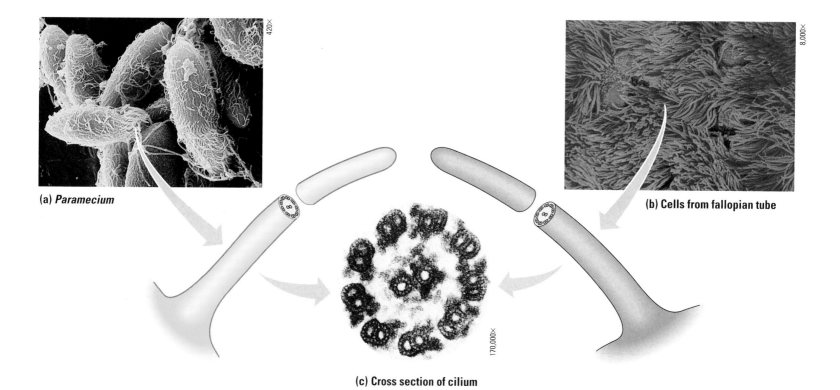

(a) *Paramecium*

(b) Cells from fallopian tube

(c) Cross section of cilium

Figure 1.9 **An example of unity underlying the diversity of life: the universal architecture of eukaryotic cilia.** Cilia (singular, *cilium*), extensions of cells that function in locomotion, occur in eukaryotic organisms as diverse as the protozoan *Paramecium* and the human. **(a)** The cilia of *Paramecium* propel the cell through water. **(b)** The cells that line the fallopian tubes (oviducts) of human females are also equipped with cilia. These cilia sweep egg cells from a woman's ovary to her uterus. **(c)** A powerful instrument called an electron microscope reveals a common pattern of tubules in the cilia of *Paramecium*, humans, and a diversity of other eukaryotes. This micrograph (photo taken with a microscope) represents this universal architecture in a cross-sectional view of a cilium.

details of cell structure (**Figure 1.9**). Above the cellular level, however, organisms are so variously adapted to their ways of life that describing and classifying biological diversity remains an important goal of biology. What can account for this combination of unity and diversity in life? The scientific explanation is the biological process called evolution.

Evolution: Biology's Unifying Theme

The history of life, as documented by fossils and other evidence, is a saga of a restless Earth billions of years old, inhabited by a changing cast of living forms (**Figure 1.10**). Life evolves. Just as an individual has a family history, each species is one twig of a branching tree of life extending back in time through ancestral species more and more remote. Species that are very similar, such as the brown bear and the polar bear, share a common ancestor that represents a relatively recent branch point on the tree of life (**Figure 1.11**). But through an ancestor that lived much farther back in time, all bears are also related to squirrels, humans, and all other mammals. Hair and milk-producing mammary glands are just two of a long list of uniquely

Figure 1.10 **Digging into the past.**
In the desert of Niger, Africa, paleontologist (fossil specialist) Paul Sereno excavates the leg bones of a giant plant-eating dinosaur called *Jobaria*. The fossil record supports other evidence that life has changed dramatically over Earth's long history.

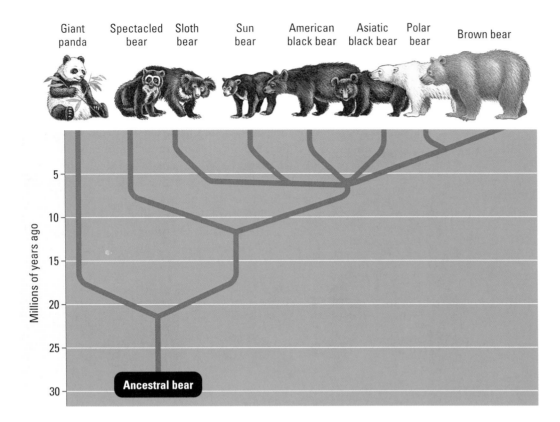

Figure 1.11
An evolutionary tree of bears.
This tree is based on both the fossil record and the evaluation of relationships among living bears by comparing genetic information.

mammalian traits. It is what we would expect if all mammals descended from a common ancestor, a prototypical mammal. And mammals, birds, reptiles, and all other vertebrates share a common ancestor even more ancient. Evidence of a still broader relationship can be found in such similarities as the matching machinery of all eukaryotic cilia, hairlike cellular projections used in locomotion (see Figure 1.9). Trace life back far enough, and there are only fossils of the primeval prokaryotes that inhabited Earth over 3 billion years ago. We can recognize some of their vestiges in our own cells. All of life is connected. And the basis for this kinship is evolution, the process that has transformed life on Earth from its earliest beginnings to the extensive diversity we see today. Evolution is the theme that unifies all of biology.

The Darwinian View of Life

The evolutionary view of life came into focus in 1859 when British biologist Charles Darwin published *The Origin of Species* (**Figure 1.12**). His book developed two main points. First, Darwin marshaled the available evidence in support of the evolutionary view that species living today descended from ancestral species (we'll examine some of the evidence for evolution in Chapter 13). Darwin called this process "descent with modification." It is an insightful euphemism for "evolution," as it captures the duality of life's unity (descent) and diversity (modification). In the Darwinian view, for example, the diversity of bears is based on different modifications of a common ancestor from which all bears descended. As the second main point in *The Origin of Species*, Darwin proposed a mechanism for descent with modification. He called this process natural selection.

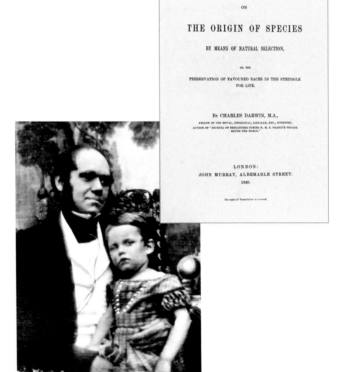

Figure 1.12 Charles Darwin (1809–1882).
Darwin and his son William posed for this photograph in 1842. Darwin was already one of Britain's most renowned biologists, but it was his publication of *The Origin of Species* 17 years later that guaranteed his immortality as the most influential scientist in the development of modern biology. He is buried next to Isaac Newton in London's Westminster Abbey.

Natural Selection

As you'll learn in Chapter 13, Charles Darwin gathered important evidence for his theories during an around-the-world voyage. He was particularly struck by the diversity of animals on the Galápagos Islands, off the coast of South America. Darwin regarded adaptation to the environment and the origin of new species as closely related processes. If some geographic barrier—an ocean separating islands, for instance—isolated two populations of a single species, the populations could diverge more and more in appearance as each adapted to local environmental conditions. Over many generations, the two populations could become dissimilar enough to be designated separate species—14 species in the case of the birds called Galápagos finches, which have beak shapes and other refinements that are adapted to their environments (**Figure 1.13**). Darwin

Activity 1E on the Web & CD
Join Darwin on his voyage to the Galápagos Islands.

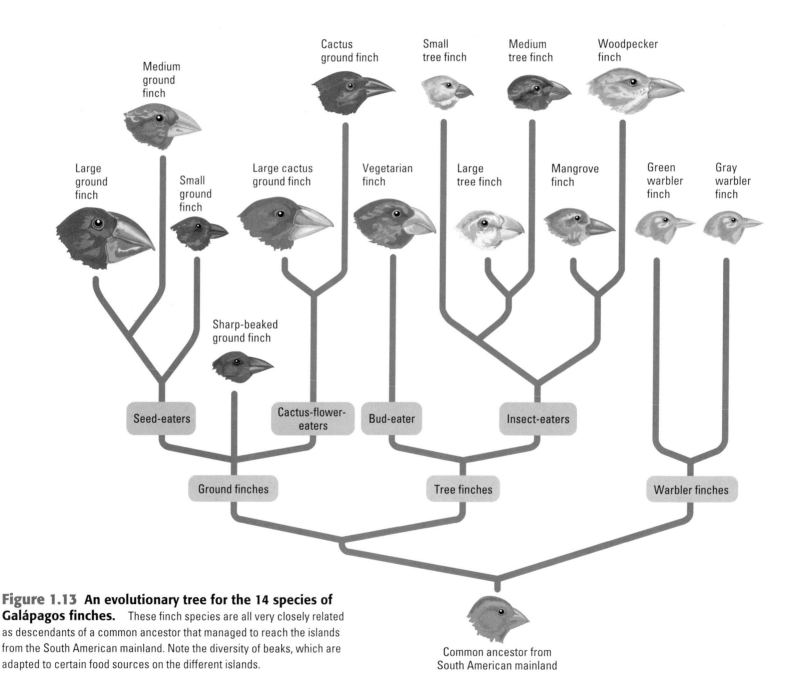

Figure 1.13 An evolutionary tree for the 14 species of Galápagos finches.   These finch species are all very closely related as descendants of a common ancestor that managed to reach the islands from the South American mainland. Note the diversity of beaks, which are adapted to certain food sources on the different islands.

realized that explaining such adaptation was the key to understanding descent with modification, or evolution. This focus on adaptation helped Darwin envision his concept of natural selection as the mechanism of evolution.

Darwin's Inescapable Conclusion   Darwin synthesized the concept of natural selection from two observations that by themselves were neither profound nor original. Others had the pieces of the puzzle, but Darwin could see how they fit together. As the late evolutionary biologist Stephen Jay Gould put it, Darwin based natural selection on "two undeniable facts and an inescapable conclusion." Let's look at his logic:

Fact 1: Overproduction and struggle for existence   Any population of a species has the potential to produce far more offspring than the environment can possibly support with resources such as food and shelter. (A population consists of individuals of the same species living in a particular location.) This overproduction leads to a struggle for existence among the varying individuals of a population.

Fact 2: Individual variation   Individuals in a population of any species vary in many heritable traits. No two individuals in a population are exactly alike. You know this variation to be true of human populations; careful observers find variation in populations of all species.

The inescapable conclusion: Unequal reproductive success   In the struggle for existence, those individuals with traits best suited to the local environment will, on average, have the greatest reproductive success: They will leave the greatest number of surviving, fertile offspring. Therefore, the very traits that enhance survival and reproductive success will be disproportionately represented in succeeding generations of a population.

❶ Population with varied inherited traits

❷ Elimination of individuals with certain traits

Case Study in the Process of Science on the Web & CD Conduct your own study of how environmental changes affect a population of leafhoppers.

It is this unequal reproductive success that Darwin called natural selection. And the product of natural selection is adaptation, the accumulation of favorable variations in a population over time. Examples are beaks well equipped for available food sources and (as we'll see later in this chapter) markings that reduce predation. As the process of adaptation, natural selection is also the mechanism of evolution (**Figure 1.14**).

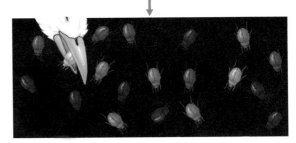
❸ Reproduction of survivors

Observing Artificial Selection   Darwin found convincing evidence for the power of unequal reproduction in examples of artificial selection, the selective breeding of domesticated plants and animals by humans. We humans have been modifying other species for centuries by selecting breeding stock

Figure 1.14 Natural selection.
❶ This imaginary beetle population has colonized a locale where the soil has been blackened by a recent brush fire. Initially, the population varies extensively in the coloration of individuals, from very light gray to charcoal. ❷ For hungry birds that prey on the beetles, it is easiest to spot the beetles that are lightest in color. ❸ The selective predation favors the survival and reproductive success of the darker beetles. Thus, genes for dark color are passed along to future generations in greater frequency than genes for light color. ❹ Generation after generation, the beetle population adapts to its environment through natural selection.

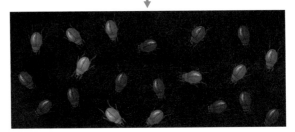
❹ Increasing frequency of traits that enhance survival and reproductive success

(a)

(b)

Figure 1.15 Examples of artificial selection.
(a) These vegetables (and more) all have a common ancestor in one species of wild mustard (inset). Humans have customized the plant by selecting different parts of the plant to accentuate as food.
(b) All modern breeds of domesticated dogs can be traced back to one population of wolves. The tremendous physical variety of modern dogs reflects thousands of years of artificial selection by humans.

with certain traits. The plants and animals we grow for food bear little resemblance to their wild ancestors (**Figure 1.15a**). The power of selective breeding is especially apparent in our pets, which have been bred more for fancy than for utility. For example, people in different cultures have customized hundreds of dog breeds as different as basset hounds and St. Bernards, all descendant from one ancestral population of wolves (**Figure 1.15b**). Darwin could see that in artificial selection, humans were substituting for the environment in screening the heritable traits of populations.

Observing Natural Selection If artificial selection could achieve so much change so rapidly, Darwin reasoned, then natural selection should be capable of considerable adaptation of species over hundreds or thousands of generations. There are many examples of natural selection in action. One example that may directly affect your life is how resistance to antibiotics by disease-causing bacteria evolves by natural selection.

Antibiotics are drugs that help cure certain infections by impairing bacteria. Antibiotics have saved millions of humans, including one of the authors of this book, who was rescued at age 13 from bacterial meningitis by injections of the antibiotic penicillin. But there's a dark side to the widespread use of antibiotics. It has caused the evolution of antibiotic-resistant populations of the very bacteria the drugs are meant to kill.

When taken, an antibiotic will probably kill most, but not all, of the infecting bacteria. An antibiotic thus selects among the varying bacteria of a population for those mutant individuals that can survive the drug. In some cases, the mechanism of resistance depends on the ability of the bacteria to destroy the antibiotic—or even use the drug as food! While the drug kills most of the bacteria, those few bacteria that are resistant multiply and quickly become the norm in the population rather than the exceptions.

The evolution of antibiotic-resistant bacteria is a huge problem in public health. For example, in New York City, there are now some strains of the tuberculosis-causing bacteria that are resistant to all three of the antibiotics that are used to treat the disease (**Figure 1.16**). Unfortunate people infected with one of these resistant strains have no better chance of surviving than did tuberculosis patients a century ago.

Figure 1.16 Natural selection in action.
Antibiotic-resistant forms of tuberculosis-causing bacteria (inset) have made the disease a threat again in the U.S. This colorized X-ray shows the lungs of a tuberculosis patient. The infection is shown in red.

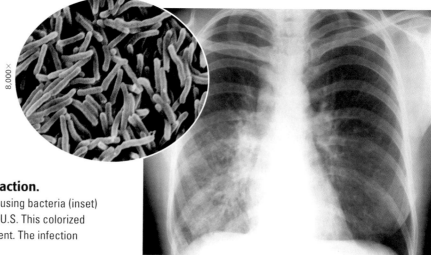

8,000×

Activity 1F on
the Web & CD
Investigate case
studies of antibiotic
resistance.

The evolution of antibiotic resistance is influencing the prescription of antibiotics by many physicians. Doctors who understand that the abuse of these drugs is speeding the evolution of resistant bacteria are less likely to prescribe the antibiotics needlessly—for instance, for a patient complaining of a common cold or flu, which are viral diseases against which the antibacterial drugs are powerless.

It is important to note that adaptation of antibiotic-resistant bacteria does not mean that the drugs created the favorable characteristics. Instead, the environment screened the heritable variations that existed among individuals of a population and favored the ones best suited to present conditions. In Chapter 13, you will learn more about how natural selection works and see several other examples of how natural selection affects your life.

Activity 1G on
the Web & CD
See a video on how
natural selection has
shaped the appear-
ance of a sea horse.

Darwin's publication of *The Origin of Species* fueled an explosion in biological research and knowledge that continues today. Over the past 140 years, a tremendous amount of research evidence has supported Darwin's theory of evolution, making it one of biology's best demonstrated, most comprehensive, and longest lasting theories. In every chapter of *Essential Biology,* we will demonstrate the connection to evolution—the unifying theme of biology.

The Process of Science

Darwin helped make biology a science by seeking natural rather than supernatural causes for the unity and diversity of life. Remember that biology is defined as the *scientific* study of life. But what *is* science? And how do we tell the difference between science and other ways we try to make sense of nature?

The word *science* is derived from a Latin verb meaning "to know." Science is a way of knowing. It developed from our curiosity about ourselves and the world around us. This basic human drive to understand is manifest in two main scientific approaches: discovery science and hypothesis-driven science. Most scientists practice a combination of these two forms of inquiry.

Discovery Science

Science seeks natural causes for natural phenomena. This limits the scope of science to the study of structures and processes that we can observe and measure, either directly or indirectly with the help of tools, such as microscopes, that extend our senses. This dependence on observations that other people can confirm demystifies nature and distinguishes science from belief in the supernatural. Science can neither prove nor disprove that angels, ghosts, deities, or spirits, both benevolent and evil, cause storms, rainbows, illnesses, and cures, for such explanations are outside the bounds of science.

Verifiable observations and measurements are the data of discovery science (**Figure 1.17**). In our quest to describe nature accurately, we discover its structure. In biology, discovery science enables us to describe life at its many levels, from ecosystems down to cells and molecules. Darwin's careful description of the diverse plants and animals he collected in South America is an example of discovery science, sometimes called descriptive science. A more recent example is the sequencing of the human genome;

Figure 1.17 Careful observation and measurement provide the raw data for science.
Cornell University's Eloy Rodriguez collects a sample of chemicals extracted from a plant. Analysis of the sample will enable Dr. Rodriguez to determine the types and amounts of different chemical substances in the plant extract.

it's not really a set of experiments, but a detailed dissection and description of the genetic material.

Discovery science can lead to important conclusions based on a type of logic called inductive reasoning. An inductive conclusion is a generalization that summarizes many concurrent observations. "All organisms are made of cells" is an example. That induction was based on two centuries of biologists discovering cells in every biological specimen they observed with microscopes. The careful observations of discovery science and the inductive conclusions they sometimes produce are fundamental to our understanding of nature.

Hypothesis-Driven Science

The observations of discovery science engage inquiring minds to ask questions and seek explanations. Ideally, such investigation consists of what is called the scientific method. As a formal process of inquiry, the scientific method consists of a series of steps, but few scientists adhere rigidly to this prescription. While it would be misleading to reduce science to a stereotyped method, we *can* identify the key element of the method that drives most modern science. It is called hypothetico-deductive reasoning or, more simply, hypothesis-driven science (**Figure 1.18**).

A hypothesis is a tentative answer to some question—an explanation on trial. It is usually an educated guess. We all use hypotheses in solving everyday problems. Let's say, for example, that your flashlight fails during a camp-out. That's an observation. The question is obvious: Why doesn't the flashlight work? A reasonable hypothesis based on past experience is that the batteries in the flashlight are dead.

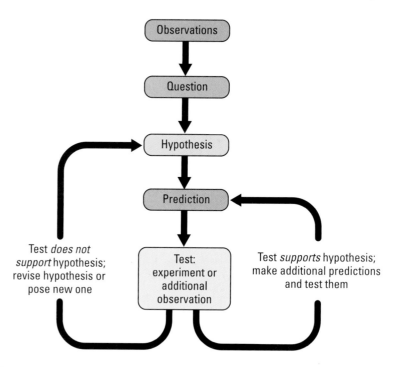

Figure 1.18 The scientific method.
Although science rarely conforms exactly to this protocol of steps, inquiry usually involves posing and testing hypotheses.

The *deductive* in hypothetico-deductive reasoning refers to the use of deductive logic to test hypotheses. Deduction contrasts with induction, which, remember, is reasoning from a set of specific observations to reach a general conclusion. In deduction, the reasoning flows in the opposite direction, from the general to the specific. From general premises, we extrapolate to the specific results we should expect if the premises are true. If all organisms are made of cells (premise 1), and humans are organisms (premise 2), then humans are composed of cells (deductive prediction about a specific case).

In the process of science, the deduction usually takes the form of predictions about what outcomes of experiments or observations we should expect *if* a particular hypothesis (premise) is correct. We then test the hypothesis by performing the experiment to see whether or not the results are as predicted. This deductive testing takes the form of "*If . . . then*" logic:

Observation: My flashlight doesn't work.

Question: What's wrong with my flashlight?

Hypothesis: The flashlight's batteries are dead.

Prediction: If this hypothesis is correct,

Experiment: and I replace the batteries with new ones,

Predicted result: then the flashlight should work.

Let's say the flashlight *still* doesn't work. We can test an alternative hypothesis if new flashlight bulbs are available (**Figure 1.19**). We could also blame the dead flashlight on campground ghosts playing tricks, but that hypothesis is untestable and therefore outside the realm of science.

A Case Study in the Process of Science

One way to learn more about how hypothesis-based science works is to examine a case study. In contrast to a made-up example, such as the flashlight problem, a case study is an in-depth examination of something that actually happened. The rest of this section is a case study of an inquiry about what harmless snakes might gain by imitating poisonous ones.

The story begins with some key observations. Many poisonous animals are brightly colored, with distinctive patterns in some species. This appearance is called warning coloration because it marks the animal as dangerous to potential predators. But there are also mimics. These imposters look like a poisonous species but are really harmless to predators. The question that follows from these observations is: What is the function of such mimicry? It is a reasonable hypothesis that such deception is an evolutionary adaptation that reduces the harmless animal's risk of being eaten.

In 2001, a team of biologists designed a simple but clever set of experiments to test the hypothesis that mimics benefit because predators confuse them with the actual harmful species. Researchers David and Karin Pfennig, along with one of their college students, tested this hypothesis by studying

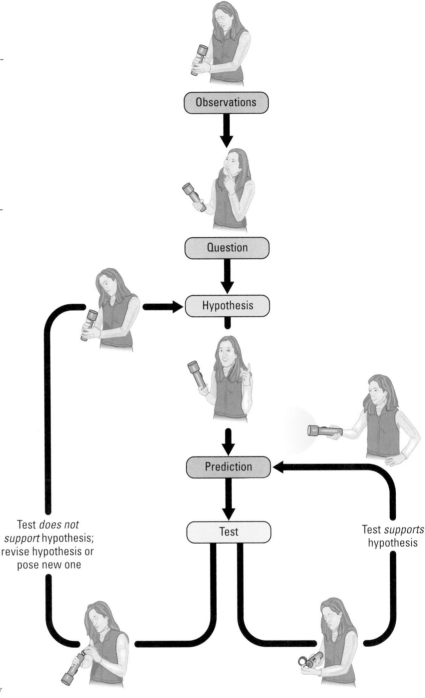

Test *does not support* hypothesis; revise hypothesis or pose new one

Test *supports* hypothesis

Figure 1.19 Applying the scientific method to a campground problem.

Figure 1.20 Mimicry in snakes.
(a) The scarlet kingsnake is a nonpoisonous species that mimics (b) the dangerous eastern coral snake.

(a) Scarlet kingsnake (nonpoisonous)

(b) Eastern coral snake (poisonous)

mimicry in snakes that live in North and South Carolina. A poisonous snake called the eastern coral snake is marked by rings of red, yellow, and black. Predators rarely attack these snakes. A nonpoisonous snake named the scarlet kingsnake mimics the ringed coloration of the coral snake (**Figure 1.20**).

What is the explanation for this case of look-alike snakes? According to the mimicry hypothesis, the coral-snakelike appearance of kingsnakes repels predators. The hypothesis predicts that predators will attack snakes with the bright rings of red, yellow, and black less frequently than they will attack snakes lacking such warning coloration. To test this prediction, the researchers made hundreds of artificial snakes out of wire and a claylike substance called plasticine. They made two types of artificial snakes: those with plain brown coloration and those with the red, black, and yellow ring pattern of scarlet kingsnakes.

The researchers placed equal numbers of the two types of artificial snakes in various field sites throughout North and South Carolina. After four weeks, the team retrieved the artificial snakes and counted how many had been attacked by looking for bite or claw marks. The bar graph in **Figure 1.21** summarizes the results of the artificial snake experiment. The data show that the plain brown artificial snakes were attacked much more frequently than the ones with colored rings.

This is an example of a controlled experiment. Such an experiment is designed to compare an experimental group (the artificial snakes with the colored rings, in this case) with a control group (plain brown artificial snakes). Ideally, a control group and an experimental group differ only in the one variable an experiment is designed to test—in our example, the effect of the snakes' coloration on the behavior of their predators. During the experiment, conditions such as light, temperature, and appetite of the predators varied, but both kinds of snakes were subject to the same variations. Thus, the control group cancelled out the effects of all variables other than the one being tested. This enabled researchers to compare the two groups and draw conclusions about the effect of the colored rings on predator behavior.

In this research, we can recognize the logic of the hypothetico-deductive process:

Observation: Both poisonous coral snakes and nonpoisonous kingsnakes have red, yellow, and black rings.

Question: What is the function of the kingsnakes' mimicry of coral snakes?

Hypothesis: Mimicry of coral snakes helps protect the kingsnakes from predators.

Figure 1.21 Effect of artificial snake coloration on attacks by predators.
Results of mimicry experiments using artificial snakes show a dramatic difference in the frequency of attacks on plain brown snakes compared to the snakes with colored rings.

Prediction: *If* predators confuse kingsnakes with coral snakes,

Experiment: and the number of attacks on ringed versus brown artificial snakes are compared,

Predicted result: *then* predators should attack fewer ringed artificial snakes than brown artificial snakes.

The experimental tests supported the hypothesis in this case.

Case Study in the Process of Science on the Web & CD Practice developing hypotheses and designing experiments on acid rain.

Our case study reinforces the important point that scientists must test their hypotheses. Without such testing, ideas about nature, such as speculations on the function of mimicry, are "just so" stories. And explaining that something is true just because "it's so" is not very convincing.

The Culture of Science

It is not unusual for several scientists to ask the same questions. Such convergence contributes to the progressive and self-correcting qualities of science. Scientists build on what has been learned from earlier research, and they pay close attention to contemporary scientists working on the same problem. Scientists share information through publications, seminars, meetings, and personal communication. The Internet has added a new medium for this exchange of ideas and data.

Both cooperation and competition characterize the scientific culture (**Figure 1.22**). Scientists working in the same research field subject one another's work to careful scrutiny. It is common for scientists to check on the conclusions of others by attempting to repeat experiments. This obsession with evidence and confirmation helps characterize the scientific style of inquiry. Scientists are generally skeptics.

We have seen that science has two key features that distinguish it from other styles of inquiry: (1) a dependence on observations and measurements that others can verify and (2) the requirement that ideas (hypotheses) are testable by experiments that others can repeat.

Theories in Science

Many people associate facts with science, but accumulating facts is not what science is primarily about. A telephone book is an impressive catalog of factual information, but it has little to do with science. It is true that facts, in the form of verifiable observations and repeatable experimental results, are the prerequisites of science. What really advances science, however, is some new theory that ties together a number of observations that previously seemed unrelated. The cornerstones of science are the explanations that apply to the greatest variety of phenomena. People like Newton, Darwin, and Einstein stand out in the history of science not because they discovered a great many facts, but because their theories had such broad explanatory power.

What is a scientific theory, and how is it different from a hypothesis? A theory is much broader in scope than a hypothesis. This is a hypothesis: "Mimicking coral snakes is an adaptation that helps protect some non-poisonous snakes from predation." But *this* is a theory: "Adaptations evolve by natural selection."

Because theories are so comprehensive, they only become widely accepted in science if they are supported by an accumulation of extensive

Figure 1.22 Science as a social process. Here, New York University plant biologist Gloria Coruzzi (left) mentors one of her students in the methods of molecular biology.

and varied evidence. This use of the term *theory* in science for a comprehensive explanation supported by abundant evidence contrasts with our everyday usage, which equates theories more with speculations or hypotheses. Natural selection qualifies as a scientific theory because of its broad application and because it has been validated by a continuum of observations and experiments. (For an example of this, you can jump ahead to the Evolution Connection at the end of Chapter 18.)

Scientific theories are not the only way of "knowing nature," of course. A comparative religion course would be a good place to learn about the diverse legends that tell of a supernatural creation of Earth and its life. Science and religion are two very different ways of trying to make sense of nature. Art is still another way. A broad education should include exposure to these different ways of viewing nature. Each of us synthesizes our worldview by integrating our life experiences and multidisciplinary education. As a science textbook and part of that broad education, *Essential Biology* showcases life in the scientific context of evolution, the one theme that continues to hold all of biology together, no matter how big and complex the subject becomes.

Evolution Connection on the Web
Examine issues surrounding the teaching of evolution.

Science, Technology, and Society

Science and technology are interdependent. New technologies, such as more powerful microscopes and computers, advance science. And scientific discoveries can lead to new technologies. In most cases, technology applies scientific discoveries to the development of new goods and services. For example, it was 50 years ago that two scientists, James Watson and Francis Crick, discovered the structure of DNA through the process of science. Their discovery eventually led to a variety of DNA technologies, including the genetic engineering of microorganisms to mass-produce human insulin and the use of DNA fingerprinting for investigating crimes (**Figure 1.23**). Perhaps Watson and Crick envisioned that their discovery would someday inform new technologies, but that probably did not motivate their research, nor could they have predicted exactly what the applications would be. The direction technology takes depends less on the curiosity that drives basic science than it does on the current needs of humans and the changing climate of culture.

Technology has improved our standard of living in many ways, but it is a double-edged sword. Technology that keeps people healthier has enabled the population to grow more than tenfold in the past three centuries, to double to 6 billion in just the past 40 years. The environmental consequences are sometimes devastating. Acid rain, deforestation, global warming, nuclear accidents, toxic wastes, and extinction of species are just a few of the repercussions of more and more people wielding more and more technology. Science can help us identify such problems and provide insight about what course of action may prevent further damage. But solutions to these problems have as much to do with politics, economics, culture, and the values of societies as with science and technology. Now that science and technology have become such powerful functions of society, every thoughtful citizen has a responsibility to develop a reasonable amount of scientific and technological literacy.

Activity 1H on the Web & CD
Weigh the pros and cons of the pesticide DDT.

We wrote this book to help students who are not biology majors develop an appreciation for the science of life and apply that understanding to evaluate such social issues as health and environmental quality. We believe that such a biological perspective is essential for any educated person, which is why we named our book *Essential Biology*. We hope it serves you well.

Figure 1.23 DNA technology and the law.
Forensic technicians can use traces of DNA extracted from a blood sample or other body tissue collected at a crime scene to produce a molecular "fingerprint." The stained bands visible in this photograph represent fragments of DNA, and the pattern of bands varies from person to person. The legal applications of DNA technology have become very public in the past decade. It is just one example of biology's prominent role in society today.

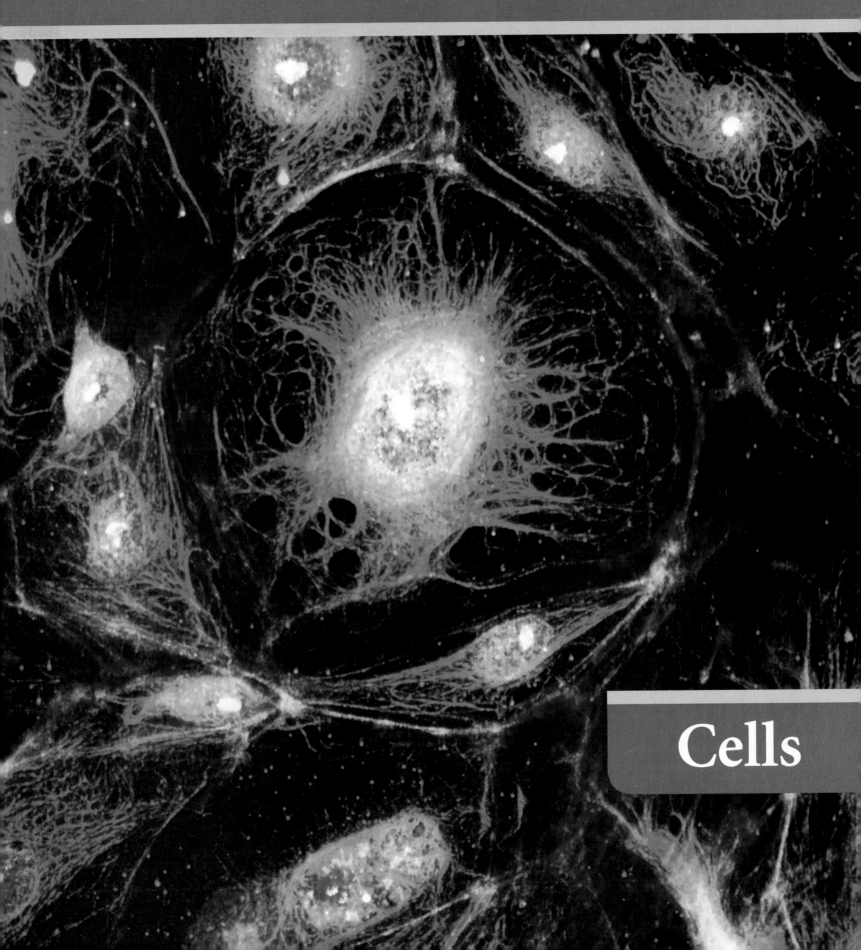

Cells

CHAPTER 2

Essential Chemistry for Biology

**Biology and Society:
Fluoride in the Water** 21

**Tracing Life Down to the
Chemical Level** 21

Some Basic Chemistry 22
Matter: Elements and Compounds
Atoms
Chemical Bonding and Molecules
Chemical Reactions

Water and Life 28
The Structure of Water
Water's Life-Supporting Properties
Acids, Bases, and pH

**Evolution Connection:
Earth Before Life** 32

Rain in the eastern United States can be more acidic than vinegar.

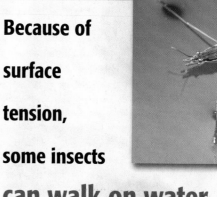

Because of surface tension, some insects can walk on water.

About 65% of your weight is oxygen atoms.

The iron in a multivitamin pill is the same element as the iron in a train or a ship.

Fluoride in the Water

One of the first things that people notice about you is your smile. But you might not realize that healthy teeth are much more common today than they were several decades ago. The main reason for the decrease in tooth decay is the addition of fluoride-containing chemicals to drinking water and dental products.

Although your teeth may seem lifeless, they are actually a dynamic, living system. Oral bacteria reside in plaque, a sticky combination of food, saliva, and dead cells that coats your teeth. As these bacteria grow, they release acids that can dissolve the dental surface, causing pain and tooth loss. Fluoride prevents cavities in two ways: by affecting the metabolism of the bacteria and by promoting the replacement of lost minerals on the tooth surface.

Fluoride is a form of the element fluorine and a common ingredient in Earth's crust. Fluoride dissolves in water as it bubbles through the ground. Frequent exposure to small amounts of fluoride is the best way to prevent tooth decay. In some areas, fluoride occurs naturally in the drinking water in sufficient concentrations to help maintain healthy teeth. But in many other areas, the water supply is fluoridated—supplemented with small amounts of fluoride as part of the municipal water treatment process.

Biology and Society on the Web
Learn what the American Dental Association has to say about the importance of fluoride.

The widespread addition of fluoride to drinking water, products containing drinking water, and toothpaste has been a major factor in the decline in tooth decay in the United States and other developed countries. This is one of many ways that society uses chemical products to improve human health. Other examples include adding iodine to salt ("iodized" salt) and adding vitamins to flour ("enriched" flour).

The use of fluoride in maintaining proper oral hygiene helps illustrate an important point: Living organisms are, at their most basic level, chemical systems. Many questions about life reduce to questions about chemicals and their interactions, making knowledge of chemistry essential to understanding biology. In this chapter, you will learn some essential chemistry that you'll be able to apply throughout your study of life.

Tracing Life Down to the Chemical Level

In Chapter 1, you learned that biology includes the study of life at many levels. In Figure 1.2, you traced some of these levels from a satellite view of Earth (the biosphere) down to the molecules that make up cells (the chemical level). This chapter focuses on life at the chemical level—a world that is mostly invisible to us because it occurs at the microscopic level of cells. So let's start with a macroscopic scene, such as the African savanna in **Figure 2.1**, and work our way down to smaller and smaller levels of biological organization.

The savanna is an example of an ecosystem. An **ecosystem** consists of all organisms living in a particular area, as well as the nonliving, physical components of the environment that affect the organisms, such as water, air, soil, and sunlight. All the organisms in the savanna (zebras, insects, grasses, bacteria, and so on) are collectively called a **community.** Within

Ecosystem
African savanna

Community
All organisms in savanna

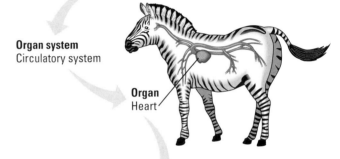

Population
Herd of zebras

Organism
Zebra

Organ system
Circulatory system

Organ
Heart

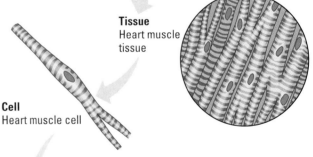

Tissue
Heart muscle tissue

Cell
Heart muscle cell

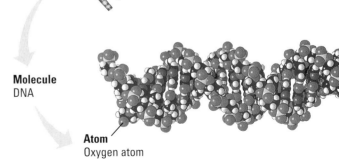

Molecule
DNA

Atom
Oxygen atom

Figure 2.1
Tracing life down to the chemical level: the hierarchy of biological organization.

communities are various **populations,** groups of interacting individuals of one species, such as a herd of zebras. Below population in the hierarchy is the **organism,** an individual living thing, such as a zebra.

Life's hierarchy continues to unfold within an individual organism. The zebra's body consists of several **organ systems,** such as the digestive and circulatory systems. Each organ system consists of **organs,** such as the heart and blood vessels of the circulatory system. As we continue downward through life's hierarchy, each organ is made up of several different **tissues,** each of which consists of a group of similar cells performing a specific function. A **cell** is life's basic unit of structure and function in all living things.

Activity 2A on the Web & CD Play solitaire while reviewing the hierarchical organization of life.

Finally, we reach the chemical level in the hierarchy. Each cell consists of an enormous number of chemicals that cooperate to give the cell the properties we recognize as life. The chemical illustrated in Figure 2.1 is DNA, the chemical of inheritance, the substance of genes. DNA is an example of a molecule, a cluster of even smaller chemical units called atoms. In the computer graphic in Figure 2.1, which illustrates a tiny segment of one DNA molecule, each sphere represents an atom.

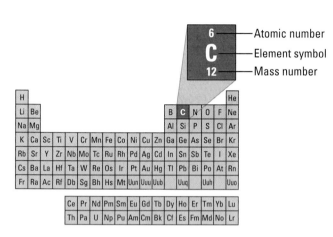

Figure 2.2 Abbreviated periodic table of the elements. The periodic table lists all of the chemical elements, both natural and human-made. In the full periodic table, each entry contains the elemental symbol in the center, with the atomic number above and the average atomic mass below, as shown here for carbon (C).

CHECKPOINT

Put these levels of biological organization in order, starting with the level that includes all others in the list: organism, molecule, cell, ecosystem, population, atom, tissue, organ, community, organ system.

Answer: ecosystem → community → population → organism → organ system → organ → tissue → cell → molecule → atom

---

Some Basic Chemistry

Take any biological system apart and you eventually end up at the chemical level. Beginning at this basic biological level, let's explore the chemistry of life. Imagination will help. Try picturing yourself small enough to actually crawl into molecules and climb around on their atoms.

Matter: Elements and Compounds

Humans and other organisms and everything around them are all made of matter, the physical "stuff" of the universe. Defined more formally, **matter** is anything that occupies space and has mass. Matter is found on Earth in three physical states: solid, liquid, and gas.

Matter is composed of chemical **elements,** substances that cannot be broken down into other substances. There are 92 naturally occurring elements on Earth; examples are carbon, oxygen, gold, and fluorine. Each element has a symbol, the first letter or two of its English, Latin, or German name. For instance, the symbol for gold, Au, is from the Latin word *aurum*. All the elements are listed in the periodic table of the elements, a familiar fixture in any chemistry or biology lab (**Figure 2.2**).

Of the 92 naturally occurring elements, 25 are essential to life. Four of these elements—oxygen (O), carbon (C), hydrogen (H), and nitrogen (N)—make up about 96% of the weight of the human body, as well as most other living matter (**Figure 2.3**). Much of the remaining 4% is accounted

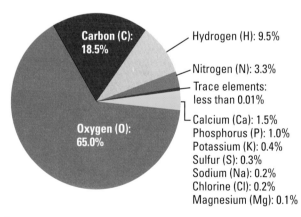

Figure 2.3 Chemical composition of the human body by weight. Note that oxygen (O), carbon (C), hydrogen (H), and nitrogen (N) make up 96.3% of the weight of the human body. (These percentages include water, H_2O.) The trace elements, which make up less than 0.01%, are boron (B), chromium (Cr), cobalt (Co), copper (Cu), fluorine (F), iodine (I), iron (Fe), manganese (Mn), molybdenum (Mo), selenium (Se), silicon (Si), tin (Sn), vanadium (V), and zinc (Zn).

Case Study in the
Process of Science
on the Web & CD
Examine rocks from
space to determine
if they contain
evidence of
extraterrestrial life.

for by 7 elements, most of which are probably familiar to you, such as calcium (Ca) and phosphorus (P). Less than 0.01% of your weight is made up of the 14 **trace elements,** listed in the legend for Figure 2.3. Trace elements are required in only very small amounts, but you cannot live without them. The average human, for example, needs only a tiny bit of iodine, about 0.15 milligram (mg) each day. An iodine deficiency in the diet, however, prevents normal functioning of the thyroid gland and results in goiter, an abnormal enlargement of the thyroid gland (**Figure 2.4**). Fluorine, discussed at the beginning of the chapter, is another trace element.

Elements can combine to form **compounds,** substances that contain two or more elements in a fixed ratio. In everyday life, compounds are much more common than pure elements. Familiar examples are relatively simple compounds such as table salt and water. Table salt is sodium chloride, NaCl, consisting of equal parts of the elements sodium (Na) and chlorine (Cl). A molecule of water, H_2O, has two atoms of hydrogen and one atom of oxygen. Most of the compounds in living organisms contain several different elements. The DNA in Figure 2.1, for example, contains carbon (shown as dark blue spheres), nitrogen (light blue), oxygen (red), hydrogen (gray), and phosphorus (yellow).

Atoms

Each element consists of one kind of atom, which is different from the atoms of other elements. An **atom,** named from a Greek word meaning "indivisible," is the smallest unit of matter that still retains the properties of an element. In other words, the smallest amount of the element carbon is one carbon atom. And that's a very small "piece" of carbon. It would take about a million carbon atoms to stretch across the period at the end of this sentence.

The Structure of Atoms Atoms are composed of subatomic particles, of which the three most important are protons, electrons, and neutrons. A **proton** is a subatomic particle with a single unit of positive electrical charge (+). An **electron** is a subatomic particle with a single unit of negative electrical charge (−). A **neutron** is electrically neutral (has no electrical charge).

Let's look at the structure of an atom of the element helium (He), the "lighter-than-air" gas that can be used to make balloons rise. Each atom of helium has 2 neutrons and 2 protons tightly packed into the **nucleus,** the atom's central core (**Figure 2.5**). Two electrons orbit the nucleus at nearly the speed of light. The attraction between the negatively charged electrons and the positively charged protons keeps the electrons in orbit. When an atom has an equal number of protons and electrons (as helium does), its net electrical charge is zero and so the atom is neutral.

Elements differ in the number of subatomic particles in their atoms. The number of protons in an atom, called the **atomic number,** determines which element it is. An atom's **mass number** is the sum of the number of protons and neutrons in its nucleus. Both the atomic number and the mass number can be read from the periodic table (see Figure 2.2). For helium, the atomic number is 2 and the mass number is 4 (2 protons + 2 neutrons).

Activity 2B on
the Web & CD
Review the structure
of the atom.

Mass is a measure of the amount of matter in an object. A proton and a neutron have nearly identical mass, called 1 atomic mass unit (amu) for convenience. An electron has very little mass—only about 1/2,000 the mass of a proton.

(a)   **(b)**

Figure 2.4 Why is salt "iodized"?

(a) Goiter, an enlargement of the thyroid gland, shown here in a Burmese woman, can occur when a person's diet does not include enough iodine, a trace element. **(b)** Goiter has been reduced in many nations by adding iodine to table salt (iodized salt). Unfortunately, goiter still affects many thousands of people in developing countries.

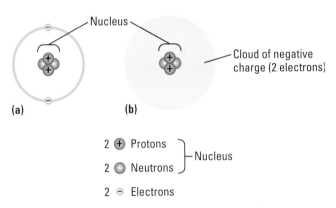

Figure 2.5 Two simplified models of a helium atom.

Both models are highly diagrammatic, and we use them only to show the basic components of atoms. The atomic nucleus consists of tightly packed neutrons and protons, 2 each in the case of helium. The electrons orbit the nucleus, held there by the electrical attraction between the negatively charged electrons and the positively charged protons in the nucleus. **(a)** This model shows the number of electrons in the atom. **(b)** This model, slightly more realistic, shows the electrons as a spherical cloud of negative charge surrounding the nucleus. Neither model is drawn to scale. In real atoms, the electron cloud is much bigger compared with the nucleus. If the electron cloud was the size of a football stadium, the nucleus would only be the size of a fly on the field.

Table 2.1	Isotopes of Carbon		
	Carbon-12	**Carbon-13**	**Carbon-14**
Protons	6 ⎫ mass	6 ⎫ mass	6 ⎫ mass
	⎬ number	⎬ number	⎬ number
Neutrons	6 ⎭ 12	7 ⎭ 13	8 ⎭ 14
Electrons	6	6	6

Isotopes Some elements have alternate mass forms called isotopes. The different **isotopes** of an element have the same numbers of protons and electrons but different numbers of neutrons. **Table 2.1** shows the numbers of subatomic particles in the three isotopes of carbon. Carbon-12 (abbreviated ^{12}C), with 6 neutrons and 6 protons (mass number 12), makes up about 99% of all naturally occurring carbon. Most of the other 1% consists of carbon-13, with 7 neutrons and 6 protons (mass number 13). A third isotope, carbon-14, with 8 neutrons and 6 protons (mass number 14), occurs in minute quantities. Notice that all three isotopes have 6 protons—otherwise, they would not be carbon. Both carbon-12 and carbon-13 are stable isotopes, meaning their nuclei remain intact more or less forever. The isotope carbon-14, on the other hand, is unstable, or radioactive. A **radioactive isotope** is one in which the nucleus decays, giving off particles and energy.

Radioactive isotopes have many uses in biological research and medicine. Living cells cannot distinguish radioactive isotopes from nonradioactive isotopes of the same element. Consequently, organisms take up and use compounds containing radioactive isotopes in the usual way. Once taken up, the location and concentration of radioactive isotopes can be detected because of the radiation they emit. This makes radioactive isotopes useful as tracers—biological spies, in effect—for monitoring the fate of atoms in living organisms.

Medical diagnosis is an important application of radioactive tracers. Certain kidney disorders, for example, can be diagnosed by injecting tiny amounts of radioactive isotopes into a patient's blood and then measuring the amount of radioactive tracer passed in the urine. Health-care professionals also use radioactive tracers in combination with sophisticated imaging instruments such as PET scanners, which can monitor chemical processes as they actually occur in the body (**Figure 2.6**).

Though radioactive isotopes have many beneficial uses, uncontrolled exposure to them can harm living organisms by damaging cellular molecules, especially DNA. The explosion of a nuclear reactor at Chernobyl, Ukraine, in 1986 released large amounts of radioactive isotopes, killing 30 people within a few weeks and exposing thousands to an increased risk of developing cancer. A 1999 nuclear accident at the Tokaimura power plant, about 150 km (95 miles) upwind from Tokyo, Japan, renewed concerns about the biological risks of radioactivity.

Natural sources of radiation can also pose a threat. Radon, a radioactive gas, may cause lung cancer. Radon can contaminate buildings in regions where underlying rocks naturally contain radioactive elements. Homeowners

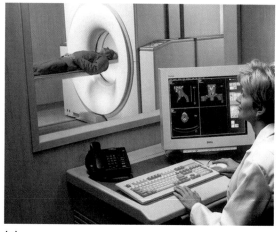

(a)

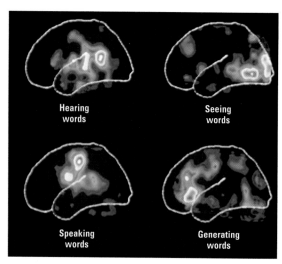

Hearing words

Seeing words

Speaking words

Generating words

(b)

Figure 2.6 PET scan, a medical application of radioactivity.
(a) PET, which stands for positron-emission tomography, detects locations of intense chemical activity in the body. The patient is first injected with a nutrient, such as sugar, that is labeled with a radioactive isotope. The isotope emits subatomic particles called positrons. The positrons collide with electrons made available by chemical reactions in the body. An instrument called the PET scanner detects the energy released by these collisions and maps the chemical "hot spots," regions of an organ that are most chemically active at the time. **(b)** A computer that is linked to the scanner translates the energy data into an anatomical image. The color of the image varies with the amount of the isotope present in an area. For example, these four PET scans of one patient's brain show chemical activity under four different conditions, all related to language. (These are side views, with the front of the brain at the left.) In addition to enabling physicians to diagnose brain disorders, PET scans can also help detect certain heart problems and cancers.

can buy a radon detector or hire a company to test their home to ensure that radon levels are safe.

Electron Arrangement and the Chemical Properties of Atoms

Of the three subatomic particles we've discussed, electrons are the ones that primarily determine how an atom behaves when it encounters other atoms. Electrons vary in the amount of energy they possess. The farther an electron is from the nucleus, the greater its energy. Electrons do not orbit an atom at just any energy level, but only at specific levels called electron shells. Depending on the number of electrons, atoms may have one, two, or more electron shells, with electrons in the outermost shell having the highest energy. Each shell can accommodate up to a specific number of electrons. The innermost shell is full with only 2 electrons, while the second and third shells can each hold up to 8 electrons.

The number of electrons in the outermost shell determines the chemical properties of an atom. Atoms whose outer shells are not full tend to interact with other atoms—that is, to participate in chemical reactions. **Figure 2.7** shows the electron shells of four biologically important elements. Because the outer shells of all four atoms are incomplete, these atoms react readily with other atoms. The hydrogen atom is highly reactive because it has only 1 electron in its single electron shell, which can accommodate 2 electrons. Atoms of carbon, nitrogen, and oxygen are also highly reactive because their outer shells, which can hold 8 electrons, are incomplete. In contrast, the helium atom in Figure 2.5 has a single, first-level shell that is full with 2 electrons. As a result, helium is chemically inert (unreactive).

Activity 2C on the Web & CD
Review electron arrangement.

Activity 2D on the Web & CD
Build your own atom.

Chemical Bonding and Molecules

Chemical reactions enable atoms to give up or acquire electrons and thereby complete their outer shells. Atoms do this by either transferring or sharing outer electrons. These interactions usually result in atoms staying close together, held by attractions called **chemical bonds.**

Ionic Bonds Table salt is an example of how the transfer of electrons can bond atoms together. The two ingredients of table salt are the elements sodium (Na) and chlorine (Cl). When a sodium atom donates 1 electron to a chlorine atom, the electron transfer results in both atoms having full outer shells of electrons (**Figure 2.8**). Before the electron transfer, each of these atoms is electrically neutral. The electron transfer moves one unit of negative charge from sodium to chlorine. The atoms are now **ions,** the term for atoms that are electrically charged as a result of gaining or losing electrons. Negatively charged ions often have names ending in "ide," like "chloride." The loss of an electron gives the sodium ion a charge of $+1$, while chlorine's gain of an electron gives it a charge of -1. The sodium ion (Na^+) and chloride ion (Cl^-) are held together by an **ionic bond,** the attraction between oppositely charged ions. Fluorine in Earth's crust is often found in the form of ionic compounds such as calcium fluoride (CaF_2), the result of bonds between calcium ions (Ca^{2+}) and fluoride ions (F^-).

Activity 2E on the Web & CD
Watch ions and ionic bonds form.

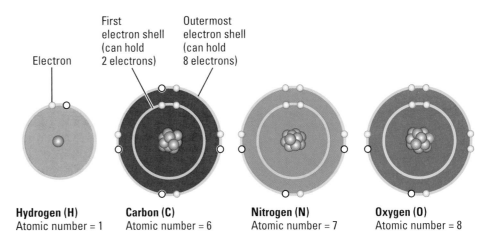

Hydrogen (H) Atomic number = 1
Carbon (C) Atomic number = 6
Nitrogen (N) Atomic number = 7
Oxygen (O) Atomic number = 8

Figure 2.7 Atoms of the four elements most abundant in living matter. All four atoms are chemically reactive because their outermost electron shells are incomplete. The small empty circles in these diagrams represent unfilled "spaces" in the outer electron shells.

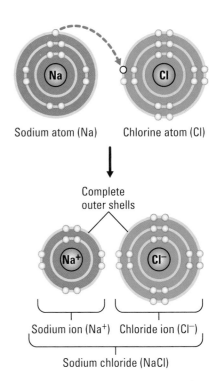

Sodium atom (Na) Chlorine atom (Cl)

Complete outer shells

Sodium ion (Na^+) Chloride ion (Cl^-)

Sodium chloride (NaCl)

Figure 2.8 Electron transfer and ionic bonding. Notice that sodium has only 1 electron in its outer shell, whereas chlorine has 7. When these atoms collide, the chlorine atom strips sodium's outer electron away. In doing so, chlorine fills its outer shell with 8 electrons. Sodium, in losing 1 electron, ends up with only two shells, the outer shell now having a full set of 8 electrons. Because electrons are negatively charged particles, the electron transfer between the two atoms moves one unit of negative charge from sodium to chlorine. The atoms are now ions, which are electrically charged atoms. Two ions with opposite charges attract each other; when the attraction holds them together, it is called an ionic bond.

Covalent Bonds In contrast to the complete transfer of electrons that leads to ionic bonds, a **covalent bond** forms when two atoms share one or more pairs of outer-shell electrons. Atoms held together by covalent bonds form a **molecule.** For example, a covalent bond connects each hydrogen atom to the carbon in the molecule CH_4, a common gas called methane (**Figure 2.9**). You can see that each of the four hydrogen atoms in a molecule of methane shares one pair of electrons with the single carbon atom.

The number of covalent bonds an atom can form is equal to the number of additional electrons needed to fill its outer shell. Note in Figure 2.9 that hydrogen (H) can form one covalent bond, oxygen (O) can form two, and carbon (C) can form four. The single covalent bond in H_2 completes the outer shells of both hydrogen atoms. In contrast, an oxygen atom needs two electrons to complete its outer shell. In an O_2 molecule, the two oxygen atoms share two pairs of electrons, forming a double covalent bond.

Though H_2 and O_2 are molecules, neither qualifies as a compound because these molecules are each composed of only one element. An example of a molecule that is also a compound is methane (CH_4). You can see in Figure 2.9 that each of the four hydrogen atoms in this molecule shares one pair of electrons with the single carbon atom. Note that water, H_2O, is also a compound.

Activity 2F on the Web & CD
See covalent bonds form.

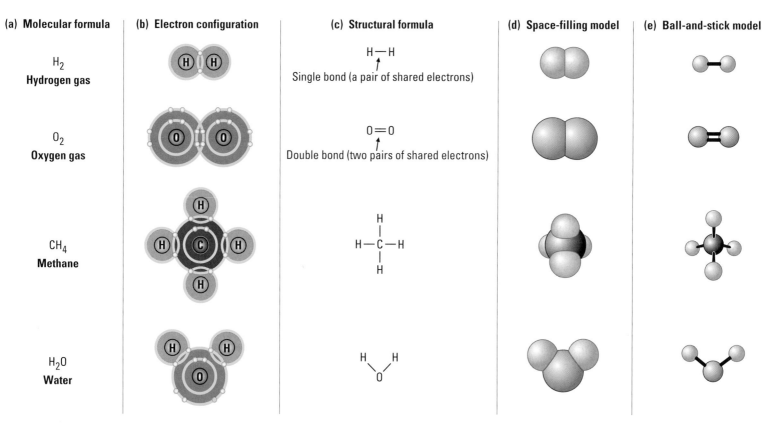

| (a) Molecular formula | (b) Electron configuration | (c) Structural formula | (d) Space-filling model | (e) Ball-and-stick model |

H_2
Hydrogen gas

H—H
Single bond (a pair of shared electrons)

O_2
Oxygen gas

O=O
Double bond (two pairs of shared electrons)

CH_4
Methane

H—C—H (with H above and below)

H_2O
Water

H H / O

Figure 2.9 Alternative ways to represent molecules.
(a) A molecular formula, such as CH_4 or O_2, tells you the number of each kind of atom in a molecule but not how they are attached together. **(b)** You can see from the electron configuration that each atom completes its outer shell by sharing electrons with its covalent partner. **(c)** In a structural formula, each line represents a covalent bond, a pair of shared electrons. Hydrogen always forms one covalent bond; oxygen, a total of two; and carbon, a total of four. The double covalent bond (double lines) in O_2 consists of two pairs of shared electrons. **(d)** A space-filling model, in which the color-coded balls symbolize atoms, shows the shape of a molecule. **(e)** In a ball-and-stick model, the "balls" represent atoms and the "sticks" represent the bonds between the atoms.

Chemical Reactions

The chemistry of life is dynamic. Your cells are constantly rearranging molecules by breaking existing chemical bonds and forming new ones. Such changes in the chemical composition of matter are called **chemical reactions.** A simple example is the reaction between oxygen gas and hydrogen gas that forms water (this is an explosive reaction, which, fortunately, does not occur in your cells):

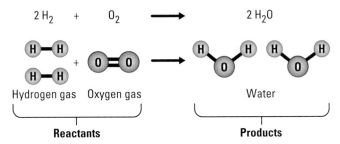

Let's translate the chemical shorthand: Two molecules of hydrogen gas (2 H_2) react with one molecule of oxygen gas (O_2) to form two molecules of water (2 H_2O). The arrow indicates the conversion of the starting materials, the **reactants** (2 H_2 and O_2), to the **products** (2 H_2O).

Notice that the same numbers of hydrogen and oxygen atoms are present in reactants and products, although they are grouped differently. Chemical reactions cannot create or destroy matter, but only rearrange it. These rearrangements usually involve the breaking of chemical bonds in reactants and the forming of new bonds in products. You can see those rearrangements in our ball-and-stick models for water formation; the "sticks" represent bonds between the atoms, which are shown as "balls."

The water molecule we have built here is a good conclusion to this section on basic chemistry. Water is a substance so important in biology that we'll take a closer look at its life-supporting properties in the next section.

CHECKPOINT

1. What four elements are most abundant in living matter?

2. Why is water (H_2O) classified as a compound but oxygen gas (O_2) is not?

3. A nitrogen atom has 7 protons, and the most common isotope of nitrogen has 7 neutrons. A radioactive isotope of nitrogen has 8 neutrons. What are the atomic numbers and mass numbers of the stable and radioactive forms of nitrogen?

4. Why are radioactive isotopes useful as tracers in research on the chemistry of life?

5. Explain what holds together the ions in a crystal of table salt (NaCl).

6. What is chemically nonsensical about this structure?

$$H-C\equiv C-H$$

Answers: 1. Oxygen, carbon, hydrogen, and nitrogen **2.** Because water consists of two different elements **3.** ^{14}N has an atomic number of 7 and a mass number of 14. ^{15}N has an atomic number of 7 and a mass number of 15. **4.** Organisms incorporate radioactive isotopes of an element into their molecules just as they do the nonradioactive isotopes, and researchers can detect the presence of the radioactive isotopes. **5.** Ionic bonds, the electrical attraction between the Na^+ ions and Cl^- ions, hold the ions together. **6.** Each carbon atom has only three covalent bonds instead of the required four.

Figure 2.10 A watery world.
Three-quarters of Earth's surface is submerged in water. Although most of this water is in liquid form, water is also present on Earth as ice and vapor (including clouds). Water is the only common substance that exists in the natural environment in all three physical states of matter: solid, liquid, and gas. These three states of water are visible in this view of Earth from space.

Water and Life

Life on Earth began in water and evolved there for 3 billion years before spreading onto land. Modern life, even land-dwelling life, is still tied to water. You've had personal experience with this dependence on water every time you seek liquids to quench your thirst and replenish your body's water content. Inside your body, your cells are surrounded by a fluid that's composed mostly of water, and your cells themselves range from 70% to 95% in water content.

The abundance of water is a major reason Earth is habitable (**Figure 2.10**). Water is so common that it's easy to overlook its extraordinary behavior. We can trace water's unique life-supporting properties to the structure and interactions of its molecules.

The Structure of Water

Studied in isolation, the water molecule is deceptively simple. Its two hydrogen atoms are joined to one oxygen atom by single covalent bonds:

However, the electrons of the covalent bonds are not shared equally between the oxygen and hydrogen. Oxygen attracts the electrons of the covalent bonds much more strongly than does hydrogen. The unequal sharing of negatively charged electrons, combined with the V shape of the molecule, makes a water molecule polar. A **polar molecule** has opposite charges on opposite ends. In the case of water, the oxygen end of the molecule has a slight negative charge, while the two hydrogen atoms are slightly positive (**Figure 2.11a**).

> **Activity 2G on the Web & CD**
> Review how water's polarity results in hydrogen bonds.

The polarity of water results in weak electrical attractions between neighboring water molecules. The molecules tend to orient such that the hydrogen atom of one molecule is near the oxygen atom of an adjacent water molecule. These weak attractions are called **hydrogen bonds** (**Figure 2.11b**).

Figure 2.11 Water, a polar molecule.
(a) The arrows indicate the stronger pull that oxygen has, compared with its hydrogen partners, on the shared electrons of covalent bonds. The roughly V-shaped water molecule is polar, its oxygen end slightly negative and its hydrogen ends slightly positive in charge. **(b)** The charged regions of the polar water molecules are attracted to oppositely charged areas of neighboring molecules. Each molecule can hydrogen-bond to a maximum of four partners. (The dashed lines represent hydrogen bonds.) At any instant in liquid water at 98.6°F (37°C, human body temperature), only about 15% of the molecules are bonded to four partners. But so many water molecules are involved in some degree of hydrogen bonding at any moment that liquid water is much more cohesive than almost any other liquid.

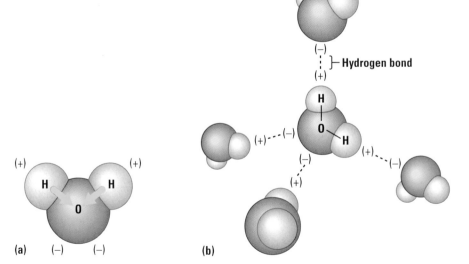

Water's Life-Supporting Properties

The polarity of water molecules and the hydrogen bonding that results explain most of water's life-supporting properties. We'll explore four of those properties here: the cohesive nature of water, the ability of water to moderate temperature, the biological significance of ice floating, and the versatility of water as a solvent.

The Cohesion of Water Water molecules stick together as a result of hydrogen bonding. Hydrogen bonds between molecules of liquid water last for only a few trillionths of a second, yet at any instant, many of the molecules are hydrogen-bonded to others. This tendency of molecules to stick together, called **cohesion,** is much stronger for water than for most other liquids. The cohesion of water is important in the living world. Trees, for example, depend on cohesion to help transport water from their roots to their leaves (**Figure 2.12**).

> **Activity 2H on the Web & CD**
> See an animation showing how water is transported long distances in trees.

Related to cohesion is surface tension, a measure of how difficult it is to stretch or break the surface of a liquid. Hydrogen bonds give water unusually high surface tension, making it behave as though it were coated with an invisible film (**Figure 2.13**).

How Water Moderates Temperature If you've ever burned your finger on a metal pot while waiting for the water in it to boil, you know that water heats up much more slowly than metal. In fact, because of hydrogen bonding, water has a stronger resistance to temperature change than most other substances.

Temperature and heat are related, but different. A swimmer crossing San Francisco Bay has a higher temperature than the water, but the bay contains far more heat because of its immense volume. **Heat** is the amount of energy associated with the movement of the atoms and molecules in a body of matter. **Temperature** measures the intensity of heat—that is, the average speed of molecules rather than the total amount of heat energy in a body of matter.

When water is heated, the heat energy first disrupts hydrogen bonds and then makes water molecules move faster. The temperature of the water doesn't go up until the water molecules start moving faster. Because heat is first used to break hydrogen bonds rather than raise the temperature, water absorbs and stores a large amount of heat while warming up only a few degrees in temperature. Conversely, when water cools, hydrogen bonds form, a process that releases heat. Thus, water can release a relatively large amount of heat to the surroundings while the water temperature drops only slightly.

Earth's giant water supply—the oceans, seas, and lakes—allows temperatures to stay within limits that permit life by storing a huge amount of heat from the sun during warm periods and giving off heat to warm the air during cold conditions. That's also why coastal areas generally have milder climates than inland regions. Water's resistance to temperature change also stabilizes ocean temperatures, creating a favorable environment for marine life.

Another way that water moderates temperature is by **evaporative cooling.** When a substance evaporates, the surface of the liquid remaining behind cools down. This occurs because the molecules with the greatest energy (the "hottest" ones) tend to vaporize first. It's as if the five fastest runners on your track team left school, lowering the average speed of the remaining team. On a global scale, surface evaporation cools tropical oceans. On the scale of individual organisms, evaporative cooling prevents some land-dwelling creatures

SEM 170×

Microscopic tubes

Figure 2.12 Cohesion and water transport in plants.
The evaporation of water from leaves pulls water upward from the roots through microscopic tubes in the trunk of the tree. Because of cohesion, the pulling force is relayed through the tubes all the way down to the roots. As a result, water rises against the force of gravity.

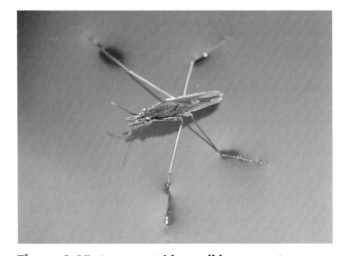

Figure 2.13 A water strider walking on water.
The cumulative strength of hydrogen bonds between water molecules allows this insect to walk on pond water without breaking the surface.

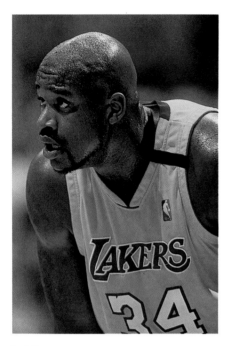

Figure 2.14
Sweating as a mechanism of evaporative cooling.

from overheating. It's why sweating helps you maintain a constant body temperature, even when exercising on a hot day (**Figure 2.14**). And the old expression "It's not the heat, it's the humidity" has its basis in the difficulty of sweating water into air that is already saturated with water vapor.

The Biological Significance of Ice Floating When most liquids get cold, their molecules move closer together. If the temperature is cold enough, the liquid freezes and becomes a solid. Water, however, behaves differently. When water molecules get cold enough, they move apart, forming ice. A chunk of ice has fewer molecules than an equal volume of liquid water, so ice floats because it is less dense than liquid water. Like water's other life-supporting properties, floating ice is a consequence of hydrogen bonding. In contrast to the short-lived hydrogen bonds in liquid water, those in solid ice last longer, with each molecule bonded to four neighbors. As a result, ice is a spacious crystal (**Figure 2.15**).

How does the fact that ice floats help support life on Earth? Imagine what would happen if ice sank. All ponds, lakes, and even the oceans would eventually freeze solid. During summer, only the upper few inches of the oceans would thaw. Instead, when a deep body of water cools, the floating ice insulates the liquid water below, allowing life to persist under the frozen surface.

Water as the Solvent of Life You know from experience that you can dissolve sugar or salt in water. This results in a mixture known as a **solution,** a liquid consisting of a homogenous mixture of two or more substances. The dissolving agent is called the **solvent,** and a substance that is dissolved is a **solute.** When water is the solvent, the result is called an **aqueous solution**.

The fluids of organisms are aqueous solutions. Water can dissolve an enormous variety of solutes necessary for life, such as the fluoride mentioned at the beginning of the chapter. Water is the solvent inside all cells, in

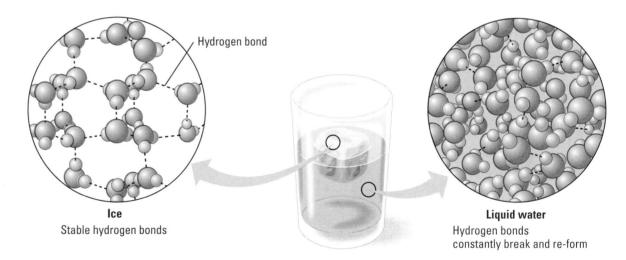

Hydrogen bond

Ice
Stable hydrogen bonds

Liquid water
Hydrogen bonds
constantly break and re-form

Figure 2.15 Why ice floats.
Compare the spaciously arranged molecules in the ice crystal with the tightly packed molecules in liquid water. The more stable hydrogen bonding in ice holds the molecules apart, resulting in ice being less dense than liquid water. The expansion of water as it freezes can crack boulders when the water is in crevices, and it can also break the water pipes of an unheated house in winter.

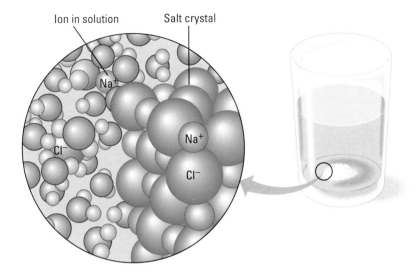

Figure 2.16 A crystal of table salt (NaCl) dissolving in water.
The sodium and chloride ions at the surface of the crystal have affinities for different parts of the water molecules. The positive Na^+ ions attract the electrically negative oxygen regions (red) of the water molecules. The negative Cl^- ions attract the positively charged hydrogen regions (gray) of water. As a result, H_2O molecules surround the ions, dissolving the crystal in the process.

blood, and in plant sap. As a solvent, it is a medium for chemical reactions. Water can dissolve ionic salts and many polar molecules, such as sugars, by orienting localized regions of positive and negative charge toward the charged regions of polar molecules (**Figure 2.16**).

Acids, Bases, and pH

In the aqueous solutions within organisms, most of the water molecules are intact. However, some of the water molecules actually break apart (dissociate) into ions. The ions formed are called hydrogen ions (H^+) and hydroxide ions (OH^-).

A chemical compound that donates H^+ ions to solutions is called an **acid.** One example of a strong acid is hydrochloric acid (HCl), the acid in your stomach. In solution, HCl dissociates completely into H^+ and Cl^- ions. A **base** (or alkali) is a compound that accepts H^+ ions and removes them from solution. Some bases, such as sodium hydroxide (NaOH), do this by releasing OH^- ions, which combine with H^+ to form H_2O.

To describe the acidity of a solution, we use the **pH scale** (pH stands for *potential hydrogen*). The scale ranges from 0 (most acidic) to 14 (most basic). Each pH unit represents a tenfold change in the concentration of H^+ (**Figure 2.17**). For example, lemon juice at pH 2 has 100 times more H^+ than an equal amount of tomato juice at pH 4. Pure water and aqueous solutions are neutral, neither acidic nor basic; they have a pH of 7. They do contain some H^+ and OH^- ions, but the concentrations of the two kinds of ions are equal. The pH of the solution inside most living cells is close to 7.

Activity 2I on the Web & CD
Explore the pHs of some common household substances.

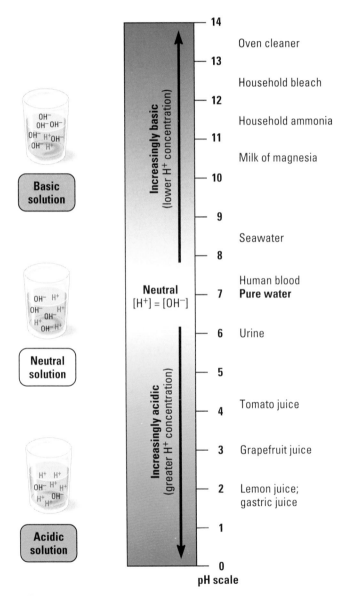

Figure 2.17 The pH scale.
A solution having a pH of 7 is neutral, meaning that its H^+ and OH^- concentrations are equal. The lower the pH below 7, the more acidic the solution, or the greater its excess of H^+ ions compared with OH^- ions. The higher the pH above 7, the more basic the solution, or the greater the deficiency of H^+ ions relative to OH^- ions.

Figure 2.18
The effects of acid precipitation on a forest.
Acid fog and acid rain may have contributed to the death of many of the fir trees in this forest in the Czech Republic. Acid precipitation results from water in the atmosphere reacting with certain pollutants spewed as exhaust from automobiles, factories, and power plants. The acid rain, snow, or fog can descend on land or lakes hundreds of miles downwind from the sources of pollution. Acid precipitation is a problem in the United States, as well; rain with a pH between 2 and 3—more acidic than vinegar—has been recorded in the eastern United States.

Even a slight change in internal pH can be harmful because a balance of H^+ ions and OH^- ions is critical for the proper functioning of chemical processes within organisms. To prevent harmful changes in pH, biological fluids contain **buffers,** substances that resist changes in pH by accepting H^+ ions when they are in excess and donating H^+ ions when they are depleted. This buffering process, however, is not foolproof. The biological damage from an unfavorable pH is apparent in the toll that acid precipitation can take on an ecosystem such as a pond or forest (**Figure 2.18**). It is a daunting reminder that the chemistry of life is linked to the chemistry of the environment. It reminds us, too, that chemistry happens on a global scale, since industrial processes in one region of the world often cause acid precipitation to fall in another part.

> **Case Study in the Process of Science on the Web & CD**
> Perform experiments to determine the effect of acid precipitation on plant life.

CHECKPOINT

1. Why is it unlikely that two neighboring water molecules would be arranged like this?

2. Explain why, if you pour very carefully, you can actually "stack" water slightly above the rim of a cup.

3. Why is it more dangerous to stay neck-deep in a 105°F (41°C) hot tub for an hour than it is to sit for an hour outside when the air temperature is 105°F (41°C)?

4. Explain why ice floats.

5. Why are blood and most other biological fluids classified as aqueous solutions?

6. Compared to a basic solution of pH 8, the same volume of an acidic solution at pH 5 has _____ times more hydrogen ions (H^+).

Answers: 1. The positively charged hydrogen regions would repel each other. **2.** Surface tension due to water's cohesion will hold the water together. **3.** Evaporative cooling in the hot tub is limited to the skin of the head and neck. **4.** Ice is less dense than liquid water because the more stable hydrogen bonds "lock" the molecules in a spacious crystal. **5.** The solvent is water. **6.** 1,000

Evolution Connection

Earth Before Life

Chemical reactions and physical processes on the early Earth created an environment that made life possible. And life, once it began, transformed the planet's chemistry. Biological and geologic histories are inseparable.

Earth began as a cold world when gravity drew together dust and ice orbiting a young sun about 4.5 billion years ago. The planet eventually melted from the heat produced by compaction, radioactive decay, and the impact of meteorites. Molten material sorted into layers of varying density. Most of the iron and nickel sank to the center and formed a dense core. Less dense material became concentrated in a layer called the mantle, which surrounds the core. And the least dense material solidified to form a thin crust. The present continents, including North America, are attached to plates of crust that float on the flexible mantle.

The first atmosphere, which was probably composed mostly of hot hydrogen gas (H_2), escaped. The gravity of Earth was not strong enough to hold such small molecules. Volcanoes belched gases that formed a new atmosphere (**Figure 2.19**). Based on analysis of gases vented by modern volcanoes, scientists have speculated that the second early atmosphere consisted mostly of water vapor (H_2O), carbon monoxide (CO), carbon dioxide (CO_2), nitrogen (N_2), methane (CH_4), and ammonia (NH_3). The first seas formed from torrential rains that began when Earth had cooled enough for water in the atmosphere to condense. In addition to an atmosphere very different from the one we know, lightning, volcanic activity, and ultraviolet radiation were much more intense when Earth was young. In such a seemingly inhospitable world, life began about 3.5–4.0 billion years ago. In Chapter 15, we'll examine some of the hypotheses and experiments of scientists who investigate the origins of the first life on Earth.

Evolution Connection on the Web
Learn more about what Earth was like before life appeared.

Figure 2.19 The gaseous exhaust of a volcano. Such volcanic activity was probably much more common when Earth was young.

Chapter Review

Summary of Key Concepts

For study help, go to the Essential Biology Website (www.essentialbiology.com) or CD-ROM to explore the Activities and Case Studies in the Process of Science.

Tracing Life Down to the Chemical Level

• The levels of biological order, beginning with the largest, are: ecosystems → communities → populations → organisms → organ systems → organs → tissues → cells → molecules → atoms.

Activity 2A The Levels of Life Card Game

Some Basic Chemistry

• **Matter: Elements and Compounds** Matter consists of elements and compounds, which are combinations of two or more elements. Of the 25 elements essential for life, oxygen, carbon, hydrogen, and nitrogen are the most abundant in living matter.

Case Study in the Process of Science How are space rocks analyzed for signs of life?

• **Atoms**

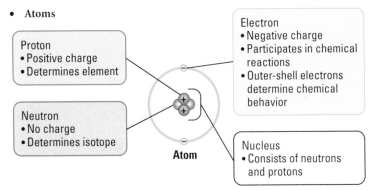

Proton
• Positive charge
• Determines element

Neutron
• No charge
• Determines isotope

Atom

Electron
• Negative charge
• Participates in chemical reactions
• Outer-shell electrons determine chemical behavior

Nucleus
• Consists of neutrons and protons

Activity 2B The Structure of Atoms
Activity 2C Electron Arrangement
Activity 2D Build an Atom

• **Chemical Bonding and Molecules** Electron transfers that complete outer electron shells result in charged atoms, or ions. Oppositely charged ions attract one another in what are called ionic bonds. Salts are ionic compounds. In covalent bonds, atoms complete their outer electron shells by sharing electrons. A molecule consists of two or more atoms connected by covalent bonds.

Activity 2E Ionic Bonds

Activity 2F Covalent Bonds

• **Chemical Reactions** By breaking bonds in reactants and forming new bonds in products, chemical reactions rearrange matter.

Water and Life

• **The Structure of Water**

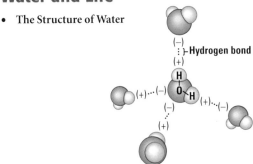

Activity 2G The Structure of Water

• **Water's Life-Supporting Properties** The ability of leaves to pull water up microscopic tubes without the water columns breaking is an example of how water's cohesion supports life. Water moderates temperature by absorbing heat in warm environments and releasing heat in cold environments. Evaporative cooling also helps stabilize the temperatures of oceans and organisms. The fact that ice floats because it is less dense than liquid water prevents the oceans from freezing solid. Blood and other biological fluids are aqueous solutions with a diversity of solutes dissolved in water, a versatile solvent.

Activity 2H *The Cohesion of Water in Trees*

- **Acids, Bases, and pH**

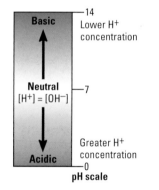

Activity 2I *Acids, Bases, and pH*

Case Study in the Process of Science *How Does Acid Rain Affect Trees?*

Self-Quiz

1. Which of the following elements is least abundant in your body?
   - a. carbon
   - b. oxygen
   - c. phosphorus
   - d. iron

2. Which of the following are compounds? $MgCl_2$, H_2, Fe, C_2H_6.

3. A chemical compound is to a(an) _____ as a body organ is to a tissue.

4. An atom can be changed into an ion by adding or removing _____. An atom can be changed into a different isotope by adding or removing _____. But if you change the number of _____, the atom becomes a different element.

5. A sulfur atom has 6 electrons in its third shell, which is its outer shell. As a result, it forms _____ covalent bonds with other atoms.

6. Explain the difference between an ionic bond and a covalent bond in terms of what happens to the electrons in the outer shell of the participating atoms.

7. Which of the following is not a chemical reaction?
   - a. Sugar ($C_6H_{12}O_6$) and oxygen gas (O_2) combine to form carbon dioxide (CO_2) and water (H_2O).
   - b. Sodium metal and chlorine gas unite to form sodium chloride.
   - c. Hydrogen gas combines with oxygen gas to form water.
   - d. Ice melts to form liquid water.

8. Some people in your study group say they don't understand what a polar molecule is. You explain that a polar molecule
   - a. is slightly negative at one end and slightly positive at the other end.
   - b. has an extra electron, giving it a positive charge.
   - c. has an extra electron, giving it a negative charge.
   - d. has covalent bonds.

9. Explain how the unique properties of water result from the fact that water is a polar molecule.

10. A can of cola consists mostly of sugar dissolved in water, with some carbon dioxide gas that makes it fizzy and makes the pH less than 7. Describe the cola using the following terms: solute, solvent, acidic, aqueous solution.

Answers to the Self-Quiz questions can be found in Appendix B.

Go to the website or CD-ROM for more Self-Quiz questions.

The Process of Science

1. Plants use a process known as photosynthesis to incorporate the carbon from atmospheric carbon dioxide gas into sugar. Outline an experiment using radioactive carbon to test the hypothesis that carbon assimilated by the leaves during photosynthesis can turn up in protein molecules stored in seeds.

2. The following diagram shows the arrangement of electrons around the nucleus of a fluorine atom (left) and a potassium atom (right). Predict what would happen if a fluorine atom and a potassium atom came into contact. What kind of bond do you think they would form? What kind of compound?

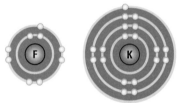

Fluorine atom   Potassium atom

Case Study in the Process of Science on the Web & CD *Examine rocks from space to determine if they contain extraterrestrial life.*

Case Study in the Process of Science on the Web & CD *Perform experiments to determine the effect of acid precipitation on plant life.*

Biology and Society

1. Evaluate this statement: "It's paranoid and ignorant to worry about industry or agriculture contaminating the environment with chemical wastes; this stuff is just made of the same atoms that were already present in our environment."

2. One solution to the problem of acid precipitation caused by emissions from power plants is to use nuclear power to produce energy. The proponents of nuclear power contend that it is the only way that the United States can increase its energy production while reducing air pollution, because nuclear power plants emit little or no acid-precipitation-causing pollutants. What are some of the benefits of nuclear power? What are the possible costs and dangers? Do you think we ought to increase our use of nuclear power to generate electricity? Why or why not? If a new power plant were to be built near your home, would you prefer it to be a coal-burning plant or a nuclear plant? Why?

Biology and Society on the Web *Learn what the American Dental Association has to say about the importance of fluoride.*

CHAPTER 3

The Molecules of Life

Biology and Society: Got Lactose? 36

Organic Molecules 36

Carbon Chemistry

Giant Molecules from Smaller Building Blocks

Biological Molecules 39

Carbohydrates

Lipids

Proteins

Nucleic Acids

Evolution Connection: DNA and Proteins as Evolutionary Tape Measures 49

Americans consume an average of 140 pounds of sugar per person per year.

Cellulose, found in plant cell walls, is the most abundant organic compound on Earth.

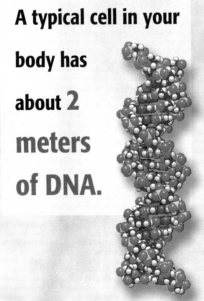

A typical cell in your body has about 2 meters of DNA.

A typical cow produces over 200 pounds of methane gas each year.

Figure 3.1 **Lactose intolerance.** People who are lactose intolerant can eat foods that contain lactose (such as ice cream) by taking pills that contain the enzyme lactase (inset).

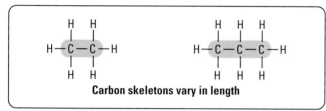

Carbon skeletons vary in length

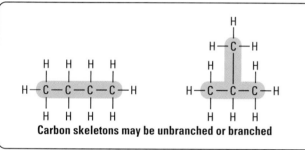

Carbon skeletons may be unbranched or branched

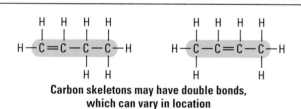

Carbon skeletons may have double bonds, which can vary in location

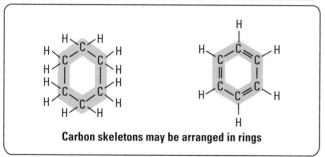

Carbon skeletons may be arranged in rings

Figure 3.2 **Variations in carbon skeletons.**
All of these examples are hydrocarbons, organic molecules consisting only of carbon and hydrogen. Notice that each carbon atom forms four bonds, and each hydrogen atom forms one bond. Remember that one line represents a single bond (sharing one pair of electrons) and two lines represent a double bond (sharing two pairs of electrons).

Got Lactose?

A milk moustache is meant to represent good health. Indeed, milk is among the healthier foods you can eat—rich in protein, minerals, and vitamins. But for some people, a glass of milk or a serving of milk-containing foods (such as ice cream or cheese) can cause bloating, gas, and other uncomfortable symptoms. Such people suffer from lactose intolerance—the inability to properly digest lactose, the main sugar found in milk.

In the digestive system, cells of the small intestine produce a molecule called lactase that breaks lactose down. Lactase is an example of an enzyme, a protein that helps drive chemical reactions (in this case, the breakdown of lactose). The cells of lactose-intolerant people produce insufficient amounts of this enzyme. As a result, lactose is not properly broken down and absorbed. This disturbs the water balance of the intestine, causing uncomfortable symptoms.

What options are there for people with lactose intolerance? The underlying cause, the underproduction of lactase, cannot be treated. However, the symptoms can be controlled through changes in diet. The first option is avoiding lactose-containing foods. Many substitutes are available, such as milk and cheese made from soy. Other lactose-free products are made from milk that has been pretreated with lactase. Alternatively, the enzyme lactase is available in pill form. Ingesting these pills along with food can artificially provide the enzyme that the body naturally lacks (**Figure 3.1**).

Biology and Society on the Web
Learn more about lactose intolerance.

Understanding the cause of lactose intolerance illustrates the importance of biological molecules to human health. Lactose, a sugar, is digested by lactase, a protein, which is encoded by a gene, made of DNA, a nucleic acid. In this chapter, we'll explore the structure and function of these and other carbon-containing molecules that are essential to life.

Organic Molecules

A cell is mostly water, but the rest of it consists mostly of carbon-based molecules. Carbon is unparalleled in its ability to form the large, complex, diverse molecules that are necessary for life functions. The study of carbon compounds is called **organic chemistry.**

Carbon Chemistry

Why are carbon atoms so versatile as molecular ingredients? Remember that an atom's bonding ability is related to the number of electrons it must share to complete its outer shell. A carbon atom has 4 electrons in an outer shell that holds 8 (see Figure 2.7). Carbon completes its outer shell by sharing electrons with other atoms in four covalent bonds. Each carbon thus acts as an intersection from which an organic molecule can branch off in up to four directions. And because carbon can use one or more of its bonds to attach to other carbon atoms, it is possible to construct an endless diversity of **carbon skeletons** varying in size and branching pattern (**Figure 3.2**). The carbon atoms of organic molecules can also use one or more of their bonds to partner with other elements, most commonly hydrogen, oxygen, and nitrogen.

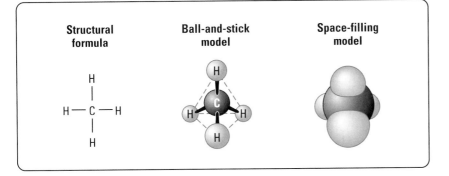

Structural formula   Ball-and-stick model   Space-filling model

Figure 3.3 Methane, the simplest hydrocarbon.
In the ball-and-stick model, notice that the four single bonds of carbon point to the corners of a tetrahedron.

Activity 3A on the Web & CD
Explore the diversity of carbon compounds and build a hydrocarbon.

In terms of chemical composition, the simplest organic compounds are **hydrocarbons,** organic molecules containing only carbon and hydrogen atoms (see Figure 3.2). And the simplest hydrocarbon is methane, a single carbon atom bonded to four hydrogen atoms (**Figure 3.3**). Methane is one of the most abundant hydrocarbons in natural gas and is also produced by prokaryotes that live in swamps and in the digestive tracts of grazing animals, such as cows. Larger hydrocarbons (such as octane, with eight carbons) are the main molecules in the gasoline we burn in cars and other machines (**Figure 3.4**). Hydrocarbons are also important fuels in your body; the energy-rich parts of fat molecules have a hydrocarbon structure.

Each type of organic molecule has a unique three-dimensional shape. Notice in Figure 3.3 that carbon's four bonds point to the corners of an imaginary tetrahedron (an object with four triangular sides). This geometric pattern occurs at each carbon "intersection" where there are four covalent bonds, and thus large organic molecules can have very elaborate shapes. A recurring theme in biology is how the molecules of your body recognize one another based on their shapes. Just one example is the chemical signaling your brain cells use to "talk" to each other (**Figure 3.5**).

Case Study in the Process of Science on the Web & CD
Perform experiments to determine the effectiveness of drugs with slightly different shapes.

The unique properties of an organic compound depend not only on its carbon skeleton but also on the atoms attached to the skeleton. In an organic molecule, the groups of atoms that usually participate in chemical reactions

Figure 3.4 Hydrocarbons as fuel.
Energy-rich hydrocarbons provide fuel for machines and, in the form of the hydrocarbon content of fats, the body's cells.

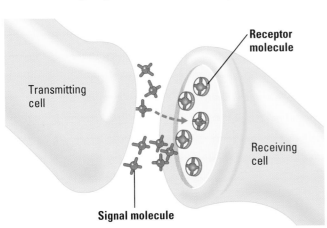

Figure 3.5 The importance of molecular shape to brain chemistry.
One nerve cell in your brain "talks" to another by releasing signal molecules with a shape that fits receptor molecules located on the surface of the receiving cell. The signal molecules cross the tiny gap between cells and bind to the receptors, stimulating the receiving cell. (The actual shapes of the signal and receptor molecules are much more complex than shown here.)

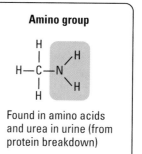

Hydroxyl group

Found in alcohols and sugars

Carbonyl group

Found in sugars

Amino group

Found in amino acids and urea in urine (from protein breakdown)

Carboxyl group

Found in amino acids, fatty acids, and some vitamins

Figure 3.6 Some common functional groups.

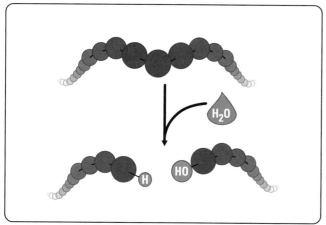

(a) Dehydration synthesis of a polymer

Short polymer   Monomer

H_2O

Longer polymer

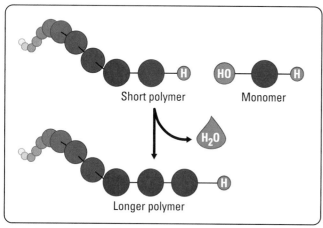

(b) Hydrolysis of a polymer

H_2O

Figure 3.7 Synthesis and digestion of polymers.
(a) The only atoms shown in these diagrams are hydrogens and hydroxyl groups (—OH or HO—, depending on orientation) in strategic locations on the monomers. A polymer grows in length when an incoming monomer and the monomer at the end of the existing chain each contribute to the formation of a water molecule. The monomers replace those lost covalent bonds with a bond to each other. **(b)** Hydrolysis reverses the process by breaking down the polymer with the addition of water molecules, breaking the bonds between monomers.

Activity 3B on the Web & CD
Test your knowledge of functional groups.

are called **functional groups. Figure 3.6** shows four of the functional groups important in the chemistry of life. Though the examples in the figure each contain only one functional group, many biological molecules have two or more. We are now ready to see how your cells make such large molecules out of smaller organic molecules.

Giant Molecules from Smaller Building Blocks

On a molecular scale, many of life's molecules are gigantic; in fact, biologists call them **macromolecules.** Examples are proteins, DNA, and carbohydrates called polysaccharides. Even though they are quite large, the structure of most macromolecules can be easily understood because they are **polymers,** large molecules made by stringing together many smaller molecules called **monomers.** A polymer is like a pearl necklace made by joining together many pearl monomers.

Cells link monomers together to form polymers by a process called **dehydration synthesis (Figure 3.7a).** For each monomer added to a chain, a water molecule (H_2O) is formed by the release of two hydrogen atoms and one oxygen atom from the monomers (hence the term *dehydration synthesis;* the monomers lose water). This same basic process of dehydration synthesis occurs regardless of the specific monomers and the type of polymer the cell is producing.

Organisms not only make macromolecules, but also have to break them down. For example, many of the molecules in your food are macromolecules. You must digest these giant molecules to make their monomers available to your cells for assimilation into your own brand of macromolecules. That digestion occurs by a process called **hydrolysis (Figure 3.7b).** Hydrolysis means to break (*lyse*) with water (*hydro*). Cells break bonds between monomers by adding water to them, a process essentially the reverse of dehydration synthesis.

Activity 3C on the Web & CD
View animations of dehydration synthesis and hydrolysis.

CHECKPOINT

1. Draw a structural formula for C_2H_4. Remember that each carbon has four bonds; each hydrogen has one.

2. When two molecules of glucose ($C_6H_{12}O_6$) are joined together in a dehydration synthesis reaction, what are the formulas of the two products? (Remember that chemical reactions involve the rearrangement of atoms and that no atoms are gained or lost.)

Answers: 1. H $\quad H$ $C=C$ H $\quad H$ **2.** $C_6H_{12}O_6 + C_6H_{12}O_6 \rightarrow C_{12}H_{22}O_{11} + H_2O$

Biological Molecules

In the remainder of the chapter, we'll explore the four categories of large molecules in cells: carbohydrates, lipids, proteins, and nucleic acids, the category that includes DNA. For each category, you'll learn about the structure and function of the molecules by first learning about the monomers used to build them.

Carbohydrates

Athletes know them as "carbs." **Carbohydrates** include the small sugar molecules dissolved in soft drinks, as well as the long starch molecules we consume in pasta and potatoes.

Monosaccharides Simple sugars, or **monosaccharides** (from the Greek *mono*, single, and *sacchar*, sugar), include glucose, found in sports drinks, and fructose, found in fruit. Both of these simple sugars are found in honey (**Figure 3.8**). Generally, monosaccharides have molecular formulas that are some multiple of CH_2O. For example, the formula for glucose is $C_6H_{12}O_6$. Fructose has the same molecular formula as glucose, $C_6H_{12}O_6$, but its atoms are arranged differently (**Figure 3.9**). Glucose and fructose are examples of **isomers**, molecules that have the same molecular formula but different structures. Because shape is so important, seemingly minor differences in chemistry give isomers different properties. In this case, the rearrangement of chemical groups makes fructose taste considerably sweeter than glucose.

Activity 3D on the Web & CD
See three different models of the ring form of glucose.

It is convenient to draw sugars as if their carbon skeletons were linear. However, in aqueous solutions, many monosaccharides form rings, as shown for glucose in **Figure 3.10**.

Monosaccharides, particularly glucose, are the main fuel molecules for cellular work. Analogous to an automobile engine consuming gasoline, your cells break down glucose molecules and extract their stored energy, giving off carbon dioxide as "exhaust." The rapid conversion of glucose to cellular energy is why intravenous dextrose (a form of glucose) is administered to sick or injured patients in an ambulance or emergency room; an aqueous solution of glucose is injected into the bloodstream to provide an immediate energy source to tissues in need of repair. Cells also use the carbon skeletons of monosaccharides as raw material for manufacturing other kinds of organic molecules.

Figure 3.8
Honey, a mixture of two simple sugars.
The sweet taste of honey comes from its main ingredients: the monosaccharides glucose and fructose.

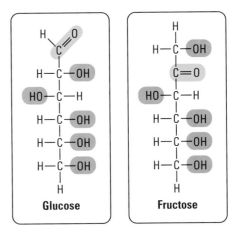

Glucose **Fructose**

Figure 3.9 Monosaccharides (simple sugars).
These molecules have the two trademarks of sugars: several hydroxyl groups (—OH) and a carbonyl group ($>C=O$). Glucose and fructose are isomers of each other: They have identical molecular formulas ($C_6H_{12}O_6$), but their structures differ because the atoms are not arranged the same way. The difference in this case is the location of the carbonyl group.

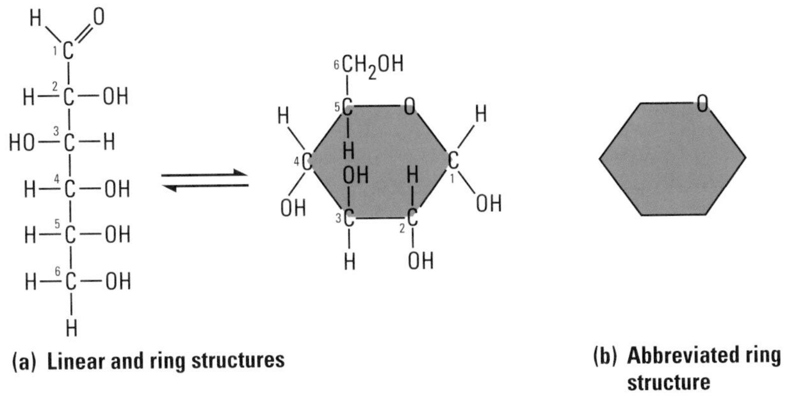

(a) Linear and ring structures

(b) Abbreviated ring structure

Figure 3.10 The ring structure of glucose.
(a) Dissolved in water, one part of a glucose molecule can bond to another part to form a ring. The carbon atoms are numbered here so that you can relate the linear and ring versions of the molecule. As the double arrows indicate, ring formation is a reversible process, but at any instant in an aqueous solution, most glucose molecules are rings. **(b)** From now on, we'll use this abbreviated ring symbol for glucose.

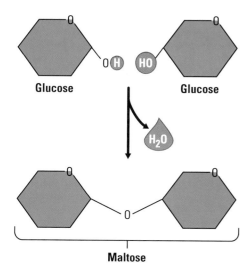

Figure 3.11
Disaccharide (double sugar) formation.
To form a disaccharide, two simple sugars are joined by dehydration synthesis, in this case forming a bond between two glucose monomers to make the double sugar maltose.

Figure 3.12 A year's supply of sugar.
Americans consume an average of 64 kg (140 pounds) of sweetener per person per year—over a third of a pound per day. The 64 kg of sucrose (table sugar) in this photograph will give you some idea of that level of consumption. We actually get much of our sugar in the form of high-fructose corn syrup (HFCS), which is used to sweeten such products as soft drinks and candy. Reading food labels will make you aware of the amount of sugar in processed foods.

Disaccharides A **disaccharide,** or double sugar, is constructed from two monosaccharides by the process of dehydration synthesis. An example of a disaccharide is maltose, also called malt sugar, which consists of two glucose monomers (**Figure 3.11**). Maltose, common in germinating seeds, is used in making beer, malted milk shakes, and malted milk ball candy. Lactose is another disaccharide, made from the monosaccharides glucose and galactose.

The most common disaccharide is sucrose, common table sugar, which consists of a glucose linked to a fructose. Sucrose is the main carbohydrate in plant sap, and it nourishes all the parts of the plant. Sucrose is extracted from the stems of sugarcane or, more commonly in the United States, the roots of sugar beets.

The history of sugar refining and consumption by humans is a story with important health and economic consequences. Before the 1980s, refined sugar was processed from cane and beets. Occupying only a small part of the sweetener market was corn syrup, which contains glucose. Because glucose is only half as sweet as sucrose, corn syrup was not a serious rival to sucrose.

The balance among sweeteners was drastically changed in the 1980s, when corn syrup producers developed a commercial method for converting much of the glucose in corn syrup to much sweeter fructose. The resulting high-fructose corn syrup (HFCS), which is about half fructose, is inexpensive and has replaced sucrose in many prepared foods. Soft drink manufacturers, once the largest commercial users of sucrose in the world, have almost completely replaced sucrose with HFCS. If you read the label on any soft drink can or bottle, you're likely to find that HFCS is the first or second ingredient listed. Because corn syrup is produced primarily in developed countries, the changeover has hurt the developing economies of tropical countries, where sugarcane is a major crop.

The United States is one of the world's leading markets for sweeteners, with the average American consuming about 64 kg (140 pounds) per year, mainly as sucrose and HFCS (**Figure 3.12**). This national "sweet tooth" persists in spite of our growing awareness about how sugar can negatively affect our health. Sugar is a major cause of tooth decay. High sugar consumption also tends to replace eating more varied and nutritious foods. The description of sugars as "empty calories" is accurate in the sense that most sweeteners contain only negligible amounts of nutrients other than carbohydrates. For good health, we also require proteins, fats, vitamins, and minerals. And we need to include substantial amounts of complex carbohydrates—that is, polysaccharides—in our diet.

Polysaccharides Complex carbohydrates, or **polysaccharides,** are long chains of sugar units—polymers of monosaccharides. One familiar example is starch, found in roots and other plant organs. **Starch** consists entirely of many glucose monomers strung together (**Figure 3.13a**). Plant cells store starch in granules, where it is available as a sugar stockpile that can be broken down as needed to provide energy and raw material for building other molecules. Potatoes and grains, such as wheat, corn, and rice, are the major sources of starch in the human diet. Humans and most other animals are able to use plant starch as food by hydrolyzing the bonds between glucose monomers within their digestive systems.

Animals store excess sugar in the form of a polysaccharide called **glycogen.** Glycogen is similar in structure to starch in that it is also a polymer of glucose monomers, but glycogen is more extensively branched (**Figure 3.13b**). Most of our glycogen is stored as granules in our liver and muscle cells, which hydrolyze the glycogen to release glucose when it is needed for energy. This is the basis for "carbo loading," the consumption of large

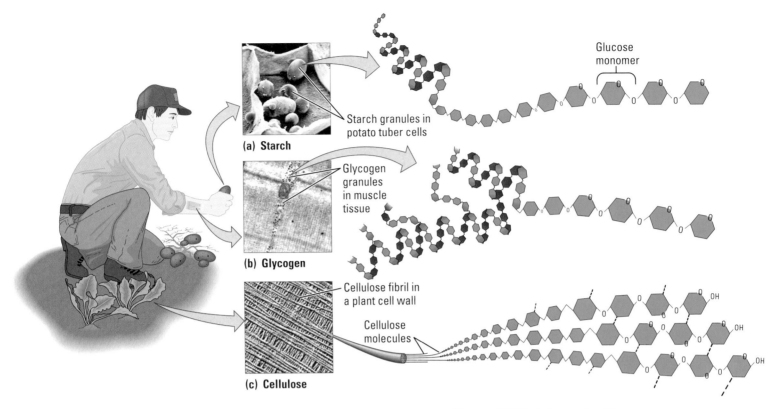

Starch granules in potato tuber cells

(a) Starch

Glycogen granules in muscle tissue

(b) Glycogen

Cellulose fibril in a plant cell wall

Cellulose molecules

(c) Cellulose

Glucose monomer

Figure 3.13 Polysaccharides.
(a) Plants store glucose by polymerizing it to form starch.
(b) Animals also store glucose, but in the form of glycogen, a polysaccharide more extensively branched than plant starch.
(c) The cellulose of plant cell walls is an example of a structural polysaccharide. The cellulose molecules are assembled into fibrils that make up the main fabric of the walls. Wood, a cell wall material consisting of such cellulose fibrils along with other polymers, is strong enough to support trees hundreds of feet high. We take advantage of that structural strength in our use of lumber as a building material.

amounts of starchy foods the night before an athletic event. The starch is converted into glycogen, which is then available for rapid use during physical activity the next day.

In addition to playing an important role in nutrition, certain polysaccharides serve as structural components. **Cellulose,** the most abundant organic compound on Earth, forms cable-like fibrils in the tough walls that enclose plant cells and is a major component of wood (**Figure 3.13c**). Cellulose resembles starch and glycogen in being a polymer of glucose, but its glucose monomers are linked together in a different orientation. Unlike the glucose linkages in starch and glycogen, those in cellulose cannot be broken by most animals. The cellulose in plant foods, which passes unchanged through our digestive tract, is commonly known as dietary "fiber" or "roughage." Because it remains undigested, fiber does not serve as a nutrient, although it does appear to help keep our digestive system healthy. Most Americans do not get the recommended levels of fiber in their diet. Foods rich in fiber include fruits and vegetables, whole grains, bran, and beans. Grazing animals and wood-eating insects such as termites have prokaryotes inhabiting their digestive tracts that break down cellulose (**Figure 3.14**).

Simple sugars (such as glucose) and double sugars (such as sucrose) dissolve readily in water, forming sugary solutions, including soft drinks. In contrast, cellulose and some forms of starch are such large molecules that they do not dissolve in water. If they did, then your cotton bath towels, which are mostly cellulose, would dissolve the first time you put them in a washing machine. In spite of this difference, almost all carbohydrates are **hydrophilic,** which literally means "water-loving." Hydrophilic molecules adhere water to their surface. It is the hydrophilic quality of cellulose that makes a fluffy bath towel so water absorbent.

Activity 3E on the Web & CD
Review carbohydrates and watch animations of disaccharides and polysaccharides forming.

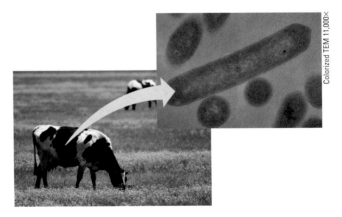

Colorized TEM 11,000×

Figure 3.14 Prokaryotes in grazing animals.
Animals lack the ability to break down the polymer cellulose into the glucose monomers that make it up. Within the digestive tract of grazing animals such as cows, prokaryotes break down consumed cellulose, converting it into a form that can be digested by the animal. A by-product of this reaction is large amounts of methane gas.

Lipids

In contrast to carbohydrates and most other biological molecules, **lipids** are **hydrophobic,** which means that they do not mix with water (from the Greek *hydro*, water, and *phobos*, fearing). You have probably observed this chemical behavior in an unshaken bottle of salad dressing: The oil, which is a type of lipid, separates from the vinegar, which is mostly water. If you shake the bottle, you can force a temporary mixture long enough to douse your salad with dressing, but what remains in the bottle will quickly separate again once you stop shaking it. Lipids are a diverse set of molecules; two examples are fats and steroids.

Fats Dietary **fat** consists largely of the molecule triglyceride, which is made of a molecule called glycerol joined with three molecules called fatty acids via dehydration synthesis (**Figure 3.15a**).

Despite the generally unfavorable attitude that most people have toward fats, these molecules actually perform essential functions within the body. The major portion of a fatty acid is a long hydrocarbon, which, like the hydrocarbons of gasoline, stores much energy. In fact, a pound of fat packs more than twice as much energy as a pound of carbohydrate such as starch. This compact energy storage enables a mobile animal such as a human to get around much better than if the animal had to lug its stored energy around in the bulkier form of carbohydrate. The downside to this energy efficiency is that it is very difficult for a person trying to lose weight to "burn off" excess body fat. It is important to understand that a reasonable amount of body fat is both normal and healthy as a fuel reserve. We stock these long-term food stores in specialized reservoirs called adipose cells, which swell and shrink when we deposit and withdraw fat from them. In addition to storing energy, adipose tissue cushions vital organs and insulates us, helping maintain a warm body temperature even when the outside air is cold.

"Saturated" versus "unsaturated" fats is a comparison you have probably encountered in advertisements for foods. What's that all about? Notice in **Figure 3.15b** that one of the fatty acids bends where there is a double bond in the carbon skeleton. That fatty acid is said to be **unsaturated** because it has less than the maximum number of hydrogens at the location of the double bond. The other two fatty acids in the fat molecule of Figure 3.15b lack double bonds in their hydrocarbon portions. Those fatty acids are **saturated,** meaning that they are bonded to the maximum number of hydrogen atoms. A saturated fat is one with all three of its fatty acids saturated. If one or more of the fatty acids is unsaturated, then it's an unsaturated fat, such as the one in Figure 3.15b. A polyunsaturated fat has several double bonds within its fatty acids.

Most animal fats, such as lard and butter, have a relatively high proportion of saturated fatty acids. The linear shape of saturated fatty acids allows them to stack easily, making saturated fats solid at room temperature. Diets rich in saturated fats may contribute to cardiovascular disease by promoting

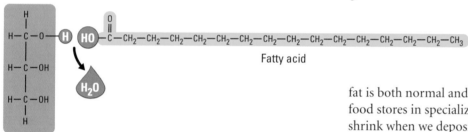

Fatty acid

(a) Dehydration synthesis linking a fatty acid to glycerol

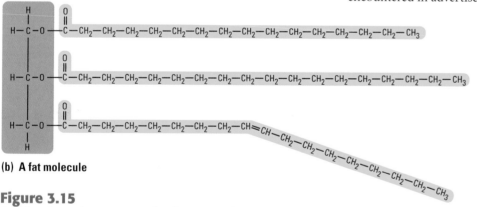

(b) A fat molecule

Figure 3.15
Synthesis and structure of a fat, or triglyceride.
(a) This diagram shows the first of three fatty acids that will attach to glycerol by dehydration synthesis. **(b)** The finished fat has a glycerol "head" and three fatty acid "tails." The fatty acids consist mainly of energy-rich hydrocarbon.

atherosclerosis. In this condition, lipid-containing deposits called plaques build up within the walls of blood vessels, reducing blood flow and increasing risk of heart attacks and strokes. In contrast, plant and fish fats are relatively high in unsaturated fatty acids. The bent shape of unsaturated fatty acids makes them less likely to form solids, so unsaturated fats are usually liquid at room temperature. Vegetable oils (such as corn and canola oil) and fish oils (such as cod liver oil) are examples.

While plant oils tend to be low in saturated fat, tropical plant oils are an exception. Cocoa butter, a main ingredient in chocolate, contains a mix of saturated and unsaturated fat that gives it a melting point near body temperature. Thus, chocolate stays solid at room temperature but melts in the mouth. This pleasing "mouth feel" is one of the reasons chocolate is so appealing.

Sometimes, such as when producing margarine and peanut butter, a food manufacturer wishes to use a vegetable oil but needs the food product to be solid. To achieve this, the manufacturer can convert unsaturated fats to saturated fats by adding hydrogen, a process called hydrogenation. Unfortunately, hydrogenation also creates trans fats, a form of fat that recent research suggests is very unhealthy. To avoid trans fats in your diet, buy foods that are labeled "trans fat free" and avoid foods with hydrogenated (or partially hydrogenated) oils listed among the ingredients.

Steroids   Classified as lipids because they are hydrophobic, **steroids** are very different from fats in structure and function. The carbon skeleton of a steroid is bent to form four fused rings. Cholesterol, which gets a lot of bad press because of its association with cardiovascular disease, is a steroid (**Figure 3.16**). But cholesterol is also an essential molecule in your body. As we will see in Chapter 4, cholesterol and another type of fat called phospholipid are present in the membranes that surround your cells. Cholesterol is also the "base steroid" from which your body produces other steroids—and that includes estrogen and testosterone, the steroids that function as sex hormones (see Figure 3.16).

> **Activity 3F on the Web & CD**
> Review lipids and build a triglyceride.

The controversial drugs called synthetic **anabolic steroids** are variants of testosterone, the male sex hormone. Testosterone causes a general buildup in muscle and bone mass during puberty in males and maintains masculine traits throughout life. Because anabolic steroids structurally resemble testosterone, they also mimic some of its effects. Some athletes use anabolic steroids to build up their muscles quickly and enhance their performance. Anabolic steroids, along with many other drugs, are banned by most athletic organizations. Nonetheless, many professional athletes admit to using them heavily.

Using anabolic steroids is indeed a fast way to increase body size beyond what hard work alone can produce (**Figure 3.17**). But at what cost? Although medical researchers still debate the extent of health risks from steroid abuse, there is evidence that these substances can cause serious physical and mental problems, including violent mood swings, deep depression, liver damage, high cholesterol, a reduced sex drive, and infertility. Why would using a mimic of a sex hormone decrease fertility? The reason is that anabolic steroids often cause the body to reduce its normal output of sex hormones. The many potential health hazards of anabolic steroids, coupled with the unfairness of an artificial advantage, strongly support the argument for banning their use in athletics.

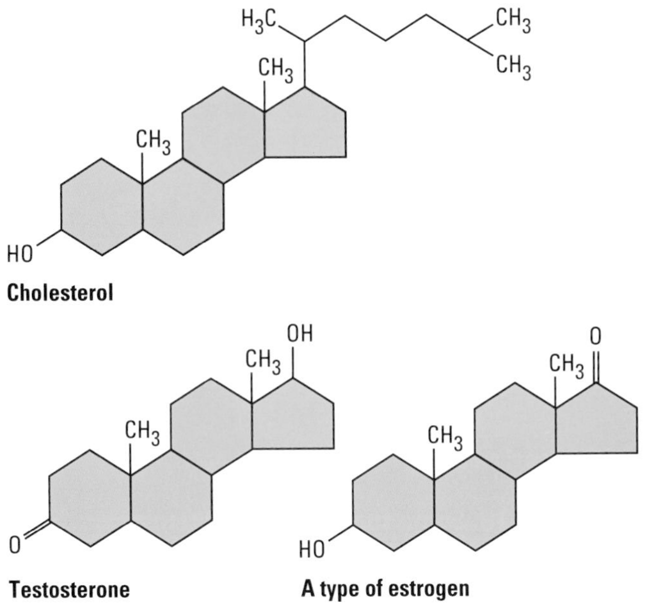

Figure 3.16 Examples of steroids.
All steroids have a carbon skeleton consisting of four fused rings, abbreviated here with all the atoms of the rings omitted. Different steroids vary in the functional groups attached to this core set of rings. For example, the subtle contrast between testosterone and estrogen influences the development of the anatomical and physiological differences between male and female mammals, including humans.

Cholesterol

Testosterone

A type of estrogen

Figure 3.17 Worth the risk?
Steroids can build muscle mass, but they can also cause serious health problems.

(b) Storage proteins

(d) Transport proteins

Figure 3.18 Some functions of proteins.
(a) Structural proteins provide support. Examples are the proteins found in hair, horns, feathers, spider webs, and connective tissues, such as tendons and ligaments. (b) Storage proteins, found in seeds and eggs, provide a source of amino acids for developing plants and animals. (c) Muscles are rich in contractile proteins. (d) Transport proteins include hemoglobin, the iron-containing protein in blood that conveys oxygen from your lungs to other parts of the body. The red blood cells in this photograph contain hemoglobin.

(a) Structural proteins

(c) Contractile proteins

Proteins

Proteins are the most elaborate of life's molecules. A **protein** is a polymer constructed from amino acid monomers. Every human body has tens of thousands of different kinds of proteins, and each protein has a unique, three-dimensional structure that corresponds to a specific function. Proteins perform most of the tasks the body needs to function. **Figure 3.18** surveys the functions of four types of proteins: structural proteins, storage proteins, contractile proteins, and transport proteins. Other types of proteins include defensive proteins, such as antibodies of the immune system, and signal proteins, which convey messages from one cell to another. Enzymes (including lactase), another important protein category, change the rate of a chemical reaction without being changed in the process. We'll discuss enzymes in more detail in Chapter 5. Now let's take a look at the architecture of proteins.

> **Activity 3G on the Web & CD**
> Click on proteins to see them in action.

The Monomers: Amino Acids All proteins are constructed from a common set of 20 kinds of amino acids. Each **amino acid** consists of a central carbon atom bonded to four covalent partners (carbon, remember, always forms four covalent bonds). Three of those attachments are common to all 20 amino acids: a carboxyl group (—COOH), an amino group (—NH$_2$), and a hydrogen atom. The variable component of amino acids, called the side group, is attached to the fourth bond of the central carbon. Each type of amino acid has a unique side group, giving that amino acid its special chemical properties (**Figure 3.19**).

Amino group   Carboxyl group

(a)   Side group

(a)

Leucine   Serine
(b) (hydrophobic)   (hydrophilic)

←—Side groups—→

Figure 3.19 Amino acids.
(a) The general structure of an amino acid. (b) The 20 amino acids vary only in their side groups, which give these monomers their unique properties. For example, the side group of the amino acid leucine is pure hydrocarbon. That region of leucine is hydrophobic, because hydrocarbons don't mix with water. In contrast, the side group of the amino acid serine has a hydroxyl (—OH) group, which is hydrophilic.

Proteins as Polymers Cells link amino acids together by—you guessed it—dehydration synthesis. The resulting bond between adjacent amino acids is called a **peptide bond** (**Figure 3.20**). Proteins usually consist of 100 or more amino acids, forming a chain called a **polypeptide.**

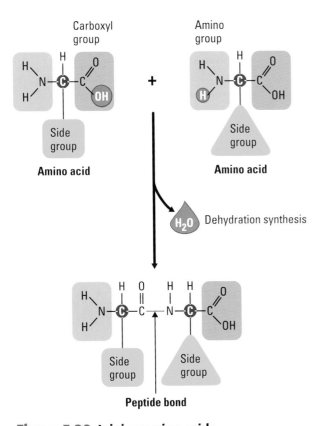

Figure 3.20 Joining amino acids.
Dehydration synthesis links adjacent amino acids by a peptide bond.

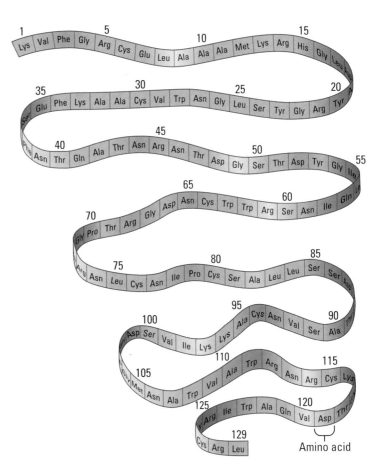

Figure 3.21 The primary structure of a protein.
This is the unique amino acid sequence, or primary structure, of a protein called lysozyme. (The chain was drawn in serpentine fashion so that it would fit on the page. The actual shape of lysozyme is much more complex.) The names of the amino acids are given as their three-letter abbreviations, with the positions of lysozyme's 129 amino acids numbered along the chain.

Your body has tens of thousands of different kinds of protein. How is it possible to make such a huge variety of proteins from just 20 kinds of amino acids? The answer is arrangement. You know that you can make many different words by varying the sequence of just 26 letters. Though the protein alphabet is slightly smaller (just 20 "letters"), the "words" are much longer, with a typical polypeptide being at least 100 amino acids in length. Just as each word is constructed from a unique succession of letters, each protein has a unique linear sequence of amino acids. This specific amino acid sequence is called the protein's **primary structure** (**Figure 3.21**).

Just as changing a single letter can drastically affect the meaning of a word ("tasty" versus "nasty," for instance), even a slight change in primary structure can affect a protein's ability to function. Consider, for example, the substitution of one amino acid for another at a particular position in hemoglobin, the blood protein that carries oxygen. Such an amino acid swap is the cause of sickle-cell disease, an inherited blood disorder (**Figure 3.22**).

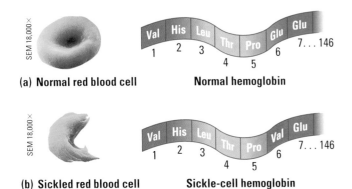

(a) Normal red blood cell   **Normal hemoglobin**

(b) Sickled red blood cell   **Sickle-cell hemoglobin**

Figure 3.22 A single amino acid substitution in a protein causes sickle-cell disease. **(a)** Red blood cells of humans are normally disk-shaped, as seen in this microscopic view. Each cell contains millions of molecules of the protein hemoglobin, which transports oxygen from the lungs to other organs of the body. Next to the photograph, you can see the first 7 of the 146 amino acids in a polypeptide chain of hemoglobin. **(b)** A slight change in the primary structure of hemoglobin—an inherited substitution of one amino acid—causes sickle-cell disease. The substitution—valine in place of the amino acid glutamic acid—occurs in the number 6 position of the polymer. The abnormal hemoglobin molecules tend to crystallize, deforming some of the cells into a sickle shape. The life of someone with the disease is characterized by dangerous episodes when the angular cells clog tiny blood vessels, impeding blood flow.

Protein Shape At this point, you might be thinking that a polypeptide chain is the same thing as a protein, but that's not quite true. The distinction between the two is analogous to the relationship between a long strand of yarn and a sweater of particular size and shape that you could knit from the yarn. A functional protein is not just a polypeptide chain, but one or more polypeptides precisely twisted, folded, and coiled into a molecule of unique shape. If we dissect the overall shape of a protein, we

> **Activity 3H on the Web & CD**
> See animations of the levels of protein structure.

can recognize at least three levels of structure: primary, secondary, and tertiary. Proteins with more than one polypeptide chain have a fourth level: quaternary structure. You can examine these levels of protein structure in **Figure 3.23**.

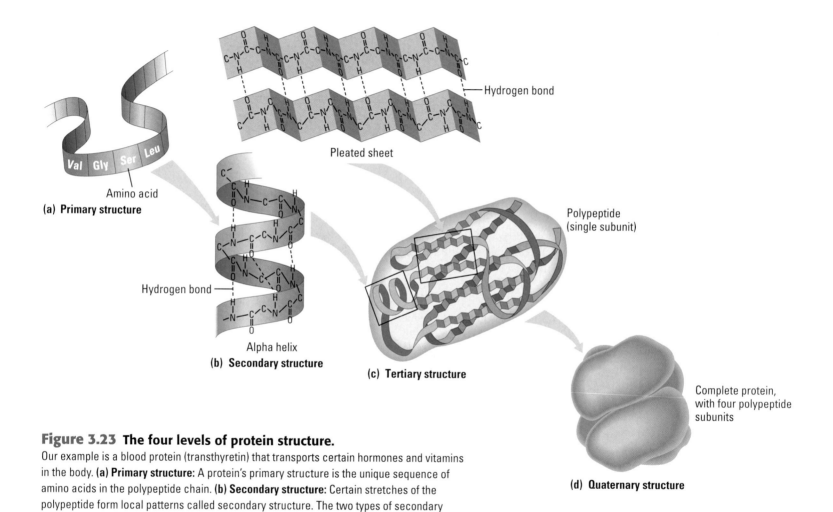

Figure 3.23 The four levels of protein structure.
Our example is a blood protein (transthyretin) that transports certain hormones and vitamins in the body. **(a) Primary structure:** A protein's primary structure is the unique sequence of amino acids in the polypeptide chain. **(b) Secondary structure:** Certain stretches of the polypeptide form local patterns called secondary structure. The two types of secondary structure are named alpha helix and pleated sheet. Secondary structure is reinforced by hydrogen bonds along the polypeptide backbone, similar to the weak hydrogen bonds between water molecules (see Figure 2.11). (Dashed lines represent the hydrogen bonds in this diagram.) The structure illustrated here is simplified, showing only the atoms of the polypeptide backbone, not those of the amino acid side groups. **(c) Tertiary structure:** The overall three-dimensional shape of the protein is called tertiary structure. It is reinforced by chemical bonds (not shown here) between the side groups of amino acids in different regions of the polypeptide chain. **(d) Quaternary structure:** Some proteins consist of two or more polypeptide chains. For example, this blood protein is constructed from four polypeptide chains. Such proteins have a quaternary structure, which results from bonding between the polypeptide chains.

When a cell makes a polypeptide, the chain usually folds spontaneously to form the functional shape for that protein. It is a protein's three-dimensional shape that enables the molecule to carry out its specific function in a cell. In almost every case, a protein's function depends on its ability to recognize and bind to some other molecule. For example, the receptors on the brain cell in Figure 3.5 are actually proteins that recognize certain chemical signals from other cells. If the protein receptor's shape were to be altered, then it would not be able to perform this recognition function. With proteins, function follows form—that is, what a protein does is a consequence of its shape.

What Determines Protein Structure? A protein's shape is sensitive to the surrounding environment. An unfavorable change in temperature, pH, or some other quality of the aqueous environment can cause a protein to unravel and lose its normal shape. This is called **denaturation** of the protein. If you cook an egg, the transformation of the egg white from clear to opaque is caused by proteins in the egg white denaturing. The denatured proteins become insoluble in water and form a white solid.

Given an environment suitable for that protein (so that it doesn't denature), the primary structure of a protein causes it to fold into its functional shape. Each kind of protein has a unique primary structure and therefore a unique shape that enables it to do a certain job in a cell. But what determines primary structure, a protein's specific amino acid sequence? Each polypeptide chain has a sequence specified by an inherited gene. And that relationship between genes and proteins brings us to this chapter's last category of molecules.

Nucleic Acids

Nucleic acids are information storage molecules that provide the directions for building proteins. The name *nucleic* comes from their location in the nuclei of eukaryotic cells. There are actually two types of nucleic acids: **DNA** (those most famous of chemical initials, which stand for *deoxyribonucleic acid*) and **RNA** (for *ribonucleic acid*). The genetic material that humans and other organisms inherit from their parents consists of giant molecules of DNA. Within the DNA are genes, specific stretches of DNA that program the amino acid sequences (primary structure) of proteins. Those programmed instructions, however, are written in a kind of chemical code that must be translated from "nucleic acid language" to "protein language." A cell's RNA molecules help translate. You'll learn more about how DNA and RNA work in Chapter 10.

Nucleic acids are polymers of monomers called **nucleotides** (**Figure 3.24**). Each nucleotide is itself a complex organic molecule with three parts. At the center of each nucleotide is a five-carbon sugar, deoxyribose in DNA and ribose in RNA. Attached to the sugar is a negatively charged phosphate group containing a phosphorus atom bonded to oxygen atoms (PO_4^-). Also attached to the sugar is a nitrogen-containing base (**nitrogenous base**) made of one or two rings. (It is called a base because it behaves like a base, or alkali, in aqueous solutions.) The sugar and phosphate are the same in all nucleotides; only the base varies. Each DNA nucleotide has one of the following four bases: adenine (abbreviated A), guanine (G), cytosine (C), or thymine (T). Thus, all genetic information is written in a four-letter alphabet—A, G, C, T—the bases that distinguish the four nucleotides that make up DNA (**Figure 3.25**).

Activity 3I on the Web & CD
Learn more about the roles of DNA and RNA in the manufacture of proteins.

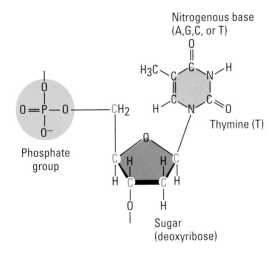
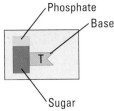

Figure 3.24 DNA Nucleotides.
Each nucleotide monomer consists of three parts: a sugar (deoxyribose, in the case of DNA nucleotides); a phosphate (a phosphorus atom bonded to oxygen); and a nitrogenous base. (In the bottom of this figure, the three parts of a nucleotide are symbolized with shapes and colors rather than actual chemical structures.)

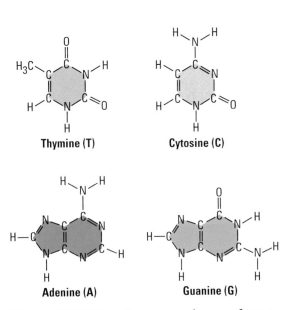

Figure 3.25 The nitrogenous bases of DNA.
Thymine and cytosine have single-ring structures. Adenine and guanine have double-ring structures.

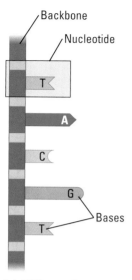

Backbone
Nucleotide
T
A
C
G
T
Bases

(a) DNA strand

Figure 3.26
The structure of DNA.

(a) A DNA strand is a polymer of nucleotides linked into a backbone, with appendages consisting of the bases. A strand has a specific sequence of the four bases, abbreviated A, G, C, and T. **(b)** A double helix consists of two DNA strands held together by bonds between bases. The bonds are individually weak—they are hydrogen bonds, like those between water molecules—but they zip the two strands together with a cumulative strength that gives the double helix its stability. The base pairing is specific: A always pairs with T; G always pairs with C.

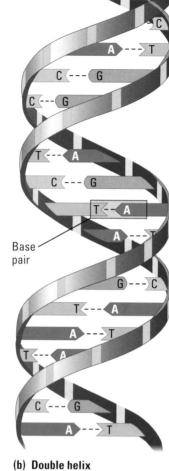

C
A --- T
C --- G
C --- G
T <-- A
C --- G
T <-- A
A --- T
G --- C
T <-- A
A --- T
T <-- A
C --- G
A --- T

Base pair

(b) Double helix

Nucleotide monomers are linked into long chains called polynucleotides, or DNA strands (**Figure 3.26a**). Nucleotides are joined together by covalent bonds between the sugar of one nucleotide and the phosphate of the next. This results in a **sugar-phosphate backbone,** a repeating pattern of sugar-phosphate-sugar-phosphate, with the bases hanging off the backbone like appendages. Polynucleotides vary in length from long to very long, so the number of possible polynucleotide sequences is very great. One long DNA strand contains many genes, each a specific series of hundreds or thousands of nucleotides. And each of these genes stores information in its unique sequence of nucleotide bases. In fact, it is this information that cells translate into an amino acid sequence to make a specific protein.

How can the cells of parents copy their genes to pass along to offspring? Inheritance is based on DNA actually being double-stranded, with the two DNA strands wrapped around each other to form a **double helix** (**Figure 3.26b**). In the central core of the helix, the bases along one DNA strand hydrogen-bond to bases along the other strand. This base pairing is specific: The base A can pair only with T, and G can pair only with C. Thus, if you know the sequence of bases along one DNA strand, you also know the sequence along the complementary strand in the double helix. As we will see in Chapter 10, this unique base pairing is the basis of DNA's ability to act as the molecule of inheritance.

What about RNA? As its name—ribonucleic acid—implies, its sugar is ribose rather than deoxyribose. By comparing the RNA nucleotide in

> **Activity 3J on the Web & CD**
> Review DNA and RNA structure.

Figure 3.27 with the DNA nucleotide in Figure 3.24, you can see that the RNA ribose sugar has an extra —OH group compared with the DNA deoxyribose sugar (*deoxy* means "without an oxygen"). Another difference between RNA and DNA is that instead of the base thymine, RNA has a similar but distinct base called uracil (U). Except for the presence of ribose and uracil, an RNA polynucleotide chain is identical to a DNA polynucleotide chain. However, RNA is usually found only in a single-stranded form, while DNA usually exists as a double helix.

Nitrogenous base
(A,G,C, or U)

Uracil (U)

Phosphate group

Sugar (ribose)

Figure 3.27 An RNA nucleotide.   Notice that this RNA nucleotide differs from the DNA nucleotide in Figure 3.24 in two ways. The RNA sugar is ribose rather than deoxyribose, and the base is uracil (U) instead of thymine (T). The other three kinds of RNA nucleotides have the bases A, C, and G, as in DNA.

CHECKPOINT

1. _____ is to polysaccharide as amino acid is to _____.
2. The formula for glucose is $C_6H_{12}O_6$. What would be the formula for a sugar having four carbons instead of six?
3. How do manufacturers produce the high-fructose corn syrup listed as an ingredient on a soft drink bottle? Why is this profitable?
4. Why can't you digest wood?
5. On a food package label, what is the meaning of "unsaturated fats"?
6. In classifying the molecules of life, what do fats and human sex hormones have in common?
7. Indicate which of the following is *not* a protein: an enzyme, hemoglobin, a gene.
8. In what way is the production of a polypeptide similar to the production of a polysaccharide?
9. Why does a denatured protein no longer function?
10. In a double helix, a region along one DNA strand has the sequence GAATGC. What is the base sequence along the complementary region of the other strand of the double helix?

Evolution Connection

DNA and Proteins as Evolutionary Tape Measures

Genes (DNA) and their products (proteins) are historical documents. These information-rich molecules are the records of an organism's hereditary background. The linear sequences of nucleotides in DNA molecules are passed from parents to offspring, and these DNA sequences determine the amino acid sequences of proteins in the offspring. The DNA and proteins of siblings are more similar than the DNA and proteins of unrelated individuals of the same species. This concept of molecular genealogy also extends to relationships between species.

Testable hypotheses are at the heart of science, and analysis of DNA and protein sequences adds a new tool for testing evolutionary hypotheses. For example, fossil evidence and anatomical similarity support the hypothesis that humans and gorillas are closely related animals. This hypothesis is testable—we can use it to make predictions about what to expect if the hypothesis is correct. For example, if humans and gorillas are closely related, then they should share a greater proportion of their inherited DNA and protein sequences than they do with more distantly related species. If molecular analysis did not confirm this prediction, then the test would cast doubt on the hypothesis of a close evolutionary relationship. In fact, however, molecular analysis of DNA and protein sequences in humans and gorillas supports the hypothesis that these two species are very closely related (**Figure 3.28**). Molecular biology has added a new tape measure to the toolkit biologists use to assess evolutionary relationships.

> **Evolution Connection on the Web**
> Learn why mice can be used as models for medical research.

Figure 3.28
Evolutionary relationships among six vertebrates, based on amino acid sequences of a protein.
Inherited DNA determines the amino acid sequences of proteins. In this case, the sequences are for one of the polypeptides making up hemoglobin, the oxygen-transporting protein in the blood of vertebrates (animals with backbones). The times marking the evolutionary branch points are based on the fossil history of these animals.

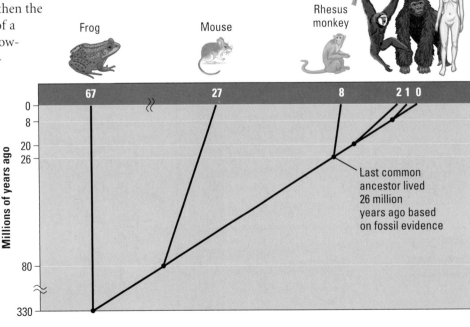

Chapter Review

For study help, go to the Essential Biology Website (www.essentialbiology.com) or CD-ROM to explore the Activities and Case Studies in the Process of Science.

Organic Molecules

• **Carbon Chemistry** Carbon atoms can form large, complex, diverse molecules by bonding to four partners, including other carbon atoms. In addition to variations in the size and shape of carbon skeletons, organic molecules also vary in the presence and locations of different functional groups.

Activity 3A *Diversity of Carbon-Based Molecules*

Case Study in the Process of Science *What Factors Determine the Effectiveness of Drugs?*

Activity 3B *Functional Groups*

• **Giant Molecules from Smaller Building Blocks**

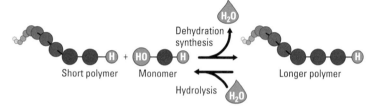

Activity 3C *Making and Breaking Polymers*

Biological Molecules

Biological macromolecule	Function	Monomer	Examples
Carbohydrates	Dietary energy; storage; plant structure	Monosaccharide	Monosaccharides: glucose, fructose. Disaccharides: lactose, sucrose. Polysaccharides: starch, cellulose.
Lipids	Long-term energy storage (for fats); hormones (for steroids)	Components of a fat molecule	Fats, oils, steroids
Proteins	Enzymes, structure, storage, contraction, transport, etc.	Amino acid	Lactase (an enzyme), hemoglobin
Nucleic acids	Information storage	Nucleotide	DNA, RNA

• **Carbohydrates** Simple sugars (monosaccharides) provide cells with energy and carbon skeletons for building other organic molecules. Double sugars (disaccharides), such as sucrose (table sugar), consist of two monosaccharides joined by dehydration synthesis. Polysaccharides are macromolecules, long polymers of sugar monomers. Starch and glycogen are storage polysaccharides in plants and animals, respectively. The cellulose of plant cell walls is an example of a structural polysaccharide.

Activity 3D *Models of Glucose*

Activity 3E *Carbohydrates*

• **Lipids** Along with other kinds of lipids, fats are hydrophobic. Fats are the major form of long-term energy storage in animals. A fat, or triglyceride, consists of three fatty acids joined to a glycerol by dehydration synthesis. Most animal fats are saturated, meaning that their fatty acids have the maximum amount of hydrogen. Plant oils are mostly unsaturated fats, having less hydrogen in the fatty acids because of double bonding in the carbon skeletons. Steroids, including cholesterol and the sex hormones, are also lipids.

Activity 3F *Lipids*

• **Proteins** The monomers of proteins, amino acids, come in 20 varieties. They are linked by dehydration synthesis to form polymers called polypeptides. A protein consists of one or more polypeptides folded into a specific three-dimensional shape. Contributing to this shape are four levels of structure: the protein's amino acid sequence (primary structure), localized coiling or pleating in certain regions of the protein (secondary structure), other contortions that reinforce the overall shape of the protein (tertiary structure), and an association of subunits in proteins with more than one polypeptide (quaternary structure). Shape is sensitive to environment, and if a protein loses its shape because of an unfavorable environment (denaturation), the protein's function is also disrupted.

Activity 3G *Protein Functions*

Activity 3H *Protein Structure*

• **Nucleic Acids** The chemical of inheritance is a nucleic acid called DNA. It takes the form of a double helix, two DNA strands (polymers of nucleotides) held together by bonds between nucleotide components called bases. There are four kinds of bases: adenine (A), guanine (G), thymine (T), and cytosine (C). A always pairs with T, and G always pairs with C. These base-pairing rules enable DNA to act as the molecule of inheritance.

Activity 3I *Nucleic Acid Functions*

Activity 3J *Nucleic Acid Structure*

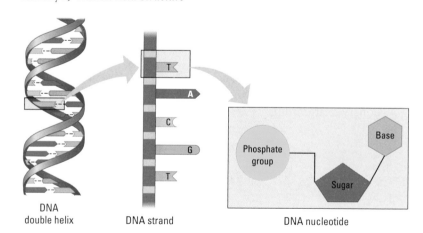

1. Monomers are joined together to form larger polymers through the chemical reaction called _____. Polymers are broken down into the monomers that make them up through the chemical reaction called _____.

2. Which of the following terms includes all the others in the list?
   a. polysaccharide
   b. carbohydrate
   c. monosaccharide
   d. disaccharide

3. One molecule of fat is made by joining three molecules of _____ to one molecule of _____.

4. Which of the following statements about saturated fats is true?
   a. Saturated fats contain one or more double bonds along the tail.
   b. Saturated fats contain the maximum number of hydrogens along the tail.
   c. Saturated fats make up the majority of most plant oils.
   d. Saturated fats are healthier for you than unsaturated fats.

5. What chemical element is present in proteins and nucleic acids, but not in sugars and fats?

6. Changing one amino acid within a protein could change what about a protein?
   a. the primary structure
   b. the overall shape of the protein
   c. the function of the protein
   d. all of the above

7. Most proteins can easily dissolve in water. Knowing that, where within the tertiary structure of a protein would you most likely find hydrophobic amino acids such as leucine?

8. A shortage of phosphorus in the soil would make it especially difficult for a plant to manufacture
   a. DNA.
   b. proteins.
   b. cellulose.
   d. fatty acids.
   e. sucrose.

9. A glucose molecule is to _____ as a _____ is to a nucleic acid.

10. Name three similarities between DNA and RNA. Name three differences.

Answers to the Self-Quiz questions can be found in Appendix B.

Go to the website or CD-ROM for more Self-Quiz questions.

The Process of Science

1. A food manufacturer is advertising a new cake mix as fat-free. Scientists at the U.S. Food and Drug Administration (FDA) are testing the product to see if it truly lacks fat. Hydrolysis of the cake mix yields glucose, fructose, glycerol, a number of amino acids, and several kinds of molecules with long hydrocarbon chains. Further analysis shows that most of the hydrocarbon chains have a carboxyl group at one end. What would you tell the food manufacturer if you were the spokesperson for the FDA?

2. As described in the Biology and Society section at the start of the chapter, lactase is an enzyme that breaks down the disaccharide lactose into the monosaccharides glucose and galactose. Imagine that you have produced several mutant versions of lactase, each of which differs from normal lactase by a single amino acid. Describe a test that could indirectly determine which of the mutations significantly alters the three-dimensional shape of the protein.

Case Study in the Process of Science on the Web & CD *Perform experiments to determine the effectiveness of drugs with slightly different shapes.*

Biology and Society

1. Some amateur and professional athletes take anabolic steroids to help them "bulk up" or build strength. The health risks of this practice are extensively documented. Apart from these health issues, what is your opinion about the ethics of athletes using chemicals to enhance performance? Is this a form of cheating, or is it just part of the preparation required to stay competitive in a sport where anabolic steroids are commonly used? Defend your opinion.

2. Recent court rulings have found that tobacco companies can be held liable for the health consequences of their products. While lung cancer does kill many people every year, heart disease kills many more. Imagine you're a juror sitting on a trial where a fast-food manufacturer is being sued for producing a harmful product. To what extent do you think manufacturers of unhealthy foods should be held responsible for the health consequences of their products? In what ways is this situation similar to the tobacco lawsuits? In what ways is it different? As a jury member, how would you vote?

3. Each year, industrial chemists develop and test thousands of new organic compounds for use as pesticides, such as insecticides, fungicides, and weed killers. In what ways are these chemicals useful and important to us? In what ways can they be harmful? Is your general opinion of pesticides positive or negative? What influences have shaped your feelings about these chemicals?

Biology and Society on the Web *Learn more about lactose intolerance.*

CHAPTER 4

A Tour of the Cell

Biology and Society:
Drugs That Target Cells 53

The Microscopic World
of Cells 53
Microscopes as Windows to Cells
The Two Major Categories of Cells
A Panoramic View of Eukaryotic Cells

Membrane Structure
and Function 58
A Fluid Mosaic of Lipids and Proteins
Selective Permeability

The Nucleus and Ribosomes:
Genetic Control of the Cell 59
Structure and Function of the Nucleus
Ribosomes
How DNA Controls the Cell

The Endomembrane System:
Manufacturing and Distributing
Cellular Products 61
The Endoplasmic Reticulum
The Golgi Apparatus
Lysosomes
Vacuoles

Chloroplasts and Mitochondria:
Energy Conversion 64
Chloroplasts
Mitochondria

The Cytoskeleton:
Cell Shape and Movement 65
Maintaining Cell Shape
Cilia and Flagella

Cell Surfaces: Protection,
Support, and Cell-Cell
Interactions 67
Plant Cell Walls and Cell Junctions
Animal Cell Surfaces and Cell
Junctions

Evolution Connection:
The Origin of Membranes 69

 **If you stacked up 8,000 cell membranes, they** would only be as thick as a page in this book.

 **An electron microscope can visualize objects a million times smaller than the head of a pin.**

Every second, your body produces about 2 million red blood cells.

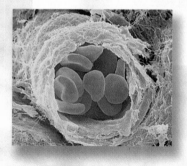

 **The cells of a whale are about the same size as the cells of a mouse.**

Drugs That Target Cells

Antibiotics are one of the great marvels of modern medicine. Antibiotics are drugs that disable or kill infectious bacteria. Most antibiotics are naturally occurring chemicals derived from other microorganisms. Penicillin, for example, was originally isolated from mold in 1928 and has been widely prescribed since the 1940s (**Figure 4.1**). A revolution in human health rapidly followed its introduction. The fatality rates of many diseases (such as bacterial pneumonia and surgical infections), were drastically reduced, saving millions of lives.

Biology and Society on the Web
Learn about the downside of antibiotic use.

The goal of treatment with antibiotic drugs is to kill invading bacteria while causing minimal harm to the human host. But how does the antibiotic recognize its bacterial target among trillions of human cells? Most antibiotics achieve such precision by binding to structures that are found only in bacterial cells. Erythromycin, for example, binds to part of the bacterial ribosome, a cellular organelle. The ribosomes of humans and bacteria are sufficiently different that erythromycin will only bind to bacterial ribosomes, leaving human ribosomes unaffected. The antibiotics streptomycin, tetracycline, and chloramphenicol also target bacterial ribosomes. The antibiotic ciprofloxacin has received a lot of attention recently because it is the drug of choice to use against anthrax-causing bacteria. It works by targeting an enzyme that bacteria need to maintain their chromosome structure. Human cells can survive just fine without this enzyme because their chromosomes have a sufficiently different makeup. Penicillin, ampicillin, and bacitracin work by disrupting the synthesis of cell walls, a structural feature of many bacterial cells that is entirely absent in human cells.

To understand how life on Earth works—including your own body—you first need to learn about cells. All organisms are made of cells. Cells are the building blocks of all life, which makes them as fundamental to biology as atoms are to chemistry. Moreover, the cell is the smallest entity that exhibits all the characteristics of life. In this chapter, we'll take a tour of cells and explore their structure and function.

Colorized SEM 3,500×

Figure 4.1 The mold that produces penicillin.
The first antibiotic discovered was penicillin, which is made by the common mold *Penicillium notatum*, shown here growing on bread and in a close-up view. Alexander Fleming, who discovered penicillin in 1928, did not set out to discover an antibiotic. He was trying to grow bacteria, but the pesky mold *Penicillium notatum* had contaminated his bacterial cultures. Fleming almost tossed out the contaminated cultures, but noticed that bacteria did not grow near the mold. Fortunately, he recognized the value of an agent that could inhibit bacterial growth, and the age of antibiotics was born.

The Microscopic World of Cells

Each cell in the human body is a miniature marvel of great complexity. Cells must be tiny for materials to move into and out of them fast enough to meet the cell's metabolic needs. But even though they are so small, cells are also very complex. If a complicated machine with millions of parts—say, a jumbo jet—were reduced to microscopic size, it would still seem simple and crude compared to the complexity of a living cell.

Organisms are either single-celled, such as most bacteria and protists, or multicelled, such as plants, animals, and most fungi. Your own body is a cooperative society of trillions of cells of many different specialized types. Three examples are the muscle cells that move your arms and legs, the nerve cells that control your muscles, and the red blood cells that carry oxygen to your muscles and nerves. Everything you do—every action and every thought—reflects processes occurring at the cellular level. For exploring this world of cells, our main tools are microscopes.

Microscopes as Windows to Cells

Our understanding of nature often parallels the invention and refinement of instruments that extend human senses to new limits. The development of microscopes, for example, provided increasingly clear windows to the world of cells.

The type of microscope used by Renaissance scientists, as well as the microscope you will use if your biology course includes a lab, is called a **light microscope (LM).** Visible light passes through the specimen, such as human white blood cells (**Figure 4.2a**). Glass lenses then enlarge the image and project it into a human eye or a camera.

Two important values of microscopes are magnification and resolving power. **Magnification** is an increase in the object's apparent size compared to its actual size. The clarity of that magnified image depends on **resolving power,** which is the ability of an optical instrument to show two objects as separate. For example, what appears to the unaided eye as one star in the sky may be resolved as two stars with a telescope. Each optical instrument—be it an eye, a telescope, or a microscope—has a limit to its resolving power. The human eye can resolve images that are as close as 1/10 millimeter (mm), which equals 10^{-4} meter (m). For light microscopes, the resolving power is limited to about 0.2 micrometer (μm), the size of a small bacterial cell (1 μm = 1/1,000 mm, or 10^{-3} mm, which equals 10^{-6} m). This limits the useful magnification to about 1,000×. With greater magnification, the image is blurry.

Activity 4A on the Web & CD
Review the metric system.

Figure 4.2 Different views of human white blood cells.

These photographs were taken with different types of microscopes: **(a)** light microscope (LM); **(b)** scanning electron microscope (SEM); and **(c)** transmission electron microscope (TEM). Such photographs taken with microscopes are called micrographs. Throughout this textbook, each micrograph will have a notation along its side. For example, "LM 1,350×" indicates that the micrograph was taken with a light microscope and the objects are magnified to 1,350 times their original size.

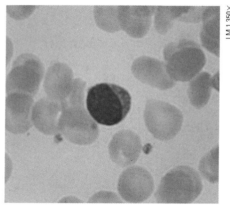

LM 1,350×

(a) Light micrograph (LM) of a white blood cell (stained purple) surrounded by red blood cells

(c) Transmission electron micrograph (TEM) of a white blood cell

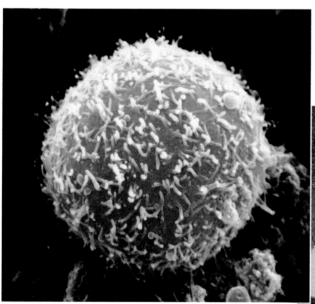

Colorized SEM 45,000×

(b) Scanning electron micrograph (SEM) of a white blood cell

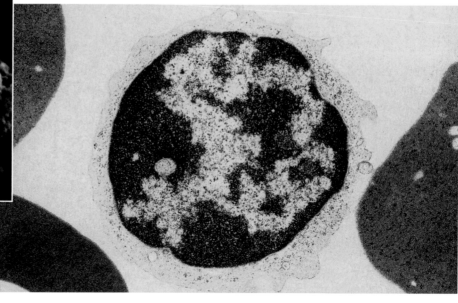

Colorized TEM 51,000×

A Panoramic View of Eukaryotic Cells

Figure 4.6 provides a panoramic view of two idealized eukaryotic cells, an animal cell and a plant cell. The similarities are more obvious than the differences. Both cells have a very thin outer membrane, the **plasma membrane,** which regulates the traffic of molecules between the cells and their surroundings. Each cell also has a prominent **nucleus,** a membrane-enclosed organelle that contains genes.

The entire region of the cell between the nucleus and plasma membrane is called the **cytoplasm.** It consists of various organelles suspended in a fluid, the **cytosol.** The structure of each organelle has become adapted during evolution to perform specific functions. Most of the organelles are enclosed by membranes, but some are not.

As you can see in Figure 4.6, most of the organelles are found in both animal and plant cells.

One important difference is the presence of chloroplasts in plant cells but not in animal cells. Chloroplasts are the organelles that convert light energy to the chemical energy of food. Also notice that plant cells have walls exterior to their plasma membranes. Animal cells lack this protective cell wall. (This is why cell walls are a good target for antibiotics like penicillin.) We'll see other differences and similarities between plant and animal cells as we now take a closer look at the architecture of eukaryotic cells, beginning with the plasma membrane.

Activity 4C on the Web & CD
Test your understanding of different types of cells.

Activity 4D on the Web & CD
Build an animal cell and a plant cell.

Figure 4.6 A Panoramic view of an idealized animal cell and plant cell. For now, the labels on the drawings are just words, but these organelles will come to life as we take a closer look at how each part of the cell functions. To keep from getting lost on our tour of the cell, we'll carry miniature versions of these overview diagrams as our roadmaps, with the structure we're interested in highlighted.

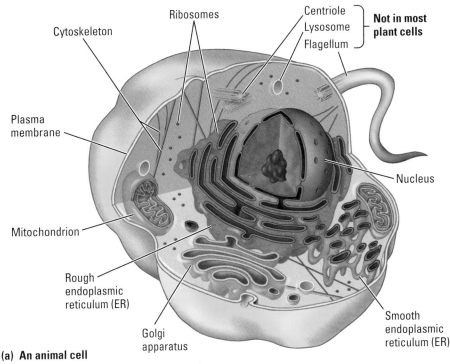

(a) An animal cell

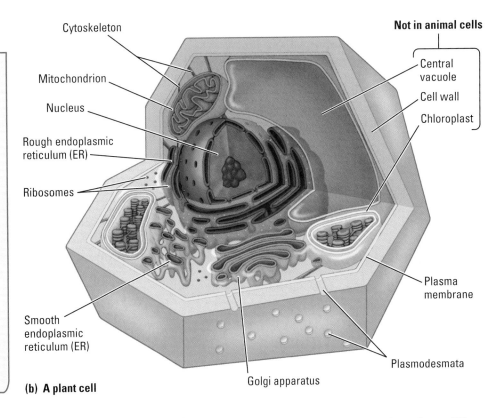

(b) A plant cell

CHECKPOINT

1. Which type of microscope would you use to study (a) the changes in shape of a living human white blood cell; (b) the finest details of surface texture of a human hair; (c) the detailed structure of an organelle in the cytoplasm of a human liver cell?

2. Using a light microscope to examine a thin section of a large spherical cell, you find that the cell is 0.3 mm in diameter. The nucleus is about one-fourth as wide. What is the diameter of the nucleus in micrometers?

3. How is the nucleoid region of a prokaryotic cell unlike the nucleus of a eukaryotic cell?

4. Name two structures in plant cells that animal cells lack.

Answers: 1. (a) light microscope; (b) scanning electron microscope; (c) transmission electron microscope **2.** About 0.075 mm, which equals 75 μm, or 7.5 x 10^{-5} m **3.** There is no membrane enclosing the DNA of the nucleoid region. **4.** Chloroplasts and a cell wall

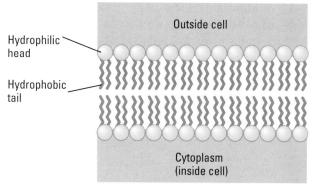

Hydrophilic head

Hydrophobic tail

Outside cell

Cytoplasm (inside cell)

(a) Phospholipid bilayer of membrane

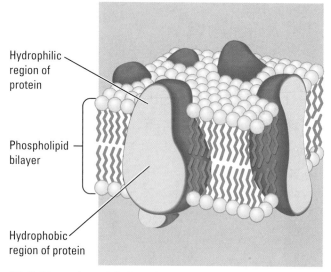

Hydrophilic region of protein

Phospholipid bilayer

Hydrophobic region of protein

(b) Fluid mosaic model of membrane

Figure 4.7 Plasma membrane structure.

(a) At the interface between two aqueous compartments, phospholipids arrange themselves into a bilayer. The symbol for phospholipids that we'll use throughout this book looks like a lollipop with two wavy sticks. The "head" of the lollipop is the end with the phosphate group, which is hydrophilic. The two sticks, the hydrocarbon tails of the phospholipid, are hydrophobic. Notice how the bilayer arrangement keeps the heads exposed to water while keeping the tails in the dry interior of the membrane. **(b)** Membrane proteins, like the phospholipids, have both hydrophilic and hydrophobic regions. The membrane behaves as a fluid mosaic because its phospholipids and many of its diverse proteins drift within the membrane.

Membrane Structure and Function

The plasma membrane is the edge of life, the boundary that separates the living cell from its nonliving surroundings. The plasma membrane is a remarkable film, so thin that you would have to stack 8,000 of them to equal the thickness of the page you're reading. Yet the plasma membrane can regulate the traffic of chemicals into and out of the cell. The key to how a membrane works is its structure.

A Fluid Mosaic of Lipids and Proteins

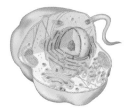

The plasma membrane and other membranes of the cell are composed mostly of lipids and proteins. The lipids belong to a special category called **phospholipids.** They are related to dietary fats but have only two fatty acids instead of three (see Figure 3.15). A phospholipid has a phosphate group (a combination of phosphorus and oxygen) in place of the third fatty acid. The phosphate group is electrically charged, which makes it hydrophilic ("water-loving"). The rest of the phospholipid, however, consisting of the two fatty acids, is hydrophobic ("water-fearing"). Thus, phospholipids have a kind of chemical ambivalence in their interactions with water. The phosphate group "head" mixes with water, while the fatty acid "tails" avoid it. This makes phospholipids good membrane material. By forming a two-layered membrane, or **phospholipid bilayer,** the hydrophobic parts of the molecules hide from water, while the hydrophilic portions are wet (**Figure 4.7a**). Most membranes have specific proteins embedded in the phospholipid bilayer (**Figure 4.7b**). **Figure 4.8** surveys some of the functions of these membrane proteins.

Activity 4E on the Web & CD Use what you have learned to label a diagram of a membrane.

Membranes are not static sheets of molecules locked rigidly in place. The phospholipids and most of the proteins are free to drift about in the plane of the membrane. This behavior is captured in the description of a membrane as a **fluid mosaic**—fluid because the molecules can move freely past one another and mosaic because of the diversity of proteins that float like icebergs in the phospholipid sea.

Selective Permeability

The plasma membrane and other membranes of the cell are **selectively permeable.** That means that a membrane allows some substances to cross more easily than others and blocks the passage of some substances altogether. For example, the plasma membrane allows the cell to take up such materials as oxygen and nutrients and to dispose of wastes such as carbon dioxide. The traffic of some substances can only occur through avenues called **transport proteins,** which are among the specialized proteins built into membranes. The sugar glucose, for example, requires a transport protein to move it into the cell. Controlling the passage of materials across membranes is an important function of all cells, and we'll explore it in detail in Chapter 5.

Activity 4F on the Web & CD See how different molecules get through the plasma membrane.

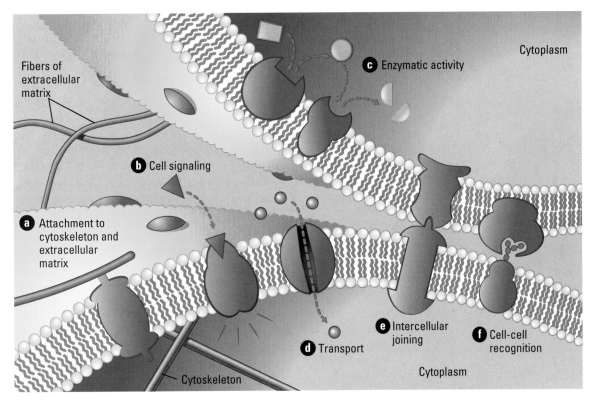

Figure 4.8 Primary functions of membrane proteins.

a **Attachment to the cytoskeleton and extracellular matrix.** Elements of the cytoskeleton may be bonded to membrane proteins, a function that helps maintain cell shape and fixes the location of certain membrane proteins. Proteins that adhere to the fibers of the extracellular matrix can coordinate extracellular and intracellular changes.

b **Cell signaling.** A membrane protein may have a binding site with a specific shape that fits the shape of a chemical messenger, such as a hormone. The external messenger (signal) may cause a change in the protein that relays the message to the inside of the cell.

c **Enzymatic activity.** A protein built into the membrane may be an enzyme with its active site exposed to substances in the adjacent solution. In some cases, several enzymes in a membrane are organized as a team that carries out the sequential steps of a metabolic pathway.

d **Transport.** A protein that spans the membrane may provide a channel across the membrane that is selective for a particular solute.

e **Intercellular joining.** Membrane proteins of adjacent cells may be hooked together to form various kinds of junctions.

f **Cell-cell recognition.** Some proteins with short chains of sugars serve as identification tags that are specifically recognized by other cells.

CHECKPOINT

1. Why do phospholipids tend to organize into a bilayer in an aqueous solution?

2. Explain how each word in the term *fluid mosaic* describes the structure of a membrane.

Answers: 1. The structure shields the hydrophobic tails of the phospholipids from water while exposing the hydrophilic heads to water. **2.** A membrane is fluid because its components are not locked into place. A membrane is mosaic because it contains a variety of different proteins embedded within it.

The Nucleus and Ribosomes: Genetic Control of the Cell

If we think of the cell as a factory, then the nucleus is its executive boardroom. The top managers are the genes, the inherited DNA molecules that direct almost all the business of the cell. Genes store the information necessary

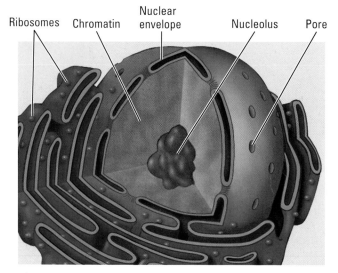

Ribosomes   Chromatin   Nuclear envelope   Nucleolus   Pore

Figure 4.9 **The nucleus.**

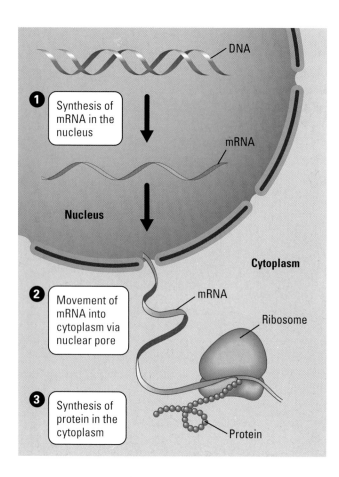

DNA

1 Synthesis of mRNA in the nucleus

mRNA

Nucleus

Cytoplasm

2 Movement of mRNA into cytoplasm via nuclear pore

mRNA

Ribosome

3 Synthesis of protein in the cytoplasm

Protein

to produce proteins, which then do most of the actual work of the cell. A **gene** is a specific stretch of DNA that contains the code for the structure of a specific protein.

Structure and Function of the Nucleus

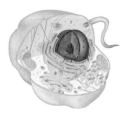

The nucleus is bordered by a double membrane called the **nuclear envelope** (**Figure 4.9**). Each membrane of the nuclear envelope is similar in structure to the plasma membrane. Pores through the envelope allow the passage of material between the nucleus and the cytoplasm. Within the nucleus, long DNA molecules and associated proteins form long fibers called **chromatin.** Each long fiber constitutes one **chromosome.** The number of chromosomes in a cell depends on the species; for example, each human body cell has 46 chromosomes, while rice cells have 24, and porcupine cells have 34. No matter their number, each individual chromosome is made of one long strand of chromatin, a combination of DNA and proteins.

In addition to chromosomes, the nucleus contains a ball-like mass of fibers and granules called the **nucleolus.** It produces the component parts of ribosomes.

Ribosomes

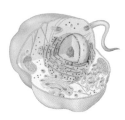

The small dots in the cells in Figure 4.6 and outside the nucleus in Figure 4.9 are **ribosomes.** Although they are assembled from components made in the nucleus, ribosomes do not begin to work until they move to the cytoplasm. Ribosomal parts travel from the nucleus to the cytoplasm via the pores in the nuclear envelope. In the cytoplasm, ribosomes build all the cell's proteins. Some ribosomes are suspended in the cytosol, the fluid of the cytoplasm. They make proteins that will remain dissolved in the cytosol. Other ribosomes are attached to the outside of a membranous organelle called the endoplasmic reticulum. These ribosomes make proteins destined to be incorporated into membranes or secreted by the cell.

How DNA Controls the Cell

Activity 4G on the Web & CD
Get an overview of protein synthesis.

How do the DNA "executives" in the nucleus direct the "workers" in the cytoplasm? DNA does this by transferring its coded information to a portable molecule called messenger RNA (mRNA) (**Figure 4.10**). These genetic messages exit the nucleus through the pores in the envelope. Like a middle manager, an RNA molecule carries the order to "build this type of protein" from the nucleus to the cytoplasm. You'll learn how the message is actually translated into a specific protein in Chapter 10.

Figure 4.10
DNA → RNA → Protein: How genes in the nucleus control the cell.
In a eukaryotic cell, DNA in the nucleus programs protein production in the cytoplasm by dictating **1** the synthesis of messenger RNA (mRNA). **2** The mRNA travels to the cytoplasm and binds to ribosomes. As a ribosome (disproportionately large in this drawing) moves along the mRNA, **3** the genetic message is translated into a protein of specific amino acid sequence. The structure and behavior of a cell reflect the kinds of proteins it has, but it is the inherited DNA that determines a cell's protein composition.

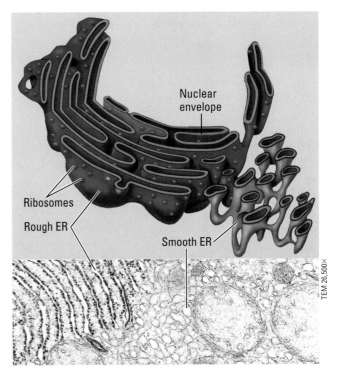

Figure 4.11 Endoplasmic reticulum (ER).
The flattened sacs of rough ER and the tubes of smooth ER are continuous, though different in structure and function. Notice that the ER is also continuous with the nuclear envelope, which is actually part of the endomembrane system.

The Endomembrane System: Manufacturing and Distributing Cellular Products

Notice again in Figure 4.6 that the cytoplasm of a eukaryotic cell is partitioned by membranes. Many of the membranous organelles belong to the **endomembrane system.** This system includes the endoplasmic reticulum, the Golgi apparatus, lysosomes, and vacuoles.

The Endoplasmic Reticulum

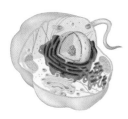

To continue our factory metaphor of the cell, the **endoplasmic reticulum (ER)** is one of the main manufacturing facilities. It produces an enormous variety of molecules. The ER is a membranous labyrinth of tubes and sacs running throughout the cytoplasm. The ER membrane separates its internal compartment from the surrounding cytosol (**Figure 4.11**). There are two distinct types of ER: rough ER and smooth ER. These two ER components are physically connected, but they differ in structure and function.

Rough ER The "rough" in **rough ER** refers to the appearance of this organelle in electron micrographs (see Figure 4.11). The roughness is due to ribosomes that stud the outside of the ER membrane. These ribosomes produce two main types of proteins: membrane proteins and secretory proteins. Some newly manufactured membrane proteins are embedded right in the ER membrane. Thus, one function of rough ER is the production of new membrane. Secretory proteins are those the cell will actually export (secrete). Cells that secrete a lot of protein, such as the cells of your salivary glands that secrete an enzyme into your mouth, are especially rich in rough ER. Some of the products manufactured by rough ER are dispatched to other locations in the cell via **transport vesicles,** membranous spheres that bud from the ER (**Figure 4.12**).

Smooth ER The "smooth" in **smooth ER** refers to the fact that this organelle lacks the ribosomes that populate the surface of rough ER (see Figure 4.11). A diversity of enzymes built into the smooth ER membrane enables this organelle to perform many functions. One is the synthesis of lipids, including steroids (see Figure 3.16). For example, the cells in your ovaries or testes that produce sex hormones, which are steroids, are enriched with smooth ER.

In liver cells, the functions of smooth ER include the detoxification of drugs and other poisons that might be present in the bloodstream. For example, certain ER enzymes detoxify sedatives such as barbiturates, stimulants such

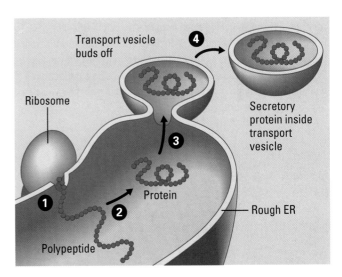

Figure 4.12 How rough ER manufactures and packages secretory proteins. A secretory protein is a protein that a cell exports to the fluid outside the cell. **1** A ribosome links amino acids to form a polypeptide chain with a unique amino acid sequence (see Chapter 3). The chain threads through the membrane and into the cavity of the ER. **2** Secretory proteins are often modified in the ER. **3** Secretory proteins depart in transport vesicles that **4** bud off from the ER. If the proteins require no further refinement in other organelles, they are secreted from the cell when the vesicles fuse with the plasma membrane, the outer membrane of the cell.

as amphetamines, and some antibiotics (which is why they don't persist in the bloodstream after combating an infection). As liver cells are exposed to the chemicals in a drug, the amounts of smooth ER and its detoxifying enzymes increase. This can strengthen the body's tolerance for the drug, meaning that higher doses will be required in the future to achieve the desired effect, such as sedation. The growth of smooth ER in response to one drug can also increase tolerance to other drugs, including important medicines. Barbiturate use, for example, may decrease the effectiveness of certain antibiotics by accelerating their breakdown in the liver. Furthermore, increasing tolerance to drugs is one of the hallmarks of addiction—a potentially serious consequence of drug use.

The Golgi Apparatus

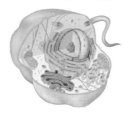

The **Golgi apparatus,** an organelle named for its discoverer, Italian scientist Camillo Golgi, is a refinery, warehouse, and shipping center. Working in close partnership with the ER, the Golgi apparatus receives, refines, stores, and distributes chemical products of the cell (**Figure 4.13**). For example, the Golgi is believed to chemically tag protein products to mark their final destination within the cell. Products made in the ER reach the Golgi in transport vesicles. Enzymes of the Golgi modify many of these products, and vesicles that bud from the Golgi then distribute the finished molecules to other organelles or to the plasma membrane. Vesicles that bind with the plasma membrane secrete finished chemical products to the outside of the cell.

Lysosomes

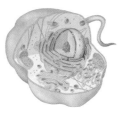

The name **lysosome** ("breakdown body") is a good description of how these organelles function in animal cells (they are absent from most plant cells). A lysosome is a membrane-enclosed sac of digestive enzymes. These enzymes can break down macromolecules such as proteins, polysaccharides, fats, and nucleic acids. The lysosome provides a compartment where the cell can digest macromolecules safely, without committing suicide by unleashing these digestive enzymes on the cell itself.

Figure 4.13 **The Golgi apparatus.**

This component of the endomembrane system consists of flattened sacs arranged something like a stack of pita bread. A cell may contain just a few Golgi stacks or hundreds of them. The Golgi apparatus receives, refines, and distributes products of the ER. One side of a Golgi stack serves as a receiving dock for vesicles transported from the ER. The ER products are often modified during their stay in the Golgi. The "shipping" side of a Golgi stack serves as a depot from which the finished products can be dispatched in transport vesicles to other organelles or to the plasma membrane.

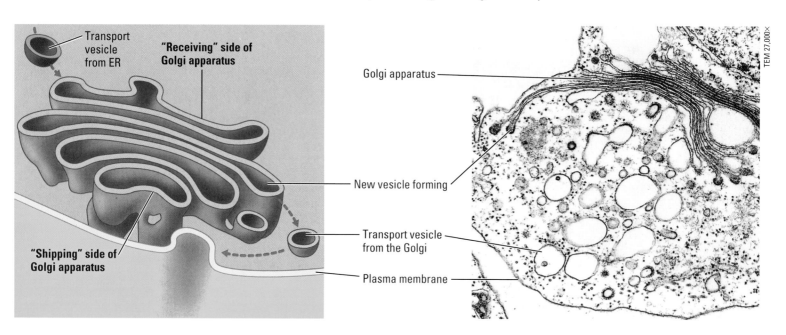

Lysosomes have several types of digestive functions. Many cells engulf nutrients into tiny cytoplasmic sacs called **food vacuoles.** Lysosomes fuse with the food vacuoles, exposing the food to enzymes that digest it (**Figure 4.14a**). Small molecules that result from this digestion, such as amino acids, leave the lysosome and nourish the cell. Lysosomes also help destroy harmful bacteria. Our white blood cells ingest bacteria into vacuoles, and lysosomal enzymes that are emptied into these vacuoles rupture the bacterial cell walls. Lysosomes also serve as recycling centers for damaged organelles. Without harming the cell, a lysosome can engulf and digest parts of another organelle, making its molecules available for the construction of new organelles (**Figure 4.14b**). Lysosomes also have sculpturing functions in embryonic development. For example, lysosomal enzymes destroy cells of the webbing that joins the fingers of early human embryos. In this case, the lysosomes act as "suicide packs," breaking open and causing the programmed death of whole cells.

Abnormal lysosomes are associated with hereditary disorders called **lysosomal storage diseases.** A person with such a disease is missing one of the digestive enzymes of lysosomes. The abnormal lysosomes become engorged with indigestible substances, and this eventually interferes with other cellular functions. Most of these diseases are fatal in early childhood. An example is Tay-Sachs disease, which ravages the nervous system. In this disorder, lysosomes lack a lipid-digesting enzyme, and nerve cells in the brain are damaged as they accumulate excess lipids.

Vacuoles

Vacuoles are membranous sacs that bud from the ER, Golgi, or plasma membrane. Vacuoles come in different sizes and have a variety of functions. Figure 4.14a shows a food vacuole budding from the plasma membrane. Certain freshwater protists have contractile vacuoles that function as pumps to expel excess water that flows into the cell from the outside environment (**Figure 4.15a**). Another type of vacuole is a plant cell's **central vacuole,** which can account for more than half the volume of a mature cell (**Figure 4.15b**).

The plant cell vacuole is a versatile compartment. It is the place where the plant stores organic nutrients. For example, proteins are stockpiled in the vacuoles of cells in seeds, such as beans and peas. Central vacuoles also

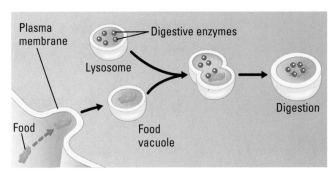

(a) Lysosome digesting food

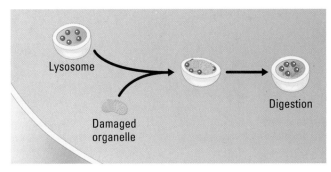

(b) Lysosome breaking down damaged organelle

Figure 4.14 The functions of lysosomes.
As sacs of digestive enzymes, lysosomes **(a)** digest food and **(b)** help recycle the molecules of the cell itself by breaking down damaged organelles.

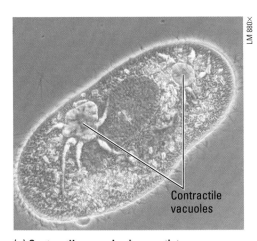

LM 880×

(a) Contractile vacuoles in a protist

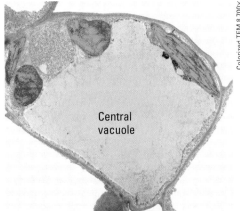

Colorized TEM 8,700×

Central vacuole

(b) Central vacuole in a plant cell

Figure 4.15 Two types of vacuoles.
(a) This single-celled organism, a protist named *Paramecium,* has two contractile vacuoles.
(b) The central vacuole is often the largest organelle in a mature plant cell.

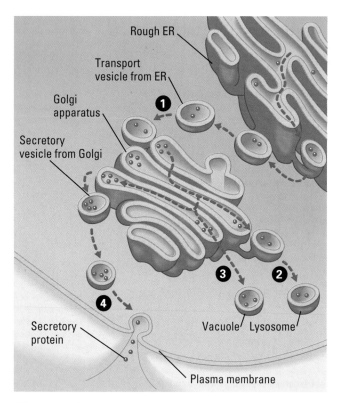

Figure 4.16 Review of the endomembrane system.
The arrows show some of the pathways of cell product distribution and membrane migration via transport vesicles. For example, ❶ digestive enzymes made by the rough ER are transported via vesicles to the Golgi for processing. ❷ Lysosomes containing the processed digestive enzymes bud off from the Golgi. ❸ Other cell products end up in vacuoles for storage. ❹ Still other cell products are secreted from the cell.

contribute to plant growth by absorbing water and causing cells to expand. Central vacuoles in flower petals may contain pigments that attract pollinating insects. Central vacuoles can also contain poisons that protect against plant-eating animals.

Figure 4.16 will help you review how all the organelles of the endomembrane system are related. Note that it is possible for a product made in one part of the endomembrane system to eventually exit the cell or become part of another organelle without ever crossing a membrane. Also note that membrane originally fabricated by the ER can eventually turn up as part of the plasma membrane through the fusion of secretory vesicles. In this way, even the plasma membrane is related to the endomembrane system.

Activity 4H on the Web & CD
Trace the movement of a protein through the endomembrane system.

CHECKPOINT

1. Which structure includes all the others in the list: rough ER, smooth ER, endomembrane system, the Golgi apparatus?

2. What makes rough ER rough?

3. What is the relationship between the Golgi apparatus and the ER in a protein-secreting cell?

4. How can defective lysosomes result in excess accumulation of a particular chemical in a cell?

Answers: 1. Endomembrane system **2.** Ribosomes attached to the membranes **3.** The Golgi receives transport vesicles that bud from the ER and that contain proteins synthesized in the ER. The Golgi finishes processing the proteins and then dispatches transport vesicles that secrete the proteins to the outside of the cell. **4.** If the lysosomes lack an enzyme needed to break down the compound, the cell will accumulate an excess of that compound.

Chloroplasts and Mitochondria: Energy Conversion

A cell requires a continuous energy supply to do all the work of life. The two types of cellular power stations are the organelles called chloroplasts and mitochondria.

Chloroplasts

Most of the living world runs on the energy provided by photosynthesis, the conversion of light energy from the sun to the chemical energy of sugar and other organic molecules. **Chloroplasts** are the organelles of plants and protists that perform photosynthesis.

Befitting an organelle that carries out complex, multistep processes, internal membranes partition the chloroplast into three major compartments (**Figure 4.17**). One compartment is the space between the two membranes that envelop the chloroplast. The **stroma,** the thick fluid within the chloroplast, is the second compartment. Suspended in

Figure 4.17 The chloroplast: site of photosynthesis.

Inner and outer membranes of envelope · Space between membranes · Stroma (fluid in chloroplast) · Granum

that fluid, the interior of a network of membrane-enclosed tubes and disks forms the third compartment. Notice in Figure 4.17 that the disks occur in stacks called **grana** (singular, *granum*). The grana are the chloroplast's solar power packs, the structures that actually trap light energy and convert it to chemical energy. You'll learn in Chapter 7 how the chloroplast functions.

Mitochondria

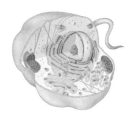

Mitochondria (singular, *mitochondrion*) are the sites of cellular respiration. This process harvests energy from sugars and other food molecules and converts it to another form of chemical energy called ATP. Cells use molecules of ATP as the direct energy source for most of their work. In contrast to chloroplasts, which are unique to the photosynthetic cells of plants and protists, mitochondria are found in almost all eukaryotic cells, including your own.

An envelope of two membranes encloses the mitochondrion, which contains a thick fluid (**Figure 4.18**). The inner membrane of the envelope has numerous infoldings called **cristae**. Many of the enzymes and other molecules that function in cellular respiration are built into the inner membrane. By increasing the surface area of this membrane, the cristae maximize ATP output. In Chapter 6, you'll learn more about how mitochondria convert food energy to ATP energy.

Activity 4I on the Web & CD
Build a chloroplast and a mitochondrion.

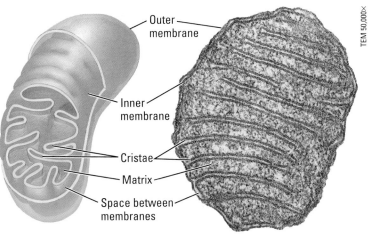

TEM 50,000×

Outer membrane
Inner membrane
Cristae
Matrix
Space between membranes

Figure 4.18
The mitochondrion: site of cellular respiration.

The Cytoskeleton: Cell Shape and Movement

If someone asked you to describe a house, you would most likely mention the number of rooms and their location. You probably would not think to mention the beams and floor joists that support the house. Yet these structures perform an extremely important function. Similarly, cells have an infrastructure called the **cytoskeleton,** a network of fibers extending throughout the cytoplasm. The cytoskeleton serves as both skeleton and "muscles" for the cell, functioning in both support and movement.

Maintaining Cell Shape

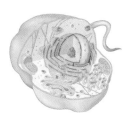

One function of the cytoskeleton is to give mechanical support to the cell and maintain its shape. This is especially important for animal cells, which lack rigid cell walls. The cytoskeleton contains several types of fibers made from different types of protein. One of the most important types of fibers

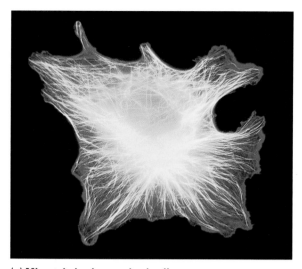

(a) Microtubules in an animal cell

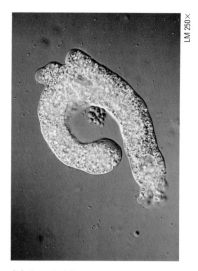

LM 250×

(b) Amoeboid movement

Figure 4.19 The cytoskeleton.
(a) In this micrograph, the microtubules of the cytoskeleton have been labeled with a fluorescent dye. **(b)** Rapid degradation and rebuilding of microtubules is responsible for the crawling movement of organisms like the protist *Amoeba.*

is **microtubules** (**Figure 4.19a**). Microtubules are straight, hollow tubes composed of globular proteins called tubulins.

Just as the bony skeleton of your body helps fix the positions of your organs, the cytoskeleton provides anchorage and reinforcement for many organelles in a cell. For instance, the nucleus is often held in place by a cytoskeletal cage of filaments. Other organelles move along tracks made from microtubules. For example, a lysosome might reach a food vacuole by moving along a microtubule. Microtubules also guide the movement of chromosomes when cells divide.

Although providing support like an animal's skeleton, the cell's cytoskeleton is more dynamic. It can quickly dismantle in one part of the cell by removing protein subunits and re-form in a new location by reattaching the subunits. Such rearrangement can provide rigidity in a new location, change the shape of the cell, or even cause the whole cell or some of its parts to move. This process contributes to the amoeboid (crawling) motions of the protist *Amoeba* and some of our white blood cells (**Figure 4.19b**).

Cilia and Flagella

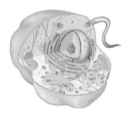

In some eukaryotic cells, a specialized arrangement of microtubules functions in the beating of flagella and cilia. Cilia and flagella are motile appendages—extensions from a cell that aid in locomotion. **Flagella** (singular, *flagellum*) propel the cell by an undulating whiplike motion. They often occur singly, such as in the sperm cells of humans and other animals (**Figure 4.20a**). **Cilia** (singular, *cilium*) are generally shorter and more numerous than flagella and promote movement by a coordinated back-and-forth motion, like the rhythmic oars of a galley ship. Both cilia and flagella propel various protists through water (**Figure 4.20b**). Though different in length, number per cell, and beating pattern, cilia and flagella have the same basic architecture with a core of microtubules wrapped in an extension of the plasma membrane (see Figure 1.9).

Activity 4J on the Web & CD
See an animation of a flagellum moving.

Some cilia or flagella extend from nonmoving cells that are part of a tissue layer. There they function to move fluid over the surface of the tissue. For example, the ciliated lining of your windpipe helps cleanse your respiratory system by sweeping mucus with trapped debris out of your lungs (**Figure 4.20c**). Tobacco smoke—whether inhaled first- or secondhand—irritates these ciliated cells, inhibiting or destroying the cilia. This interferes with the normal cleansing mechanisms and allows more toxin-laden smoke particles to reach the lungs. Frequent coughing—common in heavy smokers—then becomes the body's attempt to cleanse the respiratory system.

Because human sperm rely on flagella for movement, it's easy to understand why problems with flagella can lead to male infertility. Sperm with malfunctioning flagella are unable to travel up the female reproductive tract to fertilize the ovum (egg). Interestingly, some men with a certain type of

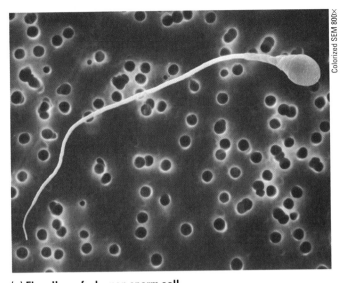

(a) Flagellum of a human sperm cell

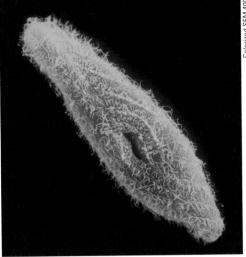

(b) Cilia on a protist

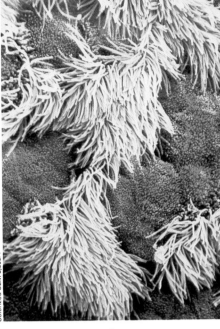

(c) Cilia lining the respiratory tract

Figure 4.20 Flagella and cilia.
(a) A flagellum usually undulates, its snakelike motion driving a cell such as this sperm cell. (b) Cilia are shorter and more numerous and move with a back-and-forth motion. A dense nap of beating cilia covers this *Paramecium*, a freshwater protist that can dart rapidly through its watery home. (c) The cilia lining your respiratory tract sweep mucus with trapped debris out of your lungs.

hereditary sterility also suffer from respiratory problems. The explanation for this lies in the similarities between flagella (found in sperm) and cilia (found lining the respiratory tract). Due to a defect in the structure of their flagella and cilia, their sperm do not swim and their cilia do not sweep mucus out of their lungs.

CHECKPOINT

1. Name two functions that the cytoskeleton performs within the cell.

2. Compare and contrast cilia and flagella.

Answers: 1. The cytoskeleton serves as an anchor onto which organelles can attach and provides a track along which organelles can move. **2.** Cilia are short and numerous and move back and forth. Flagella are longer, often occurring singly, and they undulate. Cilia and flagella have the same basic structure and help move cells or move fluid over cells.

Cell Surfaces: Protection, Support, and Cell-Cell Interactions

Having crisscrossed the interior of the cell to explore various organelles, our tour now takes us to the surface of this microscopic world. Most cells secrete materials for coats of one kind or another that are external to the plasma membrane. These extracellular coats help protect and support cells and facilitate certain interactions between cellular neighbors in tissues.

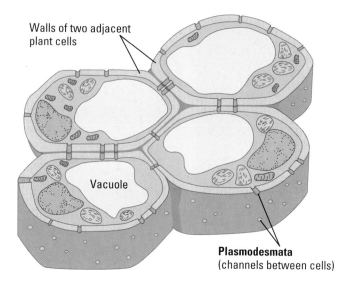

Walls of two adjacent plant cells

Vacuole

Plasmodesmata (channels between cells)

Figure 4.21
Plant cell walls with plasmodesmata.

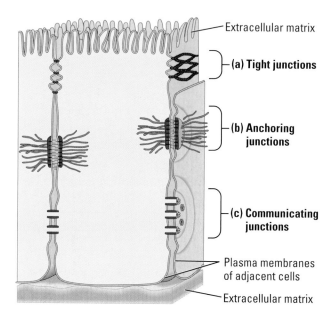

Extracellular matrix

(a) Tight junctions

(b) Anchoring junctions

(c) Communicating junctions

Plasma membranes of adjacent cells

Extracellular matrix

Figure 4.22
Animal cell surfaces and junctions.
Three types of cell-cell junctions integrate the cells of a tissue.
(a) Tight junctions bind cells together, forming a leakproof sheet. Such a sheet of tissue lines the digestive tract, preventing the contents of the tract from leaking into surrounding tissues.
(b) Anchoring junctions attach adjacent cells to each other or to the extracellular matrix. These junctions rivet cells together with cytoskeletal fibers but still allow materials to pass along the spaces between cells. **(c)** Communicating junctions are channels that allow water and other small molecules to flow between neighboring cells. These junctions are especially common in animal embryos, where chemical communication between cells is essential for development.

Plant Cell Walls and Cell Junctions

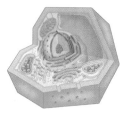

Plant cells are encased by **cell walls.** The wall protects the cell, maintains its shape, and restrains the cell from absorbing so much water that it would burst. Cell walls do not totally isolate plant cells from each other. To function in a coordinated way as part of a tissue, cells must have **cell junctions,** structures that connect them to one another. In plants, channels called **plasmodesmata** (singular, *plasmodesma*) pass through the cell walls, connecting the cytoplasm of each cell to its neighbors. These channels allow water and other small molecules to move between cells, integrating the activities of a tissue (**Figure 4.21**).

Plants do not have bones, but the walls of their cells are strong enough to hold even the tallest trees up against the pull of gravity. Much thicker and stronger than the plasma membrane, the plant cell wall is made from cellulose fibrils embedded in a matrix of other molecules (see Figure 3.13). This combination of materials, strong fibers in a "ground substance" (matrix), is the same basic architectural design found in steel-reinforced concrete and in fiberglass. It is a composite that combines the benefits of strength and flexibility. You've witnessed how this works if you've ever watched a tall tree flex back and forth in a strong wind without breaking. It is cell wall material in the tree trunk that makes that possible. We exploit this combination of strength and flexibility by using lumber cut from trees as building material.

Animal Cell Surfaces and Cell Junctions

Activity 4K on the Web & CD
Watch animations of plant and animal cell junctions.

Animal cells lack the cell walls found in plants, but most animal cells secrete a sticky coat called the **extracellular matrix.** This layer helps hold cells together in tissues, and it can also have protective and supportive functions.

Adjacent cells in many animal tissues connect by various types of cell junctions (**Figure 4.22**). **Tight junctions** bind cells together tightly, forming a leakproof sheet. **Anchoring junctions** attach cells with fibers but still allow materials to pass along the spaces between cells. **Communicating junctions** are channels that allow water and other small molecules to flow between neighboring cells.

Activities 4L and 4M on the Web & CD
Review animal and plant cell structure and function.

Evolution Connection

The Origin of Membranes

The plasma membrane is generally regarded as the boundary of the living cell. It is therefore logical to assume that one of the earliest episodes in the evolution of life on Earth may have been the formation of membranes. The ability of cells to regulate their chemical exchanges with the environment is a basic requirement for life. A selectively permeable membrane can enclose a solution that is different in composition from the surrounding solution while still permitting the uptake of nutrients and the elimination of waste products. Phospholipids, the key ingredients of membranes, were probably among the organic molecules that formed from chemical reactions on the early Earth before the emergence of life. And when mixed with water, phospholipids spontaneously self-assemble into membranes (**Figure 4.23**). This requires no genes or other information except the unique characteristics of the phospholipids themselves. The origin of membranes on a primordial Earth may have been an early step in the evolution of the first cells and the formation of the first life. We will evaluate this hypothesis in more detail in Chapter 15.

Evolution Connection on the Web
Learn what pre-life membranes may have been like.

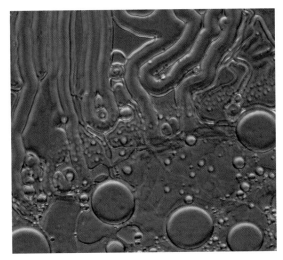

Figure 4.23 The spontaneous formation of membranes: a key step in the origin of life. When phospholipids are mixed with water, they self-assemble to form films. Agitation breaks the films into spheres. Even these primitive membranes have some ability to control the traffic of substances between the contents of a sphere and the aqueous environment outside.

Chapter Review

Summary of Key Concepts

For study help, go to the Essential Biology Website (www.essentialbiology.com) or CD-ROM to explore the Activities and Case Studies in the Process of Science.

The Microscopic World of Cells

- **Microscopes as Windows to Cells** Using early microscopes, biologists discovered that all organisms are made of cells. Resolving power limits the useful magnification of microscopes. A light microscope (LM) has useful magnifications of up to about 1,000×. Electron microscopes, both scanning (SEM) and transmission (TEM), are much more powerful.

Activity 4A *Metric System Review*

Case Study in the Process of Science *What is the Size and Scale of Our World?*

- **The Two Major Categories of Cells**

Prokaryotes	Eukaryotes
• Smaller	• Larger
• Simpler	• More complex
• Most do not have membrane-enclosed organelles	• Membrane-enclosed organelles
• Bacteria and archaea	• Protists, plants, fungi, animals

Activity 4B *Prokaryotic Cell Structure and Function*

- **A Panoramic View of Eukaryotic Cells** Many cellular functions are partitioned by membranes in the complex organization of eukaryotic cells. The largest organelle is usually the nucleus. Other organelles are located in the cytoplasm, the region between the nucleus and the plasma membrane.

Activity 4C *Comparing Cells*

Activity 4D *Build an Animal Cell and a Plant Cell*

Membrane Structure and Function

- **A Fluid Mosaic of Lipids and Proteins**

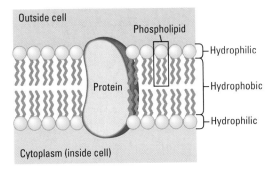

Activity 4E *Membrane Structure*

- **Selective Permeability** Membranes are selective in transporting molecules into and out of the cell.

Activity 4F *Selective Permeability of Membranes*

The Nucleus and Ribosomes: Genetic Control of the Cell

- **Structure and Function of the Nucleus** An envelope consisting of two membranes encloses the nucleus. Within the nucleus, DNA and proteins make up chromatin fibers, and each fiber is a chromosome. The nucleus also contains the nucleolus, which produces components of ribosomes.

- **Ribosomes** Ribosomes produce proteins in the cytoplasm.

- **How DNA Controls the Cell** Genetic messages are transmitted to the ribosomes via messenger RNA, which travels from the nucleus to the cytoplasm.

Activity 4G Overview of Protein Synthesis

The Endomembrane System: Manufacturing and Distributing Cellular Products

- **The Endoplasmic Reticulum** The ER consists of membrane-enclosed tubes and sacs within the cytoplasm. Rough ER, named for the ribosomes attached to its surface, makes membranes and secretory proteins. The functions of smooth ER include lipid synthesis and detoxification.

- **The Golgi Apparatus** The Golgi refines certain ER products and packages them in transport vesicles targeted for other organelles or export from the cell.

- **Lysosomes** Lysosomes, sacs containing digestive enzymes, function in digestion within the cell and chemical recycling.

- **Vacuoles** These membrane-enclosed organelles include the contractile vacuoles that expel water from certain freshwater protists and the large, multifunctional central vacuoles of plant cells.

Activity 4H The Endomembrane System

Chloroplasts and Mitochondria: Energy Conversion

- **Chloroplasts** The sites of photosynthesis in plant cells, chloroplasts convert light energy to the chemical energy of food. Grana, stacks of membranous sacs within the chloroplasts, trap the light energy.

- **Mitochondria** These are the sites of cellular respiration, which converts food energy to ATP energy. ATP drives most cellular work.

Activity 4I Build a Chloroplast and a Mitochondrion

The Cytoskeleton: Cell Shape and Movement

- **Maintaining Cell Shape** Straight, hollow microtubules are an important component of the cytoskeleton, an organelle that gives support to and maintains the shape of cells.

- **Cilia and Flagella** Cilia and flagella are both motile appendages made primarily of microtubules. Cilia are short, numerous, and move by coordinated beating. Flagella are long, often occur singly, and propel a cell through whiplike movements.

Activity 4J Cilia and Flagella

Cell Surfaces: Protection, Support, and Cell-Cell Interactions

- **Plant Cell Walls and Cell Junctions** The walls that encase plant cells consist of cellulose fibers embedded in other materials. Cell walls support plants against the pull of gravity and also prevent cells from absorbing too much water. Channels called plasmodesmata connect the cytoplasm of adjacent cells.

- **Animal Cell Surfaces and Cell Junctions** Animal cells are coated by a sticky extracellular matrix. Junctions between animal cells help hold them together and also integrate the cells of a tissue.

Activity 4K Cell Junctions

Activity 4L Review: Animal Cell Structure and Function

Activity 4M Review: Plant Cell Structure and Function

Self-Quiz

1. Emily would like to film the movement of chromosomes during cell division. Her best choice for a microscope would be a
   a. light microscope, because of its magnifying power.
   b. transmission electron microscope, because of its resolving power.
   c. scanning electron microscope, because the chromosomes are on the cell surface.
   d. light microscope, because the specimen must be kept alive.

2. You look into a light microscope and view an unknown cell. What might you see that would tell you whether the cell is prokaryotic or eukaryotic?
   a. a rigid cell wall
   b. a nucleus
   c. a plasma membrane
   d. ribosomes

3. Prokaryotic cells are characteristic of _____ and _____.

4. Which best describes the structure of the plasma membrane?
   a. proteins sandwiched between two layers of phospholipid
   b. proteins embedded in two layers of phospholipid
   c. a layer of protein coating a layer of phospholipid
   d. phospholipids embedded in two layers of protein

5. The ER has two distinct regions that differ in structure and function. Lipids are synthesized within the _____, and proteins are synthesized within the _____.

6. A type of cell called a lymphocyte makes proteins that are exported from the cell. You can track the path of these proteins within the cell from production through export by labeling them with radioactive isotopes. Which of the following organelles do you expect would be radioactively labeled in your experiment: chloroplasts, Golgi, plasma membrane, smooth ER, rough ER, nucleus, mitochondria. In what order would you expect the label to appear within your chosen organelles?

7. Name two similarities in the structure or function of chloroplasts and mitochondria. Name two differences.

8. Match the following organelles with their functions:
   a. ribosomes   1. movement
   b. microtubules   2. photosynthesis
   c. mitochondria   3. protein synthesis
   d. chloroplasts   4. digestion
   e. lysosomes   5. cellular respiration

9. DNA controls the cell by transmitting genetic messages that result in protein production. Place the following organelles in the order that represents the flow of genetic information from the DNA through the cell: nuclear pores, ribosomes, nucleus, rough ER, Golgi.

10. Match the following cell junctions with their functions:

a. anchoring junctions	1. bind animal cells together tightly
b. plasmodesmata	2. allow materials to pass from one plant cell to another
c. tight junctions	3. allow materials to pass from one animal cell to another
d. communicating junctions	4. attach animal cells but allow materials to pass along spaces between cells

Answers to the Self-Quiz questions can be found in Appendix B.

Go to the website or CD-ROM for more Self-Quiz questions.

The Process of Science

1. The cells of plant seeds store oils in the form of droplets enclosed by membranes. Unlike the membranes you learned about in this chapter, the oil droplet membrane consists of a single layer of phospholipids rather than a bilayer. Draw a model for a membrane around an oil droplet. Explain why this arrangement is more stable than a bilayer.

2. You've isolated a new prokaryotic organism, and now you want to determine whether it is photosynthetic. It is too small to visualize with the microscopes you have available. Design an experiment to determine whether the new cell actively uses photosynthesis.

Case Study in the Process of Science on the Web & CD *Explore the relative sizes of objects from the Earth-sized to the subcellular.*

Biology and Society

1. Doctors at a university medical center removed John Moore's spleen, which is a standard treatment for his type of leukemia. The disease did not recur. Researchers kept the spleen cells alive in a nutrient medium. They found that some of the cells produced a blood protein that showed promise as a treatment for cancer and AIDS. The researchers patented the cells. Moore sued, claiming a share in profits from any products derived from his cells. The U.S. Supreme Court ruled against Moore, stating that his lawsuit "threatens to destroy the economic incentive to conduct important medical research." Moore argued that the ruling left patients "vulnerable to exploitation at the hands of the state." Do you think Moore was treated fairly? Is there anything else you would like to know about this case that might help you decide?

2. Livestock producers add antibiotics to animal feed in order to keep the animals healthy, thereby promoting growth. As a result, much of the packaged meat for sale in supermarkets contains bacteria that are resistant to standard antibiotics. Why do you think this happens? Do you think all products derived from antibiotic-fed livestock should be labeled as such? Would you pay extra for a product from antibiotic-free livestock?

Biology and Society on the Web *Learn about the downside of antibiotic use.*

CHAPTER 5

The Working Cell

Biology and Society: Stonewashing Without the Stones 73

Some Basic Energy Concepts 73
Conservation of Energy
Entropy
Chemical Energy
Food Calories

ATP and Cellular Work 76
The Structure of ATP
Phosphate Transfer
The ATP Cycle

Enzymes 78
Activation Energy
Induced Fit
Enzyme Inhibitors

Membrane Transport 80
Passive Transport: Diffusion Across Membranes
Osmosis and Water Balance in Cells
Active Transport: The Pumping of Molecules Across Membranes
Exocytosis and Endocytosis: Traffic of Large Molecules
The Role of Membranes in Cell Signaling

Evolution Connection: Evolving Enzymes 84

A handful of peanuts contains enough energy to boil a quart of water.

About 75% of the energy generated by a car's engine is lost as heat.

It takes about 10 million ATP molecules per second to power an active muscle cell.

You'd have to run about 14 miles to burn the calories from a pepperoni pizza.

Stonewashing Without the Stones

When Levi Strauss began selling blue jeans in the 1870s, they were designed as work wear that could withstand the rugged demands of miners and other physical laborers who flocked to California during the Gold Rush. "Levi's" were constructed from denim, a sturdy cotton fabric, held together with rivets, and dyed a distinctive indigo color. Today, blue jeans are still worn by people in physically demanding jobs, but they are also a popular fashion choice. Their original purpose—toughness—has taken a back seat to leisure and style.

Accordingly, the clothing industry has introduced new types of denim fabric that place a greater emphasis on look and comfort. One style is known as stonewashed (**Figure 5.1**). In the process of stonewashing, rolls of fabric and pumice stones are tumbled inside huge machines. Although this process produces good-looking and comfortable jeans, it has several drawbacks: it can damage the fabric, it requires the wasteful mining of stones, and it punishes the equipment.

Recently, better results have been achieved using cellulase, a protein that breaks down the polysaccharide cellulose, the main component of cotton and other natural fibers (see Figure 3.13c).

Biology and Society on the Web
Learn more about the commercial application of enzymes.

Cellulase is an enzyme—a catalyst that speeds up chemical reactions—used by bacteria and fungi to break down plant material. By washing denim fabric in cellulase, some of the cellulose fibers in the material are broken down, releasing indigo dye. Once the desired softness and color are reached, the cellulase is removed by rinsing. The textile industry calls this process "biostoning." The enzyme is friendlier to the environment, the equipment, and the fabric, and this process allows manufacturers to closely control the final color.

Enzymes such as cellulase help living organisms control their chemical environments. In this chapter, we'll explore how cells are able to maintain this vital control—over energy, over chemical reactions, and over the movement of materials.

Figure 5.1 Stonewashed versus regular jeans.

Some Basic Energy Concepts

Energy makes the world go round—both the cellular world and the larger world outside. But what exactly is energy? Our first step in understanding the working cell is to learn a few basic concepts about energy.

Conservation of Energy

Energy is defined as the capacity to perform work. Work is performed whenever an object is moved against an opposing force. In other words, work moves things in directions in which they would not move if left alone.

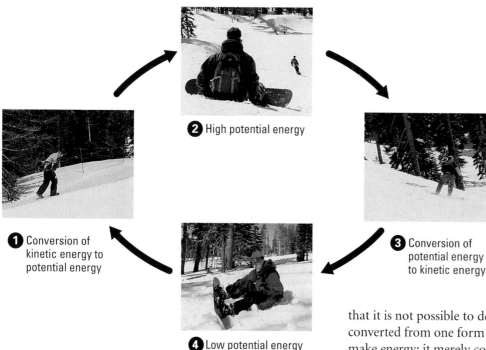

② High potential energy

① Conversion of kinetic energy to potential energy

③ Conversion of potential energy to kinetic energy

④ Low potential energy

Figure 5.2
Energy conversions while snowboarding.

As an example, imagine a snowboarder climbing up and sliding down a hill (**Figure 5.2**). To get to the top of the hill, the snowboarder must perform leg work powered by food energy to overcome the opposing force of gravity. In the process of walking uphill, food energy is converted to **kinetic energy,** the energy of motion.

What happens to the kinetic energy once the snowboarder reaches the top of the hill? Has the energy disappeared? The answer is no. You may be familiar with the principle of conservation of matter, which states that matter cannot be created nor destroyed but can only be converted from one form to another. A similar principle is known as **conservation of energy.** It states that it is not possible to destroy or create energy; energy, too, can only be converted from one form to another. A power plant, for example, does not make energy; it merely converts it to a convenient form. That's exactly what happened in climbing the hill. The kinetic energy of the snowboarder's climb is now stored in a form called potential energy. **Potential energy** is energy that an object has because of its location or arrangement, such as the energy contained by water behind a dam or by a compressed spring. In our example, the snowboarder's body has potential energy because of its elevated location. That potential energy is converted back to kinetic energy as the snowboarder slides down the hill. Life depends on the conversion |of energy from one form to another.

Entropy

If energy cannot be destroyed, where has it gone once the snowboarder reaches the bottom of the hill again? It has been converted to **heat,** a type of kinetic energy contained in the random motion of atoms and molecules. The friction between the snowboard and the snow generated heat in the snowboard, in the surrounding air, and in the snow. In fact, some of the energy released by snowboarding actually melted a bit of the snow.

All energy conversions generate some heat. While that does not destroy energy, it does make it less useful. Heat, of all energy forms, is the most difficult to "tame"—the most difficult to harness for useful work. It is energy in its most randomized form, the energy of aimless molecular movement.

Scientists use the term **entropy** as a measure of disorder, or randomness. Everything that happens in nature increases the entropy of the universe. The energy conversions during the climb up and slide down the hill increased entropy because all of the snowboarder's energy was lost to the atmosphere as heat. To climb up the hill for another run, the snowboarder must use additional stored food energy.

Chemical Energy

How can molecules derived from the food we eat provide energy for our working cells? The molecules of food, gasoline, and other fuels have a special form of potential energy called **chemical energy** that arises from the arrangement of atoms. Carbohydrates, fats, and gasoline have structures that make them especially rich in chemical energy.

Living cells and automobile engines use the same basic process to make the chemical energy stored in their fuels available for work. In both cases, this process breaks the organic fuel into smaller waste molecules that have much less chemical energy than the fuel molecules did, thereby liberating energy that can be used to perform work.

For example, the engine of an automobile mixes oxygen with gasoline in an explosive chemical reaction that breaks down the fuel molecules and pushes the pistons that eventually move the wheels. The waste products emitted from the exhaust pipe of the car are mostly carbon dioxide and water (**Figure 5.3a**). About 25% of the energy an automobile engine extracts from its fuel is converted to the kinetic energy of the car's movement. The rest is converted to heat—so much, in fact, that the engine would melt if it weren't for the car's radiator and fan dispersing excess heat into the atmosphere.

Your cells also use oxygen to help harvest chemical energy (**Figure 5.3b**). And as in a car engine, the "exhaust" is mostly carbon dioxide and water. The "combustion" of fuel in cells is called cellular respiration, a slow, gradual, more efficient "burn" of fuel compared to the explosive combustion in an automobile engine. **Cellular respiration** is the energy-releasing chemical breakdown of fuel molecules and the storage of that energy in a form the cell can use to perform work. Cellular respiration, performed within the mitochondria of cells (see Figure 4.18), will be discussed in detail in the next chapter. You convert about 40% of your food energy to useful work, such as the contraction of your muscles. The other 60% of the energy released by the breakdown of fuel molecules generates body heat. Humans and other animals can use this heat to keep the body at an almost constant temperature (37°C, or 98.6°F, in the case of humans), even when the surrounding air is much colder. The liberation of heat energy also explains why you feel so hot after vigorous exercise. Sweating and other cooling mechanisms enable your body to lose the excess heat, much as a car's radiator keeps the engine from overheating.

Activity 5A on the Web & CD
Use archery to review energy concepts.

Food Calories

Read any packaged food label and you'll find the number of calories in each serving of that food. Calories are units of energy. A **calorie** is the amount of energy that raises the temperature of 1 gram (g) of water by 1 degree Celsius. You could actually measure the caloric content of a peanut by burning it under a container of water to convert all of the stored chemical energy to heat and then measuring the temperature increase of the water.

Calories are tiny units of energy, so using them to describe the fuel content of foods is not practical. Instead, it's conventional to use kilocalories (kcal), units of 1,000 calories. In fact, the "calories" on a food package are

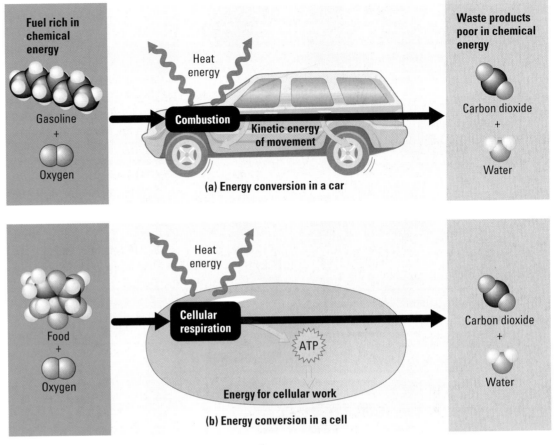

Fuel rich in chemical energy

Gasoline + Oxygen

Heat energy

Combustion

Kinetic energy of movement

(a) Energy conversion in a car

Waste products poor in chemical energy

Carbon dioxide + Water

Heat energy

Food + Oxygen

Cellular respiration

ATP

Energy for cellular work

(b) Energy conversion in a cell

Carbon dioxide + Water

Figure 5.3
Energy conversions in (a) a car and (b) a cell.
Like an automobile engine, cellular respiration uses oxygen to harvest the chemical energy of organic fuel molecules. Cellular respiration breaks sugars and other organic molecules down to smaller waste products. This chemical breakdown releases energy that was stored in the fuel molecules. Cells use this energy for their work.

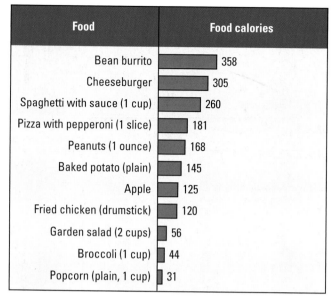

Food	Food calories
Bean burrito	358
Cheeseburger	305
Spaghetti with sauce (1 cup)	260
Pizza with pepperoni (1 slice)	181
Peanuts (1 ounce)	168
Baked potato (plain)	145
Apple	125
Fried chicken (drumstick)	120
Garden salad (2 cups)	56
Broccoli (1 cup)	44
Popcorn (plain, 1 cup)	31

(a) Food calories (kilocalories) in various foods

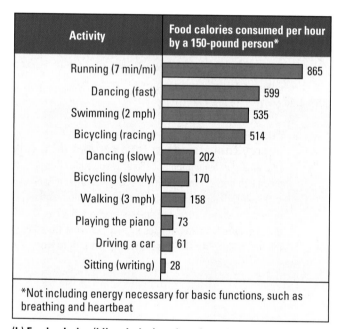

Activity	Food calories consumed per hour by a 150-pound person*
Running (7 min/mi)	865
Dancing (fast)	599
Swimming (2 mph)	535
Bicycling (racing)	514
Dancing (slow)	202
Bicycling (slowly)	170
Walking (3 mph)	158
Playing the piano	73
Driving a car	61
Sitting (writing)	28

*Not including energy necessary for basic functions, such as breathing and heartbeat

(b) Food calories (kilocalories) we burn in various activities

Figure 5.4 Some caloric accounting.

actually kilocalories. That's a lot of energy. For example, just one peanut has about 5 food calories (5 kcal). That's enough energy to increase the temperature of 1 kg (a little more than a quart) of water by 5°C in our peanut-burning experiment. And just a handful of peanuts packs enough food calories, if converted to heat, to boil 1 kg of water. In living organisms, food isn't used to boil water, of course, but to fuel the activities of life. **Figure 5.4** shows the number of food calories in several foods and how many food calories are burned off by some typical activities.

ATP and Cellular Work

The carbohydrates, fats, and other fuel molecules we obtain from food do not drive the work in our cells directly. Instead, the chemical energy released by the breakdown of organic molecules during cellular respiration within the mitochondria is used to generate molecules of ATP. These molecules of ATP then power cellular work.

The Structure of ATP

The initials ATP stand for adenosine triphosphate. **ATP** consists of an organic molecule called adenosine plus a tail of three phosphate groups (**Figure 5.5**).

The triphosphate tail is the "business" end of ATP, the part that provides energy for cellular work. Each phosphate group is negatively charged. Negative charges repel each other. The crowding of negative charge in the triphosphate tail contributes to the potential energy of ATP. It's analogous to storing energy by compressing a spring; if you release the spring, it will "relax," and you can use that springiness to do some useful work. For ATP power, it is release of the phosphate at the tip of

Activity 5B on the Web & CD Make and break ATP.

Figure 5.5 ATP power.   Each P in the triphosphate tail of ATP represents a phosphate group, a phosphorus bonded to oxygen atoms. The triphosphate tail is unstable, partly because of repulsion between the phosphate groups, which are negatively charged. There is a tendency for the phosphate group at the end of the triphosphate tail to "break away" from ATP and bond to other molecules instead. This phosphate transfer, catalyzed by enzymes, energizes cellular work. The leftover molecule is ADP (adenosine diphosphate).

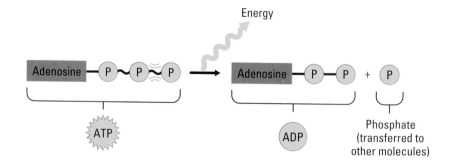

the triphosphate tail that makes energy available to working cells. What is left behind is now called **ADP,** adenosine diphosphate (two phosphate groups instead of three; see Figure 5.5).

Phosphate Transfer

When ATP drives work in cells, phosphate groups don't just fly off into space. ATP energizes other molecules in cells by transferring phosphate groups to those molecules (**Figure 5.6**). This helps cells perform three main kinds of work: mechanical work, transport work, and chemical work.

Imagine another climb up a snowboarding hill. In the muscle cells of the snowboarder's legs, ATP transfers phosphate groups to special motor proteins. The proteins change their shape, causing the muscle cells to contract and perform mechanical work (Figure 5.6a). As an example of transport work, ATP enables brain cells to pump ions across their membranes (Figure 5.6b). This prepares the brain cells to transmit signals. And ATP drives the chemical work of making a cell's giant molecules. An example is the linking of amino acids to make a protein (Figure 5.6c). In Chapter 3, we saw that dehydration synthesis links amino acids; it is ATP that provides the energy for this process. Notice again in Figure 5.6 that all these types of work occur when target molecules accept phosphate from ATP. Special enzymes catalyze these phosphate transfers that energize the working parts of cells.

The ATP Cycle

Your cells spend your ATP continuously. Fortunately, it is a renewable resource. ATP can be restored by adding a phosphate group back to ADP. That takes energy, like recompressing a spring. And that's where food reenters the story. The chemical energy that cellular respiration harvests from sugars and other organic fuels is put to work regenerating a cell's supply of ATP. Cells thus have an ATP cycle: Cellular work spends ATP, which is recycled from ADP and phosphate via cellular respiration (**Figure 5.7**). Thus, ATP functions in what is called **energy coupling:** the transfer of energy from processes that yield energy, such as the breakdown of organic fuels, to processes that consume energy, such as muscle contraction and other types of cellular work. The third phosphate group acts as an energy shuttle within the ATP cycle.

The ATP cycle runs at an astonishing pace. A working muscle cell recycles all of its ATP about once each minute. That amounts to about 10 million ATP molecules spent and regenerated per second per cell.

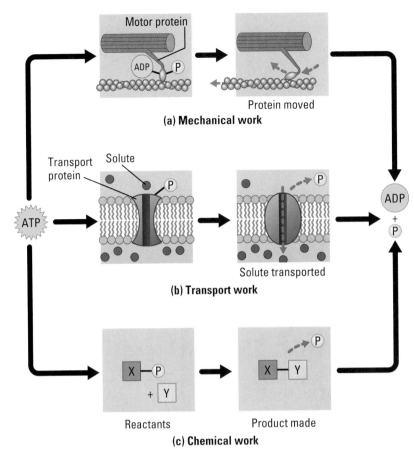

(a) Mechanical work

(b) Transport work

(c) Chemical work

Figure 5.6 How ATP drives cellular work.
Each type of work is powered when enzymes transfer phosphate from ATP to a recipient molecule: **(a)** a motor protein (mechanical work), **(b)** a transport protein (transport work), or **(c)** a chemical reactant (chemical work).

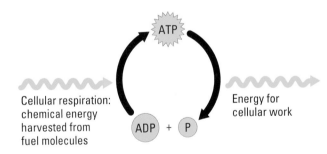

Cellular respiration: chemical energy harvested from fuel molecules

Energy for cellular work

Figure 5.7 The ATP cycle.

CHECKPOINT

1. Explain how ATP powers cellular work.

2. What is the source of energy for regenerating ATP from ADP? Where in the cell does this take place?

Answers: 1. ATP transfers phosphate groups to other molecules, increasing the energy content of those molecules. **2.** Chemical energy harvested from sugars and other organic fuels via cellular respiration in the mitochondria

Enzymes

The many chemical reactions that occur in organisms are collectively called **metabolism.** Few metabolic reactions would occur without the assistance of **enzymes,** specialized proteins that speed up chemical reactions.

Activation Energy

To initiate a chemical reaction, chemical bonds in the reactant molecules must be broken. This process requires that the molecules absorb energy from their surroundings, usually in the form of heat. This energy is called **activation energy** because it activates the reactants and triggers a chemical reaction.

One way to speed up a chemical reaction is to increase the supply of energy by heating the mixture; that's why Bunsen burners are used in chemistry labs. Boiling your cells, however, is not an option for speeding up your metabolism. Enzymes enable metabolism to occur at cooler temperatures by reducing the amount of activation energy required to break the bonds of reactant molecules. If you think of the requirement for activation energy as a barrier to a chemical reaction, then the function of an enzyme is to lower that barrier (**Figure 5.8**). An enzyme does this by binding to reactant molecules and putting them under some kind of physical or chemical stress, making it easier to break their bonds and start a reaction.

Induced Fit

Each enzyme is very selective in the reaction it catalyzes. This specificity is based on the ability of the enzyme to recognize the shape of a certain reactant molecule, which is termed the enzyme's **substrate.** And the ability of the enzyme to recognize and bind to its specific substrate depends on the enzyme's shape. A special region of the enzyme, called the **active site,** has a shape and chemistry that fits it to the substrate molecule. When a substrate molecule slips into this docking station, the active site changes shape slightly to embrace the substrate and catalyze the reaction. This interaction is called **induced fit,** because the entry of the substrate induces the enzyme to change its shape slightly and make the fit between substrate and active site even snugger. Think of a handshake: As your hand makes contact with your friend's, it changes shape slightly to make a better fit.

After the products are released from the active site, the enzyme is once again available to accept another molecule of its specific substrate. In fact, the ability to function over and over again is a key characteristic of enzymes. **Figure 5.9** follows the action of a specific enzyme called sucrase, which hydrolyzes the disaccharide sucrose (the substrate). Like sucrase, many enzymes are named for their substrates, but with an *-ase* ending.

Activity 5C on the Web & CD
See enzymes in action.

Case Study in the Process of Science on the Web & CD
Investigate the effect of enzymes on reaction rate.

Enzyme Inhibitors

Certain molecules can inhibit a metabolic reaction by binding to an enzyme and disrupting its function (**Figure 5.10**). Some of these **enzyme inhibitors** are actually substrate imposters that plug up the active site. Other inhibitors bind to the enzyme at some site remote to the active site, but the binding changes the shape of the enzyme so that its active site is no

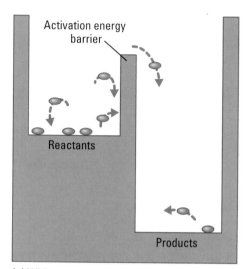

(a) Without enzyme

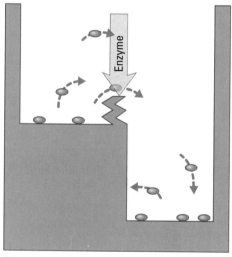

(b) With enzyme

Figure 5.8
Activation energy: a jumping-bean analogy.
(a) A chemical reaction requires activation energy to break the bonds of the reactant molecules. The jumping beans in this analogy represent reactant molecules that must overcome the barrier of activation energy before they can reach the "product" side. **(b)** An enzyme speeds the process by lowering the barrier of activation energy.

longer receptive to the substrate. In some cases, the binding is reversible, enabling certain inhibitors to regulate metabolism. For example, if metabolism is producing more of a certain product than a cell needs, that product may inhibit an enzyme required for its production. This **feedback regulation** keeps the cell from wasting resources that could be put to better use.

Some enzyme inhibitors act as poisons that block metabolic processes essential to the survival of an organism. For example, an insecticide called malathion inhibits an enzyme required for normal functioning of the insect nervous system. Many antibiotics that kill disease-causing bacteria are also enzyme inhibitors. For instance, penicillin inhibits an enzyme that many bacteria use to make their cell walls. Because humans do not have this enzyme, the antibiotic can help us combat an infection without destroying our own cells.

CHECKPOINT

1. What effect does an enzyme have on the activation energy of a chemical reaction?

2. How does an enzyme recognize its substrate?

3. How does the antibiotic penicillin work?

Answers: 1. An enzyme lowers the activation energy of a chemical reaction. **2.** The substrate and the enzyme's active site are complementary in shape and chemical nature. **3.** It inhibits an enzyme that certain bacteria use to make their cell walls.

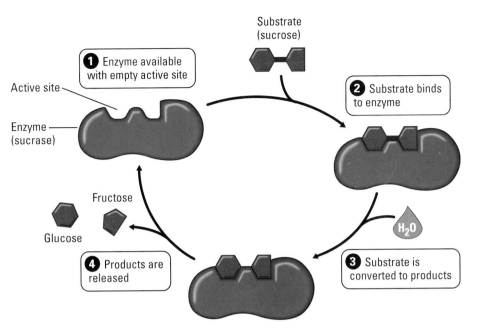

Figure 5.9 How an enzyme works.
Our example is the enzyme sucrase, named for its substrate, sucrose. **1** With its active site empty, sucrase is receptive to a molecule of its substrate. **2** The substrate has a shape that fits the shape of the active site. Sucrose enters the active site, and **3** the enzyme catalyzes the chemical reaction—in this case, the hydrolysis of sucrose. **4** The products—glucose and fructose, in this example—exit the active site, and the sucrase is available to receive another molecule of its substrate.

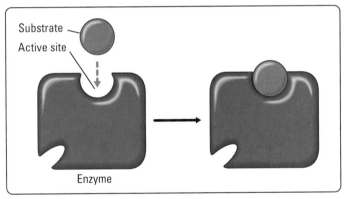

(a) **Normal enzyme action**

Figure 5.10 Enzyme inhibitors.

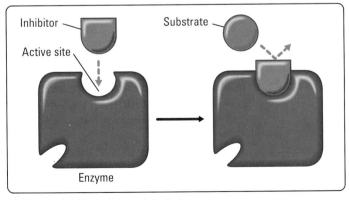

(b) **Enzyme inhibition by a substrate imposter**

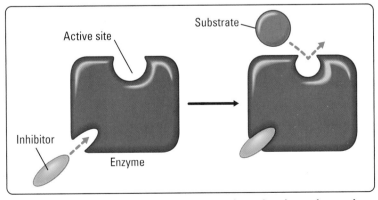

(c) **Enzyme inhibition by a molecule that causes the active site to change shape**

Membrane Transport

Activity 5D on the Web & CD
Review membrane structure.

Activity 5E on the Web & CD
See transport proteins in action.

Besides controlling the flow of energy and chemical reactions, working cells must also control the flow of materials. In Chapter 4, you learned about the structure of the plasma membrane (see Figure 4.7). Its primary function is to regulate the passage of materials into and out of the cell. **Transport proteins** embedded within the phospholipid bilayer help with this task. In this section, you'll learn about the most important mechanisms of passage across membranes.

Passive Transport: Diffusion Across Membranes

Molecules are restless. The heat energy they contain makes them vibrate and wander randomly. One result of this motion is **diffusion,** the tendency for molecules of any substance to spread out into the available space. Each molecule moves randomly, and yet diffusion of a population of molecules may be directional. For example, imagine many molecules of perfume isolated inside a bottle. If you remove the bottle top, every molecule of perfume will move randomly about, but the net direction of the perfume molecules will be out of the bottle to fill the room. You could, with great effort, return the perfume to its bottle, but this would require an expenditure of energy.

For an example closer to a living cell, imagine a membrane separating pure water from a solution of a dye dissolved in water (**Figure 5.11**). Assume that this membrane is permeable to the dye molecules—meaning that the membrane allows the dye molecules to pass through. Although each dye molecule moves randomly, there will be a net migration across the membrane to the side that began as pure water. The spreading of the dye across the membrane will continue until both solutions have equal concentrations of the dye. Once that point is reached, there will be a dynamic equilibrium, with as many dye molecules moving per second across the membrane in one direction as the other.

Activity 5F on the Web & CD
Watch an animation of diffusion.

These examples illustrate a simple rule of diffusion: Each substance will diffuse from where it is more concentrated to where it is less concentrated. Put another way, a substance tends to diffuse down its concentration gradient.

Diffusion across a membrane is called **passive transport**—passive because the cell does not expend any energy for it to happen. The membrane does, however, play a regulatory role by being selectively permeable. For example, small molecules generally pass through more readily than large molecules such as proteins (otherwise, the cell would lose its macromolecules). However, the membrane is relatively impermeable to even some very small substances, such as hydrogen (H^+) and other inorganic ions, which are too hydrophilic to pass through the phospholipid bilayer of the membrane. Such substances can be transported via **facilitated diffusion** by specific transport proteins that act as selective corridors.

Activity 5G on the Web & CD
See an animation of facilitated diffusion.

Passive transport is extremely important to all cells. In our lungs, for example, passive transport along concentration gradients is the sole means by which oxygen (O_2), essential for metabolism, enters red blood cells and carbon dioxide (CO_2), a metabolic waste, passes out of them. One of the most important substances that crosses membranes by passive transport is water.

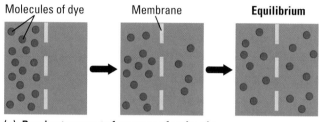

(a) Passive transport of one type of molecule

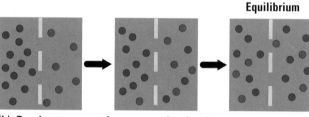

(b) Passive transport of two types of molecules

Figure 5.11
Passive transport: diffusion across a membrane.
(a) The membrane is permeable to these dye molecules, which diffuse down their concentration gradient. At equilibrium, the molecules are still restless, but the rate of transport is equal in both directions. **(b)** If solutions have two or more solutes, each will diffuse down its own concentration gradient.

Osmosis and Water Balance in Cells

The passive transport of water across a selectively permeable membrane is called **osmosis** (**Figure 5.12**). Consider the case of a membrane separating two solutions with different concentrations of a solute—say, the sugar sucrose. The membrane is permeable to water but not to the solute. The solution with a higher concentration of solute is said to be **hypertonic.** The solution with the lower solute concentration is **hypotonic.** Note that the hypotonic solution, by having the lower solute concentration, has the higher water concentration. Therefore, water will diffuse across the membrane along its concentration gradient from an area of higher water concentration (hypotonic solution) to one of lower water concentration (hypertonic solution). This reduces the difference in solute concentrations and changes the volumes of the two solutions. Once the total solute concentrations become the same on both sides of the membrane, water molecules will move at the same rate in both directions, so there will be no net change in solute concentration. Solutions of equal solute concentration are said to be **isotonic.**

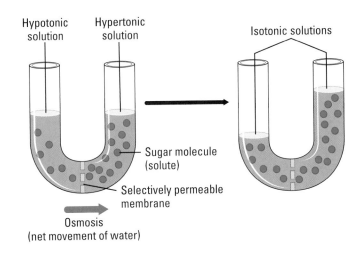

Figure 5.12 Osmosis.
A membrane separates two solutions with different sugar concentrations. Water molecules can pass through the membrane, but sugar molecules cannot. Osmosis, the passive transport of water across the membrane, reduces the difference in sugar concentrations and changes the volumes of the two solutions.

Water Balance in Animal Cells The survival of a cell depends on its ability to balance water uptake and loss. When an animal cell, such as a red blood cell, is immersed in an isotonic solution, the cell's volume remains constant because the cell gains water at the same rate that it loses water (**Figure 5.13a**, top). By definition, the cell is isotonic to its surroundings because the two solutions have the same total concentration of solutes. Many marine animals, such as sea stars and crabs, are isotonic to seawater.

Activity 5H on the Web & CD
Run experiments demonstrating osmosis.

What happens if an animal cell finds itself in a hypotonic solution, which has a lower solute concentration than the cell? The cell gains water, swells, and may burst (lyse) like an overfilled water balloon (**Figure 5.13b**, top). A hypertonic environment is also harsh on an animal cell; the cell shrivels and can die from water loss (**Figure 5.13c**, top).

For an animal to survive if its cells are exposed to a hypotonic or hypertonic environment, the animal must have a way to balance the excessive uptake or excessive loss of water. The control of water balance is called **osmoregulation.** For example, a freshwater fish, which lives in a hypotonic environment, has kidneys and gills that work constantly to prevent an excessive buildup of water in the body. And if you look at Figure 4.15a, you'll see *Paramecium*'s contractile vacuole, which bails out the excess water that continuously enters the cell from the hypotonic pond water.

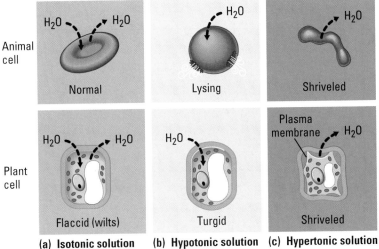

(a) Isotonic solution **(b) Hypotonic solution** **(c) Hypertonic solution**

Figure 5.13 The behavior of animal and plant cells in different osmotic environments.

Water Balance in Plant Cells Water balance problems are somewhat different for plant cells because of their rigid cell walls. A plant cell immersed in an isotonic solution is flaccid (floppy), and a plant wilts in this situation (Figure 5.13a, bottom). In contrast, a plant cell is turgid (firm) and healthiest in a hypotonic environment, with a net inflow of water (Figure 5.13b, bottom). Although the elastic cell wall expands a bit, the back pressure it exerts prevents the cell from taking in too much water and bursting, as an animal cell would in this environment. Turgor is necessary for plants to retain their upright posture and the extended state of their leaves (**Figure 5.14**). However, in a hypertonic environment, a plant cell is no better

Case Study in the Process of Science on the Web & CD
Investigate the effect of salt concentrations on living cells.

Figure 5.14 Plant turgor.
An underwatered plant wilts due to a drop in turgor.

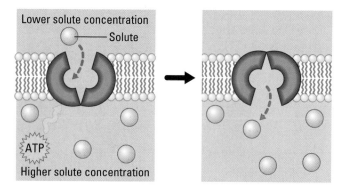

Figure 5.15 Active transport. Like enzymes, transport proteins are specific in their recognition of molecules. This transport protein has a binding site that accepts only a certain solute. Using energy from ATP, the protein pumps the solute against its concentration gradient.

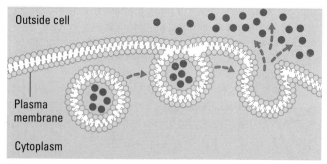

(a) Exocytosis

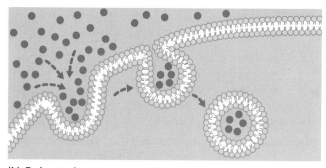

(b) Endocytosis

Figure 5.16 Exocytosis and endocytosis.

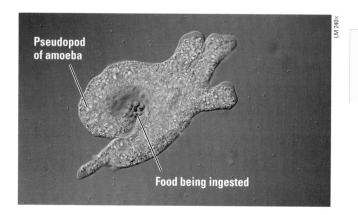

Figure 5.17 Phagocytosis.
An amoeba uses a cellular extension called a pseudopod to engulf food and package it in a food vacuole. A similar process is used by the white blood cells of your immune system to destroy invaders.

off than an animal cell. As a plant cell loses water, it shrivels, and its plasma membrane pulls away from the cell wall (Figure 5.13c, bottom). This process, called **plasmolysis,** usually kills the cell.

Active Transport: The Pumping of Molecules Across Membranes

In contrast to passive transport, **active transport** requires that a cell expend energy to move molecules across a membrane. In active transport, a specific transport protein actively pumps a solute across a membrane against the solute's concentration gradient—that is, away from the side where it is less concentrated (**Figure 5.15**). Membrane proteins usually use ATP as their energy source for active transport.

Active transport enables cells to maintain internal concentrations of small molecules that differ from environmental concentrations. For example, compared to its surroundings, an animal cell has a much higher concentration of potassium ions and a much lower concentration of sodium ions. The plasma membrane helps maintain these steep gradients by pumping sodium out of the cell and potassium into the cell. This particular case of active transport (called the sodium-potassium pump) is vital in the propagation of nerve signals.

Activity 5I on the Web & CD
Test your knowledge of active transport.

Exocytosis and Endocytosis: Traffic of Large Molecules

So far, we've focused on how water and small solutes enter and leave cells by moving through the plasma membrane. The story is different for large molecules such as proteins, which are much too big to fit through the membrane itself. Their traffic into and out of the cell depends on being packaged in vesicles. You have already seen an example in the packaging and secretion of proteins. During protein production by the cell, secretory proteins exit the cell in transport vesicles that fuse with the plasma membrane, spilling the contents outside the cell (see Figures 4.12 and 4.16). That process is called **exocytosis** (**Figure 5.16a**). When you cry, for example, cells in your tear glands use exocytosis to export the salty tears. The reverse process, **endocytosis,** takes material into the cell within vesicles that bud inward from the plasma membrane (**Figure 5.16b**).

There are three types of endocytosis. In **phagocytosis** ("cellular eating"), a cell engulfs a particle and packages it within a food vacuole (**Figure 5.17**). In **pinocytosis** ("cellular drinking"), the cell "gulps" droplets of fluid by forming tiny vesicles. Because any and all solutes dissolved in the droplet are taken into the cell, pinocytosis is unspecific in the substances it transports. In contrast, **receptor-mediated endocytosis** is very specific: It is triggered by the binding of certain external molecules to specific receptor proteins built into the plasma membrane. This binding causes the membrane protein to transport the specific substance into the cell. An example

Activity 5J on the Web & CD
See animations of exocytosis and endocytosis.

of receptor-mediated endocytosis is the mechanism human liver cells use to take up cholesterol particles that circulate in the blood (**Figure 5.18**).

The Role of Membranes in Cell Signaling

The cells of your body talk to each other by chemical signaling across their plasma membranes. Communication begins with the reception of an extra-cellular signal, such as a hormone, by a specific receptor protein built into the plasma membrane (**Figure 5.19**). This signal triggers a chain reaction in one or more molecules that function in transduction (passing the signal along). The proteins and other molecules of the **signal transduction pathway** relay the signal and convert it to chemical forms that work within the cell. This leads to chemical responses, such as the activation of certain metabolic functions, and structural responses, such as rearrangements of the cytoskeleton. Communication between cells is another example of how the plasma membrane serves as a cell's interface with its surroundings.

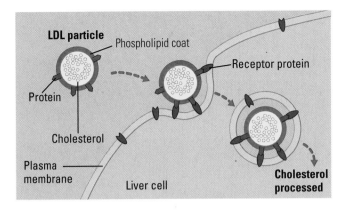

Figure 5.18 Cholesterol uptake by human cells. Normal liver cells remove excess cholesterol from the blood by receptor-mediated endocytosis. This mechanism prevents the excessive cholesterol levels in the blood that contribute to cardio-vascular disease. Cholesterol circulates in the blood mainly in particles called low-density lipoproteins, or LDLs. An LDL is a cho-lesterol globule coated by phospholipid. Proteins embedded in the phospholipid coat attach to receptor proteins built into liver cell membranes. This enables liver cells to take up LDLs and process the cholesterol. Faulty liver cell membranes can overload the blood with cholesterol. For example, in one type of hereditary disorder, LDL receptors on liver cell membranes are missing or reduced in number. This defect results in a heavy load of blood cholesterol, which accumulates on artery walls and causes early onset of cardiovascular disease.

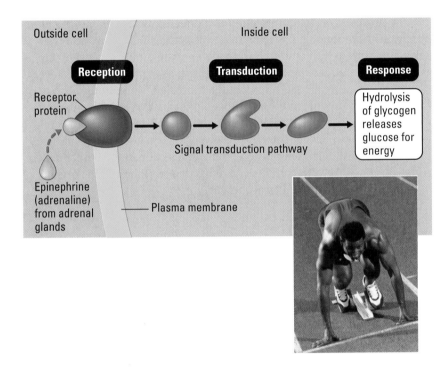

Figure 5.19 An example of cell signaling. When a person gets "psyched up" for an athletic contest, cells in adrenal glands call muscle cells to action. The adrenal cells secrete a hormone called epinephrine (also called adrenaline) into the blood-stream. When that signal reaches muscle cells, it is recognized by receptor proteins built into the plasma membrane. This triggers responses within the muscle cells without the hormone even entering. One of the responses is the hydrolysis of glycogen (a storage polysac-charide), which makes the sugar glucose available as an energy source for the muscle cells. This chain of events is part of the "fight-or-flight" response that enables you to attack or run when you're in danger—or keep your edge during an intense competition.

CHECKPOINT

1. Explain why it is not enough just to say that a solution is hypertonic.

2. What is the energy source for active transport?

3. What is the primary difference between passive and active transport in terms of concentration gradients?

Answers: 1. Hypertonic and hypotonic are relative terms. A solution that is hypertonic to tap water could be hypotonic to seawater. In using these terms, you must provide a comparison, as in "The so-lution is hypertonic to the cell." **2.** ATP **3.** Passive transport moves molecules along the concentra-tion gradient (from higher to lower concentration), while active transport moves molecules against the concentration gradient.

Evolution Connection

Evolving Enzymes

Organisms use a great variety of enzymes. How could such diversity arise? Recall that most enzymes are proteins, and proteins are encoded by genes. Data gathered from the analysis of genetic sequences suggest that some of our genes arose through molecular evolution: One ancestral gene randomly duplicated, and the two copies of that gene diverged over time via genetic mutation, eventually becoming two distinct genes that produce two distinct enzymes.

Enzyme evolution can be greatly accelerated in the laboratory. For example, one group of researchers sought to modify an enzyme that *E. coli* bacteria use to split the disaccharide lactose into its two component monosaccharides. In a process called directed evolution, many copies of the gene for the original enzyme were randomly mutated. The mutated enzymes that could best perform a new function (in this case, to split a bond in a different substrate molecule) were artificially selected using a screening test. Those enzymes were then subjected to another round of duplication, mutation, and selection. After seven rounds, this process produced a novel enzyme that recognized a new substrate.

The processes of natural selection and directed evolution both result in the production of new enzymes with new functions. There are, however, several important differences between the two schemes. The most obvious is the amount of time involved: many, many years for the former, just weeks for the latter. But the most significant difference between the two processes is mentioned in the name: The laboratory experiment was *directed;* the researchers engineered the evolutionary outcome to suit their chosen purpose. Natural selection, on the other hand, selects enzyme variants that work best in the organism's natural environment.

Evolution Connection on the Web
Enzymes are very important to life. How did life and enzymes get together?

Chapter Review

Summary of Key Concepts

For study help, go to the Essential Biology Website (www.essentialbiology.com) or CD-ROM to explore the Activities and Case Studies in the Process of Science.

Some Basic Energy Concepts

- **Conservation of Energy** Machines and organisms can transform kinetic energy (energy of motion) to potential energy (stored energy) and vice versa. In all such energy transformations, total energy is conserved. Energy cannot be created or destroyed.

- **Entropy** Every energy conversion releases some randomized energy in the form of heat. This is an example of the tendency for the entropy, or disorder, of the universe to increase.

- **Chemical Energy** Molecules store varying amounts of potential energy in the arrangement of their atoms. Organic molecules are relatively rich in such chemical energy.

Activity 5A *Energy Concepts*

- **Food Calories** Actually kilocalories, food calories are units for the amount of energy in our foods and also for the amount of energy we expend in various activities.

ATP and Cellular Work

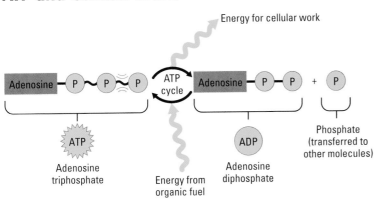

Activity 5B *The Structure of ATP*

Enzymes

- **Activation Energy**   Enzymes are biological catalysts that speed up metabolic reactions by lowering the activation energy required to break the bonds of reactant molecules.

- **Induced Fit**   The entry of a substrate molecule into the active site of an enzyme causes the enzyme to change shape slightly, allowing for a better fit.

Activity 5C *How Enzymes Work*

Case Study in the Process of Science *How Is the Rate of Enzyme Catalysis Measured?*

- **Enzyme Inhibitors**   Enzyme inhibitors are molecules that can disrupt metabolic reactions by binding to enzymes.

Membrane Transport

- **Passive Transport, Osmosis, and Active Transport**

Requires no energy		Requires energy
Passive transport	**Osmosis**	**Active transport**
Higher solute concentration	Higher water concentration (hypotonic)	Higher solute concentration
Lower solute concentration	Lower water concentration (hypertonic)	Lower solute concentration

Most animal cells require an isotonic environment. Plant cells need a hypotonic environment, which keeps the walled cells turgid.

Activity 5D *Membrane Structure*

Activity 5E *Selective Permeability of Membranes*

Activity 5F *Diffusion*

Activity 5G *Facilitated Diffusion*

Activity 5H *Osmosis and Water Balance in Cells*

Case Study in the Process of Science *How Does Osmosis Affect Cells?*

Activity 5I *Active Transport*

- **Exocytosis and Endocytosis: Traffic of Large Molecules**

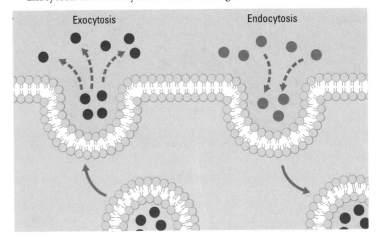

The three kinds of endocytosis are phagocytosis ("cellular eating"); pinocytosis ("cellular drinking"); and receptor-mediated endocytosis, which enables the cell to take in specific large molecules.

Activity 5J *Exocytosis and Endocytosis*

- **The Role of Membranes in Cell Signaling**   Receptors on the cell surface trigger signal transduction pathways that control processes within the cell.

Activity 5K *Cell Signaling*

Case Study in the Process of Science *How Do Cells Communicate with Each Other?*

Self-Quiz

1. You have to expend considerable energy to carry a heavy sofa up a flight of stairs. Since energy cannot be destroyed, what happens to all the energy you spend?

2. _____ is the capacity to perform work, while _____ is a measure of randomness.

3. The label on a candy bar says that it contains 150 calories. If you were able to convert all of that energy to heat, you could raise the temperature of how much water by 15°C?

4. Why does removing a phosphate group from the triphosphate tail in a molecule of ATP release energy?

5. Your digestive system is equipped with a diversity of enzymes that break the polymers in your food down to monomers that your cells can assimilate. A generic name for a digestive enzyme is hydrolase. What is the chemical basis for that name?

6. Explain how an inhibitor can disrupt the action of an enzyme even though it does not bind to the enzyme's active site.

7. If someone sitting at the other end of a restaurant smokes a cigarette, you may still breathe in some of that smoke. The movement of smoke through the air of the restaurant is an example of what type of transport?

    a. osmosis   b. diffusion

    c. facilitated diffusion   d. active transport

8. The total solute concentration in a red blood cell is about 2%. Sucrose cannot pass through a red blood cell's plasma membrane, but water and urea can. Osmosis will cause such a cell to shrink the most when the cell is immersed in which of the following?

    a. a hypertonic sucrose solution

    b. a hypotonic sucrose solution

    c. a hypertonic urea solution

    d. a hypotonic urea solution

9. Which of the following types of cellular transport require(s) an expenditure of energy?

    a. facilitated diffusion   b. active transport

    c. osmosis   d. a and b

10. A _____ is a process that links the reception of cell signals to responses within the cell.

Answers to the Self-Quiz questions can be found in Appendix B.

Go to the website or CD-ROM for more Self-Quiz questions.

The Process of Science

1. HIV, the virus that causes AIDS, depends on an enzyme called reverse transcriptase in order to multiply. Reverse transcriptase reads a molecule of RNA and creates a molecule of DNA from it. A molecule of AZT, the first drug approved to treat AIDS, has a shape very similar to that of the DNA base thymine. Propose a model for how AZT is able to inhibit HIV.

2. Gaining and losing weight are matters of caloric accounting: calories in via the food you eat minus calories you spend in activity. One pound of human body fat contains approximately 3,500 calories. Using Figure 5.4, compare various ways you could burn off that many calories. How far would you have to run, swim, or walk to burn the equivalent of 1 pound of fat? For how much time would you have to do each activity? Which method of burning calories appeals the most to you? The least?

Case Study in the Process of Science on the Web & CD *Investigate the effect of enzymes on reaction rate.*

Case Study in the Process of Science on the Web & CD *Investigate the effect of salt concentrations on living cells.*

Case Study in the Process of Science on the Web & CD *Investigate how cells communicate with each other?*

Biology and Society

1. Obesity is a serious health problem for many Americans. It seems that there are as many fad diets as there are people trying to lose weight. The best-selling diet book in history is *Dr. Atkins' New Diet Revolution* which proposes a low-carbohydrate diet. Most people who follow the Atkins diet compensate for the reduced carbohydrates in their diet by increasing their intake of protein and fat. What advantages are there to such a diet? What disadvantages? Do you think that the government should regulate the claims of diet books? How would you propose that the claims be tested? Do you think diet proponents should be forced to obtain and publish data before they can make claims? Have you ever tried a fad diet?

2. Lead acts as an enzyme inhibitor, and it can interfere with the development of the nervous system. One manufacturer of lead-acid batteries instituted a "fetal protection policy" that banned female employees of childbearing age from working in areas where they might be exposed to high levels of lead. Under this policy, women were involuntarily transferred to lower-paying jobs in lower-risk areas. A group of employees challenged the policy in court, claiming that it deprived women of job opportunities available to men. The U.S. Supreme Court ruled the policy illegal. Nonetheless, many people are uncomfortable about the "right" to work in an unsafe environment. What rights and responsibilities of employers, employees, and government agencies are in conflict in this situation? Whose responsibility should it be to determine what makes a safe environment and who should or should not work there? What criteria should be used to decide?

Biology and Society on the Web *Learn more about the commercial application of enzymes.*

CHAPTER 6

Cellular Respiration: Harvesting Chemical Energy

Biology and Society: Feeling the "Burn" 88

Energy Flow and Chemical Cycling in the Biosphere 88

Producers and Consumers

Chemical Cycling Between Photosynthesis and Cellular Respiration

Cellular Respiration: Aerobic Harvest of Food Energy 90

The Relationship Between Cellular Respiration and Breathing

The Overall Equation for Cellular Respiration

The Role of Oxygen in Cellular Respiration

The Metabolic Pathway of Cellular Respiration

Fermentation: Anaerobic Harvest of Food Energy 98

Fermentation in Human Muscle Cells

Fermentation in Microorganisms

Evolution Connection: Life on an Anaerobic Earth 99

Bacteria are used to produce yogurt, sour cream, pepperoni, and cheese.

All the energy in all the food you eat can be traced back to sunlight.

If you exercise too hard, your muscles shut down from a lack of oxygen.

Both carbon monoxide and cyanide kill by disrupting cellular respiration.

Feeling the "Burn"

When you exercise, your muscles need energy in order to perform work. Your muscle cells use oxygen to help release energy from the sugar glucose. To keep moving, your muscles need a steady supply of oxygen.

When there is enough oxygen reaching your cells to support their energy needs, metabolism is said to be aerobic. As your muscles work harder, you breathe faster and deeper. If you continue to pick up the pace, you will approach your aerobic capacity, the maximum rate at which oxygen can be taken in and used by muscle cells and therefore the highest level of exercise that your body can maintain aerobically.

If you exceed your aerobic capacity, the demand for oxygen in your muscles outstrips the body's ability to deliver it; metabolism now becomes anaerobic. With insufficient oxygen, your muscle cells switch to an "emergency mode" in which they break down glucose very inefficiently and produce lactic acid as a by-product. As lactic acid accumulates, it impairs muscle activity and causes the "burn" associated with heavy exercise ("Feel the burn!—it's a *good* burn!"). The problem is that your muscles can work under these conditions for only a few minutes. If too much lactic acid builds up, your muscles give out (**Figure 6.1**).

With physical conditioning, the body can increase its ability to deliver oxygen to the muscles. Furthermore, top athletes learn exactly how hard they can work without exceeding their aerobic capacity. Well-trained athletes are thus able to push their bodies to provide the maximum amount of work. Short-distance sprinters don't have to worry about building up too much lactic acid before the end of the race. However, athletes involved in longer competitions—particularly runners, bikers, and rowers—have to be sure to stay within their aerobic capacity until the final sprint, at which time they can crank it up to anaerobic levels. Among well-trained atheletes, collapsing at the end of the race is a sign that the switch to an anaerobic pace was well timed.

Biology and Society on the Web
Learn about the proper ways to exercise.

The metabolic pathways that provide energy to an athlete are used by nonathletes as well. In fact, we need energy to walk, talk, and think—in short, to stay alive. The human body has trillions of cells, all hard at work, all demanding fuel continuously. We all know how it feels to be hungry at certain times of the day; that "hunger drive" is an evolutionary adaptation that reminds us that it's time for a fuel stop—for a meal. In this chapter, you'll learn how cells harvest food energy and put it to work.

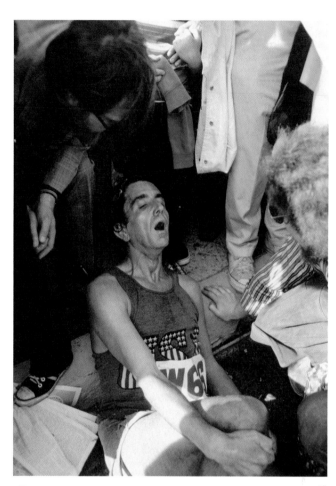

Figure 6.1 Feeling the "burn."
A build-up of lactic acid can cause muscles to give out and "burn."

Energy Flow and Chemical Cycling in the Biosphere

In an indirect way, the fuel molecules in food represent solar energy. We can trace the energy stored in all our food to the sun. Humans and other animals depend on plants to convert the energy of sunlight to the chemical energy of sugars and other organic molecules. And we depend on plants for more than our food. You're probably wearing clothing made of another product of photosynthesis—cotton. Most of our homes are framed with lumber, which is wood produced by photosynthetic trees. Even the text you are now reading is printed on paper, still another material that can be traced to photosynthesis in plants. But for animals, photosynthesis is primarily about feeding the biosphere.

Producers and Consumers

The star we call the sun is at the center of our solar system, almost 100 million miles away. A giant thermonuclear reactor, the sun converts matter to energy, radiating some of it as visible light. A miniscule portion of that light reaches planet Earth. And a tiny fraction of the light that illuminates Earth powers life. The process that makes this possible is called **photosynthesis.** *Photo* means "light," and *to synthesize* means "to make something." Putting this together, photosynthesis uses light energy from the sun to power a chemical process that makes organic molecules.

Here on land, photosynthesis occurs mainly in green cells within the leaves of plants. Leaves owe their greenness to chlorophyll, the pigment contained in chloroplasts. Chloroplasts are the organelles that house the equipment for photosynthesis (see Figure 4.17). Chloroplasts trap light energy and use it to produce sugars and other energy-rich organic molecules. (You'll learn more about photosynthesis in Chapter 7.)

Plants and other **autotrophs** ("self-feeders") are organisms that make all their own organic matter—including carbohydrates, lipids, proteins, and nucleic acids—from inorganic nutrients. The term is a bit misleading in its implication that plants do not require nutrients—they do. But those nutrients are entirely inorganic: carbon dioxide from the air, and water and minerals from the soil. In contrast, humans and other animals are **heterotrophs** ("other-feeders"), organisms that cannot make organic molecules from inorganic ones. That's why we must eat—to get our nutrients from food. Heterotrophs depend on autotrophs for their organic fuel and material for growth and repair.

Most ecosystems depend entirely on photosynthesis for food. For this reason, biologists refer to plants and other autotrophs as **producers.** Heterotrophs, in contrast, are **consumers,** because they obtain their food by eating plants or by eating animals that have eaten plants (**Figure 6.2**). We animals depend on food not only for fuel, but also for the raw organic materials we need to build our cells and tissues.

Chemical Cycling Between Photosynthesis and Cellular Respiration

The ingredients for photosynthesis are carbon dioxide (CO_2) and water (H_2O). Carbon dioxide, a gas in the surrounding air, enters plants through tiny pores in the surface of leaves. Water is absorbed from the damp soil by the plant's roots, and it moves up the plant's veins to the leaves (see Figure 2.12). Chloroplasts rearrange the atoms of these ingredients to produce

Figure 6.2 Producer and consumer. Plants and other photosynthetic organisms use light energy to drive the production of their own organic material. Animals and other consumers depend on this photosynthetic product for energy and building material. In this photo, a porcupine (consumer) eats fruit produced by a photosynthetic plant (producer).

sugars, most importantly glucose ($C_6H_{12}O_6$), and other organic molecules. A by-product of photosynthesis is oxygen gas (O_2) (**Figure 6.3**).

Plants have chloroplasts and so are capable of producing fuel from sunlight. Animals and plants use these organic compounds to obtain energy. A chemical process called cellular respiration harvests energy that is stored in sugars and other organic molecules. Cellular respiration uses O_2 to help convert energy extracted from organic fuel to another form of chemical energy called ATP. Cells spend ATP for almost all their work. In both plants and animals, the production of ATP during cellular respiration occurs mainly in the organelles called mitochondria (see Figure 4.18).

Notice in Figure 6.3 that the waste products of cellular respiration are CO_2 and H_2O—the very same chemical ingredients used for photosynthesis. Plants store chemical energy via photosynthesis and then harvest this energy via cellular respiration. However, plants usually make more organic molecules than they need for fuel. This photosynthetic surplus provides the organic material for the plant to grow. It is also the source of food for humans and other consumers. Analyze nearly any food chain, and you can trace the energy and raw materials for growth back to solar-powered photosynthesis.

> **Activity 6A on the Web & CD**
> Build your own chemical cycling system.

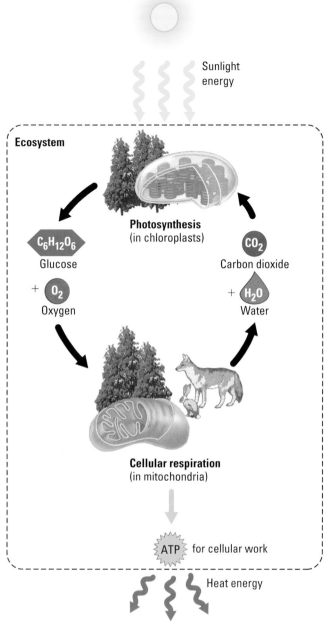

Figure 6.3
Energy flow and chemical cycling in ecosystems.
Energy enters a forest or other ecosystem as sunlight and exits in the form of heat. Organisms temporarily trap the energy for their work. Photosynthesis in the chloroplasts of plants converts light energy to the chemical energy of sugars, such as glucose, and other organic molecules. Cellular respiration in the mitochondria of plants, animals, and other eukaryotes harvests the food energy to generate ATP. These molecules of ATP directly drive most cellular work. Chemical elements essential for life recycle between cellular respiration and photosynthesis.

CHECKPOINT

1. Although they are "self-feeders," photosynthetic autotrophs are not totally self-sufficient. What chemical ingredients do they require from the environment in order to synthesize sugar?

2. Why are plants called producers? Why are animals called consumers?

3. What is misleading about the following statement? "Plants perform photosynthesis and animals perform cellular respiration."

Answers: **1.** Light, CO_2, and H_2O. (Plants also require soil minerals.) **2.** Plants produce organic molecules by photosynthesis. Consumers must acquire organic material by consuming it rather than making it. **3.** It implies that cellular respiration does not occur in plants. It does.

Cellular Respiration: Aerobic Harvest of Food Energy

Internal combustion engines, like the ones found in cars, use O_2 (via the air intakes) to break down gasoline. A cell also requires O_2 to break down its fuel. Cellular respiration—a living version of internal combustion—is the main way that chemical energy is harvested from food and converted to ATP energy. Cellular respiration is an **aerobic** process, which is just another way of saying that it requires oxygen. Putting all this together, we can now define **cellular respiration** as the aerobic harvesting of chemical energy from organic fuel molecules.

The Relationship Between Cellular Respiration and Breathing

We sometimes use the word *respiration* to mean breathing. While respiration on the organismal level should not be confused with cellular respiration, the two processes are closely related. Cellular respiration requires a cell to exchange two gases with its surroundings. The cell takes in oxygen in the form of the gas O₂. It gets rid of waste in the form of the gas carbon dioxide, or CO_2. Breathing exchanges these same gases between your blood and the outside air. Oxygen present in the air you inhale diffuses across the lining of your lungs and into your bloodstream. And the CO_2 in your bloodstream diffuses across the lining of your lungs and exits when you exhale (**Figure 6.4**).

The Overall Equation for Cellular Respiration

A common fuel molecule for cellular respiration is glucose, a six-carbon sugar with the formula $C_6H_{12}O_6$ (see Figure 3.9). Here is the overall equation for what happens to glucose during cellular respiration:

$C_6H_{12}O_6$ + 6 O₂ → 6 CO_2 + 6 H_2O + ATP
Glucose   Oxygen   Carbon dioxide   Water   Energy

The series of arrows indicates that cellular respiration consists of many chemical steps, not just a single chemical reaction. Remember, the main function of cellular respiration is to generate ATP for cellular work. In fact, the process can produce up to 38 ATP molecules for each glucose molecule consumed.

Notice that cellular respiration also transfers hydrogen atoms from glucose to oxygen, forming water. That hydrogen transfer turns out to be the key to why oxygen is so vital to the harvest of energy during cellular respiration.

The Role of Oxygen in Cellular Respiration

In tracking the transfer of hydrogen from sugar to oxygen, we are also following the transfer of electrons. The atoms of sugar and other molecules are bonded together by their sharing of electrons (see Figure 2.8). During cellular respiration, hydrogen and its bonding electrons change partners, from sugar to oxygen, forming water as a product.

Redox Reactions Chemical reactions that transfer electrons from one substance to another are called oxidation-reduction reactions, or **redox reactions** for short. The loss of electrons during a redox reaction is called **oxidation.** Glucose is oxidized during cellular respiration, losing electrons to oxygen. The acceptance of electrons during a redox reaction is called **reduction.** Oxygen is reduced during cellular respiration, accepting electrons (and hydrogen) lost from glucose:

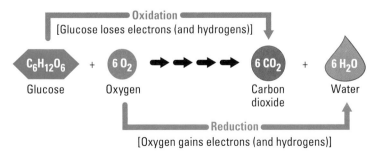

Oxidation
[Glucose loses electrons (and hydrogens)]

$C_6H_{12}O_6$ + 6 O₂ → 6 CO_2 + 6 H_2O
Glucose   Oxygen   Carbon dioxide   Water

Reduction
[Oxygen gains electrons (and hydrogens)]

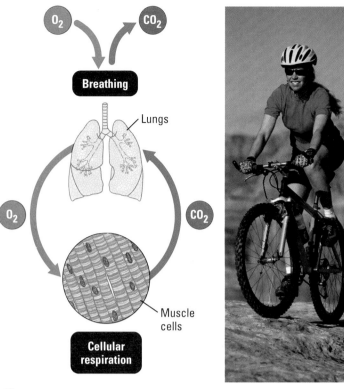

Figure 6.4
How breathing is related to cellular respiration.
When you inhale, you breathe in O₂, which diffuses across the lining of your lungs and into your bloodstream. O₂ is delivered to your cells, where it is used in cellular respiration. Carbon dioxide, a waste product of cellular respiration, diffuses from your cells to your blood and travels to your lungs. CO_2 diffuses across the lining of your lungs and exits when you exhale.

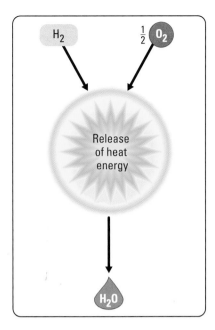

Figure 6.5 A rapid electron "fall."
An all-at-once redox reaction, such as the reaction of hydrogen and oxygen to form water, releases a burst of energy. It is difficult to use such an explosion for productive work.

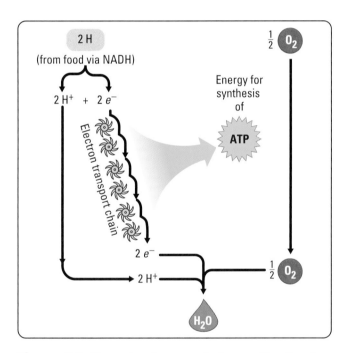

Figure 6.6 The role of oxygen in harvesting food energy.   Like the explosive formation of water in Figure 6.5, cellular respiration combines oxygen and hydrogen to produce water. However, respiration breaks up the "fall" of electrons from food to oxygen into steps. NADH transfers electrons (e^-) from food to an electron transport chain. Oxygen pulls the electrons down the chain. Cells use the step-wise release of energy to make ATP. The oxygen combines with the electrons and hydrogen from food to produce water. The oxygen molecules (O_2) we inhale are the source of the oxygen atoms we use during cellular respiration, written here as $\frac{1}{2} O_2$.

When hydrogen and its bonding electrons change partners, from sugar to oxygen, energy is released.

Why does electron transfer to oxygen release energy? In redox reactions, oxygen is an "electron grabber." An oxygen atom attracts electrons more than almost any other type of atom. When electrons move (along with hydrogen) from glucose to oxygen, it is as though they were falling. They are not really falling in the sense of an apple dropping from a tree. However, in the case of both falling objects and electron transfer to oxygen, potential energy is unlocked. Instead of gravity, it is the attraction of electrons to oxygen that causes the "fall" and energy release during cellular respiration.

A very rapid electron "fall" generates an explosive release of energy in the form of heat and light. For example, a spark will trigger a reaction between hydrogen gas (H_2) and oxygen gas (O_2) that produces water (H_2O). The reaction also releases a large amount of energy as the electrons of the hydrogen "fall" into their new bonds with oxygen (**Figure 6.5**). As another example, some fancy drinks and desserts are topped with a flaming cube of sugar. During this dramatic display, the hydrogen and electrons from the sugar "fall" to the oxygen in the air. The flame represents energy released as heat and light.

It would be difficult to capture the burst of energy released from such an explosive reaction and put it to useful work. Cellular respiration is a more controlled "fall" of electrons—more like a stepwise cascade of electrons down an energy staircase. Instead of liberating food energy in a burst of flame, cellular respiration unlocks chemical energy in smaller amounts that cells can put to productive use.

NADH and Electron Transport Chains Let's take a closer look at the path that electrons take on their way down from glucose to oxygen (**Figure 6.6**). The first stop is an electron acceptor called NAD^+ (an abbreviation for a positively charged ion with a very long chemical name: <u>n</u>icotinamide <u>a</u>denine <u>d</u>inucleotide). The transfer of electrons from organic fuel to NAD^+ reduces the NAD^+ to **NADH** (the H represents the transfer of hydrogen along with the electrons). In our staircase analogy, the electrons have taken one baby step down in their trip from glucose to oxygen. The rest of the staircase consists of an **electron transport chain.** Each link in an electron transport chain is actually a molecule, usually a protein. In a series of redox reactions, each member of the chain can first accept and then donate electrons. The electrons give up a small amount of energy with each transfer. The energy given up can be used to generate ATP from ADP. At the "uphill" end, the first molecule of the chain accepts electrons from NADH. Thus, NADH carries electrons from glucose and other fuel molecules and deposits them at the top of an electron transport chain. The electrons then cascade down the chain, from molecule to molecule. The molecule at the bottom of the chain finally "drops" the electrons to oxygen. The oxygen also picks up hydrogen to form water.

The overall effect of all this electron traffic during cellular respiration is a downhill trip for electrons from glucose to oxygen via NADH and electron transport chains. During the stepwise freeing of chemical energy during electron transport, our cells make most of their ATP from ADP. Oxygen, the "electron grabber," makes it all possible. By pulling electrons down the transport chain from fuel molecules, oxygen functions somewhat like gravity pulling objects downhill. This is how the oxygen we breathe functions in our cells and why we cannot survive more than a few minutes without it.

The Metabolic Pathway of Cellular Respiration

Cellular respiration is an example of **metabolism,** the general term for all the chemical processes that occur in cells. More specifically, cellular respiration is a metabolic pathway. That means that it is not a single chemical reaction, but a series of reactions. A specific enzyme catalyzes each reaction in a metabolic pathway. More than two dozen reactions are involved in cellular respiration. We can group them into three main metabolic stages: glycolysis, the Krebs cycle, and electron transport (which you've already encountered). Let's see how these stages cooperate to harvest food energy.

A Road Map for Cellular Respiration **Figure 6.7** is a map that will help you follow glucose through the metabolic pathway of cellular respiration. The map also shows you where the three stages of respiration occur in your cells.

During **glycolysis,** a molecule of glucose is split into two molecules of a compound called pyruvic acid. The enzymes for glycolysis are located outside the mitochondria, dissolved in the cytosol. The **Krebs cycle** completes the breakdown of sugar all the way to CO_2, the waste product of cellular respiration. The enzymes for the Krebs cycle are dissolved in the fluid within mitochondria. Glycolysis and the Krebs cycle generate a small amount of ATP directly. They generate much more ATP indirectly, via redox reactions that transfer electrons from fuel molecules to NAD^+. The third stage of cellular respiration is electron transport. Electrons captured from food by NADH "fall" down electron transport chains to oxygen. The proteins and other molecules that make up electron transport chains reside within mitochondria. Electron transport from NADH to oxygen releases the energy your cells use to make most of their ATP.

Activity 6B on the Web & CD
Get the big picture of cellular respiration.

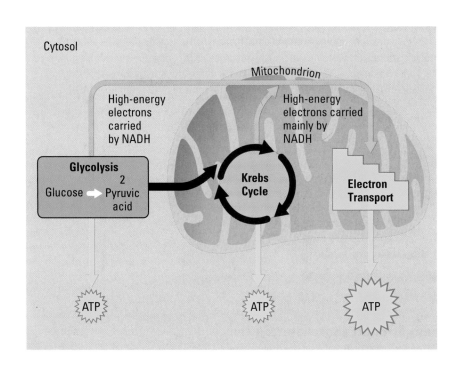

Figure 6.7
A road map for cellular respiration.
The three main stages of cellular respiration—glycolysis, the Krebs cycle, and electron transport—are color-coded in this diagram. We'll carry a smaller version of this map with us so you can keep the overall process of cellular respiration in plain view as we take a closer look at its three stages.

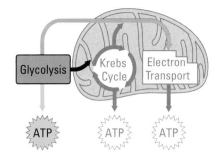

Stage 1: Glycolysis The word *glycolysis* means "splitting of sugar." That is exactly what happens (**Figure 6.8**). Glycolysis breaks a six-carbon glucose molecule in half, forming two three-carbon molecules. These molecules then donate high-energy electrons to NAD$^+$, the electron carrier. Glycolysis also makes some ATP directly when enzymes transfer phosphate groups from fuel molecules to ADP (**Figure 6.9**). What remains of the fractured glucose at the end of glycolysis are two molecules of pyruvic acid. The pyruvic acid still holds most of the energy of glucose, and that energy is harvested in the Krebs cycle.

Activity 6C on the Web & CD
Watch the process of glycolysis.

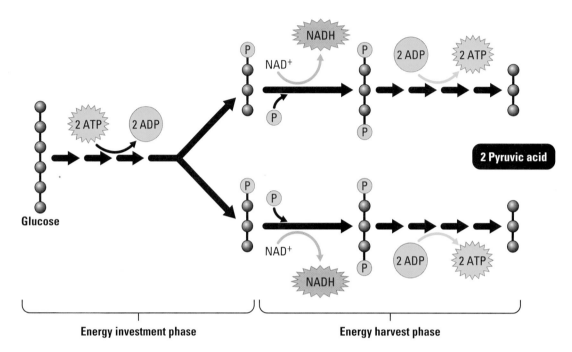

Energy investment phase   Energy harvest phase

Figure 6.8 Glycolysis.

Each gray ball in this diagram represents a carbon atom. A team of enzymes splits glucose, eventually forming two molecules of pyruvic acid. Along the way, energy is stored as ATP and NADH. Note that the cell actually invests some ATP to get glycolysis started. Enzymes attach phosphate groups (P) to the fuel molecules during this energy investment phase. That investment is paid back with dividends during the energy harvest phase. Glycolysis generates some ATP directly, but it also donates high-energy electrons to NAD$^+$, forming NADH.

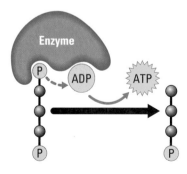

Figure 6.9 ATP synthesis by direct phosphate transfer.

Glycolysis generates ATP when enzymes transfer phosphate groups directly from fuel molecules to ADP.

Stage 2: The Krebs Cycle Pyruvic acid, the fuel that remains after glycolysis, is not quite ready for the Krebs cycle. First, the fuel must be "prepped"—converted to a form the Krebs cycle can use (**Figure 6.10**). The actual fuel consumed by the Krebs cycle is a two-carbon compound called acetic acid. It must enter the Krebs cycle in the form of acetyl-CoA, in which the acetic acid is bonded to a carrier molecule called coenzyme A (the *CoA* in acetyl-CoA).

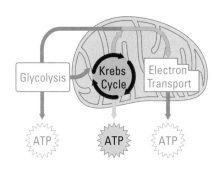

Activity 6D on the Web & CD See the Krebs cycle in action.

The Krebs cycle finishes extracting the energy of sugar by breaking the acetic acid molecules (two per glucose) all the way down to CO_2 (**Figure 6.11**). The cycle uses some of this energy to make ATP by the direct method (see Figure 6.9). However, the Krebs cycle captures much more energy in the form of NADH and a second electron carrier, $FADH_2$. Electron transport then converts NADH and $FADH_2$ energy to ATP energy.

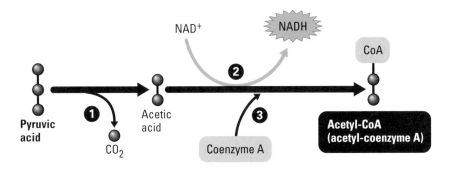

Figure 6.10 The link between glycolysis and the Krebs cycle: the conversion of pyruvic acid to acetyl-CoA. ❶ Each pyruvic acid (there are two per starting glucose, remember) loses a carbon as CO_2. This is the first of this waste product we've seen so far in the breakdown of glucose. The remaining fuel molecules, each with only two carbons left, are called acetic acid (the same acid that's in vinegar). ❷ Oxidation of the fuel generates NADH. ❸ Finally, the acetic acid is attached to a molecule called coenzyme A (CoA) to form acetyl-CoA. The CoA escorts the acetic acid into the first reaction of the Krebs cycle.

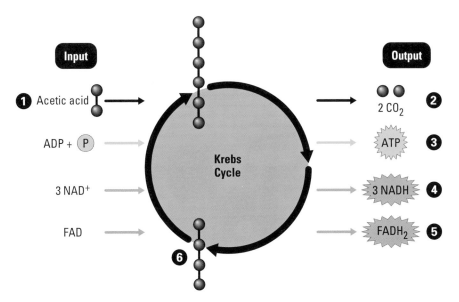

Figure 6.11 The Krebs cycle. This simplified version of the Krebs cycle emphasizes inputs and outputs. ❶ Input includes fuel in the form of acetic acid from acetyl-CoA. It joins a four-carbon acceptor molecule to form a six-carbon product. For every acetic acid molecule that enters the cycle as fuel, ❷ two CO_2 molecules eventually exit as exhaust (waste product). Along the way, the Krebs cycle harvests energy from the fuel. ❸ Some of the energy is used to produce ATP directly. ❹ Most of the energy is trapped by NADH. ❺ $FADH_2$ is another electron carrier that works something like NADH. ❻ All the carbon atoms that entered the cycle as fuel are accounted for as CO_2 exhaust, and the four-carbon acceptor molecule is recycled. We have tracked only one acetic acid molecule through the Krebs cycle here. But recall that glycolysis splits glucose in two (see Figure 6.8). So the Krebs cycle actually turns twice for each glucose molecule that fuels a cell.

Stage 3: Electron Transport The molecules of electron transport chains are built into the inner membranes of mitochondria. An entire electron transport chain functions as a chemical machine that uses the energy released by the "fall" of electrons to pump hydrogen ions (H^+) across the inner mitochondrial membrane. Remember from Chapter 2 that H^+ ions are dissolved as solutes in varying amounts in biological fluids. By pumping H^+ ions, electron transport chains store potential energy by making the ions more concentrated on one side of the membrane than on the other.

The energy stored by electron transport behaves something like the elevated reservoir of water behind a dam. There is a tendency for the H^+ ions

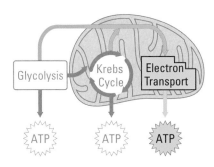

to gush back to where they are less concentrated, just as there is a tendency for water to flow downhill. The membrane, analogous to the dam, temporarily restrains the H^+ ions.

The energy of dammed water can be harnessed to do work. Gates in the dam allow the water to rush downhill, turning giant wheels called turbines as it goes. The spinning turbines can do work—generate electricity, for example. Your mitochondria have structures that act like turbines. Each of these miniature machines, called an **ATP synthase,** is constructed from several proteins built into the inner mitochondrial membrane, the same membrane where electron transport chains are located (**Figure 6.12**). Electron transport provides energy for operating the ATP synthase machines, but indirectly. Hydrogen ions pumped by electron transport rush back "downhill" through an ATP synthase. This action spins a component of the ATP synthase, just as water turns the turbines in a dam. The rotation activates catalytic sites in the synthase that attach phosphate groups to ADP molecules to generate ATP.

> **Activity 6E on the Web & CD**
> Produce ATP using electron transport.

Some of the deadliest poisons do their damage by disrupting electron transport in mitochondria. For example, carbon monoxide and cyanide both kill by blocking the transfer of electrons from electron transport chains to oxygen. With its energy-harvesting mechanism shut down, the mitochondrial membrane can no longer convert food energy to ATP energy. Cells stop working, and the organism dies.

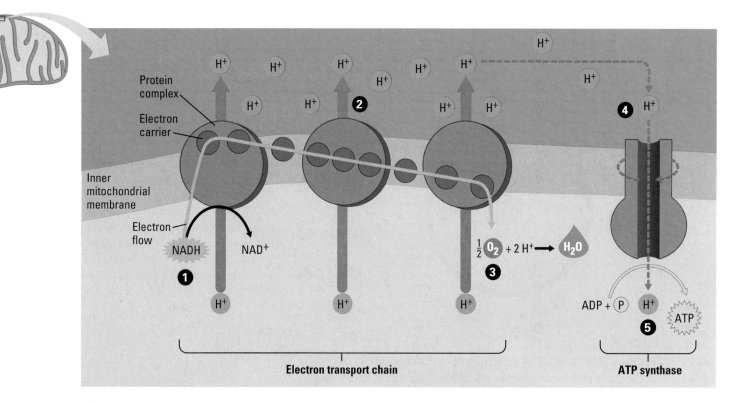

Figure 6.12 How electron transport drives ATP synthase machines.

This figure is a simplified view of how the energy stored in NADH can be used to generate molecules of ATP. ❶ NADH transfers electrons from food to electron transport chains. ❷ Electron transport chains use this energy supply to pump H^+ ions across the inner membrane of the mitochondrion. The infoldings of the inner membrane increase surface area, maximizing the number of electron transport chains built into the membrane. ❸ Again, notice that the oxygen you breathe pulls electrons down the transport chain. ❹ The H^+ ions flow back through an ATP synthase. This spins a part of the synthase, much like water turns a turbine when it flows through the gates in a dam. ❺ The ATP synthase uses the energy of the H^+ gradient to generate ATP from ADP.

The Versatility of Cellular Respiration We have seen so far that food provides the energy to make the ATP our cells use for all their work. We have concentrated on the sugar glucose as the fuel that is broken down in cellular respiration, but respiration is a versatile metabolic furnace that can "burn" many other kinds of food molecules. **Figure 6.13** diagrams some metabolic routes for the use of diverse carbohydrates, fats, and proteins as fuel for cellular respiration.

Adding Up the ATP from Cellular Respiration Taking cellular respiration apart to see how all the molecular nuts and bolts of its metabolic machinery work, it's easy to lose sight of the overall function: generating ATP. **Figure 6.14** will help you add up the ATP molecules a cell can make for each glucose molecule it consumes as fuel. Notice that most of that ATP production is powered by electron transport. And electron transport depends on the presence of oxygen. Next, we'll see what happens when cells harvest food energy without the help of oxygen.

> **Case Study in the Process of Science on the Web & CD** Perform experiments that measure the rate of cellular respiration.

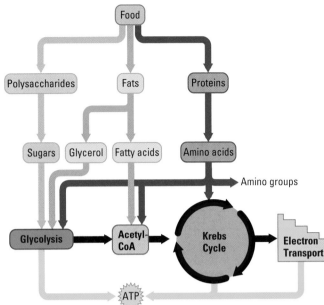

Figure 6.13 Energy from food. Carbohydrates, fats, and proteins can all serve as fuel for cellular respiration. Digestion hydrolyzes the large food molecules to their monomers (see Chapter 3), which are distributed to cells by the circulatory system. The monomers "feed" into glycolysis and the Krebs cycle at various points.

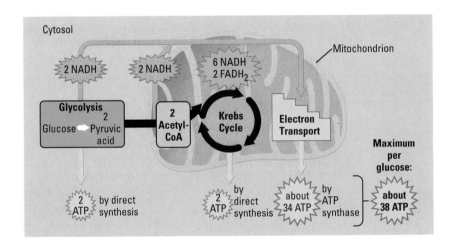

Figure 6.14 A summary of ATP yield during cellular respiration. A cell can convert the energy of each glucose molecule to as many as 38 ATP molecules. Glycolysis and the Krebs cycle each contribute 2 ATPs by direct synthesis (see Figure 6.9). The other 34 ATP molecules are produced by the ATP synthase machines. The "fall" of electrons from food to oxygen powers ATP synthase. The electrons are carried from the organic fuel to electron transport chains by NADH and $FADH_2$. Each electron pair "dropped" down a transport chain from NADH can power the synthesis of up to 3 ATPs. Each electron pair transferred to an electron transport chain from FADH is worth up to 2 ATPs.

CHECKPOINT

1. How is your breathing related to your cellular respiration?

2. At the "downhill" end of the electron transport chain, when electrons from NADH are finally passed to oxygen, what waste product of cellular respiration is produced? (*Hint*: Review Figure 6.6.)

3. What is the potential energy source that directly drives ATP production by ATP synthase?

4. Of the three main stages of cellular respiration represented in Figure 6.7, which one uses oxygen directly to extract chemical energy from organic compounds?

5. The oxidation of acetic acid by NAD^+ extracts some chemical energy from the acetic acid. How can the cell harness that energy to make ATP?

6. Of the three stages of cellular respiration, which occurs in the cytosol, outside mitochondria?

Answers: 1. In breathing, your lungs exchange CO_2 and O_2 between your body and the atmosphere. In cellular respiration, your cells consume the O_2 in extracting energy from food and release CO_2 as a waste product. **2.** Water (H_2O) **3.** A concentration gradient of H^+ ions across the inner membrane of a mitochondrion **4.** Electron transport **5.** The NADH can supply electrons to the electron transport chain, which generates a hydrogen ion gradient that drives ATP synthesis. **6.** Glycolysis

Fermentation: Anaerobic Harvest of Food Energy

Although you must breathe to stay alive, some of your cells can actually work for short periods without O_2. Your muscle cells are a good example. They can produce some ATP under conditions that are **anaerobic** ("without oxygen"). This anaerobic harvest of food energy is called **fermentation**.

Fermentation in Human Muscle Cells

When you walk between classes, ATP powers the muscle cells of your legs. Cellular respiration regenerates the ATP supply to keep you going. Blood provides your muscle cells with enough O_2 to keep electrons "falling" down transport chains in your mitochondria. But if you run because you're late for class, your muscles are forced to keep working under anaerobic conditions. That's because they are spending ATP at a rate that outpaces your bloodstream's delivery of O_2 from your lungs to your muscles.

Muscle cells have enough ATP to support such an anaerobic activity for about 5 seconds. A secondary supply of phosphate bond energy, called creatine phosphate, can keep you going another 10 seconds or so. But what if you still haven't made it to class? To keep running, your muscles must generate ATP by the anaerobic process of fermentation.

The metabolic pathway that provides ATP during fermentation is glycolysis, the same pathway that functions as the first stage of cellular respiration. Glycolysis does not require O_2 (see Figure 6.8). And glycolysis, remember, does produce a small amount of ATP directly: 2 ATPs for each glucose molecule broken down to pyruvic acid. That isn't very efficient compared with the 38 ATPs each glucose generates during cellular respiration, but it can energize your leg muscles fast enough and long enough for you to make it to class. However, your cells will have to consume more glucose fuel per second, since so much less ATP per glucose molecule is generated under anaerobic conditions.

There is more to fermentation than just glycolysis. To harvest food energy during glycolysis, NAD^+ must be present as an electron acceptor (see Figure 6.8). This is no problem under aerobic conditions, because the cell regenerates its NAD^+ when NADH drops its electron cargo down electron transport chains to O_2 (see Figure 6.6). However, this productive recycling of NAD^+ cannot occur under anaerobic conditions because there is no O_2 to accept the electron. Instead, NADH disposes of electrons by adding them to the pyruvic acid produced by glycolysis (**Figure 6.15**). This restores NAD^+ and keeps glycolysis working.

The reduction (addition of electrons) of pyruvic acid produces a waste product called lactic acid. A temporary accumulation of lactic acid in muscle cells probably contributes to the soreness or burning you feel after an exhausting run or other anaerobic spurt of activity. The lactic acid is eventually transported in the blood from muscles to the liver, where liver cells convert it back to pyruvic acid. That metabolic process requires O_2, which is why you breathe hard for some time even after you've stopped vigorous activity. If lactic acid accumulates to a critical level, extreme fatigue ("hitting the wall," as discussed in the chapter's opening section) may cause the muscles to temporarily shut down.

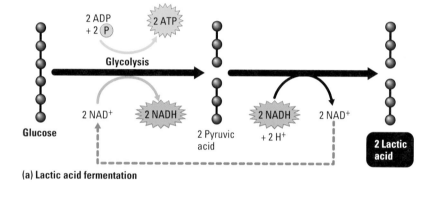

(a) Lactic acid fermentation

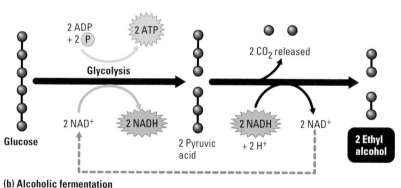

(b) Alcoholic fermentation

Figure 6.15 Fermentation Glycolysis produces ATP without the help of O_2. This process requires a continuous supply of NAD^+ to accept electrons from glucose. The NAD^+ is regenerated when NADH transfers the electrons it removed from food to pyruvic acid, thereby producing **(a)** lactic acid, **(b)** ethyl alcohol, or other waste products, depending on the species of organism.

Fermentation in Microorganisms

Like human muscle cells, certain fungi and bacteria produce lactic acid as their waste product during fermentation. We have domesticated such microbes to transform milk into cheese, sour cream, and yogurt. These foods owe their sharp or sour flavor mainly to lactic acid. The food industry also uses fermentation to produce soy sauce from soybeans, to pickle cucumbers, olives, and cabbage, and to produce meat products like sausage, pepperoni, and salami.

Yeast, a microscopic fungus, is capable of both cellular respiration and fermentation. If you keep yeast cells in an anaerobic environment, they are forced to ferment sugars and other foods to stay alive. In contrast to muscle cells, the yeast style of fermentation produces ethyl alcohol as a waste product instead of lactic acid. This alcoholic fermentation also releases CO_2 (see Figure 6.15b). For thousands of years, humans have put yeast to work producing alcoholic beverages such as beer and wine (**Figure 6.16a**). And as every baker knows, the CO_2 bubbles from fermenting yeast also cause bread dough to rise (**Figure 6.16b**).

> **Activity 6F on the Web & CD**
> Compare lactic acid fermentation and alcoholic fermentation.

Yeast is an example of what is called a **facultative anaerobe**, an organism with the metabolic versatility to harvest food energy by either cellular respiration or fermentation. In contrast are **obligate anaerobes**, organisms that are actually poisoned by oxygen, such as certain bacteria that live in stagnant ponds or deep in the soil. At the cellular level, our muscle cells, like yeast cells, behave as facultative anaerobes. But as whole organisms, we are best described, in metabolic terms, as **obligate aerobes** because of our dependence on oxygen and cellular respiration to stay alive.

(a)

(b)

Figure 6.16 Using fermentation to make food.
(a) A fermentation tank at a brewery. **(b)** The air bubbles in bread are produced by fermenting yeast.

CHECKPOINT

1. How many molecules of ATP are generated per molecule of glucose during fermentation? How many are generated during cellular respiration?

2. _____ acid is to human muscle cells as ethyl _____ is to yeast.

Answers: 1. 2; 38 **2.** Lactic; alcohol

Evolution Connection

Life on an Anaerobic Earth

The Krebs cycle and the electron transport chain operate only under aerobic conditions, in the presence of O_2. Glycolysis, on the other hand, occurs under both aerobic and anaerobic conditions. The role of glycolysis in both cellular respiration and fermentation has an evolutionary basis. Ancient bacteria probably used glycolysis to make ATP before O_2 was present in Earth's atmosphere. The oldest known fossils of bacteria date back over 3.5 billion years, but O_2 did not accumulate in the atmosphere until about 2.5 billion years ago. For a billion years, bacteria must have generated ATP exclusively from glycolysis, which does not require O_2. Glycolysis is the most widespread metabolic pathway, which also suggests that it evolved very early in ancestors common to all kingdoms of life. And the fact that glycolysis occurs in the cytosol not in mitochondria implies great antiquity, too; mitochondria evolved long after the origin of prokaryotic life. Glycolysis is a metabolic heirloom from the earliest cells that continues to function today in the harvest of food energy.

> **Evolution Connection on the Web**
> Learn about the problems life faced when oxygen first began to accumulate in the atmosphere.

Summary of Key Concepts

For study help, go to the Essential Biology Website (www.essentialbiology.com) or CD-ROM to explore the Activities and Case Studies in the Process of Science.

Energy Flow and Chemical Cycling in the Biosphere

- **Producers and Consumers** Autotrophs (producers) make organic molecules from inorganic nutrients via photosynthesis. Heterotrophs (consumers) must consume organic material and obtain energy via cellular respiration.

- **Chemical Cycling Between Photosynthesis and Cellular Respiration**

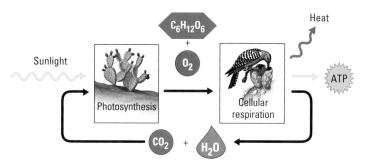

- Activity 6A *Build a Chemical Cycling System*

Cellular Respiration: Aerobic Harvest of Food Energy

- **The Relationship Between Cellular Respiration and Breathing** The bloodstream distributes O_2 from the lungs to all the cells of the body and transports CO_2 waste from the cells to the lungs for disposal.

- **The Overall Equation for Cellular Respiration**

$C_6H_{12}O_6$ (Glucose) + $6 O_2$ (Oxygen) → → → $6 CO_2$ + $6 H_2O$ (Water) + ATP (Energy)

- **The Role of Oxygen in Cellular Respiration**

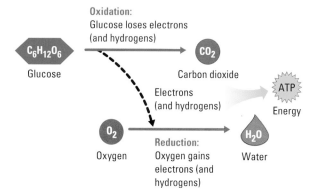

Redox reactions transfer electrons from food molecules to an electron acceptor called NAD^+, forming NADH. The NADH then passes the high-energy electrons to an electron transport chain that eventually "drops" them to O_2.

The energy released during this electron transport is used to regenerate ATP from ADP. The affinity of oxygen for electrons keeps the redox reactions of cellular respiration working.

- **The Metabolic Pathway of Cellular Respiration** You can follow the flow of molecules through the process of cellular respiration in the following diagram:

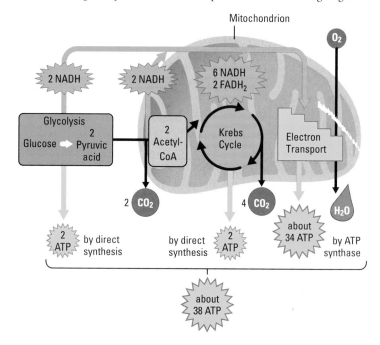

The electron transport chains pump H^+ across the membrane as electrons flow stepwise from NADH to oxygen. Backflow of H^+ across the membrane powers the ATP synthases, which attach phosphate to ADP to make ATP.

Activity 6B *Overview of Cellular Respiration*

Activity 6C *Glycolysis*

Activity 6D *The Krebs Cycle*

Activity 6E *Electron Transport*

Case Study in the Process of Science *How Is the Rate of Cellular Respiration Measured?*

Fermentation: Anaerobic Harvest of Food Energy

- **Fermentation in Human Muscle Cells** When muscle cells consume ATP faster than O_2 can be supplied for cellular respiration, they regenerate ATP by fermentation. The waste product under these anaerobic conditions is lactic acid. The ATP yield per glucose is about 19 times less during fermentation than during cellular respiration.

- **Fermentation in Microorganisms** Yeast and other facultative anaerobes can survive with or without O_2. Wastes from fermentation can be ethyl alcohol, lactic acid, or other compounds, depending on the species. Some microorganisms are obligate anaerobes, which are poisoned by O_2.

Activity 6F *Fermentation*

Self-Quiz

1. Which of the following statements is a correct distinction between autotrophs and heterotrophs?
    a. Only heterotrophs require chemical compounds from the environment.
    b. Cellular respiration is unique to heterotrophs.
    c. Only heterotrophs have mitochondria.
    d. Only autotrophs can live on nutrients that are entirely inorganic.

2. Of the three metabolic pathways that contribute to cellular respiration, which produces the most ATP molecules per glucose?

3. In glycolysis, _____ is oxidized and _____ is reduced.

4. The final electron acceptor of electron transport chains in mitochondria is _____.

5. The poison cyanide acts by blocking a key step in the electron transport chain. Knowing this, explain why cyanide kills so quickly.

6. Cells can harvest the most total chemical energy from which of the following?
    a. an NADH molecule
    b. a glucose molecule
    c. six carbon dioxide molecules
    d. two pyruvic acid molecules

7. Cellular respiration that consumes oxygen occurs in all of the following except
    a. aerobic consumers.
    b. plants.
    c. human muscle cells.
    d. obligate anaerobes.

8. _____ is a metabolic pathway common to both fermentation and cellular respiration.

9. Sports physiologists at an Olympic training center wanted to monitor athletes to determine at what point their muscles were functioning anaerobically. They could do this by checking for a buildup of
    a. ADP.
    b. lactic acid.
    c. carbon dioxide.
    d. oxygen.

10. A glucose-fed yeast cell is moved from an aerobic environment to an anaerobic one. For the cell to continue to generate ATP at the same rate, how much glucose must it consume in the anaerobic environment compared to the aerobic environment?

Answers to the Self-Quiz questions can be found in Appendix B.

Go to the website or CD-ROM for more Self-Quiz questions.

The Process of Science

Your body makes NAD^+ from two B vitamins, niacin and riboflavin. You need only tiny amounts of these vitamins. The U.S. Food and Drug Administration's recommended dietary allowances are 20 milligrams (mg) daily for niacin and 1.7 mg daily for riboflavin. These amounts are thousands of times less than the amount of glucose your body needs each day to fuel its energy requirements. How many NAD^+ molecules are needed for the breakdown of each glucose molecule? Why do you think your daily requirement for these substances is so small?

Case Study in the Process of Science on the Web & CD *Perform experiments that measure the rate of cellular respiration.*

Biology and Society

1. Nearly all human societies use fermentation to produce alcoholic drinks such as beer and wine. The technology dates back to the earliest civilizations. Suggest a hypothesis for how humans first discovered fermentation. In pre-industrial cultures, why do you think wine was a more practical beverage than the grape juice from which it was made?

2. The consumption of alcohol by a pregnant woman can cause a series of birth defects called fetal alcohol syndrome (FAS). Symptoms of FAS include head and facial irregularities, heart defects, mental retardation, and behavioral problems. The U.S. Surgeon General's Office recommends that pregnant women abstain from drinking alcohol, and the government has mandated that a warning label be placed on liquor bottles. Imagine you are a server in a restaurant. An obviously pregnant woman orders a strawberry daiquiri. How would you respond? Is it the woman's right to make those decisions about her unborn child's health? Do you bear any responsibility in the matter? Is a restaurant responsible for monitoring the dietary habits of its customers?

Biology and Society on the Web *Learn about the proper ways to exercise.*

CHAPTER 7

Photosynthesis: Converting Light Energy to Chemical Energy

Biology and Society:
Plant Power 103

The Basics of
Photosynthesis 103

Chloroplasts: Sites of Photosynthesis

The Overall Equation for
Photosynthesis

A Photosynthesis Road Map

The Light Reactions:
Converting Solar Energy to
Chemical Energy 106

The Nature of Sunlight

Chloroplast Pigments

How Photosystems Harvest Light
Energy

How the Light Reactions Generate
ATP and NADPH

The Calvin Cycle: Making Sugar
from Carbon Dioxide 111

Water-Saving Adaptations of C_4 and
CAM Plants

The Environmental Impact of
Photosynthesis 113

How Photosynthesis Moderates the
Greenhouse Effect

Evolution Connection:
The Oxygen Revolution 114

Life on Earth is solar powered.

Without the greenhouse **effect** of the atmosphere, the average air temperature on Earth would be about 10°C (18°F) colder.

Photosynthesis produces 160 billion metric tons of carbohydrates each year.

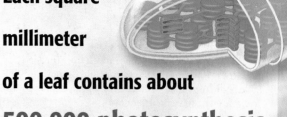

Each square millimeter of a leaf contains about 500,000 photosynthesis factories called chloroplasts.

Plant Power

On a global scale, the productivity of photosynthesis is astounding. By converting the energy of sunlight to chemical energy, Earth's plants and other photosynthetic organisms make about 160 billion metric tons of organic material per year (a metric ton is 1,000 kg, about 1.1 tons). That much organic material is equivalent in weight to about 25 stacks of this book reaching from Earth to the sun!

All of the food consumed by humans can be traced back to photosynthetic plants. But plants provide much more than just nourishment. They also supply many of the raw materials we need to survive. For example, during the vast majority of human history, burning plant material was the only nonsolar source of heat, light, and fuel for cooking. Over the last century, wood has been largely displaced as an energy source by fossil fuels such as coal, gas, and oil (which can also be traced back to photosynthesis sources, albeit ancient ones). But recently, the benefits of plant matter as an energy source have regained attention.

Figure 7.1 shows an "energy plantation" in upstate New York. There, willow trees are grown to test these plants as a renewable energy source. The trees are cut once every three years, and the harvested wood is sent to power plants to generate electricity. Because they grow exceptionally fast and resprout after cutting, willows are a renewable resource. Other tree species, including sycamore, eucalyptus, and black locust, are being tested elsewhere.

Burning wood for energy has several advantages over burning fossil fuels. Wood has very little of the sulfur impurities that cause acid rain. Energy plantations also provide habitat for wildlife, reduce erosion, renew the soil, and help farmers diversify. Perhaps most significantly, vigorously growing young plants remove a lot of carbon dioxide from the air. Substituting energy crops for fossil fuels could reduce the levels of gases in the atmosphere that cause global warming. Today, so-called biomass energy accounts for only about 4% of all energy consumed in the United States. But this renewable, low-emissions resource may play a significant role in satisfying our future energy needs.

In this chapter, you'll learn how plants convert solar energy to chemical energy. Although photosynthesis is a complex process, some basic concepts will keep us oriented.

Biology and Society on the Web
Learn more about the potential of biomass energy.

Activity 7A on the Web & CD
Discover more ways we use plants.

Figure 7.1
One-year-old willow trees growing in an "energy plantation" in upstate New York.

The Basics of Photosynthesis

Almost all plants are photosynthetic autotrophs—meaning that they generate their own organic matter from inorganic ingredients via photosynthesis—and so are certain groups of protists and bacteria (**Figure 7.2**, see page 104). This section presents an overview of the process of photosynthesis. In later sections, we'll take an even closer look.

Chloroplasts: Sites of Photosynthesis

In Chapter 4, you learned that photosynthesis occurs within **chloroplasts,** organelles present in certain plant cells. All green parts of a plant have chloroplasts and can carry out photosynthesis. However, in most plants,

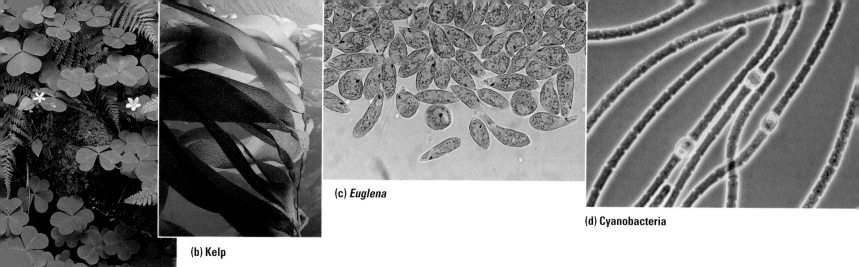

(c) *Euglena*

(d) Cyanobacteria

(b) Kelp

(a) Mosses, ferns, and flowering plants

Figure 7.2 Photosynthetic autotrophs: producers for most ecosystems.
(a) On land, plants are the predominant producers of food. Three major groups of plants—mosses, ferns, and flowering plants—are represented in this scene. In oceans, lakes, ponds, streams, and other aquatic habitats, photosynthetic organisms include **(b)** large algae, such as this kelp, **(c)** certain microscopic protists, such as *Euglena*, and **(d)** bacteria called cyanobacteria.

the leaves have the most chloroplasts and are the major sites of photosynthesis. More specifically, the green color in plants is from chlorophyll pigments in the chloroplasts. Chlorophyll absorbs the light energy that the chloroplast puts to work in making food.

Chloroplasts are concentrated in the cells of the **mesophyll,** the green tissue in the interior of the leaf (**Figure 7.3**). Carbon dioxide (CO_2) enters, and oxygen (O_2) exits, by way of tiny pores called **stomata** (singular, *stoma,* meaning "mouth"). Most stomata are found on the lower epidermis of leaves. In addition to carbon dioxide, photosynthesis also requires water as an inorganic ingredient. This needed water is mainly absorbed by the plant's roots, then travels via veins to the leaves.

Activity 7B on the Web & CD
Review the location and structure of chloroplasts.

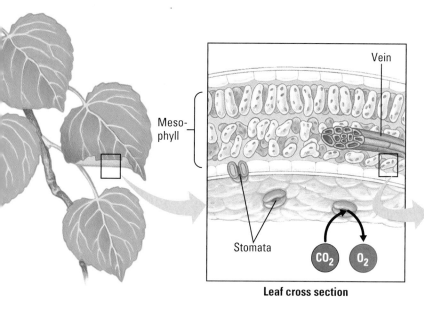

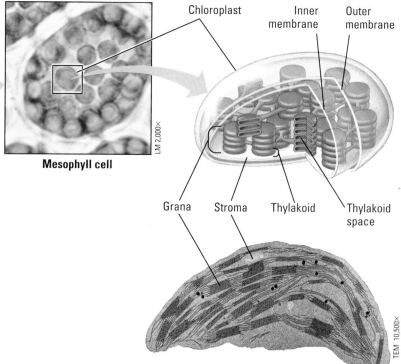

Figure 7.3 Journey into a leaf.
This series of blowups takes you into the mesophyll of a leaf, then into a cell, and finally into a chloroplast, the site of photosynthesis.

Membranes in the chloroplast form the apparatus where many of the reactions of photosynthesis occur. Like the mitochondrion, the chloroplast has a double-membrane envelope (see Figures 4.17 and 4.18). The chloroplast's inner membrane encloses a compartment filled with **stroma,** a thick fluid. Suspended in the stroma is an elaborate system of disklike membranous sacs called **thylakoids.** The thylakoids are concentrated in stacks called **grana** (singular, *granum*). The chlorophyll molecules that capture light energy are built into the thylakoid membranes. The structure of chloroplasts—with stacks of disks—aids its function by providing a large surface area for the reactions of photosynthesis.

The Overall Equation for Photosynthesis

The following chemical equation provides an overall look at the reactants and products of photosynthesis:

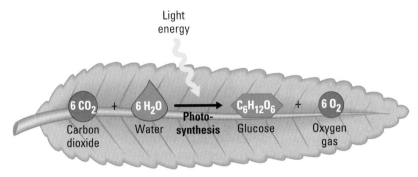

Light energy

$6 CO_2$ + $6 H_2O$ → Photosynthesis → $C_6H_{12}O_6$ + $6 O_2$
Carbon dioxide   Water   Glucose   Oxygen gas

Notice that the reactants of photosynthesis, carbon dioxide (CO_2) and water (H_2O), are also the waste products of cellular respiration (see Figure 6.2). And photosynthesis produces what respiration uses, namely glucose ($C_6H_{12}O_6$) and oxygen (O_2). In other words, photosynthesis takes the "exhaust" of cellular respiration and rearranges its atoms to produce food and oxygen. It's a chemical transformation that requires much energy, and sunlight absorbed by chlorophyll provides that energy.

You learned in Chapter 6 that cellular respiration is a process of electron transfer, or reduction and oxidation (redox). A "fall" of electrons from food molecules to oxygen to form water releases the energy that mitochondria can use to make ATP (see Figure 6.5). The opposite occurs in photosynthesis: Electrons are boosted "uphill" and added to carbon dioxide to produce sugar. Hydrogen is moved along with the electrons, so the redox process takes the form of hydrogen transfer from water to carbon dioxide. This requires the chloroplast to actually split water molecules into hydrogen and oxygen. The chloroplast transfers the hydrogen along with electrons to carbon dioxide to form sugar. The oxygen from the splitting of water escapes through stomata into the atmosphere as O_2, a waste product of photosynthesis. It takes a lot of energy to split water, which is a very stable molecule. Again, it is sunlight that provides the energy.

A Photosynthesis Road Map

The equation for photosynthesis is a deceptively simple summary of a very complex process. Actually, photosynthesis is not a single process, but two processes, each with many steps. These two stages of photosynthesis are called the light reactions and the Calvin cycle (**Figure 7.4**).

The **light reactions** convert solar energy to chemical energy. They use light energy to drive the synthesis of two molecules: ATP and NADPH. We

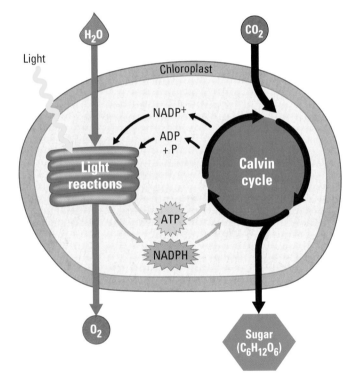

Figure 7.4 A road map for photosynthesis.
Thylakoids, the membranous sacs stacked as grana in chloroplasts, are the sites of the light reactions. Chlorophyll built into the thylakoid membranes absorbs light energy. The thylakoids convert the light energy to the chemical energy of ATP and NADPH. The Calvin cycle uses these two products of the light reactions to power the production of sugar from CO_2. ATP provides energy, and NADPH is a source of high-energy electrons to convert CO_2 to sugar. The enzymes for the Calvin cycle are dissolved in the stroma, the thick fluid within the chloroplast. We'll carry a smaller version of this road map for orientation as we take a closer look at the light reactions and the Calvin cycle.

have already met ATP as the molecule that drives most cellular work. **NADPH,** chemical cousin of the NADH that appeared in Chapter 6, is an electron carrier. In cellular respiration, remember, NADH carries electrons from food molecules. In photosynthesis, light drives electrons from water to $NADP^+$ (the oxidized form of the carrier) to form NADPH (the reduced form of the carrier). Although the light reactions convert light energy to the chemical energy of ATP and NADPH, this stage of photosynthesis does not produce sugar.

Activity 7C on the Web & CD
See simple animations of the overall process of photosynthesis.

It is the **Calvin cycle** that actually makes sugar from carbon dioxide. The ATP generated by the light reactions provides the energy for sugar synthesis. And the NADPH produced by the light reactions provides the high-energy electrons for the reduction of carbon dioxide to glucose. Thus, the Calvin cycle depends on light, but only indirectly; it depends on the supply of ATP and NADPH produced by the light reactions.

The road map in Figure 7.4 will help you keep oriented in this energy traffic from the light reactions to the Calvin cycle as we now take a closer look at how these two stages of photosynthesis work.

CHECKPOINT

1. For chloroplasts to produce sugar from carbon dioxide in the dark, they would require an artificial supply of the molecules _____ and _____.

2. What are the primary inputs and outputs of the Calvin cycle?

Answers: 1. ATP; NADPH **2.** Inputs: CO₂, ATP, NADPH; output: glucose

---

The Light Reactions: Converting Solar Energy to Chemical Energy

Chloroplasts are chemical factories powered by the sun, an energy source over 90 million miles from Earth. In this section, we'll track sunlight into a chloroplast to see how it is converted to the chemical energy of ATP and NADPH.

The Nature of Sunlight

Sunlight is a type of energy called radiation, or electromagnetic energy. Electromagnetic energy travels through space as rhythmic waves analogous to those made by a pebble dropped in a puddle of water. The distance between the crests of two adjacent waves is called a **wavelength.** The full range of radiation, from the very short wavelengths of gamma rays to the very long wavelengths of radio signals, is called the **electromagnetic spectrum.** Visible light composes only a small fraction of the spectrum. It consists of those wavelengths that our eyes see as different colors (**Figure 7.5**).

When sunlight shines on a pigmented material, certain wavelengths (colors) of the visible light are absorbed and disappear from the light that is reflected by the material. For example, we see a pair of jeans as blue because the pigments in the material absorb the other colors, leaving only light in the

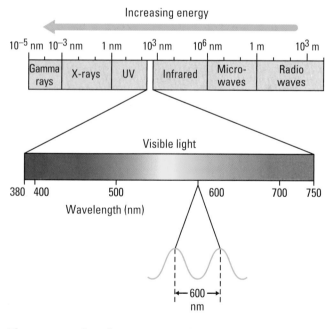

Figure 7.5 The electromagnetic spectrum.
This diagram expands the thin slice of the spectrum that is visible to us as different colors of light. Visible light ranges from about 380 nanometers (nm) to about 750 nm in wavelength. The bottom of the figure shows electromagnetic waves of one particular wavelength of visible light.

blue part of the spectrum to be reflected from the material to our eyes. And the leaves of plants appear green because pigments in their chloroplasts absorb mainly blue-violet and red-orange wavelengths (**Figure 7.6**). Of course, energy cannot be destroyed. If a pigment has subtracted certain wavelengths of light from the spectrum through absorption, that energy has been converted to other forms. Chloroplasts are able to convert some of the solar energy they absorb into chemical energy.

Chloroplast Pigments

Different pigments absorb light of different wavelengths, and chloroplasts contain several kinds of pigments. One, **chlorophyll a,** absorbs mainly blue-violet and red light. It looks grass-green because it reflects mainly green light. Chlorophyll *a* is the pigment that participates directly in the light reactions. A very similar molecule, chlorophyll *b*, absorbs mainly blue and orange light and reflects (appears) yellow-green. Chlorophyll *b* does not participate directly in the light reactions, but it broadens the range of light that a plant can use by conveying absorbed energy to chlorophyll *a*, which then puts the energy to work in the light reactions.

Activity 7D on the Web & CD
Review the nature of light and chloroplast pigments.

Chloroplasts also contain a family of yellow-orange pigments called carotenoids, which absorb mainly blue-green light. Some may pass energy to chlorophyll *a*. Other carotenoids have a protective function: They absorb and dissipate excessive light energy that would otherwise damage chlorophyll. (Similar carotenoids, which we obtain from carrots and certain other plants, may help protect our eyes from very bright light.) The spectacular colors of fall foliage in certain parts of the world are due partly to decreases in green chlorophyll, allowing the yellow-orange hues of longer-lasting carotenoids to show through (**Figure 7.7**).

Case Study in the Process of Science on the Web & CD
Learn how to separate plant pigments by chromatography.

All of these chloroplast pigments are built into the thylakoid membranes, especially where those membranes are stacked as grana (see Figure 7.3). There the pigments are part of light-harvesting complexes called photosystems.

How Photosystems Harvest Light Energy

The theory of light as waves explains most of light's properties. However, light also behaves as discrete packets of energy called photons. A **photon** is a fixed quantity of light energy. The shorter the wavelength of light, the greater

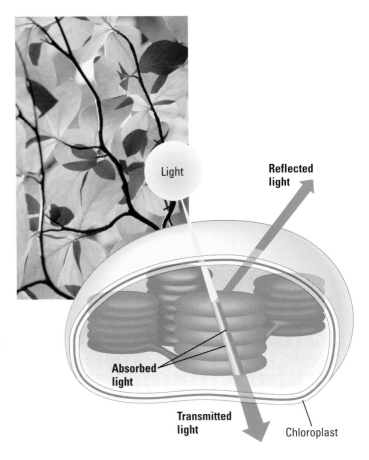

Figure 7.6 Why are leaves green?

Chlorophyll and the other pigments built into the membranes of grana mainly absorb light in the blue-violet and red-orange part of the electromagnetic spectrum. The pigments do not absorb much green light, which is reflected to our eyes.

Figure 7.7 Photosynthetic pigments.

Falling autumn temperatures cause a decrease in the levels of green chlorophyll within the leaves of deciduous trees. This causes the colors of the carotenoids to become prominent.

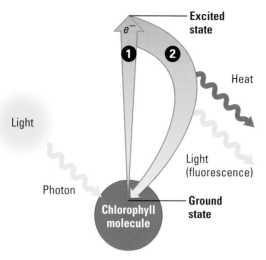

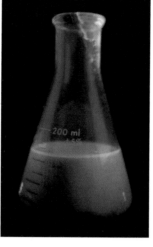

(a) Absorption of a photon

(b) Fluorescence of isolated chlorophyll in solution

(c) Fluorescence of a glow stick

Figure 7.8 The behavior of isolated chlorophyll.
(a) ❶ The absorption of a photon drives an electron (e^-) from its ground state to an excited state. **❷** In a billionth of a second, the excited electron falls back to its ground state, releasing heat and light. **(b)** The afterglow from chlorophyll is called fluorescence. This flask contains a solution of chlorophyll illuminated with ultraviolet (UV) light. The red-orange wavelengths of the fluorescence are less energetic than the UV light that excited the chlorophyll. Remember that energy cannot be destroyed, however; heat makes up the difference. **(c)** A glow stick contains chemicals that, when mixed, excite electrons within a fluorescent dye. As the electrons fall from their excited state to the ground state, the excess energy is emitted as glowing light.

Figure 7.9 A photosystem.   The pigment cluster of the photosystem functions as a light-gathering antenna that focuses energy onto the reaction center. At the reaction center of the photosystem, a chlorophyll *a* molecule transfers its light-excited electron (e^-) to a primary electron acceptor.

the energy of a photon. For example, a photon of violet light packs nearly twice as much energy as a photon of red light (see Figure 7.5).

When a pigment molecule absorbs a photon, one of the pigment's electrons gains energy, and we say that the electron has been raised from a ground state to an excited state. The excited state is very unstable, and generally the electron loses the excess energy and falls back to its ground state almost immediately (**Figure 7.8a**). Most pigments merely release heat energy as their light-excited electrons fall back to their ground state. (That's why a dark surface, such as a black automobile roof, gets so hot on a sunny day.) Some pigments, including isolated chlorophyll that has been extracted from chloroplasts, emit light as well as heat after absorbing photons (**Figure 7.8b**). The fluorescent light emitted by a glow stick (**Figure 7.8c**) is caused by a similar chemical process. When you bend a glow stick, you break an internal glass vial, causing two chemicals to combine. This starts a chemical reaction that excites electrons within a fluorescent dye. The electrons quickly fall back down to their ground state, releasing energy in the form of fluorescence.

Light-excited chlorophyll behaves very differently in an intact chloroplast than it does in isolation. In its native habitat of the thylakoid membrane, chlorophyll is organized with other molecules into photosystems. Each **photosystem** has a cluster of a few hundred pigment molecules, including chlorophylls *a* and *b* and carotenoids (**Figure 7.9**). This cluster of

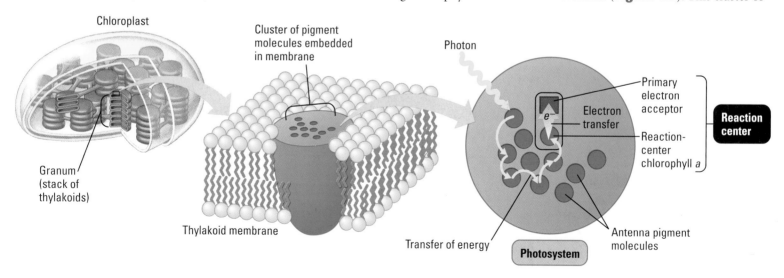

Chloroplast

Granum (stack of thylakoids)

Cluster of pigment molecules embedded in membrane

Thylakoid membrane

Photon

Transfer of energy

Photosystem

Primary electron acceptor

Electron transfer

e^-

Reaction-center chlorophyll *a*

Antenna pigment molecules

Reaction center

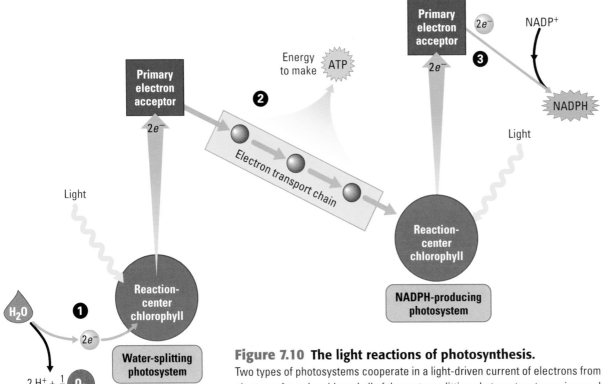

Figure 7.10 The light reactions of photosynthesis.
Two types of photosystems cooperate in a light-driven current of electrons from water to NADP$^+$. Photons excite electrons from the chlorophyll of the water-splitting photosystem to a primary electron acceptor. **❶** The water-splitting photosystem replaces its light-excited electrons by extracting electrons from H$_2$O. This is the step that releases O$_2$ during photosynthesis. **❷** Energized electrons from the water-splitting photosystem pass down an electron transport chain to the NADPH-producing photosystem. The chloroplast uses the energy released by this electron "fall" to make ATP. **❸** The NADPH-producing photosystem transfers its light-excited electrons to NADP$^+$, reducing it to NADPH. The electron transport chain replaces the electrons lost from the photosystem's chlorophyll.

pigment molecules functions as a light-gathering antenna. When a photon strikes one pigment molecule, the energy jumps from pigment to pigment until it arrives at the **reaction center** of the photosystem. The reaction center consists of a chlorophyll *a* molecule that sits next to another molecule called a **primary electron acceptor.** This acceptor traps the light-excited electron from the reaction-center chlorophyll. Another team of molecules built into the thylakoid membrane then uses that trapped energy to make ATP and NADPH.

How the Light Reactions Generate ATP and NADPH

Two types of photosystems cooperate in the light reactions. One is a water-splitting photosystem, which uses light energy to extract electrons from H$_2$O (**Figure 7.10**). This is the process that releases O$_2$ as a waste product of photosynthesis. The second photosystem is an NADPH-producing photosystem. It produces NADPH by transferring light-excited electrons from chlorophyll to NADP$^+$. An electron transport chain connecting the two photosystems releases energy that the chloroplast uses to make ATP. The traffic of electrons through the two photosystems is analogous to the cartoon in **Figure 7.11.**

Case Study in the Process of Science on the Web & CD Perform experiments to measure the rate of the light reactions.

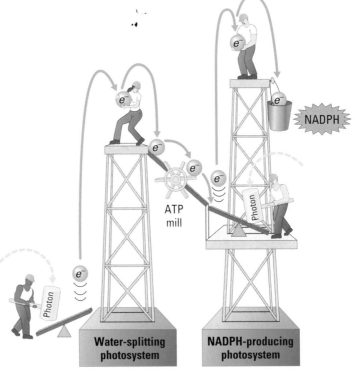

Figure 7.11
A hard-hat analogy for the light reactions.

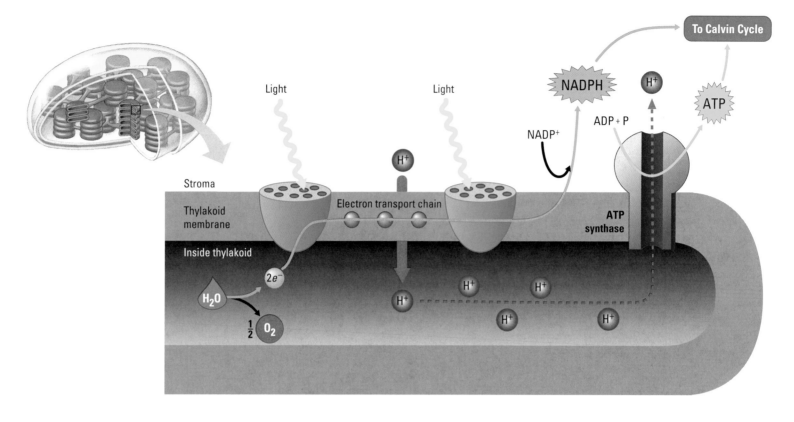

Figure 7.12
How the thylakoid membrane converts light energy to the chemical energy of NADPH and ATP.
The two photosystems and the electron transport chain that connects them transfer electrons from H_2O to $NADP^+$. The electron transport chain also functions as a hydrogen ion (H^+) pump. ATP synthase molecules, much like the ones in mitochondria, use the energy of the H^+ gradient to make ATP.

Figure 7.12 places the light reactions in the thylakoid membrane. Notice that the mechanism of ATP production during the light reactions is

Activity 7E on the Web & CD
See animations of the light reactions.

very similar to the ATP production we saw in cellular respiration (see Chapter 6). In both cases, an electron transport chain pumps hydrogen ions (H^+) across a membrane—the inner mitochondrial membrane in the case of respiration and the thylakoid membrane in photosynthesis. And in both cases, ATP synthases use the energy stored by the H^+ gradient to make ATP. The main difference is that food provides the high-energy electrons in cellular respiration, while it is light-excited electrons that flow down the transport chain during photosynthesis.

We have seen how the light reactions convert solar energy to the chemical energy of ATP and NADPH. Notice again, however, that the light reactions produce no sugar. That's the job of the Calvin cycle, which spends the ATP and NADPH produced by the light reactions—as we'll see next.

CHECKPOINT

1. Compared to a solution of isolated chlorophyll, why do intact chloroplasts release less heat and fluorescence when illuminated?

2. Why is water required as a reactant in photosynthesis?

3. In addition to conveying electrons from the water-splitting photosystem to the NADPH-producing photosystem, the electron transport chains of chloroplasts also provide the energy for the synthesis of _____.

Answers: 1. In the chloroplasts, the light-excited electrons are trapped by a primary electron acceptor rather than immediately giving up all their energy as heat and light. **2.** It is the splitting of water that provides electrons for converting CO_2 to sugar (via electron transfer by NADPH). **3.** ATP

The Calvin Cycle: Making Sugar from Carbon Dioxide

The Calvin cycle functions like a sugar factory within a chloroplast. It is called a cycle because, like the Krebs cycle in cellular respiration, the starting material is regenerated with each turn of the cycle (**Figure 7.13**). And with each turn, there are chemical inputs and outputs. The inputs are CO_2 from the air and ATP and NADPH produced by the light reactions. Using carbon from CO_2, energy from ATP, and high-energy electrons from NADPH, the Calvin cycle constructs an energy-rich sugar molecule. That sugar is not glucose, but a smaller sugar named glyceraldehyde 3-phosphate (G3P). The plant cell can then use G3P as the raw material to make the glucose and other organic molecules it needs.

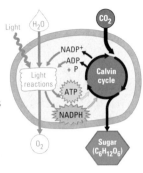

Activity 7F on the Web & CD
Watch animations of the Calvin cycle.

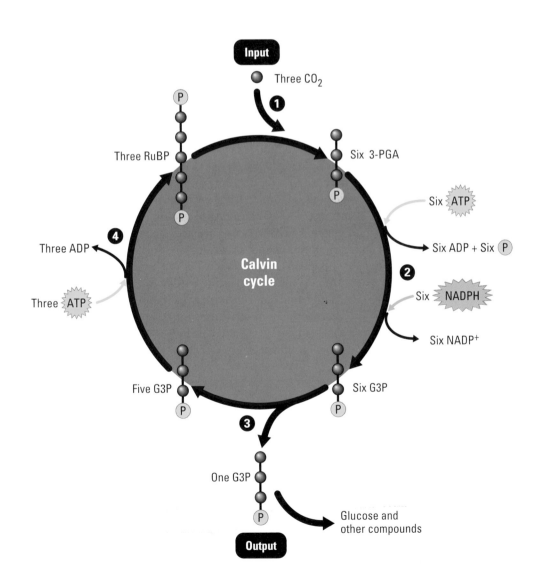

Figure 7.13 The Calvin cycle.
The gray balls in this diagram symbolize carbon atoms. The cycle produces one G3P sugar molecule for every three CO_2 molecules that enter the cycle. ❶ Carbon enters the cycle as CO_2. An enzyme adds the CO_2 to RuBP (ribulose bisphosphate), a five-carbon sugar already present in the chloroplast as a product of the cycle. (The Ⓟ s in the diagram are phosphate groups.) The product then breaks into a three-carbon compound called 3-PGA (3-phospho-glyceric acid). ❷ ATP and NADPH from the light reactions provide energy and electrons. Enzymes use the ATP energy and high-energy electrons from NADPH to convert the 3-PGA to a three-carbon sugar, G3P (glyceraldehyde 3-phosphate). ❸ Carbon exits the cycle as sugar. The cycle has converted three CO_2 molecules to one molecule of the sugar G3P. This is the direct product of photosynthesis, but plant cells can use the G3P to make glucose and other organic compounds for growth and fuel. ❹ The cycle regenerates its starting material. Note that of the six G3P molecules produced in step 3, only one of them represents net sugar output. That's because we started with a total of 15 sugar carbons in the three RuBP molecules that accepted CO_2 back in step 1. Enzymes now regenerate the RuBP by rearranging the five G3P molecules that are left after one of those sugars exits the cycle.

(a) Sugarcane (b) Pineapple

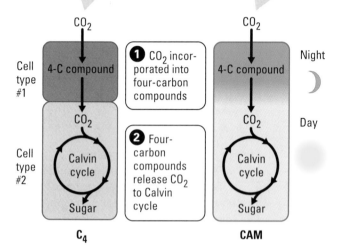

Water-Saving Adaptations of C₄ and CAM Plants

During the long evolution of plants in diverse environments, natural selection has refined photosynthetic adaptations that enable certain plants to continue producing food even in arid conditions.

Plants in which the Calvin cycle uses CO_2 directly from the air are called **C₃ plants** because the first organic compound produced is the three-carbon compound 3-PGA (see Figure 7.13). C_3 plants are common and widely distributed, and some of them, such as soybeans, oats, wheat, and rice, are important in agriculture. One of the problems that farmers face in growing C_3 plants, however, is that dry weather can reduce the rate of photosynthesis and decrease crop productivity. On a hot, dry day, plants close their stomata, the pores in the undersurface of a leaf. Closing stomata is an adaptation that reduces water loss, but it also prevents CO_2 from entering the leaf. As a result, CO_2 levels can get very low in the leaf, and sugar production ceases, at least in C_3 plants.

In contrast to C_3 plants, so-called **C₄ plants** have special adaptations that save water without shutting down photosynthesis. When the weather is hot and dry, a C_4 plant keeps its stomata closed most of the time, thus conserving water. At the same time, it continues making sugars by photosynthesis, using the route shown in **Figure 7.14a**. A C_4 plant has an enzyme that incorporates carbon from CO_2 into a four-carbon (4-C) compound instead of into 3-PGA. This enzyme has an intense affinity for CO_2 and can continue to mine it from the air spaces of the leaf even when the stomata are closed. The four-carbon compound the enzyme produces acts as a carbon shuttle; it donates the CO_2 to the Calvin cycle in a nearby cell, which therefore keeps on making sugars even though the plant's stomata are closed. Corn, sorghum, and sugarcane are examples of agriculturally important C_4 plants. All three evolved in hot regions of the tropics where there are frequent dry seasons.

Another photosynthetic adaptation that conserves water evolved in pineapples, many cacti, and most of the so-called succulent plants (those with very juicy tissues), such as aloe and jade plants. Collectively called **CAM plants,** most such species are adapted to very dry climates. CAM stands for *crassulacean acid metabolism,* after the plant family Crassulaceae (jade plants and others), in which this important water-saving adaptation was first discovered. A CAM plant conserves water by opening its stomata and admitting CO_2 mainly at night (**Figure 7.14b**). When CO_2 enters the leaves, it is incorporated into a four-carbon compound, as in C_4 plants. The four-carbon compound in a CAM plant banks CO_2 at night and releases it to the Calvin cycle during the day. This keeps photosynthesis operating during the day, even though the leaf admits no more CO_2 because the stomata are closed. Note that in all plants—C_3, C_4, and CAM types—it is the Calvin cycle that is ultimately responsible for the synthesis of sugar.

Figure 7.14 C₄ and CAM photosynthesis compared.
Both adaptations are characterized by ❶ the preliminary incorporation of CO₂ into four-carbon compounds, followed by ❷ the transfer of the CO₂ to the Calvin cycle. **(a)** In C₄ plants, such as sugarcane, these two steps are separated spatially; they are segregated into two cell types. **(b)** In CAM plants, such as pineapple, the two steps are separated in time; carbon incorporation into four-carbon compounds occurs at night, and the Calvin cycle operates during the day. The C₄ and CAM pathways are two evolutionary solutions to the problem of maintaining photosynthesis with stomata partially or completely closed on hot, dry days.

Activity 7G on the Web & CD
Learn more about how plants in dry climates photosynthesize.

CHECKPOINT

1. In terms of the spatial organization of photosynthesis within the chloroplast, what is the advantage of the light reactions producing NADPH and ATP on the stroma side of the thylakoid membrane?

2. What is the function of NADPH in the Calvin cycle?

3. How do special enzymes enable C₄ and CAM plants to conserve water during photosynthesis?

Answers: 1. The Calvin cycle, which consumes the NADPH and ATP, occurs in the stroma. **2.** It provides the high-energy electrons for the reduction of CO₂ to form sugar. **3.** By allowing photosynthesis to continue even when stomata are closed during dry conditions

The Environmental Impact of Photosynthesis

Figure 7.15 reviews how the light reactions and the Calvin cycle cooperate in converting light energy to the chemical energy of food. What the diagram doesn't show is the transfer of this organic material from the producers (plants and other photosynthetic organisms) to consumers (such as the animals that eat plants). Even the energy we acquire when we eat meat was originally captured by photosynthesis. The energy in a hamburger, for instance, came from sunlight that was originally converted to chemical energy in chloroplasts in grasses eaten by cattle.

In addition to producing food, photosynthesis also has an enormous impact on the atmosphere by swapping O_2 for CO_2. This O_2, of course, sustains cellular respiration in all organisms. But gas exchange by plants also helps moderate temperatures on Earth, as you'll see next.

How Photosynthesis Moderates the Greenhouse Effect

Old-growth forests are the focus of a long-standing controversy between the timber industry and conservationists (**Figure 7.16**). The giant trees of these forests contain a lot of marketable lumber, and economic arguments can be made for harvesting them. On the other side of the controversy, conservationists argue that old-growth forests are home to many species of plants and wildlife that can survive nowhere else and that we should save these remnants of our ancient forests for future generations.

The photosynthesis carried out by old-growth forests has direct bearing on the controversy. At center stage is CO_2, the gas that plants use to make sugars in photosynthesis and that all organisms give off as waste from cellular respiration. Carbon dioxide normally makes up about 0.03% of the air we breathe. This amount of CO_2 in the atmosphere provides plants with plenty of carbon. It also helps moderate world climates, because CO_2 retains heat from the sun that would otherwise radiate from Earth back into space. Warming induced by CO_2 is called the **greenhouse effect** because atmospheric CO_2 traps heat and warms the air just as clear glass does in a greenhouse (**Figure 7.17**). This natural effect is highly beneficial. Without it, Earth would be about 10°C (18°F) colder and much less hospitable to life.

Ironically, planet Earth may now be overheating from the greenhouse effect. Since the start of the Industrial Revolution, the atmospheric concentration of

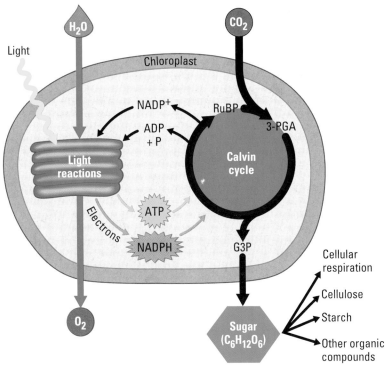

Figure 7.15 **A review of photosynthesis.**

Figure 7.16
An old-growth forest in the Pacific Northwest.
"Old growth" refers to ancient forests that have never been seriously disturbed by humans. Many of the trees—mainly Douglas fir here—are thousands of years old. The forest they dominate is one of a few remaining undisturbed areas that contain harvestable timber in the United States.

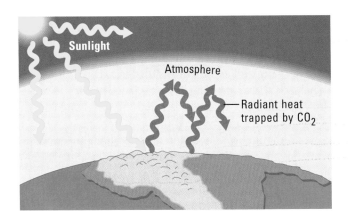

Figure 7.17
The greenhouse effect, a result of CO_2 in the atmosphere.

CO_2 has increased about 30%, predominantly from the combustion of carbon-based fossil fuels. Increasing concentrations of greenhouse gases is the most likely cause of **global warming,** a slow but steady rise in Earth's surface temperature. Photosynthesis consumes CO_2, tending to counteract the greenhouse effect, but the global rate of photosynthesis may decline as we clear huge tracts of forest for logging, farming, and urban expansion. Extensive deforestation continues, especially in the tropics, the northwestern and southeastern United States, Canada, and Siberia.

Activity 7H on the Web & CD
Perform experiments to see how various factors affect global temperature.

The greenhouse effect has come up on both sides of the old-growth forest controversy. Those who favor saving the old-growth trees argue that these large plants remove a lot of potentially harmful CO_2 from the atmosphere. Those favoring harvesting argue that replacing the old trees (which have a lot of nonphotosynthetic wood tissue) with seedlings would actually increase photosynthesis and reduce atmospheric CO_2.

Relative to their size and weight, young, rapidly growing trees do, in fact, take up CO_2 at a faster rate than old trees. However, when old trees are harvested, much less than half their bulk becomes lumber. All the roots and many branches are left behind to decompose. Most of the wood itself is turned into paper, sawdust, or fuel—products that usually decompose or are burned within a few years. Decomposition and burning turn the carbon compounds that were in the trees into CO_2, returning a lot of it to the atmosphere.

CHECKPOINT

How might the combustion of fossil fuels and wood be contributing to global warming?

Answer: By raising concentrations of atmospheric CO_2 and increasing the greenhouse effect

Evolution Connection

The Oxygen Revolution

The atmospheric oxygen we breathe and use for cellular respiration is a by-product of the water-splitting step of photosynthesis. The first photosynthetic organisms with the metabolic equipment to split water were prokaryotes called cyanobacteria (see Figure 7.2d). They changed Earth forever by adding O_2 to the atmosphere.

Cyanobacteria evolved between 2.5 and 3.4 billion years ago. As their numbers grew, the gradual accumulation of O_2 in the atmosphere created a crisis for other ancient forms of life, because oxygen attacks the bonds of organic molecules. The corrosive oxygen-containing atmosphere probably caused the extinction of many prokaryotic forms unable to cope. Other species survived in habitats that remained anaerobic (such as deep in the soil), where we find their descendants living today. These organisms die if they are exposed to oxygen. Still other species adapted to the changing environment, actually putting the oxygen to use in extracting energy from food, the key process of cellular respiration. The "oxygen revolution" was a major episode in the history of life on Earth.

Evolution Connection on the Web
Learn more about the organisms that first released large amounts of oxygen into Earth's atmosphere.

Chapter Review

Summary of Key Concepts

For study help, go to the Essential Biology Website (www.essentialbiology.com) or CD-ROM to explore the Activities and Case Studies in the Process of Science.

The Basics of Photosynthesis

• **Chloroplasts: Sites of Photosynthesis** Chloroplasts contain a thick fluid called stroma surrounding a network of membranes called thylakoids.

Activity 7A *Plants in Our Lives*

Activity 7B *The Sites of Photosynthesis*

• **The Overall Equation for Photosynthesis**

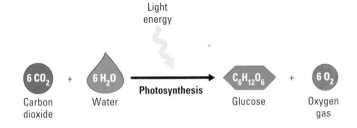

Light energy

$6\ CO_2$ + $6\ H_2O$ → **Photosynthesis** → $C_6H_{12}O_6$ + $6\ O_2$

Carbon dioxide | Water | Glucose | Oxygen gas

• **A Photosynthesis Road Map**

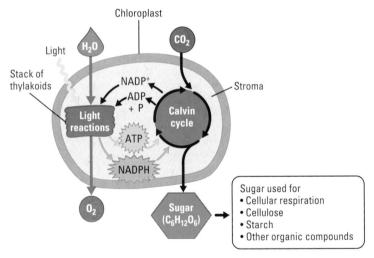

Chloroplast

Light

H_2O | CO_2

Stack of thylakoids

$NADP^+$
$ADP + P$
Light reactions
Calvin cycle
ATP
NADPH

Stroma

O_2

Sugar ($C_6H_{12}O_6$)

Sugar used for
• Cellular respiration
• Cellulose
• Starch
• Other organic compounds

Activity 7C *Overview of Photosynthesis*

The Light Reactions: Converting Solar Energy to Chemical Energy

• **The Nature of Sunlight** Visible light is part of the electromagnetic spectrum of radiation. It travels through space as waves.

• **Chloroplast Pigments** Pigments absorb light energy of certain wavelengths and reflect other wavelengths. We see the reflected wavelengths as the color of the pigment. Several chloroplast pigments absorb light of various wavelengths, but it is the green pigment chlorophyll *a* that participates directly in the light reactions.

Activity 7D *Light Energy and Pigments*

Case Study in the Process of Science *How Does Paper Chromatography Separate Plant Pigments?*

• **How Photosystems Harvest Light Energy; How the Light Reactions Generate ATP and NADPH**

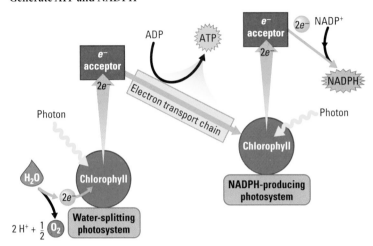

e^- acceptor
ADP
ATP
e^- acceptor
$2e^-$ | $NADP^+$
$2e^-$
NADPH
$2e^-$
Electron transport chain
Photon
Photon
Chlorophyll
H_2O
$2e^-$
Chlorophyll
NADPH-producing photosystem
$2\ H^+ + \frac{1}{2}\ O_2$
Water-splitting photosystem

Case Study in the Process of Science *How Is the Rate of Photosynthesis Measured?*

Activity 7E *The Light Reactions*

The Calvin Cycle: Making Sugar from Carbon Dioxide

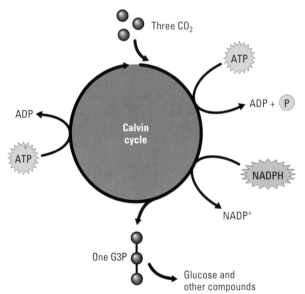

Three CO_2
ATP
ADP + P
ADP
Calvin cycle
ATP
NADPH
$NADP^+$
One G3P
Glucose and other compounds

Activity 7F *The Calvin Cycle*

• **Water-Saving Adaptations of C₄ and CAM Plants** Photosynthetic adaptations of C₄ and CAM plants enable sugar production to continue even when stomata are closed, thereby reducing water loss in arid environments.

Activity 7G *Photosynthesis in Dry Climates*

The Environmental Impact of Photosynthesis

• The process of photosynthesis provides organic material and chemical energy for life on Earth. Photosynthesis also swaps O_2 for CO_2 in the atmosphere.

- **How Photosynthesis Moderates the Greenhouse Effect** Atmospheric CO_2 traps heat and contributes to our planet's relatively warm temperature. Photosynthesis makes use of some of this CO_2. Levels of CO_2 in the atmosphere are increasing, raising concerns about global warming. Deforestation and the burning of fossil fuels may be contributing to these global changes in atmosphere and climate.

Activity 7H *The Greenhouse Effect*

Self-Quiz

1. The light reactions take place in the region of the chloroplast called the _____, while the Calvin cycle takes place in the _____.

2. Which of the following are inputs to photosynthesis? Which are outputs?
   - a. CO_2
   - b. O_2
   - c. sugar
   - d. H_2O
   - e. light

3. What color of light is the least effective in driving photosynthesis? Why?

4. When light strikes chlorophyll molecules, they lose electrons, which are ultimately replaced by splitting a molecule of _____.

5. Which of the following are produced by reactions that take place in the thylakoids and are consumed by reactions in the stroma?
   - a. CO_2 and H_2O
   - b. $NADP^+$ and ADP
   - c. ATP and NADPH
   - d. glucose and O_2

6. The reactions of the Calvin cycle are not directly dependent on light, and yet they usually do not occur at night. Why?

7. Why is it difficult for most plants to carry out photosynthesis in very hot, dry environments, such as deserts?

8. What is the primary advantage offered by the C_4 and CAM pathways?

9. Of the following metabolic processes, which one is common to photosynthesis and cellular respiration?
   - a. reactions that convert light energy to chemical energy
   - b. reactions that split H_2O molecules and release O_2
   - c. reactions that store energy by pumping H^+ across membranes
   - d. reactions that convert CO_2 to sugar

10. The combustion of fossil fuels may be contributing to global warming mainly by raising atmospheric concentrations of _____.

Answers to the Self-Quiz questions can be found in Appendix B.

Go to the website or CD-ROM for more Self-Quiz questions.

The Process of Science

1. Tropical rain forests cover only about 3% of Earth's surface, but they are estimated to be responsible for more than 20% of global photosynthesis. For this reason, rain forests are often referred to as the "lungs" of the planet, providing O_2 for life all over Earth. However, most experts believe that rain forests make little or no net contribution to global O_2 production. From your knowledge of photosynthesis and cellular respiration, can you explain why they might think this? (*Hint:* What happens to the food produced by a rain forest tree when it is eaten by animals or when the tree dies?)

2. Suppose you wanted to discover whether the oxygen atoms in the glucose produced by photosynthesis come from H_2O or CO_2. Explain how you could use a radioactive isotope to find out.

Case Study in the Process of Science on the Web & CD *Learn how to separate plant pigments by chromatography.*

Case Study in the Process of Science on the Web & CD *Perform experiments to measure the rate of the light reactions.*

Biology and Society

1. There is growing evidence that Earth is getting warmer owing to an intensified greenhouse effect resulting from increased CO_2 emissions from industry, vehicles, and the burning of forests. Global warming could influence agriculture and perhaps even melt polar ice and flood coastal regions. In response, 178 countries accepted the Kyoto agreement, which calls for mandatory reductions of greenhouse gas emissions in 30 developed nations by 2012. But in 2002, the Bush administration rejected the Kyoto agreement, instead proposing a more modest set of voluntary goals, allowing businesses to decide whether they wish to participate or not and providing a series of tax incentives to encourage them to do so. President Bush stated that the primary reasons for rejecting the agreement were that it would hurt the American economy and that some developing countries (such as India) were exempted from it, even though they produce a lot of pollution. Do you agree with the administration's decision? In what ways might efforts to reduce greenhouse gases hurt the economy? How can those costs be weighed against the costs of global warming? Do you think that poorer, less developed nations should carry an equal burden to reduce their emissions?

2. The use of biomass energy avoids many of the problems associated with gathering, refining, transporting, and burning fossil fuels. Yet biomass energy is not without its own set of problems. What challenges do you think would arise from a large-scale conversion to biomass energy? How do these challenges compare with those encountered with fossil fuels? Which set of challenges do you think is more likely to be eventually overcome? Do you think any one type of energy has more benefits and fewer costs than the others? Which one, and why?

Biology and Society on the Web *Learn more about the potential of biomass energy.*

Genetics

CHAPTER 8

The Cellular Basis of Reproduction and Inheritance

Biology and Society:
A $50,000 Egg! 119

What Cell Reproduction
Accomplishes 119

Passing On the Genes from
Cell to Cell

The Reproduction of Organisms

The Cell Cycle and Mitosis 121

Eukaryotic Chromosomes

The Cell Cycle

Mitosis and Cytokinesis

Cancer Cells: Growing Out of Control

Meiosis, the Basis of Sexual
Reproduction 128

Homologous Chromosomes

Gametes and the Life Cycle
of a Sexual Organism

The Process of Meiosis

Review: Comparing Mitosis
and Meiosis

The Origins of Genetic Variation

When Meiosis Goes Awry

Evolution Connection:
New Species from Errors
in Cell Division 137

You began life as a single cell, but there are now more cells in your body than **stars in the Milky Way.**

The **dance of the chromosomes** in a dividing cell is so precise that only one error occurs in 100,000 cell divisions.

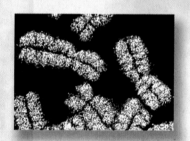

Just in the past second, **millions of your cells** have divided in two.

Each **sperm or egg** produced in your reproductive organs carries one of over 8 million possible combinations of parental chromosomes.

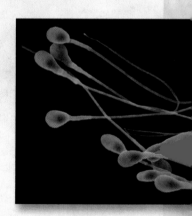

Biology and Society

A $50,000 Egg!

A few years ago, an advertisement appeared in several Ivy League college newspapers with this heading: "Egg Donor Needed—Large Financial Incentive." The ad was seeking a woman between the ages of 21 and 32 who was tall, athletic, and healthy and who had scored over 1400 on her SATs. If such a woman were willing to "donate" some of her eggs, the couple who placed the ad would pay her $50,000. The ad caught the attention of the media because of the large sum of money involved ("Sterile Couple Seeks Ivy Dream Girl's Eggs," said one tabloid headline). While the amount is unusual, this type of transaction is actually becoming fairly common in our society.

Infertility, the inability to produce children through normal sexual means after one year of trying, affects one in ten American couples. Many infertile couples turn to in vitro fertilization (IVF), a laboratory procedure that involves joining sperm and egg in a petri dish, allowing the fertilized egg to grow into an eight-cell embryo, and then implanting the embryo into the woman's uterus (**Figure 8.1**). A woman may provide her own eggs for an IVF procedure, but if she cannot, eggs must be obtained from an egg donor. Some women have a friend or relative willing to provide eggs, but most don't, and they therefore have to obtain them from a stranger. Unfortunately, the medical procedure for removing eggs involves pain and risk for the donor. Therefore, the demand for donated eggs far outstrips the supply. Since the 1990s, increasingly large sums of money have been offered to potential donors in an attempt to increase the supply of available eggs.

The economics of egg "donation" raise a series of ethical and legal questions that our society must face. Are we comfortable with the idea of buying and selling human eggs? How much money is enough—or too much? Because IVF is an expensive procedure, couples who seek eggs are usually financially secure. The women paid to donate eggs, on the other hand, are usually not. Do we wish to encourage a system in which poor women provide eggs to rich women?

Biology and Society on the Web
Learn more about controversial reproductive technologies.

As we have seen, a multicellular organism begins as a single cell that divides into two cells, which divides into four cells, then eight, and so on. The perpetuation of life depends on creating new cells. In this chapter, we'll learn how individual cells reproduce, and then turn our attention to the cellular basis of sexual reproduction.

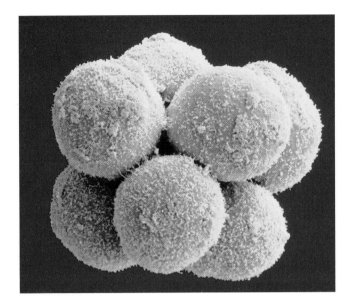

Figure 8.1 A human embryo at the eight-cell stage, the result of in vitro fertilization.   This human embryo was created through in vitro fertilization, a process that joins egg and sperm in a petri dish to produce an embryo. The original cell of the embryo has divided three times to form eight cells. The eight cells have identical sets of genes, as will the cells they give rise to. Cells continue to divide aand then specialize, developing into a baby, which eventually grows to be an adult.

What Cell Reproduction Accomplishes

When you hear the word *reproduction,* you probably think of the birth of new organisms. But reproduction actually occurs much more often at the cellular level. Consider the skin on your arm. The surface is a protective layer of dead cells, but underneath are layers of living cells busy carrying out the chemical reactions you studied in Unit One. The living cells are also engaged in another vital activity: They are reproducing themselves. The new cells are moving outward toward the skin's surface, replacing dead cells that have rubbed off. This renewal of your skin goes on throughout your life. And when your skin is injured, additional cell reproduction helps heal the wound.

The replacement of lost or damaged cells is just one of the important roles that cell reproduction, or **cell division,** plays in your life. Another is growth. All of the trillions of cells in your body result from repeated cell divisions that began in your mother's body with a single fertilized egg cell.

Passing On the Genes from Cell to Cell

When a cell divides, the two "daughter" cells that result are ordinarily genetically identical to each other and to the original "parent" cell. (By convention, biologists use the word *daughter* in this context; it does not imply gender.) Before the parent cell splits into two, it duplicates its **chromosomes,** the DNA-containing structures that carry the organism's genes. Then, during the division process, the two sets of chromosomes are distributed to the daughter cells. As a rule, the daughter cells receive identical sets of chromosomes, with identical genes.

The Reproduction of Organisms

Some organisms reproduce by simple cell division. Single-celled organisms, such as amoebas, reproduce this way, and the offspring are genetic replicas of the parent (**Figure 8.2a**). Because it does not involve fertilization of an egg by a sperm, this type of reproduction is called **asexual reproduction.** Offspring produced by asexual reproduction inherit all their chromosomes from a single parent. Many multicellular organisms can reproduce asexually as well. For example, some sea stars can divide into two pieces that regrow into two whole new individuals (**Figure 8.2b**). And if you've ever grown an African violet or spider plant from a clipping, you've observed asexual

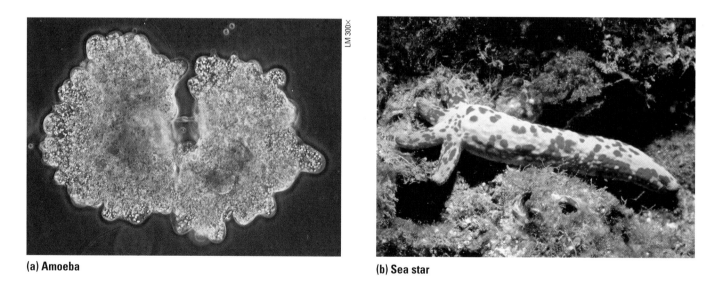

LM 300×

(a) Amoeba

(b) Sea star

Figure 8.2 Asexual reproduction. (a) This single-celled amoeba is reproducing by dividing in half. Its chromosomes have been duplicated, and the two identical sets of chromosomes have been allocated to opposite sides of the parent. When division is complete, the two daughter amoebas will be genetically identical to each other and to their parent. (b) If a sea star is divided into two pieces, each piece may regrow into a whole new organism. If this happens, both resulting organisms will be genetically identical to the original.

reproduction in plants. In asexual reproduction, there is one simple principle of inheritance: The lone parent and each of its offspring have identical genes.

Sexual reproduction is different; it requires fertilization of an egg by a sperm. The production of egg and sperm cells involves a special type of cell division, called meiosis, that occurs only in reproductive organs (such as testes and ovaries). As we'll discuss later, a sperm or egg cell has only half as many chromosomes as the parent cell that gave rise to it. So two kinds of cell division—meiosis and ordinary cell division—are involved in the lives of sexually reproducing organisms.

Activity 8A on the Web & CD Learn more about asexual and sexual reproduction.

The remainder of the chapter is divided into two main sections. The first section deals with the cell cycle and mitosis, the type of cell division responsible for asexual reproduction and for the growth and maintenance of multicellular organisms. The second section focuses on meiosis, the special type of cell division that produces sperm and eggs for sexual reproduction.

CHECKPOINT

Ordinary cell division produces two daughter cells that are genetically identical. Name three functions of this type of cell division.

Answer: Cell replacement, growth of an organism, asexual reproduction of an organism

---

The Cell Cycle and Mitosis

A **genome** is a complete set of an organism's genes; in a human, that's around 35,000 genes. In eukaryotic cells, the vast majority of the genome is located on chromosomes in the cell nucleus. (The main exceptions are genes on small DNA molecules found in mitochondria and chloroplasts.) As leading players in cell division, the chromosomes deserve a little more of our attention before we broaden our focus to the cell as a whole.

Eukaryotic Chromosomes

Each eukaryotic chromosome contains a single long DNA molecule, typically bearing thousands of genes. The number of chromosomes in a eukaryotic cell, like the number of genes, depends on the species. For example, human body cells generally have 46 chromosomes, while those of a dog have 78, and those of a mouse have 40. Chromosomes are made up of a material called **chromatin,** a combination of DNA and protein molecules. The protein molecules help organize the chromatin and help control the activity of its genes.

Most of the time, the chromosomes exist as a diffuse mass of very long fibers that are much longer than the diameter of the nucleus they are stored in. In fact, the total DNA in a single human cell's 46 chromosomes could stretch for over 2 meters! As a cell prepares to divide, its chromatin fibers coil up, forming compact chromosomes. When they are in this state, chromosomes are clearly visible under the light microscope, as shown in the plant cell in **Figure 8.3** (each dark purple thread is an individual chromosome). When a cell is not dividing, the chromosomes are too thin to be clearly seen in a light micrograph.

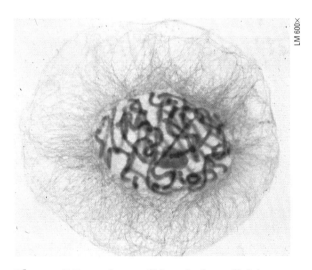

LM 600×

Figure 8.3 A plant cell just before division. The purple threads are the chromosomes (the colors result from staining). The thinner red threads in the surrounding cytoplasm are the cytoskeleton.

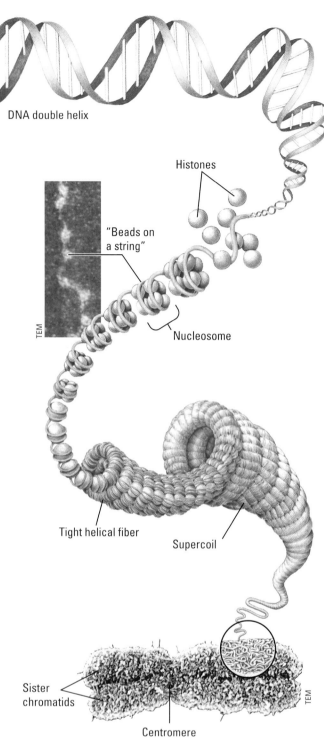

DNA double helix

Histones

"Beads on a string"

TEM

Nucleosome

Tight helical fiber

Supercoil

Sister chromatids

Centromere

Figure 8.4
DNA packing in a eukaryotic chromosome. Successive levels of coiling of DNA and associated proteins ultimately results in highly condensed chromosomes. When not dividing, the DNA of active genes is only lightly packed, in the "beads on a string" arrangement. At the bottom of the figure, you can see a highly compacted chromosome from a cell preparing to divide. This chromosome has already duplicated, and the two copies are called sister chromatids. The constricted region is the centromere. The fuzzy appearance comes from the intricate twists and folds of the chromatin fibers.

Such long molecules of DNA can fit into the nucleus because within each chromosome the DNA is packed into an elaborate, multilevel system of coiling and folding. A crucial aspect of DNA packing is the association of the DNA with small proteins called **histones,** found only in eukaryotes. (Bacteria have analogous proteins, but they lack the degree of DNA packing found in eukaryotes.)

Figure 8.4 presents a simplified model for the main levels of DNA packing. At the first level of packing, shown near the top, histones attach to the DNA. In electron micrographs, the combination of DNA and histones has the appearance of beads on a string. Each "bead," called a **nucleosome,** consists of DNA wound around a protein core of eight histone molecules. At the next level of packing, the beaded string is wrapped into a tight helical fiber. Then this fiber coils further into a thick supercoil. Looping and folding can further compact the DNA, as you can see in the chromosome at the bottom of the figure. Viewed as a whole, Figure 8.4 gives a sense of how successive levels of coiling and folding enable a huge amount of DNA to fit into a relatively small cell nucleus.

Before a cell begins the division process, it duplicates all of its chromosomes. The DNA molecule of each chromosome is copied through the process of DNA replication (see Chapter 10), and new protein molecules attach as needed. The result is that each chromosome now consists of two copies called **sister chromatids,** which contain identical genes. At the bottom of Figure 8.4, you can see an electron micrograph of a human chromosome that has been duplicated. The two chromatids are joined together especially tightly at a region called the **centromere.**

When the cell divides, the sister chromatids of a duplicated chromosome separate from each other, as shown in the simple diagram in **Figure 8.5**. Once separated from its sister, each chromatid is considered a full-fledged chromosome, and it is identical to the original chromosome. One of the new chromosomes goes to one daughter cell, and the other goes to the other daughter cell. In this way, each daughter cell receives a complete and identical set of chromosomes. A dividing human skin cell, for example, has 46 duplicated chromosomes, and each of the two daughter cells that results from it has 46 single chromosomes.

Let's now review the roles of cell division: In single-celled organisms, it is a means of asexual reproduction. In multicellular organisms, it enables the organism to grow from a single cell. Cell division also replaces worn-out or damaged cells, keeping the total cell number in a mature individual relatively constant. In your body, for example, millions of cells must divide every second to maintain the total number of about 60 trillion cells.

The Cell Cycle

How do chromosome duplication and cell division fit into the life of a cell? The rate at which a cell divides depends on its role within the organism's body. Some cells divide once a day, others less often, and highly specialized cells, such as mature muscle cells, not at all. Eukaryotic cells that do divide undergo a **cell cycle,** an orderly sequence of events that extends from the time a cell first arises until it itself divides.

As **Figure 8.6** shows, most of the cell cycle is spent in **interphase.** This is a time when a cell performs its normal functions within the organism. For example, a cell in your stomach lining might make and release enzyme molecules that aid in digesting the food you eat. During interphase, a cell roughly doubles everything in its cytoplasm. It increases its supply of proteins, increases the number of many of its organelles (such as mitochondria and ribosomes), and grows in size. Typically, interphase lasts for at least 90% of the cell cycle.

From the standpoint of cell reproduction, the most important event of interphase is chromosome duplication, when the DNA in the nucleus is precisely doubled. This occurs approximately in the middle of interphase, and the period when it is occurring is called the S phase (for DNA *synthesis*). The interphase periods before and after the S phase are called the G_1 and G_2 phases, respectively (G stands for *gap*). During G_2, each chromosome in the cell consists of two identical sister chromatids, and the cell is preparing to divide.

The part of the cell cycle when the cell is actually dividing is called the **mitotic phase** (M phase). It includes two overlapping processes, mitosis and cytokinesis. In **mitosis,** the nucleus and its contents, notably the duplicated chromosomes, divide and are evenly distributed, forming two daughter nuclei. In **cytokinesis,** the cytoplasm is divided in two. Cytokinesis usually begins before mitosis is completed. The combination of mitosis and cytokinesis produces two genetically identical daughter cells, each with a single nucleus, surrounding cytoplasm, and a plasma membrane.

Activity 8B on the Web & CD
Test your knowledge of the cell cycle.

Mitosis is a remarkably accurate mechanism for allocating identical copies of a large amount of genetic material to two daughter cells. Experiments with yeast cells, for example, indicate that an error in chromosome distribution occurs only once in about 100,000 cell divisions. Mitosis is unique to eukaryotes. Prokaryotes have only a single small chromosome (see Chapter 10) and use a simpler mechanism for allocating DNA to daughter cells.

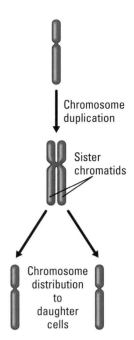

Figure 8.5
Chromosome duplication and distribution.
During cell reproduction, the cell duplicates each chromosome and distributes the two copies to the daughter cells. This diagram focuses on a single chromosome.

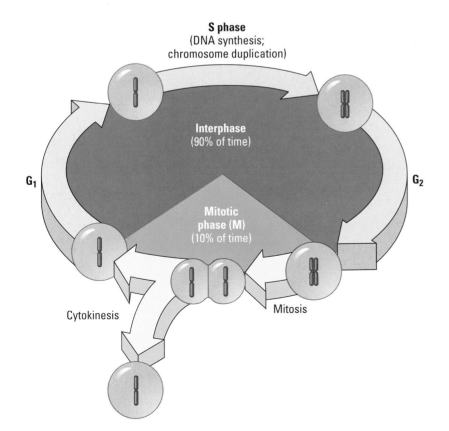

Figure 8.6 The eukaryotic cell cycle. The cell cycle extends from the "birth" of a cell, resulting from cell reproduction, to the time the cell itself divides in two. The cell spends most of the cycle in interphase. A key event of interphase is the duplication of the chromosomes; the period during which this occurs is called the S phase (for DNA synthesis). Before the S phase, the cell is said to be in the G_1 phase; after the S phase, the cell is in G_2. The cell metabolizes and grows throughout interphase. The actual division process occurs during the mitotic phase (M phase), which includes mitosis (the division of the cell's nucleus) and cytokinesis (the division of the cytoplasm). During interphase, the chromosomes are diffuse masses of thin fibers; they do not actually appear in the rodlike form you see here.

Mitosis and Cytokinesis

Figure 8.7 Cell reproduction: A dance of the chromosomes. After the chromatin doubles during interphase, the elaborately choreographed stages of mitosis—prophase, metaphase, anaphase, and telophase—distribute the duplicate sets of chromosomes to two separate nuclei. Cytokinesis then divides the cytoplasm, yielding two genetically identical daughter cells. The micrographs here show cells from a newt. The drawings include details not visible in the micrographs. For simplicity, only four chromosomes appear in the drawings.

The light micrographs in **Figure 8.7** show the cell cycle for an animal cell, with most of the figure devoted to the mitotic phase. With the onset of mitosis, striking changes are visible in the nucleus and other cellular structures. The text under the figure describes the events occurring at each stage. Mitosis is a continuum, but biologists distinguish four main stages: **prophase, metaphase, anaphase,** and **telophase.**

The chromosomes are the stars of the mitotic drama, and their movements depend on the **mitotic spindle,** a football-shaped structure of

Activities 8C & 8D on the Web & CD
See an animation and a video of mitosis and cytokinesis.

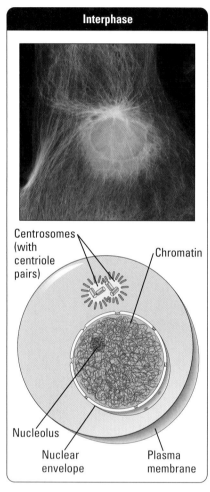

Interphase

Interphase is the period of cell growth, when the cell makes new molecules and organelles. At the point shown here, late interphase (G$_2$), the cytoplasm contains two centrosomes. Within the nucleus, the chromosomes are duplicated, but they cannot be distinguished individually because they are still in the form of loosely packed chromatin fibers. The prominent nucleolus is an indication that the cell is making ribosomes.

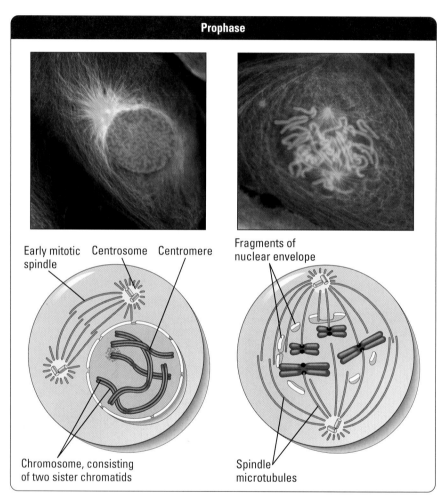

Prophase

During prophase, changes occur in both nucleus and cytoplasm. In the nucleus, the chromatin fibers coil, so that the chromosomes become thick enough to be seen with the light microscope. The nucleoli disappear. Each chromosome appears as two identical sister chromatids joined together, with a narrow "waist" at the centromere. In the cytoplasm, the mitotic spindle begins to form as microtubules grow out from the centrosomes, which are moving away from each other.

Late in prophase, the nuclear envelope breaks up. The spindle microtubules can now reach the chromosomes, which are thick and have a protein structure (black dot) at their centromeres. Some of the spindle microtubules capture chromosomes by attaching to these structures, throwing the chromosomes into agitated motion. Other microtubules make contact with microtubules coming from the opposite spindle pole. The spindle moves the chromosomes toward the center of the cell.

microtubules that guides the separation of the two sets of daughter chromosomes. The spindle microtubules grow from two **centrosomes,** clouds of cytoplasmic material that in animal cells contain centrioles. (Centrioles are can-shaped structures made of microtubules, which were introduced in Chapter 4.)

Case Study in the Process of Science on the Web & CD Investigate how much time cells spend in each phase of mitosis.

Cytokinesis, the actual division of the cytoplasm into two cells, typically occurs during telophase. In animal cells, the cytokinesis process is known as cleavage. The first sign of cleavage is the appearance of a **cleavage furrow,** an indentation at the equator of the cell. A ring of microfilaments in the cytoplasm just under the plasma membrane is responsible for the cleavage

Metaphase	Anaphase	Telophase and Cytokinesis
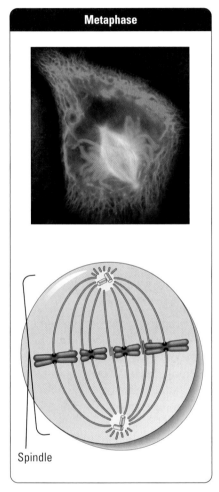	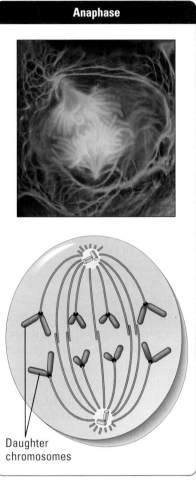	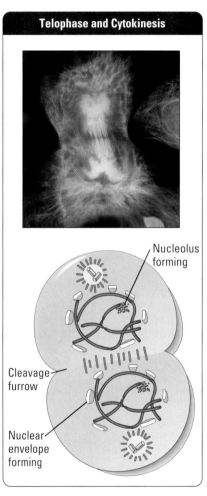

Metaphase

The mitotic spindle is now fully formed. The chromosomes convene on an imaginary plate equidistant from the two poles of the spindle. The centromeres of all the chromosomes are lined up at this plate. For each chromosome, the spindle microtubules attached to the two sister chromatids pull toward opposite poles. This tug of war keeps the chromosomes in the middle of the cell.

Anaphase

Anaphase begins suddenly, when the sister chromatids of each chromosome separate. Each is now considered a full-fledged (daughter) chromosome. Motor proteins at the centromeres "walk" the daughter chromosomes along their microtubules toward opposite poles of the cell (see motor proteins in Figure 5.6a). Meanwhile, these microtubules shorten. However, the microtubules *not* attached to chromosomes lengthen, pushing the poles farther apart and elongating the cell.

Telophase and Cytokinesis

Telophase begins when the two groups of chromosomes have reached the cell poles. Telophase is the reverse of prophase: Nuclear envelopes form, the chromosomes uncoil, nucleoli reappear, and the spindle disappears. Mitosis, the division of one nucleus into two genetically identical daughter nuclei, is now finished.

Cytokinesis, the division of the cytoplasm, usually occurs with telophase. In animals, a cleavage furrow pinches the cell in two, producing two daughter cells.

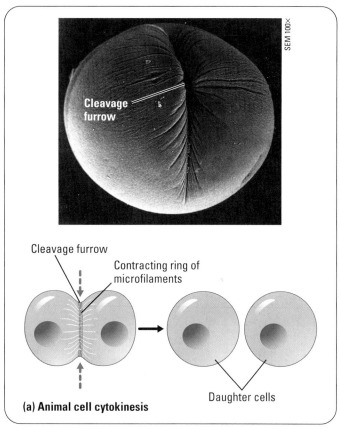

Cleavage furrow

Contracting ring of microfilaments

Daughter cells

(a) Animal cell cytokinesis

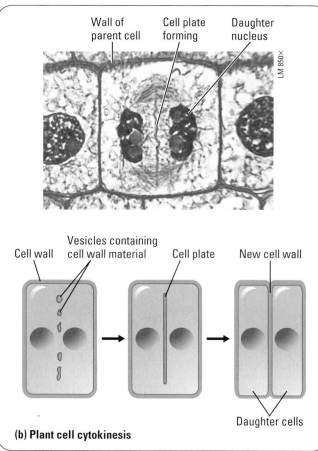

Wall of parent cell

Cell plate forming

Daughter nucleus

Cell wall

Vesicles containing cell wall material

Cell plate

New cell wall

Daughter cells

(b) Plant cell cytokinesis

Figure 8.8 Cytokinesis in animal and plant cells.

furrow. The ring contracts like the pulling of a drawstring, deepening the furrow and pinching the parent cell in two (**Figure 8.8a**). Microfilaments are made of actin, a protein that also enables muscle cells to contract.

Cytokinesis in a plant cell occurs differently. Membrane-enclosed vesicles containing cell wall material collect at the middle of the cell. The vesicles gradually fuse, forming a membranous disk called the **cell plate.** The cell plate grows outward, accumulating more cell wall material as more vesicles join it. Eventually, the membrane of the cell plate fuses with the plasma membrane, and the cell plate's contents join the parental cell wall, resulting in two daughter cells (**Figure 8.8b**).

Cancer Cells: Growing Out of Control

For a plant or animal to grow and maintain its tissues normally, it must be able to control the timing of cell division. The sequential events of the cell cycle are directed by a **cell cycle control system** that consists of special proteins within the cell. These proteins integrate information from the environment and from other body cells and send "stop" and "go-ahead" signals at certain key points during the cell cycle using signal transduction pathways (see Chapter 5). For example, the cell cycle normally halts at the G_1 phase of interphase unless the cell receives a go-ahead signal via certain cell cycle control system proteins. If that signal never arrives, the cell will switch into a permanently nondividing state called G_0. Some of our nerve and muscle cells, for example, are arrested at G_0. If the go-ahead signal is received and the G_1 checkpoint is passed, the cell will usually complete the rest of the cycle.

If the cell cycle control system malfunctions, cells may reproduce at the wrong time or in the wrong place. The result may be a tumor, an abnormally growing mass of body cells. A **benign tumor** is one that remains at its original site in the body. Benign tumors can cause problems if they grow in certain organs, such as the brain, but often they can be completely removed by surgery.

What Is Cancer? Cancer, which currently claims the lives of one out of every five people in the United States and other developed countries, is caused by a breakdown in the control of the cell cycle. Unlike normal cells of the body, **cancer cells** have a severely deranged cell cycle control system; not only do they divide excessively, but they also exhibit other kinds of bizarre behavior. A lump resulting from the reproduction of a cancer cell is called a **malignant tumor.**

The most dangerous attribute of malignant cancer cells is their ability to spread into neighboring tissues and often to other parts of the body. Like a benign tumor, a malignant tumor displaces normal tissue as it grows (**Figure 8.9**). But if a malignant tumor is not killed or removed, it can spread into surrounding tissues. More alarming still, cells may split off from the tumor, invade the circulatory system (lymph vessels and blood vessels), and travel to new locations, where they can form new tumors. The spread of cancer cells beyond their original site is called **metastasis.** Malignant cancer cells may continue to divide and spread until the host organism dies.

Cancers are named according to where they originate. Liver cancer, for example, always begins in liver tissue and may or may not spread from there. Cancers are grouped into four categories based on their sites of origin. **Carcinomas** are cancers that originate in the external or internal coverings of the body, such as the skin or the lining of the intestine. **Sarcomas** arise in tissues that support the body, such as bone and muscle. Cancers of blood-forming tissues, such as bone marrow and lymph nodes, are called **leukemias** and **lymphomas.**

Cancer Treatment Once a tumor starts growing in the body, how can it be treated? The three main types of cancer treatment are sometimes referred to as "slash, burn, and poison." Surgery to remove a tumor ("slash") is usually the first step. "Burn" and "poison" refer to treatments that attempt to stop cancer cells from dividing. In **radiation therapy** ("burn"), parts of the body that have cancerous tumors are exposed to high-energy radiation, which disrupts cell division. Because cancer cells divide more often than most normal cells, they are more likely to be dividing at any given time. Therefore, radiation can often destroy cancer cells without seriously injuring the normal cells of the body. However, there is sometimes enough damage to normal body cells to produce bad side effects. For example, side effects of radiation therapy can include nausea and hair loss. Additionally, damage to cells of the ovaries or testes can cause sterility.

Chemotherapy ("poison") uses the same basic strategy as radiation; in this case, drugs are administered that disrupt cell division. These drugs work in a variety of ways. Some, called antimitotic drugs, prevent cell division by interfering with the mitotic spindle. One antimitotic drug, vinblastine, prevents the spindle from forming in the first place; another, paclitaxel (trade name Taxol), freezes the spindle after it forms, keeping it from functioning. Vinblastine was first obtained from the periwinkle, a flowering plant native to tropical rain forests in Madagascar. Taxol is made from a chemical found in the bark of the Pacific yew, a tree found mainly in the northwestern United States. Taxol has fewer side effects than many anticancer drugs and seems to be effective against some hard-to-treat cancers of the ovary and breast.

In the laboratory, researchers can grow cancer cells in culture. The cells are placed in a glass container, and nutrients are provided by an artificial liquid medium (**Figure 8.10**). Normal mammalian cells grow in culture for only about 50 cell generations, after which they cease to divide. But cancer cells are "immortal"—they can continue to divide indefinitely, as long as they have a supply of nutrients. It is by studying cancer cells in culture that researchers are learning about the molecular changes that make a cell cancerous. We will return to the topic of cancer in Chapter 11, after learning more about genes.

Cancer Prevention and Survival Although cancer can strike anyone, there are certain lifestyle changes you can make to reduce your chances of dying from cancer. Not smoking, getting adequate exercise, avoiding over exposure to the sun, and eating a high-fiber, low-fat diet can all help prevent cancer. Regular visits to the doctor can help identify tumors early, thereby significantly increasing the possibility of successful treatment. Seven types of cancer can be easily detected: skin and oral (via physical exam), breast (via self-exams and mammograms for higher risk women), prostate (via rectal exam), cervical (via Pap smear), testicular (via self-exam), and colon (via colonoscopy).

Activity 8E on the Web & CD
Learn more about the causes of cancer.

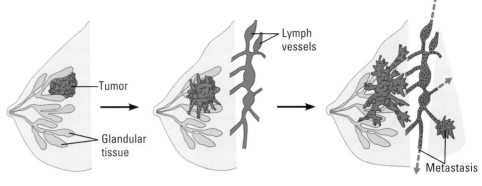

A tumor grows from a single cancer cell.

Cancer cells invade neighboring tissue.

Cancer cells spread through lymph and blood vessels to other parts of the body.

Lymph vessels

Tumor

Glandular tissue

Metastasis

Figure 8.9 The growth and metastasis of a malignant (cancerous) tumor of the breast.

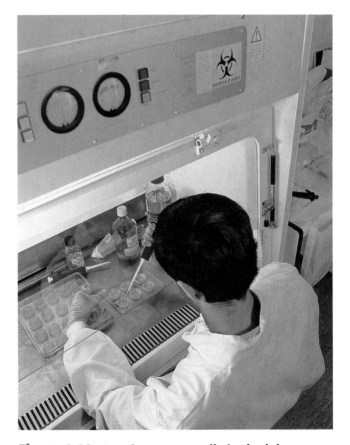

Figure 8.10 Growing cancer cells in the lab.
This researcher is working under a special hood that helps maintain sterile conditions for his experiment.

Figure 8.11 The varied products of sexual reproduction. A multi-ethnic family poses for a snapshot. Each child has inherited a unique combination of genes from the parents and displays a unique combination of traits.

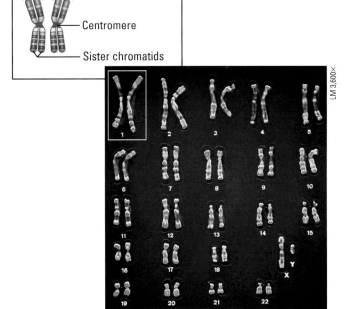

Figure 8.12 Pairs of homologous chromosomes. To make this karyotype (chromosome display) of a man, a scientist broke open a cell in metaphase of mitosis, stained the released chromosomes with special dyes, made a micrograph, and then arranged the chromosome images in matching pairs. The result: 22 well-matched pairs (autosomes) and a twenty-third pair that consists of an X chromosome and a Y chromosome (sex chromosomes). Each chromosome consists of two sister chromatids closely attached all along their lengths. Notice that with the exception of X and Y, the homologous chromosomes of each pair match in size, centromere position, and staining pattern.

Meiosis, the Basis of Sexual Reproduction

Only maple trees produce more maple trees, only goldfish make more goldfish, and only people make more people. These simple facts of life have been recognized for thousands of years and are reflected in the age-old saying, "Like begets like." But in a strict sense, "Like begets like" applies only to asexual reproduction, such as the reproduction of the amoeba in Figure 8.2a. In that case, because offspring inherit all their DNA from a single parent, they are exact genetic replicas of that one parent and of each other, and their appearances are very similar.

The family photo in **Figure 8.11** makes the point that in a sexually reproducing species, like does not exactly beget like. You probably resemble your parents more closely than you resemble a stranger, but you do not look exactly like your parents or your siblings. Each offspring of sexual reproduction inherits a unique combination of genes from its two parents, and this combined set of genes programs a unique combination of traits. As a result, sexual reproduction can produce great variation among offspring.

Sexual reproduction depends on the cellular processes of meiosis and fertilization. But before discussing these processes, we return to chromosomes and their role in the life cycles of sexually reproducing organisms.

Homologous Chromosomes

If we examine a number of cells from any individual organism, we discover that virtually all of them have the same number and types of chromosomes. Likewise, if we examine cells from different individuals of a single species—sticking to one gender, for now—we find that they have the same number and types of chromosomes. Viewed with a microscope, your chromosomes would look just like those of Madonna (if you're a woman) or Michael Jordan (if you're a man).

A typical body cell, called a **somatic cell,** has 46 chromosomes in humans. If we break open a human cell in metaphase of mitosis, make a micrograph of the chromosomes, and arrange the chromosome images in an orderly array, we produce a display called a **karyotype** (**Figure 8.12**). Every (or almost every) chromosome has a twin that resembles it in size and shape. The two chromosomes of such a matching pair, called **homologous chromosomes,** carry the same sequence of genes controlling the same inherited characteristics. If a gene influencing eye color is located at a particular place on one chromosome—for example, within the yellow band in the Figure 8.12 drawing—then the homologous chromosome has a similar gene for eye color there. However, the two genes may be slightly different versions. Altogether, we humans have 23 homologous pairs of chromosomes. Other species have different numbers of chromosomes, but these, too, usually match in pairs. For example, the number of pairs of chromosomes in the genomes of mammals ranges from 3 (an Indian muntjac deer) to 67 (a black rhinoceros).

For a human female, the 46 chromosomes fall neatly into 23 homologous pairs, with the members of each pair essentially identical in appearance. For a male, however, one pair of chromosomes do *not* look alike. This nonmatching pair is the male's sex chromosomes. **Sex chromosomes** determine a person's gender. Like all mammals, human males have one X chromosome and one Y chromosome (see Figure 8.12). Females have two X chromosomes. (While X and Y sex chromosomes determine gender in mammals, other organisms have different systems; in this chapter, we focus on humans.) The remaining chromosomes, found in both males and females, are called **autosomes.** For both autosomes and sex chromosomes, we inherit one chromosome of each pair from our mother and the other from our father.

Gametes and the Life Cycle of a Sexual Organism

The **life cycle** of a multicellular organism is the sequence of stages leading from the adults of one generation to the adults of the next. Having two sets of chromosomes, one inherited from each parent, is a key factor in the human life cycle, outlined in **Figure 8.13**, and in the life cycles of all other species that reproduce sexually. Let's follow the chromosomes through the human life cycle.

Humans (as well as most plants and animals) are said to be **diploid** organisms because all body cells contain two homologous sets of chromosomes. The total number of chromosomes, 46 in humans, is the diploid number, represented as $2n$. The exceptions are the egg and sperm cells, known as **gametes.**

Activity 8F on the Web & CD
Review the human life cycle.

Made by meiosis in an ovary or testis, each gamete has a single set of chromosomes: 22 autosomes plus a sex chromosome, X or Y. A cell with a single chromosome set is called a **haploid** cell; it has only one member of each homologous pair. For humans, the haploid number, n, is 23.

In the human life cycle, sexual intercourse allows a haploid sperm cell from the father to reach and fuse with a haploid egg cell of the mother in the process known as **fertilization.** The resulting fertilized egg, called a **zygote,** is diploid. It has two homologous sets of chromosomes, one set from each parent. The life cycle is completed as a sexually mature adult develops from the zygote. Mitotic cell division ensures that all somatic cells of the human body receive a copy of all of the zygote's 46 chromosomes. Thus, every one of the trillions of cells in your body can trace its ancestry back through mitotic divisions to the single zygote cell produced when your father's sperm and mother's egg fused.

All sexual life cycles involve an alternation of diploid and haploid stages. Producing haploid gametes by meiosis keeps the chromosome number from doubling in every generation (**Figure 8.14**).

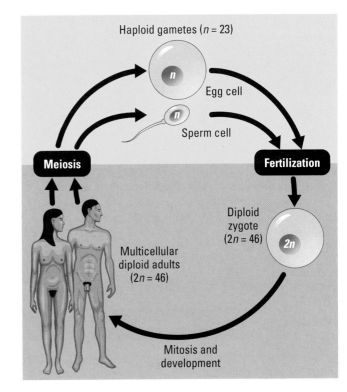

Figure 8.13 The human life cycle. In each generation, the doubling of chromosome number that results from fertilization is offset by the halving of chromosome number that occurs in meiosis. For humans, the number of chromosomes in a haploid cell (sperm or egg) is 23 (that is, $n = 23$). The number of chromosomes in the diploid zygote and all somatic cells arising from it is 46 ($2n = 46$).

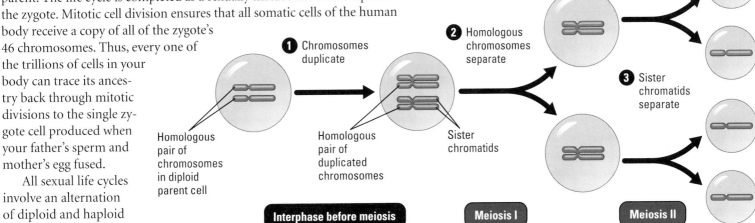

Figure 8.14 How meiosis halves the chromosome number. This simplified diagram tracks one pair of homologous chromosomes. ❶ Each of the chromosomes is duplicated during the preceding interphase. ❷ The first division, meiosis I, segregates the two chromosomes of the homologous pair, packaging them in separate (haploid) daughter cells. But each chromosome is still doubled. ❸ Meiosis II separates the sister chromatids. Each of the four daughter cells is haploid and contains only one single chromosome from the homologous pair.

The Process of Meiosis

Meiosis, the process that produces haploid daughter cells in diploid organisms, resembles mitosis, but with two special features. The first is the halving of the number of chromosomes. In meiosis, a cell that has duplicated its chromosomes undergoes *two consecutive divisions,* called *meiosis I* and *meiosis II.* Because the two divisions of meiosis are preceded by only one duplication of the chromosomes, each of the four daughter cells resulting from meiosis has only half as many chromosomes as the starting cell—a haploid set of chromosomes.

The second special feature of meiosis is an exchange of genetic material—pieces of chromosomes—between homologous chromosomes. This exchange, called **crossing over,** occurs during the first prophase of meiosis. We'll look more closely at crossing over later. For now, study **Figure 8.15,** including the text below it, which describes the stages of meiosis in detail.

Figure 8.15 The stages of meiosis. The drawings here show the two cell divisions of meiosis, starting with a diploid animal cell containing four chromosomes. Each homologous pair consists of a red chromosome and a blue chromosome of the same size. The colors remind us that the members of a homologous pair were inherited from different parents and carry different versions of some genes.

Meiosis I: Homologous chromosomes separate

Interphase	Prophase I	Metaphase I	Anaphase I
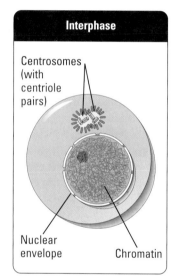	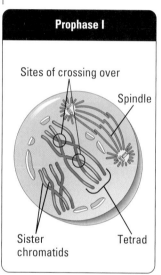	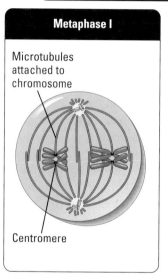	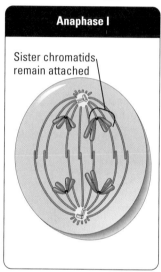
Centrosomes (with centriole pairs); Nuclear envelope; Chromatin	Sites of crossing over; Spindle; Sister chromatids; Tetrad	Microtubules attached to chromosome; Centromere	Sister chromatids remain attached
Chromosomes duplicate	**Homologous chromosomes pair and exchange segments**	**Tetrads line up**	**Pairs of homologous chromosomes split up**

Interphase

Like mitosis, meiosis is preceded by an interphase during which the chromosomes duplicate. Each chromosome then consists of two identical sister chromatids.

Meiosis I

Prophase I Prophase I is the most complicated stage of meiosis. As the chromatin condenses, special proteins cause the homologous chromosomes to stick together in pairs. The resulting structure has four chromatids and is called a **tetrad.** Within each tetrad, chromatids of the homologous chromosomes exchange corresponding segments—they "cross over." Because the versions of the genes on a chromosome (or one of its chromatids) may be different from those on its homologue, crossing over rearranges genetic information.

As prophase I continues, the chromosomes condense further, a spindle forms, and the tetrads are moved toward the center of the cell.

Metaphase I At metaphase I, the tetrads are aligned in the middle of the cell. The sister chromatids of each chromosome are still attached at their centromeres, where they are anchored to spindle microtubules. Notice that for each tetrad, the spindle microtubules attached to one homologous chromosome come from one pole of the cell, and the microtubules attached to the other chromosome come from the opposite pole. With this arrangement, the homologous chromosomes of each tetrad are poised to move toward opposite poles of the cell.

Anaphase I As in anaphase of mitosis, chromosomes now migrate toward the poles of the cell. But in contrast to mitosis, the sister chromatids migrate as a pair instead of splitting up. They are separated not from each other, but from their homologous partners. So in the drawing, you see two still-doubled chromosomes moving toward each pole.

As you go through Figure 8.15, keep in mind the difference between homologous chromosomes and sister chromatids: The two chromosomes of a homologous pair are individual chromosomes that were inherited from different parents, one from the mother and one from the father. Homologues appear alike in the microscope, but they have different versions of some of their genes (for example, a gene for freckles on one chromosome and a gene for the absence of freckles at the same place on the homologue). The homologues in Figure 8.15 (and later figures) are colored red and blue to remind you that they differ in this way. In the interphase just before meiosis, each homologue replicates to form sister chromatids that remain together until anaphase of meiosis II. Before crossing over occurs, sister chromatids are identical and carry the same versions of all their genes.

Meiosis II: Sister chromatids separate

Telophase I and Cytokinesis	Prophase II	Metaphase II	Anaphase II	Telophase II and Cytokinesis

Cleavage furrow

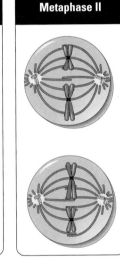

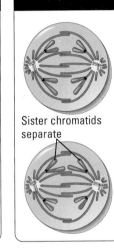

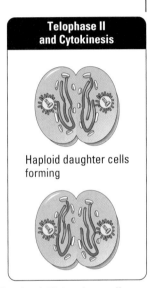

Sister chromatids separate

Haploid daughter cells forming

Two haploid cells form; chromosomes are still double

During another round of cell division, the sister chromatids finally separate; four haploid daughter cells result, containing single chromosomes

Telophase I and Cytokinesis

In telophase I, the chromosomes arrive at the poles of the cell. When they finish their journey, each pole has a haploid chromosome set, although each chromosome is still in duplicate form. Usually, cytokinesis occurs along with telophase I, and two haploid daughter cells are formed.

Depending on the species, the nuclei may or may not return to an interphase state. But in either case, there is no further chromosome duplication.

Meiosis II

Meiosis II is essentially the same as mitosis. The important difference is that meiosis II starts with a haploid cell.

During prophase II, a spindle forms and moves the chromosomes toward the middle of the cell. During metaphase II, the chromosomes are aligned as they are in mitosis, with the microtubules attached to the sister chromatids of each chromosome coming from opposite poles. In anaphase II, the centromeres of sister chromatids finally separate, and the sister chromatids of each pair, now individual daughter chromosomes, move toward opposite poles of the cell. In telophase II, nuclei form at the cell poles, and cytokinesis occurs at the same time. There are now four daughter cells, each with the haploid number of single chromosomes.

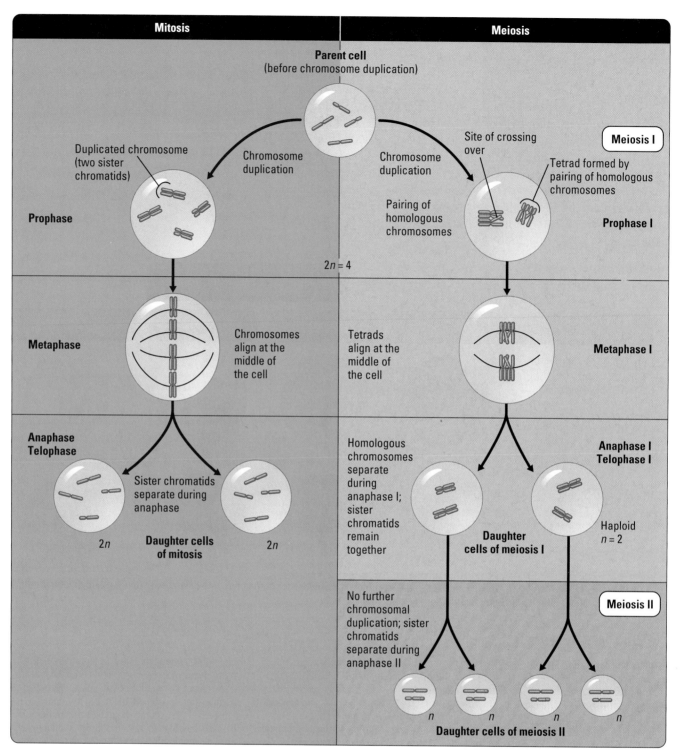

Figure 8.16

Comparing mitosis and meiosis.
The events unique to meiosis occur during meiosis I: In prophase I, duplicated homologous chromosomes pair to form tetrads, and crossing over occurs between homologous (nonsister) chromatids. In metaphase I, tetrads (rather than individual chromosomes) are aligned at the center of the cell. During anaphase I, sister chromatids of each chromosome stay together and go to the same pole of the cell as homologous chromosomes separate. At the end of meiosis I, there are two haploid cells, but each chromosome still has two sister chromatids. Meiosis II is virtually identical to mitosis and separates sister chromatids. But unlike mitosis, meiosis II yields daughter cells with a haploid set of chromosomes.

Review: Comparing Mitosis and Meiosis

We have now described the two ways that cells of eukaryotic organisms divide. Mitosis, which provides for growth, tissue repair, and asexual reproduction, produces daughter cells genetically identical to the parent cell. Meiosis, needed for sexual reproduction, yields genetically unique haploid daughter cells—cells with only one member of each homologous chromosome pair.

For both mitosis and meiosis, the chromosomes duplicate only once, in the preceding interphase. Mitosis involves one division of the nucleus, and it is usually accompanied by cytokinesis, producing two diploid cells. Meiosis entails two nuclear and cytoplasmic divisions, yielding four haploid cells.

Figure 8.16 compares mitosis and meiosis, tracing these two processes for a diploid parent cell with four chromosomes. As before, homologous chromosomes are those matching in size. Notice that all the events unique to meiosis occur during meiosis I.

The Origins of Genetic Variation

As we discussed earlier, offspring that result from sexual reproduction are genetically different from their parents and from one another. When we discuss evolution in Unit Three, we'll see that this genetic variety in offspring is the raw material for natural selection. For now, let's take another look at meiosis and fertilization to see how genetic variety arises.

Independent Assortment of Chromosomes Figure 8.17 illustrates one way in which meiosis contributes to genetic variety. The figure shows how the arrangement of homologous chromosome pairs at metaphase of meiosis I affects the resulting gametes. Once again, our example is from an organism with a diploid chromosome number of 4, with red and blue used to differentiate homologous chromosomes.

The orientation of the homologous pairs of chromosomes (tetrads) at metaphase I is a matter of chance, like the flip of a coin. Thus, in this example, there are two possible ways that the two tetrads can align during metaphase I. In possibility 1, the tetrads are oriented with both red chromosomes on the same side. In this case, each of the gametes produced at the end of meiosis II has only red or only blue chromosomes (combinations a and b). In possibility 2, the tetrads are oriented differently. This arrangement produces gametes that each have one red and one blue chromosome. Furthermore, half the gametes have a big blue chromosome and a small red one (combination c), and half have a big red one and a small blue one (combination d). Thus, in this example, the organism will produce gametes with four different combinations of chromosomes. For a species with more than two pairs of chromosomes, such as the human, every chromosome pair orients independently of all the others at metaphase I. (Chromosomes X and Y behave as a homologous pair in meiosis.)

For any species, the total number of chromosome combinations that can appear in gametes is 2^n, where n is the haploid number. For the organism in this figure, $n = 2$, so the number of chromosome combinations is 2^2, or 4. For a human ($n = 23$), there are 2^{23}, or about 8 million, possible chromosome combinations! This means that every gamete a human produces contains one of about 8 million possible combinations of maternal and paternal chromosomes.

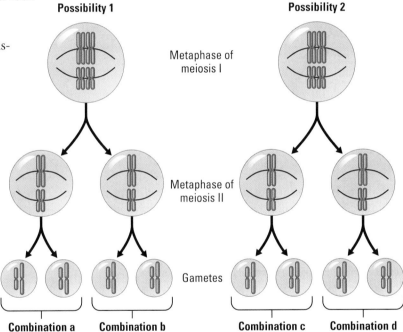

Figure 8.17 Results of alternative arrangements of chromosomes at metaphase of meiosis I.
In this figure, we consider the consequences of meiosis in a diploid organism with four chromosomes (two homologous pairs). The positioning of each homologous pair of chromosomes (tetrad) at metaphase of meiosis I is random; the two red chromosomes can be on the same side (possibility 1) or on opposite sides (possibility 2). The arrangement of chromosomes at metaphase I determines which chromosomes will be packaged together in the haploid gametes. Because possibilities 1 and 2 are equally likely, the four possible types of gametes will be made in approximately equal numbers.

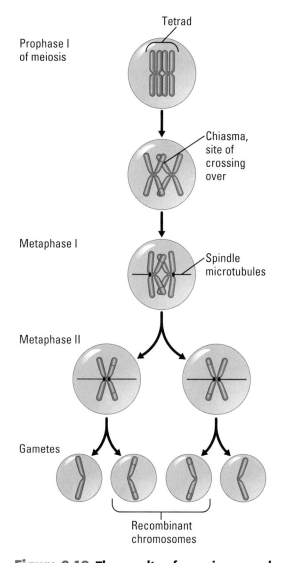

Prophase I
of meiosis

Tetrad

Chiasma,
site of
crossing
over

Metaphase I

Spindle
microtubules

Metaphase II

Gametes

Recombinant
chromosomes

Figure 8.18 The results of crossing over during meiosis. This diagram focuses on a single pair of homologous chromosomes (a tetrad). Early in prophase I of meiosis, homologous (*non*sister) chromatids exchange corresponding segments, remaining attached at the crossover points. Sister chromatids are joined at their centromeres. Following these chromosomes through the rest of meiosis, we see that crossing over gives rise to *recombinant* chromosomes—individual chromosomes that combine genetic information originally derived from different parents. With multiple pairs of homologous chromosomes, the result is a huge variety of gametes.

Random Fertilization How many possibilities are there when a gamete from one individual unites with a gamete from another individual during fertilization? A human egg cell, representing one of about 8 million possibilities, is fertilized at random by one sperm cell, representing one of about 8 million other possibilities. By multiplying 8 million by 8 million, we find that a man and a woman can produce a diploid zygote with any of 64 trillion combinations of chromosomes! So we see that the random nature of fertilization adds a huge amount of potential variability to the offspring of sexual reproduction. And as we'll see next, the sources of variety described thus far are only part of the picture.

Crossing Over So far, we have focused on genetic variability in gametes and zygotes at the whole-chromosome level. We'll now take a closer look at crossing over, the exchange of corresponding segments between two homologous chromosomes, which occurs during prophase I of meiosis. **Figure 8.18** shows crossing over between two homologous chromosomes and the results in the gametes. At the time that crossing over begins, homologous chromosomes are closely paired all along their lengths, with a precise gene-by-gene alignment. The sites of crossing over appear as X-shaped regions; each is called a **chiasma** (plural, *chiasmata*). The homologous chromatids remain attached to each other at chiasmata until anaphase I.

The exchange of segments by homologous chromatids adds to the genetic variety resulting from sexual reproduction. In Figure 8.18, if there were no crossing over, meiosis could produce only two types of gametes. These would be the ones ending up with the "parental" types of chromosomes, either all blue or all red (as in Figure 8.17). With crossing over, gametes arise that have chromosomes that are part red and part blue. These chromosomes are called "recombinant" because they result from **genetic recombination,** the production of gene combinations different from those carried by the parental chromosomes.

Because most chromosomes contain thousands of genes, a single crossover event can affect many genes. When we also consider that multiple crossovers can occur in each tetrad, it's not surprising that gametes and the offspring that result from them can be so varied.

Case Study in the Process of Science on the Web & CD Estimate the frequency of crossing over in a fungus.

Activity 8H on the Web & CD Review the origins of genetic variation.

When Meiosis Goes Awry

So far, our discussion of meiosis has focused on the process as it normally and correctly occurs. But what happens when an error occurs in the process?

Down Syndrome: An Extra Chromosome 21 Figure 8.12 showed a normal human complement of 23 pairs of chromosomes. Compare it with **Figure 8.19**; besides having two X chromosomes (because it's from a female), the karyotype in Figure 8.19 has *three* number 21 chromosomes. This condition is called **trisomy 21.**

In most cases, a human embryo with an abnormal number of chromosomes is spontaneously aborted (miscarried) long before birth. However, some aberrations in chromosome number, including trisomy 21, seem to upset the genetic balance less drastically, and individuals carrying them survive. These people usually have a characteristic set of symptoms, called a syndrome. A person with trisomy 21, for instance, is said to have **Down syndrome** (named after John Langdon Down, who described it in 1866).

Affecting about one out of every 700 children born, trisomy 21 is the most common chromosome number abnormality and the most common serious birth defect in the United States. Chromosome 21 is one of our smallest chromosomes, but an extra copy produces a number of effects. Down syndrome includes characteristic facial features—frequently a fold of skin at the inner corner of the eye (epicanthic fold), a round face, a flattened nose bridge, and small, irregular teeth—as well as short stature, heart defects, and susceptibility to respiratory infection, leukemia, and Alzheimer disease.

People with Down syndrome usually have a life span shorter than normal. They also exhibit varying degrees of mental retardation. However, individuals with the syndrome may live to middle age or beyond, and many are socially adept and able to hold a job. A few women with Down syndrome have had children, though most people with the syndrome are sexually underdeveloped and sterile. Half the eggs produced by a woman with Down syndrome will have an extra chromosome 21, so there is a 50% chance that she will transmit the disorder to her child.

As indicated in **Figure 8.20**, the incidence of Down syndrome in the offspring of normal parents increases markedly with the age of the mother. Down syndrome affects less than 0.05% of children (fewer than one in 2,000) born to women under age 30. The risk climbs to 1% for mothers in their late 30s and is even higher for older mothers. Because of this relatively high risk, pregnant women over 35 are candidates for fetal testing for trisomy 21 and other chromosomal abnormalities (see Chapter 9).

What causes trisomy 21? We address that question next.

How Accidents During Meiosis Can Alter Chromosome Number

Within the human body, meiosis occurs repeatedly as the testes or ovaries produce gametes. Almost always, the meiotic spindle distributes chromosomes to daughter cells without error. But occasionally there is an accident, called a **nondisjunction,** in which the members of a chromosome pair fail to separate at anaphase. Nondisjunction can occur during meiosis I or II (**Figure 8.21**). In either case, gametes with abnormal numbers of chromosomes are the result.

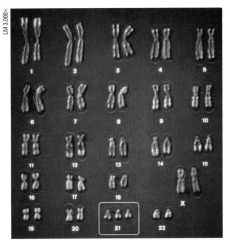

Figure 8.19 Trisomy 21 and Down syndrome. The child displays the characteristic facial features of Down syndrome. The karyotype (left) shows trisomy 21; notice the three copies of chromosome 21.

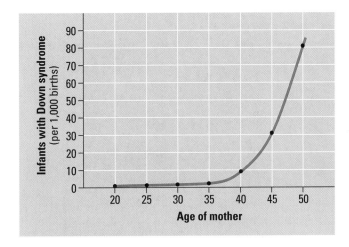

Figure 8.20 Maternal age and Down syndrome. The chance of having a baby with Down syndrome rises with the age of the mother.

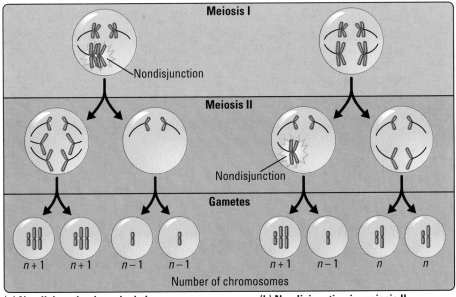

(a) Nondisjunction in meiosis I   **(b) Nondisjunction in meiosis II**

Figure 8.21 Two types of nondisjunction. In both parts of the figure, the cell at the top is diploid (2*n*), with two pairs of homologous chromosomes. **(a)** A pair of homologous chromosomes fails to separate during anaphase of meiosis I, even though the rest of meiosis occurs normally. In this case, all the resulting gametes end up with abnormal numbers of chromosomes. **(b)** Meiosis I is normal, but a pair of sister chromatids fail to move apart in one of the cells during anaphase of meiosis II. In this case, two gametes have the normal complement of two chromosomes each, but the other two gametes are abnormal.

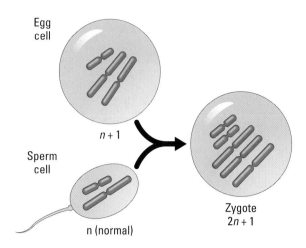

Egg cell

n + 1

Sperm cell

n (normal)

Zygote
2*n* + 1

Figure 8.22 Fertilization after nondisjunction in the mother. Assuming that the organism has a diploid number of 4 (2*n* = 4), the sperm is a normal haploid cell (*n* = 2). The egg cell, however, contains an extra copy of the larger chromosome as a result of nondisjunction in meiosis; it has a total of *n* + 1 = 3 chromosomes. When the sperm and egg fuse during fertilization, the result is an abnormal zygote with an extra chromosome; it has 2*n* + 1 = 5 chromosomes.

Figure 8.22 shows what can happen when an abnormal gamete produced by nondisjunction unites with a normal gamete in fertilization. When a normal sperm fertilizes an egg cell with an extra chromosome, the result is a zygote with a total of 2*n* + 1 chromosomes. Mitosis then transmits the abnormality to all embryonic cells. If the organism survives, it will have an abnormal karyotype and probably a syndrome of disorders caused by the abnormal number of genes.

Nondisjunction explains how abnormal chromosome numbers come about, but what causes nondisjunction in the first place? We do not know the answer, nor do we fully understand why offspring with trisomy 21 are more likely to be born as a woman ages. We do know, however, that meiosis begins in a woman's ovaries before she is born but is not completed until years later, at the time of an ovulation. Because only one egg cell usually matures each month, a cell might remain arrested in the middle of meiosis for decades. Perhaps damage to the cell during this time leads to meiotic errors. It seems that the longer the time lag, the greater the chance that there will be errors such as nondisjunction when meiosis is completed.

Abnormal Numbers of Sex Chromosomes Nondisjunction in meiosis does not affect just autosomes, such as chromosome 21. It can also lead to abnormal numbers of sex chromosomes (X and Y). Unusual numbers of sex chromosomes seem to upset the genetic balance less than unusual numbers of autosomes. This may be because the Y chromosome is very small and carries fewer genes than other chromosomes. Also, most of the genes on the Y chromosome affect maleness but not functions that are essential to the person's survival. A peculiarity of X chromosomes in humans and other mammals also helps an individual tolerate unusual numbers of X chromosomes: In mammals, the cells normally operate with only one functioning X chromosome because any other copies of the chromosome become inactivated in each cell (see Chapter 11).

Table 8.1 lists the most common sex chromosome abnormalities. An extra X chromosome in a male, making him XXY, produces a condition called Klinefelter syndrome. Men with this disorder have male sex organs, but the testes are abnormally small and the individual is sterile. The extra X chromosome often includes breast enlargement and other feminine body contours (**Figure 8.23a**). The extra X chromosome does not seem to affect intelligence. Klinefelter syndrome is also found in individuals with more than one additional sex chromosome, such as XXYY, XXXY, or XXXXY. These abnormal numbers of sex chromosomes probably result from multiple nondisjunctions. Such men are more likely to be mentally retarded than XY or XXY individuals.

Human males with a single extra Y chromosome (XYY) do not have any well-defined syndrome, although they tend to be taller than average.

Table 8.1	Abnormalities of Sex Chromosome Number in Humans		
Sex Chromosomes	**Syndrome**	**Origins of Nondisjunction**	**Frequency in Population**
XXY	Klinefelter syndrome (male)	Meiosis in egg or sperm formation	$\frac{1}{2,000}$
XYY	None (normal male)	Meiosis in sperm formation	$\frac{1}{2,000}$
XXX	Metafemale	Meiosis in egg or sperm formation	$\frac{1}{1,000}$
XO	Turner syndrome (female)	Meiosis in egg or sperm formation	$\frac{1}{5,000}$

Females with an extra X chromosome (XXX) are called metafemales. They have limited fertility but are otherwise apparently normal.

Females who are lacking an X chromosome are designated XO; the O indicates the absence of a second sex chromosome. These women have Turner syndrome. They have a characteristic appearance, including short stature and often a web of skin extending between the neck and shoulders (**Figure 8.23b**). Women with Turner syndrome are sterile because their sex organs do not fully mature at adolescence, and they have poor development of breasts and other secondary sex characteristics. However, they are usually of normal intelligence.

The sex chromosome abnormalities described here illustrate the crucial role of the Y chromosome in determining a person's sex. In general, a single Y chromosome is enough to produce "maleness," regardless of the number of X chromosomes. The absence of a Y chromosome results in "femaleness."

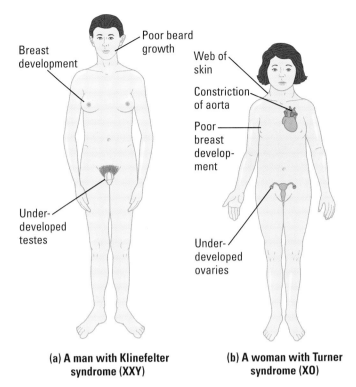

(a) A man with Klinefelter syndrome (XXY)

(b) A woman with Turner syndrome (XO)

Figure 8.23 Syndromes associated with unusual numbers of sex chromosomes.

CHECKPOINT

1. _____ is to somatic cells as haploid is to _____.

2. If a single diploid cell with 18 chromosomes undergoes meiosis to produces sperm, the result will be _____ sperm, each with _____ chromosomes.

3. Explain how mitosis conserves chromosome number while meiosis reduces the number by half.

4. In what important way is anaphase of meiosis I different from anaphase of mitosis?

5. Name two events during meiosis that contribute to genetic variety among gametes. During what stages of meiosis does each occur?

6. How does the karyotype of a human female differ from that of a male?

7. What is the chromosomal basis of Down syndrome?

8. Explain how nondisjunction in meiosis could result in a diploid gamete.

Answers: 1. Diploid; gametes **2.** 4; 9 **3.** In mitosis, a single replication of the chromosomes is followed by one division of the cell. In meiosis, a single replication of the chromosomes is followed by two cell divisions. **4.** In anaphase of mitosis, chromosomes line up single file and sister chromatids separate. In anaphase of meiosis I, homologous pairs of chromosomes line up and separate. **5.** Crossing over between homologous chromosomes during prophase I and independent orientation of tetrads at metaphase I **6.** A female has two X chromosomes; a male has an X and a Y. **7.** Three copies of chromosome 21 (trisomy 21) **8.** A diploid gamete would result if there were nondisjunction of *all* the chromosomes during meiosis I or II.

Evolution Connection

New Species from Errors in Cell Division

Errors in meiosis or mitosis do not always lead to problems. In fact, biologists believe that such errors have been instrumental in the evolution of many species. Numerous plant species, in particular, seem to have originated from accidents during cell division that resulted in extra sets of chromosomes. The new species is **polyploid,** meaning that it has more than two sets of homologous chromosomes in each somatic cell. At least half of all species of flowering plants are polyploid, including such useful ones as wheat, potatoes, apples, and cotton.

Figure 8.24

Chock full of chromosomes—a tetraploid mammal?
The somatic cells of this red viscacha rat from Argentina have about twice as many chromosomes as those of closely related species. (The head of its sperm is unusually large, presumably a necessity for holding all that genetic material.) Scientists think that this rat is a tetraploid species that arose when an ancestor somehow doubled its chromosome number, probably by errors in mitosis or meiosis within the animal's reproductive organs. Researchers are studying the rat's chromosomes to verify that it actually has four homologous sets.

Activity 8I on the Web & CD
See how polyploid plants come about.

Evolution Connection on the Web
Learn more about the role of polyploidy in the evolution of plants.

Let's consider one scenario by which a diploid (2n) plant species might generate a tetraploid (4n) plant. Imagine that, like many plants, our diploid plant produces both sperm and egg cells and can self-fertilize. If meiosis fails to occur in the plant's reproductive organs and gametes are instead produced by mitosis, the gametes will be diploid. The union of a diploid (2n) sperm with a diploid (2n) egg in self-fertilization will produce a tetraploid (4n) zygote, which may develop into a mature tetraploid plant that can itself reproduce by self-fertilization. The tetraploid plants will constitute a new species, one that has evolved in just one generation.

Although polyploid animal species are less common than polyploid plants, they are known to occur among the fishes and amphibians. Moreover, researchers in Chile have identified the first candidate for polyploidy among the mammals, a rat whose cells seem to be tetraploid (**Figure 8.24**). Tetraploid organisms are sometimes strikingly different from their recent diploid ancestors—larger, for example. Scientists don't yet understand exactly how polyploidy brings about such differences.

You'll learn more about the evolution of polyploid species in Chapter 14. In Chapter 9, we continue our study of genetic principles by looking at the rules governing the inheritance of biological traits and the connection between these traits and the organism's chromosomes.

Chapter Review

Summary of Key Concepts

For study help, go to the Essential Biology Website (www.essentialbiology.com) or CD-ROM to explore the Activities and Case Studies in the Process of Science.

What Cell Reproduction Accomplishes

• **Passing On the Genes from Cell to Cell** Cell reproduction, usually called cell division, involves the duplication of all the chromosomes, followed by the distribution of the two identical sets of chromosomes to two "daughter" cells when the cell divides in two. The daughter cells are genetically identical.

• **The Reproduction of Organisms** Some organisms use ordinary cell division to reproduce. Their offspring are therefore genetically identical to the lone parent and to each other. Cell division also enables multicellular organisms to grow and develop and to replace damaged or lost cells. Organisms that reproduce sexually, by the union of a sperm with an egg cell, carry out meiosis, a type of cell division that yields sperm and egg cells with only half as many chromosomes as body (somatic) cells.

Activity 8A Asexual and Sexual Reproduction

The Cell Cycle and Mitosis

• **Eukaryotic Chromosomes** The many genes of a eukaryotic genome are grouped into multiple chromosomes in the nucleus. Each chromosome contains one very long DNA molecule, with thousands of genes, that is tightly packed around histone proteins. Individual chromosomes are visible with a light microscope only when the cell is in the process of dividing; otherwise, they are in the form of thin, loosely packed chromatin fibers. Before a cell starts dividing, the chromosomes duplicate, producing sister chromatids (containing identical DNA) joined together at the centromere.

• **The Cell Cycle**

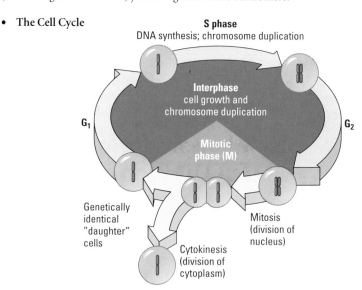

Activity 8B The Cell Cycle

- **Mitosis and Cytokinesis**   At the start of mitosis, the chromosomes coil up, the nuclear envelope breaks down, and a mitotic spindle made of microtubules moves the chromosomes to the middle of the cell. The sister chromatids then separate and are moved to opposite poles of the cell, where two new nuclei form. Cytokinesis overlaps the end of mitosis. Mitosis and cytokinesis produce genetically identical cells. In animals, cytokinesis occurs by cleavage, which pinches the cell apart. In plants, a membranous cell plate splits the cell in two.

Activity 8C   Mitosis and Cytokinesis Animation

Activity 8D   Mitosis and Cytokinesis Video

Case Study in the Process of Science   How Much Time Do Cells Spend in Each Phase of Mitosis?

- **Cancer Cells: Growing Out of Control**   When the cell cycle control system malfunctions, a cell may divide excessively and form a tumor. Cancer cells have highly abnormal cell cycles. They can grow to form malignant tumors, invade other tissues (metastasize), and even kill the organism. Surgery can remove tumor cells, and radiation and chemotherapy are effective as treatments because they interfere with cell division. You can protect yourself against cancer through lifestyle changes and regular screenings.

Activity 8D   Mitosis and Cytokinesis Video

Meiosis, the Basis of Sexual Reproduction

- **Homologous Chromosomes**   The somatic cells (body cells) of each species contain a specific number of chromosomes; for example, human cells have 46, made up of 23 pairs (two sets) of homologous chromosomes. The chromosomes of a homologous pair carry genes for the same characteristics at the same places. In mammalian males, one pair of chromosomes are only partially homologous: the sex chromosomes X and Y. Females have two X chromosomes, while males have one X and one Y.

- **Gametes and the Life Cycle of a Sexual Organism**

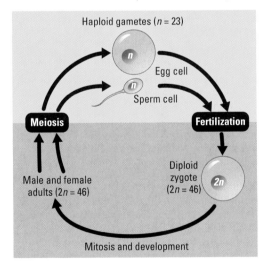

Activity 8F   Human Life Cycle

- **The Process of Meiosis**   Meiosis, like mitosis, is preceded by chromosome duplication. But in meiosis, the cell divides twice to form four daughter cells. The first division, meiosis I, starts with the pairing of homologous chromosomes. In crossing over, homologous chromosomes exchange corresponding

segments. Meiosis I separates the members of the homologous pairs and produces two daughter cells, each with one set of (duplicated) chromosomes. Meiosis II is essentially the same as mitosis; in each of the cells, the sister chromatids of each chromosome separate.

Activity 8G   Meiosis Animation

Case Study in the Process of Science   How Can the Frequency of Crossing Over Be Estimated?

- **Review: Comparing Mitosis and Meiosis**

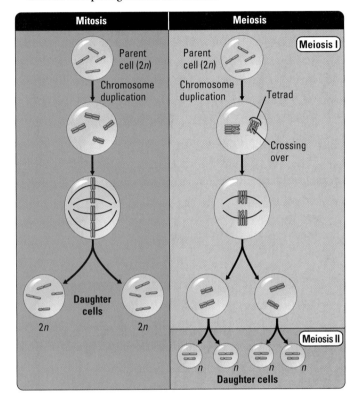

- **The Origins of Genetic Variation**   Because the chromosomes of a homologous pair come from different parents, they carry different versions of many of their genes. The large number of possible arrangements of chromosome pairs at metaphase of meiosis I leads to many different combinations of chromosomes in eggs and sperm. This is one source of the variation in offspring that results from sexual reproduction. Random fertilization of eggs by sperm greatly increases the variation. Crossing over during prophase of meiosis I increases variation still further.

Activity 8H   Origins of Genetic Variation

- **When Meiosis Goes Awry**   Sometimes a person has an abnormal number of chromosomes, which causes problems. Down syndrome is caused by an extra copy of chromosome 21. The abnormal chromosome count is a product of nondisjunction, the failure of a homologous pair of chromosomes to separate during meiosis I or of sister chromatids to separate during meiosis II. Nondisjunction can also produce gametes with extra or missing sex chromosomes, which lead to varying degrees of malfunction in humans but do not usually affect survival.

Activity 8I   Polyploid Plants

Self-Quiz

1. Which of the following is not a function of mitosis in humans?

   a. repair of wounds

   b. growth

   c. production of gametes from diploid cells

   d. replacement of lost or damaged cells

2. Why is it difficult to observe individual chromosomes during interphase?

3. A biochemist measures the amount of DNA in cells growing in the laboratory. The quantity of DNA in a cell would be seen to double

   a. between prophase and anaphase of mitosis.

   b. between the G_1 and G_2 phases of the cell cycle.

   c. during the M phase of the cell cycle.

   d. between prophase I and prophase II of meiosis.

4. Which two phases of mitosis are essentially opposites in terms of changes in the nucleus?

5. A chemical that disrupts microfilament formation would interfere with

   a. DNA replication.

   b. formation of the mitotic spindle.

   c. cleavage.

   d. crossing over.

6. If an intestinal cell in a dog contains 78 chromosomes, a dog sperm cell would contain _____ chromosomes.

7. A micrograph of a dividing cell from a mouse shows 19 chromosomes, each consisting of two sister chromatids. During which stage of meiosis could this picture have been taken? (Explain your answer.)

8. Tumors that remain at their site of origin are called _____, while tumors from which cells migrate to other body tissues are called _____.

9. A fruit fly somatic cell contains eight chromosomes. This means that _____ different combinations of chromosomes are possible in its gametes.

10. Although nondisjunction is a random event, there are many more individuals with an extra chromosome 21, which causes Down syndrome, than individuals with an extra chromosome 3 or chromosome 16. Propose an explanation for this.

Answers to the Self-Quiz questions can be found in Appendix B.

Go to the website or CD-ROM for more Self-Quiz questions.

The Process of Science

1. A mule is the offspring of a horse and a donkey. A donkey sperm contains 31 chromosomes and a horse egg 32 chromosomes, so the zygote contains a total of 63 chromosomes. The zygote develops normally. The combined set of chromosomes is not a problem in mitosis, and the mule combines some of the best characteristics of horses and donkeys. However, a mule is sterile; meiosis cannot occur normally in its testes or ovaries. Explain why mitosis is normal in cells containing both horse and donkey chromosomes but the mixed set of chromosomes interferes with meiosis.

2. You prepare a slide with a thin slice of an onion root tip. You see the following view in a light microscope. Identify the stage of mitosis for each of the outlined cells, a–d.

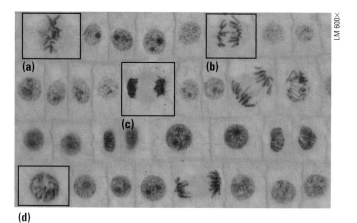

Case Study in the Process of Science on the Web & CD *Investigate how much time cells spend in each phase of mitosis.*

Case Study in the Process of Science on the Web & CD *Estimate the frequency of crossing over in a fungus.*

Biology and Society

1. Every year, about a million Americans are diagnosed with cancer. This means that about 75 million Americans now living will eventually have cancer, and one in five will die of the disease. There are many kinds of cancers and many causes of the disease. For example, smoking causes most lung cancers. Overexposure to ultraviolet rays in sunlight causes most skin cancers. There is evidence that a high-fat, low-fiber diet is a factor in breast, colon, and prostate cancers. And agents in the workplace, such as asbestos and vinyl chloride, are also implicated as causes of cancer. Hundreds of millions of dollars are spent each year in the search for effective treatments for cancer, yet far less money is spent on preventing cancer. Why might this be true? What kinds of lifestyle changes could we make to help prevent cancer? What kinds of prevention programs could be initiated or strengthened to encourage these changes? What factors might impede such changes and programs? Should we devote more of our resources to treating cancer or preventing it? Why?

2. The practice of buying and selling gametes, particularly eggs from fertile women, is becoming increasingly common in the United States and other developed countries. The story that opened this chapter described a couple who was willing to pay $50,000 for the eggs of a woman with particular physical and mental traits. Do you have any objections to this transaction? Would you be willing to sell your gametes? Whether you are willing to do so or not, do you think that other people should be restricted from doing so? Why do you think the couple sought out a woman with those specific traits? Do you agree with their reasoning?

Biology and Society on the Web *Learn more about controversial reproductive technologies.*

CHAPTER 9

Patterns of Inheritance

Biology and Society:
Testing Your Baby   142

Heritable Variation and Patterns
of Inheritance   142

In an Abbey Garden

Mendel's Principle of Segregation

Mendel's Principle of Independent
Assortment

Using a Testcross to Determine
an Unknown Genotype

The Rules of Probability

Family Pedigrees

Human Disorders Controlled
by a Single Gene

Beyond Mendel   154

Incomplete Dominance in Plants
and People

Multiple Alleles and Blood Type

Pleiotropy and Sickle-Cell Disease

Polygenic Inheritance

The Role of Environment

The Chromosomal Basis
of Inheritance   159

Gene Linkage

Genetic Recombination:
Crossing Over and Linkage Maps

Sex Chromosomes and
Sex-Linked Genes   163

Sex Determination in Humans
and Fruit Flies

Sex-Linked Genes

Sex-Linked Disorders in Humans

Evolution Connection:
The Telltale Y Chromosome   166

The first genetics research "lab" was a monk's abbey garden.

In certain isolated communities, one out of 100 boys is born with the genetic defect that causes **Duchenne muscular dystrophy.**

The same genetic defect that causes sickle-cell disease can also **protect you against malaria.**

Intermarriage caused the disease **hemophilia** to spread through the royal families of Europe.

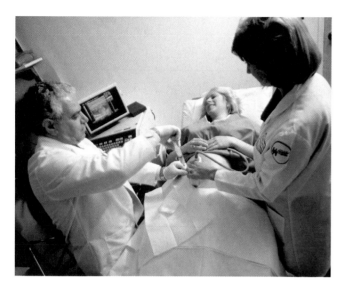

Figure 9.1 Amniocentesis. A physician inserts a needle into the mother's uterus and extracts 10 mL (about 2 teaspoonsful) of amniotic fluid. Cells collected from the fluid can be genetically tested.

Testing Your Baby

Few events in life are as exciting as the news that you're having a baby. So many decisions to make! How should you prepare the nursery? What about names? Modern technology has added another question: Should the fetus be genetically tested? Some parents are aware that they have an increased risk of having a baby with a genetic disease. For example, pregnant women over 35 have an increased risk of bearing children with Down syndrome, and other couples may be aware that certain genetic diseases run in their families. These prospective parents may want to learn more about their baby's genetic legacy.

Genetic testing before birth requires the collection of fetal cells. In a procedure called amniocentesis, performed between weeks 14 and 20 of pregnancy, a physician carefully inserts a needle through the mother's abdomen into her uterus, avoiding the fetus (**Figure 9.1**). The physician extracts 10 milliliters (mL), about 2 teaspoonsful, of the amniotic fluid that surrounds the developing fetus. Cells from the fetus (mostly shed skin cells) can then be isolated from this fluid.

In another procedure, called chorionic villus sampling (CVS), a physician inserts a narrow, flexible tube through the mother's vagina and into her uterus. A small amount of fetal tissue is removed by suction. CVS can be performed early, at 10 to 12 weeks of pregnancy.

Once cells are obtained, they can be screened for genetic diseases. Some genetic disorders, such as Tay-Sachs disease, can be detected by the presence of certain chemicals in the amniotic fluid. Karyotyping can be used to detect chromosomal abnormalities, such as Down syndrome.

Unfortunately, both of these methods of sampling fetal DNA involve some risk of complications, such as maternal bleeding, miscarriage, or premature birth. Complication rates for CVS and amniocentesis are about 2% and 1%, respectively. Because of the risks, these techniques are usually reserved for situations in which the possibility of a genetic disease or other type of birth defect is significant.

If fetal tests reveal a genetic disease that cannot be helped by routine surgery or other therapy, the parents must make a choice—either to terminate the pregnancy or to prepare themselves for a baby with a serious birth defect. CVS

Biology and Society on the Web
Read about socioeconomic factors related to fetal testing.

gives parents a chance to make a decision while the fetus is still quite young. For some, an abortion for any reason is unthinkable. For others, it is a regrettable but permissible way to prevent suffering. Even if abortion is not a consideration, identifying a genetic disease early can give families time to prepare—emotionally, medically, and financially.

Fetal testing is just one example of how genetics, the science of heredity, affects our lives. In this chapter, you will learn the basic rules of genetics and how the behavior of chromosomes accounts for these rules. Along the way, you'll see many examples of the effects of gene mutations on human health.

Figure 9.2 Heritable variation in budgies.
In nature, budgies (also called parakeets) are most often green and yellow, like the second from the right in this photograph. This is the wild-type coloration. Birds of other colors are rarer in nature, but a wide variety have been bred in captivity.

Heritable Variation and Patterns of Inheritance

Budgies, also known as parakeets, are small, long-tailed parrots native to Australia that are kept as pets throughout the world. Wild budgies typically have green underparts and yellow upperparts barred with black, like the bird second from the right in **Figure 9.2**. Geneticists call such traits—the ones found most commonly in nature—the **wild type.** The white and blue birds in the photo are uncommon in nature, but large numbers of these and other color variants are raised in captivity and are available in pet stores. A variant called sky-blue is especially popular.

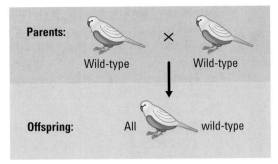

Parents: Wild-type × Wild-type

Offspring: All wild-type

(a) Offspring from the mating of two wild-type birds

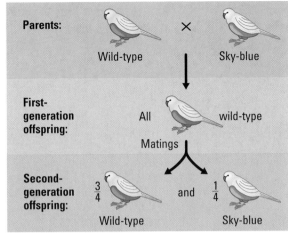

Parents: Wild-type × Sky-blue

First-generation offspring: All wild-type

Matings

Second-generation offspring: $\frac{3}{4}$ Wild-type and $\frac{1}{4}$ Sky-blue

(b) Two generations of offspring from the mating of a wild-type with a sky-blue bird

Figure 9.3 Patterns of inheritance in budgies.
(a) In nature, when two wild-type (green and yellow) budgies mate, all the offspring usually look like the parents. (The cross symbol indicates a mating.) **(b)** When a wild-type bird mates with a sky-blue bird, all their offspring will also be wild-type in appearance. But when these first-generation offspring mate with each other, about one-quarter of their offspring, the second generation, will be sky-blue.

Feather color in budgies is an inherited characteristic, and pet breeders can predict which color variants will result from particular parents. For instance, if two wild-type (green and yellow) budgies are mated (**Figure 9.3a**), it's likely that all their offspring will be green and yellow. The same is true when a wild-type bird and a sky-blue are mated (**Figure 9.3b**). But if two offspring of a wild-type budgie and a sky-blue are mated, their offspring, the second generation, will most likely include both wild-type (green and yellow) birds and sky-blues. Knowing how these feather color traits are inherited—their *patterns of inheritance*—a breeder would predict that about one in four second-generation birds would be sky-blue.

Over a century ago, a monk named Gregor Mendel discovered that the inheritance of many genetic characteristics follows a few simple rules. These genetic rules, or principles, apply to budgies, humans, and all other sexually reproducing organisms.

Gregor Mendel was the first to analyze patterns of inheritance in a systematic, scientific way. During the 1860s, using the garden of an abbey in Brunn, Austria (now Brno, in the Czech Republic), Mendel deduced the fundamental principles of genetics by breeding garden peas (**Figure 9.4**). His work is a classic in the history of biology. Strongly influenced by his study of physics, mathematics, and chemistry at the University of Vienna, his research was both experimental and mathematically rigorous, and these qualities were largely responsible for his success.

In a paper published in 1866, Mendel correctly argued that parents pass on to their offspring discrete heritable factors that are responsible for inherited traits. (It is interesting to note that Mendel's publication came just seven years after Darwin's 1859 publication of *The Origin of Species*, making the 1860s a banner decade in the development of modern biology.) In his paper, Mendel stressed that the heritable factors (today called genes) retain their individuality generation after generation. In other words, genes are like marbles of different colors: Just as marbles retain their colors permanently, no matter how they are mixed up or temporarily covered up, genes permanently retain their identities.

Figure 9.4 Gregor Mendel in his garden.
This painting shows Mendel examining some garden peas, the plant he chose for his experiments.

In an Abbey Garden

Mendel probably chose to study garden peas because he was familiar with them from his rural upbringing, they were easy to grow, and they came in many readily distinguishable varieties. And, importantly, Mendel was able to exercise strict control over the matings of his pea plants. The petals of the pea flower almost completely enclose the female and male parts—the carpel and stamens, respectively (**Figure 9.5**). Consequently, in

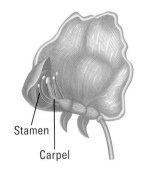

Stamen

Carpel

Figure 9.5 The structure of a pea flower.
To reveal the reproductive organs, the stamens and carpel, one of the petals has been removed in this drawing. The carpel produces egg cells; the stamens make pollen, which carries sperm.

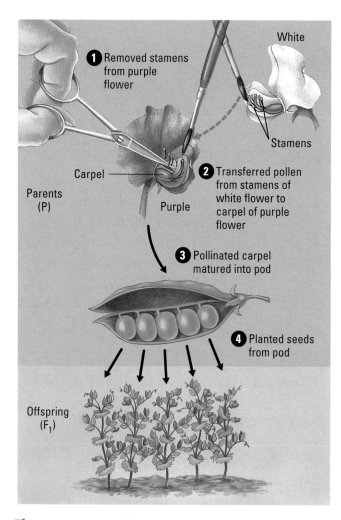

Figure 9.6 Mendel's technique for cross-fertilizing pea plants. ❶ To prevent self-fertilization, he cut off the stamens from an immature flower before they produced pollen. This stamenless plant would be the female parent in the experiment. ❷ To cross-fertilize this female, he dusted its carpel with pollen from another plant. After pollination, ❸ the carpel developed into a pod, containing seeds (peas). ❹ He planted the seeds, and they grew into offspring plants.

nature, pea plants usually **self-fertilize** because sperm-carrying pollen grains released from the stamens land on the tip of the egg-containing carpel of the same flower. Mendel could ensure self-fertilization by covering a flower with a small bag so that no pollen from another plant could reach the carpel. When he wanted **cross-fertilization** (fertilization of one plant by pollen from a different plant), he pollinated the plants by hand, as shown in **Figure 9.6**. Thus, whether Mendel let a pea plant self-fertilize or cross-fertilized it with a known source of pollen, he could always be sure of the parentage of new plants.

Mendel's success was due not only to his experimental approach and choice of organism, but also to his selection of characteristics to study. Each of the characteristics he chose to study, such as flower color, occurs in two distinct forms. Mendel worked with his plants until he was sure he had **true-breeding** varieties—that is, varieties for which self-fertilization produced offspring all identical to the parent. For instance, he identified a purple-flowered variety that, when self-fertilized, produced offspring plants that all had purple flowers.

Now Mendel was ready to ask what would happen when he crossed his different true-breeding varieties with each other. For example, what offspring would result if plants with purple flowers and plants with white flowers were cross-fertilized as shown in Figure 9.6? In the language of breeders and geneticists, the offspring of two different true-breeding varieties are called **hybrids,** and the cross-fertilization itself is referred to as a hybridization, or simply a **cross.** The parental plants are called the **P generation** (P for parental), and their hybrid offspring are the **F_1 generation** (F for *filial,* from the Latin for son or daughter). When F_1 plants self-fertilize or fertilize each other, their offspring are the **F_2 generation.**

Mendel's Principle of Segregation

Mendel performed many experiments in which he tracked the inheritance of characteristics, such as flower color, that occur as two alternatives (**Figure 9.7**). The results led him to formulate several hypotheses about inheritance. Let's look at some of his experiments and follow the reasoning that led to his hypotheses.

Monohybrid Crosses Figure 9.8a starts with a cross between a pea plant with purple flowers and one with white flowers. This is called a **monohybrid cross** because the parent plants differ in only one characteristic. Mendel saw that the F_1 plants (monohybrids) all had purple flowers. Was the heritable factor for white flowers now lost as a result of the hybridization? By mating the F_1 plants, Mendel found the answer to be no. Of the F_2 plants, one-fourth had white flowers. Mendel concluded that the heritable factor for white flowers did not disappear in the F_1 plants, but that only the purple-flower factor was affecting F_1 flower color. He also deduced that the F_1 plants must have carried two factors for the flower color characteristic, one for purple and one for white. From these results and others, Mendel developed four hypotheses. Using modern terminology (including "gene" instead of "heritable factor"), here are his hypotheses:

1. There are alternative forms of genes, the units that determine heritable traits. For example, the gene for flower color in pea plants exists in one form for purple and another for white. We now call alternative forms of genes **alleles.**
2. For each inherited characteristic, an organism has two genes, one from each parent. These genes may be the same allele or different alleles.

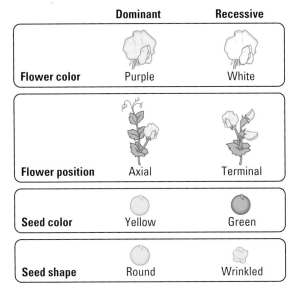

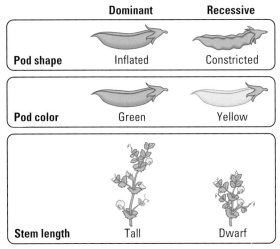

Figure 9.7 The seven characteristics of pea plants studied by Mendel. Each characteristic comes in the two alternatives shown here. The alternative on the left in each pair (such as purple flower color) is dominant; the one on the right (such as white flower color) is recessive. You will learn about dominant and recessive genes shortly.

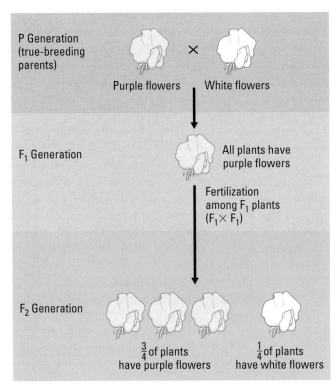

(a) Mendel's crosses tracking one characteristic (flower color)

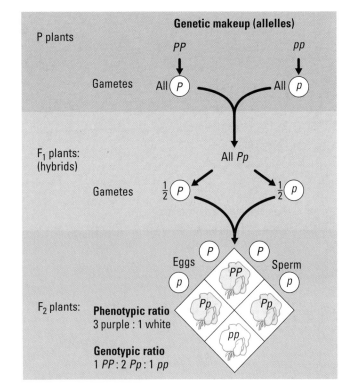

(b) Explanation of the results in part (a)

Figure 9.8 The principle of segregation. (a) Crossing true-breeding purple-flowered plants with true-breeding white-flowered plants produced F_1 plants with purple flowers. Mendel cross-pollinated F_1 plants with other F_1 plants (or allowed F_1 plants to self-pollinate) to produce 929 F_2 plants. Of these, 705, about three-quarters, had purple flowers, while 224, about one-quarter, had white flowers. In other words, there was a ratio of about 3:1 of the two varieties in the F_2 generation. (b) This diagram illustrates Mendel's explanation for the inheritance pattern shown in part (a) using modern terminology. Uppercase and lowercase letters distinguish dominant from recessive alleles, with P representing the dominant allele, for purple flowers, and p standing for the recessive allele, for white flowers. At the top of the diagram are the alleles carried by the parental plants. Both plants were true-breeding, so one parental variety must have had two

alleles for purple flowers (PP), while the other variety had two alleles for white flowers (pp). Gametes, symbolized with circles, each contained only one allele for the flower color gene. Union of the parental gametes produced F_1 hybrids having a Pp combination. Because the purple-flower allele is dominant, all these hybrids had purple flowers. When the hybrid plants produced gametes, the two alleles segregated, half the gametes receiving the P allele and the other half receiving the p allele. Random combinations of these gametes resulted in the 3:1 ratio that Mendel observed in the F_2 generation. The box at the bottom of the figure is a Punnett square, a useful tool for showing all possible combinations of alleles in offspring. Each square represents an equally probable product of fertilization. For example, the box at the right corner of the Punnett square shows the genetic combination resulting from a p sperm fertilizing a P egg.

3. A sperm or egg carries only one allele for each inherited characteristic, because allele pairs segregate (separate) during the production of gametes. Then when sperm and egg unite at fertilization, each contributes its allele, restoring the paired condition in the offspring.

4. When the two genes of a pair are different alleles and one is fully expressed while the other has no noticeable effect on the organism, the alleles are called the **dominant allele** and the **recessive allele,** respectively. Dominant alleles are represented by uppercase letters (in our example, *P*, the purple-flower allele) and recessive alleles by lowercase letters (*p*, the white-flower allele).

Do Mendel's hypotheses account for the 3:1 ratio he observed in the F_2 generation? His hypotheses predict that when alleles segregate during gamete formation in the F_1 plants, half the gametes will receive a purple-flower allele (*P*) and the other half a white-flower allele (*p*). During pollination among the F_1 plants, the gametes unite randomly. An egg with a purple-flower allele has an equal chance of being fertilized by a sperm with a purple-flower allele or one with a white-flower allele. Since the same is true for an egg with a white-flower allele, there are a total of four equally likely combinations of sperm and egg. **Figure 9.8b** (see page 145) illustrates these combinations using a diagram called a **Punnett square,** a handy tool for predicting the results of a genetic cross.

Activity 9A on the Web & CD
Try a monohybrid cross with MendAliens.

What will be the physical appearance of these F_2 offspring? One-fourth of the plants have two alleles specifying purple flowers (*PP*); clearly, these plants will have purple flowers. One-half (two-fourths) of the F_2 offspring have inherited one allele for purple flowers and one allele for white flowers (*Pp*); like the F_1 plants, these plants will also have purple flowers, the dominant trait. (Note that *Pp* and *pP* are equivalent and always written as the former.) Finally, one-fourth of the F_2 plants have inherited two alleles specifying white flowers (*pp*) and will express this recessive trait. Thus, Mendel's model accounts for the 3:1 ratio that he observed in the F_2 generation.

Because an organism's appearance does not always reveal its genetic composition, geneticists distinguish between an organism's expressed, or physical, traits, called its **phenotype** (such as purple or white flowers), and its genetic makeup, its **genotype** (in our example, *PP*, *Pp*, or *pp*). Now we can see that Figure 9.8a shows the phenotypes and Figure 9.8b the genotypes in our sample crosses. For the F_2 plants, the ratio of plants with purple flowers to those with white flowers (3:1) is called the phenotypic ratio. The genotypic ratio is 1(*PP*):2(*Pp*):1(*pp*).

Mendel found that each of the seven characteristics he studied had the same inheritance pattern: One parental trait disappeared in the F_1 generation, only to reappear in one-fourth of the F_2 offspring. The underlying mechanism is stated by **Mendel's principle of segregation:** *Pairs of alleles segregate (separate) during gamete formation; the fusion of gametes at fertilization creates allele pairs again.* Research since Mendel's day has established that the principle of segregation applies to all sexually reproducing organisms.

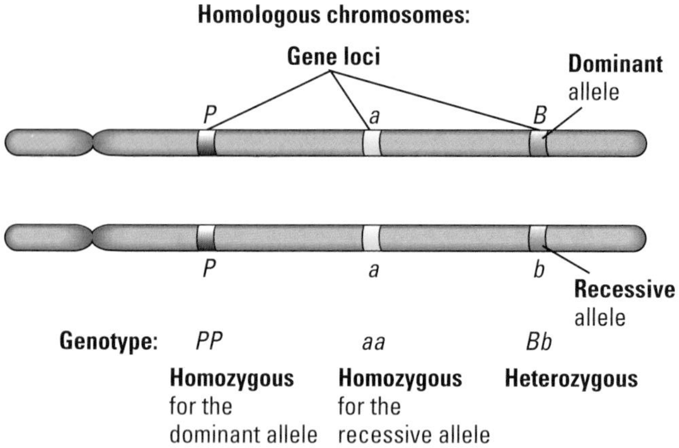

Genetic Alleles and Homologous Chromosomes Before continuing with Mendel's experiments, let's see how some of the concepts we discussed in Chapter 8 fit with what we've said about genetics so far. The diagram in **Figure 9.9** shows a pair of homologous chromosomes. Recall from Chapter 8 that every diploid individual, whether pea plant or human, has two sets of homologous chromosomes. One set comes from the organism's female parent, the other from the male parent. In the diagram, three gene

Figure 9.9 The relationship between alleles and homologous chromosomes. The two chromosomes shown here are a homologous pair. The labeled bands on the chromosomes represent three gene loci, specific locations of genes along the chromosome. The matching colors of corresponding loci on the two homologues highlight the fact that homologous chromosomes carry genes for the same characteristics at the same positions along their lengths. However, as the uppercase and lowercase gene labels indicate, the two chromosomes may bear either identical alleles or different ones at any one locus. In other words, the organisms may be homozygous or heterozygous for the gene at that locus. We use uppercase for dominant alleles, lowercase for recessive alleles.

loci (singular, *locus*) are indicated on the chromosomes. These are the specific locations of three genes. You can see the connection between Mendel's principle of segregation and homologous chromosomes: Alleles (alternative forms) of a gene reside at the same locus on homologous chromosomes.

Notice that the homologous chromosomes may bear either the same alleles or different ones at a given locus. When an organism has a pair of identical alleles for a gene, it is said to be **homozygous** for that gene. (True-breeding organisms are homozygous.) When the two alleles are different, it is **heterozygous.** We will return to the chromosomal basis of Mendel's principles later in the chapter.

Mendel's Principle of Independent Assortment

Two of the seven characteristics Mendel studied were seed shape and seed color. Mendel's seeds were either round or wrinkled in shape and either yellow or green in color. From tracking these characteristics one at a time in monohybrid crosses, Mendel knew that the allele for round shape (designated *R*) was dominant to the allele for wrinkled shape (*r*) and that the allele for yellow seed color (*Y*) was dominant to the allele for green seed color (*y*). What would result from a **dihybrid cross,** the mating of parental varieties differing in two characteristics? Mendel crossed homozygous plants having round yellow seeds (genotype *RRYY*) with plants having wrinkled green seeds (*rryy*). As shown in **Figure 9.10**, the union of

> **Activity 9B on the Web & CD**
> See the results of a dihybrid cross with MendAliens.

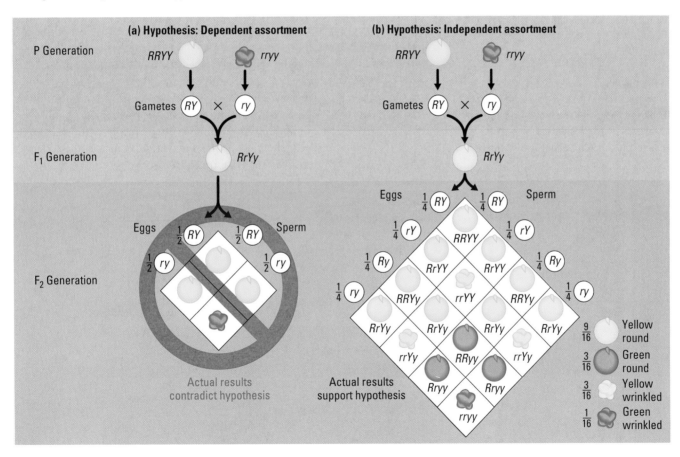

Figure 9.10 Testing two hypotheses for gene assortment in a dihybrid cross.
(a) One hypothesis leads to the prediction that every plant in the F₂ generation will have seeds exactly like one of the parents, either round and yellow or wrinkled and green. **(b)** The other hypothesis, independent assortment, predicts F₂ plants with four different seed phenotypes. The results of experiments support this hypothesis.

RY and *ry* gametes from the P generation yielded hybrids heterozygous for both characteristics (*RrYy*)—that is, dihybrids. As we would expect, all of these offspring, the F₁ generation, had round yellow seeds. But were the two characteristics transmitted from parents to offspring as a package, or was each characteristic inherited independently of the other?

The question was answered when Mendel allowed fertilization to occur among the F₁ plants. If the genes for the two characteristics were inherited together (Figure 9.10a), then the F₁ hybrids would produce only the same two kinds (genotypes) of gametes that they received from their parents. In that case, the F₂ generation would show a 3:1 phenotypic ratio (three plants with round yellow seeds for every one with wrinkled green seeds), as in the Punnett square in Figure 9.10a. If, however, the two seed characteristics segregated independently, then the F₁ generation would produce four gamete genotypes—*RY*, *rY*, *Ry*, and *ry*—in equal quantities. The Punnett square in Figure 9.10b shows all possible combinations of alleles that can result in the F₂ generation from the union of four kinds of sperm with four kinds of eggs. If you study the Punnett square, you'll see that it predicts nine different genotypes in the F₂ generation. These will produce four different phenotypes in a ratio of 9:3:3:1.

The right-hand Punnett square in Figure 9.10 also reveals that a dihybrid cross is equivalent to two monohybrid crosses occurring simultaneously. From the 9:3:3:1 ratio, we can see that there are 12 plants with round seeds to 4 with wrinkled seeds and also 12 yellow-seeded plants to 4 green-seeded ones. These 12:4 ratios each reduce to 3:1, which is the F₂ ratio for a monohybrid cross. Mendel tried his seven pea characteristics in various dihybrid combinations and always observed a 9:3:3:1 ratio (or two simultaneous 3:1 ratios) of phenotypes in the F₂ generation. These results supported hypothesis (b) in Figure 9.10, now known as **Mendel's principle of independent assortment:** *Each pair of alleles segregates independently of the other pairs during gamete formation.* In other words, the inheritance of one characteristic has no effect on the inheritance of another. For another application of this principle, examine **Figure 9.11**, showing the independent inheritance of two characteristics in Labrador retrievers.

Figure 9.11 Independent assortment of genes in Labrador retrievers. (a) The inheritance of two hereditary characteristics in Labrador retrievers, black versus chocolate coat color and normal vision versus the eye disorder progressive retinal atrophy (PRA), is controlled by separate genes. Black Labs have at least one copy of an allele called *B*, which gives their hairs densely packed granules of a dark pigment. The *B* allele is dominant to *b*, which leads to a less tightly packed distribution of pigment granules. As a result, the coats of dogs with genotype *bb* are chocolate in color. The allele that causes PRA, called *n*, is recessive to allele *N*, which is necessary for normal vision. Thus, only dogs of genotype *nn* become blind from PRA. (Blanks in the genotypes indicate alleles that can be dominant or recessive.) (b) If you mate two doubly heterozygous (*BnNn*) Labs, the phenotypic ratio of the offspring (F₂) is 9:3:3:1. These results resemble the F₂ results in Figure 9.10, demonstrating that the coat color and PRA genes are inherited independently.

Activity 9C on the Web & CD
Perform experiments on pea plants like those carried out by Gregor Mendel.

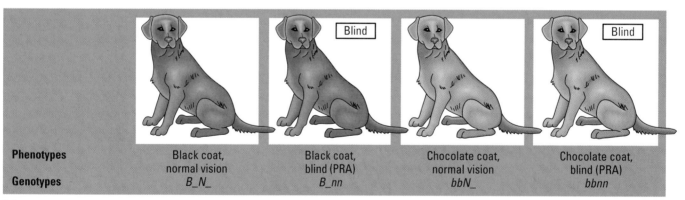

Phenotypes	Black coat, normal vision	Black coat, blind (PRA)	Chocolate coat, normal vision	Chocolate coat, blind (PRA)
Genotypes	*B_N_*	*B_nn*	*bbN_*	*bbnn*

(a)

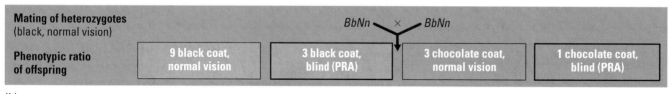

Mating of heterozygotes (black, normal vision) *BbNn* × *BbNn*

Phenotypic ratio of offspring	9 black coat, normal vision	3 black coat, blind (PRA)	3 chocolate coat, normal vision	1 chocolate coat, blind (PRA)

(b)

Using a Testcross to Determine an Unknown Genotype

Suppose you have a pea plant with white flowers. You know its genotype: it must be *pp*, because that's the only combination of alleles that produces the white-flower phenotype. But what if you have a pea plant with purple flowers? It could have one of two possible genotypes—*PP* or *Pp*—and there is no way to tell by looking at the flower which one it is. To determine the genotype of this purple flower, you need to perform what geneticists call a **testcross,** a mating between an individual of unknown genotype (your pea plant with purple flowers) and a homozygous recessive individual—in this case, a *pp* pea plant with white flowers.

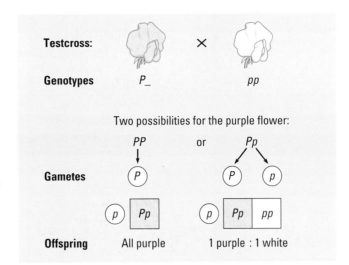

Testcross:

Genotypes   *P_*   *pp*

Two possibilities for the purple flower:

PP   or   *Pp*

Gametes

Offspring   All purple   1 purple : 1 white

Figure 9.12
A flower color testcross.
To determine the genotype of a purple flower, it is crossed with a white flower (homozygous recessive, *pp*). If all the offspring are purple, the purple parent must have genotype *PP*. If half the offspring are white, the purple parent must be heterozygous (*Pp*).

Figure 9.12 shows the offspring that could result from such a mating. If, as shown on the left, the purple parent's genotype is *PP*, we would expect all the offspring to have purple flowers, because a cross between genotypes *PP* and *pp* can produce only *Pp* offspring. On the other hand, if the purple parent is *Pp*, we would expect both purple (*Pp*) and white (*pp*) offspring. Thus, the appearance of the offspring reveals the purple-flowered plant's genotype. The figure also shows that we would expect the white and purple offspring of a *Pp* × *pp* cross to exhibit a 1:1 (1 purple to 1 white) phenotypic ratio.

Mendel used testcrosses to determine whether he had true-breeding varieties of plants. The testcross continues to be an important tool of geneticists for determining genotypes.

The Rules of Probability

Mendel's strong background in mathematics served him well in his studies of inheritance. He understood, for instance, that the segregation of allele pairs during gamete formation and the re-forming of pairs at fertilization obey the rules of probability—the same rules that apply to the tossing of coins, the rolling of dice, and the drawing of cards. Mendel also appreciated the statistical nature of inheritance. He knew that he needed to obtain large samples—count many offspring from his crosses—before he could begin to interpret inheritance patterns.

An important lesson we can learn from coin tossing is that for each and every toss of the coin, the probability of heads is $\frac{1}{2}$. In other words, the outcome of any particular toss is unaffected by what has happened on previous attempts. Each toss is an independent event.

If two coins are tossed simultaneously, the outcome for each coin is an independent event, unaffected by the other coin. What is the chance that both coins will land heads-up? The probability of such a compound event is the product of the separate probabilities of the independent events—for the coins, $\frac{1}{2} \times \frac{1}{2} = \frac{1}{4}$. This is called the **rule of multiplication,** and it holds true for independent events that occur in genetics as well as coin tosses, as shown in **Figure 9.13**. In our dihybrid cross of Labradors (see Figure 9.11), the genotype of the F₁ dogs for coat color was *Bb*. What is the probability that a particular F₂ dog will have the *bb* genotype? To produce a *bb* offspring, both egg and sperm must carry the *b* allele. The probability that an egg will have the *b* allele is $\frac{1}{2}$, and the probability that a sperm will have the *b* allele is also $\frac{1}{2}$.

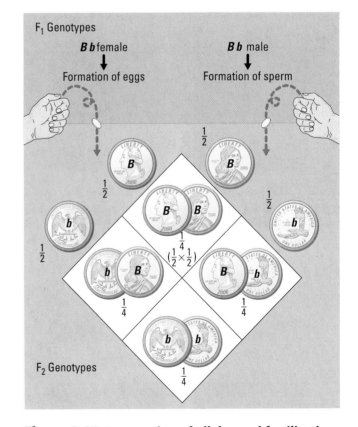

F₁ Genotypes

Bb female   *Bb* male

Formation of eggs   Formation of sperm

F₂ Genotypes

Figure 9.13 Segregation of alleles and fertilization as chance events. When a heterozygote (*Bb*) forms gametes, segregation of alleles is like the toss of a coin. An egg cell has a 50% chance of receiving the dominant allele and a 50% chance of receiving the recessive allele. The same odds apply to a sperm cell. Like two separately tossed coins, segregation during sperm and egg formation occurs as two independent events. To determine the probability that an individual offspring will inherit the dominant allele from both parents, we multiply the probabilities of each required event: $\frac{1}{2} \times \frac{1}{2} = \frac{1}{4}$.

By the rule of multiplication, the probability that two *b* alleles will come together at fertilization is $\frac{1}{2} \times \frac{1}{2} = \frac{1}{4}$. This is exactly the answer given by the Punnett square. If we know the genotypes of the parents, we can predict the probability for any genotype among the offspring.

By applying the rules of probability to segregation and independent assortment, we can solve some rather complex genetics problems. For instance, we can predict the results of trihybrid crosses, in which three different characteristics are involved. Consider a cross between two organisms that both have the genotype *AaBbCc*. What is the probability that an offspring from this cross will be a recessive homozygote for all three genes (*aabbcc*)? Because each allele pair assorts independently, we can treat this trihybrid cross as three separate monohybrid crosses:

Aa × *Aa*: Probability of *aa* offspring $= \frac{1}{4}$

Bb × *Bb*: Probability of *bb* offspring $= \frac{1}{4}$

Cc × *Cc*: Probability of *cc* offspring $= \frac{1}{4}$

Because the segregation of each allele pair is an independent event, we use the rule of multiplication to calculate the probability that the offspring will be *aabbcc*:

$$\tfrac{1}{4}\,aa \times \tfrac{1}{4}\,bb \times \tfrac{1}{4}\,cc = \tfrac{1}{64}$$

We could reach the same conclusion by constructing a 64-section Punnett square, but that would take a lot of time—and also a lot of space!

Family Pedigrees

Mendel's principles apply to the inheritance of many human traits. **Figure 9.14** illustrates alternative forms of three human characteristics that are each thought to be determined by simple dominant-recessive inheritance at one gene locus. (The genetic basis of many of the human characteristics that might interest you the most, such as eye color and hair color, are not well understood.) If we call the dominant allele of any such gene *A*, the dominant phenotype results from either the homozygous genotype *AA* or the heterozygous genotype *Aa*. Recessive phenotypes always result from the homozygous genotype *aa*. In genetics, the word *dominant* does not imply that a phenotype is either normal or more common than a recessive phenotype; wild-type traits (those prevailing in nature) are not necessarily specified by dominant alleles. In genetics, dominance means that a heterozygote (*Aa*), carrying only a single copy of a dominant allele, displays the dominant phenotype. By contrast, the phenotype of the corresponding recessive allele is seen only in a homozygote (*aa*). Recessive traits are often more common in the population than dominant ones. For example, the absence of freckles is more common than their presence.

How do we know how particular human traits are inherited? Researchers working with pea plants or budgies can perform testcrosses. But geneticists who study humans obviously cannot control the mating of their subjects. Instead, they must analyze the results of matings that have already occurred. First, the geneticist collects as much information as possible about a family's history for the trait. Then the researcher assembles this information into a family tree—the family **pedigree.** Finally, to analyze the pedigree, the geneticist uses Mendel's concept of dominant and recessive alleles and his principle of segregation.

Let's apply this approach to the example in **Figure 9.15**, which shows part of a real pedigree. This pedigree is from a family that lived on Martha's Vineyard, an island off the coast of Massachusetts, where a particular kind

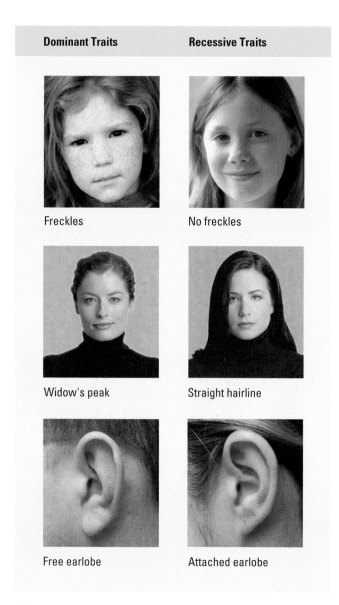

Dominant Traits	Recessive Traits
Freckles	No freckles
Widow's peak	Straight hairline
Free earlobe	Attached earlobe

Figure 9.14 Examples of inherited traits in humans.

of inherited deafness was once prevalent. In the pedigree, squares represent males, circles represent females, and colored symbols indicate deafness. The earliest generation studied is at the top of the pedigree. Notice that deafness did not appear in this generation and that it showed up in only two of the seven children in the third generation. By applying Mendel's principles, we can deduce that the deafness allele is recessive; that's the only way that Jonathan Lambert could be deaf while both of his parents were not. Once we know that, we can label all the deaf individuals as homozygous recessive (*dd*). Mendel's principles also enable us to deduce the other genotypes that are shown. Jonathan and Elizabeth's hearing children, for example, must be heterozygous (*Dd*) because they all inherited one copy of the recessive (*d*) gene from their father. People such as these, who have one copy of the allele for a recessive disorder and do not exhibit symptoms, are called **carriers** of the disorder.

Human Disorders Controlled by a Single Gene

The hereditary deafness found on Martha's Vineyard is just one of over a thousand human genetic disorders currently known to be inherited as dominant or recessive traits controlled by a single gene locus; ten more examples are listed in **Table 9.1**. These disorders show simple inheritance patterns like the traits Mendel studied in pea plants. The genes involved are all located on **autosomes,** chromosomes other than the sex chromosomes X and Y.

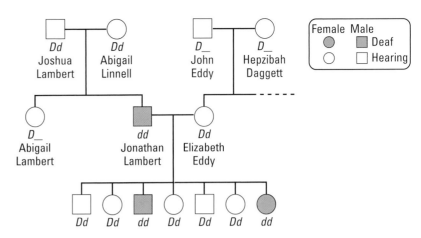

Figure 9.15 A family pedigree showing the inheritance of deafness. The letter *D* stands for the hearing allele, and *d* symbolizes the recessive allele for deafness. The first deaf individual to appear in this pedigree is Jonathan Lambert. Because only people who are homozygous for the recessive allele are deaf, his genotype must have been *dd*. Therefore, both his parents must have carried a *d* allele along with the *D* allele that gave them normal hearing. Two of Jonathan's children were deaf (*dd*), so his wife, who had normal hearing, must also have carried a *d* allele. Likewise, all of the couple's normal children must have been heterozygous (*Dd*). What are the genotypes of Elizabeth Eddy's parents and Jonathan's sister Abigail? These three people had normal hearing, so they must have carried at least one *D* allele. And at least one of Elizabeth's parents must have had a *d* allele. But more than that we cannot say without additional information.

Table 9.1	Some Autosomal Disorders in Humans	
Disorder	**Major Symptoms**	**Incidence**
Recessive disorders		
Albinism	Lack of pigment in skin, hair, and eyes	$\frac{1}{22,000}$
Cystic fibrosis	Excess mucus in lungs, digestive tract, liver; increased susceptibility to infections; death in infancy unless treated	$\frac{1}{1,800}$ European Americans
Galactosemia	Accumulation of galactose in tissues; mental retardation; eye and liver damage	$\frac{1}{100,000}$
Phenylketonuria (PKU)	Accumulation of phenylalanine in blood; lack of normal skin pigment; mental retardation unless treated	$\frac{1}{10,000}$ in U.S. and Europe
Sickle-cell disease (homozygous)	Sickled red blood cells; damage to many tissues	$\frac{1}{500}$ African Americans
Tay-Sachs disease	Lipid accumulation in brain cells; mental deficiency; blindness; death in childhood	$\frac{1}{3,500}$ Ashkenazi Jews
Dominant disorders		
Achondroplasia	Dwarfism	$\frac{1}{25,000}$
Alzheimer's disease (one type)	Mental deterioration; usually strikes late in life	Not known
Huntington's disease	Mental deterioration and uncontrollable movements; strikes in middle age	$\frac{1}{25,000}$
Hypercholesterolemia	Excess cholesterol in blood; heart disease	$\frac{1}{500}$ is heterozygous

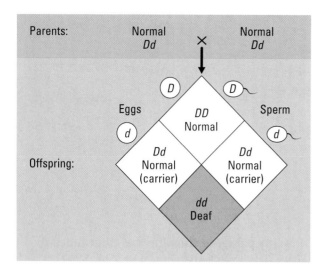

Parents: Normal *Dd* × Normal *Dd*

Eggs

Sperm

D *D*

d *d*

DD
Normal

Dd
Normal
(carrier)

Dd
Normal
(carrier)

dd
Deaf

Offspring:

Figure 9.16 Predicted offspring when both parents are carriers for a recessive disorder. Carriers are heterozygotes who have one copy of a potentially harmful recessive allele. Because carriers also have the dominant allele, they do not display the disorder. As in the last figure, *d* stands for a recessive allele associated with deafness and *D* for the dominant allele. As you can see in the Punnett square, one-fourth of the offspring of two carriers would be expected to be homozygous recessive and therefore deaf.

Recessive Disorders Most human genetic disorders are recessive. They range in severity from relatively harmless conditions, such as albinism (lack of pigmentation), to deadly diseases. The vast majority of people afflicted with recessive disorders are born to normal parents who are both heterozygotes—that is, who are carriers of the recessive allele for the disorder but are phenotypically normal.

Using Mendel's principles, we can predict the fraction of affected offspring likely to result from a marriage between two carriers. Suppose one of Jonathan Lambert's hearing sons (*Dd*) married a hearing woman whose pedigree indicated that her genotype was also *Dd*. What is the probability that they would have a deaf child? As the Punnett square in **Figure 9.16** shows, each child of two carriers has a $\frac{1}{4}$ chance of inheriting two recessive alleles. Thus, we can say that about one-fourth of the children of this marriage are likely to be deaf. We can also say that a hearing ("normal") child from such a marriage has a $\frac{2}{3}$ chance of being a carrier (that is, two out of three of the offspring with the hearing phenotype are likely to be carriers). We can apply this same method of pedigree analysis and prediction to any genetic trait controlled by a single gene locus.

The most common lethal genetic disease in the United States is cystic fibrosis. Though the disease affects only about one in 17,000 African Americans and about one in 90,000 Asian Americans, it occurs in approximately one out of every 1,800 European Americans. The cystic fibrosis allele is recessive and is carried by about one in every 25 European Americans. A person with two copies of this allele has cystic fibrosis, which is characterized by an excessive secretion of very thick mucus from the lungs, pancreas, and other organs. This mucus can interfere with breathing, digestion, and liver function and makes the person vulnerable to pneumonia and other infections. Untreated, most children with cystic fibrosis die by the time they are 5 years old. However, a special diet, antibiotics to prevent infection, frequent pounding of the chest and back to clear the lungs, and other treatments can prolong life to adulthood, although there is no cure for this fatal disease.

Like cystic fibrosis, most genetic disorders are not evenly distributed across all ethnic groups. Such uneven distribution is the result of prolonged geographic isolation of certain populations. For example, the isolated lives of the Martha's Vineyard inhabitants between 1700 and 1900 fostered marriage between close relatives. Consequently, the frequency of deafness remained high, and the deafness allele was only rarely transmitted to outsiders.

With the increased mobility in most societies today, it is relatively unlikely that two carriers of a rare, harmful allele will meet and mate. However, the probability increases greatly if close relatives marry and have children. People with recent common ancestors are more likely to carry the same recessive alleles than are unrelated people. Therefore, a mating of close relatives, called **inbreeding,** is more likely to produce offspring homozygous for a harmful recessive trait.

Most societies have taboos and laws forbidding marriage between very close relatives. These rules may have arisen out of the observation that stillbirths and birth defects are more common when parents are closely related. Such effects can also be observed in many types of inbred animals. For example, the type of hereditary blindness called progressive retinal atrophy (PRA) is common among Labrador retrievers and other purebred dogs (see Figure 9.11). The detrimental effects of inbreeding are may also affect some endangered species, such as cheetahs (see Figure 13.22).

Although inbreeding is clearly dangerous, geneticists debate the extent to which human inbreeding increases the risk of inherited diseases. Many harmful mutations have such severe effects that a homozygous embryo spontaneously aborts long before birth—perhaps so early that the miscarriage goes undetected. Furthermore, as some geneticists argue, marriage between relatives is just as likely to concentrate favorable alleles as harmful ones. There are, in fact, some human populations, such as the Tamils of India, in which marriage between first cousins has long been common and has not produced observable ill effects.

Dominant Disorders Although most harmful alleles are recessive, a number of human disorders are caused by dominant alleles. Some conditions are nonlethal, such as extra fingers and toes, or fingers and toes that are webbed.

One serious disorder caused by a dominant allele is **achondroplasia,** a form of dwarfism. Among people with this disorder, the head and torso of the body develop normally, but the arms and legs are short (**Figure 9.17**). About one out of 25,000 people have achondroplasia. Only heterozygotes, individuals with a single copy of the defective allele, have the disease. The homozygous dominant genotype causes the death of the embryo. Therefore, all those who do not have achondroplasia, more than 99.99% of the population, are homozygous for the normal, recessive allele. This example makes it clear that a dominant allele is not necessarily more plentiful in a population than the corresponding recessive allele. In this case, a person with achondroplasia has a 50% chance of passing the dominant allele for the condition on to any children.

Dominant alleles that are lethal are, in fact, much less common than lethal recessives. One reason for this difference is that the dominant lethal allele cannot be carried by heterozygotes without affecting them. Many lethal dominant alleles result from mutations in a sperm or egg that subsequently kill the embryo. And if the afflicted individual is born but does not survive long enough to reproduce, he or she will not pass on the lethal allele. This is in contrast to lethal recessive mutations, which are perpetuated from generation to generation by the reproduction of heterozygous carriers.

A lethal dominant allele can escape elimination, however, if it does not cause death until a relatively advanced age. The allele that causes **Huntington disease,** a degeneration of the nervous system that usually does not begin until middle age, is one example. As the disease progresses, it produces uncontrollable movements in all parts of the body. The loss of brain cells leads to memory loss and impaired judgment, and contributes to depression. Diminished motor skills eventually affect the ability to swallow and speak. Death usually ensues 10 to 20 years after the onset of symptoms. Unfortunately, by the time the symptoms of Huntington disease become evident, the afflicted individual may have had children, half of whom (on average) will have received the lethal dominant allele for the disease. This example demonstrates that a dominant allele is not necessarily "better" than the corresponding recessive allele.

Figure 9.17 Achondroplasia, a dominant trait.
The late David Rappaport, an actor, had achondroplasia, a form of dwarfism in which the head and torso of the body develop normally, but the arms and legs are short.

1. What is a wild-type trait?

2. What is meant by "self-fertilization" of a plant?

3. How can two plants that have different genotypes for flower color be identical in phenotype?

4. Which of Mendel's principles is represented by each statement?
   a. Each pair of alleles segregates independently during gamete formation.
   b. Alleles segregate during gamete formation; fertilization creates pairs of alleles once again.

5. An imaginary creature called a glump has blue skin, controlled by the allele *B*, which is dominant to allele *b*. Glumps of genotype *bb* are albino (white). The independently inherited characteristic of eye color is specified by alleles *E* and *e*. Glumps of genotype *EE* have orange eyes; glumps that are homozygous recessives for the eye color gene have pink eyes.
   a. If a homozygous blue-skinned glump mates with an albino glump, what will be the phenotypes of their offspring?
   b. You use a testcross to determine the genotype of an orange-eyed glump. Half of the offspring of the testcross have orange eyes and half have pink eyes. What is the genotype of the orange-eyed parent?
   c. A glump of genotype *BbEe* is crossed with one that is *bbEe*. What is the probability of their producing an albino baby glump with pink eyes?

6. A man and a woman who are both carriers of cystic fibrosis allele *c*, which is recessive to the normal allele *C*, have had three children without cystic fibrosis. If the couple has a fourth child, what is the probability that the child will have the disorder?

7. Peter is a 28-year-old man whose father died of Huntington disease. Neither Peter's much older sister, who is 44, nor his 60-year-old mother shows any signs of the disease. What is the probability that Peter has inherited Huntington disease?

Answers: 1. A trait that is the prevailing one in nature **2.** An egg cell in the plant's carpel is fertilized by a sperm-carrying pollen grain from a stamen of the same plant. **3.** One could be homozygous for the dominant allele, while the other is heterozygous. **4.** Statement **a** is the principle of independent assortment. Statement **b** is the principle of segregation. **5. a.** All will have blue skin (*BB* × *bb* → *Bb*). **b.** *Ee* **c.** $\frac{1}{8}$ (that is, $\frac{1}{2} \times \frac{1}{4}$) **6.** $\frac{1}{4}$ (The genotypes and phenotypes of their other children are irrelevant.) **7.** $\frac{1}{2}$

Beyond Mendel

Mendel's two principles explain inheritance in terms of discrete factors—genes—that are passed along from generation to generation according to simple rules of probability. Mendel's principles are valid for all sexually reproducing organisms, including garden peas, Labrador retrievers, and human beings. But just as the basic rules of musical harmony cannot account for all the rich sounds of a symphony, Mendel's principles stop short of explaining some patterns of genetic inheritance. In particular, they do not explain the many cases of inherited characteristics that exist in more than two clear-cut variants—such as flower color that can be red, white, or pink, or human skin color in all its range of shades. In fact, for most sexually reproducing organisms, cases where Mendel's rules can strictly account for the patterns of inheritance are relatively rare. More often, the observed inheritance patterns are more complex. We will now add extensions to Mendel's principles that account for this complexity.

Incomplete Dominance in Plants and People

The F_1 offspring of Mendel's crosses always looked like one of the two parent plants because the dominant alleles had the same effect in one or two copies. But for some characteristics, the F_1 hybrids have an appearance in between the phenotypes of the two parents, an effect called **incomplete dominance.** For instance, when red snapdragons are crossed with white snapdragons, all the F_1 hybrids have pink flowers (**Figure 9.18**). And in the F_2 generation, the genotypic ratio and the phenotypic ratio are the same, 1:2:1.

Activity 9D on the Web & CD Investigate incomplete dominance in MendAliens.

We also see examples of incomplete dominance in humans. One case involves the recessive allele (h) responsible for **hypercholesterolemia,** a condition characterized by dangerously high levels of cholesterol in the blood. Normal individuals are HH. Heterozygotes (Hh; about one in 500 people) have blood cholesterol levels about twice normal. They are unusually prone to atherosclerosis, the blockage of arteries by cholesterol buildup in artery walls, and they may have heart attacks from blocked heart arteries by their mid-30s. Hypercholesterolemia is even more serious in homozygous individuals (hh; about one in a million people). These homozygotes have about five times the normal amount of blood cholesterol and may have heart attacks as early as age 2.

If we look at the molecular basis for hypercholesterolemia, we can understand the intermediate phenotype of heterozygotes (**Figure 9.19**). The H allele specifies a cell surface receptor protein that certain cells use to mop up excess cholesterol from the blood. With only half as many receptors as HH individuals, heterozygotes can remove much less excess cholesterol.

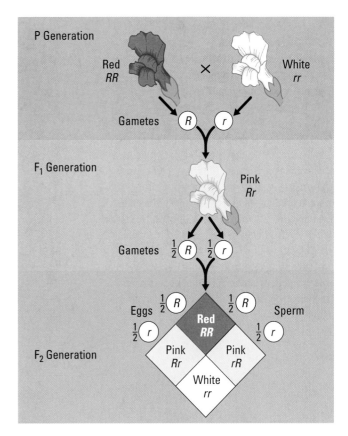

Figure 9.18 Incomplete dominance in snapdragons. In incomplete dominance, a heterozygote has a phenotype in between the phenotypes of the two kinds of homozygotes. The phenotypic ratios in the F_2 generation are the same as the genotypic ratios, 1:2:1. Compare this diagram with Figure 9.8, where one of the alleles displays complete dominance.

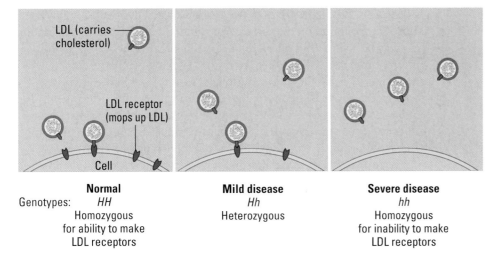

Figure 9.19 Incomplete dominance in human hypercholesterolemia. The dominant allele, which normal individuals carry in duplicate (HH), specifies a cell surface protein called an LDL receptor. LDLs, or low-density lipoproteins, are cholesterol-containing particles in the blood. The LDL receptors pick up LDL particles from the blood and promote their uptake by cells that break down the cholesterol (see Figure 5.18). This process helps prevent the accumulation of cholesterol in the arteries. Heterozygotes (Hh) have only half the normal number of LDL receptors, and homozygotes (hh) have none, thereby allowing dangerous levels of LDL to build up in the blood.

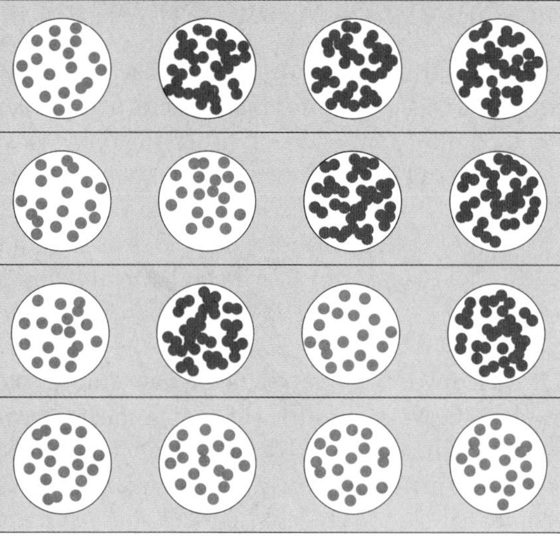

Figure 9.20 Multiple alleles for the ABO blood groups. For the gene that controls a person's ABO blood type, three alleles exist in the human population: I^A, I^B, and i. Because each person carries two alleles, six genotypes are possible. Whenever the I^A or I^B allele is present, the corresponding carbohydrate (A or B, respectively) is present on the surface of red blood cells. Both of these alleles, which are codominant, are dominant to the i allele, which does not specify any surface carbohydrate. A person produces antibodies against foreign blood carbohydrates. When these antibodies meet up with red blood cells carrying the corresponding carbohydrate—anti-A with A, for example—the cells clump. This is the basis of blood-typing and of the adverse reaction that occurs when someone receives a transfusion of incompatible blood.

Blood Group (Phenotype)	Genotypes	Antibodies Present in Blood	Reaction When Blood from Groups Below Is Mixed with Antibodies from Groups at Left			
			O	A	B	AB
O	ii	Anti-A Anti-B				
A	$I^A I^A$ or $I^A i$	Anti-B				
B	$I^B I^B$ or $I^B i$	Anti-A				
AB	$I^A I^B$	—				

Multiple Alleles and Blood Type

So far, we have discussed inheritance patterns involving only two alleles per gene. But many genes have multiple alleles. Although each individual carries, at most, two different alleles for a particular gene, in cases of multiple alleles, more than two possible alleles exist in the population. The **ABO blood groups** in humans are an example of multiple alleles. There are three common alleles for the characteristic of ABO blood type, which in various combinations produce four phenotypes. A person's blood type may be O, A, B, or AB. These letters refer to two carbohydrates, designated A and B, which may be found on the surface of red blood cells. A person's red blood cells may be coated with one substance or the other (type A or B), with both (type AB), or with neither (type O). Matching compatible blood groups is critical for blood transfusions. If a donor's blood cells have a carbohydrate (A or B) that is foreign to the recipient, then proteins called antibodies in the recipient's blood bind specifically to the foreign carbohydrates and cause the donated blood cells to clump together. This clumping can kill the recipient. It is also the basis of blood-typing (**Figure 9.20**).

The four blood types result from various combinations of the three different alleles, symbolized as I^A (for the ability to make substance A), I^B (for B), and i (for neither A nor B). Each person inherits one of these alleles from each parent. Because there are three alleles, there are six possible genotypes, as listed in the figure. Both the I^A and I^B alleles are dominant to the i allele. Thus, $I^A I^A$ and $I^A i$ people have type A blood, and $I^B I^B$ and $I^B i$ people have type B. Recessive homozygotes (ii) have type O blood; they make neither the A nor the B substance. The I^A and I^B alleles exhibit **codominance,** meaning that both alleles are expressed in heterozygous individuals ($I^A I^B$), who have type AB blood. Codominance (the expression of both alleles) is different from incomplete dominance (the expression of one intermediate trait).

Pleiotropy and Sickle-Cell Disease

Most of our genetic examples to this point have been cases in which each gene specified only one hereditary characteristic. But in many cases, one gene influences several characteristics. The impact of a single gene on more than one characteristic is called **pleiotropy.**

An example of pleiotropy in humans is **sickle-cell disease,** a disorder characterized by the diverse symptoms shown in **Figure 9.21.** All of these possible phenotypic effects result from the action of a single kind of allele when it is present on both homologous chromosomes. The direct effect of the sickle-cell allele is to make red blood cells produce abnormal hemoglobin molecules. As mentioned in Chapter 3, these abnormal molecules tend to link together and crystallize, especially when the oxygen content of the blood is lower than usual because of high altitude, overexertion, or respiratory ailments. As the hemoglobin crystallizes, the normally disk-shaped red blood cells deform to a sickle shape with jagged edges. Sickling of the cells, in turn, can lead to the cascade of symptoms in Figure 9.21. Blood transfusions and certain drugs may relieve some of the symptoms, but there is no cure, and sickle-cell disease kills about 100,000 people in the world annually.

In most cases, only people who are homozygous for the sickle-cell allele suffer from the disease. Heterozygotes, who have one sickle-cell allele and one nonsickle allele, are usually healthy, although in rare cases they may experience some pleiotropic effects when oxygen in the blood is severely reduced, such as at very high altitudes. These effects may occur because the nonsickle and sickle-cell alleles are actually codominant: Both alleles are expressed in heterozygous individuals, and their red blood cells contain both normal and abnormal hemoglobin. A simple blood test can identify homozygotes and heterozygotes.

Sickle-cell disease is by far the most common inherited illness among black people, striking one in 500 African American children born in the United States. About one in ten African Americans is a heterozygote. Among Americans of other ancestries, the sickle-cell allele is very rare.

One in ten is an unusually high frequency of heterozygotes for an allele with such harmful effects in homozygotes. We might expect that the frequency of the sickle-cell allele in the population would be much lower because many homozygotes die before passing their genes to the next generation. The high frequency appears to be a vestige of the roots of African Americans. Sickle-cell disease is most common in tropical Africa, where the deadly disease malaria is also prevalent. The microorganism that causes malaria spends part of its life cycle inside red blood cells. When it enters the red blood cells of a person with the sickle-cell allele, it triggers sickling. The body destroys most of the sickled cells, and the microbe does not grow well in the cells that remain. Consequently, in many parts of Africa, sickle-cell carriers (heterozygotes) who contract malaria live longer and have more offspring than noncarriers with malaria. In this way, malaria has kept the frequency of the sickle-cell allele relatively high in much of the African continent. To put it in evolutionary terms, as long as malaria is a danger,

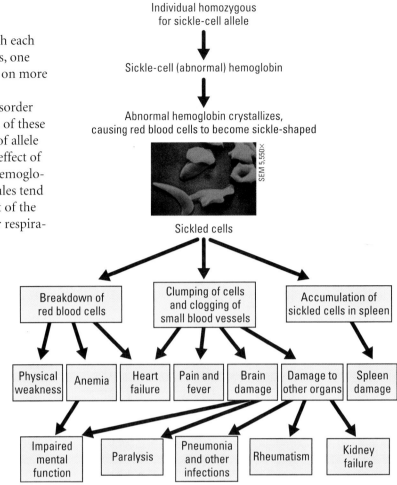

Figure 9.21 Sickle-cell disease, multiple effects of a single human gene. People who are homozygous for the sickle-cell allele have sickle-cell disease. They produce abnormal hemoglobin molecules, which cause their red blood cells, normally smooth disks, to become sickle-shaped. Sickled cells are destroyed rapidly by the body, and their destruction may cause anemia and general weakening of the body (hence the alternative name for the disorder, sickle-cell anemia). Also, because of their angular shape, sickled cells do not flow smoothly in the blood and tend to accumulate and clog tiny blood vessels. Blood flow to body parts is reduced, resulting in periodic fever, pain, and damage to various organs, including the heart, brain, and kidneys. Sickled cells accumulate in the spleen, damaging that organ.

individuals with the sickle-cell allele have a selective advantage. (You'll learn more about this topic in Chapters 10 and 13.)

Polygenic Inheritance

Mendel studied genetic characteristics that could be classified on an either-or basis, such as purple or white flower color. However, many characteristics, such as human skin color and height, vary along a continuum in a population. Many such features result from **polygenic inheritance,** the additive effects of two or more genes on a single phenotypic characteristic. (This is the converse of pleiotropy, in which a single gene affects several characteristics.)

Let's consider a hypothetical example. Assume that skin pigmentation in humans is controlled by three genes that are inherited separately, like Mendel's pea genes. The dark-skin allele for each gene (*A*, *B*, and *C*) contributes one "unit" of darkness to the phenotype and is incompletely dominant to the other alleles (*a*, *b*, and *c*). A person who is *AABBCC* would be very dark, while an *aabbcc* individual would be very light. An *AaBbCc* person would have skin of an intermediate shade. Because the alleles have an additive effect, the genotype *AaBbCc* would produce the same skin color as any other genotype with just three dark-skin alleles, such as *AABbcc*. **Figure 9.22** shows how inheritance of these three genes could lead to a range of skin color in a population. Seven levels of pigmentation would arise at the frequencies indicated by the bars in the graph.

The Role of Environment

If we examine a real human population for the skin color phenotype, we would see more shades than just seven. The true range might be similar to the entire spectrum of color under the bell-shaped curve in Figure 9.22. In fact, no matter how carefully we characterize the genes for skin color, a purely genetic description will always be incomplete. This is because some intermediate shades of skin color result from the effects of environmental factors, such as exposure to the sun.

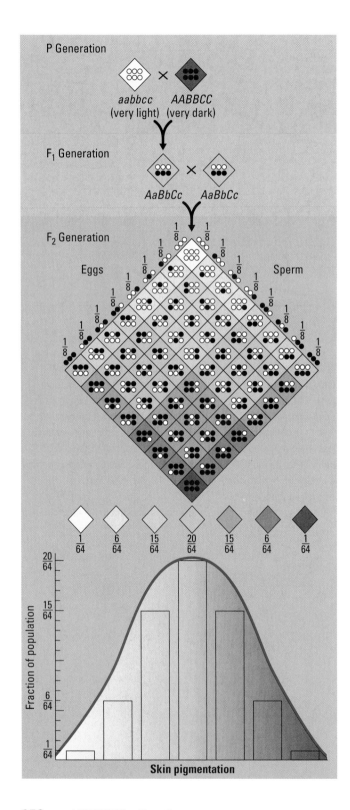

Figure 9.22 A model for polygenic inheritance of skin color.
According to this model, three separately inherited genes affect the darkness of skin. For each gene, there is an allele for dark skin (*A*, *B*, or *C*) that is incompletely dominant to an allele for light skin (*a*, *b*, or *c*). Each dominant allele brings one "unit" of skin pigmentation. An individual who is heterozygous for all three genes (*AaBbCc*), or has some other combination of three dominant alleles and three recessive alleles, has three units of darkness, indicated in the figure by three black dots in the diamonds. Thus, diamonds representing the triply heterozygous F₁ individuals have three black dots. The Punnett square shows all possible genotypes of F₂ offspring. The row of diamonds below the Punnett square shows the seven skin pigmentation phenotypes that would theoretically result. The seven bars in the graph at the bottom of the figure depict the relative numbers of each of the phenotypes in the F₂ generation. The bell-shaped curve indicates the distribution of an even greater variety of skin shades in the population that might result from the combination of heredity and environmental effects, such as sun-tanning.

Many human characteristics result from a combination of heredity and environment. Some characteristics, such as eye color, appear to be almost entirely genetically determined. Other characteristics, such as height, clearly have a large environmental component—health and diet during adolescence, for example. And some features, such as taste in music, nail length, and body piercings, probably have no genetic component and derive entirely from environmental influences. As geneticists learn more and more about our genes, it's becoming clear that many human characteristics—such as a person's susceptibility to heart disease, cancer, alcoholism, and schizophrenia—are influenced by both genes and environment.

Simply spending time with identical twins will convince anyone that environment, and not just genes, affects a person's traits. However, it is important to realize that although many individual features of organisms arise from a combination of genetic and environmental factors, only genetic influences are heritable. Any effects of the environment are not passed on to the next generation through biological means.

CHECKPOINT

1. Why is a testcross unnecessary to determine whether a snapdragon with red flowers is homozygous or heterozygous?

2. Maria has type O blood, and her sister has type AB blood. What are the genotypes of the girls' parents?

3. How does sickle-cell disease exemplify the concept of pleiotropy?

4. Based on the skin tone model in Figure 9.22, put the following individuals in order by skin tone, from lightest to darkest: *AAbbCC, aaBBcc, AabBCc, Aabbcc, AaBBCC.*

Answers: 1. Only plants homozygous for the dominant allele have red flowers; heterozygotes have pink flowers. **2.** One parent is $I^A i$, and the other parent is $I^B i$. **3.** Homozygotes for the sickle-cell allele have abnormal hemoglobin, and its effect on the shape of red blood cells leads to a cascade of symptoms affecting many organs of the body. **4.** *Aabbcc, aaBBcc, AabBCc, AAbbCC, AaBBCC*

The Chromosomal Basis of Inheritance

Mendel published his results in 1866, but not until long after he died did biologists understand the significance of his work. Cell biologists worked out the processes of mitosis and meiosis in the late 1800s (see Chapter 8 to review these processes). Then, around 1900, researchers began to notice parallels between the behavior of chromosomes and the behavior of Mendel's heritable factors. One of biology's most important concepts—the chromosome theory of inheritance—began to emerge.

The **chromosome theory of inheritance** states that genes are located on chromosomes and that the behavior of chromosomes during meiosis and fertilization accounts for inheritance patterns. Indeed, it is chromosomes that undergo segregation and independent assortment during meiosis and thus account for Mendel's principles. You can see the chromosomal basis of Mendel's principles by following the fates of two genes during

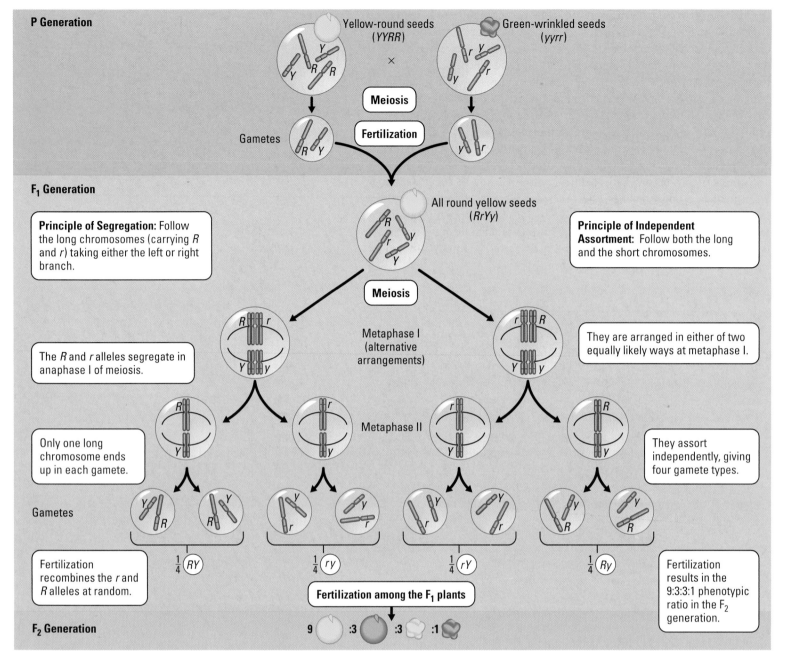

P Generation

Yellow-round seeds (*YYRR*) × Green-wrinkled seeds (*yyrr*)

Meiosis

Gametes — Fertilization

F₁ Generation

All round yellow seeds (*RrYy*)

Principle of Segregation: Follow the long chromosomes (carrying *R* and *r*) taking either the left or right branch.

Principle of Independent Assortment: Follow both the long and the short chromosomes.

Meiosis

Metaphase I (alternative arrangements)

The *R* and *r* alleles segregate in anaphase I of meiosis.

They are arranged in either of two equally likely ways at metaphase I.

Metaphase II

Only one long chromosome ends up in each gamete.

They assort independently, giving four gamete types.

Gametes

$\frac{1}{4}$ *RY* $\frac{1}{4}$ *rY* $\frac{1}{4}$ *rY* $\frac{1}{4}$ *Ry*

Fertilization recombines the *r* and *R* alleles at random.

Fertilization results in the 9:3:3:1 phenotypic ratio in the F₂ generation.

Fertilization among the F₁ plants

F₂ Generation   9   :3   :3   :1

Figure 9.23 The chromosomal basis of Mendel's principles.   Here we correlate the results of one of Mendel's dihybrid crosses (see Figure 9.10) with the behavior of chromosomes. Starting with two true-breeding parental plants, the diagram follows two genes through the F₁ and F₂ generations. The two genes specify seed shape (allelles *R* and *r*) and seed color (alleles *Y* and *y*) and are on different chromosomes.

meiosis and fertilization in pea plants (**Figure 9.23**). Recasting Mendel's principles in terms of the interactions of chromosomes provides us with a greater understanding of the underlying principles of genetics and shows us how Mendel's work can be extended in several important ways.

Gene Linkage

In 1908, British biologists William Bateson and Reginald Punnett (originator of the Punnett square) discovered an inheritance pattern that seemed totally inconsistent with Mendelian principles. Bateson and Punnett were working with two characteristics in sweet-pea plants, flower color and

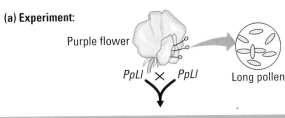

(a) Experiment:

Purple flower

$PpLl$ × $PpLl$ Long pollen

Phenotypes	Observed offspring	Prediction (9:3:3:1)
Purple long	284	215
Purple round	21	71
Red long	21	71
Red round	55	24

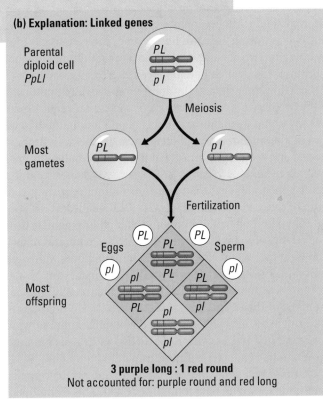

(b) Explanation: Linked genes

Parental diploid cell $PpLl$

Meiosis

Most gametes

Fertilization

Eggs   Sperm

Most offspring

3 purple long : 1 red round
Not accounted for: purple round and red long

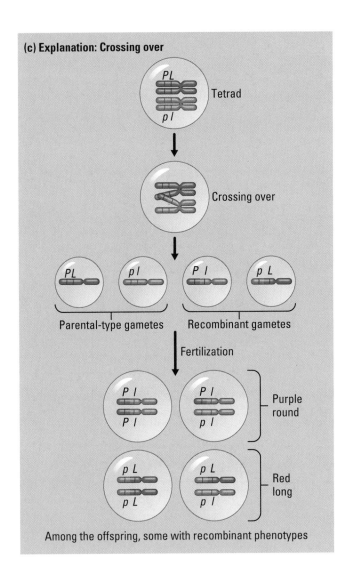

(c) Explanation: Crossing over

Tetrad

Crossing over

Parental-type gametes   Recombinant gametes

Fertilization

Purple round

Red long

Among the offspring, some with recombinant phenotypes

Figure 9.24 Linked genes and crossing over. **(a)** In a famous experiment, Bateson and Punnett looked at the offspring of sweet-pea plants heterozygous for flower color (Pp) and pollen shape (Ll). The parent plants had purple flowers (expression of the dominant P allele) and long pollen grains (expression of the L allele). The corresponding recessive traits are red flowers (in pp plants) and round pollen (in ll plants). When the researchers tabulated the phenotypes of their 381 new plants, the results were consistent with Mendel's principle of segregation but not with Mendel's principle of assortment. They did not see the 9:3:3:1 ratio predicted for a dihybrid cross. Instead, as shown in the table, they found a disproportionately large number of plants with either purple flowers and long pollen (284 of 381) or red flowers and round pollen (55 of 381). **(b)** The disproportionately large numbers of two of the phenotypes among the offspring are explained by gene linkage, the fact that the gene loci for the two characteristics are on the same chromosome and tend to remain together during meiosis and fertilization. In the starting plants in this experiment, alleles P and L were together on one chromosome and p and l on its homologue. **(c)** The small numbers of the other two offspring phenotypes are explained by crossing over during gamete formation. Some of the gametes end up with recombinant chromosomes, carrying new combinations of alleles, either P and l or p and L. When the recombinant gametes participate in fertilization, recombinant offspring can result.

pollen shape. In an experiment that at first seemed just like one of Mendel's dihybrid crosses, they bred plants that were heterozygous for both characteristics. These plants exhibited the dominant traits of purple flowers and long pollen grains, as expected. But when the biologists crossed these (F_1) heterozygotes, the offspring (F_2) plants gave them a surprise. The plants did not show the 9:3:3:1 ratio predicted for a dihybrid cross. Instead, there were too many plants in two of the phenotypic categories and too few plants in the other two categories (**Figure 9.24a**).

These results were explained several years later, when other studies revealed that the sweet-pea genes for flower color and pollen shape are on the same chromosome (**Figure 9.24b**). Alleles that start out together on the same chromosome—such as P with L and p with l—would be expected to travel together during meiosis and fertilization. In general, genes that are located close together on a chromosome, called **linked genes,** do tend to be inherited together. They do not follow Mendel's principle of independent assortment. In the Bateson-Punnett experiment, meiosis in the heterozygous sweet-pea plant yielded mostly two genotypes of gametes (PL and pl), rather

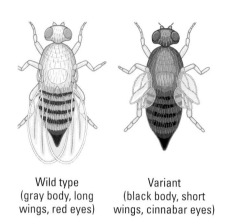

Wild type
(gray body, long
wings, red eyes)

Variant
(black body, short
wings, cinnabar eyes)

Figure 9.25 The fruit fly *Drosophila melanogaster*.

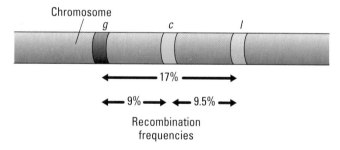

Figure 9.26 Using crossover data to map genes.
This diagram represents a part of the *Drosophila* chromosome that carries three linked genes. Their recessive alleles specify black body (*g*), cinnabar eyes (*c*), and short wings (*l*). The corresponding dominant alleles specify the wild-type traits of gray body, red eyes, and long wings. Under the chromosome are the actual recombination frequencies (crossover frequencies) between these genes, taken two at a time: 17% between *g* and *l*, 9% between *g* and *c*, and 9.5% between *l* and *c*. Geneticists reasoned that these values represent the relative distances between the genes. Because the recombination frequencies between *g* and *c* and between *l* and *c* are approximately half that between *g* and *l*, gene *c* must lie roughly midway between *g* and *l*. Thus, the sequence of these genes on their chromosome must be *g-c-l* (or the equivalent *l-c-g*).

than equal numbers of the four types of gametes that would result if the flower color and pollen shape genes were not linked. The large numbers of plants with purple long and red round traits in the experiment resulted from fertilization among these two types of gametes.

Genetic Recombination: Crossing Over and Linkage Maps

But what about the smaller numbers of plants with purple round and red long traits? Crossing over accounts for these offspring. In Chapter 8, we saw that during meiosis, crossing over between homologous chromosomes shuffles chromosome segments in the haploid daughter cells (see Figure 8.18), thereby producing new combinations of alleles. Figure 9.24c reviews crossing over, showing how two linked genes can give rise to four different gamete genotypes. Two of the gamete genotypes reflect the presence of parental-type chromosomes, which have not been altered by crossing over. In contrast, the other two gamete genotypes are recombinant. The chromosomes of these gametes carry new combinations of alleles that result from the exchange of chromosome segments in crossing over. Thus, in the Bateson-Punnett experiment, the small fraction of offspring plants with recombinant (purple round and red long) phenotypes must have resulted from fertilization involving recombinant gametes. The percentage of recombinant offspring among the total is called the **recombination frequency.** In this case, it is the sum of the recombinant plants, 21 + 21, divided by 381, or 11%.

The discovery of how crossing over adds to gamete diversity confirmed the relationship between chromosome behavior and inheritance. Some of the most important early studies of crossing over were performed in the laboratory of American embryologist Thomas Hunt Morgan in the early 1900s. Morgan and his colleagues used the fruit fly *Drosophila melanogaster* in many of their experiments (**Figure 9.25**). Often seen flying around overripe fruit, *Drosophila* is a good research animal for studies of inheritance because it is easily and inexpensively grown, and can produce several generations in a matter of months.

Using *Drosophila,* one of Morgan's students developed a method for analyzing crossover data to map gene loci. This technique is based on the assumption that the farther apart two genes are on a chromosome, the higher the probability that a crossover will occur between them. In other words, the greater the distance between two genes, the more points there are between them where crossing over can occur. (This assumption is not entirely accurate, but it is good enough to provide useful data.) With this principle in mind, geneticists can use recombination data to assign genes to relative positions on chromosomes—that is, to map genes. **Figure 9.26** shows some of the data used to map three genes that reside on one of the *Drosophila* chromosomes. The result is called a **linkage map.**

The linkage mapping method has proved extremely valuable in establishing the relative positions of many genes in many organisms. The real beauty of the technique is that a wealth of information about genes can be learned simply by breeding and observing the organisms; no fancy equipment is required. Today, with DNA technology, geneticists can determine the physical distances in nucleotides between linked genes. However, linkage mapping still played a critical role in the mapping of the human genome, which we'll discuss in Chapter 12.

Activity 9E on the Web & CD
Explore linkage and crossing over in MendAliens.

Sex Chromosomes and Sex-Linked Genes

The patterns of inheritance we've discussed so far have involved only genes located on autosomes, not on the sex chromosomes. We're now ready to look at the role of sex chromosomes in humans and fruit flies and the inheritance patterns exhibited by the characteristics they control.

Sex Determination in Humans and Fruit Flies

A number of organisms, including fruit flies and all mammals, have a pair of **sex chromosomes,** designated X and Y, that determine an individual's sex. **Figure 9.27** reviews what you learned in Chapter 8 about sex determination in humans. Individuals with one X chromosome and one Y chromosome are males; XX individuals are females. Human males and females both have 44 autosomes (chromosomes other than sex chromosomes). As a result of chromosome segregation during meiosis, each gamete contains one sex chromosome and a haploid set of autosomes (22 in humans). All eggs contain a single X chromosome. Of the sperm cells, half contain an X chromosome and half contain a Y chromosome. An offspring's sex depends on whether the sperm cell that fertilizes the egg bears an X or a Y. The same is true for *Drosophila*.

The genetic basis of sex determination in humans is not yet completely understood, but one gene on the Y chromosome plays a crucial role. This gene, called *SRY*, triggers the development of the testes. In the absence of a functioning version of *SRY*, an individual develops ovaries rather than testes. Other genes on the Y chromosome are also necessary for normal sperm production. The X-Y system in other mammals is similar to that in humans. In the fruit fly's X-Y system, some genetic details are different, although a Y chromosome is still essential for sperm formation.

Sex-Linked Genes

Besides bearing genes that determine sex, the so-called sex chromosomes also contain genes for characteristics unrelated to maleness or femaleness. Any gene located on a sex chromosome is called a **sex-linked gene.** (Note that the use of "linked" here is somewhat different from its use in the term *linked genes*.) The X chromosome contains many more genes than the Y; therefore, most sex-linked genes unrelated to sex determination are found on the X chromosome.

Activity 9F on the Web & CD
Study sex-linkage in MendAliens.

Sex linkage was discovered by T. H. Morgan, while he was studying the inheritance of white eye color in fruit flies. Wild-type fruit flies have red

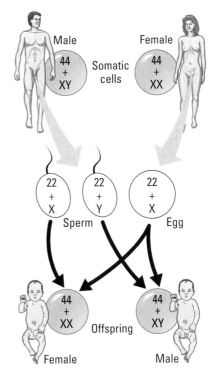

Figure 9.27 Sex determination in humans: a review.

Figure 9.28 Fruit fly eye color.

(a) The wild-type color for fruit fly eyes is red.
(b) White eye color is a rare variant. Study of the inheritance patterns of the white-eye trait led to the conclusion that the gene controlling it was carried on the fly's X chromosome. Such a gene and trait are said to be sex-linked.

(a)

(b)

eyes; white eyes are very rare (**Figure 9.28**). White eye color turned out to be a recessive trait whose gene is on the fly's X chromosome; the gene and trait are said to be X-linked (or sex-linked).

Figure 9.29a shows what happens when a white-eyed male fly is mated with a homozygous red-eyed female. All the offspring have red eyes, suggesting that the wild type is dominant. When those offspring are bred to each other, the classical 3:1 phenotypic ratio of wild-type to white-eyed appears among the offspring (**Figure 9.29b**). However, there is a surprising twist: The white-eye trait shows up only in males. All the females have red eyes, while half the males have red eyes and half have white eyes. This is

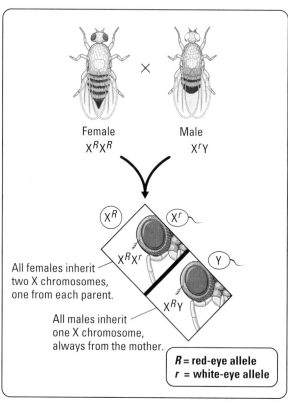

All females inherit two X chromosomes, one from each parent.

All males inherit one X chromosome, always from the mother.

R = red-eye allele
r = white-eye allele

(a) Homozygous red-eyed female × white-eyed male

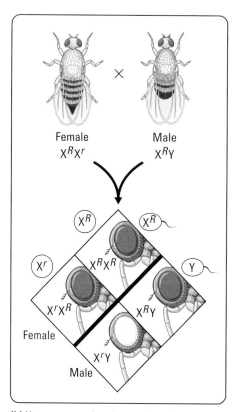

(b) Heterozygous female × red-eyed male

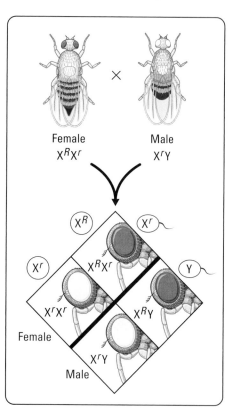

(c) Heterozygous female × white-eyed male

Figure 9.29 The inheritance of white eye color, a sex-linked trait.

We use the uppercase letter *R* for the dominant, wild-type, red-eye allele and *r* for the recessive, white-eye allele. To indicate that these alleles are on the X chromosome, we show them as superscripts to the letter X. Thus, red-eyed male fruit flies have the genotype $X^R Y$, and white-eyed males are $X^r Y$. The Y chromosome does not have a gene locus for eye color; therefore, the male's phenotype results entirely from his single X-linked gene. In the female, $X^R X^R$ and $X^R X^r$ flies have red eyes, and $X^r X^r$ flies have white eyes.

because the gene involved in this inheritance pattern is located exclusively on the X chromosome; there is no corresponding eye color locus on the Y. Thus,

females (XX) carry two copies of the gene for this characteristic, while males (XY) carry only one. Because the white-eye allele is recessive, a female will have white eyes only if she receives that allele on both X chromosomes (**Figure 9.29c**). For a male, however, a single copy of the white-eye allele confers white eyes. Since a male has only one X chromosome, there can be no wild-type allele present to offset the recessive allele.

Sex-Linked Disorders in Humans

A number of human conditions, including red-green color blindness, hemophilia, and a type of muscular dystrophy, result from sex-linked (X-linked) recessive alleles that are inherited in the same way as the white-eye allele in fruit flies. The fruit fly model also shows us why recessive sex-linked traits are expressed much more frequently in men than in women. Like a male fruit fly, if a man inherits only one sex-linked recessive allele—from his mother—the allele will be expressed. In contrast, a woman has to inherit two such alleles—one from each parent—to exhibit the trait.

Red-green color blindness is a common sex-linked disorder characterized by a malfunction of light-sensitive cells in the eyes. It is actually a class of disorders involving several X-linked genes. A person with normal color vision can see more than 150 colors. In contrast, someone with red-green color blindness can see fewer than 25. For some affected people, red hues appear gray; others see gray instead of green; still others are green-weak or red-weak, tending to confuse shades of these colors. Mostly males are affected, but heterozygous females have some defects. (If you have red-green color blindness, you probably cannot see the numeral in **Figure 9.30**.)

Hemophilia is a sex-linked recessive trait with a long, well-documented history. Hemophiliacs bleed excessively when injured because they have inherited an abnormal allele for a factor involved in blood clotting. The most seriously affected individuals may bleed to death after relatively minor bruises or cuts. A high incidence of hemophilia has plagued the royal families of Europe. The first royal hemophiliac seems to have been a son of Queen Victoria (1819–1901) of England. It is likely that the hemophilia allele arose through a mutation in one of the gametes of Victoria's mother or father, making Victoria a carrier of the deadly allele. Hemophilia was eventually introduced into the royal families of Prussia, Russia, and Spain through the marriages of two of Victoria's daughters who were carriers. In this way, the age-old practice of strengthening international alliances by marriage effectively spread hemophilia through the royal families of several nations (**Figure 9.31**).

Another sex-linked recessive disorder is **Duchenne muscular dystrophy,** a condition characterized by a progressive weakening and loss of muscle tissue. Almost all cases are males, and the first symptoms appear in early childhood, when the child begins to have difficulty standing up. He is inevitably wheelchair-bound by age 12. Eventually, he becomes severely weakened, and normal breathing becomes difficult. Death usually occurs by age 20. For such a severe disease, Duchenne muscular dystrophy is relatively common. In the general U.S. population, about one in 3,500 male babies is affected, and the disease is even more common in some inbred populations. In one Amish community in Indiana, for instance, one out of every 100 males is born with the disease. With the help of DNA technology (discussed in Chapter 12), the gene whose defectiveness causes Duchenne muscular dystrophy has been mapped at a particular point on the X chromosome. The gene's wild-type allele codes for a protein that is present in normal muscle but missing in Duchenne patients.

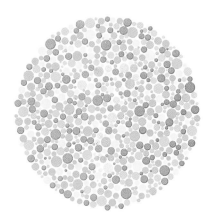

Figure 9.30 A test for red-green color blindness. Can you see a green numeral 7 against the reddish background? If not, you probably have some form of red-green color blindness, an X-linked trait.

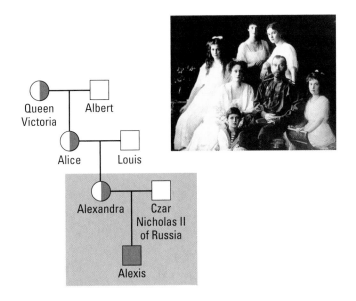

Figure 9.31 Hemophilia in the royal family of Russia. The photograph shows Queen Victoria's granddaughter Alexandra, her husband Nicholas, who was the last czar of Russia, their son Alexis, and their daughters. The pedigree uses half-colored symbols to represent heterozygous carriers of the hemophilia allele. As you can see in the pedigree, Alexandra, like her mother and grandmother, was a carrier, and Alexis had the disease.

Evolution Connection

The Telltale Y Chromosome

The Y chromosome of human males is only about one-third the size of the X chromosome and carries only $\frac{1}{100}$ as many genes. As mentioned earlier, most of the Y genes seem to function in maleness and male fertility and are not present on the X. In prophase I of meiosis, only two tiny regions of the X and Y chromosomes can cross over (recombine). Crossing over requires that the DNA in the recombining regions line up and match very closely, and for the human X and Y chromosomes, this can only happen at their tips.

Nevertheless, biologists believe that X and Y were once a fully homologous pair, having evolved from a pair of autosomes about 300 million years ago. Since that time, four major episodes of change, the most recent about 40 million years ago, have rearranged pieces of the Y chromosome in a way that prevents the matching required for recombination with the X. Over millions of years, many Y genes have disappeared, shrinking the chromosome. Meanwhile, the lack of substantial exchange between X and Y chromosomes has prevented male-determining genes from migrating to the X chromosome—which could have had the disastrous consequence of making XX individuals male.

Because most of the DNA of the Y chromosome passes more or less intact from father to son—the main changes are rare mutations—the Y chromosome provides a window to the ancestry of the male lines of humanity. Recently, researchers used comparisons of Y DNA to confirm the claim by an Lemba people of southern Africa, that they are descended from ancient Jews (**Figure 9.32**). Certain Y DNA sequences had previously been shown to be distinctive of Jews, particularly of the priestly caste called Cohanim (descendants of Moses' brother Aaron, according to the Bible). These same sequences were found at equally high frequencies among the Lemba. And it was a study of Y DNA that supported the oral tradition among the descendants of the slave Sally Hemings that Thomas Jefferson was their ancestor. On a grander scale, Y chromosome studies are providing evidence bearing on the question of when and where fully modern humans first evolved.

Evolution Connection on the Web
Learn more about how the Y chromosome is used to study ancestry.

Figure 9.32 A Lemba man.
DNA sequences from the Y chromosomes of Lemba men suggest that the Lemba are descended from ancient Jews.

Chapter Review

Summary of Key Concepts

For study help, go to the Essential Biology Website (www.essentialbiology.com) or CD-ROM to explore the Activities and Case Studies in the Process of Science.

Heritable Variation and Patterns of Inheritance

- For traits that vary within a population, the one most commonly found in nature is called the wild type. Breeders of budgies and other animals can predict the traits of offspring by using the genetic principles discovered by Gregor Mendel. Gregor Mendel was the first to analyze patterns of inheritance in a scientific manner. In his 1866 paper, he emphasized that heritable factors (genes) retain permanent identities.

- **In an Abbey Garden** Mendel started with true-breeding varieties of pea plants representing two alternative variants of hereditary characteristics, such as flower color. He then crossed the different varieties and traced the inheritance of traits from generation to generation.

- **Mendel's Principle of Segregation** Pairs of alleles separate during gamete formation, and fertilization restores the pairs.

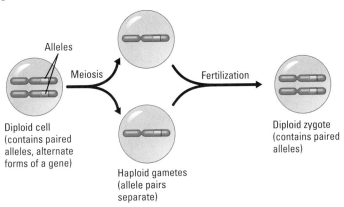

If an individual's genotype (genetic makeup) has two different alleles for a gene and only one influences the organism's phenotype (appearance), that allele is said to be dominant and the other allele recessive. Alleles of a gene reside at the same locus, or position, on homologous chromosomes. Where the allele pair match, the organism is homozygous; where they're different, the organism is heterozygous.

Activity 9A Monohybrid Cross

- **Mendel's Principle of Independent Assortment** By following two characteristics at once, Mendel found that the alleles of a pair segregate independently of other allele pairs during gamete formation.

Activity 9B Dihybrid Cross

Activity 9C Gregor's Garden

- **Using a Testcross to Determine an Unknown Genotype**

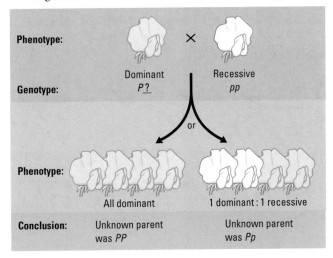

- **The Rules of Probability** Inheritance follows the rules of probability. The chance of inheriting a recessive allele from a heterozygous parent is $\frac{1}{2}$. The chance of inheriting it from both of two heterozygous parents is $\frac{1}{2} \times \frac{1}{2} = \frac{1}{4}$, illustrating the rule of multiplication for calculating the probability of two independent events.

- **Family Pedigrees** The inheritance of many human traits, from freckles to genetic diseases, follows Mendel's principles and the rules of probability. Geneticists use family pedigrees to determine patterns of inheritance and individual genotypes among humans.

- **Human Disorders Controlled by a Single Gene** Many inherited disorders in humans are controlled by a single gene (represented by two alleles). Most such disorders, such as cystic fibrosis, are caused by autosomal recessive alleles. A few, such as Huntington disease, are caused by dominant alleles.

Beyond Mendel

- **Incomplete Dominance in Plants and People**

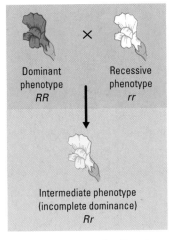

Activity 9D Incomplete Dominance

- **Multiple Alleles and Blood Type**   Within a population, there are often multiple kinds of alleles for a characteristic, such as the three alleles for the ABO blood groups. The alleles determining the A and B blood factors are codominant; that is, both are expressed in a heterozygote.

- **Pleiotropy and Sickle-Cell Disease**

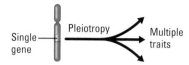

The presence of two copies of the sickle-cell allele at a single gene locus brings about the many symptoms of sickle-cell disease. But having just one copy of the sickle-cell allele may be beneficial because it provides some protection against the disease malaria.

- **Polygenic Inheritance**

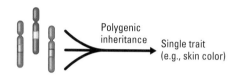

Multiple genes

- **The Role of Environment**   Many human characteristics result from a combination of genetic and environmental effects, but only genetic influences are biologically heritable.

The Chromosomal Basis of Inheritance

- Genes are located on chromosomes, whose behavior during meiosis and fertilization accounts for inheritance patterns (see Figure 9.23).

- **Gene Linkage**   Certain genes are linked: They tend to be inherited together because they lie close together on the same chromosome.

- **Genetic Recombination: Crossing Over and Linkage Maps**   Crossing over can separate linked alleles, producing gametes with recombinant chromosomes and offspring with recombinant phenotypes. The fact that crossing over between linked genes is more likely to occur between genes that are farther apart enables geneticists to map the relative positions of genes on chromosomes.

Activity 9E *Linked Genes and Crossing Over*

Sex Chromosomes and Sex-Linked Genes

- **Sex Determination in Humans and Fruit Flies**   In humans, the Y chromosome has a gene that triggers the development of testes; an absence of this gene triggers the development of ovaries.

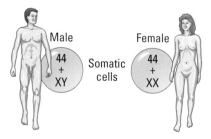

- **Sex-Linked Genes**   Genes on the sex chromosomes are said to be sex-linked. In both fruit flies and humans, the X chromosome carries many genes unrelated to sex. Their inheritance pattern reflects the fact that females have two homologous X chromosomes, but males have only one.

Activity 9F *Sex-Linked Genes*

Case Study in the Process of Science *What Can Fruit Flies Reveal about Inheritance?*

- **Sex-Linked Disorders in Humans**   Most sex-linked human disorders, such as red-green color blindness and hemophilia, are due to recessive alleles and are seen mostly in males. A male receiving a single X-linked recessive allele from his mother will have the disorder; a female has to receive the allele from both parents to be affected.

Self-Quiz

1. Most genes come in alternative forms called _____. A diploid individual with two identical versions of a gene is said to be _____, while an individual with two different versions of a gene is said to be _____.

2. The genetic makeup of an organism is called its _____, while the physical traits of an organism are called its _____.

3. Edward was found to be heterozygous (*Ss*) for the sickle-cell trait. The alleles represented by the letters *S* and *s* are
    a. on the X and Y chromosomes.
    b. linked.
    c. on homologous chromosomes.
    d. both present in each of Edward's sperm cells.

4. Whether an allele is dominant or recessive depends on
    a. how common the allele is, relative to other alleles.
    b. whether it is inherited from the mother or the father.
    c. whether it or another allele determines the phenotype when both are present.
    d. whether or not it is linked to other genes.

5. Two fruit flies with eyes of the usual red color are crossed, and their offspring are as follows: 77 red-eyed males, 71 ruby-eyed males, 152 red-eyed females. The gene that controls whether eyes are red or ruby is _____, and the allele for ruby eyes is _____.
    a. autosomal (carried on an autosome); dominant
    b. autosomal; recessive
    c. sex-linked; dominant
    d. sex-linked; recessive

6. All the offspring of a white hen and a black rooster are gray. The simplest explanation for this pattern of inheritance is
    a. pleiotropy.
    b. sex linkage.
    c. codominance.
    d. incomplete dominance.

7. A man who has type B blood and a woman who has type A blood could have children of which of the following phenotypes?

   a. A, B, or O

   b. AB only

   c. AB or O

   d. A, B, AB, or O

Answers to the Self-Quiz questions can be found in Appendix B.

Go to the website or CD-ROM for more Self-Quiz questions.

More Genetics Problems

1. In fruit flies, the genes for wing shape and body stripes are linked. In a fly whose genotype is *WwSs*, *W* is linked to *S*, and *w* is linked to *s*. Show how this fly can produce gametes containing four different combinations of alleles. Which are parental-type gametes? Which are recombinant gametes? What process produces recombinant gametes?

2. Adult height in humans is at least partially hereditary; tall parents tend to have tall children. But humans come in a range of sizes, not just tall or short. What extension of Mendel's model could produce this variation in height?

3. A true-breeding brown mouse is repeatedly mated with a true-breeding white mouse, and all their offspring are brown. If two of these brown offspring are mated, what fraction of the F_2 mice will be brown?

4. How could you determine the genotype of one of the brown F_2 mice in problem 3? How would you know whether a brown mouse is homozygous? Heterozygous?

5. Tim and Jan both have freckles (a dominant trait), but their son Michael does not. Show with a Punnett square how this is possible. If Tim and Jan have two more children, what is the probability that both of them will have freckles?

6. Incomplete dominance is seen in the inheritance of hypercholesterolemia. Mack and Toni are both heterozygous for this characteristic, and both have elevated levels of cholesterol. Their daughter Zoe has a cholesterol level six times normal; she is apparently homozygous, *hh*. What fraction of Mack and Toni's children are likely to have elevated but not extreme levels of cholesterol, like their parents? If Mack and Toni have one more child, what is the probability that the child will suffer from the more serious form of hypercholesterolemia seen in Zoe?

7. A female fruit fly with forked bristles on her body is mated with a male fly with normal bristles. Their offspring are 121 females with normal bristles and 138 males with forked bristles. Explain the inheritance pattern for this trait.

8. A couple are both phenotypically normal, but their son suffers from hemophilia, a sex-linked recessive disorder. Draw a pedigree that shows the genotypes of the three individuals. What fraction of the couple's children are likely to suffer from hemophilia? What fraction are likely to be carriers?

9. Heather was surprised to discover that she suffered from red-green color blindness. She told her biology professor, who said, "Your father is color-blind too, right?" How did her professor know this? Why did her professor not say the same thing to the color-blind males in the class?

Answers to More Genetic Problems can be found in Appendix B.

The Process of Science

1. In 1981, a stray cat with unusual curled-back ears was adopted by a family in Lakewood, California. Hundreds of descendants of this cat have since been born, and cat fanciers hope to develop the "curl" cat into a show breed. The curl allele is apparently dominant and carried on an autosome. Suppose you owned the first curl cat and wanted to develop a true-breeding variety. Describe tests that would determine whether the curl gene is dominant or recessive and whether it is autosomal or sex-linked.

2. Imagine that you had a large collection of fruit flies divided into ten different strains. Each strain is true-breeding (homozygous) and differs from wild-type flies in just one characteristic. The only special equipment available is a magnifying glass that lets you determine the sex and traits of any fly, a large number of bottles to perform controlled matings, and an anesthetic liquid that enables you to examine and sort live flies. Using only this equipment, how much could you learn about the genetic makeup of the flies? Describe a series of experiments that would give you knowledge about fruit fly genetics.

Case Study in the Process of Science on the Web & CD *Investigate inheritance in fruit flies.*

Biology and Society

1. Gregor Mendel never saw a gene, yet he concluded that "heritable factors" were responsible for the patterns of inheritance he observed in peas. Similarly, maps of *Drosophila* chromosomes (and the very idea that genes are carried on chromosomes) were conceived by observing the patterns of inheritance of linked genes, not by observing the genes directly. Is it legitimate for biologists to claim the existence of objects and processes they cannot actually see? How do scientists know whether an explanation is correct?

2. Many infertile couples turn to in vitro fertilization to try to have a baby. In this technique, sperm and ova are collected and used to create eight-cell embryos for implantation into a woman's uterus. At the eight-cell stage, one of the fetal cells can be removed without causing harm to the developing baby. Once removed, the cell can be genetically tested. Some couples may know that a particular genetic disease runs in their family. They might wish to avoid implanting any embryos with the disease-causing genes. Do you think this is an acceptable use of genetic testing? What if a couple wanted to use genetic testing to select embryos for traits unrelated to disease, such as freckles? Do you think that couples undergoing in vitro fertilization should be allowed to perform whatever genetic tests they wish? Or do you think there should be limits on what tests can be performed? How do you draw the line between genetic tests that are acceptable and those that are not?

Biology and Society on the Web *Read about socioeconomic factors related to fetal testing.*

CHAPTER 10

Molecular Biology of the Gene

Biology and Society:
Sabotaging HIV 171

The Structure and
Replication of DNA 171

DNA and RNA:
Polymers of Nucleotides

Watson and Crick's Discovery of
the Double Helix

DNA Replication

The Flow of Genetic
Information from DNA to
RNA to Protein 176

How an Organism's DNA Genotype
Produces Its Phenotype

From Nucleotide Sequence to Amino
Acid Sequence: An Overview

The Genetic Code

Transcription: From DNA to RNA

The Processing of Eukaryotic RNA

Translation: The Players

Translation: The Process

Review: DNA → RNA → Protein

Mutations

Viruses: Genes in Packages 188

Bacteriophages

Plant Viruses

Animal Viruses

HIV, the AIDS Virus

Evolution Connection:
Emerging Viruses 192

In 2002, **5 million people worldwide** were newly infected with HIV.

Because all organisms use the same genetic code, scientists can make a plant glow like a firefly.

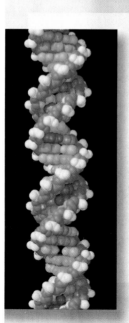

The loss of a single nucleotide from a **1,000-nucleotide gene** can completely destroy the gene's function.

Many viruses have genes that are not made of DNA.

Biology and Society

Sabotaging HIV

AIDS, acquired immuneodeficiency syndrome, is one of the most significant health challenges facing the world today. The cause of AIDS is infection by HIV, the human immunodeficiency virus. AIDS spreads when an infected person transmits HIV to an uninfected person through the transfer of bodily fluids. For example, an HIV-infected mother can pass the virus to her baby through her blood during childbirth or through her milk during breastfeeding.

In the United States, only a few hundred HIV-infected babies are born each year. In contrast, over half a million HIV-infected babies are born in developing nations annually. Why are there so many more AIDS babies born in these countries than in the United States?

One reason for the disparity is the difference in availability of anti-HIV drugs. In particular, the drug AZT, first approved for the treatment of AIDS in 1987, is very effective at preventing the spread of HIV during childbirth. In the United States, AZT has been widely available to HIV-infected mothers for over a decade. However, even though treatment with AZT costs only a few dollars, it is still too expensive for economically disadvantaged people and governments.

To understand how AZT prevents the spread of AIDS, we have to think on the molecular level. After infecting a human cell, HIV depends on a special viral enzyme to make DNA. The shape of AZT is very similar to part of the T (thymine) nucleotide, one of the building blocks of DNA (**Figure 10.1**). In fact, AZT looks so similar to the T nucleotide that it binds to the viral enzyme instead of T, acting as a molecular saboteur that interferes with the synthesis of HIV DNA. Because this synthesis is an essential step in the reproductive cycle of HIV, AZT effectively blocks the spread of the virus.

AZT is a good example of how a detailed understanding of biological molecules can help improve human health. In this chapter, you will learn about molecular biology, the study of heredity at the molecular level. You will learn in detail how the DNA of genes exerts its effects on the cell and on the whole organism. You'll learn, too, how DNA replicates—the molecular basis for the similarities between parents and offspring—and how it can mutate. You will also learn about viruses that infect bacteria, plants, and animals.

Biology and Society on the Web
Learn about other drugs that combat HIV.

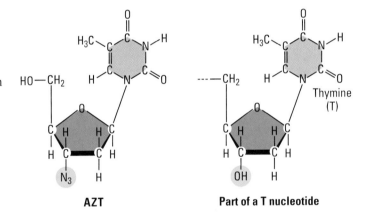

Figure 10.1 AZT and the T nucleotide.
The anti-HIV drug AZT (left) has a chemical shape very similar to part of the T (thymine) nucleotide of DNA.

The Structure and Replication of DNA

DNA was known as a chemical in cells by the end of the nineteenth century, but Mendel and other early geneticists did all their work without any knowledge of DNA's role in heredity. By the late 1930s, experimental studies had convinced most biologists that a specific kind of molecule, rather than some complex chemical mixture, was the basis of inheritance. Attention focused on chromosomes, which were already known to carry genes. By the 1940s, scientists knew that chromosomes consisted of two types of chemicals: DNA and protein. And by the early 1950s, a series of discoveries had convinced the scientific world that DNA was the hereditary material.

Activity 10A on the Web & CD
Find out how Alfred Hershey and Martha Chase demonstrated that DNA, not protein, is the genetic material.

What came next was one of the most celebrated quests in the history of science—the effort to figure out the structure of DNA. A good deal was already known about DNA. Scientists had identified all its atoms and knew how they were covalently bonded to one another. What was not understood was the specific arrangement that gave DNA its unique properties—the capacity to store genetic information, copy it, and pass it from generation to generation. A race was on to discover how the structure of this molecule could account for its role in heredity. We will describe that momentous discovery shortly. First, let's look at the underlying chemical structure of DNA and its chemical cousin RNA.

DNA and RNA: Polymers of Nucleotides

Recall from Chapter 3 that both DNA and RNA are nucleic acids, which consist of long chains (polymers) of chemical units (monomers) called **nucleotides.** (This would be a good time to review the information on nucleic acids in Chapter 3, particularly Figures 3.25–3.27.) A very simple diagram of a nucleotide polymer, or **polynucleotide,** is shown on the left in **Figure 10.2**. This sample polynucleotide chain shows only one possible arrangement of the four different types of nucleotides (abbreviated A, C, T, and G) that make up DNA. Polynucleotides tend to be very long and can have any sequence of nucleotides, so a great number of polynucleotide chains are possible.

The nucleotides are joined to one another by covalent bonds between the sugar of one nucleotide and the phosphate of the next. This results in a

Figure 10.2 **The structure of a DNA polynucleotide.**
A molecule of DNA contains two polynucleotides, each a chain of nucleotides. Each nucleotide consists of a nitrogenous base, a sugar (blue), and a phosphate group (gold). The nucleotides are linked, the sugar of one connected to the phosphate of the next, forming a sugar-phosphate backbone, with the bases protruding from the sugars. The chemical structure at the right shows the details of a DNA nucleotide. The sugar has five carbon atoms (shown in red for emphasis) and is called deoxyribose. The phosphate group has given up an H+ ion, acting as an acid. Hence, the full name for DNA is deoxyribonucleic acid.

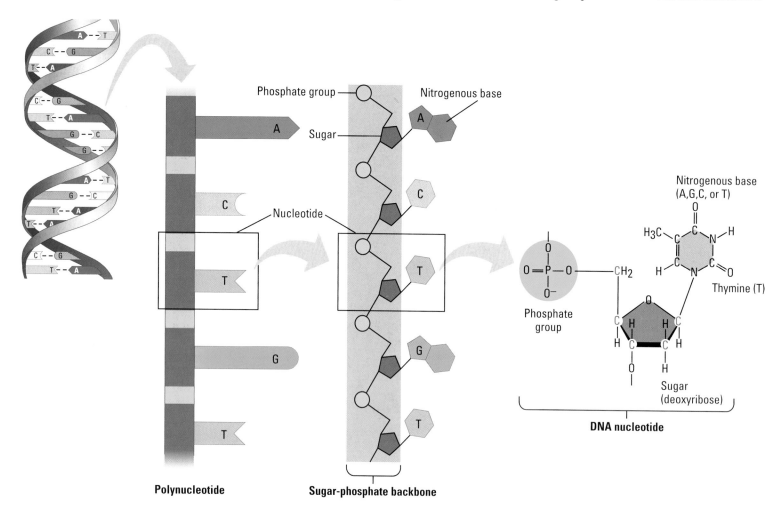

sugar-phosphate backbone, a repeating pattern of sugar-phosphate-sugar-phosphate. The nitrogenous bases are arranged like appendages along this backbone. Zooming in on our polynucleotide in Figure 10.2, we see that each nucleotide consists of three components: a nitrogenous base, a sugar (blue), and a phosphate group (gold). Examining a single nucleotide even more closely (Figure 10.2, right), we see the chemical structure of its three components. The phosphate group, with a phosphorus atom (P) at its center, is the source of the *acid* in nucleic acid. (The phosphate has given up a hydrogen ion, H^+, leaving a negative charge on one of its oxygen atoms.) The sugar has five carbon atoms: four in its ring and one extending above the ring. The ring also includes an oxygen atom. The sugar is called *deoxyribose* because, compared to the sugar ribose, it is missing an oxygen atom. The full name for DNA is *deoxyribonucleic acid,* with the *nucleic* part coming from DNA's location in the nuclei of eukaryotic cells. The nitrogenous base (thymine, in our example) has a ring of nitrogen and carbon atoms with various functional groups attached. In contrast to the acidic phosphate group, nitrogenous bases are basic; hence their name.

The four nucleotides found in DNA differ only in their nitrogenous bases (see Figure 3.25 for a review). At this point, the structural details are not as important as the fact that the bases are of two types. **Thymine (T)** and **cytosine (C)** are single-ring structures. **Adenine (A)** and **guanine (G)** are larger, double-ring structures. (The one-letter abbreviations can be used for either the bases alone or for the nucleotides containing them.) Recall from Chapter 3 that RNA has the nitrogenous base **uracil (U)** instead of thymine (uracil is very similar to thymine). RNA also contains a slightly different sugar than DNA (ribose instead of deoxyribose). Other than that, RNA and DNA polynucleotides have the same chemical structure.

Activity 10B on the Web & CD
Review the structure of DNA and RNA.

Watson and Crick's Discovery of the Double Helix

The celebrated partnership that resulted in the determination of the physical structure of DNA began soon after a 23-year-old American named James D. Watson journeyed to Cambridge University, where Englishman Francis Crick was studying protein structure with a technique called X-ray crystallography (**Figure 10.3a**). While visiting the laboratory of Maurice Wilkins at King's College in London, Watson saw an X-ray crystallographic photograph of DNA, produced by Wilkins's colleague Rosalind Franklin (**Figure 10.3b**). The photograph clearly revealed the basic shape of DNA to be a helix (spiral). On the basis of Watson's later recollection of the photo, he and Crick deduced that the diameter of the helix was uniform. The thickness of the helix suggested that it was made up of two polynucleotide strands—in other words, a **double helix.**

Using wire models, Watson and Crick began trying to construct a double helix that would conform both to Franklin's data and to what was then known about the chemistry of DNA. After failing to make a satisfactory model that placed the sugar-phosphate backbones inside the double helix, Watson tried putting the backbones on the outside and forcing the nitrogenous bases to swivel to the interior of the molecule. It occurred to him that the four kinds of bases might pair in a specific way. This idea of specific base pairing was a flash of inspiration that enabled Watson and Crick to solve the DNA puzzle.

At first, Watson imagined that the bases paired like with like—for example, A with A, C with C. But that kind of pairing did not fit with the fact

(a) James Watson and Francis Crick

(b) Rosalind Franklin

Figure 10.3 Discoverers of the double helix.
(a) James Watson (left) and Francis Crick deduced that the DNA helix consisted of two intertwined polynucleotide chains, with hydrogen-bonded base pairs on the inside of the molecule. Watson and Crick are shown in 1953 with their model of DNA. **(b)** Rosalind Franklin performed the X-ray crystallography that revealed the helical nature of the DNA molecule.

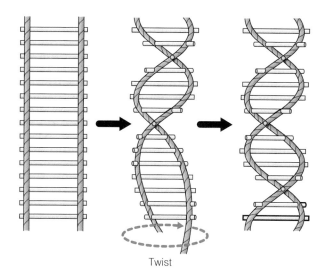

Figure 10.4 A rope-ladder model of a double helix.
The ropes at the sides represent the sugar-phosphate backbones. Each wooden rung stands for a pair of bases connected by hydrogen bonds.

that the DNA molecule has a *uniform* diameter. An AA pair (made of double-ringed bases) would be almost twice as wide as a CC pair (made of single-ringed bases), causing bulges in the molecule. It soon became apparent that a double-ringed base on one strand must always be paired with a single-ringed base on the opposite strand. Moreover, Watson and Crick realized that the individual structures of the bases dictated the pairings even more specifically. Each base has chemical side groups that can best form hydrogen bonds with one appropriate partner (to review the hydrogen bond, see Figure 2.11). Adenine can best form hydrogen bonds with thymine, and guanine with cytosine. In the biologist's shorthand, A pairs with T, and G pairs with C. A is also said to be "complementary" to T, and G to C.

You can picture the model of the DNA double helix proposed by Watson and Crick as a rope ladder having rigid, wooden rungs, with the ladder twisted into a spiral (**Figure 10.4**). **Figure 10.5** shows three more detailed representations of the double helix. The ribbonlike diagram in **Figure 10.5a** symbolizes the bases with shapes that emphasize their complementarity. **Figure 10.5b** is a more chemically precise version showing only four base pairs, with the helix untwisted

Activity 10C on the Web & CD
See rotating 3D models of the DNA double helix.

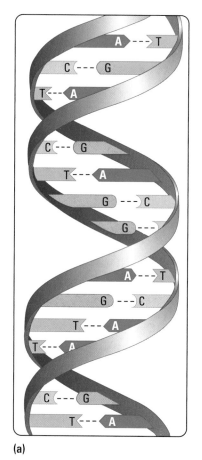

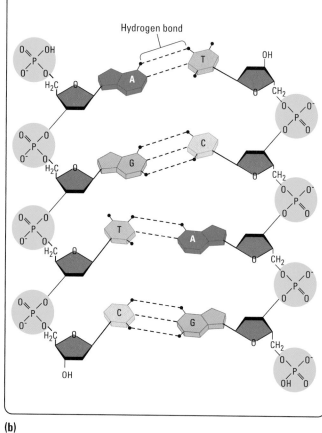

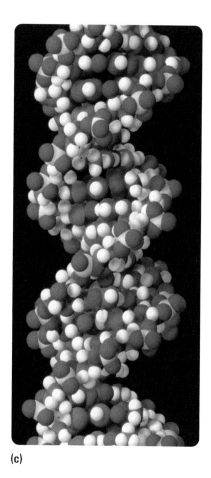

(a)   (b)   (c)

Figure 10.5 Three representations of DNA.
(a) In this model, the sugar-phosphate backbones are blue ribbons, and the bases are complementary shapes in shades of green and orange. (b) In this more chemically detailed structure, you can see the individual hydrogen bonds (dashed lines). You can also see that the strands run in opposite directions; notice that the sugars on the two strands are upside down with respect to each other. (c) In this computer graphic of a DNA double helix, each atom is shown as a sphere, creating a space-filling model.

and the individual hydrogen bonds specified by dashed lines; you can see that the double helix has an antiparallel arrangement—that is, the two sugar-phosphate backbones are oriented in opposite directions. **Figure 10.5c** is a computer graphic showing every atom of part of a double helix.

Although the base-pairing rules dictate the side-by-side combinations of nitrogenous bases that form the rungs of the double helix, they place no restrictions on the *sequence* of nucleotides along the length of a DNA strand. In fact, the sequence of bases can vary in countless ways.

In April 1953, Watson and Crick shook the scientific world with a succinct, two-page announcement of their molecular model for DNA in the journal *Nature*. Few milestones in the history of biology have had as broad an impact as their double helix, with its AT and CG base pairing. In 1962, Watson, Crick, and Wilkins received the Nobel Prize for their work. (Franklin may have received the prize as well, had she not died from cancer in 1958.)

In their 1953 paper, Watson and Crick wrote that the structure they proposed "immediately suggests a possible copying mechanism for the genetic material." In other words, the structure of DNA also points toward a molecular explanation for life's unique properties of reproduction and inheritance, as we see next.

DNA Replication

When a cell or a whole organism reproduces, a complete set of genetic instructions must pass from one generation to the next. For this to occur, there must be a means of copying the instructions. Watson and Crick's model for DNA structure immediately suggested to them that DNA replicates by a template mechanism—each DNA strand can serve as a mold, or template, to guide reproduction of the other strand. The logic behind the Watson-Crick proposal for how DNA is copied is quite simple. If you know the sequence of bases in one strand of the double helix, you can very easily determine the sequence of bases in the other strand by applying the base-pairing rules: A pairs with T (and T with A), and G pairs with C (and C with G). For example, if one polynucleotide has the sequence ATCG, then the complementary polynucleotide in that DNA molecule must have the sequence TAGC.

Figure 10.6 shows how the template model can account for the direct copying of a piece of DNA. The two strands of parental DNA separate, and each becomes a template for the assembly of a complementary strand from a supply of free nucleotides. The nucleotides are lined up one at a time along the template strand in accordance with the base-pairing rules. Enzymes link the nucleotides to form the new DNA strands. The completed new molecules, identical to the parental molecule, are known as daughter DNA molecules (no gender should be inferred from this name).

Although the general mechanism of DNA replication is conceptually simple, the actual process is complex and requires the cooperation of more than a dozen enzymes and other proteins. The enzymes that actually make the covalent bonds between the nucleotides of a

Case Study in the Process of Science on the Web & CD Virtually reenact famous experiments that tested different models of DNA replication.

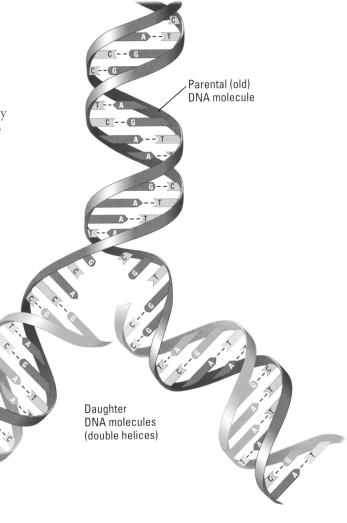

Figure 10.6 DNA replication. The two strands of the original (parental) DNA molecule (blue) serve as templates for making new (daughter) strands (orange). Replication results in two daughter DNA molecules, each consisting of one old strand and one new strand. The parental DNA untwists as its strands separate, and the daughter DNA rewinds as it forms.

Parental (old) DNA molecule

Daughter (new) strand

Daughter DNA molecules (double helices)

Figure 10.7 Damage to DNA by ultraviolet light.
The ultraviolet (UV) radiation in sunlight can damage the DNA in skin cells. Fortunately, the cells can repair some of the damage, using enzymes that include some that catalyze DNA replication. You can protect yourself from UV radiation by wearing protective clothing and sunscreen.

new DNA strand are called **DNA polymerases.** As an incoming nucleotide base-pairs with its complement on the template strand, a DNA polymerase adds it to the end of the growing daughter strand (polymer). The process is both fast and amazingly accurate; typically, DNA replication proceeds at a rate of 50 nucleotides per second, with only about one in a billion incorrectly paired. In addition to their roles in DNA replication, DNA polymerases and some of the associated proteins are also involved in repairing damaged DNA. DNA can be harmed by toxic chemicals in the environment or by high-energy radiation, such as X-rays and ultraviolet light (**Figure 10.7**).

DNA replication begins at specific sites on a double helix, called origins of replication. It then proceeds in both directions, creating what are called replication "bubbles" (**Figure 10.8**). The parental DNA strands open up as daughter strands elongate on both sides of each bubble. The DNA molecule of a eukaryotic chromosome has many origins where replication can start simultaneously, shortening the total time needed for the process. Eventually, all the bubbles merge, yielding two completed double-stranded daughter DNA molecules.

Activity 10D on the Web & CD
Build your own DNA molecule.

DNA replication ensures that all the somatic cells in a multicellular organism carry the same genetic information. It is also the means by which genetic instructions are copied for the next generation of the organism.

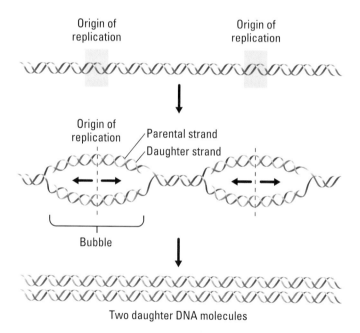

Figure 10.8 Multiple "bubbles" in replicating DNA.
DNA replication begins at sites called origins of replication and proceeds in both directions, producing the "bubbles" shown here. The long DNA molecules of eukaryotic chromosomes each have many origins, leading to multiple bubbles as they replicate. The bubbles merge, producing two completed daughter DNA molecules.

CHECKPOINT

1. Compare and contrast the chemical components of DNA and RNA.
2. Along one strand of a DNA double helix is the nucleotide sequence GGCATAGGT. What is the sequence for the other DNA strand?
3. How does complementary base pairing make the replication of DNA possible?
4. What is the function of DNA polymerase in DNA replication?

Answers: 1. Both are polymers of nucleotides. A nucleotide consists of a sugar + a nitrogenous base + a phosphate group. In RNA, the sugar is ribose; in DNA, it is deoxyribose. Both RNA and DNA have the bases A, G, and C; for a fourth base, DNA has T and RNA has U. **2.** CCGTATCCA **3.** When the two strands of the double helix separate, each serves as a template on which nucleotides can be arranged by specific base pairing into new complementary strands. **4.** This enzyme covalently connects nucleotides one at a time to one end of a growing daughter strand as the nucleotides line up along a template strand according to the base-pairing rules.

The Flow of Genetic Information from DNA to RNA to Protein

We are now ready to address the question of how DNA functions as the inherited directions for a cell and for the organism as a whole. What exactly are the instructions carried by the DNA, and how are these instructions carried out? You learned the general answers to these questions in Chapter 4, but here we will explore them in more detail.

How an Organism's DNA Genotype Produces Its Phenotype

Knowing the structure of DNA, we can now define genotype and phenotype more precisely than we did in Chapter 9. An organism's *genotype*, its genetic makeup, is the sequence of nucleotide bases in its DNA. The *phenotype* is the organism's specific traits. The molecular basis of the phenotype lies in proteins with a variety of functions. For example, structural proteins help make up the body of an organism, and enzymes catalyze its metabolic activities.

What is the connection between the genotype and the protein molecules that more directly determine the phenotype? Recall from Chapter 4 that

Activity 10E on the Web & CD
Get an overview of protein synthesis.

DNA specifies the synthesis of proteins. A gene does not build a protein directly, but rather dispatches instructions in the form of RNA, which in turn programs protein synthesis. This central concept in biology is summarized in **Figure 10.9**. The chain of command is from DNA in the nucleus of the cell to RNA to protein synthesis in the cytoplasm. The two main stages are **transcription,** the transfer of genetic information from DNA into an RNA molecule, and **translation,** the transfer of the information in the RNA into a protein.

The relationship between genes and proteins was first proposed in 1909, when English physician Archibald Garrod suggested that genes dictate phenotypes through enzymes, the proteins that catalyze chemical processes. Garrod's idea came from his observations of inherited diseases. He hypothesized that an inherited disease reflects a person's inability to make a particular enzyme, and he referred to such diseases as "inborn errors of metabolism." He gave as one example the hereditary condition called alkaptonuria, in which the urine appears dark red because it contains a chemical called alkapton. Garrod reasoned that normal individuals have an enzyme that breaks down alkapton, whereas alkaptonuric individuals lack the enzyme. Garrod's hypothesis was ahead of its time, but research conducted decades later proved him right. In the intervening years, biochemists accumulated evidence that cells make and break down biologically important molecules via metabolic pathways, as in the synthesis of an amino acid or the breakdown of a sugar. As you learned in Chapter 5, each step in a metabolic pathway is catalyzed by a specific enzyme. If a person lacks one of the enzymes, the pathway cannot be completed.

The major breakthrough in demonstrating the relationship between genes and enzymes came in the 1940s from the work of American geneticists George Beadle and Edward Tatum with the orange bread mold *Neurospora crassa*. Beadle and Tatum studied strains of the mold that

Case Study in the Process of Science on the Web & CD
Virtually reenact George Beadle and Edward Tatum's experiments.

were unable to grow on the usual simple growth medium. Each of these strains turned out to lack an enzyme in a metabolic pathway that produced some molecule the mold needed, such as an amino acid. Beadle and Tatum also showed that each mutant was defective in a single gene. Accordingly, they formulated the one gene–one enzyme hypothesis, which states that the function of an individual gene is to dictate the production of a specific enzyme.

The one gene–one enzyme hypothesis has been amply confirmed, but with some important modifications. First it was extended beyond enzymes to include *all* types of proteins. For example, alpha-keratin, the structural protein of your hair, is the product of a gene. So biologists soon began to think in terms of one gene–one *protein*. Then it was discovered that many proteins have two or more different polypeptide chains, and each polypeptide

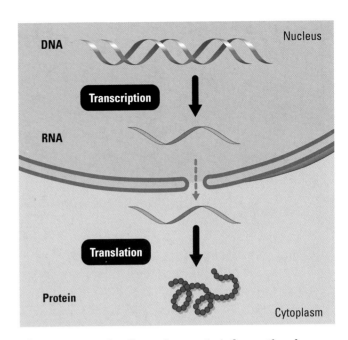

Figure 10.9 The flow of genetic information in a eukaryotic cell: a review. A sequence of nucleotides in the DNA is transcribed into a molecule of RNA in the cell's nucleus (purple area). The RNA travels to the cytoplasm (blue-green area), where it is translated into the specific amino acid sequence of a protein.

is specified by its own gene. Thus, Beadle and Tatum's hypothesis has come to be restated as one gene–one *polypeptide*.

From Nucleotide Sequence to Amino Acid Sequence: An Overview

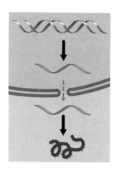

Stating that genetic information in DNA is transcribed into RNA and then translated into polypeptides does not explain *how* these processes occur. Transcription and translation are linguistic terms, and it is useful to think of nucleic acids and polypeptides as having languages, too. To understand how genetic information passes from genotype to phenotype, we need to see how the chemical language of DNA is translated into the different chemical language of polypeptides.

What exactly is the language of nucleic acids? Both DNA and RNA are polymers made of monomers in specific sequences that carry information, much as specific sequences of letters carry information in English. In DNA, the monomers are the four types of nucleotides, which differ in their nitrogenous bases (A, T, C, and G). The same is true for RNA, although it has the base U instead of T.

DNA's language is written as a linear sequence of nucleotide bases, a sequence such as the one you see on the enlarged DNA strand in **Figure 10.10**. Specific sequences of bases, each with a beginning and an end, make up the genes on a DNA strand. A typical gene consists of thousands of nucleotides, and a DNA molecule may contain thousands of genes.

When DNA is transcribed, the result is an RNA molecule. The process is called transcription because the nucleic acid language of DNA has simply been rewritten (transcribed) as a sequence of bases of RNA; the language is still that of nucleic acids. The nucleotide bases of the RNA molecule are complementary to those on the DNA strand. As you will soon see, this is because the RNA was synthesized using the DNA as a template.

Translation is the conversion of the nucleic acid language into the polypeptide language. Like nucleic acids, polypeptides are polymers, but the monomers that make them up—the letters of the polypeptide alphabet—are the 20 amino acids common to all organisms. Again, the language is written in a linear sequence, and the sequence of nucleotides of the RNA molecule dictates the sequence of amino acids of the polypeptide. But remember, RNA is only a messenger; the genetic information that dictates the amino acid sequence is based in DNA.

What are the rules for translating the RNA message into a polypeptide? In other words, what is the correspondence between the nucleotides of an RNA molecule and the amino acids of a polypeptide? Keep in mind that there are only four different kinds of nucleotides in DNA (A, G, C, T) and RNA (A, G, C, U). In translation, these four must somehow specify 20 amino acids. If each nucleotide base coded for one amino acid, only 4 of the 20 amino acids could be accounted for. What if the language consisted of two-letter code words? If we read the bases of a gene two at a time, AG, for example, could specify one amino acid, while AT could designate a different amino acid. However, when the four bases are taken two by two, there are only 16 (that is, 4^2) possible arrangements—still not enough to specify all 20 amino acids.

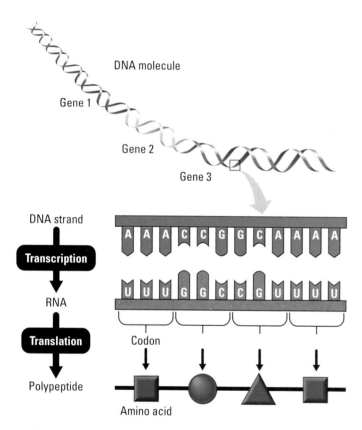

Figure 10.10 Transcription and translation of codons. This figure focuses on a small region of one of the genes carried by a DNA molecule. The enlarged segment from one strand of gene 3 shows its specific sequence of bases. The red strand underneath represents the results of transcription: an RNA molecule. Its base sequence is complementary to that of the DNA. The purple chain represents the results of translation: a polypeptide. The brackets indicate that *three* RNA nucleotides code for each amino acid.

Triplets of bases are the smallest "words" of uniform length that can specify all the amino acids. There can be 64 (that is, 4^3) possible code words of this type—more than enough to specify the 20 amino acids. Indeed, there are enough triplets to allow more than one coding for each amino acid. For example, the base triplets AAT and AAC could both code for the same amino acid—and, in fact, they do.

Experiments have verified that the flow of information from gene to protein is based on a *triplet* code. The genetic instructions for the amino acid sequence of a polypeptide chain are written in DNA and RNA as a series of three-base words called **codons.** Three-base codons in the DNA are transcribed into complementary three-base codons in the RNA, and then the RNA codons are translated into amino acids that form a polypeptide. Next we turn to the codons themselves.

The Genetic Code

In 1799, a large stone tablet was found in Rosetta, Egypt, carrying the same lengthy inscription in three ancient scripts: Greek, Egyptian hieroglyphics, and Egyptian written in a simplified script. The Rosetta stone provided the key that enabled scholars to crack the previously indecipherable hieroglyphic code.

In cracking the **genetic code,** the set of rules relating nucleotide sequence to amino acid sequence, scientists wrote their own Rosetta stone. It was based on a series of elegant experiments that revealed the amino acid translations of each of the nucleotide-triplet code words. The first codon was deciphered in 1961 by American biochemist Marshall Nirenberg. He synthesized an artificial RNA molecule by linking together identical RNA nucleotides having uracil as their base. No matter where this message started or stopped, it could contain only one type of triplet codon: UUU. Nirenberg added this "poly U" to a test-tube mixture containing ribosomes and the other ingredients required for polypeptide synthesis. This mixture translated the poly U into a polypeptide containing a single kind of amino acid, phenylalanine. In this way, Nirenberg learned that the RNA codon UUU specifies the amino acid phenylalanine (Phe). By variations on this method, the amino acids specified by all the codons were determined.

As **Figure 10.11** shows, 61 of the 64 triplets code for amino acids. The triplet AUG has a dual function: It not only codes for the amino acid methionine (Met) but can also provide a signal for the start of a polypeptide chain. Three of the other codons do not designate amino acids. They are the stop codons that instruct the ribosomes to end the polypeptide.

Notice in Figure 10.11 that there is redundancy in the code but no ambiguity. For example, although codons UUU and UUC both specify

Second base

First base	U	C	A	G	Third base
U	UUU Phenylalanine (Phe)	UCU Serine (Ser)	UAU Tyrosine (Tyr)	UGU Cysteine (Cys)	U
	UUC	UCC	UAC	UGC	C
	UUA Leucine (Leu)	UCA	UAA Stop	UGA Stop	A
	UUG	UCG	UAG Stop	UGG Tryptophan (Trp)	G
C	CUU Leucine (Leu)	CCU Proline (Pro)	CAU Histidine (His)	CGU Arginine (Arg)	U
	CUC	CCC	CAC	CGC	C
	CUA	CCA	CAA Glutamine (Gln)	CGA	A
	CUG	CCG	CAG	CGG	G
A	AUU Isoleucine (Ile)	ACU Threonine (Thr)	AAU Asparagine (Asn)	AGU Serine (Ser)	U
	AUC	ACC	AAC	AGC	C
	AUA	ACA	AAA Lysine (Lys)	AGA Arginine (Arg)	A
	AUG Met or start	ACG	AAG	AGG	G
G	GUU Valine (Val)	GCU Alanine (Ala)	GAU Aspartic acid (Asp)	GGU Glycine (Gly)	U
	GUC	GCC	GAC	GGC	C
	GUA	GCA	GAA Glutamic acid (Glu)	GGA	A
	GUG	GCG	GAG	GGG	G

Figure 10.11 The dictionary of the genetic code, listed by RNA codons. The three bases of an RNA codon are designated here as the first, second, and third bases. Practice using this dictionary by finding the codon UGG. This is the only codon for the amino acid tryptophan (Trp), but most amino acids are specified by two or more codons. For example, both UUU and UUC stand for the amino acid phenylalanine (Phe). Notice that the codon AUG not only stands for the amino acid methionine (Met) but also functions as a signal to "start" translating the RNA at that place. Three of the 64 codons function as "stop" signals. Any one of these termination codons marks the end of a genetic message.

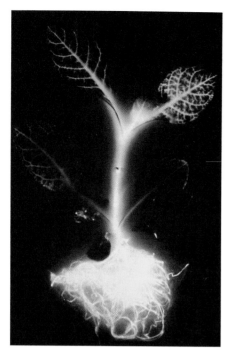

Figure 10.12 A tobacco plant expressing a firefly gene.   Because diverse organisms share a common genetic code, it is possible to program one species to produce a protein characteristic of another species by transplanting DNA. This photo shows the results of an experiment in which researchers incorporated a gene from a firefly into the DNA of a tobacco plant. The gene codes for the firefly enzyme that causes a glow.

phenylalanine (redundancy), neither of them ever represents any other amino acid (no ambiguity). The codons in the figure are the triplets found in RNA. They have a straightforward, complementary relationship to the codons in DNA. The nucleotides making up the codons occur in a linear order along the DNA and RNA, with no gaps or "punctuation" separating the codons.

Almost all of the genetic code is shared by all organisms, from the simplest bacteria to the most complex plants and animals. The universality of the genetic vocabulary suggests that it arose very early in evolution and was passed on over the eons to all the organisms living on Earth today. As you will learn in Chapter 12, such universality is extremely important to modern DNA technologies. Because the code is the same in different species, genes can be transcribed and translated after transfer from one species to another, even when the organisms are as different as a bacterium and a human, or a firefly and a tobacco plant (**Figure 10.12**). This allows scientists to mix and match genes from various species—a procedure with many useful applications.

Transcription: From DNA to RNA

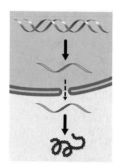

Let's look more closely at transcription, the transfer of genetic information from DNA to RNA. An RNA molecule is transcribed from a DNA template by a process that resembles the synthesis of a DNA strand during DNA replication. **Figure 10.13a** is a close-up view of this process. As with replication, the two DNA strands must first separate at the place where the process will start. In transcription, however, only one of the DNA strands serves as a template for the newly forming molecule. The nucleotides that make up the new RNA molecule take their places one at a time along the DNA template strand by forming hydrogen bonds with the nucleotide bases there. Notice that the RNA nucleotides follow the same base-pairing rules that govern DNA replication, except that U, rather than T, pairs with A. The RNA nucleotides are linked by the transcription enzyme **RNA polymerase.**

Figure 10.13b is an overview of the transcription of an entire gene. Special sequences of DNA nucleotides tell the RNA polymerase where to start and where to stop the transcribing process.

Initiation of Transcription   The "start transcribing" signal is a nucleotide sequence called a **promoter,** which is located in the DNA at the beginning of the gene. A promoter is a specific place where RNA polymerase attaches. The first phase of transcription, called initiation, is the attachment of RNA polymerase to the promoter and the start of RNA synthesis. For any gene, the promoter dictates which of the two DNA strands is to be transcribed (the particular strand varying from gene to gene).

RNA Elongation   During the second phase of transcription, elongation, the RNA grows longer. As RNA synthesis continues, the RNA strand peels away from its DNA template, allowing the two separated DNA strands to come back together in the region already transcribed.

Termination of Transcription   In the third phase, termination, the RNA polymerase reaches a special sequence of bases in the DNA template called

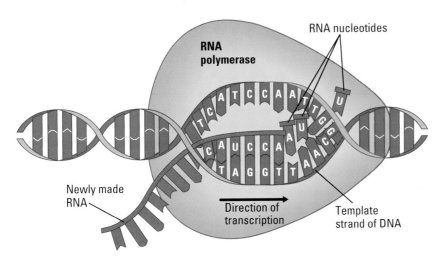

(a) A close-up view of transcription

Figure 10.13 Transcription. **(a)** As RNA nucleotides base-pair one by one with DNA bases on one DNA strand (called the template strand), the enzyme RNA polymerase links the RNA nucleotides into an RNA chain. The orange shape in the background is the RNA polymerase. **(b)** The transcription of an entire gene occurs in three phases: initiation, elongation, and termination of the RNA. The section of DNA where the RNA polymerase starts is called the promoter; the place where it stops is called the terminator.

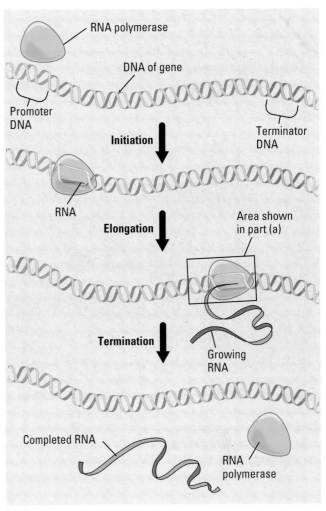

(b) Transcription of a gene

a **terminator.** This sequence signals the end of the gene. At this point, the polymerase molecule detaches from the RNA molecule and the gene.

In addition to producing RNA that encodes amino acid sequences, transcription makes two other kinds of RNA that are involved in building polypeptides. We discuss these kinds of RNA a little later.

The Processing of Eukaryotic RNA

In prokaryotic cells, which lack nuclei, the RNA transcribed from a gene immediately functions as the messenger molecule that is translated, called **messenger RNA (mRNA).** But this is not the case in eukaryotic cells. The eukaryotic cell not only localizes transcription in the nucleus but also modifies, or *processes,* the RNA transcripts there before they move to the cytoplasm for translation by the ribosomes.

One kind of RNA processing is the addition of extra nucleotides to the ends of the RNA transcript. These additions, called the **cap** and **tail,** protect the RNA from attack by cellular enzymes and help ribosomes recognize the RNA as mRNA.

Another type of RNA processing is made necessary in eukaryotes by noncoding stretches of nucleotides that interrupt the nucleotides that actually code for amino acids. It is as if unintelligible sequences of letters were randomly interspersed in an otherwise intelligible document. Most genes of plants and animals, it turns out, include such internal noncoding regions, which are called **introns.** (The functions of introns, if any, and how introns evolved remain a mystery.) The coding regions—

Activity 10F on the Web & CD
See transcription and RNA processing in action.

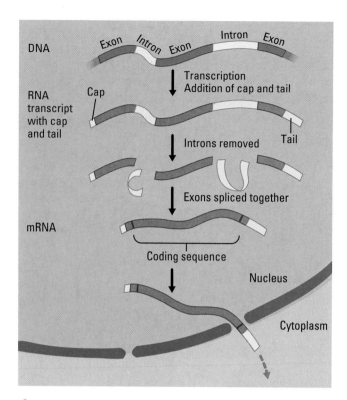

RNA transcript with cap and tail

DNA

Exon Intron Exon   Intron Exon

Transcription
Addition of cap and tail

Cap

Introns removed   Tail

Exons spliced together

mRNA

Coding sequence

Nucleus

Cytoplasm

Figure 10.14 The production of messenger RNA in a eukaryotic cell. Both exons and introns are transcribed from the DNA. Additional nucleotides, making up the cap and tail, are attached at the ends of the RNA transcript. The exons are spliced together. The product, a molecule of messenger RNA (mRNA), then travels to the cytoplasm of the cell. There the coding sequence will be translated.

the parts of a gene that are expressed—are called **exons.** As **Figure 10.14** illustrates, both exons and introns are transcribed from DNA into RNA. However, before the RNA leaves the nucleus, the introns are removed, and the exons are joined to produce an mRNA molecule with a continuous coding sequence. This process is called **RNA splicing.** RNA splicing is believed to play a significant role in humans in allowing our approximately 35,000 genes to produce many thousands more polypeptides. This is accomplished by varying the exons that are included in the final mRNA.

With capping, tailing, and splicing completed, the "final draft" of eukaryotic mRNA is ready for translation.

Translation: The Players

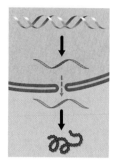

As we have already discussed, translation is a conversion between different languages—from the nucleic acid language to the protein language—and it involves more elaborate machinery than transcription.

Messenger RNA (mRNA) The first important ingredient required for translation is the mRNA produced by transcription. Once it is present, the machinery used to translate mRNA requires enzymes and sources of chemical energy, such as ATP. In addition, translation requires two heavy-duty components: ribosomes and a kind of RNA called transfer RNA.

Transfer RNA (tRNA) Translation of any language into another language requires an interpreter, a person or device that can recognize the words of one language and convert them into the other. Translation of the genetic message carried in mRNA into the amino acid language of proteins also requires an interpreter. To convert the three-letter words (codons) of nucleic acids to the one-letter, amino acid words of proteins, a cell uses a molecular interpreter, a type of RNA called **transfer RNA,** abbreviated **tRNA** (**Figure 10.15**).

A cell that is ready to have some of its genetic information translated into polypeptides has in its cytoplasm a supply of amino acids, either obtained from food or made from other chemicals. The amino acids themselves cannot recognize the codons arranged in sequence along messenger RNA. It is up to

Figure 10.15 The structure of tRNA. **(a)** The RNA polynucleotide is a "rope" whose appendages are the nitrogenous bases. Dashed lines are hydrogen bonds, which connect some of the bases. The site where an amino acid will attach is a three-nucleotide segment at one end (purple). Note the three-base anticodon at the bottom of the molecule (dark green). The overall shape of a tRNA molecule is like the letter L. **(b)** This is the representation of tRNA that we use in later diagrams.

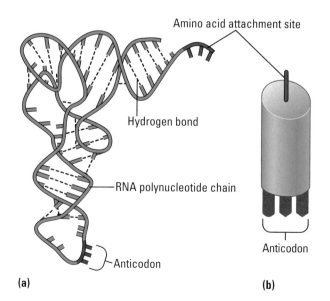

Amino acid attachment site

Hydrogen bond

RNA polynucleotide chain

Anticodon

Anticodon

(a)   (b)

the cell's molecular interpreters, tRNA molecules, to match amino acids to the appropriate codons to form the new polypeptide. To perform this task, tRNA molecules must carry out two distinct functions: (1) to pick up the appropriate amino acids, and (2) to recognize the appropriate codons in the mRNA. The unique structure of tRNA molecules enables them to perform both tasks.

As shown in Figure 10.15a, a tRNA molecule is made of a single strand of RNA—one polynucleotide chain—consisting of about 80 nucleotides. The chain twists and folds upon itself, forming several double-stranded regions in which short stretches of RNA base-pair with other stretches. At one end of the folded molecule is a special triplet of bases called an **anticodon.** The anticodon triplet is complementary to a codon triplet on mRNA. During translation, the anticodon on tRNA recognizes a particular codon on mRNA by using base-pairing rules. At the other end of the tRNA molecule is a site where an amino acid can attach. Although all tRNA molecules are similar, there is a slightly different version of tRNA for each amino acid.

Ribosomes Ribosomes are the organelles that coordinate the functioning of the mRNA and tRNA and actually make polypeptides. As you can see in **Figure 10.16a,** a ribosome consists of two subunits. Each subunit is made up of proteins and a considerable amount of yet another kind of RNA, **ribosomal RNA (rRNA).** A fully assembled ribosome has a binding site for mRNA on its small subunit and binding sites for tRNA on its large subunit. **Figure 10.16b** shows how two tRNA molecules get together with an mRNA molecule on a ribosome. One of the tRNA binding sites, the P site, holds the tRNA carrying the growing polypeptide chain, while another, the A site, holds a tRNA carrying the next amino acid to be added to the chain. The anticodon on each tRNA base-pairs with a codon on mRNA. The subunits of the ribosome act like a vise, holding the tRNA and mRNA molecules close together. The ribosome can then connect the amino acid from the A site tRNA to the growing polypeptide.

Now let's examine translation in more detail, starting at the beginning.

Translation: The Process

Translation can be divided into the same three phases as transcription: initiation, elongation, and termination.

Initiation This first phase brings together the mRNA, the first amino acid with its attached tRNA, and the two subunits of a ribosome. An mRNA molecule, even after splicing, is longer than the genetic message it carries (**Figure 10.17**). Nucleotide sequences at either end of the molecule are not part of the message, but along with the cap and tail in eukaryotes, they help the mRNA bind to the ribosome. The initiation process determines

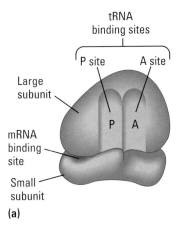

(a)

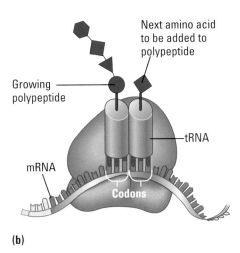

(b)

Figure 10.16 The ribosome. (a) A simplified diagram of a ribosome, showing its two subunits and sites where mRNA and tRNA molecules bind. (b) When functioning in polypeptide synthesis, a ribosome holds one molecule of mRNA and two molecules of tRNA. The growing polypeptide is attached to one of the tRNAs.

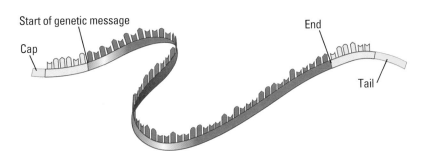

Figure 10.17 A molecule of mRNA. The pink ends are nucleotides that are not part of the message; that is, they are not translated. These nucleotides, along with the cap and tail (yellow), help the mRNA attach to the ribosome.

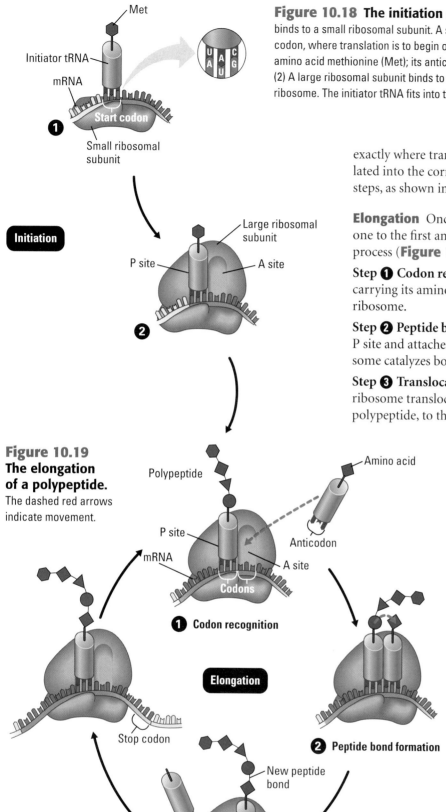

Figure 10.18 The initiation of translation. (1) An mRNA molecule binds to a small ribosomal subunit. A special initiator tRNA then binds to the start codon, where translation is to begin on the mRNA. The initiator tRNA carries the amino acid methionine (Met); its anticodon, UAC, binds to the start codon, AUG. (2) A large ribosomal subunit binds to the small one, creating a functional ribosome. The initiator tRNA fits into the P site on the ribosome.

Initiation

**Figure 10.19
The elongation
of a polypeptide.**
The dashed red arrows
indicate movement.

① Codon recognition

Elongation

② Peptide bond formation

③ Translocation

exactly where translation will begin so that the mRNA codons will be translated into the correct sequence of amino acids. Initiation occurs in two steps, as shown in **Figure 10.18**.

Elongation Once initiation is complete, amino acids are added one by one to the first amino acid. Each addition occurs in a three-step elongation process (**Figure 10.19**).

Step ① Codon recognition. The anticodon of an incoming tRNA molecule, carrying its amino acid, pairs with the mRNA codon in the A site of the ribosome.

Step ② Peptide bond formation. The polypeptide leaves the tRNA in the P site and attaches to the amino acid on the tRNA in the A site. The ribosome catalyzes bond formation. Now the chain has one more amino acid.

Step ③ Translocation. The P site tRNA now leaves the ribosome, and the ribosome translocates (moves) the remaining tRNA, carrying the growing polypeptide, to the P site. The mRNA and tRNA move as a unit. This movement brings into the A site the next mRNA codon to be translated, and the process can start again with step 1.

Termination Elongation continues until a **stop codon** reaches the ribosome's A site. Stop codons— UAA, UAG, and UGA—do not code for amino acids but instead tell translation to stop. The completed polypeptide, typically several hundred amino acids long, is freed, and the ribosome splits into its subunits.

**Activity 10G on
the Web & CD**
Test your skills by
translating mRNA
into a polypeptide.

Review: DNA → RNA → Protein

Figure 10.20 reviews the flow of genetic information in the cell, from DNA to RNA to protein. In eukaryotic cells, transcription—the stage from DNA to RNA—occurs in the nucleus, and the RNA is processed before it enters the cytoplasm. Translation is rapid; a single ribosome can make an average-sized polypeptide in less than a minute. As it is made, a polypeptide coils and folds, assuming a three-dimensional shape, its tertiary structure. Several polypeptides may come together, forming a protein with quaternary structure (see Figure 3.23).

What is the overall significance of transcription and translation? These are the processes whereby genes control the structures and activities of cells— or, more broadly, the way the genotype produces

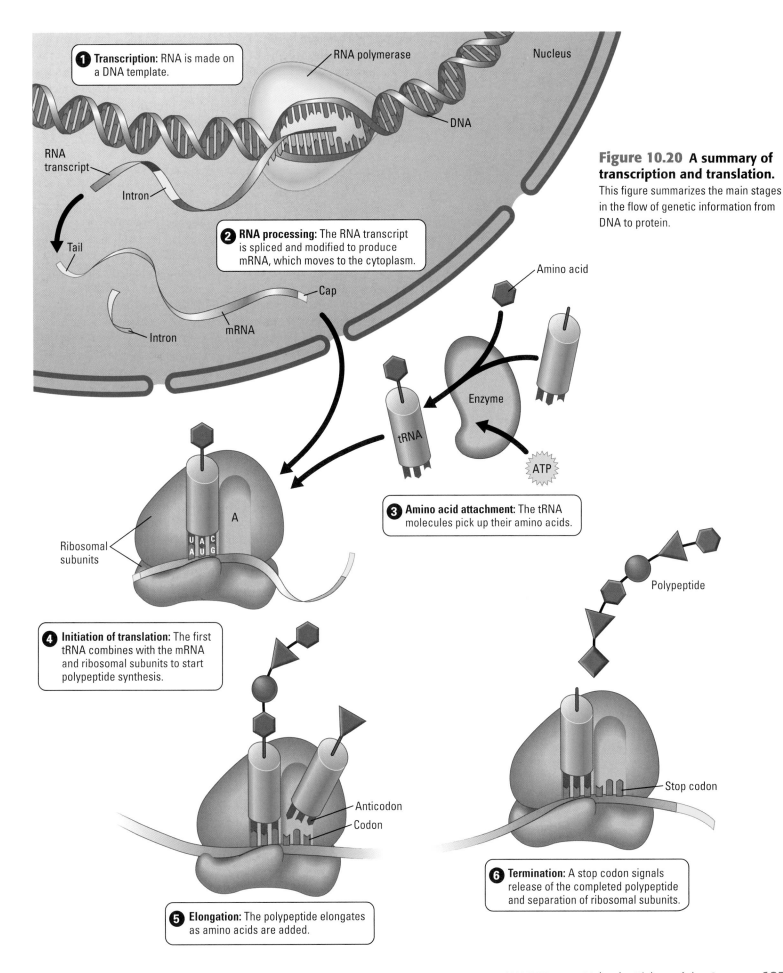

1 Transcription: RNA is made on a DNA template.

RNA polymerase

Nucleus

DNA

RNA transcript

Intron

Tail

2 RNA processing: The RNA transcript is spliced and modified to produce mRNA, which moves to the cytoplasm.

Cap

Intron

mRNA

Figure 10.20 A summary of transcription and translation. This figure summarizes the main stages in the flow of genetic information from DNA to protein.

Amino acid

Enzyme

tRNA

ATP

3 Amino acid attachment: The tRNA molecules pick up their amino acids.

A

Ribosomal subunits

Polypeptide

4 Initiation of translation: The first tRNA combines with the mRNA and ribosomal subunits to start polypeptide synthesis.

Anticodon

Codon

Stop codon

6 Termination: A stop codon signals release of the completed polypeptide and separation of ribosomal subunits.

5 Elongation: The polypeptide elongates as amino acids are added.

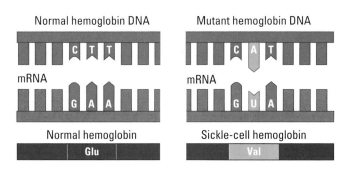

Normal hemoglobin DNA　　　　Mutant hemoglobin DNA

C T T　　　　C A T

mRNA　　　　　　　　　　　mRNA

G A A　　　　G U A

Normal hemoglobin　　　　　Sickle-cell hemoglobin
Glu　　　　　　　　　　　　Val

Figure 10.21 The molecular basis of sickle-cell disease. The sickle-cell allele differs from its normal counterpart, a gene for hemoglobin, by only one nucleotide. This difference changes the mRNA codon from one that codes for the amino acid glutamic acid (Glu) to one that codes for valine (Val).

Normal gene

mRNA　A U G A A G U U U G G C G C A
Protein　Met｜Lys｜Phe｜Gly｜Ala

(a) Base substitution

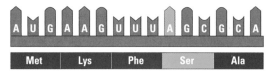

A U G A A G U U U A G C G C A
Met｜Lys｜Phe｜Ser｜Ala

(b) Nucleotide deletion

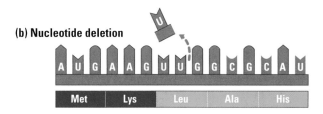

A U G A A G U U G G C G C A U
Met｜Lys｜Leu｜Ala｜His

Figure 10.22 Two types of mutations and their effects. Mutations are changes in DNA, but they are represented here as reflected in mRNA and its polypeptide product. **(a)** In the base substitution shown here, an A replaces a G in the fourth codon of the mRNA. The result in the polypeptide is a serine (Ser) instead of a glycine (Gly). This amino acid substitution may or may not affect the protein's function. **(b)** When a nucleotide is deleted (or inserted), the reading frame is altered, so that all the codons from that point on are misread. The resulting polypeptide is likely to be completely nonfunctional.

the phenotype. The chain of command originates with the information in a gene, a specific linear sequence of nucleotides in DNA. The gene serves as a template, dictating the transcription of a complementary sequence of nucleotides in mRNA. In turn, mRNA specifies the linear sequence in which amino acids appear in a polypeptide. Finally, the proteins that form from the polypeptides determine the appearance and capabilities of the cell and organism.

Mutations

Since discovering how genes are translated into proteins, scientists have been able to describe many heritable differences in molecular terms. For instance, when a child is born with sickle-cell disease (see Figure 9.21), the condition can be traced back through a difference in a protein to one tiny change in a gene. In one of the polypeptides in the hemoglobin protein, the sickle-cell child has a single different amino acid, as you may recall from Chapter 3. This difference is caused by a single nucleotide difference in the coding strand of DNA (**Figure 10.21**). In the double helix, a base pair is changed.

The sickle-cell allele is not a unique case. We now know that the various alleles of many genes result from changes in single base pairs in DNA. Any change in the nucleotide sequence of DNA is called a **mutation.** Mutations can involve large regions of a chromosome or just a single nucleotide pair, as in the sickle-cell allele. Let's consider how mutations involving only one or a few nucleotide pairs can affect gene translation.

Types of Mutations Mutations within a gene can be divided into two general categories: base substitutions and base insertions or deletions (**Figure 10.22**). A base substitution is the replacement of one base, or nucleotide, by another. Depending on how a base substitution is translated, it can result in no change in the protein, in an insignificant change, or in a change that might be crucial to the life of the organism. Because of the redundancy of the genetic code, some substitution mutations have no effect. For example, if a mutation causes an mRNA codon to change from GAA to GAG, no change in the protein product would result, because GAA and GAG both code for the same amino acid (Glu). Such a change is called a silent mutation.

Other changes of a single nucleotide do change the amino acid coding. Such mutations are called missense mutations. For example, if a mutation causes an mRNA codon to change from GGC to AGC, the resulting protein will have a serine (Ser) instead of a glycine (Gly) at this position (see Figure 10.22a). Some missense mutations have little or no effect on the resulting protein, but others, as we saw in the sickle-cell case, cause changes in the protein that prevent it from performing normally.

Occasionally, a base substitution leads to an improved protein or one with new capabilities that enhance the success of the mutant organism and its descendants. Much more often, though, mutations are harmful. Some base substitutions, called nonsense mutations, change an amino acid codon into a stop codon. For example, if an AGA (Arg) codon is mutated to a UGA (stop) codon, the result will be a prematurely terminated protein, which may not function properly.

Case Study in the Process of Science on the Web & CD Act as a medical sleuth to investigate a patient's condition.

Mutations involving the insertion or deletion of one or more nucleotides in a gene often have disastrous effects. Because mRNA is read as a series of nucleotide triplets during translation, adding or subtracting nucleotides may alter the **reading frame** (triplet grouping) of the genetic message. All the nucleotides that are "downstream" of the insertion or deletion will be regrouped into different codons. For example, consider an mRNA molecule containing the sequence AAG-UUU-GGC-GCA; this codes for Lys-Phe-Gly-Ala. If a U is deleted in the second codon, the resulting sequence will be AAG-UUG-GCG-CAU, which codes for Lys-Leu-Ala-His (see Figure 10.22b). The altered polypeptide is likely to be nonfunctional. Inserting one or two mRNA nucleotides would have a similarly large effect.

Mutagens What causes mutations? Mutagenesis, the creation of mutations, can occur in a number of ways. Mutations resulting from errors during DNA replication or recombination are known as spontaneous mutations, as are other mutations of unknown cause. Other sources of mutation are physical and chemical agents called **mutagens.** The most common physical mutagen is high-energy radiation, such as X-rays and ultraviolet (UV) light. Chemical mutagens are of various types. One type, for example, consists of chemicals that are similar to normal DNA bases but that base-pair incorrectly when incorporated into DNA.

Many mutagens can act as carcinogens, agents that cause cancer. What can you do to avoid exposure to mutagens? Several lifestyle practices can

Activity 10H on the Web & CD Learn more about agents that cause cancer.

help, including wearing protective clothing and sunscreen to minimize direct exposure to the sun's UV rays and not smoking. But such precautions are not foolproof; for example, you cannot entirely avoid UV radiation.

Although mutations are often harmful, they are also extremely useful, both in nature and in the laboratory. Mutations are the source of the rich diversity of genes in the living world, a diversity that makes evolution by natural selection possible (**Figure 10.23**). Mutations are also essential tools for geneticists. Whether naturally occurring or created in the laboratory, mutations are responsible for the different alleles needed for genetic research.

Figure 10.23 Mutations and diversity.
Mutations are the ultimate source of the diversity of life in this underwater scene near the Fiji Islands.

CHECKPOINT

1. What are transcription and translation?

2. How many nucleotides are necessary to code for a polypeptide that is 100 amino acids long?

3. An mRNA molecule contains the nucleotide sequence CCAUUUACG. Using Figure 10.11, translate this sequence into the corresponding amino acid sequence.

4. How does RNA polymerase "know" where to start transcribing a gene?

5. What is an anticodon?

6. What is the function of the ribosome in protein synthesis?

7. What would happen if a mutation changed a start codon to some other codon?

8. Once polypeptide synthesis starts, what are the three main steps by which it grows (elongates)?

9. Which of the following does not participate directly in translation: ribosomes, transfer RNA, messenger RNA, DNA, enzymes, ATP?

10. What happens when one nucleotide is lost from the middle of a gene?

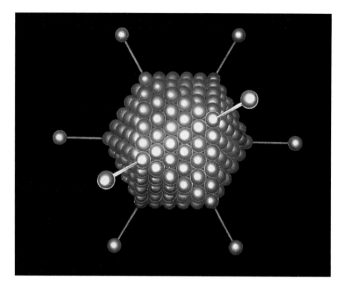

Figure 10.24 The adenovirus. A virus that infects the human respiratory system, an adenovirus consists of DNA enclosed in a protein shell shaped like a 20-sided polyhedron, shown here in a computer-generated model. At each vertex of the polyhedron is a protein spike, which helps the virus attach to a susceptible cell.

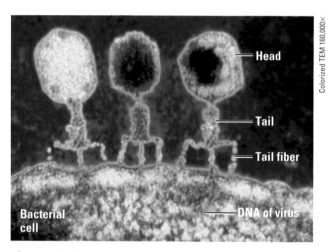

Colorized TEM 160,000×

Head

Tail

Tail fiber

Bacterial cell

DNA of virus

Figure 10.25
Bacteriophages infecting a bacterial cell.
The "landing crafts" settling on the surface of the "planet" in this electron micrograph are actually viruses in the process of infecting a bacterium. The viruses are called bacteriophage T4, and the bacterium is *E. coli*. The phage consists of a molecule of DNA enclosed within an elaborate structure made of proteins. The "legs" of the phage (the tail fibers) bend when they touch the cell surface. The tail is a hollow rod enclosed in a springlike sheath. As the legs bend, the spring compresses, the bottom of the rod punctures the cell membrane, and the viral DNA passes from inside the head of the virus into the cell. Once inside, the viral DNA takes over the molecular machinery of the cell and directs the manufacture of hundreds of new viruses, which are released when the cell bursts open. Each phage is only about 200 nm tall (2 ten-thousandths of a millimeter).

Viruses: Genes in Packages

Viruses sit on the fence between life and nonlife. A virus is lifelike in having genes and a highly organized structure, but it differs from a living organism in not being cellular or able to reproduce on its own. In many cases, a virus is nothing more than packaged genes, a bit of nucleic acid wrapped in a protein

Activity 10I on the Web & CD
Get an overview of how viruses reproduce.

coat (**Figure 10.24**). A virus can survive only by infecting a living cell with genetic material that directs the cell's molecular machinery to make more viruses. In this section, we're going to take a look at the relationship between viral structure and the processes of nucleic acid replication, transcription, and translation. We'll consider viruses that infect different types of host organisms, starting with bacteria.

Bacteriophages

Activity 10J on the Web & CD
See an animation of the lytic cycle of a phage.

Viruses that attack bacteria are called **bacteriophages** ("bacteria-eaters"), or **phages** for short (**Figure 10.25**). Once they infect a bacterium, most phages enter a reproductive cycle called the **lytic cycle.** The lytic cycle gets its name from the fact that, after many copies of the phage are produced within the bacterial cell, the bacterium lyses (breaks open). Some viruses can also reproduce by an alternative route—the **lysogenic cycle.** During a lysogenic cycle, viral DNA replication occurs without phage production or the death of the cell.

Figure 10.26 illustrates the two kinds of cycles for a phage named lambda that can infect *E.coli* bacteria. Lambda has a head (containing DNA) and a tail. Before embarking on one of the two cycles, ❶ lambda binds to the

Activity 10K on the Web & CD
Watch animations of a phage lysogenic cycle and lytic cycle.

outside of a bacterium and injects its DNA inside. ❷ The injected lambda DNA forms a circle. In the lytic cycle, this DNA immediately turns the cell into a virus-producing factory. ❸ The cell's own machinery for DNA replication, transcription, and translation is hijacked by the virus and used to produce copies of the virus. ❹ The cell lyses, releasing the new phages.

In the lysogenic cycle, ❺ the DNA inserts by genetic recombination into the bacterium's DNA. Called a **bacterial chromosome,** this DNA is a single circular molecule. Once inserted into the bacterial chromosome, the phage DNA is referred to as a **prophage,** and most of its genes are inactive. Survival of the prophage depends on the reproduction of the cell where it resides. ❻ The host cell replicates the prophage DNA along with its cellular DNA and

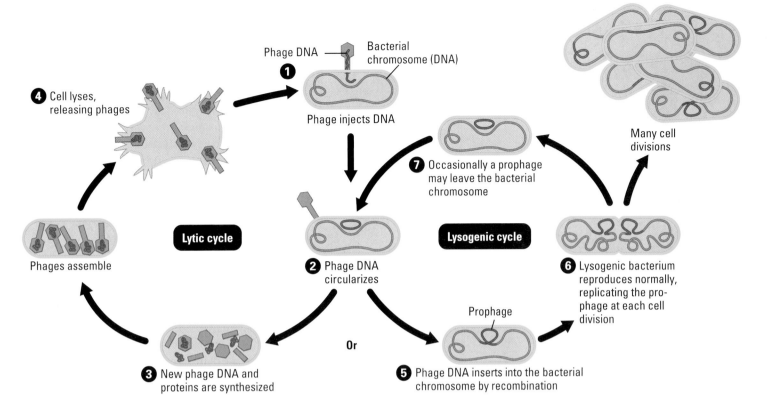

① Phage injects DNA

Phage DNA — Bacterial chromosome (DNA)

④ Cell lyses, releasing phages

Lytic cycle

Phages assemble

② Phage DNA circularizes

③ New phage DNA and proteins are synthesized

Or

⑦ Occasionally a prophage may leave the bacterial chromosome

Lysogenic cycle

Many cell divisions

⑥ Lysogenic bacterium reproduces normally, replicating the prophage at each cell division

Prophage

⑤ Phage DNA inserts into the bacterial chromosome by recombination

then, upon dividing, passes on both the prophage and the cellular DNA to its two daughter cells. A single infected bacterium can quickly give rise to a large population of bacteria that all carry prophages. The prophages may remain in the bacterial cells indefinitely. ⑦ Occasionally, however, one leaves its chromosome; this event may be triggered by environmental conditions such as exposure to a mutagen. Once separate, the lambda DNA usually switches to the lytic cycle, which results in the production of many copies of the virus and bursting of the host cell.

Sometimes the few prophage genes active in a lysogenic bacterial cell can cause medical problems. For example, the bacteria that cause diphtheria, botulism, and scarlet fever would be harmless to humans if it were not for the prophage genes they carry. Certain of these genes direct the bacteria to produce the toxins responsible for making people ill.

Plant Viruses

Viruses that infect plant cells can stunt plant growth and diminish crop yields. Most plant viruses discovered to date have RNA rather than DNA as their genetic material. Many of them, like the tobacco mosaic virus in **Figure 10.27**, are rod-shaped with a spiral arrangement of proteins surrounding the nucleic acid.

To infect a plant, a virus must first get past the plant's outer protective layer of cells (the epidermis). For this reason, a plant damaged by wind, chilling, injury, or insects is more susceptible to infection than a healthy plant. Some insects also carry and transmit plant viruses. Farmers and gardeners, too, may spread plant viruses through the use of pruning shears and other tools. And infected plants may pass viruses to their offspring.

Once a virus enters a plant cell and begins reproducing, it can spread throughout the entire plant through plasmodesmata, the channels that connect the cytoplasms of adjacent plant cells (see Figure 4.21). As with animal

Figure 10.26
Alternative types of phage reproductive cycles.
Certain phages, like lambda, can undergo alternative reproductive cycles. After entering the bacterial cell, the phage DNA can either integrate into the bacterial chromosome (lysogenic cycle) or immediately start the production of progeny phages (lytic cycle), destroying the cell. In most cases, the phage follows the lytic pathway, but once it enters a lysogenic cycle, the prophage (the phage's DNA) may be carried in the host cell's chromosome for many generations. The bacterial chromosome is a single DNA molecule in the form of a closed loop.

Protein   RNA

Figure 10.27 The tobacco mosaic virus. The photo shows the mottling of leaves in tobacco mosaic disease. The rod-shaped virus causing the disease has RNA as its genetic material.

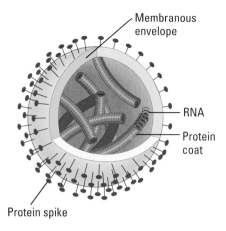

Figure 10.28 An influenza virus. The genetic material of this virus consists of eight separate molecules of RNA, each wrapped in a protein coat. Around the outside of the virus is an envelope made of membrane, studded with protein spikes.

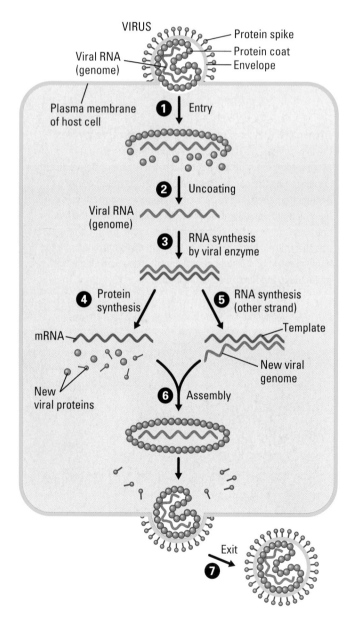

Figure 10.29 The reproductive cycle of an enveloped virus. This virus is the one that causes mumps. Like the flu virus, it has a membranous envelope with protein spikes, but its genome is a single molecule of RNA.

viruses, there is no cure for most viral diseases of plants, and agricultural scientists focus on reducing the number of plants that are infected and on breeding varieties of crop plants that resist viral infection. For example, strains of tomato, squash, and cantaloupe have been bred to be resistant to the tobacco mosaic virus, which can infect a wide variety of agricultural crops.

More recently, genetic engineering methods (discussed in detail in Chapter 12) have been used to create plant breeds that are resistant to some plant viruses. For example, this has been done with the papaya, Hawaii's second largest crop. The spread of papaya ringspot potyvirus (PRSV) by aphids had wiped out the papaya in certain island regions. But since 1998, farmers have been able to plant a newly engineered PRSV-resistant strain of papaya, and papayas are now being reintroduced into their old habitats.

Animal Viruses

Viruses that infect animal cells are common causes of disease. We have all suffered from viral infections. **Figure 10.28** shows the structure of an influenza (flu) virus. Like many animal viruses, this one has an outer envelope made of phospholipid membrane, with projecting spikes of protein. The envelope helps the virus enter and leave a cell. Also, like many other animal viruses, flu viruses have RNA rather than DNA as their genetic material. Other RNA viruses include those that cause the common cold, measles, and mumps, as well as ones that cause more serious human diseases, such as AIDS and polio. Diseases caused by DNA viruses include hepatitis, chicken pox, and herpes infections.

Figure 10.29 shows the reproductive cycle of an enveloped RNA virus (the mumps virus). When the virus contacts a susceptible cell, protein spikes on its outer surface attach to receptor proteins on the cell's plasma membrane. The viral envelope fuses with the cell's membrane, allowing the protein-coated RNA to ❶ enter the cytoplasm. ❷ Enzymes then remove the protein coat. ❸ An enzyme that entered the cell as part of the virus uses the virus's RNA genome as a template for making complementary strands of RNA. The new strands have two functions: ❹ They serve as mRNA for the synthesis of new viral proteins, and ❺ they serve as templates for synthesizing new viral genome RNA. ❻ The new coat proteins assemble around the new viral RNA. ❼ Finally, the viruses leave the cell by cloaking themselves in plasma membrane. In other words, the virus obtains its envelope from the cell, leaving the cell without necessarily lysing it.

Not all animal viruses reproduce in the cytoplasm. For example, the viruses called herpesviruses—which cause chicken pox, shingles, cold sores, and genital herpes—are enveloped DNA viruses that reproduce in a cell's nucleus, and they get their envelopes from the cell's nuclear membranes. Copies of the herpesvirus DNA usually remain behind as mini-chromosomes in the nuclei of certain nerve cells. There they remain latent until some sort of physical stress, such as a cold or sunburn, or emotional stress triggers the herpesvirus DNA to begin producing the virus, resulting in unpleasant symptoms. Once acquired, herpes infections may flare up repeatedly throughout a person's life.

The amount of damage a virus causes the body depends partly on how quickly the immune system responds to fight the infection and partly on the

the ability of the infected tissue to repair itself. We usually recover completely from colds because our respiratory tract tissue can efficiently replace damaged cells by mitosis. In contrast, the poliovirus attacks nerve cells, which do not usually divide. The damage to such cells by polio, unfortunately, is permanent. In such cases, we try to prevent the disease with vaccines. The antibiotic drugs that help us recover from bacterial infections are powerless against viruses. The development of antiviral drugs has been slow because it is difficult to find ways to kill a virus without killing the host cells.

HIV, the AIDS Virus

The devastating disease AIDS is caused by a type of RNA virus with some special twists. In outward appearance, the AIDS virus (**Figure 10.30a**) somewhat resembles the flu or mumps virus. Its envelope enables HIV to enter and leave a cell much the way the mumps virus does. But HIV has a different mode of reproduction. It is a **retrovirus,** an RNA virus that reproduces by means of a DNA molecule. Retroviruses are so named because they reverse the usual DNA → RNA flow of genetic information. They carry molecules of an enzyme called **reverse transcriptase,** which catalyzes reverse transcription, the synthesis of DNA on an RNA template.

Figure 10.30b illustrates what happens after HIV RNA is uncoated in the cytoplasm of a cell. The reverse transcriptase (green) ❶ uses the RNA as a template to make a DNA strand and then ❷ adds a second, complementary DNA strand. ❸ The resulting double-stranded viral DNA then enters the cell nucleus and inserts itself into the chromosomal DNA, becoming a **provirus.** Occasionally, the provirus is ❹ transcribed into RNA and ❺ translated into viral proteins. ❻ New viruses assembled from these components eventually leave the cell and can then infect other cells. This is the standard reproductive cycle for retroviruses.

> **Activity 10L on the Web & CD**
> See an animation of HIV reproducing.

AIDS stands for acquired immuneodeficiency syndrome, and **HIV** for human immunodeficiency virus; these terms describe the main effect of the virus on the body. HIV infects and eventually kills several kinds of white blood cells that are important in the body's immune system (**Figure 10.30c**). The loss of such cells causes the body to become susceptible to other infections that it would normally be able to fight off. Such secondary infections cause the syndrome (a collection of symptoms) that eventually kills AIDS patients.

> **Case Study in the Process of Science on the Web & CD**
> Learn about the infections that affect AIDS patients.

Two main types of anti-HIV drugs are currently used in treating AIDS. Both interfere with the reproduction of the virus. The first type inhibits the action of the HIV enzyme reverse transcriptase; AZT, mentioned at the start of this chapter, is the most widely used drug of this type. The second type inhibits the action of enzymes called proteases, which are needed for making HIV proteins. Many HIV-infected people in the United States and other developed countries take both reverse-transcriptase inhibitors and protease inhibitors, and the combination seems to be much more effective than the individual drugs in keeping the virus at bay and extending the patients'

> **Case Study in the Process of Science on the Web & CD**
> Find out why AIDS rates differ across the United States.

lives. However, even in combination, the drugs do not completely rid the body of the virus. Typically, HIV reproduction and the symptoms of AIDS return if a patient discontinues the medications. AIDS has no cure yet—leaving prevention (by avoiding unprotected sex and needle sharing) as the only healthy option.

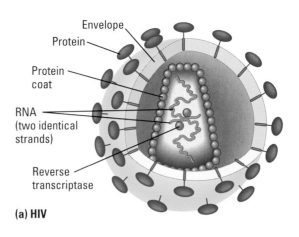

(a) HIV

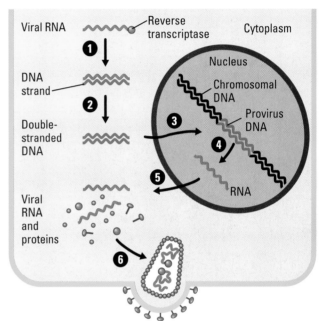

(b) The behavior of HIV nucleic acid in an infected cell

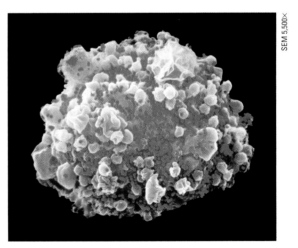

SEM 5,500×

(c) HIV infecting a white blood cell

Figure 10.30 HIV, the AIDS virus. **(a)** The structure of HIV, the human immunodeficiency virus. **(b)** Once the virus enters a cell and its RNA is unwrapped, it can produce new viruses as shown here. **(c)** HIV (red spots) in the process of infecting a white blood cell.

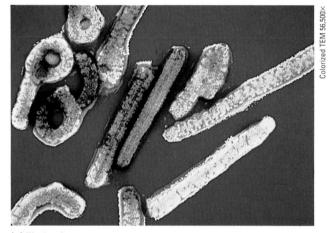

(a) **Ebola virus**

Colorized TEM 56,500×

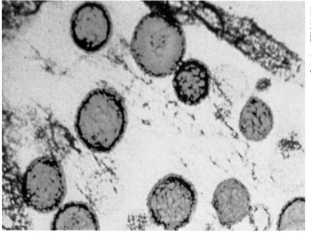

(b) **Hantavirus**

Colorized TEM 26,000×

Figure 10.31 Emerging viruses. Emerging viruses are viruses that have recently appeared or recently come to the attention of medical scientists. **(a)** Ebola virus, an enveloped thread of protein-coated RNA. **(b)** Hantavirus, another enveloped RNA virus.

Evolution Connection

Emerging Viruses

Acknowledging the persistent threat that viruses pose to human health, geneticist and Nobel Prize winner Joshua Lederberg once warned, "We live in evolutionary competition with microbes. There is no guarantee that we will be the survivors." Lederberg cited the AIDS epidemic and recurrent flu epidemics as examples of the human population's vulnerability to viral attacks.

The AIDS virus (HIV), which seemed to arise abruptly in the early 1980s, and the new varieties of flu virus that frequently appear, are not the only examples of newly dangerous viruses. A deadly virus called the Ebola virus menaces central African nations periodically, and many biologists fear its emergence as a global threat (**Figure 10.31a**). Another newly dangerous virus was responsible for killing dozens of people in the southwestern United States in 1993. It turned out to be the hantavirus, known to Western medicine since the Korean War and named for a river in Korea (**Figure 10.31b**). In 2003, a deadly new disease called SARS (severe acute respiratory syndrome) appeared in China and soon spread throughout the world (**Figure 10.32**). SARS is caused by a previously unknown member of the coronavirus family, viruses that usually cause only mild cold-like symptoms in humans.

Although the viruses mentioned here may not have originated as recently as once thought, they still raise intriguing questions: How do new viruses arise, and what causes certain viruses to emerge as major threats? Most biologists favor the hypothesis that the very first viruses arose from fragments of cellular nucleic acid that had some means of moving from one cell to another. Consistent with this idea, viral nucleic acid usually has more in common with its host cell's DNA than with the nucleic acid of viruses that infect other hosts. Indeed, some viral genes are virtually identical to genes of their host. The earliest viruses may have been naked bits of nucleic acid that traveled from one cell to another via injured cell surfaces. Genes coding for viral coat proteins may have evolved later.

Evolution Connection on the Web
Discover why you can get the flu year after year after year . . .

Today, the mutation of existing viruses is a major source of new viral diseases. The RNA viruses tend to have an unusually high rate of mutation because the replication of their nucleic acid does not involve proofreading steps, as does DNA replication. Some mutations may enable existing viruses to evolve into new genetic varieties that can cause disease in individuals who had developed immunity to the ancestral virus. Flu epidemics are caused by viruses that are genetically different enough from earlier years' viruses that people have little immunity to them.

Another source of new viral diseases is the spread of existing viruses to a new host species. For example, hantavirus is common in rodents, especially deer mice. The population of deer mice in the southwestern United States exploded in 1993 after unusually wet weather increased the rodents' food supply. Humans acquired hantavirus when they inhaled dust containing traces of urine and feces from infected mice.

Finally, a viral disease may start out in a small, isolated population and then rather suddenly become widespread. AIDS, for example, went unnamed and virtually unnoticed for decades before starting to spread around the world. In this case, technological and social factors, including affordable international travel, blood transfusion technology, sexual promiscuity, and the abuse of intravenous drugs, allowed a previously rare human disease to become a global epidemic. It is likely that when we do find the means to control HIV and other deadly viruses, genetic research—in particular, molecular biology—will be responsible for the discovery.

Figure 10.32 SARS virus. Young girls in Hong Kong attempt to protect themselves from exposure to the SARS virus.

Chapter Review

Summary of Key Concepts

For study help, go to the Essential Biology Website (www.essentialbiology.com) or CD-ROM to explore the Activities and Case Studies in the Process of Science.

The Structure and Replication of DNA

Activity 10A The Hershey-Chase Experiment

- **DNA and RNA: Polymers of Nucleotides**

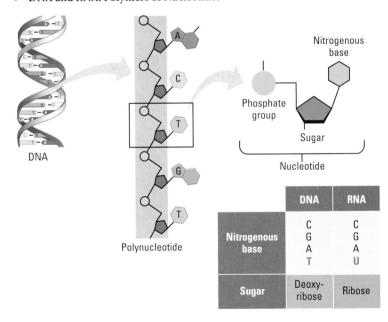

Activity 10B DNA and RNA Structure

- **Watson and Crick's Discovery of the Double Helix**   Watson and Crick worked out the three-dimensional structure of DNA: two polynucleotide strands wrapped around each other in a double helix. Hydrogen bonds between bases hold the strands together. Each base pairs with a complementary partner: A with T, and G with C.

Activity 10C DNA Double Helix

- **DNA Replication**

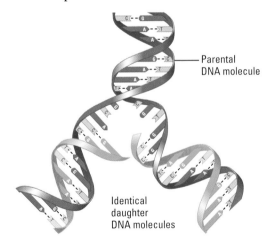

Case Study in the Process of Science *What Is the Correct Model for DNA Replication?*

Activity 10D DNA Replication

The Flow of Genetic Information from DNA to RNA to Protein

- **How an Organism's DNA Genotype Produces Its Phenotype**
The information constituting an organism's genotype is carried in the sequence of its DNA bases. Studies of inherited metabolic defects first suggested that phenotype is expressed through proteins. A particular gene—a linear sequence of many nucleotides—specifies a polypeptide. The DNA of the gene is transcribed into RNA, which is translated into the polypeptide.

Activity 10E Overview of Protein Synthesis

Case Study in the Process of Science How Is a Metabolic Pathway Analyzed?

- **From Nucleotide Sequence to Amino Acid Sequence: An Overview**
The DNA of a gene is transcribed into RNA using the usual base-pairing rules, except that an A in DNA pairs with U in RNA. In the translation of a genetic message, each triplet of nucleotide bases in the RNA, called a codon, specifies one amino acid in the polypeptide.

- **The Genetic Code**   In addition to codons that specify amino acids, the genetic code has one that is a start signal and three that are stop signals for translation. The genetic code is redundant: There is more than one codon for each amino acid.

- **Transcription: From DNA to RNA**   In transcription, RNA polymerase binds to the promoter of a gene, opens the DNA double helix there, and catalyzes the synthesis of an RNA molecule using one DNA strand as a template. As the single-stranded RNA transcript peels away from the gene, the DNA strands rejoin.

- **The Processing of Eukaryotic RNA**   The RNA transcribed from a eukaryotic gene is processed before leaving the nucleus to serve as messenger RNA (mRNA). Introns are spliced out, and a cap and tail are added.

Activity 10F Transcription and RNA Processing

- **Translation: The Players**

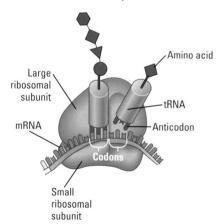

- **Translation: The Process**   In initiation, a ribosome assembles with the mRNA and the initiator tRNA bearing the first amino acid. One by one, the codons of the mRNA are recognized by tRNAs bearing succeeding amino acids. The ribosome bonds the amino acids together. With each addition, the mRNA translocates by one codon through the ribosome. When a stop codon is reached, the completed polypeptide is released.

Activity 10G Translation

- **Review: DNA → RNA → Protein**   Figure 10.20 summarizes transcription, RNA processing, and translation. The sequence of codons in DNA, via the sequence of codons in mRNA, spells out the primary structure of a polypeptide.

- **Mutations**   Mutations are changes in the DNA base sequence, caused by errors in DNA replication or by mutagens. Substituting, inserting, or deleting nucleotides in a gene has varying effects on the polypeptide and organism.

Case Study in the Process of Science How Do You Diagnose a Genetic Disorder?

Activity 10H Causes of Cancer

Viruses: Genes in Packages

Activity 10I Simplified Reproductive Cycle of a Virus

- **Bacteriophages**   Viruses can be regarded simply as genes packaged in protein. When phage DNA enters a lytic cycle inside a bacterium, it is replicated, transcribed, and translated. The new viral DNA and protein molecules then assemble into new phages, which burst from the cell. In the lysogenic cycle, phage DNA inserts into the cell's chromosome and is passed on to generations of daughter cells. Much later, it may initiate phage production.

Activity 10J Phage Lytic Cycle

Activity 10K Phage Lysogenic and Lytic Cycles

- **Plant Viruses**   Viruses that infect plants can be a serious agricultural problem. Most have RNA genomes. They enter plants via breaks in the plant's outer layers.

- **Animal Viruses**   Many animal viruses, such as flu viruses, have RNA genomes; others, such as hepatitis viruses, have DNA. Some animal viruses "steal" a bit of cell membrane as a protective envelope. Some insert their DNA into a cellular chromosome and remain latent for long periods.

- **HIV, the AIDS Virus**   HIV is a retrovirus. Inside a cell it uses its RNA as a template for making DNA, which is then inserted into a chromosome.

Activity 10L HIV Reproductive Cycle

Case Study in the Process of Science What Causes Infections in AIDS Patients?

Case Study in the Process of Science Why Do AIDS Rates Differ Across the United States?

Self-Quiz

1. A molecule of DNA contains two polymer strands called _____, made by bonding many monomers called _____ together. Each monomer contains three parts: _____, _____, and _____.

2. Which of the following correctly ranks nucleic acid structures in order of size, from largest to smallest?

   a. gene, chromosome, nucleotide, codon

   b. chromosome, gene, codon, nucleotide

   c. nucleotide, chromosome, gene, codon

   d. chromosome, nucleotide, gene, codon

3. A scientist inserts a radioactively labeled DNA molecule into a bacterium. The bacterium replicates this DNA molecule and distributes one daughter molecule (double helix) to each of two daughter cells. How much radioactivity will the DNA in each of the two daughter cells contain? Why?

4. The nucleotide sequence of a DNA codon is GTA. In an mRNA molecule transcribed from this DNA, the codon has the sequence _____. In the process of protein synthesis, a transfer RNA pairs with the mRNA codon. The nucleotide sequence of the tRNA anticodon is _____. The amino acid attached to the tRNA is _____. (See Figure 10.11.)

5. Describe the process by which the information in a gene is transcribed and translated into a protein. Correctly use these terms in your description: tRNA, amino acid, start codon, transcription, mRNA, gene, codon, RNA polymerase, ribosome, translation, anticodon, peptide bond, stop codon.

6. Match the following molecules with the cellular process or processes in which they are primarily involved.

a. ribosomes   1. DNA replication

b. tRNA   2. transcription

c. DNA polymerases   3. translation

d. RNA polymerase

e. mRNA

7. A geneticist found that a particular mutation had no effect on the polypeptide encoded by the gene. This mutation probably involved

a. deletion of one nucleotide.

b. alteration of the start codon.

c. insertion of one nucleotide.

d. substitution of one nucleotide.

8. Scientists have discovered how to put together a bacteriophage with the protein coat of phage A and the DNA of phage B. If this composite phage were allowed to infect a bacterium, the phages produced in the cell would have

a. the protein of A and the DNA of B.

b. the protein of B and the DNA of A.

c. the protein and DNA of A.

d. the protein and DNA of B.

9. HIV requires an enzyme called _____ to convert its RNA genome into a DNA version.

10. Why is reverse transcriptase a particularly good target for anti-AIDS drugs? (*Hint:* Would you expect such a protocol to harm the human host?)

Answers to the Self-Quiz questions can be found in Appendix B.

Go to the website or CD-ROM for more Self-Quiz questions.

The Process of Science

1. A cell containing a single chromosome is placed in a medium containing radioactive phosphate, making any new DNA strands formed by DNA replication radioactive. The cell replicates its DNA and divides. Then the daughter cells (still in the radioactive medium) replicate their DNA and divide, resulting in a total of four cells. Sketch the DNA molecules in all four cells, showing

a normal (nonradioactive) DNA strand as a solid line and a radioactive DNA strand as a dashed line.

2. In a classic 1952 experiment, biologists Alfred Hershey and Martha Chase labeled two batches of bacteriophage, one with radioactive sulfur (which only tags protein) and the other with radioactive phosphorus (which only tags DNA). In separate test tubes, they allowed each batch of phage to bind to non-radioactive bacteria and inject its DNA. After a few minutes, they separated the bacterial cells from the viral parts that remained outside the bacterial cells and measured the radioactivity of both portions. What results do you think they obtained? How would these results help them to determine which viral component—DNA or protein—was the infectious portion?

Case Study in the Process of Science on the Web & CD *Virtually reenact famous experiments that tested different models of DNA replication.*

Case Study in the Process of Science on the Web & CD *Virtually reenact George Beadle and Edward Tatum's experiments.*

Case Study in the Process of Science on the Web & CD *Act as a medical sleuth to investigate a patient's condition.*

Case Study in the Process of Science on the Web & CD *Learn about the infections that affect AIDS patients.*

Case Study in the Process of Science on the Web & CD *Find out why AIDS rates differ across the United States.*

Biology and Society

1. Scientists at the National Institutes of Health (NIH) have worked out thousands of sequences of genes and the proteins they encode, and similar analysis is being carried out at universities and private companies. Knowledge of the nucleotide sequences of genes might be used to treat genetic defects or produce lifesaving medicines. The NIH and some United States biotechnology companies have applied for patents on their discoveries. In Britain, the courts have ruled that a naturally occurring gene cannot be patented. Do you think individuals and companies should be able to patent genes and gene products? Before answering, consider the following: What are the purposes of a patent? How might the discoverer of a gene benefit from a patent? How might the public benefit? What negative effects might result from patenting genes?

2. AZT and other anti-HIV drugs have the potential to prevent the transmission of HIV to the children of infected mothers. But anti-AIDS drugs are too expensive for most people in developing nations. Some researchers have proposed a series of trials in which HIV-infected pregnant women are given minimal doses of anti-HIV medications. Such research has the potential to determine the smallest effective dose of medication required to prevent transmission, and this could save many lives by reducing the cost of drug treatment. But such a trial would also result in many HIV-infected babies being born—a situation that could potentially have been prevented. Do you think such trials should be permitted? What are the potential benefits? The potential costs? How do you measure the relative value of each?

Biology and Society on the Web *Learn about other drugs that combat HIV.*

CHAPTER 11

Gene Regulation

Biology and Society:
Baby's First Bank Account   197

From Egg to Organism:
How and Why Genes Are
Regulated   197

Patterns of Gene Expression in
Differentiated Cells

DNA Microarrays:
Visualizing Gene Expression

The Genetic Potential of Cells

Reproductive Cloning of Animals

Therapeutic Cloning and Stem Cells

The Regulation of
Gene Expression   203

Gene Regulation in Bacteria

Gene Regulation in the
Nucleus of Eukaryotic Cells

Regulation in the Cytoplasm

Cell Signaling

The Genetic Basis
of Cancer   208

Genes That Cause Cancer

The Effects of Lifestyle on Cancer Risk

Evolution Connection:
Homeotic Genes   212

Shortly after you drink a milk shake, bacteria living in your large intestine turn on certain genes.

Cancer-causing genes were first discovered in a chicken virus.

If a salamander loses a leg, it can grow a new one.

Lung cancer causes more deaths than any other kind of cancer.

Biology and Society

Baby's First Bank Account

In recent years, many new parents have opened savings accounts for their newborns. But these are not the traditional college tuition plans. In fact, these accounts don't contain any money. Instead, they contain frozen cells collected from their child's umbilical cord and placenta at birth. To open one of these accounts, a physician inserts a needle into the umbilical cord 10 minutes after birth and extracts ¼ to ½ cup of blood. After collection at the hospital, the blood is sent to a storage facility and frozen in liquid nitrogen (**Figure 11.1**).

Why would you save such a thing? The reason is that umbilical cord blood is rich in special cells, known as stem cells, that can give rise to many of the body's other types of cells. Stem cells have many potential medical applications. For example, some children stricken by leukemia appear to have been successfully cured after treatment with their own stem cells.

As of 2003, there are about 150,000 cord blood samples stored worldwide, at an initial cost of about $1,500 each, plus an annual storage fee of about $100. (These fees are not usually covered by medical insurance.) The American Academy of Pediatrics recommends cord blood banking only for babies born into families with a known genetic risk. Realistically, the chance of needing stored cells and successfully using them is miniscule, but anxious parents are willing to spend the money just in case. So far, the promise of cord blood banking vastly exceeds the accomplishments. Parents should be wary of the claims of some banking companies—that the stored cells can be used to cure paralysis or grow a new heart, for example—that are, at this time, wholly unsubstantiated.

The power of stem cells lies in their ability to develop into a variety of different body cells. Most cells of the adult body lack this ability. The cellular process that underlies this potential is gene regulation, the turning on and off of genes. How genes are controlled and the role of gene regulation in the lives of organisms are the subject of this chapter. As you'll see, the subject of gene regulation touches on some of the most interesting topics in all of biology, including cloning, stem cell research, and cancer.

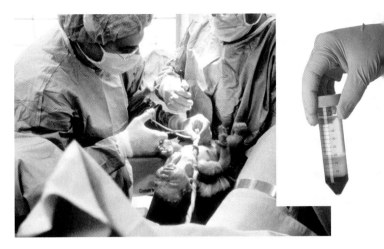

Figure 11.1 Umbilical cord blood banking.
Just after birth, a doctor may remove blood from a newborn's umbilical cord. The umbilical cord blood (inset)—rich in stem cells—is stored for possible future use.

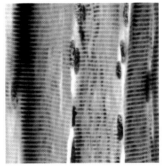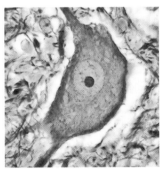

(a) Three muscle cells (partial)   **(b) A nerve cell (partial)**

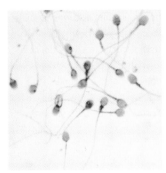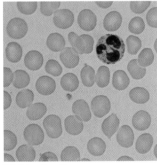

(c) Sperm cells   **(d) Blood cells**

From Egg to Organism: How and Why Genes Are Regulated

The micrographs in **Figure 11.2** show four of the many different types of human cells. Each type of cell has a unique structure appropriate for carrying out its function. For example, a sperm cell has a powerful flagellum that

Figure 11.2 Four types of human cells. All four micrographs are shown at the same magnification (750×). **(a)** This micrograph shows short segments of three muscle cells, which are long fibers with multiple nuclei (dark vertical rods here). The horizontal stripes result from the arrangement of the proteins that enable the cells to contract. **(b)** The large orange shape is part of a nerve cell, with its nucleus appearing as a lighter orange disk. (The dark spot in the nucleus is the nucleolus.) You can see only a little of this cell's three long extensions, which carry nerve signals from one part of the body to another. **(c)** Sperm cells have long flagella that can propel them to meet an egg cell. **(d)** A single white blood cell (with a three-part darkly stained nucleus) is surrounded by red blood cells. The small size and disklike shape of red blood cells help them efficiently transport oxygen molecules through the narrowest of blood vessels.

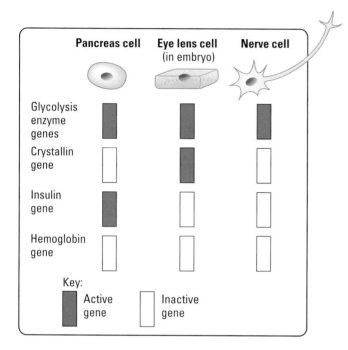

Figure 11.3 Patterns of gene expression.
Shown here are the activity states of four genes in three types of human cells. Different types of cells express different combinations of genes. The genes for the enzymes of glycolysis are "housekeeping" genes; they are active in all metabolizing cells. The specialized proteins represented here are the transparent protein crystallin, which forms the lens of the eye; insulin, a hormone made in the pancreas; and the oxygen transport protein hemoglobin. Notice that the hemoglobin genes are not active in any of the cell types shown here. They are turned on only in cells that are developing into red blood cells. Insulin genes are activated only in the cells of the pancreas that produce that hormone. Nerve cells express genes for other specialized proteins not indicated here.

can propel it through a woman's reproductive tract to meet an egg, and a nerve cell has long extensions for conducting nerve signals between widely separated parts of the body. Together, the various types of human cells enable a person to function as a whole.

What makes each of these types of cells different? Do they have different genes? Remember that every somatic cell in your body (that is, every cell except your gametes) was produced by repeated rounds of mitosis that started with a zygote. Since mitosis duplicates the genome exactly (or very nearly so), each of the many different kinds of cells in your body has the same DNA as the zygote.

How, then, do cells become different from one another? The only way that cells with the same genetic information can develop into cells with different structures and functions is if gene activity is regulated. In some way, control mechanisms must determine that certain genes will be turned on while others remain turned off in a particular cell. In other words, individual cells must undergo **cellular differentiation**—that is, they must become specialized in structure and function. It is the regulation of genes that leads to this specialization.

Patterns of Gene Expression in Differentiated Cells

What does it mean to say that genes are active or inactive, turned on or off? As we discussed in Chapter 10, genes determine the nucleotide sequences of specific mRNA molecules, and mRNA in turn determines the sequences of amino acids in protein molecules. A gene that is turned on is being transcribed into RNA, and that message is being translated into specific protein molecules. The overall process by which genetic information flows from genes to proteins—that is, from genotype to phenotype—is called **gene expression.**

Because all the differentiated cells in an individual organism contain the same genes, and all the genes have the potential of being expressed, the great differences among cells in an organism must result from the selective expression of genes. The regulation of gene expression within cells and across cell boundaries plays a central role in the development of a unicellular zygote into a multicellular organism. During embryonic development, groups of cells follow diverging developmental pathways, and each group develops into a particular kind of tissue. In the mature organism, each cell type—nerve or pancreas, for instance—has a different pattern of turned-on genes.

For example, consider the patterns of gene expression for four genes in three different specialized cells of an adult human (**Figure 11.3**). The genes for the enzymes of the metabolic pathway of glycolysis are on in all the cells. However, the genes for specialized proteins are expressed only by particular kinds of cells.

DNA Microarrays: Visualizing Gene Expression

The selective expression of genes can be visualized using a recently developed technology called a **DNA microarray** (**Figure 11.4a**). A DNA microarray is a glass slide carrying thousands of different kinds of single-stranded DNA fragments arranged in an array (grid). Each DNA fragment is obtained from a particular gene; a single microarray thus carries DNA from thousands of genes.

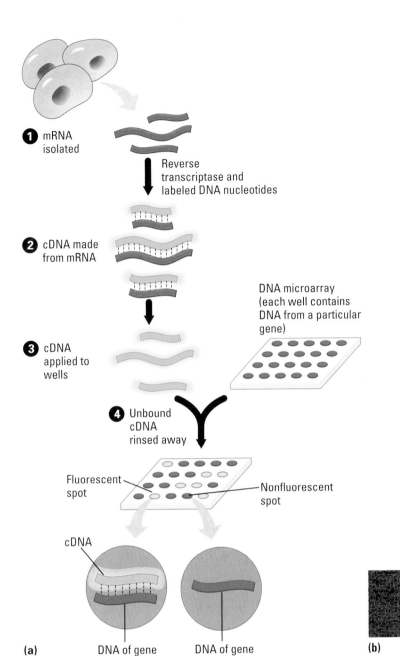

(a)

Figure 11.4 A DNA microarray.
(a) To visualize gene expression, researchers use a collection of mRNA from one type of cell to make fluorescently labeled complementary DNA (cDNA). The cDNA is applied to thousands of single-stranded DNA fragments from a large number of different genes fixed to the DNA microarray. After unbound cDNA is rinsed away, any remaining fluorescent spots represent active genes in the original cells. **(b)** Microarrays are used for testing the activity of many genes at one time. Shown actual size, this microarray contains tiny samples of 6,400 human genes arranged in an orderly pattern.

DNA microarray, actual size (6,400 genes)

(b)

To use a microarray, ❶ a researcher collects all of the mRNA transcribed from genes in a particular type of cell. This mRNA is mixed with reverse transcriptase, a viral enzyme (see Chapter 10) that ❷ catalyzes the synthesis of DNA molecules that are complementary to each mRNA sequence. This cDNA (complementary DNA) is produced in the presence of nucleotides that have been modified to fluoresce (glow). ❸ A small amount of the fluorescently labeled cDNA mixture is added to each of the thousands of kinds of single-stranded DNA fragments in the microarray. If a molecule in the cDNA mixture is complementary to a DNA fragment at a particular location, the cDNA molecule binds to it, becoming fixed there. ❹ After nonbinding cDNA is rinsed away, the remaining cDNA produces a detectable glow in the microarray (**Figure 11.4b**). The pattern of glowing spots enables the researcher to determine which genes are turned on or off in the starting cells.

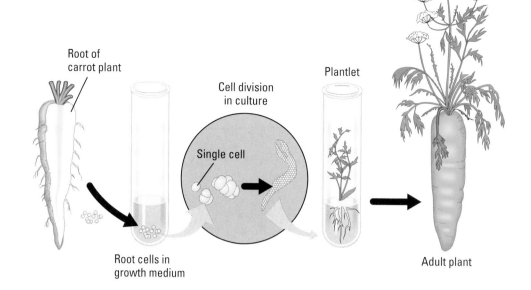

Figure 11.5
Test-tube cloning of a carrot plant.
An entire carrot plant can be grown from a differentiated root cell. The new plant is a genetic duplicate of the parent plant. This process proves that mature plant cells can dedifferentiate and then redifferentiate into all the specialized cells of an adult plant.

Root of
carrot plant

Cell division
in culture

Plantlet

Single cell

Root cells in
growth medium

Adult plant

Researchers can use microarrays to learn which genes are active in different tissues or in tissues from individuals in different states of health. This powerful new tool is giving researchers new insight into gene regulation.

The Genetic Potential of Cells

If different cells with specific functions contain a complete set of DNA, then every cell has the potential to act like every other cell if its pattern of gene expression is altered. If you have ever grown a plant from a small cutting, you've seen evidence that differentiated plant cells have the ability to develop into a whole new organism. **Figure 11.5** shows this process using a carrot plant. A single cell removed from the carrot's root and placed in a growth medium may begin dividing and eventually grow into an adult plant. This technique can be used to produce hundreds or thousands of genetically identical organisms—**clones**—from the somatic cells of a single plant. In this way, growers can propagate large numbers of plants that have desirable traits, such as high fruit yield and resistance to disease. The success of plant cloning shows that a mature plant cell can dedifferentiate— that is, lose its differentiation—and then redifferentiate to give rise to all the specialized cells of a new plant. This ability indicates that differentiation does not cause irreversible changes in the DNA.

In animals, an example of naturally occurring dedifferentiation makes possible **regeneration,** the regrowth of lost body parts. When a salamander loses a leg, for example, certain cells in the leg stump reverse their differentiated state, divide, and then redifferentiate to give rise to a new leg.

Reproductive Cloning of Animals

Cells from animals that do not naturally regenerate major body parts also retain their full genetic potential. This was demonstrated in the 1950s when researchers replaced the nuclei of frog egg cells with nuclei from tadpole intestinal cells. Some of the resulting embryos developed into tadpoles and then frogs. The replacement of an egg's nucleus with one from a differentiated cell, called **nuclear transplantation,** has been used to clone a variety of animals (**Figure 11.6**).

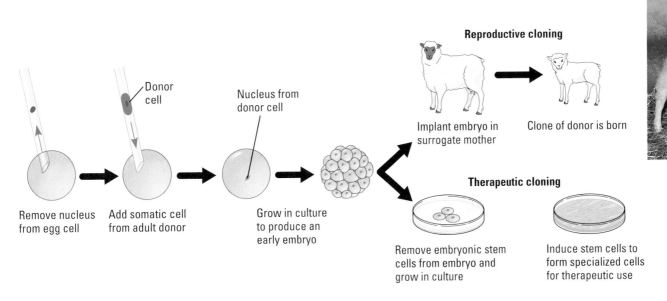

Reproductive cloning

Implant embryo in surrogate mother

Clone of donor is born

Therapeutic cloning

Remove embryonic stem cells from embryo and grow in culture

Induce stem cells to form specialized cells for therapeutic use

Donor cell

Nucleus from donor cell

Remove nucleus from egg cell

Add somatic cell from adult donor

Grow in culture to produce an early embryo

Figure 11.6 Cloning by nuclear transplantation.
Nuclear transplantation involves the injection of an adult somatic nucleus into a nucleus-free egg cell. The resulting embryo may then be used to produce a new organism (reproductive cloning, shown in the upper branch) or to provide stem cells (therapeutic cloning, lower branch). The lamb in the photograph is the famous Dolly, shown with her surrogate mother. Dolly died in 2003 of causes unrelated to her unusual origin.

(a) Piglets

(b) Banteng

A breakthrough in animal cloning came in 1997 with the first clone of a mammal using a nucleus from an adult cell (the upper branch of Figure 11.6). In this case, Scottish researchers used an electric shock to fuse specially treated sheep udder cells with 277 eggs from which they had removed the nuclei. After several days of growth, 29 of the resulting embryos were implanted in the uteruses of unrelated ewes (surrogate mothers). One of the embryos developed into the celebrated Dolly. As expected, Dolly resembled her genetic parent, the udder cell donor, not the egg donor or the surrogate ewe. Within a few years, the nuclear transplantation procedure was used to clone several other species, including mice, cows, goats, pigs, and cats.

The procedure depicted in the upper branch of Figure 11.6 is called **reproductive cloning** because it results in the creation of a cloned animal. Why would anyone want to do this? In agriculture, farm animals with specific sets of desirable traits might be cloned to produce herds of animals with these traits. In research, genetically identical animals may help researchers identify the roles of specific genes. The pharmaceutical industry is experimenting with cloning animals that have potential medical use. For example, the pigs in **Figure 11.7a** are clones that lack a gene for a protein that can cause immune system rejection in humans. Wildlife biologists hope that reproductive cloning can be used to restock populations of endangered animals. For example, in 2003 an American biotechnology company cloned a banteng, an endangered wild ox, using a cow as a surrogate mother (**Figure 11.7b**).

Therapeutic Cloning and Stem Cells

The lower branch of Figure 11.6 depicts an alternative cloning outcome called therapeutic cloning. The purpose of **therapeutic cloning** is not to produce a viable organism but to produce special cells called embryonic stem cells.

Figure 11.7 Reproductive cloning of mammals.
(a) These piglets are clones of a pig that was genetically modified to lack a protein that causes transplant rejection in humans.
(b) This baby banteng, a type of ox, was produced by cloning in 2003 using frozen skin cells from a male banteng that died at the San Diego Zoo in 1980. The nuclei from the banteng cells were inserted in cow eggs.

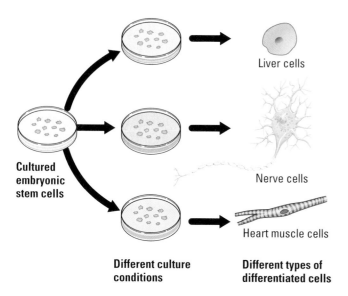

Cultured embryonic stem cells

Liver cells

Nerve cells

Heart muscle cells

Different culture conditions

Different types of differentiated cells

Figure 11.8 Differentiation of embryonic stem cells in culture. Scientists hope to discover growth conditions that will stimulate cultured embryonic stem cells to differentiate into specialized cells of various types.

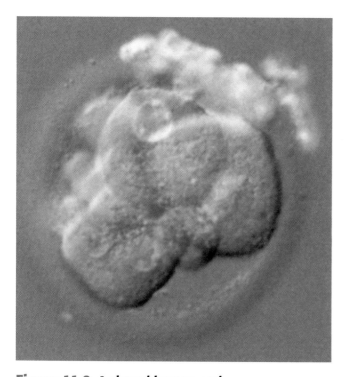

Figure 11.9 A cloned human embryo.
This cloned human embryo, created in 2001 by a biotechnology company, stopped growing at the six-cell stage. It was created for the purpose of therapeutic cloning, which is the production of embryonic stem cells for medical use.

Embryonic stem cells (ES cells) are cells in an early animal embryo that differentiate during development to give rise to all the specialized cells in the body. When grown in laboratory culture, ES cells can divide indefinitely. The right conditions—such as the presence of certain growth factors—can induce changes in gene expression that cause differentiation into a particular cell type (**Figure 11.8**). If scientists can discover the right conditions, they will be able to grow cells for the repair of irreversibly injured or diseased organs. In the future, embryos may be created using a cell nucleus from a patient so that ES cells can be harvested and induced to develop into whole organs. Presumably, the patient's body would readily accept such a transplant because it would be a genetic match.

Adult stem cells are cells present in adult tissues that generate replacements for nondividing differentiated cells. Unlike ES cells, adult stem cells are partway along the road to differentiation; in the body, they usually give rise to only a few related types of specialized cells. For example, stem cells in bone marrow generate the different kinds of blood cells. Adult stem cells are much more difficult than ES cells to grow in culture, but researchers have had some success. Another difference between adult and embryonic stem cells is that adult stem cells can be harvested from adult patients, whereas ES cells can only be obtained from an embryo (or from the culture of cells originally taken from an embryo). Because no embryonic tissue is involved, adult stem cells may provide an ethically less problematic route for human tissue and organ replacement.

In 2001, a biotechnology company in Massachusetts named Advanced Cell Technology (ACT) announced that it had created the first human embryo by cloning (**Figure 11.9**). Although their most successful embryo stopped growing at the six-cell stage, this announcement heightened debate about whether human cloning—reproductive or therapeutic—should be allowed to proceed. As critics point out, there are many practical and ethical obstacles. Mammalian cloning is currently very inefficient; for example, the success of Dolly was preceded by 276 failures, and 17 human eggs failed to produce an embryo before ACT's first "success." Even with improvements, the procedure would likely be too expensive to use for individual patients. More troubling to many are the ethical questions raised by creating human embryos for medical purposes. A public consensus on the moral status of the human embryo is unlikely to be reached any time soon, and debate on the use of stem cells continues.

CHECKPOINT

1. What is cellular differentiation?

2. If your nerve cells and skin cells have exactly the same genes, how can they be so different?

3. What evidence supports the view that differentiation is based on regulating gene expression rather than on irreversible changes in the genome?

4. What is the difference between reproductive and therapeutic cloning?

Answers: **1.** The structural and functional specialization of cell types during development of an organism **2.** Each cell type must be expressing certain genes that are not expressed in the other cell type. **3.** The nuclear transplantation experiments in animals and the cloning of plants from differentiated cells **4.** Reproductive cloning results in the production of a live individual, whereas therapeutic cloning produces stem cells.

The Regulation of Gene Expression

We've seen that different cell types arise from selective gene expression. But exactly how is gene expression regulated in a cell? The pathway from gene to functioning protein in eukaryotic cells is a long one, providing a number of points where the process can be regulated: turned on or off, speeded up or slowed down. Picture the series of pipes that carry water from your local water supply to a faucet in your home. At various points, valves control the flow of water. We use this model in **Figure 11.10** to illustrate the flow of genetic information from a eukaryotic chromosome—a reservoir of genetic information—to an active protein that has been made in the cell's cytoplasm. The multiple mechanisms that control gene expression are analogous to the control valves in your water pipes. In the figure, a control knob indicates a gene expression "valve." All these knobs represent possible control points, although only one or a few control points are likely to be important for a typical protein.

Gene Regulation in Bacteria

Before discussing further how eukaryotes control gene expression, let's see how simpler organisms do it. Picture an *Escherichia coli* bacterium living in your intestine. It will be bathed in varying nutrients, depending on what you eat. If you drink a milk shake, for example, there will be a sudden rush of lactose. In response, *E. coli* will express three genes that produce enzymes to absorb and digest this sugar. Once the lactose is gone, *E. coli* does not waste its energy continuing to produce these enzymes. Thus, the bacterium is able to adjust its gene expression to changes in the environment.

How does the presence or absence of lactose influence the activity of the genes responsible for lactose metabolism? The key is the way the three genes are organized: They are adjacent in the DNA and regulated as a single unit. This regulation is achieved through control sequences, stretches of DNA that help turn all three genes on and off, coordinating their expression. Such a cluster of genes with related functions, along with the control sequences, is called an **operon.** The operon considered here, the *lac* (short for *lactose*) operon, was first described in the 1960s by French biologists François Jacob and Jacques Monod. The *lac* operon illustrates principles of gene regulation that apply to a wide variety of prokaryotic genes.

How are DNA control sequences able to cause genes to turn on or off? One control sequence, called a **promoter,** is the site where the transcription enzyme, RNA polymerase, attaches and initiates transcription—in our example, transcription of the three lactose enzyme genes. Between the promoter and the enzyme genes, a DNA segment called an **operator** acts as a switch that is turned on or off, depending on whether a specific protein is bound there. The operator and protein together determine whether RNA polymerase can attach to the promoter and start transcribing the genes. In the *lac* operon, when the operator switch is turned on, all the enzymes needed to metabolize lactose are made at once.

Figure 11.10 **The gene expression "pipeline" in a eukaryotic cell.**
Each valve in the pipeline represents a stage at which the pathway from chromosome to functioning protein can be regulated. Most regulation of gene expression, in both eukaryotes and prokaryotes, occurs at the transcription stage (large yellow knob). Regulating the pathway at this early stage allows the cell to avoid wasting valuable resources.

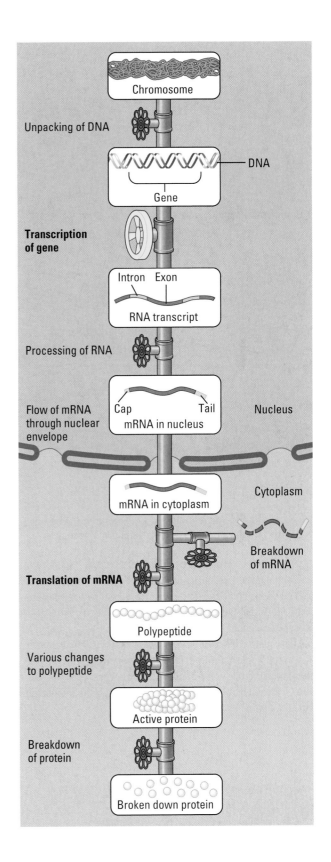

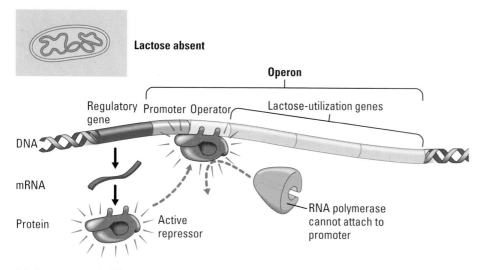

Lactose absent

Operon

Regulatory gene · Promoter · Operator · Lactose-utilization genes

DNA

mRNA

Protein

Active repressor

RNA polymerase cannot attach to promoter

(a) Operon turned off (default state when no lactose is present)

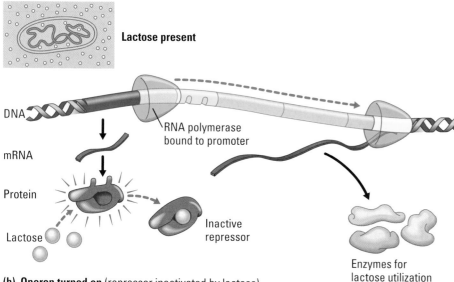

Lactose present

DNA

RNA polymerase bound to promoter

mRNA

Protein

Lactose

Inactive repressor

Enzymes for lactose utilization

(b) Operon turned on (repressor inactivated by lactose)

Figure 11.11 The *lac* operon of *E. coli.* In the DNA represented here, a small segment of the bacterium's chromosome, the three genes that code for the lactose-utilization enzymes are next to each other (light blue). These genes and two control sequences, a promoter and an operator, make up the *lac* operon. A regulatory gene coding for a repressor is nearby. **(a)** In the absence of lactose, the repressor (red) binds to the operator (yellow), preventing transcription of the operon. **(b)** Lactose *de*represses the operon by inactivating the repressor. The result is the production of the three enzymes for lactose utilization.

Figure 11.11a shows the *lac* operon in "off" mode, its status when there is no lactose around. Transcription is turned off when a protein called a **repressor** (shown in red) binds to the operator and physically blocks the attachment of RNA polymerase to the promoter.

Figure 11.11b shows the operon in "on" mode, when lactose is present. Lactose interferes with the attachment of the *lac* repressor to the operator by binding to the repressor and changing its shape. In its new shape, the repressor is inactive; it cannot bind to the operator, and the operator switch remains on. RNA polymerase can now bind to the promoter and from there transcribe the genes of the operon into mRNA. Translation produces all three enzymes needed for lactose utilization.

Many operons have been identified in bacteria. Some are quite similar in structure to the *lac* operon, while others have somewhat different mechanisms of control. For example, operons that control amino acid synthesis cause bacteria to stop making these molecules when they are already present in the environment, saving materials and energy for the cells. In these cases, the amino acid activates the repressor, rather than inactivating it, as for the *lac* operon. Armed with a variety of operons, *E. coli* and other prokaryotes can thrive in frequently changing environments.

Gene Regulation in the Nucleus of Eukaryotic Cells

While operons are a common method of gene regulation in bacteria, they generally do not exist in eukaryotes. Eukaryotic cells have more elaborate mechanisms than bacteria for regulating the expression of their genes. This is not surprising because, as a single cell, a prokaryote does not require the elaborate regulation of gene expression that leads to cell specialization in multicellular organisms. Using a reduced version of Figure 11.10 as a guide to the flow of genetic information through the cell, we will explore several ways that eukaryotes can control gene expression, starting with what goes on in the cell nucleus.

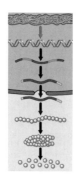

The Regulation of DNA Packing Recall from Chapter 8 that eukaryotic chromosomes may be in a more or less condensed state, with the DNA and accompanying proteins more or less tightly wrapped together. DNA packing tends to prevent gene expression, presumably by preventing RNA polymerase and other transcription proteins from contacting the DNA. Highly compacted chromatin, which is found in mitotic chromosomes and also in some regions of interphase chromosomes, is generally not expressed at all. For a gene to be transcribed, DNA packing must be loosened.

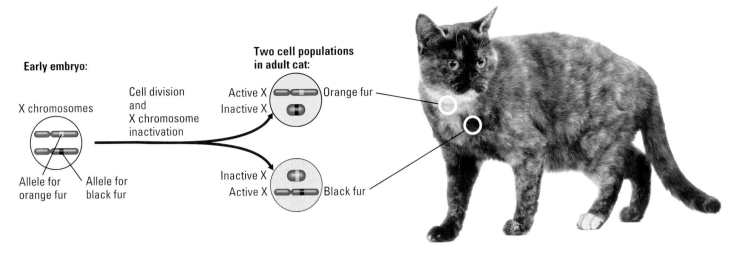

Early embryo:

X chromosomes

Allele for orange fur   Allele for black fur

Cell division and X chromosome inactivation

Two cell populations in adult cat:

Active X
Inactive X

Orange fur

Inactive X
Active X

Black fur

Cells may use DNA packing for the long-term inactivation of genes. One intriguing case is seen in female mammals, where one entire X chromosome in each somatic cell is highly compacted and almost entirely inactive. This **X chromosome inactivation** first takes place early in embryonic development, when one of the two X chromosomes in each cell is inactivated at random. The inactivation is subsequently inherited by a cell's descendants. Consequently, a female heterozygous for genes on the X chromosome has populations of cells that express different X-linked alleles. A striking effect of X chromosome inactivation is the tortoiseshell cat, which has orange and black patches of fur (**Figure 11.12**).

Figure 11.12 X chromosome inactivation: the tortoiseshell pattern on a cat. The tortoiseshell gene is on the X chromosome, and the tortoiseshell phenotype requires the presence of two different alleles, one for orange fur and one for nonorange (black) fur. Normally, only females can have both alleles, because only they have two X chromosomes. If a female is heterozygous for the tortoiseshell gene, she is tortoiseshell. Orange patches are formed by populations of cells in which the X chromosome with the orange allele is active; black patches have cells in which the X chromosome with the nonorange allele is active. (Calico cats also have white areas, which are determined by another gene.)

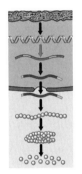

The Initiation of Transcription The most important stage for regulating gene expression is transcription. The eukaryotic control mechanisms involve regulatory proteins that, like prokaryotic repressors, bind to specific segments of DNA. A model for the eukaryotic regulation of transcription is shown in **Figure 11.13**. This model features an intricate array of regulatory proteins that interact with DNA and with one another to turn genes on or off.

In contrast to the genes of bacterial operons, each eukaryotic gene usually has its own promoter and other control sequences. Moreover, repressors are less common than **activators,** proteins that turn genes *on* by binding to DNA. (Activators act by making it easier for RNA polymerase to bind to the promoter.) The use of activators is efficient because a typical animal or plant cell needs to turn on (transcribe) only a small percentage of its genes, those required for the cell's specialized structure and function. The "default" state for most genes in multicellular eukaryotes seems to be "off," with the exception of "housekeeping" genes for routine activities such as glucose metabolism.

Case Study in the Process of Science on the Web & CD Experiment with a eukaryotic promoter.

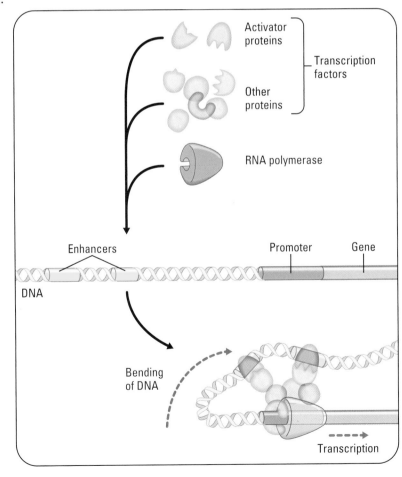

Activator proteins

Other proteins

Transcription factors

RNA polymerase

Enhancers

Promoter   Gene

DNA

Bending of DNA

Transcription

Figure 11.13 A model for the turning on of a eukaryotic gene. A large assembly of proteins and several control sequences in the DNA are involved in initiating the transcription of a eukaryotic gene.

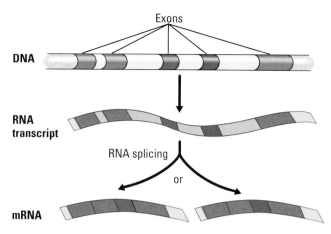

DNA

Exons

RNA transcript

RNA splicing

or

mRNA

Figure 11.14 Producing two different mRNAs from the same gene. By recognizing different parts of a gene as exons, two different cells can use a given DNA gene to synthesize somewhat different mRNAs and proteins. The cells might be cells of different tissues or the same type of cell at different times in the organism's life. Alternative RNA splicing is a means by which the cell can exert differential control over the expression of certain genes.

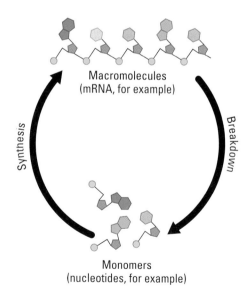

Macromolecules
(mRNA, for example)

Synthesis

Breakdown

Monomers
(nucleotides, for example)

Figure 11.15 Recycling of macromolecules in the cell. The breakdown of mRNA and proteins after varying lengths of time are modes of regulation that occur in the cytoplasm. The components of these macromolecules can be recycled and used for making new macromolecules.

Turning on a eukaryotic gene involves regulatory proteins called **transcription factors** (activators are one type), in addition to RNA polymerase. In the model depicted in Figure 11.13, the first step in initiating gene transcription is the binding of activator proteins to DNA sequences called **enhancers.** With bending of the DNA, the bound activators interact with other transcription factors, which then bind together at the gene's promoter. This large assembly of proteins facilitates the correct attachment of RNA polymerase to the promoter and the initiation of transcription. As shown in the figure, several enhancers and activators may be involved. Genes coding for related enzymes, such as those in a metabolic pathway, may share a specific enhancer (or collection of enhancers), allowing these genes to be activated at the same time. Not shown are repressor proteins that may bind to DNA sequences called **silencers,** inhibiting the start of transcription.

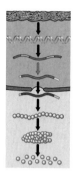

RNA Processing The eukaryotic cell localizes transcription in the nucleus and also (as you learned in Chapter 10) modifies, or processes, the RNA transcripts there, before they move to the cytoplasm for translation by the ribosomes. This processing includes the addition of a cap and a tail to the RNA, as well as the removal of any introns—noncoding DNA segments that interrupt the genetic message—and the splicing together of the remaining exons. Such processing creates additional opportunities for regulating gene expression.

In some cases, a cell can carry out exon splicing in more than one way, generating different mRNA molecules from the same RNA transcript. Notice in **Figure 11.14**, for example, that one mRNA molecule ends up with the green exon and the other with the brown exon. With this sort of **alternative RNA splicing,** an organism can get more than one type of polypeptide from a single gene. Results from the Human Genome Project suggest that alternative RNA splicing is very common in humans. In one case, the mRNA transcript from one gene can be spliced to encode seven different versions of a cellular protein. Each of the seven is made in a different type of cell.

Another potential control point in the eukaryotic gene expression pathway is the transport of mRNA from the nucleus to the cytoplasm. By facilitating or blocking this transport, the cell can regulate the timing and amount of protein synthesis from a molecule of mRNA.

Regulation in the Cytoplasm

After eukaryotic mRNA is transported to the cytoplasm, there are additional opportunities for regulation.

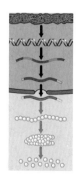

The Breakdown of mRNA The lifetime of mRNA molecules is an important factor regulating the amounts of various proteins the cell produces. Eukaryotic mRNAs can have lifetimes of hours or even weeks. Long-lived mRNAs get translated into many more protein molecules than do short-lived ones. But eventually, all mRNAs are broken down and their parts recycled (**Figure 11.15**).

The Regulation of Translation The process of translation also offers opportunities for regulation. Among the molecules involved in translation are a great many proteins

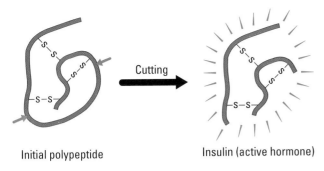

Initial polypeptide   Insulin (active hormone)

Figure 11.16 The formation of an active insulin molecule. One type of post-translational control is the cutting of a polypeptide to convert it to its active form. The hormone insulin, for example, is synthesized as one polypeptide. A special enzyme then removes an interior section. The remaining polypeptide chains are held together by covalent bonds between the sulfur atoms of sulfur-containing amino acids. Only in this form does the protein act as a hormone.

that have regulatory functions. Red blood cells, for instance, have a protein that prevents the translation of hemoglobin mRNA unless the cell has a supply of heme, an iron-containing chemical group essential for hemoglobin function.

Protein Alterations The final opportunities for regulating gene expression occur after translation. Post-translational control mechanisms in eukaryotes often involve cutting polypeptides into smaller, active final products. The hormone insulin, for example, is synthesized in pancreatic cells as one long inactive polypeptide (**Figure 11.16**). After translation, a large center portion is cut away, leaving two shorter chains that constitute the active insulin molecule.

Protein Breakdown Another control mechanism operating after translation is the selective breakdown of proteins. Some of the proteins that trigger metabolic changes in cells are broken down within a few minutes or hours. This regulation allows a cell to adjust the kinds and amounts of its proteins in response to changes in its environment. In addition, when proteins are damaged, they are usually broken down right away and replaced by new ones that function properly.

Cell Signaling

In a multicellular organism, the process of gene regulation can cross cell boundaries. A cell can produce and secrete chemicals, such as hormones, that induce a neighboring or distant cell to be regulated in a certain way.

Cell-to-cell signaling, with proteins or other kinds of molecules carrying messages from signaling cells to receiving (target) cells, is a key mechanism in the development of a multicellular organism from a fertilized egg, as well as in the coordination of cellular activities in the mature organism. As you saw in Figure 5.18, a signal molecule usually acts by binding to a receptor protein in the plasma membrane of the target cell and initiating a signal-transduction pathway, a series of molecular changes that converts a signal received on a target cell's surface into a

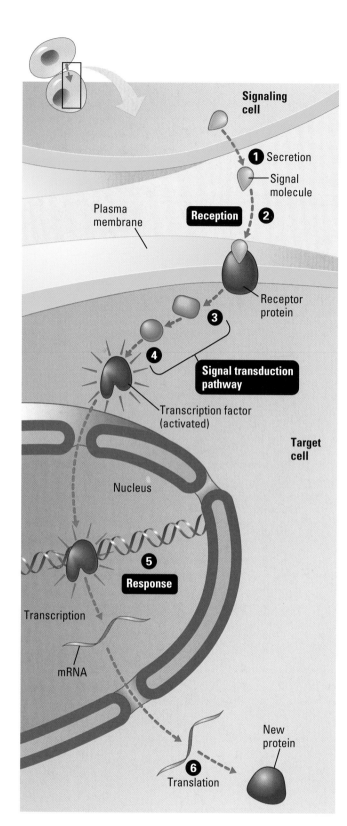

Signaling cell

1 Secretion

Signal molecule

Plasma membrane

Reception **2**

Receptor protein

3

4

Signal transduction pathway

Transcription factor (activated)

Target cell

Nucleus

5

Response

Transcription

mRNA

New protein

6

Translation

Figure 11.17 A cell-signaling pathway that turns on a gene.

The coordination of cellular activities in a multicellular organism depends on cell-to-cell signaling that helps to regulate genes. **1** First, the signaling cell secretes the signal molecule. **2** This molecule binds to a specific receptor protein embedded in the target cell's plasma membrane—the reception step. **3** The binding activates the first in a series of relay proteins (green) within the target cell. Each relay molecule activates another. This is a signal transduction pathway. **4** The last relay molecule in the series activates a transcription factor that **5** triggers the transcription of a specific gene, the response to the signal. **6** Translation of the mRNA produces a protein.

specific response inside the cell. **Figure 11.17** shows the main elements of a signal transduction pathway in which the target cell's response is the transcription (turning on) of a gene.

The Genetic Basis of Cancer

In Chapter 8, we introduced cancer as a variety of diseases in which cells escape from the control mechanisms that normally limit their growth and division. In recent years, scientists have learned that this escape from normal controls is due to changes in some of the cells' genes or to changes in the way certain genes are expressed.

Genes That Cause Cancer

The abnormal behavior of cancer cells was observed years before anything was known about the cell cycle, its control, or the role genes play in making cells cancerous. One of the earliest clues to the cancer puzzle was the discovery, in 1911, of a virus that causes cancer in chickens. Recall that viruses are simply "packaged genes," molecules of DNA or RNA surrounded by protein and in some cases a membranous envelope. Viruses that cause cancer can become permanent residents in host cells by inserting their nucleic acid into the DNA of host chromosomes. It is now known that a number of viruses that can produce cancer carry specific cancer-causing genes in their nucleic acid. When inserted into a host cell, these genes can make the host cell cancerous. A gene that causes cancer is called an **oncogene** ("tumor gene").

Oncogenes and Tumor-Suppressor Genes

In 1976, American molecular biologists J. Michael Bishop, Harold Varmus, and their colleagues made a startling discovery. They found that a virus that causes cancer in chickens contains an oncogene that is an altered version of a normal chicken gene. Subsequent research has shown that the chromosomes of many animals, including humans, contain genes that can be converted to oncogenes. A normal gene with the potential to become an oncogene is called a **proto-oncogene.** (These terms can be confusing, so they bear repeating: a *proto-oncogene* is a normal gene that, if changed, can become a cancer-causing *oncogene.*) A cell can acquire an oncogene from a virus or from the conversion of one of its own proto-oncogenes.

How can a change in a gene cause cancer? Searching for the normal roles of proto-oncogenes in the cell, researchers found that many of these genes code for **growth factors**—proteins that stimulate cell division—or for other proteins that affect the cell cycle. When all these proteins are functioning normally, in the right amounts at the right times, they help keep the rate of cell division at an appropriate level. When they malfunction, cancer—uncontrolled cell growth—may result.

For a proto-oncogene to become an oncogene, a mutation must occur in the cell's DNA. The mutations that produce most cancers occur in somatic cells, those not involved in gamete formation. **Figure 11.18** illustrates three kinds of changes in somatic cell DNA that can produce active oncogenes. In all three cases, normal gene expression is changed, and the cell is stimulated to divide excessively.

Changes in genes whose products *inhibit* cell division are also involved in cancer. These genes are called **tumor-suppressor genes** because the proteins they encode normally help prevent uncontrolled cell growth (**Figure 11.19**). Any mutation that keeps a normal tumor-suppressor protein from being made or from functioning may contribute to the development of cancer.

The Effects of Cancer Genes on Cell-Signaling Pathways

Researchers are now working out the details of how oncogenes and defective tumor-suppressor genes contribute to uncontrolled cell growth. Normal proto-oncogenes and tumor-suppressor genes often code for proteins involved in signal-transduction pathways leading to gene expression, pathways similar to the one shown in Figure 11.17. Let's consider the case of a proto-oncogene named *ras,* which in mutant form is an oncogene involved in many kinds of cancer. The protein encoded by *ras* is a component of signal-transduction pathways leading to growth-stimulating responses. Therefore,

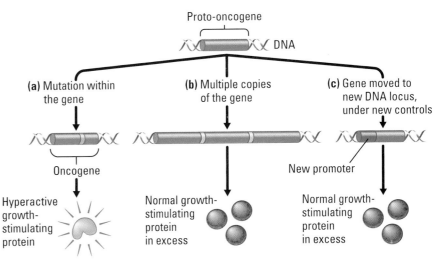

Figure 11.18 Alternative ways to make an oncogene from a proto-oncogene. Let's assume that the starting proto-oncogene codes for a protein that stimulates cell division. **(a)** A mutation (green) in the proto-oncogene itself may create an oncogene that codes for a hyperactive protein, one whose stimulating effect is stronger than normal. **(b)** An error in DNA replication or recombination may generate multiple copies of the gene, which are all transcribed and translated. The result would be an excess of the normal stimulatory protein. **(c)** The proto-oncogene may be moved from its normal location in the cell's DNA to another location. At its new site, the gene may be under the control of a different promoter and other transcriptional control sequences that cause it to be transcribed more often than normal. In that case, the normal protein would be made in excess.

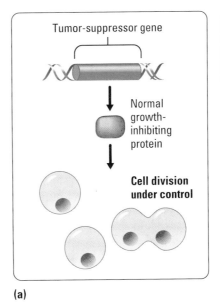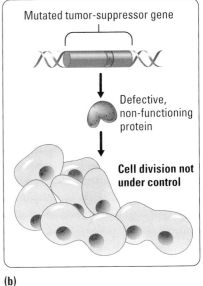

(a)   (b)

Figure 11.19 Tumor-suppressor genes.
(a) A tumor-suppressor gene normally codes for a protein that inhibits cell growth and division. In this way, the gene helps prevent cancerous tumors from arising. **(b)** When a mutation in a tumor-suppressor gene makes its protein defective, cells that are usually under the control of the normal protein may divide excessively, eventually forming a tumor.

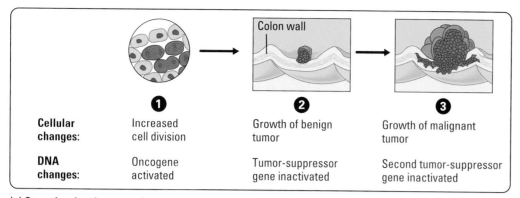

(a) Stepwise development of a typical colon cancer

Cellular changes:	Increased cell division	Growth of benign tumor	Growth of malignant tumor
DNA changes:	Oncogene activated	Tumor-suppressor gene inactivated	Second tumor-suppressor gene inactivated

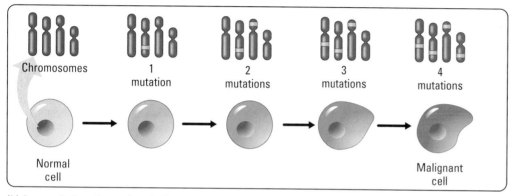

(b) Accumulation of mutations in the development of a cancer cell

Figure 11.20 The development of colon cancer.

Like most cancers, colon cancer usually results from a series of mutations. **(a) ❶** Colon cancer begins as the unusually frequent division of apparently normal cells in the colon lining. **❷** Later, a benign tumor appears in the colon wall. **❸** Eventually, a malignant tumor develops. These cellular changes parallel changes at the DNA level, including the activation of a cellular oncogene and the inactivation of two tumor-suppressor genes. **(b)** Mutations leading to cancer accumulate in a lineage of somatic cells. Colors distinguish the normal cell from cells with one or more mutations, leading to increased cell division and cancer. Once a cancer-promoting mutation occurs (orange band on chromosome), it is passed to all the descendants of the cell carrying it. In our example, the first two mutations make the cells divide more rapidly; otherwise, the cells appear normal. The third mutation increases the rate of cell division even more and also causes some changes in the cells' appearance. Finally, a cell accumulates a fourth cancer-promoting mutation and begins dividing uncontrollably. The structure of this cell and its descendants is grossly altered.

if an oncogene-creating mutation causes the *ras* protein to be hyperactive, for example, the pathway may be overstimulated, and excessive cell division may result. To produce cancer, however, further abnormalities are usually needed, as we discuss next.

The Progression of a Cancer Nearly 150,000 Americans are predicted to be stricken by cancer of the colon (large intestine) or rectum in 2003. One of the best-understood types of human cancer, colon cancer illustrates an important principle about how cancer develops: *More than one somatic mutation is needed to produce a full-fledged cancer cell.* As in many cancers, the development of colon cancer that metastasizes (spreads) is gradual.

As shown in **Figure 11.20a**, colon cancer begins as an unusually frequent division of normal-looking cells in the colon lining. This and later cellular changes parallel changes at the DNA level, including the activation of a cellular oncogene and the inactivation of two tumor-suppressor genes. These genetic changes (mutations) result in altered signal-transduction pathways. The requirement for several mutations—the total number is usually four or more—explains why cancers can take a long time to develop. **Figure 11.20b** shows how mutations leading to cancer may accumulate in a lineage of somatic cells. By the time a fourth mutation has occurred, the structure of this cell and its descendants is grossly altered.

"Inherited" Cancer Cancer is always a genetic disease in the sense that it always results from changes in DNA. Most mutations that lead to cancer arise in the organ where the cancer starts—the colon, for example. Because these mutations do not affect the cells that give rise to eggs or sperm, they are not passed from parent to child. In some families, however, there are inherited mutations in one or more of these same genes. Such a mutation is passed on and predisposes the recipient to cancer. We say that this cancer is familial or "inherited," even though it doesn't appear unless the person acquires additional somatic mutations.

One example of inherited cancer genes is associated with breast cancer, a disease that strikes one out of every ten American women. The vast majority of breast cancer cases seem to have nothing to do with inherited mutations. But in some families, breast cancer appears frequently, suggesting that an inherited trait might be involved. In 1994, cancer researchers identified a gene called *BRCA1* that is mutated in many families with familial breast cancer. Some mutations in gene *BRCA1* put a woman at high risk for breast cancer (and ovarian cancer as well)—a more than 80% risk of developing cancer during her lifetime. Research suggests that the protein encoded by the normal version of *BRCA1* acts as a tumor suppressor.

In recent years, clinical tests have been developed for the presence of mutations affecting *BRCA1* and *BRCA2*, another breast cancer gene. These tests offer women, particularly those with a family history of breast cancer, the option of being tested. Unfortunately, this knowledge is of limited use, because surgical removal of the breasts and/or ovaries is the only preventive option currently available to women who carry the mutant genes.

The Effects of Lifestyle on Cancer Risk

Cancer is now one of the leading causes of death in most developed countries, including the United States. Death rates due to certain forms of cancer (including stomach, cervical, and uterine cancers) have decreased, but the overall cancer death rate is still on the rise, currently increasing at about 1% per decade.

Cancer-causing agents are called **carcinogens.** Most mutagens are carcinogens, agents capable of bringing about cancer-causing DNA changes like the ones in Figure 11.20. In many cases, these changes result from decades of exposure to the mutagenic effects of carcinogens. One of the most potent carcinogens (and mutagens) is ultraviolet (UV) radiation. Excessive exposure to UV radiation from the sun can cause skin cancer, including a deadly type called melanoma.

The one substance known to cause more cases and types of cancer than any other single agent is tobacco. In 1900, lung cancer was a rare disease. Since then, largely because of an increase in cigarette smoking, rates of lung cancer have steadily increased, and today more people die of lung cancer (over 150,000 Americans in 2003) than any other form of cancer. Most tobacco-related cancers come from smoking, but the passive inhalation of secondhand smoke also poses a risk. As **Table 11.1** indicates, tobacco use,

Table 11.1	Cancer in the United States		
Cancer	**Known or Likely Carcinogen or Factor**	**Estimated Cases (2003)**	**Estimated Deaths (2003)**
Prostate	Testosterone; possibly dietary fat	220,900	28,900
Breast	Estrogen; possibly dietary fat	212,600	40,200
Lung	Cigarette smoke	171,900	157,200
Colon and rectum	High dietary fat; low dietary fiber	147,500	57,100
Lymphomas	Viruses (for some types)	61,000	24,700
Melanoma of skin	Ultraviolet light	58,800	9,800
Bladder	Cigarette smoke	57,400	12,500
Uterus	Estrogen	40,100	6,800
Kidney	Cigarette smoke	31,900	11,900
Pancreas	Cigarette smoke	30,700	30,000
Leukemias	X-rays; benzene; virus (for some types)	30,600	21,900
Ovary	Large number of ovulation cycles	25,400	14,300
Stomach	Table salt; cigarette smoke	22,400	12,100
Mouth and throat	Tobacco in various forms; alcohol	20,600	5,500
Brain and nerve	Trauma; X-rays	18,300	13,100
Liver	Alcohol; hepatitis viruses	17,300	14,400
Cervix	Viruses; cigarette smoke	12,200	4,100
All others		154,500	92,000
Total		1,334,100	556,500

Data from the American Cancer Society.

sometimes in combination with alcohol consumption, causes a number of other types of cancer in addition to lung cancer. In nearly all cases, cigarettes are the main culprit, but smokeless tobacco products (snuff and chewing tobacco) are linked to cancer of the mouth and throat. Exposure to some of the most lethal carcinogens is often a matter of individual choice. Tobacco use, the consumption of animal fat and alcohol, and excessive time spent in the sun, for example, are all behavioral factors that affect cancer risk.

Activity 11E on the Web & CD
Learn more about the causes of cancer.

Avoiding carcinogens is not the whole story. There is increasing evidence that some food choices significantly reduce a person's cancer risk. For instance, eating 20–30 g of plant fiber daily (about twice the amount the average American consumes) while eating less animal fat may help prevent colon cancer. Evidence also indicates that other substances in fruits and vegetables, including vitamins C and E and certain compounds related to vitamin A, may offer protection against a variety of cancers. Cabbage and its relatives, such as broccoli and cauliflower, are thought to be especially rich in substances that help prevent cancer, although the identities of all these substances are not yet established. Determining how diet influences cancer has become a primary focus of research.

The battle against cancer is being waged on many fronts, and there is reason for optimism in the progress being made. It is especially encouraging that we can help reduce our risk of acquiring some of the most common forms of cancer by the choices we make in daily life.

CHECKPOINT

1. How can a mutation in a tumor-suppressor gene contribute to the development of cancer?

2. Why is most breast cancer considered nonhereditary?

3. Of all known behavioral factors, which one causes the most cancer cases and deaths?

Answers: 1. A mutated tumor-suppressor gene may produce a defective protein unable to function in a pathway that normally inhibits cell division (that is, normally suppresses tumors). **2.** Most breast cancers are associated with somatic mutations, not inherited mutations that are passed from parents to offspring via gametes. **3.** Tobacco use

Evolution Connection

Homeotic Genes

Homeotic genes are master control genes that regulate batteries of other genes that actually create the anatomical identity of parts of the body during embryonic development. For example, one set of homeotic genes in fruit flies instructs cells in the embryonic head and thorax (midbody) to form antennae and legs, respectively. Elsewhere, these homeotic genes remain turned off, while others are turned on. Mutations in homeotic genes can produce bizarre effects. For example, homeotic fruit fly mutants may have extra sets of wings (**Figure 11.21**) or legs growing from their head.

Among the most exciting biological discoveries in recent years is that homeotic genes help direct embryonic development in a wide variety of organisms. Researchers studying homeotic genes in fruit flies found a common structural feature: Every homeotic gene they looked at contained a

Mutant fruit fly

Normal fruit fly

Figure 11.21 The effect of a homeotic gene.
As a result of a mutation in a homeotic (master control) gene, the fly on the left has an extra pair of wings.

common sequence of 180 nucleotides. Very similar sequences have since been found in virtually every eukaryotic organism examined so far, including yeasts, plants, earthworms, frogs, chickens, mice, and humans. These nucleotide sequences are called **homeoboxes,** and each is translated into a segment (60 amino acids long) of the protein product of the homeotic gene. The polypeptide segment encoded by the homeobox binds to specific sequences in DNA, enabling homeotic proteins that contain it to turn groups of genes on or off during development.

Figure 11.22 highlights some striking similarities in the chromosomal locations and developmental roles of homeobox-containing homeotic genes in two quite different animals. Notice that the order of genes on the fly chromosome is the same as on the four mouse chromosomes and that the gene order on the chromosomes corresponds to analogous body regions in both animals. These similarities suggest that the original version of these homeotic genes arose very early in the history of life and that the genes have remained remarkably unchanged over eons of animal evolution.

Evolution Connection on the Web You may be more like a fruit fly than you think.

By their presence in such diverse creatures, homeotic genes illustrate one of the central themes of biology: unity in diversity. The fact that these key genes are control genes underscores the importance of regulation in the lives of organisms.

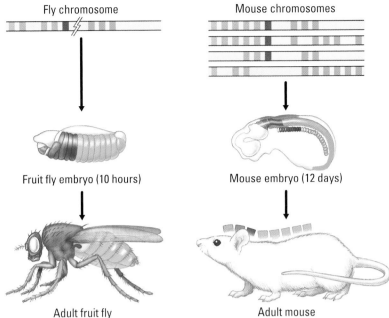

Figure 11.22 Homeotic genes in two different animals.   At the top of the figure are portions of chromosomes that carry homeotic genes in the fruit fly and the mouse. The colored boxes represent homeotic genes that are very similar (have especially close DNA sequences) in flies and mice. The same color coding identifies the parts of the animals that are affected by these genes.

Chapter Review

Summary of Key Concepts

For study help, go to the Essential Biology Website (www.essentialbiology.com) or CD-ROM to explore the Activities and Case Studies in the Process of Science.

From Egg to Organism: How and Why Genes Are Regulated

- Gene expression—the flow of information from genes to proteins—is subject to control, mainly by the turning on and off of genes.

- **Patterns of Gene Expression in Differentiated Cells**   The various cell types of a multicellular organism are different because different combinations of genes are expressed.

- **DNA Microarrays: Visualizing Gene Expression**   DNA microarrays can be used to determine which of many genes are turned on in a particular cell type through the attachment of fluorescently labeled DNA complements of mRNA to stretches of DNA from the genes.

- **The Genetic Potential of Cells**   Most differentiated cells retain a complete set of genes, so a carrot plant, for example, can be made to grow from a single carrot cell. Under special conditions, animals can also be cloned.

- **Reproductive Cloning of Animals**   Nuclear transplantation is a procedure whereby a donor cell nucleus is inserted into a nucleus-free egg. First demonstrated in frogs in the 1950s, reproductive cloning was used in 1997

to clone a sheep from an adult mammary cell and has since been used to create many other cloned animals.

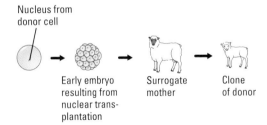

- **Therapeutic Cloning and Stem Cells**   The purpose of therapeutic cloning is to produce embryonic stem cells for medical uses. Both embryonic and adult stem cells show promise for future therapeutic uses.

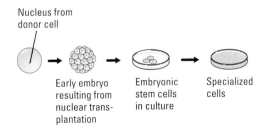

The Regulation of Gene Expression

- **Gene Regulation in Bacteria**   An operon is a cluster of genes with related functions together with sequences for controlling their expression. The *lac* operon produces enzymes that break down lactose only when lactose is present.

A typical operon

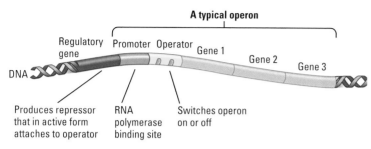

DNA

Regulatory gene — Produces repressor that in active form attaches to operator

Promoter — RNA polymerase binding site

Operator — Switches operon on or off

Gene 1
Gene 2
Gene 3

Activity 11A *The* lac *Operon in* E. coli

- **Gene Regulation in the Nucleus of Eukaryotic Cells**   In the nucleus of eukaryotic cells, there are multiple possible control points in the pathway of gene expression. DNA packing tends to block gene expression, presumably by preventing access of transcription proteins to the DNA. An extreme example is X chromosome inactivation in the cells of female mammals. The most important control point in both eukaryotes and prokaryotes is at gene transcription. A variety of regulatory proteins interact with DNA and with each other to turn the transcription of eukaryotic genes on or off. There are also opportunities for the control of eukaryotic gene expression after transcription, when introns are cut out of the RNA and a cap and tail are added.

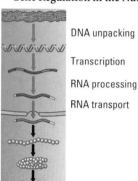

DNA unpacking

Transcription

RNA processing

RNA transport

Case Study in the Process of Science *How Do You Design a Gene Expression System?*

- **Regulation in the Cytoplasm**

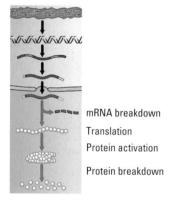

mRNA breakdown

Translation

Protein activation

Protein breakdown

The lifetime of an mRNA molecule helps determine how much protein is made, as do factors involved in translation. Finally, the cell may activate the finished protein in various ways (for instance, by cutting out portions) and later break it down.

Activity 11B *Gene Regulation in Eukaryotes*

Activity 11C *Review: Gene Regulation in Eukaryotes*

- **Cell Signaling**   Cell-to-cell signaling is key to the development and functioning of multicellular organisms. Signal-transduction pathways convert molecular messages to cell responses, often the transcription of particular genes.

Activity 11D *Signal-Transduction Pathway*

The Genetic Basis of Cancer

- **Genes That Cause Cancer**   Cancer cells, which divide uncontrollably, can result from mutations in genes whose protein products regulate the cell cycle.

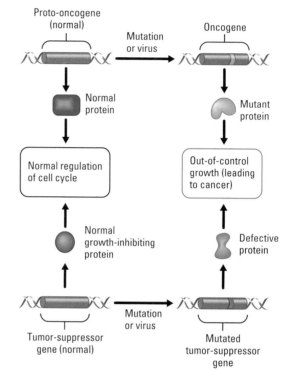

Proto-oncogene (normal) — Mutation or virus → Oncogene

Normal protein
Mutant protein

Normal regulation of cell cycle
Out-of-control growth (leading to cancer)

Normal growth-inhibiting protein
Defective protein

Tumor-suppressor gene (normal) — Mutation or virus → Mutated tumor-suppressor gene

Many proto-oncogenes and tumor-suppressor genes code for proteins active in signal-transduction pathways regulating cell division. Mutations of these genes cause malfunction of the pathways. Cancers result from a series of genetic changes in a cell lineage. Researchers have gained insight into the genetic basis of breast cancer by studying families in which a disease-predisposing mutation is inherited.

- **The Effects of Lifestyle on Cancer Risk**   Reducing exposure to carcinogens (which induce cancer-causing mutations) and other lifestyle choices can help reduce cancer risk.

Activity 11E *Causes of Cancer*

Self-Quiz

1. Your bone cells, muscle cells, and skin cells look different because
    a. different kinds of genes are present in each kind of cell.
    b. they are present in different organs.
    c. different genes are active in each kind of cell.
    d. different mutations have occurred in each kind of cell.

2. What kinds of evidence demonstrate that differentiated cells in a plant or animal retain their full genetic potential?

3. The most common procedure for cloning an animal is called _____.

4. What is learned from a DNA microarray?

5. Which of the following is a valid difference between embryonic stem cells and the stem cells found in adult tissues?

   a. In laboratory culture, only adult stem cells are immortal.

   b. In nature, only embryonic stem cells give rise to all the different types of cells in the organism.

   c. Only adult stem cells can be made to differentiate in the laboratory.

   d. Only embryonic stem cells are found in every tissue of the adult body.

6. A group of prokaryotic genes with related functions that are regulated as a single unit, along with the control sequences that perform this regulation, is called a(n) _____.

7. The regulation of gene expression must be more complex in multicellular eukaryotes than in prokaryotes because

   a. eukaryotic cells are much smaller.

   b. in a multicellular eukaryote, different cells are specialized for different functions.

   c. prokaryotes are restricted to stable environments.

   d. eukaryotes have fewer genes, so each gene must do several jobs.

8. A eukaryotic gene was inserted into the DNA of a bacterium. The bacterium then transcribed this gene into mRNA and translated the mRNA into protein. The protein produced was useless; it contained many more amino acids than the protein made by the eukaryotic cell. Why?

   a. The mRNA was not spliced as it is in eukaryotes.

   b. Eukaryotes and prokaryotes use different genetic codes.

   c. Repressor proteins interfered with transcription and translation.

   d. Ribosomes were not able to bind to tRNA.

9. What is the difference between oncogenes and proto-oncogenes? How does one turn into the other? What purpose do proto-oncogenes serve?

10. A mutation in a single gene may cause a major change in the body of a fruit fly, such as an extra pair of legs or wings. Yet it takes the combined action of many genes to produce a wing or leg. How can a change in just one gene cause such a big change in the body? What are such genes called?

Answers to the Self-Quiz questions can be found in Appendix B.

Go to the website or CD-ROM for more Self-Quiz questions.

The Process of Science

1. Study the depiction of the *lac* operon in Figure 11.11. Normally, the genes are turned off when lactose is not present. Lactose activates the genes, which code for enzymes that enable the cell to use lactose. Mutations can alter the function of this operon; in fact, it was the effects of various mutations that enabled Jacob and Monod to figure out how the operon works. Predict how the following mutations would affect the function of the operon in the presence and absence of lactose:

   a. Mutation of regulatory gene; repressor will not bind to lactose.

   b. Mutation of operator; repressor will not bind to operator.

   c. Mutation of regulatory gene; repressor will not bind to operator.

   d. Mutation of promoter; RNA polymerase will not attach to promoter.

2. One of the surprising results from the Human Genome Project is that there are many fewer genes than there are proteins in the human body. This result seems to highlight the importance of alternative RNA splicing, which allows several different mRNAs to be made from a single gene. Suppose you have samples of two types of adult cells from one person. These cell types are different in structure and function. Design an experiment using microarrays to determine whether or not the different gene expression in the two cell types is due to alternative RNA splicing.

3. Because a cat must have both orange and nonorange alleles to be tortoiseshell (see Figure 11.12), we would expect only female cats, which have two X chromosomes, to be tortoiseshell. Normal male cats (XY) can carry only one of the two alleles. Male tortoiseshell cats are occasionally seen, although they are usually sterile. What might you guess their genotype to be? (*Hint:* Some of the abnormal human genotypes described in Figure 8.23 can also be found in cats.)

Case Study in the Process of Science on the Web & CD *Experiment with a eukaryotic promoter.*

Biology and Society

1. A chemical called dioxin is produced as a by-product of certain chemical manufacturing processes. Trace amounts of this substance were present in Agent Orange, a defoliant sprayed on vegetation during the Vietnam War. There has been a continuing controversy over its effects on soldiers exposed to Agent Orange during the war. Animal tests have suggested that dioxin can cause cancer, liver and thymus damage, immune system suppression, and birth defects; at high dosage it can be lethal. But such animal tests are inconclusive; a hamster is not affected by a dose that can kill a much larger guinea pig, for example. Researchers have discovered that dioxin enters a cell and binds to a protein that in turn attaches to the cell's DNA. How might this mechanism help explain the variety of dioxin's effects on different body systems and in different animals? How might you determine whether a particular individual became ill as a result of exposure to dioxin? Do you think this information is relevant in the lawsuits of soldiers suing over exposure to Agent Orange? Why or why not?

2. There are genetic tests available for several types of "inherited cancer." The results from these tests cannot usually predict that someone *will* get cancer within a particular amount of time. Rather, they indicate only that a person has an increased risk of developing cancer. For many of the cancers involved, there are no lifestyle changes that can decrease a person's risk of actually getting the disease. Some people feel that this makes the tests useless. If your close family had a history of cancer and a test was available, would you want to get screened? Why or why not? What would you do with this information? If a sibling decided to get screened, would you want to know the results?

Biology and Society on the Web *Learn about the use of umbilical cord blood in the treatment of disease.*

CHAPTER 12

DNA Technology

Biology and Society: Hunting for Genes 217

Recombinant DNA Technology 217

From Humulin to Genetically Modified Foods

Recombinant DNA Techniques

DNA Fingerprinting and Forensic Science 224

Murder, Paternity, and Ancient DNA

DNA Fingerprinting Techniques

Genomics 229

The Human Genome Project

Tracking the Anthrax Killer

Genome-Mapping Techniques

Human Gene Therapy 234

Treating Severe Combined Immunodeficiency

Safety and Ethical Issues 235

The Controversy Over Genetically Modified Foods

Ethical Questions Raised by DNA Technology

Evolution Connection: Genomes Hold Clues to Evolution 237

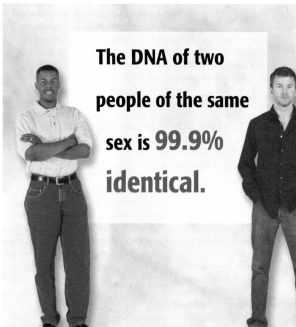

The DNA of two people of the same sex is **99.9%** identical.

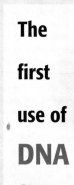

The first use of **DNA fingerprinting** in a murder case proved one man innocent and another guilty.

Animals, plants, and even **bacteria** can be genetically modified to produce **human** proteins.

Genetically modified strains account for half of the U.S. corn crop.

Hunting for Genes

DNA technology, a set of methods for studying and manipulating genetic material, has brought about some of the most remarkable scientific advances in recent years: Corn has been genetically modified to produce its own insecticide; DNA fingerprints have been used to solve crimes and study the origins of ancient peoples; and significant advances have been made toward curing fatal genetic diseases. Perhaps the most exciting use of DNA technology in basic biological research began in 1990 with the launch of the Human Genome Project. The goal of the Human Genome Project is to determine the nucleotide sequence of all DNA in the human genome and to identify the location and function of every gene. In 2003, the Project announced completion of 99% of the sequencing work. This ambitious project is revealing the genetic basis of what it means to be human.

Biology and Society on the Web
Learn more about the Human Genome Project.

The potential benefits of having a complete map of the human genome are enormous. For instance, hundreds of disease-associated genes have already been identified. One example is a gene that is mutated in an inherited type of Parkinson disease, a debilitating brain disorder that causes tremors of increasing severity. Half a million Americans suffer from this disorder, including actor Michael J. Fox and boxer Muhammad Ali (**Figure 12.1**). Until recently, Parkinson disease was thought to have an environmental, rather than a hereditary, basis. But data from the Human Genome Project mapped some cases of Parkinson disease to a specific gene. Interestingly, an altered version of the protein encoded by this gene has also been tied to Alzheimer disease, suggesting a link between these two brain disorders. The same gene is also found in rats, where it plays a role in the sense of smell, and in zebra finches, where it is thought to be involved in song learning. Cross-species comparisons such as these may uncover clues about the role played by the normal version of the protein in the brain. And such knowledge could eventually lead to treatments for the half million Americans with Parkinson disease.

In this chapter, you'll learn about four different ways that DNA technology plays a significant role in society: the use of recombinant DNA to produce useful products; the application of DNA fingerprinting in forensic science; the comparison of whole genomes as an investigative tool; and the use of human gene therapy for the treatment of diseases. For each of these applications, you'll learn about the specific techniques involved. You'll also consider some of the social, legal, and ethical issues that our society must face as we apply this emerging technology in new ways.

Figure 12.1 The fight against Parkinson Disease.
One goal of the Human Genome Project is to identify genes for inherited human diseases. Mutations in one such gene cause some forms of Parkinson disease, a neurological disorder that affects half a million Americans. In 2002, actor Michael J. Fox and boxer Muhammad Ali testified before the Senate on the status of federal funding for research into this disease.

Recombinant DNA Technology

In 1946, American geneticists Joshua Lederberg and Edward Tatum performed a series of experiments with *E. coli* that demonstrated that two individual bacteria can combine genes—a phenomenon that was previously thought to be limited to sexually reproducing eukaryotic organisms. With this work, they pioneered bacterial genetics, a field that within 20 years made *E. coli* the most thoroughly understood organism at the molecular level. In the 1970s, research on *E. coli* led to the development of

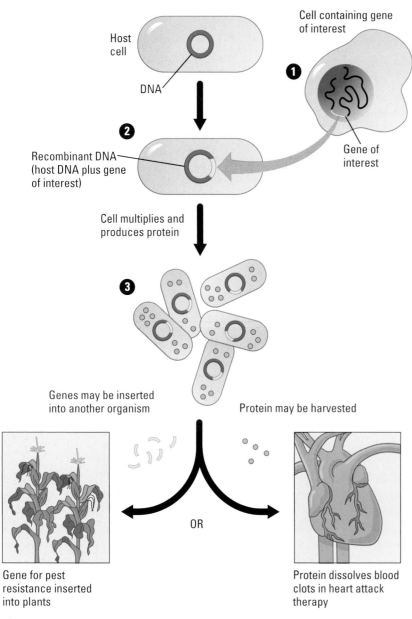

Figure 12.2 illustrates how recombinant DNA technology can be used to produce useful products. ❶ DNA carrying a gene of interest is taken from a cell of one organism and inserted into the DNA of a host cell. The gene of interest can be a human gene encoding a protein of medical value or perhaps a plant gene conferring resistance to pest insects. ❷ The host cell now contains **recombinant DNA,** a molecule carrying DNA from more than one source. An organism that carries recombinant DNA is called a **genetically modified (GM) organism.** A host that carries DNA acquired from a different species is called a **transgenic organism.** ❸ As a transgenic cell reproduces and copies its genome, it also copies the gene of interest. Furthermore, the host cell may express the foreign gene to produce the desired protein.

DNA technology has revolutionized biotechnology. As shown at the bottom of Figure 12.2, these methods can be used to create many useful products. In the example on the left, copies of the gene itself are the immediate product. In the example on the right, the protein product of the gene is harvested.

recombinant DNA technology, a set of laboratory techniques for combining genes from different sources—even different species—into a single DNA molecule.

Today, recombinant DNA technology is widely used to alter the genes of many types of cells for practical purposes. Scientists have genetically engineered bacteria to mass-produce many useful chemicals, from cancer drugs to pesticides. Genes have also been transferred from bacteria into plants and from humans into farm animals. These applications are the latest developments in **biotechnology,** the use of organisms to perform practical tasks. Biotechnology actually goes back thousands of years to the first uses of yeast to make bread and wine.

Activity 12A on the Web & CD Explore some applications of DNA technology.

Figure 12.2 Using recombinant DNA technology to produce useful products.

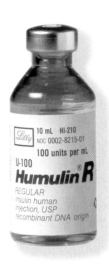

Figure 12.3 Humulin. Humulin, human insulin produced by genetically modified bacteria, was the first recombinant DNA drug approved for use in humans.

From Humulin to Genetically Modified Foods

By transferring the gene for a desired protein product into a bacterium, yeast, cultured mammalian cell, or other kind of cell that is easy to grow, proteins that are present naturally in only small amounts can be produced in large quantities. In this section, you'll learn about a few specific examples of this application of recombinant DNA technology.

Making Humulin In 1982, the biotechnology company Genentech began mass production of the world's first genetically engineered pharmaceutical product: Humulin, human insulin produced by genetically modified bacteria (**Figure 12.3**). In humans, insulin is a protein normally produced by the pancreas. Functioning as a hormone, insulin regulates the level of glucose in the blood. If the body fails to produce enough insulin, the result is type I diabetes. Because there is no cure, people with this disease must administer daily doses of insulin throughout their lives.

Since the 1920s, diabetes has been treated using insulin extracted from cows and pigs. This treatment is not without problems, however. Because the chemical structure of pig and cow insulins are not exactly the same as that of human insulin, they can cause allergic reactions in people using them. In addition, by the 1970s, the U.S. supply of beef and pork pancreas available for insulin extraction could not keep up with the demand. A new source was needed. With the advent of recombinant DNA technology, the artificial production of human insulin became an obvious goal.

In 1978, scientists working with Genentech took on the challenge. They began by chemically synthesizing the genes for human insulin. Since the amino acid sequence of the human insulin protein was already known, it was easy to use the genetic code (see Figure 10.11) to determine a nucleotide sequence that would code for the protein. Researchers arduously synthesized individual pieces of DNA and linked them together to form the desired insulin genes. In 1979, the researchers succeeded in inserting these artificial human genes into *E.coli* host cells. Under proper growing conditions, each transgenic bacterium cranked out large quantities of the human protein.

In 1982, Humulin became the first recombinant DNA drug approved by the Food and Drug Administration (FDA). Today, Humulin is produced in gigantic fermentation vats, four stories high and filled with bacteria, that operate 24 hours a day. Every day, more than 4 million people with diabetes use the insulin collected, purified, and packaged at such facilities.

Insulin is just one of many human proteins that have been produced by transgenic bacteria. Another success story is human growth hormone (HGH). Abnormally low levels of this hormone during childhood and adolescence can cause dwarfism. Because growth hormones from other animals are not effective in humans, HGH was an early target of genetic engineers. Before genetically engineered HGH became available in 1985, children with an HGH deficiency could only be treated with scarce and expensive supplies of HGH obtained from human cadavers.

Besides bacteria, mammalian cells can also be used to produce medically valuable human proteins. For example, transgenic mammalian cells growing in laboratory cultures are currently used to produce erythropoietin (EPO), a treatment for anemia.

DNA technology is also helping medical researchers develop vaccines. A **vaccine** is a harmless variant or derivative of a pathogen (a disease-causing organism, such as a bacterium or virus) that is used to prevent an infectious disease. When a person—a potential host for the pathogen—is inoculated, the vaccine stimulates the immune system to develop lasting defenses against the pathogen. For the many viral diseases for which there is no effective drug treatment, prevention by vaccination is virtually the only medical way to fight the disease. One approach to vaccine production is to use genetically engineered cells to make large amounts of a protein molecule that is found on the pathogen's outside surface. **Figure 12.4** shows a tank for growing transgenic yeast that produces a vaccine against hepatitis B, a disabling and sometimes fatal liver disease.

Genetically Modified (GM) Foods

Since ancient times, humans have selectively bred agricultural crops to make them more useful (see the discussion of artificial selection in Chapter 1). Today, DNA technology is quickly replacing traditional breeding programs as scientists try to improve the productivity of plants and animals important to agriculture. In the year 2002, roughly half of the American crops of soybeans and corn were genetically modified in some way. **Figure 12.5** shows a field of corn that has been genetically engineered to resist attack by an insect called the European

Figure 12.4 Equipment used in the production of a vaccine against hepatitis B. Yeast cells carrying genes from the hepatitis B virus are growing in the stainless steel tank.

Figure 12.5 Genetically modified corn. (a) The corn plants in this field carry a bacterial gene that enables them to resist infestation by the European corn borer. (b) The European corn borer at work and the damage that it causes.

Figure 12.6 Genetically modified rice.
"Golden rice," shown here alongside ordinary rice, has been genetically modified to produce high levels of beta-carotene, a precursor to vitamin A.

Figure 12.7 Genetically modified sheep.
These transgenic sheep carry a gene for a human blood protein, which they secrete in their milk. This protein inhibits an enzyme that contributes to lung damage in patients with cystic fibrosis and some other chronic respiratory diseases. Easily purified from the sheep's milk, the protein is currently being tested as a treatment for cystic fibrosis.

corn borer. Growing insect-resistant plants reduces the need for chemical insecticides. In another example, modified strawberries produce bacterial proteins that act as a natural antifreeze, providing protection from cold weather that can harm this delicate crop. Genetically-engineered potatoes produce harmless proteins derived from the cholera bacterium; researchers hope that these modified potatoes can serve as an edible vaccine to produce immunity against the disease cholera in people in developing nations, where cholera kills thousands of children every year. Scientists are also using genetic engineering to improve the nutritional value of crop plants. One recent development is transgenic "golden rice" that produces yellow rice grains containing beta-carotene, which our body uses to make vitamin A

Activity 12B on the Web & CD
Examine the issues surrounding "golden rice."

(**Figure 12.6**). This rice could help prevent vitamin A deficiency among people who depend on rice as their staple food—half the world's population. However, controversy surrounds the use of GM foods, as we'll discuss at the end of the chapter.

Farm Animals and "Pharm" Animals While transgenic plants are used today as commercial products, transgenic whole animals are currently only in the testing phase. **Figure 12.7** shows a herd of transgenic sheep that carry a gene for a human blood protein. The human protein can be harvested from the sheep's milk and is being tested as a treatment for cystic fibrosis. Because transgenic animals are difficult to produce, researchers may produce a single transgenic animal and then clone it. The resulting herd of genetically identical transgenic animals, all carrying a recombinant human gene, could then serve as a grazing pharmaceutical factory—"pharm" animals.

While transgenic animals are currently used to produce potentially useful proteins, none are yet found in our food supply. It is possible that DNA technology will eventually replace traditional animal breeding—for instance, to make a pig with leaner meat or a cow that will mature in less time. Someday soon, scientists might, for example, identify a gene that causes the development of larger muscles (which make up most of the meat we eat) in one variety of cattle and transfer it to other cattle or even to sheep.

Recombinant DNA technology serves many roles today and will certainly play an even larger part in our future. In the next section, you'll learn more about the methods that scientists use to create and manipulate recombinant DNA.

Recombinant DNA Techniques

Although recombinant DNA techniques can use cells from yeasts, plants, and animals, bacteria are the workhorses of modern biotechnology. To manipulate genes in the laboratory, biologists often use bacterial **plasmids,** which are small, circular DNA molecules that are separate from the much larger bacterial chromosome. In **Figure 12.8**, you can see a bacterial cell that has been ruptured, revealing one long chromosome and several smaller plasmids. Plasmids carry genes, like the main bacterial chromosome. Plasmids have several features that make them useful to biologists. For example, plasmids are small and readily taken up by bacterial cells. When they are

Case Study in the Process of Science on the Web & CD
Learn how plasmids are introduced into bacterial cells.

taken up, plasmids are acting as **vectors,** DNA carriers that move genes from one cell to another. When the bacterial cell divides, the plasmid, just like the main bacterial chromosome, is copied by the cell's replication machinery. Any foreign DNA that has been inserted will be replicated right along with the rest of the plasmid.

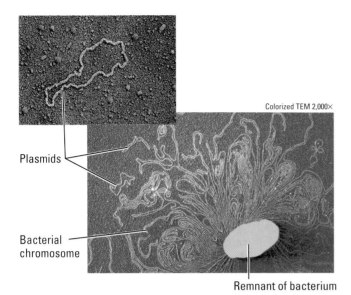

Plasmids

Bacterial
chromosome

Remnant of bacterium

Colorized TEM 2,000×

Figure 12.8 Bacterial plasmids.
The large, light blue oval shape is the remnant of a bacterium that has ruptured and released all of its DNA. Most of the DNA is the bacterial chromosome, which extends in loops from the cell. Two plasmids are also present. The inset shows an enlarged view of a single plasmid.

Making recombinant DNA in large enough quantities to be useful requires several steps. Consider a typical genetic engineering challenge: A molecular biologist at a pharmaceutical company has identified a human gene *V* that codes for a valuable product—a hypothetical substance called protein V that kills certain human viruses. The biologist wants to set up a system for manufacturing the protein on a large scale. **Figure 12.9** illustrates a way to accomplish this using recombinant DNA techniques.

❶ First, the biologist isolates two kinds of DNA: many copies of a bacterial plasmid (to serve as a vector) and human DNA containing many genes, including gene *V*, the gene of interest. ❷ The researcher cuts both the plasmids and the human DNA. The human DNA is cut into many fragments, one of which carries gene *V*. The figure shows the processing of just three human DNA fragments and three plasmids, but actually millions of plasmids and human DNA fragments (most of which do not contain gene *V*) are treated simultaneously. ❸ Next, the human DNA fragments are mixed with the cut plasmids. The plasmid and human DNA join together, resulting in recombinant DNA plasmids, some of which contain gene *V*. ❹ The recombinant plasmids are then mixed with bacteria. Under the right conditions, the bacteria take up the recombinant plasmids. ❺ Each bacterium, with its recombinant plasmid, is allowed to reproduce. This step is the actual **gene cloning,** the production of multiple copies of the gene. As the bacterium forms a clone (a group of identical cells descended from a single ancestral cell), any genes carried by the recombinant plasmid are also cloned (copied). ❻ The molecular biologist finds those bacterial clones that contain gene *V*. ❼ The transgenic bacteria with gene *V* can then be grown in large tanks, producing protein V in marketable quantities.

Activity 12C on the Web & CD
View an animation that will help you further understand gene cloning.

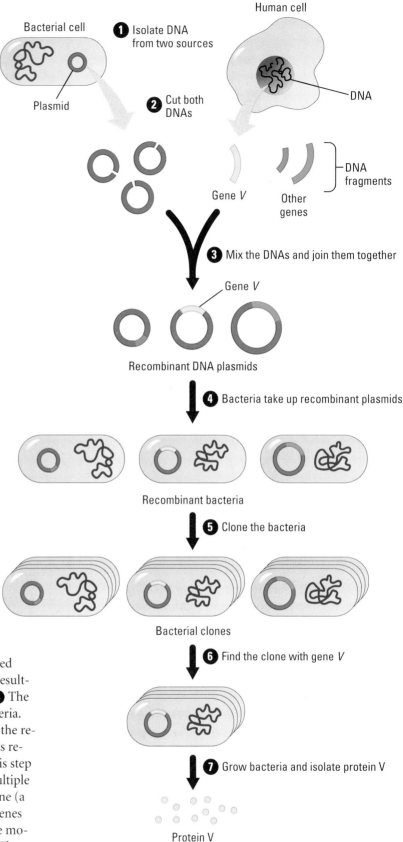

Bacterial cell ❶ Isolate DNA from two sources Human cell

Plasmid

DNA

❷ Cut both DNAs

Gene *V* Other genes DNA fragments

❸ Mix the DNAs and join them together

Gene *V*

Recombinant DNA plasmids

❹ Bacteria take up recombinant plasmids

Recombinant bacteria

❺ Clone the bacteria

Bacterial clones

❻ Find the clone with gene *V*

❼ Grow bacteria and isolate protein V

Protein V

Figure 12.9 An overview of recombinant DNA techniques.

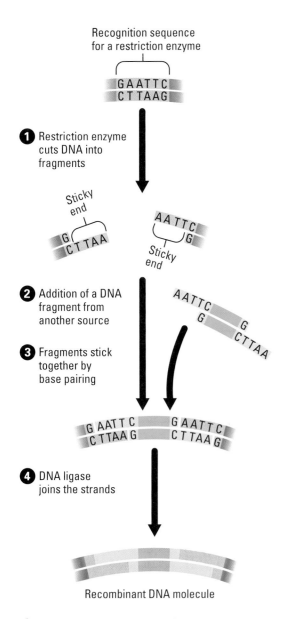

1 Restriction enzyme cuts DNA into fragments

Sticky end

Sticky end

2 Addition of a DNA fragment from another source

3 Fragments stick together by base pairing

4 DNA ligase joins the strands

Recognition sequence for a restriction enzyme

G A A T T C
C T T A A G

G
C T T A A

A A T T C
G

A A T T C
G   G
          C T T A A

G A A T T C   G A A T T C
C T T A A G   C T T A A G

Recombinant DNA molecule

Figure 12.10 Cutting and pasting DNA.
The key tools for creating recombinant DNA are two enzymes: a restriction enzyme, which cuts the original DNA molecules into pieces, and DNA ligase, which connects them together.

A Closer Look: Cutting and Pasting DNA with Restriction Enzymes

As you saw in the overview of techniques in Figure 12.9, recombinant DNA is produced by combining two ingredients: a bacterial plasmid and the gene of interest (see steps **2** and **3**). To understand this important process of cutting and pasting DNA, you need to learn about enzymes used for DNA splicing.

The cutting tools for making recombinant DNA are bacterial enzymes called **restriction enzymes.** Most restriction enzymes recognize short nucleotide sequences in DNA molecules and cut at specific points within these recognition sequences. Several hundred different restriction enzymes have been isolated. The top of **Figure 12.10** shows a piece of DNA that contains a recognition sequence for a particular restriction enzyme. **1** The restriction enzyme cuts the DNA strands between the bases A and G within the recognition sequence. (The places where DNA is cut are called **restriction sites.**) The staggered cuts yield two double-stranded DNA fragments with single-stranded ends, called "sticky ends." Sticky ends are the key to joining DNA restriction fragments originating from different sources. **2** Next, a piece of DNA (orange) from another source is added. Notice that the orange DNA has single-stranded ends identical in base sequence to the sticky ends on the blue DNA because the same restriction enzyme was used to cut both types of DNA. **3** The complementary ends on the blue and orange fragments bond together by base-pairing. **4** This union is subsequently made permanent by the "pasting" enzyme **DNA ligase.** This enzyme, which is one of the proteins the cell normally uses in DNA replication, connects the DNA pieces into continuous strands by forming covalent bonds between adjacent nucleotides. The final outcome is recombinant DNA, a molecule that contains DNA from two different sources.

> **Activity 12D on the Web & CD**
> See how different restriction enzymes cut DNA.

A Closer Look: Obtaining the Gene of Interest

The procedure shown in Figure 12.9 can yield millions of different recombinant plasmids, carrying many different segments of foreign DNA. Such a procedure is called a "shotgun" approach to gene cloning because it "hits" a huge variety of different pieces of DNA. The entire collection of cloned DNA fragments from a shotgun experiment, in which the starting material is bulk DNA from whole cells, is called a **genomic library.** A typical cloned DNA fragment is big enough to carry one or a few genes, and together the collection of fragments includes the entire genome of the organism from which the DNA was derived.

Once you've created a genomic library, you have to find the right book—that is, you must identify the bacterial clone containing a desired gene (step **6** in Figure 12.9). Methods for detecting a gene depend on base pairing between the gene and a complementary sequence on another nucleic acid molecule, either DNA or RNA. When at least part of the nucleotide sequence of a gene is already known or can be guessed, this information can be used to advantage. For example, if we know that a gene contains the sequence TAGGCT, a biochemist can use nucleotides labeled with a radioactive isotope to synthesize a short single strand of DNA with a complementary sequence (ATCCGA). This sort of labeled nucleic acid molecule is called a **nucleic acid probe** because it is used to find a specific gene or other nucleotide sequence within a mass of DNA. (In practice, a probe molecule would be considerably longer than six nucleotides.) When a radioactive DNA probe is added to the DNA of various clones, it tags the correct molecule—finds the right book in the library—by base-pairing to

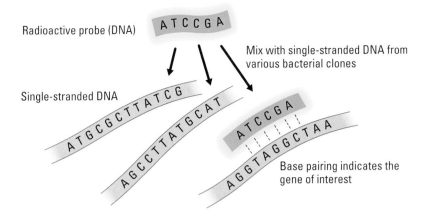

Figure 12.11 How a DNA probe tags a gene. The probe is a short, single-stranded molecule of DNA that is radioactive (it can also be RNA). When it is mixed with single-stranded DNA from a gene with a complementary base sequence, it attaches by hydrogen bonds. In this way, it labels the gene.

the complementary sequence in the gene of interest (**Figure 12.11**). Once a probe detects the desired clone within a library, the cells can be grown further and the gene of interest isolated in large amounts.

Another approach to obtain the gene of interest is to synthesize it. One method uses reverse transcriptase, a retroviral enzyme that can produce a molecule of DNA from a molecule of mRNA. **Figure 12.12** shows the steps involved. In steps ❶ and ❷, a eukaryotic cell transcribes the gene of interest and processes the transcript to produce mRNA. A researcher then isolates the mRNA and uses it as a template for synthesizing **complementary DNA (cDNA)** (steps ❸–❺). The cDNA that results from this procedure represents only those genes that are actually transcribed in the starting cells. Furthermore, because these artificial genes lack introns (see Figure 10.14), they are shorter, and therefore easier to work with, than the full version of the genes.

When the desired gene is small, it can be synthesized from scratch. In the earliest recombinant DNA experiments, including those leading to the production of Humulin, researchers laboriously synthesized genes. Today, automated DNA-synthesizing machines can accurately and rapidly produce customized DNA molecules of any sequence, in lengths up to hundreds of nucleotides long.

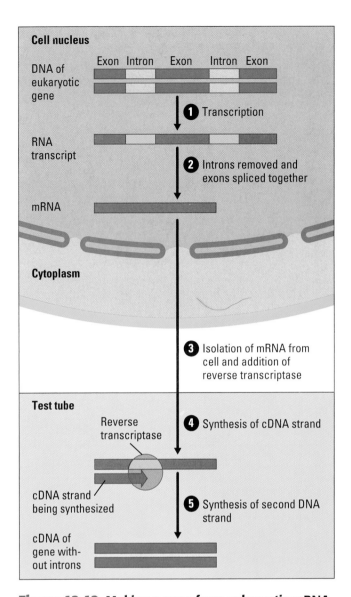

Figure 12.12 Making a gene from eukaryotic mRNA. Using the enzyme reverse transcriptase, a researcher can produce an artificial DNA gene (cDNA) from a molecule of messenger RNA (mRNA). Because the original gene's introns were spliced out during RNA processing, the artificial gene will not include introns and so the gene will be shorter than the original gene.

CHECKPOINT

1. What is recombinant DNA technology?

2. Corn that carries a bacterial gene is an example of a _____ organism.

3. Why are plasmids valuable tools for producing recombinant DNA?

4. Name three different ways that a gene of interest can be obtained.

Answers: 1. A set of methods for creating a DNA molecule that carries DNA originating from different sources **2.** transgenic or genetically modified **3.** Plasmids can carry virtually any foreign gene, are small, are easily taken up by bacteria, and are replicated by their bacterial host cells. **4.** The gene can be isolated from a genomic library created by a shotgun approach, produced from mRNA using reverse transcriptase, or synthesized from scratch.

DNA Fingerprinting and Forensic Science

Modern DNA technology has allowed biologists to make significant contributions to several fields of study. For example, DNA technology has rapidly revolutionized the field of **forensics,** the scientific analysis of evidence for crime scene investigations and other legal proceedings. Because the DNA sequence of every person is unique (except for identical twins), **DNA fingerprinting**—a procedure that analyzes a person's unique collection of DNA restriction fragments—can be used to determine whether two samples of genetic material are from the same individual.

Figure 12.13 presents an overview of a typical investigation using DNA fingerprinting. After a crime has occurred, ❶ DNA samples are collected from different sources; they may be from the present crime scene, from an old crime scene, from a suspect, or from a victim. The first problem that forensic scientists must overcome is often one of quantity; blood droplets or other biological evidence from a crime scene can be minuscule and insufficient to analyze. Therefore, before DNA forensics can begin, it may be necessary to ❷ amplify a DNA sample—that is, to copy the DNA evidence precisely to prepare a sufficient sample. Once sufficient quantities are obtained, ❸ the DNA samples are cut into fragments and then ❹ the fragments are compared. DNA fingerprinting provides data about which samples are from the same individual and which samples are unique.

Activity 12E on the Web & CD
Compare DNA fingerprints in a murder case.

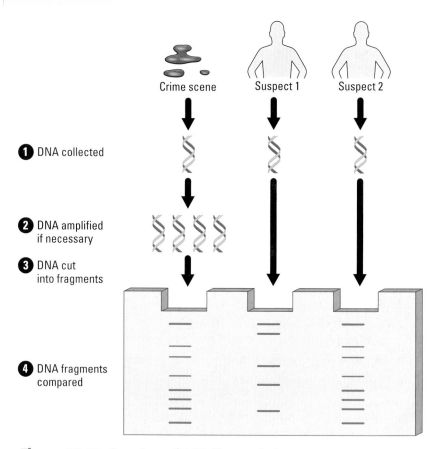

Figure 12.13 Overview of DNA fingerprinting. In this example, DNA from suspect 2 matches DNA found at the crime scene, but DNA from suspect 1 does not match.

Murder, Paternity, and Ancient DNA

The First Case On November 21, 1983, a 15-year-old girl named Lynda Mann was raped and murdered on a country lane near her home in Narborough, England. The killer left behind few clues, except for semen on the victim's body and clothes. The trail of evidence ran cold and the crime went unsolved.

On July 31, 1986, another 15-year-old girl, Dawn Ashworth, was raped and murdered less than a mile away from the first site. Police believed that a double murderer was at large. A worker from a nearby hospital was arrested and charged with both crimes. The worker confessed to the second murder, but denied committing the first.

In an attempt to link both murders to the suspect, the police turned to Alec Jeffreys, a professor at Leicester University who had recently developed the first DNA fingerprint identification system. Jeffreys compared DNA extracted from the semen samples collected at both murder scenes with DNA extracted from the suspect's blood. The DNA analysis proved that both murders had in fact been committed by the same killer. But, surprisingly, DNA from the suspect did not match either crime scene sample. For reasons unknown, the suspect had falsely confessed. On November 21, 1986, the suspect was released—and entered legal history as the first person ever exonerated by DNA evidence.

Finding themselves back at square one, the detectives decided to ask every young male from the surrounding area to voluntarily donate blood for DNA testing. Nearly 5,000 men were sampled, but none matched. The case was finally broken when a man in a pub was overheard describing how a local man named Colin Pitchfork had bullied him into submitting blood on Pitchfork's behalf. Detectives promptly arrested Pitchfork and took a blood sample. His DNA matched the samples from the two crime scenes. On January 22, 1988, the case again made history as Colin Pitchfork pleaded guilty to both crimes, closing the first murder case ever to be solved by DNA evidence.

Crimes and Other Investigations Since its introduction in 1986, DNA fingerprinting has become a standard part of law enforcement (**Figure 12.14**) and has provided crucial evidence in many famous cases. In the O. J. Simpson murder trial, DNA analysis proved that blood in Simpson's car belonged to the victims and that blood at the crime scene belonged to Simpson. (The jury in this case did not find the DNA evidence alone to be sufficient to convict the suspect, and Simpson was found not guilty.) During the investigation that led up to his impeachment, President Bill Clinton repeatedly denied that he had sexual relations with Monica Lewinsky—until DNA fingerprinting proved that his semen was on her dress. Of course, DNA evidence can prove innocence as well as guilt. The Innocence Project, headquartered in New York City, has used DNA technology and legal work to exonerate over 100 convicted criminals, including several on death row. DNA fingerprinting can also be used to identify crime victims. The largest such effort in history took place after the World Trade Center attack on September 11, 2001. Forensic scientists worked around the clock matching over 10,000 samples of victims' remains to DNA from items provided by families, such as toothbrushes, cigarette butts, and blood samples from close relatives.

The use of DNA fingerprinting extends beyond crimes. For instance, comparing the DNA of a mother, her child, and the purported father can conclusively settle a question of paternity. Sometimes paternity is of historical interest: DNA fingerprinting proved that Thomas Jefferson or a close male relative fathered at least one of the children of his slave Sally Hemings.

Figure 12.14 DNA data used to solve crimes.
The head of the Connecticut State Forensic Laboratory examines DNA data that will be stored in a state database.

DNA analysis can also be used to probe the origin of nonhuman materials. In 1998, the U.S. Fish and Wildlife Service began testing the DNA in caviar to determine if the fish eggs originated from the species claimed on the label. DNA fingerprinting can also help protect endangered species by conclusively proving the origin of contraband animal products.

Modern methods of DNA fingerprinting are so specific and powerful that the starting DNA material can even be in a partially degraded state. This allows DNA analysis to be applied in a great number of ways. In evolution research, the technique has been used to study DNA pieces recovered from an ancient mummified human, from a 40,000-year-old woolly mammoth frozen in a glacier, and from a 30-million-year-old plant fossil. One of the strangest cases of DNA fingerprinting is that of Cheddar Man, a 9,000-year-old skeleton found in a cave near Cheddar, England. DNA was extracted from his tooth and analyzed by DNA fingerprinting. The results showed that Cheddar Man was related to a present-day schoolteacher living only a half mile from the cave!

DNA Fingerprinting Techniques

In this section, you'll learn how DNA fingerprinting is carried out as we describe the methods that molecular biologists use to perform the steps in Figure 12.13.

The Polymerase Chain Reaction (PCR) The **polymerase chain reaction (PCR)** is a technique by which any segment of DNA can be amplified (cloned). Through PCR, a scientist can obtain enough DNA from even minute amounts of blood or other tissue to allow DNA fingerprinting.

In principle, PCR is surprisingly simple (**Figure 12.15**). The key to PCR is an unusual DNA polymerase that was first isolated from prokaryotes living in hot springs. Unlike most proteins, this enzyme can withstand the heat needed to separate DNA strands during the PCR procedure. The DNA sample to be amplified is heated and mixed with the special heat-tolerant DNA polymerase, nucleotide monomers, and short pieces of synthetic single-stranded DNA that serve as primers for DNA synthesis. (DNA polymerase needs primers because it can only add nucleotides to a preexisting nucleotide chain; see Chapter 10.) In this mixture, the DNA is replicated, producing two daughter DNAs; the daughter molecules, in turn, are replicated; and so on. Each time replication occurs, the amount of DNA doubles. Starting with a single DNA molecule, automated PCR can generate 100 billion copies in a few hours.

Restriction Fragment Length Polymorphism (RFLP) Analysis Once a suitable quantity of sample DNA is available, DNA fingerprinting can begin. How do you prove that two samples of DNA are the same? You could compare the complete nucleotide sequences of the genomes found in the two samples. But this approach is extremely impractical, since it requires a lot of time and money to sequence even one genome. DNA fingerprinting relies instead on indirect methods to compare samples. One method is called RFLP (pronounced "rif-lip") analysis, where RFLP stands for restriction fragment length polymorphism.

Unless you have an identical twin, the total nucleotide sequence of your DNA is unique. Geneticists can use any DNA segment that varies from person to person as a **genetic marker,** a chromosomal landmark whose inheritance can be studied. And just like a gene, a genetic marker is more likely to be an exact match to the comparable marker in a relative than to the marker in an unrelated individual.

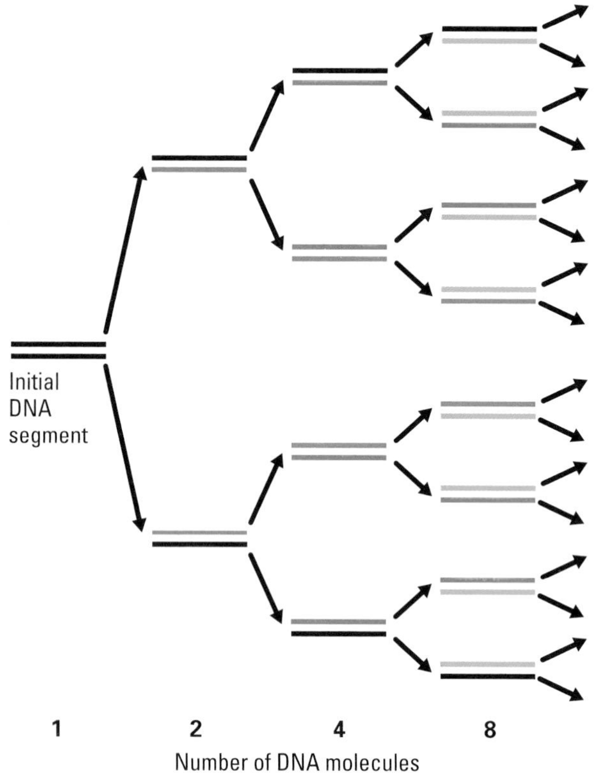

Figure 12.15 DNA amplification by PCR.
The polymerase chain reaction (PCR) is a method for making many copies of a specific segment of DNA. Each round of PCR doubles the total quantity of DNA.

Initial DNA segment

| 1 | 2 | 4 | 8 |

Number of DNA molecules

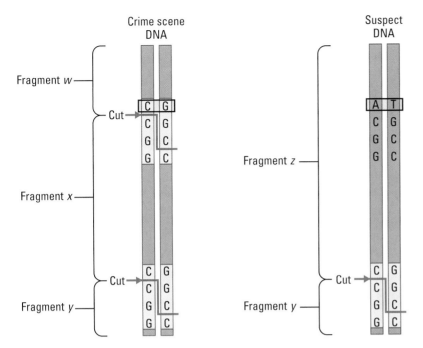

Figure 12.16 How restriction fragment formation reflects DNA sequence. The two double-stranded DNA segments shown have nucleotide sequences that differ at one base pair (yellow boxes). A particular restriction enzyme may therefore cut the segments at different places. Imagine that the first DNA sample comes from a crime scene and the second from a suspect. A certain restriction enzyme cuts the DNA from the crime scene in two places, but it only cuts the suspect's DNA in one place because of the sequence difference. These differences indicate that the DNA samples come from different individuals.

The goal of DNA fingerprinting by RFLP analysis is to determine whether samples of DNA contain identical markers or not. Consider the two samples of DNA shown in **Figure 12.16**. Imagine that the first DNA segment was extracted from DNA at a crime scene and the second from a suspect's blood. Notice that the segments differ by a single base pair (boxed in yellow). To create DNA fingerprints, a geneticist treats each DNA sample with a restriction enzyme, creating a mixture of restriction fragments. In the example shown in the figure, the restriction enzyme cuts the DNA between two cytosine (C) bases in the sequence CCGG or in its complement, GGCC. Because the DNA from the crime scene has two recognition sequences for the restriction enzyme, it is cleaved in two places, yielding three restriction fragments. The suspect's DNA, however, has only one recognition sequence and yields only two restriction fragments. Notice that the crime scene's DNA fragments differ in length and number from the suspect's DNA fragments, reflecting the different sequence of bases in the two samples of DNA. Hence the name RFLP: the *length* of *restriction fragments* differs (is *polymorphic*) among different individuals. **RFLP analysis** is the comparison of the set of restriction fragments produced by DNA from different individuals. How this comparison is made is our next topic.

Gel Electrophoresis Once a sample of DNA is chopped up by restriction enzymes, the next step of RFLP analysis is to determine the number and size of the restriction fragments. An essential tool of DNA technology, **gel electrophoresis** is a method for sorting macromolecules—proteins or nucleic acids—

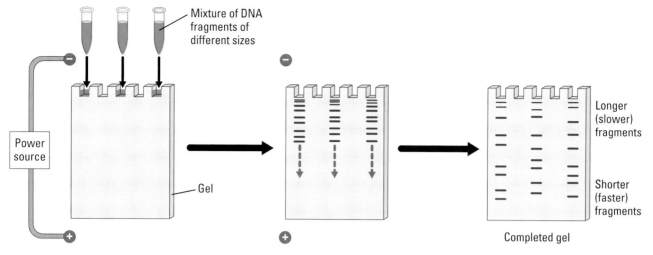

Figure 12.17
Gel electrophoresis of DNA molecules.

Activity 12F on the Web & CD
View animations of the steps of gel electrophoresis.

primarily on the basis of their electrical charge and length. **Figure 12.17** shows how gel electrophoresis can be used to separate the various DNA fragments in three different mixtures. A sample of each mixture is placed in a well at one end of a flat, rectangular gel, a thin slab of jellylike material. A negatively charged electrode is then attached to the DNA-containing end of the gel and a positive electrode to the other end. Because the DNA fragments have a negative charge owing to their phosphate (PO_4^-) groups, they move through the gel toward the positive pole. However, the longer DNA fragments are held back by the molecules of the gel, so they move more slowly than the shorter DNA fragments—and thus not as far in a given time period. When the current is turned off, a series of bands is left in each "lane" of the gel. Each band consists of DNA molecules of one size. The bands are visualized by staining or, if the DNA is radioactively labeled, by exposure onto photographic film. **Figure 12.18** shows the bands that would result from using gel electrophoresis to separate the DNA fragments from the example in Figure 12.16. The differences in the number and locations of the bands prove that the crime scene DNA did not come from the suspect.

Activity 12G on the Web & CD
Analyze DNA fragments using gel electrophoresis.

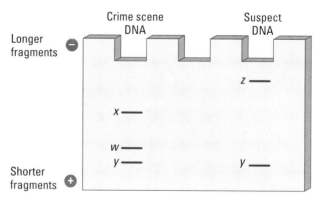

Figure 12.18 Visualizing restriction fragment patterns.
This figure shows the bands that would result from gel electrophoresis of the DNA fragments in Figure 12.16.

DNA fingerprints from an actual murder case are shown on a gel in **Figure 12.19**. The restriction fragments from the victim's DNA clearly match the fragments from the blood on the defendant's clothes. Furthermore, the markers from the defendant's DNA are clearly different. Thus, electrophoresis allows us to see similarities as well as differences between mixtures of restriction fragments, reflecting similarities as well as differences between the nucleotide sequences from two DNA samples.

Case Study in the Process of Science on the Web & CD Conduct virtual gel electrophoresis.

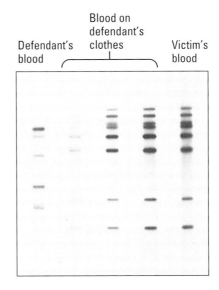

Blood on defendant's clothes

Defendant's blood

Victim's blood

Figure 12.19 DNA fingerprints from a murder case. DNA from bloodstains on the defendant's clothes matches the DNA fingerprint of the victim but differs from the DNA fingerprint of the defendant. This is evidence that the blood on the defendant's clothes came from the victim, not the defendant, and places the defendant at the scene of the crime.

CHECKPOINT

1. Why is only the slightest trace of DNA at a crime scene often sufficient for forensic analysis?

2. You use a restriction enzyme to cut a DNA molecule. The base sequence of this DNA is known, and the molecule has a total of three restriction sites clustered close together near one end. When you separate the restriction fragments by electrophoresis, how do you expect the bands to be distributed in the electrophoresis lane?

3. Put these three techniques in the order that would allow you to create a DNA fingerprint from a minuscule crime scene sample: gel electrophoresis, PCR, treatment with restriction enzymes.

Answers: 1. Because PCR can be used to produce enough molecules for analysis **2.** Three bands near the positive pole at the bottom of the gel (small fragments) and one band near the negative pole at the top of the gel (large fragment) **3.** PCR, treatment with restriction enzymes, gel electrophoresis

Genomics

By the 1980s, the nucleotide sequences of many important genes from humans and other organisms had been determined. It didn't take long for biologists to think on a larger scale. In the 1990s, a team of scientists announced that they had determined the nucleotide sequence of the entire genome of *Haemophilus influenzae,* a bacterium that can cause several human diseases, including pneumonia and meningitis. **Genomics,** the science of studying whole genomes, was born.

The first targets of genomics research were pathogenic bacteria. The genome of *H. influenzae,* for example, contains 1.8 million nucleotides and 1,709 genes. But soon, the attention of many genomics researchers turned toward more complex organisms with much larger genomes. Inevitably, biologists began to work on bagging the ultimate prize in molecular genetics: the complete sequencing of the human genome.

Figure 12.20 Completion of the Human Genome Project. Yoshiyuki Sakaki (left), director of the Japanese branch of the Human Genome Project, presents Japan's Prime Minister Junichiro Koizumi with a set of CDs that contain data from the completed project.

The Human Genome Project

In 1990, an international consortium of government-funded researchers began the Human Genome Project. Several years into the project, private companies, chiefly Celera Genomics in the United States, joined the effort. The original project deadline was 2005, but fierce competition between the public consortium and Celera spurred both groups on. On April 14, 2003, the international consortium announced the successful completion of the Human Genome Project more than two years head of schedule (**Figure 12.20**).

The 24 different kinds of chromosomes in the human genome (22 autosomes plus the X and Y sex chromosomes) contain approximately 3.2 billion nucleotide pairs of DNA and 30,000 to 40,000 genes. To try to get a sense of this quantity of DNA, imagine that its nucleotide sequence is printed in letters (A, T, C, and G) like the letters in this book. At this size, the sequence would fill a stack of books 18 stories high!

Activity 12H on the Web & CD
Look at what we've learned about human chromosome 17.

Our genome presents a major challenge not only because of its size but also because, like that of most complex eukaryotes, only a small amount of our total DNA is contained in genes that code for proteins, tRNAs, or rRNAs. Most complex eukaryotes have a huge amount of noncoding DNA—about 97% of human DNA is of this type. Some noncoding DNA is made up of gene control sequences such as promoters and enhancers (see Chapter 11). The remaining DNA has been called "junk DNA," a tongue-in-cheek way of saying that scientists don't understand its functions. This DNA includes introns (whose total length in a gene may be ten times greater than the total length of the exons) and noncoding DNA located between genes and at the ends of chromosomes.

Much of the DNA between genes consists of **repetitive DNA,** nucleotide sequences present in many copies in the genome. There are two main types of repetitive DNA. In one type, a unit of just a few nucleotide pairs is repeated many times in a row. Stretches of DNA with thousands of such repetitions are prominent at centromeres and **telomeres,** the ends of chromosomes. This suggests that repetitive DNA plays a role in chromosome structure: It may help keep the rest of the DNA properly organized during DNA replication and mitosis. In the second main type of repetitive DNA, each repeated unit is hundreds of nucleotides long, and the copies are scattered around the genome. Little is known about the functions of this DNA.

The markers most often used in DNA fingerprinting are inherited variations in the lengths of the first type of repetitive DNA. For example, one person may have the nucleotides AC repeated 65 times at one place in the genome, 118 times at a second place, and so on, whereas another person is likely to have different numbers of repeats at these places.

As of 2003, the genomes of over 100 organisms have been sequenced (**Table 12.1**). The majority are prokaryotes, including *E. coli,* a number of other bacteria (some of medical importance), and several archaea. About a dozen eukaryotic genomes have been completed. Yeast was the first eukaryote to have its full sequence determined, and the nematode *Caenorhabditis elegans,* a simple worm, was the first multicellular organism. Other sequenced animals include the fruit fly *Drosophila melanogaster* and the mouse. Plants, such as *Arabidopsis thaliana,* another important research organism, and rice, one of the world's most economically important crops, have also been completed. Recently, the genomes of the mosquito *Anopheles gambiae* and the parasite *Plasmodium falciparum* were sequenced. These organisms are important to study because together they transmit malaria, a disease that kills over 1 million children worldwide each year.

Table 12.1	Some Important Completed Genomes		
Organism	Date Completed	Size of Genome (in base pairs)	Approximate Number of Genes
Haemophilus influenzae (bacterium)	1995	1.8 million	1,700
Saccharomyces cerevisiae (yeast)	1996	12 million	6,000
Escherichia coli (bacterium)	1997	4.6 million	4,400
Caenorhabditis elegans (roundworm)	1998	97 million	19,100
Drosophila melanogaster (fruit fly)	2000	180 million	13,600
Arabidopsis thaliana (mustard plant)	2000	100 million	25,000
Homo sapiens (human)	2001	3.2 billion	30,000–40,000
Mus musculus (mouse)	2001	3 billion	35,000
Oryza sativa (rice)	2002	466 million	46,000–56,000

Tracking the Anthrax Killer

In October 2001, Bob Stevens, a 63-year-old editor at a Florida media company, died from inhalation anthrax, a disease caused by breathing spores of the bacterium *Bacillus anthracis*. As the first victim of this disease in the United States since 1976, his death was immediately suspicious. By the end of the year, four other people had died from anthrax. Law enforcement officials realized that someone was sending anthrax spores through the mail. The United States found itself in the grip of an unprecedented bioterrorist attack.

In the investigation that followed, one of the most helpful clues turned out to be the anthrax spores themselves. Examining the genomes of the spores could answer several crucial questions. Was the anthrax in all of the letters identical, indicating a lone source? Or had multiple strains been used, suggesting multiple perpetrators? Could the mailed spores be traced back to a particular strain, perhaps to one particular laboratory? To address these questions, samples were sent to genome centers for sequencing.

Investigators compared the genomes—which were 3 million nucleotides long—of the mailed anthrax spores with several laboratory strains. They quickly established that all of the mailed spores (including samples from Florida, New York, and Washington, DC) were identical, suggesting that a single perpetrator was behind all the attacks. Furthermore, they were able to match the deadly spores with a laboratory subtype called the Ames strain. The Ames strain was first isolated from a dead Texas cow in 1981. From there, it was sent to the U.S. Army Medical Research Institute in Fort Detrick, Maryland. And from there, it was sent to at least 14 other labs for use in experiments.

Using the data from the sequencing project, scientists were able to clearly differentiate the mailed strain from non-Ames strains. Unfortunately, the data were not detailed enough to tie the mailed samples to any particular laboratory. Researchers hope that future studies will be able to pinpoint the source of the bacteria used in the attacks.

The anthrax investigation is a prominent example of the new field of comparative genomics, the comparison of whole genomes. Other investigations have also taken advantage of this approach. In 1991, sequence data

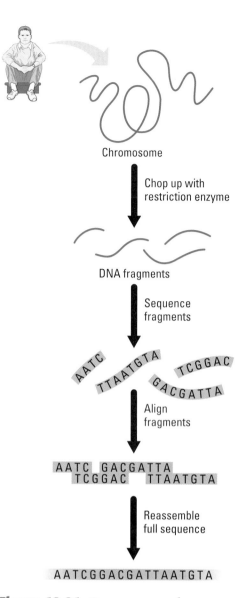

Chromosome

Chop up with
restriction enzyme

DNA fragments

Sequence
fragments

AATC TTAATGTA TCGGAC GACGATTA

Align
fragments

AATC GACGATTA
TCGGAC TTAATGTA

Reassemble
full sequence

AATCGGACGATTAATGTA

Figure 12.21 Genome mapping.

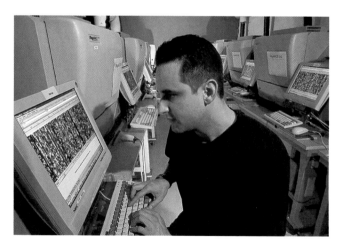

Figure 12.22 DNA sequencers.
Computers report the data obtained from the DNA-sequencing machines shown here (large gray boxes). Each machine can sequence DNA at the rate of 350,000 nucleotides per day.

from virus samples proved that a Florida dentist transmitted HIV to several patients. In 1993, after the Aum Shinrikyo cult released anthrax spores in downtown Tokyo, genomic analysis showed why their attack didn't kill anyone: They had used a harmless veterinary vaccine strain. Investigation of the West Nile virus outbreak in 1999 proved that a single natural strain of virus was infecting both birds and humans. Comparative genomics can even be used to help pinpoint what makes us human. Humans and chimpanzees share much of their DNA (estimates range between 95–99%), and plans are already under way to use the techniques of genomics to find and study the important differences.

Genome-Mapping Techniques

When it first began in 1990, the Human Genome Project was the most ambitious application of DNA technology ever attempted. Human chromosomes range in size from 50 to 300 million base pairs, and it is simply not possible to sequence that much DNA at once. Instead, chromosomes must be chopped up into much smaller fragments using restriction enzymes. Each piece is inserted into a vector (bacteria or yeast) and cloned. The DNA in each fragment is sequenced; that is, the exact order of A, T, C, and G nucleotides is determined. The collection of partial sequences from all the fragments is then aligned and reassembled in the proper order to yield the full sequence of the chromosome (**Figure 12.21**).

Although this approach is straightforward in principle, the huge size and complexity of the human genome are daunting. To deal with such a challenging task, the public consortium divided the genome project into the three stages described below. Each stage focused on the DNA in greater detail; the earlier stages provided "road markers" to help researchers find their way among the huge amounts of data generated by the last stage.

1. Genetic (Linkage) Mapping In Chapter 9, you learned how geneticists use data from genetic crosses to determine the order of linked genes on a chromosome and the relative distances between the genes. Scientists combined pedigree analysis of large families with DNA technology to map over 5,000 genetic markers. These markers included both coding regions (genes) and noncoding regions (such as stretches of repetitive DNA). The resulting low-resolution genetic map provided anchor points that enabled researchers to map other markers by testing for genetic linkage to the known markers.

2. Physical Mapping Researchers used several different restriction enzymes to break the DNA of each chromosome into a number of identifiable fragments, which they cloned. They then determined the original order of the fragments in the chromosome by overlapping the fragments and matching up their ends. They used probes to relate the fragments to the markers mapped in stage 1. The end result was a series of DNA segments that spanned the genome in a known order.

3. DNA Sequencing The most arduous part of the project was determining the nucleotide sequences of the set of DNA fragments created in stage 2. Advances in automated DNA sequencing were crucial to this endeavor. Automated sequencing machines like the one shown in **Figure 12.22** worked around the clock, feeding their results to computers that stored, analyzed, and reassembled the data. As the sequence of each cloned fragment from the physical map of stage 2 was determined, the fragments were reassembled in their proper order, producing large-scale sequences.

The Whole Genome Shotgun Method The government-sponsored Human Genome Project, begun in 1990, proceeded with the three stages just described for nearly a decade. In 1998, the private company Celera Genomics entered the race and was able to produce a draft of the human genome within three years. How did they accomplish the task so quickly?

The researchers at Celera pioneered a technique called the whole genome shotgun method. This procedure essentially skips the first two stages described and proceeds directly to the third. An entire genome is chopped by restriction enzymes into fragments that are cloned and sequenced. High-performance computers running specialized mapping software can reassemble the millions of partial sequences into an entire genome. With the human genome, the whole genome shotgun researchers relied on earlier maps created by the public consortium. But the method has also been used successfully to sequence many other genomes that did not have preliminary maps.

In 2003, when the international Human Genome Project concluded their sequencing work, 99% of the human genome was complete, leaving only a few regions (including the centromeres) unsequenced. The project researchers also bested their goal of fewer than one error per 10,000 bases.

The DNA sequences determined by the public consortium are deposited in a database that is available to researchers all over the world via the Internet. (You can browse it yourself at the website for the National Center for Biotechnology Information.) Scientists use software to scan and analyze the sequences for genes, control elements, and other features. Now comes the most exciting challenge: figuring out the functions of the genes and other sequences and how they work together to direct the structure and function of a living organism. This challenge and the applications of the new knowledge should keep scientists busy well into the twenty-first century.

One interesting question to ask about the Human Genome Project is whose genome was sequenced? The answer is no one's—or at least not any one person's. The first human genome to be sequenced by the public consortium was actually a reference genome compiled from a group of individuals. This representative sequence will serve as a standard, so that comparisons of individual differences and similarities can be made. Eventually, as the amount of sequence data multiplies, the small differences that account for individual variation within our species will come to light.

CHECKPOINT

1. How many nucleotides and genes are contained in the human genome?

2. Name three types of DNA that do not code for another molecule.

3. Why did the whole genome shotgun sequencing method allow Celera Genomics to produce a draft of the human genome so quickly?

Answers: 1. About 3.2 billion nucleotides and 30,000 to 40,000 genes **2**. Introns, repetitive DNA, and control elements (such as promoters and enhancers) **3**. The whole genome shotgun method does not require the steps involved in creating preliminary maps.

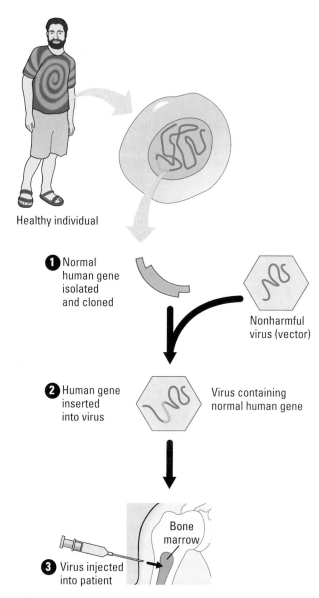

Healthy individual

1 Normal human gene isolated and cloned

Nonharmful virus (vector)

2 Human gene inserted into virus

Virus containing normal human gene

3 Virus injected into patient

Bone marrow

Figure 12.23 One approach to human gene therapy.

Human Gene Therapy

Human gene therapy is a recombinant DNA procedure that seeks to treat disease by altering an afflicted person's genes. In some cases, a mutant version of a gene may be replaced or supplemented with the normal allele. This could potentially correct a genetic disorder, perhaps permanently. In other cases, genes are inserted and expressed only long enough to treat a medical problem.

Figure 12.23 summarizes the steps in one approach to human gene therapy. **1** A gene from a normal individual is isolated and cloned by recombinant DNA techniques. **2** The gene is inserted into a vector, such as a nonharmful virus. **3** The virus is then injected into the patient. The virus inserts a copy of its genome, including the human gene, into the DNA of the patient's cells. The normal gene is then transcribed and translated within the patient's body, producing the desired protein. Ideally, the nonmutant version of the gene would be inserted into cells that multiply throughout a person's life. Bone marrow cells, which include the stem cells that give rise to all the cells of the blood and immune system, are prime candidates. If the procedure succeeds, the cells will multiply throughout the patient's life and express the normal gene. The engineered cells will supply the missing protein, and the patient will be cured.

Treating Severe Combined Immunodeficiency

Severe combined immunodeficiency (SCID) is a fatal inherited disease caused by a single defective gene. Absence of the enzyme encoded by this gene prevents the development of the immune system, requiring patients to remain isolated within protective "bubbles." Unless treated with a bone marrow transplant (effective just 60% of the time), SCID patients quickly die from infections by ever-present microbes that most of us easily fend off. The first trial of human gene therapy began at the National Institutes of Health in 1990 on a 4-year-old girl with SCID. Immune system cells were periodically removed from her blood, infected with a virus engineered to carry the normal allele of the defective gene, then reinjected into her bloodstream. The patient is still alive today and leads a normal life. However, it is unclear to what extent the gene therapy was responsible for her recovery, since her gene therapy treatment lasted a limited time and was only one of several treatments she received.

Most human gene therapy experiments to date have been preliminary, designed to test the safety and effectiveness of a procedure rather than to attempt a cure. Despite repeated hype in the news media over the past decade, it was not until April 2000 that the first scientifically strong evidence of effective gene therapy was reported. This landmark case involved the treatment of two infants suffering from SCID. Working at a Paris hospital, researchers used a procedure similar to the one in Figure 12.23 to provide the infants with functional copies of their defective gene. After treatment, the children were healthy and were able to leave their protective isolation. After this success, other children with SCID were treated with gene therapy. However, the SCID study was halted in 2002 after two patients developed leukemia-like symptoms.

Safety and Ethical Issues

As soon as scientists realized the power of DNA technology, they began to worry about potential dangers. Early concerns focused on the possibility that recombinant DNA technology might create hazardous new pathogens. What might happen, for instance, if cancer cell genes were transferred into infectious bacteria or viruses? To address such concerns, scientists developed a set of guidelines that in the United States and some other countries have become formal government regulations.

One type of safety measure is a set of strict laboratory procedures designed to protect researchers from infection by engineered microbes and to prevent the microbes from accidentally leaving the laboratory (**Figure 12.24**). In addition, strains of microorganisms to be used in recombinant DNA experiments are genetically crippled to ensure that they cannot survive outside the laboratory. Finally, certain obviously dangerous experiments have been banned. Today, most public concern about possible hazards centers not on recombinant microbes but on genetically modified (GM) foods.

The Controversy Over Genetically Modified Foods

GM strains account for a significant percentage of several agricultural crops. In 1999, controversy about the safety of these foods exploded in the United Kingdom and soon spread throughout Europe (**Figure 12.25**). In response to these concerns, the European Union suspended the introduction of new GM crops and started considering the possibility of banning the import of all GM foodstuffs. In the United States and other countries where the GM revolution has proceeded more quietly, the labeling of GM foods is now being debated but has not yet become law.

Advocates of a cautious approach fear that crops carrying genes from other species might be hazardous to human health or harm the environment. A major concern is that transgenic plants might pass their new genes to close relatives in nearby wild areas. We know that lawn and crop grasses, for example, commonly exchange genes with wild relatives via pollen transfer. If domestic plants carrying genes for resistance to herbicides, diseases, or insect pests pollinated wild plants, the offspring might become "super-weeds" that would be very difficult to control. However, researchers may be able to prevent the escape of such plant genes by engineering plants so that they cannot hybridize. The U.S. National Academy of Sciences released a study finding no scientific evidence that transgenic crops pose any special health or environmental risks. But the authors of the study also recommended more stringent long-term monitoring to watch for unanticipated environmental impacts.

Negotiators from 130 countries (including the United States) have agreed on a Biosafety Protocol that requires exporters to identify GM organisms present in bulk food shipments and allows importing countries to

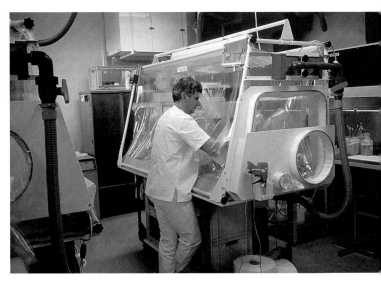

Figure 12.24 Maximum security laboratory.
A scientist in a high-containment laboratory at the Pasteur Institute in Paris uses a "glove box" for working with dangerous microorganisms. To manipulate the microbial cultures inside the box, he inserts his hands into gloves that are sealed to the edges of round holes in the box. In this way, he avoids all contact with the microbes and prevents the microbes from escaping into the environment.

Figure 12.25 Opposition to genetically modified organisms (GMO).
In Europe, there is strong opposition to GM foods. This photo shows tractors destroying a field of genetically modified rape plants in Belgium during the summer of 2002. The rape plant is a member of the mustard family and is grown as a forage crop for sheep and hogs. The seeds are used to make cooking oil and as bird feed. The destruction was ordered by the Belgian Minister for Consumer Protection. The sign reads "GMO, no thanks! Yes to biodiversity."

decide whether the shipments pose environmental or health risks. This agreement has been hailed as a breakthrough by environmentalists.

Today, governments and regulatory agencies throughout the world are grappling with how to facilitate the use of biotechnology in agriculture, industry, and medicine while ensuring that new products and procedures are safe. In the United States, all projects are evaluated for potential risks by a number of regulatory agencies, including the Food and Drug Administration, the Environmental Protection Agency, the National Institutes of Health, and the Department of Agriculture.

Ethical Questions Raised by DNA Technology

DNA technology raises many questions—moral, legal, and ethical—few of which have clear answers. Consider, for example, the child in **Figure 12.26**. She is growing at a normal rate, thanks to regular injections of human growth hormone (HGH) produced by genetically engineered bacteria. Like any new drug, this HGH was subjected to exhaustive laboratory tests before it was released for human use. However, because it is a powerful hormone that affects the body in a number of ways, this drug may be more likely than most ordinary drugs to produce unanticipated long-term side effects. Should this type of therapy therefore be reserved for treating only serious conditions? Should parents of short but hormonally normal children be able to seek HGH treatment to make their kids taller? If not, who decides which children are "tall enough" to be excluded from treatment?

Genetic engineering of gametes (sperm or ova) and zygotes has been accomplished in lab animals. It has not been attempted in humans because such a procedure would raise very difficult ethical questions. Should we try to eliminate genetic defects in our children and their descendants? Should we interfere with evolution in this way? From a long-term perspective, the elimination of unwanted versions of genes from the gene pool could backfire. Genetic variety is a necessary ingredient for the adaptation of a species as environmental conditions change with time. Genes that are damaging under some conditions may be advantageous under others (one example is the sickle-cell allele, discussed in Chapter 9). Are we willing to risk making genetic changes that could be detrimental to our species in the future? We may have to face such questions soon.

Advances in genetic fingerprinting raise privacy issues. In the future, DNA technology could prevent anyone from getting away with rape, since it is nearly impossible for someone to commit a rape without leaving DNA evidence behind. If we were to create a DNA fingerprint of every person at birth, then we could theoretically match every rape to a perpetrator. But are we, as a society, prepared to sacrifice our genetic privacy, even for such a worthwhile goal?

And what of the information being obtained in the Human Genome Project? The potential benefits to human health provide strong ethical support for the project. But there is a danger that information about disease-associated genes could be abused. A very big question is the problem of possible discrimination and stigmatization. Would you, as an employer, want to know if a potential employee had an increased risk of schizophrenia? Should you be able to find out? Should insurance companies have the right to screen applicants for disease genes? People might be coerced into taking a DNA test in order to be considered for a job or an insurance policy. How do we prevent genetic information from being used in a discriminatory manner? One

Figure 12.26 Treatment with human growth hormone. This child has been treated with human growth hormone made by bacteria.

of the missions of the Human Genome Project is to address questions like these. Indeed, up to 5% of the government-supplied budget for the project was set aside to study social, legal, and ethical implications.

A much broader ethical question is, How do we really feel about wielding one of nature's singular powers—the ability to make new microorganisms, plants, and even animals? Some might ask, Do we have any right to alter an organism's genes—or to add our new creations to an already beleaguered environment?

Such questions must be weighed against the apparent benefits to humans and the environment that can be brought about by DNA technology. For example, bacteria are being engineered to clean up mining wastes and other pollutants that threaten our soil, water, and air. These organisms may be the only feasible solutions to some of our most pressing environmental problems.

DNA technologies raise many complex issues that have no easy answers. It is up to you, as a voter and future policymaker, to educate yourself about these issues so that you can make informed choices.

CHECKPOINT

1. What is the main concern about adding genes for herbicide resistance to crop plants?

2. Why is genetically modifying a human gamete morally different from genetically modifying a human somatic cell?

Answers: 1. The possibility that the genes could escape, via cross-pollination, to weeds that are closely related to the crop species **2.** A genetically modified somatic cell will affect only the patient. Modifying a gamete will affect an unborn individual as well as all of his or her descendants.

Evolution Connection

Genomes Hold Clues to Evolution

The DNA sequences determined to date confirm the evolutionary connections between even distantly related organisms and the relevance of research on simpler organisms to understanding human biology. Yeast, for example, has a number of genes close enough to the human versions that they can substitute for them in a human cell. In fact, researchers can sometimes work out what a human disease gene does by studying its counterpart in yeast. Many genes of disparate organisms are turning out to be astonishingly similar, to the point that one researcher has joked that he now views fruit flies as "little people with wings." On a grander scale, comparisons of the completed genome sequences of bacteria, archaea, and eukaryotes strongly support the theory that these are the three fundamental domains of life—a topic we discuss further in the next unit, "Evolution and Diversity."

Evolution Connection on the Web
Investigate relationships among different life forms by examining genes.

Chapter Review

For study help, go to the Essential Biology Website (www.essentialbiology.com) or CD-ROM to explore the Activities and Case Studies in the Process of Science.

Recombinant DNA Technology

• Recombinant DNA technology is a set of laboratory procedures for combining DNA from different sources—even different species—into a single DNA molecule.

Activity 12A Applications of DNA Technology

• **From Humulin to Genetically Modified Foods** Recombinant DNA techniques have been used to create non-human cells that produce human proteins, genetically modified (GM) food crops, and transgenic farm animals.

Activity 12B DNA Technology and Golden Rice

• **Recombinant DNA Techniques** Review the steps of recombinant DNA technology in the following diagram.

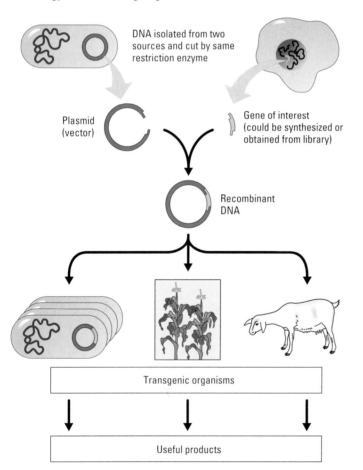

Case Study in the Process of Science *How Are Plasmids Introduced into Bacterial Cells?*

Activity 12C Cloning a Gene in Bacteria

Activity 12D Restriction Enzymes

DNA Fingerprinting and Forensic Science

• DNA fingerprinting is used to determine whether two DNA samples come from the same individual.

Activity 12E DNA Fingerprinting

• **Murder, Paternity, and Ancient DNA** DNA fingerprinting can be used to establish innocence or guilt of a criminal suspect, identify victims, determine paternity, and contribute to basic research.

• **DNA Fingerprinting Techniques** Restriction fragment length polymorphism (RFLP) analysis compares DNA fragments using restriction enzymes and gel electorophesis.

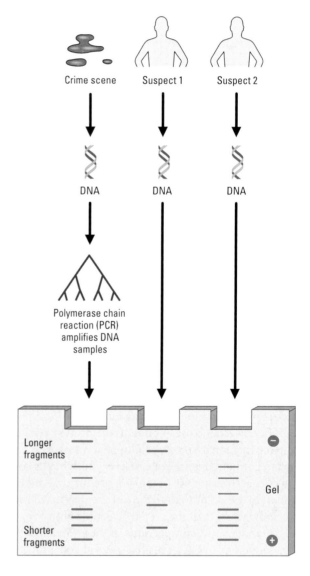

Activity 12F Gel Electrophoresis of DNA

Activity 12G Analyzing DNA Fragments Using Gel Electrophoresis

Case Study in the Process of Science *How Can Gel Electrophoresis Be Used to Analyze DNA?*

Genomics

• **The Human Genome Project**   Started in 1990 and completed in 2003, the nucleotide sequence of the human genome is providing a wealth of useful data. The 24 different chromosomes of the human genome contain about 3.2 billion nucleotide pairs and 30,000 to 40,000 genes. The majority of the genome consists of noncoding DNA, which includes repetitive nucleotide sequences.

Activity 12H *The Human Genome Project: Human Chromosome 17*

• **Tracking the Anthrax Killer**   Comparing genomes can aid criminal investigations and basic research.

• **Genome-Mapping Techniques**   The Human Genome Project proceeded in three stages: creating a genetic map, creating a physical map, and sequencing DNA. The whole genome shotgun method involves sequencing DNA fragments without first creating maps.

Human Gene Therapy

• In the following diagram, trace the steps of one approach to gene therapy.

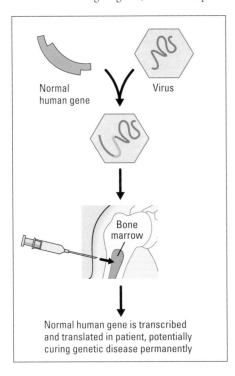

Normal
human gene

Virus

Bone
marrow

Normal human gene is transcribed
and translated in patient, potentially
curing genetic disease permanently

• **Treating Severe Combined Immunodeficiency**   Gene therapy trials have focused on SCID, an inherited immune disease.

Safety and Ethical Issues

• **The Controversy Over Genetically Modified Foods**   The debate about genetically modified crops centers on whether they might harm humans or damage the environment by transferring genes through cross-pollination with other species.

• **Ethical Questions Raised by DNA Technology**   We as a society and as individuals must become educated about DNA technologies to address the ethical questions raised by their use.

Self-Quiz

1. Suppose you wish to create a large batch of the protein lactase using recombinant DNA. Place the following steps in the order you would have to perform them.

    a. Find the clone with the gene for lactase.

    b. Insert the plasmids into bacteria and grow the bacteria into clones.

    c. Isolate the gene for lactase.

    d. Create recombinant plasmids, including one that carries the gene for lactase.

2. Why is an artificial gene made using reverse transcriptase often shorter than the natural form of the gene?

3. A carrier that moves DNA from one cell to another, such as a plasmid, is called a _____.

4. In making recombinant DNA, what is the benefit of using a restriction enzyme that cuts DNA in a staggered fashion?

5. A paleontologist has recovered a bit of organic material from the 400-year-old preserved skin of an extinct dodo. She would like to compare DNA from the sample with DNA from living birds. The most useful method for increasing the amount of dodo DNA available for testing is _____.

6. Could two pieces of DNA with different nucleotide sequences produce the same RFLP pattern? Why or why not?

7. What feature of a DNA fragment causes it to move through a gel during electrophoresis?

    a. the electrical charge of its phosphate group

    b. its nucleotide sequence

    c. the hydrogen bonds between its base pairs

    d. its double helix shape

8. The pattern of bars in a DNA fingerprint shows

    a. the order of bases in a particular gene.

    b. the presence of various-sized fragments of DNA.

    c. the order of genes along particular chromosomes.

    d. the exact location of a specific gene in a genomic library.

9. If you wanted to sequence an entire genome rapidly, you could chop it into restriction fragments, sequence each one, then reassemble the sequences in their proper order. This method is called _____.

10. Put the following steps of human gene therapy in the correct order.

    a. Virus is injected into patient.

    b. Human gene is inserted into a virus.

    c. Normal human gene is isolated and cloned.

    d. Normal human gene is transcribed and translated in the patient.

Answers to the Self-Quiz questions can be found in Appendix B.

Go to the website or CD-ROM for more Self-Quiz questions.

The Process of Science

1. A biochemist hopes to find a gene in human liver cells that codes for an important blood-clotting protein. She knows that the nucleotide sequence of a small part of the gene is CTGGACTGACA. Briefly explain how to obtain the desired gene.

2. Some scientists have joked that once the Human Genome Project is complete, "we can all go home" because there will be nothing left for genetic researchers to discover. Do you agree? Why or why not?

Case Study in the Process of Science on the Web & CD *Learn how plasmids are introduced into bacterial cells.*

Case Study in the Process of Science on the Web & CD *Conduct virtual gel electrophoresis.*

Biology and Society

1. In the not-too-distant future, gene therapy may be an option for the treatment and cure of many inherited disorders. What do you think are the most serious ethical issues that must be dealt with before human gene therapy is used on a large scale? Why do you think these issues are important?

2. Today, it is fairly easy to make transgenic plants and animals. What are some important safety and ethical issues raised by this use of recombinant DNA technology? What are some of the possible dangers of introducing genetically engineered organisms into the environment? What are some reasons for and against leaving decisions in these areas to scientists? Who do you think should make these decisions?

3. In October 2002, the government of the African nation of Zambia announced that it was refusing to distribute 15,000 tons of corn donated by the United States, enough corn to feed 2.5 million Zambians for 3 weeks. The government rejected the corn because it was likely to contain genetically modified kernels. The government made the decision after its scientific advisers concluded that the studies of the health risks posed by GM crops "are inconclusive." Do you agree with this assessment? Do you think that it is a good justification for refusing the donated corn? At the time of the government's decision, Zambia was facing food shortages, and 35,000 Zambians were expected to die from starvation in the following 6 months. In light of this, do you think it was morally acceptable for the government to refuse the food? How can the relative risks posed by GM crops be compared to the relative risks posed by starvation?

4. In 1983, a ten-year-old girl was kidnapped from her home, raped, and murdered. A jury convicted a local teenager of the crimes and sentenced him to death for the brutal killing. In 1995, DNA analysis proved that semen found near the scene could not have come from the man accused. After 12 years on death row, he was exonerated and released from prison. His case, which took place in Illinois, was far from unique. Since the 1977 imposition of the death penalty in Illinois, 12 convicts have been executed, and 13 exonerated. In 2000, the governor of Illinois declared a moratorium on all executions in his state because the death penalty system was "fraught with errors." Even though he supported the death penalty, the governor declared that the justice system was "broken." Yet, to date, no other states have followed Illinois in declaring a ban on the death penalty, even though the case described above is just one of many similar exonerations. Do you support the Illinois governor's decision? Why do you think no other states have done the same? What rights should death penalty inmates have with regards to DNA testing of old evidence? Who should pay for this additional testing?

Biology and Society on the Web *Learn more about the Human Genome Project.*

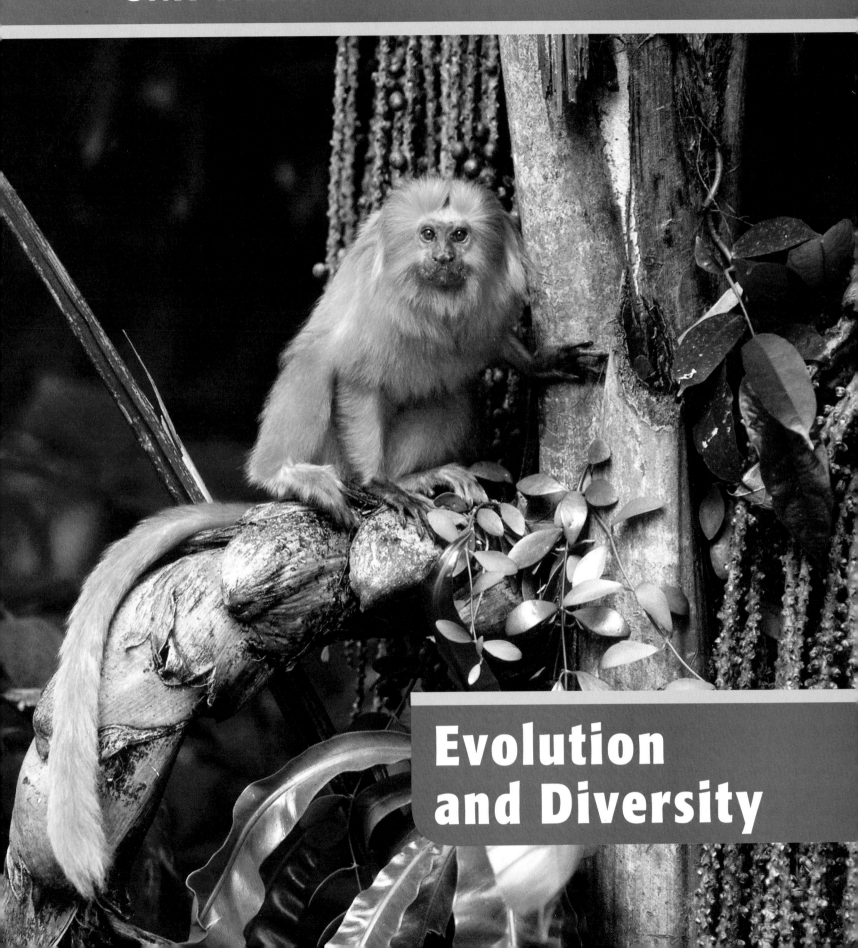

Evolution and Diversity

CHAPTER 13

How Populations Evolve

**Biology and Society:
Persistent Pests** 243

**Charles Darwin and *The Origin
of Species*** 244

*Darwin's Cultural and Scientific
Context*

Descent with Modification

Evidence of Evolution 249

The Fossil Record

Biogeography

Comparative Anatomy

Comparative Embryology

Molecular Biology

**Natural Selection and
Adaptive Evolution** 254

Darwin's Theory of Natural Selection

Natural Selection in Action

**The Modern Synthesis:
Darwinism Meets Genetics** 256

Populations as the Units of Evolution

Genetic Variation in Populations

Analyzing Gene Pools

*Population Genetics and Health
Science*

*Microevolution as Change in a
Gene Pool*

**Mechanisms of
Microevolution** 260

Genetic Drift

Gene Flow

Mutations

Natural Selection: A Closer Look

**Evolution Connection:
Population Genetics of the
Sickle-Cell Allele** 266

All humans are connected by descent from African ancestors.

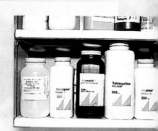

Abuse of antibiotics has hastened the evolution of antibiotic-resistant bacteria.

The same type of bones make up the forelimbs of humans, cats, whales, and bats.

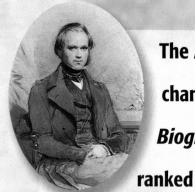

The A&E channel's *Biography* ranked Charles Darwin as the fourth most influential person of the past 1,000 years.

Persistent Pests

In the 1960s, the World Health Organization (WHO) began a campaign to eradicate the mosquitoes that transmit the disease malaria. It was a noble goal, since malaria kills an estimated 3 million people each year in the world's tropical regions, predominantly southern Africa. WHO led an effort to spray the mosquitoes' habitat with a chemical pesticide—a poison used to kill insects—called DDT. Early results were promising, and the mosquito was eliminated from the edge of its native range. The effort soon faltered, however, and the eradication plan was dropped. How could a tiny mosquito thwart the best efforts of a large group of well-funded scientists?

Situations like this one have occurred dozens of times in the last several decades. In a common scenario, whenever a new type of pesticide is used to control agricultural pests, the early results are encouraging. A relatively small amount of the poison dusted onto a crop may kill 99% of the insects. However, the relatively few survivors of the first pesticide wave are insects with genes that somehow enable them to resist the chemical attack (**Figure 13.1**). For example, the lucky few may carry genes coding for enzymes that destroy the pesticide. The poison kills most members of the insect population, leaving only the resistant individuals to reproduce. And when they do, their offspring inherit the genes for pesticide resistance. In each generation, the proportion of pesticide-resistant individuals in the insect population increases, making subsequent sprayings less and less effective.

Since the widespread use of chemical pesticides began in the 1940s, scientists have documented pesticide resistance in more than 500 species of in-

Biology and Society on the Web
Learn about the evolution of bacteria that are resistant to antibiotics.

sects. The problems such insects pose—through their impact on agriculture and medicine—are just some of the many ways that evolution has a direct connection to our daily lives. Everywhere, all the time, populations of organisms are fine-tuning adaptations to local environments through the evolutionary process of natural selection. Given the dynamics of Earth and its life, it is not surprising that even the kinds of organisms on the planet—the species—have changed over time.

But even as change characterizes life, so does continuity. All members of your family are connected by shared ancestry. In fact, all humans are connected by descent from our African ancestors. And all of life, with its dazzling diversity of millions of species, is united by descent from the first microbes that populated the primordial planet. It is this duality of life's unity and diversity that defines modern biology.

An understanding of evolution informs every field of biology, from exploring life's molecules to analyzing ecosystems. And applications of evolutionary biology are transforming medicine, agriculture, biotechnology, and conservation biology. Because evolution integrates all of biology, it is the thematic thread woven throughout this book. This unit of chapters features mechanisms of evolution and traces the history of life on Earth. This chapter starts with the story of how Darwin came to formulate his ideas. Next, some of the lines of evidence in support of evolution are presented, followed by a closer look at Darwin's theory of natural selection. The chapter then focuses on the genetic basis of evolution and the mechanisms by which it proceeds.

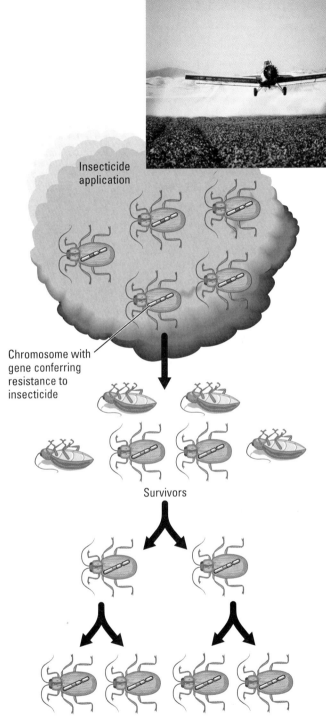

Insecticide application

Chromosome with gene conferring resistance to insecticide

Survivors

Additional applications of the same insecticide will be less effective, and the frequency of resistant insects in the population will grow

Figure 13.1 Evolution of pesticide resistance in insect populations. By spraying crops with poisons to kill insect pests, humans have unwittingly favored the reproductive success of insects with inherent resistance to the poisons.

Charles Darwin and *The Origin of Species*

Biology came of age November 24, 1859, the day Charles Darwin published *On the Origin of Species by Means of Natural Selection*. Darwin's book presented two main concepts. First, Darwin argued convincingly from several lines of evidence that contemporary species arose from a succession of ancestors through a process of "descent with modification," his phrase for evolution. Darwin's second concept in *The Origin of Species* was his theory for *how* life evolves. This proposed mechanism of evolution is called natural selection.

The basic idea of **natural selection** is that a population of organisms can change over the generations if individuals having certain heritable traits leave more offspring than other individuals. The result of natural selection is **evolutionary adaptation,** a population's increase in the frequency of traits that are suited to the environment (**Figure 13.2**). In modern terms, we would say that the genetic composition of the population had changed over time, and that is one way of defining **evolution.** But we can also use the term *evolution* on a much grander scale to mean all of biological history, from the earliest microbes to the enormous diversity of modern organisms.

Darwin's book drew a cohesive picture of life by connecting the dots of what had once seemed a bewildering array of unrelated facts. *The Origin of*

(b) A Trinidad tree mantid that mimics dead leaves

(a) A flower mantid in Malaysia

(c) A leaf mantid in Costa Rica

Figure 13.2 Camouflage as an example of evolutionary adaptation.
Related species of insects called mantids have diverse shapes and colors that evolved in different environments.

Species focused biologists' attention on the great diversity of organisms—their origins and relationships, their similarities and differences, their geographic distribution, and their adaptations to surrounding environments. Before we examine how natural selection works and how Darwin derived the idea, let's place the Darwinian revolution in its historical context.

Darwin's Cultural and Scientific Context

The view of life developed in *The Origin of Species* contrasts sharply with the view that prevailed during Darwin's lifetime. Most scientists of his day thought Earth was relatively young and populated by millions of unrelated species. *The Origin of Species* challenged that widely held notion. It was truly radical for its time. Not only did it challenge prevailing scientific views, it also shook the deepest roots of Western culture.

The Idea of Fixed Species   Like many concepts in science, the basic idea of biological evolution can be traced back to the ancient Greeks. About 2,500 years ago, the Greek philosopher Anaximander promoted the idea that life arose in water and that simpler forms of life preceded more complex ones. However, the Greek philosopher Aristotle, whose views had an enormous impact on Western culture, generally held that species are fixed, or permanent, and do not evolve. Judeo-Christian culture fortified this idea with a literal interpretation of the biblical book of Genesis, which tells the story of each form of life being individually created. The idea that all living species are static in form and inhabit an Earth that is only about 6,000 years old dominated the cultural climate of the Western world for centuries.

Lamarck and Adaptive Evolution   In the mid-1700s, the study of **fossils,** which are the imprints or remnants of organisms that lived in the past, began to take form as a branch of science. The study of fossils led French naturalist Georges Buffon to suggest that Earth might be much older than 6,000 years. He also observed some telling similarities between specific fossils and certain living animals. In 1766, Buffon proposed the possibility that a species represented by a particular fossil form could be an ancient version of a group of similar living species. Then, in the early 1800s, French naturalist Jean Baptiste Lamarck suggested that the best explanation for this relationship of fossils to current organisms is that life evolves. Lamarck explained evolution as a process of adaptation, the refinement of characteristics that equip organisms to perform successfully in their environments. An example of evolutionary adaptation is the powerful beak of a bird that feeds by cracking tough seeds.

Unfairly, we remember Lamarck today mainly for his erroneous view of how adaptations evolve. He proposed that by using or not using its body parts, an individual develops certain characteristics, which it passes on to its offspring. Lamarck's proposal is known as the inheritance of acquired characteristics. For example, this idea would view the strong beaks of seed-cracking birds as the cumulative result of ancestors exercising their beaks during feeding and passing that acquired beak power on to offspring. However, there is no evidence for inheritance of acquired characteristics. A carpenter who builds up strength and stamina through a lifetime of pounding nails with a heavy hammer will not pass enhanced biceps on to children. The incompatibility of modern genetics with Larmarck's idea that acquired characteristics can be inherited obscures the important fact that Lamarck helped set the stage for Darwin by proposing that adaptations evolve as a result of interactions between organisms and their environments.

The Voyage of the *Beagle* Charles Darwin was born in 1809 on the same day as Abraham Lincoln. (In that same year, Larmarck published some of his ideas on evolution.) Even as a boy, Darwin's consuming interest in nature was evident. When he was not reading nature books, he was in the fields and forests fishing, hunting, and collecting insects. His father, an eminent physician, could see no future for a naturalist and sent Charles to the University of Edinburgh to study medicine. But Charles, only 16 years old at the time, found medical school boring and distasteful. He left Edinburgh without a degree and then enrolled at Christ College at Cambridge University, intending to become a minister. At Cambridge, Darwin became the protégé of the Reverend John Henslow, a professor of botany. Soon after Darwin received his B.A. degree in 1831, Professor Henslow recommended the young graduate to Captain Robert FitzRoy, who was preparing the survey ship *Beagle* for a voyage around the world. It was a tour that would have a profound effect on Darwin's thinking and eventually on the thinking of the entire world.

Darwin was 22 years old when he sailed from Great Britain on HMS *Beagle* in December 1831. The main mission of the voyage was to chart poorly known stretches of the South American coastline (**Figure 13.3**). While the crew of the ship surveyed the coast, Darwin spent most of his time on shore, observing and collecting thousands of specimens of the native plants and animals of South America. As the ship worked its way around the continent, Darwin observed the various adaptations of organisms that inhabited such diverse environments as the Brazilian jungles, the grasslands of the Argentine pampas, the desolate and frigid lands of Tierra del Fuego near Antarctica, and the towering heights of the Andes.

Activity 13A on the Web & CD
Join Charles Darwin on his round-the-world voyage.

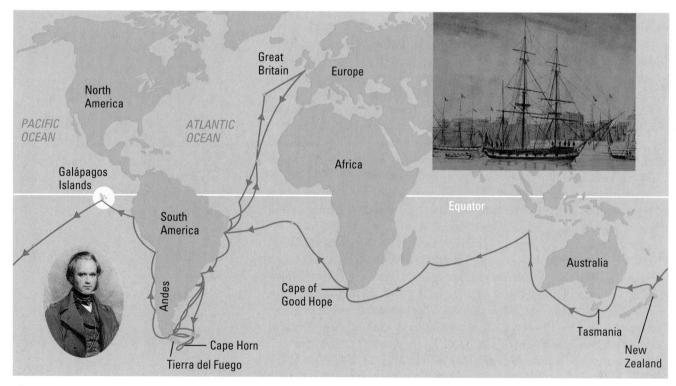

Figure 13.3 The voyage of the *Beagle*. The two insets show the ship and a young Darwin.

In spite of their unique adaptations, the plants and animals throughout the continent all had a definite South American stamp, very distinct from the life-forms of Europe. That in itself may not have surprised Darwin. But the plants and animals living in temperate regions of South America seemed more closely related to species living in tropical regions of that continent than to species living in temperate regions of Europe. And the South American fossils Darwin found, though clearly different species from modern ones, were distinctly South American in their resemblance to the living plants and animals of that continent. Despite growing up in the generally antievolutionist climate of the Victorian era, Darwin had a questioning mind. Could his observations mean that the contemporary species owed their South American features to descent from ancestral species on that continent?

Darwin was particularly intrigued by the geographic distribution of organisms on the Galápagos Islands. These are relatively young volcanic islands about 900 km (540 miles) off the Pacific coast of South America. Most of the animals of the Galápagos live nowhere else in the world, but they resemble species living on the South American mainland (**Figure 13.4**). It is as

Figure 13.4 A marine iguana, an example of the unique species inhabiting the Galápagos. These reptiles dive into the ocean to feed on algae. Partially webbed feet and a flattened tail are two of the adaptations that make marine iguanas such good swimmers.

Activity 13B on the Web & CD Visit the Galápagos Islands with Charles Darwin.

though the islands had been colonized by plants and animals that strayed from the mainland and then diversified as they adapted to environments on the different islands. Among the birds Darwin collected on the Galápagos were several types of finches. Some were unique to individual islands, while others were distributed on two or more islands that were close together. The unique adaptations of these birds included beaks modified for feeding on certain kinds of foods. Darwin did not appreciate the full significance of the finches he collected until years after returning to Britain. Since then, biologists have applied modern methods of comparing species to reconstruct the evolutionary history of Darwin's finches. The branching of this evolutionary tree traces the "descent with modification" of the 14 finch species from a common South American ancestor (see Figure 1.13).

The New Geology During the *Beagle*'s long sails between ports, Darwin, in spite of his seasickness, managed to do a lot of reading. He was strongly influenced by the recently published *Principles of Geology*, by Scottish geologist Charles Lyell. The book presented the case for an ancient Earth sculpted by gradual geologic processes that continue today. In essence, the key to the past is the present. The gradual erosion of a riverbed can add up over the millennia to a deep river-carved canyon. A mighty mountain range can be thrust up millimeter by millimeter by earth-moving earthquakes occurring sporadically over thousands or millions of years. Darwin actually experienced such an earthquake while doing field studies in the Andes Mountains of Chile. He also collected fossils of marine (sea) snails at high Andean altitudes. Perhaps, Darwin reasoned, earthquakes gradually lifted the rock bearing those marine fossils from the seafloor. In the context of such experiences, Lyell began to make sense to Darwin.

The evidence presented by Lyell and other proponents of the "new geology" pointed to two conclusions. First, Earth must be very old if it has been shaped by such slow processes as mountain building and erosion. (What you need to carve a deep river gorge at a rate of a millimeter per decade is a lot of time!) Second, the new geology explained how slow and subtle processes occurring over vast tracts of time can cause enormous change. Darwin would eventually apply this principle of gradualism to the evolution of Earth's life.

Descent with Modification

By the early 1840s, Darwin had composed a long essay describing the major features of natural selection. He realized that his evolutionary ideas and evidence would cause a social furor, however, and he delayed publication of his essay. Then, in the mid-1850s, Alfred Wallace, a British naturalist doing fieldwork in Indonesia, developed a concept of natural selection identical to Darwin's. When Wallace sent Darwin a manuscript describing his own ideas on natural selection, Darwin wrote, ". . . all my originality . . . will be smashed." However, in 1858, two of Darwin's colleagues presented Wallace's paper and excerpts from Darwin's earlier essay together to the scientific community. With the publication in 1859 of *The Origin of Species,* Darwin presented the world with an avalanche of evidence and a strong, logical argument for evolution. He also explained natural selection as the mechanism of descent with modification.

Darwin made two main points in *The Origin of Species.* First, he argued from evidence that the species of organisms inhabiting Earth today descended from ancestral species. In the first edition of his book, Darwin did not actually use the word *evolution.* He referred instead to "descent with modification." Darwin postulated that as the descendants of the earliest organisms spread into various habitats over millions of years, they accumulated different modifications, or adaptations, to diverse ways of life. In the Darwinian view, the history of life is analogous to a tree. Patterns of descent are like the branching and rebranching from a common trunk, the first organism, to the tips of millions of twigs representing the species living today. At each fork of the evolutionary tree is an ancestor common to all evolutionary branches extending from that fork. Closely related species, such as Asian and African elephants, share many characteristics because their lineage of common descent traces to a recent fork of the tree of life (**Figure 13.5**).

Darwin's second main point in *The Origin of Species* was his argument for natural selection as the mechanism for descent with modification. When biologists speak of Darwin's theory of evolution, they mean natural selection as a cause of evolution, not the phenomenon of evolution itself. In the next sections of this chapter, we will examine the evidence for evolution and then look more closely at Darwin's theory of natural selection.

CHECKPOINT

1. What is gradualism? How did Darwin apply that idea to the evolution of life?

2. What were the two main points in Darwin's *The Origin of Species*?

3. Darwin's phrase for evolution, _____ with _____, captured the idea that an ancestral species could diversify into many descendant species by the accumulation of different _____ to various environments.

Answers: 1. Gradualism is the idea that large changes on Earth can result from the accumulation of small changes over a very long time. Darwin applied this idea to suggest that species evolve through the slow accumulation of small changes over time. **2.** Descent of diverse species from common ancestors and natural selection as the mechanism of evolution **3.** descent; modification; adaptations

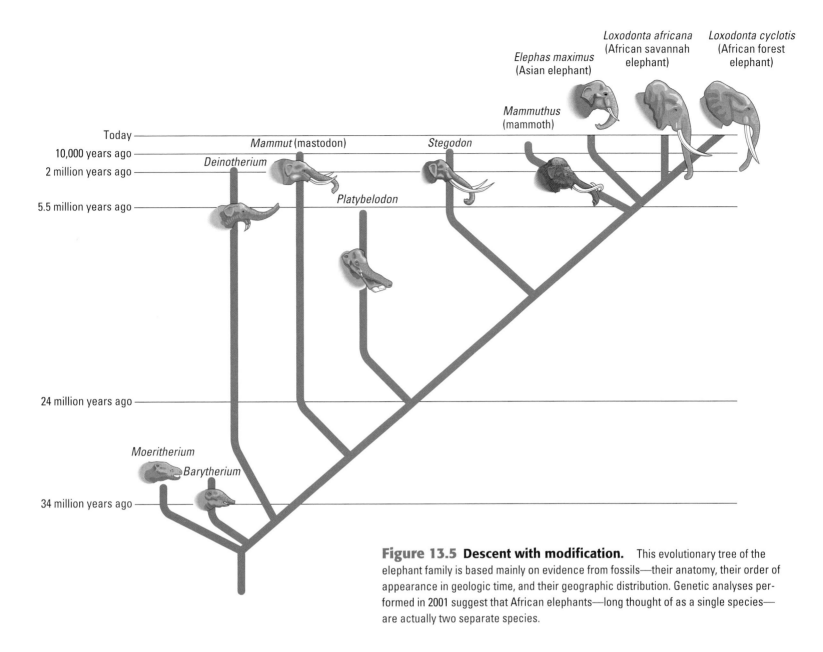

Figure 13.5 Descent with modification. This evolutionary tree of the elephant family is based mainly on evidence from fossils—their anatomy, their order of appearance in geologic time, and their geographic distribution. Genetic analyses performed in 2001 suggest that African elephants—long thought of as a single species—are actually two separate species.

Evidence of Evolution

Evolution leaves observable signs. Such clues to the past are essential to any historical science. Historians of human civilization can study written records from earlier times. But they can also piece together the evolution of societies by recognizing vestiges of the past in modern cultures. Even if we did not know from written documents that Spaniards colonized the Americas, we would deduce this from the Hispanic stamp on Latin American culture. Similarly, biological evolution has left marks.

In this section, we will examine a few of the many lines of evidence in support of evolution. One of them—fossils—is a historical record. The other three—biogeography, comparative anatomy, and comparative embryology—encompass historical vestiges of evolution evident in modern life.

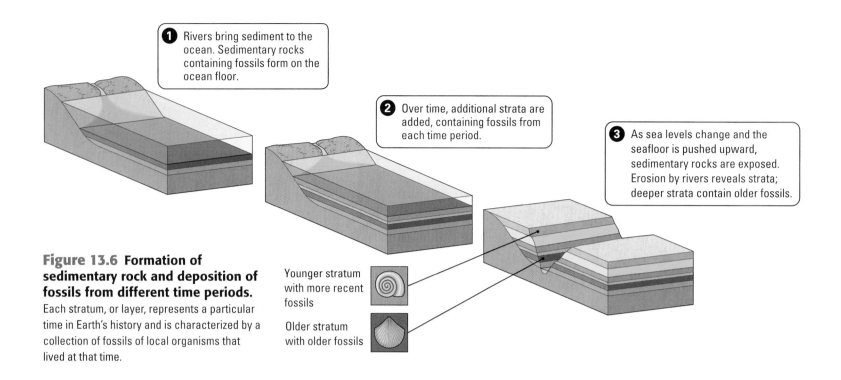

① Rivers bring sediment to the ocean. Sedimentary rocks containing fossils form on the ocean floor.

② Over time, additional strata are added, containing fossils from each time period.

③ As sea levels change and the seafloor is pushed upward, sedimentary rocks are exposed. Erosion by rivers reveals strata; deeper strata contain older fossils.

Figure 13.6 Formation of sedimentary rock and deposition of fossils from different time periods. Each stratum, or layer, represents a particular time in Earth's history and is characterized by a collection of fossils of local organisms that lived at that time.

Younger stratum with more recent fossils

Older stratum with older fossils

Figure 13.7 Strata of sedimentary rock at the Grand Canyon. The Colorado River has cut through over 2,000 m of rock, exposing sedimentary layers that are like huge pages from the book of life. Scan the canyon wall from rim to floor, and you look back through hundreds of millions of years. Each layer entombs fossils that represent some of the organisms from that period of Earth's history.

The Fossil Record

Fossils are the preserved remnants or impressions left by organisms that lived in the past. Most fossils are found in sedimentary rocks. Sand and silt eroded from the land are carried by rivers to seas and swamps, where the particles settle to the bottom. Over millions of years, deposits pile up and compress the older sediments below into rock (**Figure 13.6**). Rock strata, or layers, form when the rates of sedimentation and the types of particles that settle vary over time. When aquatic organisms die, they settle along with the sediments and may leave imprints in the rocks. Some organisms living on land may be swept into swamps and seas. Land organisms that remain in place when they die may first be covered by windblown silt and then buried in waterborne sediments when sea levels rise over them. Thus, each of the rock strata bears a unique set of fossils representing a local sampling of the organisms that lived when the sediment was deposited. Younger strata are on top of older ones. Thus, the positions of fossils in the strata reveal their relative age. The **fossil record** is this chronology of fossil appearances in the rock layers, marking the passing of geologic time (**Figure 13.7**).

The fossil record testifies that organisms have appeared in a historical sequence. The oldest known fossils, dating from about 3.5 billion years ago, are prokaryotes (bacteria and archaea). This fits with the molecular and cellular evidence that prokaryotes are the ancestors of all life. Fossils in younger layers of rock chronicle the evolution of various groups of eukaryotic organisms. One example is the successive appearance of the different classes of vertebrates (animals with backbones). Fishlike fossils are the oldest vertebrates in the fossil record. Amphibians are next, followed by reptiles, then mammals and birds.

Paleontologists (scientists who study fossils) have discovered many transitional forms that link past and present. For example, a series of fossils documents the changes in skull shape and size that occurred as mammals evolved from reptiles. Another example is a series of fossilized whales that connect these aquatic mammals to their land-dwelling ancestors (**Figure 13.8**).

Biogeography

Study of the geographic distribution of species is called **biogeography.** It was biogeography that first suggested to Darwin that today's organisms evolved from ancestral forms. Consider, for example, Darwin's visit to the Galápagos Islands, which are located about 900 km off the west coast of Ecuador (see Figure 13.3). Darwin noted that the Galápagos animals resembled species of the South American mainland more than they resembled animals on similar but distant islands. This is what we should expect if the Galápagos species evolved from South American immigrants.

There are many other examples from biogeography that seem baffling without an evolutionary perspective. Why are the tropical animals of South America more closely related to species in the South American deserts than to species in the African tropics? Why is Australia home to such an impressive diversity of pouched mammals (marsupials) but relatively few placental mammals (those in which embryonic development is completed in the uterus)? It is *not* because Australia is inhospitable to placental mammals. Humans have introduced rabbits, foxes, and many other placental mammals to Australia, where these introduced species have thrived to the point of becoming ecological and economic nuisances. The prevailing hypothesis is that the unique Australian wildlife evolved on that island continent in isolation from regions where early placental mammals diversified (**Figure 13.9**).

Biogeography makes little sense if we imagine that species were individually placed in suitable environments. In the Darwinian view, we find species where they are because they evolved from ancestors that inhabited those regions.

Comparative Anatomy

The comparison of body structures between different species is called **comparative anatomy.** Certain anatomical similarities among species bear witness to evolutionary history. For example, the same skeletal elements make up the forelimbs of humans, cats, whales, and bats, all of which are

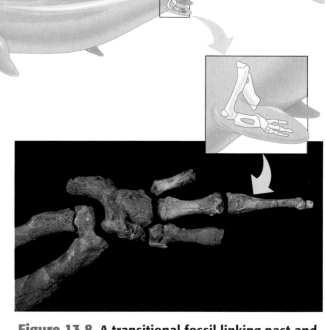

Figure 13.8 A transitional fossil linking past and present.   The hypothesis that whales evolved from terrestrial (land-dwelling) ancestors predicts a four-limbed beginning for whales. Paleontologists digging in Egypt and Pakistan have identified extinct whales that had hind limbs. Shown here are the fossilized leg bones of *Basilosaurus,* one of those ancient whales. These whales were already aquatic animals that no longer used their legs to support their weight. The leg bones of an even older fossilized whale named *Ambulocetus* are heftier. *Ambulocetus* may have been amphibious, living on land and in water.

Koala

Kangaroo

Figure 13.9
Evidence of evolution from biogeography.
The continent of Australia is home to many unique plants and animals, such as these marsupials, which evolved in relative isolation from other continents where placental mammals diversified.

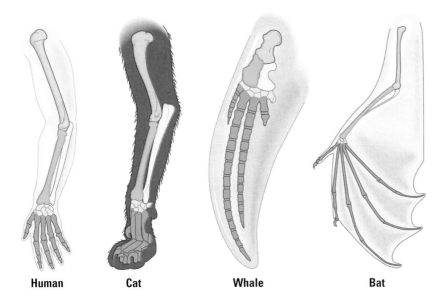

Human   Cat   Whale   Bat

Figure 13.10 Homologous structures: anatomical signs of descent with modification. The forelimbs of all mammals are constructed from the same skeletal elements. (Homologous bones in each of these four mammals are colored the same.) The hypothesis that all mammals descended from a common ancestor predicts that their forelimbs, though diversely adapted, would be variations on a common anatomical theme.

mammals (**Figure 13.10**). The functions of these forelimbs differ. A whale's flipper does not do the same job as a bat's wing. If these limbs had completely separate origins, we would expect that their basic designs would be very different. However, structural similarity would not be surprising if all mammals descended from a common ancestor with the prototype forelimb. Arms, forelegs, flippers, and wings of different mammals are variations on a common anatomical theme that has become adapted to different functions. Such similarity due to common ancestry is called **homology.** The forelimbs of diverse mammals are homologous structures.

Activity 13C on the Web & CD Reconstruct homologous forelimbs.

Comparative anatomy confirms that evolution is a remodeling process. Ancestral structures that originally functioned in one capacity become modified as they take on new functions—descent with modification. The historical constraints of this retrofitting are evident in anatomical imperfections. For example, the human spine and knee joint were derived from ancestral structures that supported four-legged mammals. Almost none of us will reach old age without experiencing knee or back problems. If these structures had first taken form specifically to support our bipedal posture, we would expect them to be less subject to sprains, spasms, and other common injuries. The anatomical remodeling that stood us up was apparently constrained by our evolutionary history.

Comparative Embryology

The comparison of structures that appear during the development of different organisms is called **comparative embryology.** Closely related organisms often have similar stages in their embryonic development. One sign that vertebrates evolved from a common ancestor is that all of them have an embryonic stage in which structures called gill pouches appear on the sides of the throat. At this stage, the embryos of fishes, frogs, snakes, birds, apes—indeed, all vertebrates—look more alike than different (**Figure 13.11**). The different classes of vertebrates take on more and more distinctive features as development progresses. In fishes, for example, most of the gill pouches develop into gills. In land vertebrates, however, these embryonic features develop into other kinds of structures, such as bones of the skull, bones supporting the tongue, and the voice box of mammals.

Figure 13.11 Evolutionary signs from comparative embryology. At this early stage of development, the kinship of vertebrates is unmistakable. Notice, for example, the gill pouches and tails in both **(a)** the bird (chick) embryo and **(b)** the human embryo. Comparative embryology helps biologists identify anatomical homology that is less apparent in adults.

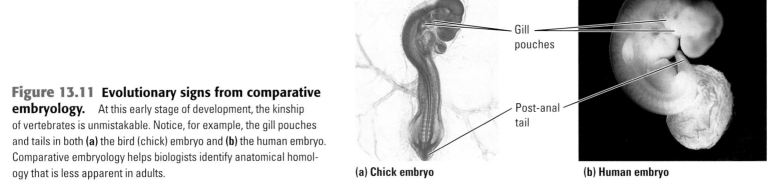

Gill pouches

Post-anal tail

(a) Chick embryo   **(b) Human embryo**

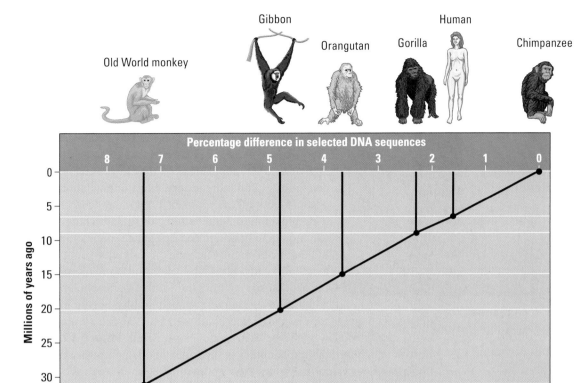

Figure 13.12
Genetic relationships among some primates. The scale across the top of this diagram gives you the percentage difference in selected DNA sequences between a chimpanzee and other primates, the animal group that includes monkeys, apes, and humans. For example, note that chimps and humans are less than 2% different in their DNA sequences. Put another way, the DNA sequences of chimps and humans are better than a 98% match within the chosen regions. In contrast, the DNA sequences of chimps and Old World monkeys (macaques, mandrills, baboons, and rhesus monkeys) differ by over 7%. The evolutionary relationships documented by these genetic comparisons are compatible with other types of evidence, including fossils. For example, the fossil record suggests that chimps and humans diverged from a common ancestor only about 6–7 million years ago (left scale on this diagram). The evolutionary split between Old World monkeys and the branch that includes humans and chimps was much earlier, over 30 million years ago.

Molecular Biology

Evolutionary relationships among species leave signs in DNA and proteins —in genes and gene products (see Evolution Connection in Chapter 3). If two species have libraries of genes and proteins with sequences of monomers that match closely, the sequences must have been copied from a common ancestor (**Figure 13.12**). By analogy, if two long paragraphs are identical except for the substitution of a letter here and there, we can be quite sure that they came from a common source.

Darwin's boldest hypothesis was that *all* forms of life are related to some extent through branching evolution from the earliest organisms. A century after Darwin's death, molecular biology began confirming the fossil record and other evidence supporting the Darwinian view of kinship among all life. The molecular signs include a common genetic code shared by all species (see Chapter 10). Evidently, the genetic language has been passed along through all branches of evolution ever since its beginnings in an early form of life. Molecular biology has added the latest chapter to the volumes of evidence validating evolution as a natural explanation for the unity and diversity of life.

CHECKPOINT

1. Why are older fossils generally in deeper rock layers than younger fossils?

2. What is homology?

Answers: 1. Sedimentation adds younger rock layers on top of older ones. **2.** Similarity between species that is due to shared ancestry

(a) Large ground finch

(b) Small tree finch

(c) Woodpecker finch

Figure 13.13 Beak adaptations of Galápagos finches. The most striking difference between these closely related species is their beaks, which are adapted for specific diets. **(a)** The large ground finch has a large beak specialized for cracking seeds that fall from plants to the ground. **(b)** The small tree finch uses its beak to grasp insects. **(c)** The woodpecker finch uses tools such as cactus spines to probe for termites and other wood-boring insects.

Figure 13.14 Overproduction of offspring.
A cloud of millions of spores is exploding from this puffball, a type of fungus. The wind will disperse the spores far and wide. Each spore, if it lands in a suitable environment, has the potential to grow and develop into a new fungus. Only a tiny fraction of the spores will actually give rise to offspring that survive and reproduce. Otherwise, it wouldn't take too many generations before puffballs filled the universe!

Natural Selection and Adaptive Evolution

Darwin perceived adaptation to the environment and the origin of new species as closely related processes. Imagine, for example, that an animal species from a mainland colonizes a chain of distant, relatively isolated islands. In the Darwinian view, populations on the different islands may diverge more and more in appearance as each population adapts to its local environment. Over many generations, the populations on different islands could become dissimilar enough to be designated as separate species. Evolution of the finches on the Galápagos Islands is an example. It is a reasonable hypothesis that the islands were colonized by finches that strayed from the South American mainland and then diversified on the different islands. Among the differences between the Galápagos finches are their beaks, which are adapted to the specific foods available on each species' home island (**Figure 13.13**). Darwin anticipated that explaining how such adaptations arise is the key to understanding evolution. And his theory of natural selection remains our best explanation for adaptive evolution.

Darwin's Theory of Natural Selection

Darwin based his theory of natural selection on two key observations. First, Darwin recognized that all species tend to produce excessive numbers of offspring (**Figure 13.14**). Darwin deduced that because natural resources are limited, the production of more individuals than the environment can support leads to a struggle for existence among the individuals of a population. In most cases, only a small percentage of offspring will survive in each generation. Many eggs are laid, young born, and seeds spread, but only a tiny fraction complete their development and leave offspring of their own. The rest are starved, eaten, frozen, diseased, unmated, or unable to reproduce for other reasons.

The second key observation that led Darwin to natural selection was his awareness of variation among individuals of a population. Just as no two people in a human population are alike, individual variation abounds in almost all species (**Figure 13.15**). Much of this variation is heritable. Siblings share more traits with each other and with their parents than they do with less closely related members of the population.

Figure 13.15
A few of the color variations in a population of Asian lady beetles.

From these two observations, Darwin arrived at the conclusion that defines natural selection: Individuals whose inherited traits are best suited to the local environment are more likely than less fit individuals to survive and reproduce. In other words, the individuals that function best tend to leave the most offspring. Darwin's genius was in connecting two observations that anyone could make and drawing an inference that could explain how adaptations evolve.

- *Observation 1: Overproduction.* Populations of many species have the potential to produce far more offspring than the environment can possibly support with food, space, and other resources. This overproduction makes a struggle for existence among individuals inevitable.

- *Observation 2: Individual variation.* Individuals in a population vary in many heritable traits.

→ *Inference: Differential reproductive success (natural selection).* Those individuals with traits best suited to the local environment generally leave a disproportionately large share of surviving, fertile offspring.

Darwin's insight was both simple and profound. The environment screens a population's inherent variability. Differential success in reproduction causes the favored traits to accumulate in the population over the generations. Natural selection causes adaptive evolution.

Natural Selection in Action

Natural selection and the adaptive evolution it causes are observable phenomena. You learned about one classic and unsettling example at the start of the chapter: the evolution of pesticide resistance in hundreds of insect species. We have used pesticides to control insects that eat our crops, transmit diseases such as malaria, or just annoy us around the house or campground. But widespread use of these poisons has led to the unintended development of pesticide-resistant insect populations. Another example of natural selection in action was described in Chapter 1: the development of antibiotic-resistant bacteria. More recently, doctors have documented an increase in drug-resistant strains of HIV, the virus that causes AIDS.

Case Study in the Process of Science on the Web & CD Investigate case studies of antibiotic resistance.

These examples highlight two key points about natural selection. First, notice that natural selection is more a process of editing than it is a creative mechanism. A pesticide does not create resistant individuals, but selects for resistant insects that were already present in the population (see Figure 13.1).

Second, note that natural selection is timely and regional. It favors those characteristics in a varying population that fit the current, local environment. Environmental factors vary from place to place and from time to time. An adaptation in one situation may be useless or even detrimental in different circumstances. For example, some genetic mutations that happen to endow houseflies with resistance to the pesticide DDT also reduce a fly's growth rate. Before DDT was introduced to environments, the gene for resistance was a handicap. But the appearance of DDT changed the rules in the environmental arena and favored pesticide-resistant individuals in the reproduction sweepstakes. Such are the dynamics of adaptive evolution by natural selection.

CHECKPOINT

1. Define natural selection.
2. Explain why the following statement is incorrect: "Pesticides have created pesticide resistance in insects."

Answers: 1. Natural selection is the differential reproductive success among a population's varying individuals. **2.** An environmental factor does not create new traits such as pesticide resistance, but rather selects among the traits that are already represented in the population.

The Modern Synthesis: Darwinism Meets Genetics

Natural selection requires hereditary processes that Darwin could not explain. How do the variations that are the raw material for natural selection arise in a population? And how are these variations passed along from parents to offspring? Darwin and Gregor Mendel lived and worked at the same time. However, Mendel's discoveries went unnoticed or unappreciated by most of the scientific community (see Chapter 9). In fact, by breeding peas in his abbey garden, Mendel illuminated the very hereditary processes required for natural selection to work. Mendelism and Darwinism finally came together in the mid-1900s, decades after both scientists were dead. This fusion of genetics with evolutionary biology came to be known as the **modern synthesis.** One of its key elements is an emphasis on the biology of populations.

Populations as the Units of Evolution

We have already used the term *population* several times in this chapter. Now it's time for a biological definition: A **population** is a group of individuals of the same species living in the same area at the same time. One population may be isolated from others of the same species. If individuals of an isolated population interbreed only rarely with those in another population, there will be little exchange of genes between the populations. Such isolation is common for populations confined to widely separated islands, unconnected lakes, or mountain ranges separated by lowlands. However, populations are not usually so isolated, and they rarely have sharp boundaries. One population center may blur into another in a region of overlap, where members of both populations are present but less numerous (**Figure 13.16**). However, individuals are more concentrated in the population centers, and they are likely to

(a)

(b)

Figure 13.16 Populations.

(a) Two dense populations of Douglas fir trees are separated by a river bottom where firs are uncommon. The two populations are not totally isolated. Interbreeding occurs when wind blows pollen between the populations. Nevertheless, trees are more likely to breed with members of the same population than with trees on the other side of the river.

(b) This nighttime satellite view of the United States shows the lights of human population centers, or cities. People move around the country, of course, and there are suburban and rural communities between cities, but people are more likely to choose mates locally.

breed with other locals. Therefore, organisms of a population are generally more closely related to one another than to members of other populations.

A population is the smallest biological unit that can evolve. A common misconception is that individual organisms evolve (in the Darwinian sense) during their lifetimes. It is true that natural selection acts on individuals. Inherited characteristics affect their survival and reproductive success. However, the evolutionary impact of this natural selection is only apparent in tracking how a population changes over time. In our example of pesticide resistance in insects, we measured evolution by the change in the relative numbers of resistant individuals over a span of generations.

A focus on populations as the evolutionary units led to a new field in science called **population genetics.** Population genetics emphasizes the extensive genetic variation within populations and tracks the genetic makeup of populations over time.

Genetic Variation in Populations

You have no trouble recognizing your friends in a crowd. Each person has a unique genome, reflected in individual variations in appearance and temperament. Individual variation abounds in populations of all species that reproduce sexually. In addition to differences we can see, most populations have a great deal of variation that can be detected only by biochemical means. For example, you cannot tell a person's ABO blood group (A, B, AB, or O) just by looking at her or him.

Not all variation in a population is heritable. Phenotype results from a combination of the genotype, which is inherited, and the many environmental influences. For instance, a strength-training program can build up your muscle mass beyond what would naturally occur from your genetic makeup. However, you would not pass this environmentally induced physique on to your offspring. Only the genetic component of variation is relevant to natural selection.

Figure 13.17 Polymorphism in a garter snake population. These four garter snakes, which belong to the same species, were all captured in one Oregon field. The behavior of each morph (form) is correlated with its coloration. When approached, spotted snakes, which blend in with their background, generally freeze. In contrast, snakes with stripes, which make it difficult to judge the speed of motion, usually flee rapidly when approached.

Many of the variable traits in a population result from the combined effect of several genes (see Chapter 9). This polygenic inheritance produces traits that vary more or less continuously—in human height, for instance, from very short individuals to very tall ones. By contrast, other features, such as human ABO blood group, are determined by a single genetic locus, with different alleles producing only distinct phenotypes; there are no in-between types. In such cases, when a population includes two or more forms of a phenotypic characteristic, the contrasting forms are called morphs. A population is said to be **polymorphic** for a characteristic if two or more morphs are present in readily noticeable numbers—that is, if neither morph is extremely rare (**Figure 13.17**).

Sources of Genetic Variation   Mutations and sexual recombination, which are both random processes, produce genetic variation.

Mutations, random changes in the genetic material, can actually create new alleles. For example, a mutation in a gene may substitute one nucleotide for another. This type of change will be harmless if it does not affect the function of the protein the DNA encodes. However, if it does affect the protein's function, the mutation will probably be harmful. An organism is a refined product of thousands of generations of past selection. A random mutation is like a shot in the dark; it is not likely to improve a genome any more than shooting a bullet through the hood of a car is likely to improve engine performance.

On rare occasions, however, a mutant allele may actually enhance reproductive success. This kind of effect is more likely when the environment is changing in such a way that alleles that were once disadvantageous are favorable under the new conditions. We already considered one example, the changing fortunes of the DDT resistance allele in housefly populations.

Organisms with very short generation spans, such as bacteria, can evolve rapidly with mutations as the only source of genetic variation. Bacteria multiply so quickly that natural selection can increase a population's frequency of a beneficial mutation in just hours or days. For most animals and plants, however, their long generation times prevent new mutations from significantly affecting overall genetic variation from one generation to the next, at least in large populations.

Activity 13D on the Web & CD
Review how sexual recombination produces genetic variation.

Animals and plants depend mainly on sexual recombination for the genetic variation that makes adaptation possible. The two sexual processes of meiosis and random fertilization shuffle alleles and deal them out to offspring in fresh combinations (see Chapter 8).

While the processes that generate genetic variation—mutations and sexual recombination—are random, natural selection (and hence evolution) is not. The environment selectively promotes the propagation of those genetic combinations that enhance survival and reproductive success.

Analyzing Gene Pools

A key concept of population genetics is the gene pool. The **gene pool** consists of all alleles (alternative forms of genes) in all the individuals making up a population. The gene pool is the reservoir from which the next generation draws its genes.

Imagine a wildflower population with two varieties (morphs) contrasting in flower color. An allele for red flowers, which we will symbolize by *R*, is dominant to an allele for white flowers, symbolized by *r*. These are the only two alleles for flower color in the gene pool of this plant population. Now, let's say that 80%, or 0.8, of all flower color loci in the gene pool have the *R* allele. We'll use the letter *p* to represent the relative frequency of the dominant allele in the population. Thus, $p = 0.8$. Because there are only two alleles in this example, the *r* allele must be present at the other 20% (0.2) of the gene pool's flower color loci. Let's use the letter *q* for the frequency of the recessive allele in the population. For the wildflower population, $q = 0.2$. And since there are only two alleles for flower color, we know that

$$p + q = 1$$

Frequency of the dominant allele   Frequency of the recessive allele

Notice that if we know the frequency of either allele in the gene pool, we can subtract it from 1 to calculate the frequency of the other allele.

From the frequencies of alleles, we can also calculate the frequencies of different genotypes in the population if the gene pool is completely stable (not evolving). In the wildflower population, what is the probability of producing an *RR* individual by "drawing" two *R* alleles from the pool of gametes? Here we apply the rule of multiplication that you learned in Chapter 9. The probability of drawing an *R* sperm multiplied by the probability of drawing an *R* egg is $p \times p = p^2$, or $0.8 \times 0.8 = 0.64$. In other words, 64% of the plants in the population will have the *RR* genotype. Applying the same math, we also know the frequency of *rr* individuals in the population: $q^2 = 0.2 \times 0.2 = 0.04$. Thus, 4% of the plants are *rr*, giving them white flowers. Calculating the frequency of heterozygous individuals, *Rr*, is trickier. That's because the heterozygous genotype can form in *two* ways, depending on whether the sperm or egg supplies the dominant allele. So the frequency of the *Rr* genotype is $2pq$, which is $2 \times 0.8 \times 0.2 = 0.32$. In our imaginary wildflower population, 32% of the plants are *Rr*, with red flowers. **Figure 13.18** reviews these calculations.

Now we can write a general formula for calculating the frequencies of genotypes in a gene pool from the frequencies of alleles, and vice versa:

$$p^2 + 2pq + q^2 = 1$$

Frequency of homozygous dominants   Frequency of heterozygotes   Frequency of homozygous recessives

Notice that the frequencies of all genotypes in the gene pool must add up to 1. This formula is called the **Hardy-Weinberg formula,** named for the two scientists who derived it in 1908.

Population Genetics and Health Science

We can use the Hardy-Weinberg formula to calculate the percentage of a human population that carries the allele for a particular inherited disease. Consider phenylketonuria (PKU), which is an inherited inability to break down the amino acid phenylalanine. If untreated, the disorder causes severe mental retardation. PKU occurs in about one out of 10,000 babies born in the United States. Newborn babies are now routinely tested for PKU, and symptoms can be prevented by following a strict diet (**Figure 13.19**).

PKU is due to a recessive allele. Thus, the frequency of individuals in the U.S. population born with PKU corresponds to the q^2 term in the Hardy-Weinberg formula. For one PKU occurrence per 10,000 births, $q^2 = 0.0001$. Therefore, *q*, the frequency of the recessive allele in the population, equals

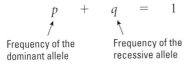

Allele frequencies $p = 0.8$ $q = 0.2$
(*R*)   (*r*)

Sperm   *R* $p = 0.8$ *RR* $p^2 = 0.64$ *R* $p = 0.8$ Eggs

r $q = 0.2$ *rR* $qp = 0.16$ *Rr* $pq = 0.16$ *r* $q = 0.2$

rr $q^2 = 0.04$

Genotype frequencies $p^2 = 0.64$ $2pq = 0.32$ $q^2 = 0.04$
(*RR*)   (*Rr*)   (*rr*)

Figure 13.18
A mathematical swim in the gene pool. Each of the four boxes in the grid corresponds to an equally probable "draw" of alleles from the gene pool.

Figure 13.19 A warning to individuals with PKU.
People with PKU (phenylketonurics) must avoid all foods with the amino acid phenylalanine. In addition to natural sources, phenylalanine is found in aspartame, a common artificial sweetener. The frequency of the PKU allele is high enough to warrant a public health program that includes warnings on foods that contain phenylalanine.

the square root of 0.0001, or 0.01. And *p*, the frequency of the dominant allele, equals 1 − *q*, or 0.99.

Now let's calculate the frequency of heterozygous individuals, who carry the PKU allele in a single dosage. These carriers are free of the disorder but may pass the PKU allele on to offspring. Carriers are represented in the Hardy-Weinberg formula by *2pq*. And that's 2 × 0.99 × 0.01, or 0.0198. Thus, the formula tells us that about 2% (actually 1.98%) of the U.S. population carries the PKU allele. Estimating the frequency of a harmful allele is essential for any public health program dealing with genetic diseases.

Microevolution as Change in a Gene Pool

How can we tell if a population is evolving? It helps, as a basis of comparison, to know what to expect if a population is *not* evolving. A nonevolving population is in genetic equilibrium, also called **Hardy-Weinberg equilibrium.** The population's gene pool remains constant over time. From generation to generation, the frequencies of alleles and genotypes are unchanged. Sexual shuffling of genes cannot by itself change a gene pool. But natural selection can. As an example, let's return to our wildflower population. Imagine the arrival of an insect species that is a vigorous pollinator of plants. If this insect is attracted to white flowers, its presence could enhance the reproductive success of white-flowered plants. Over the generations, this selection factor would increase the frequency of the *r* allele at the expense of the *R* allele. In contrast to the Hardy-Weinberg equilibrium of a nonevolving population, we now have the changing gene pool of an evolving population.

One of the products of the modern synthesis was a definition of evolution that is based on population genetics: *Evolution is a generation-to-generation change in a population's frequencies of alleles.* Because this describes evolution on the smallest scale, it is sometimes referred to more specifically as **microevolution.**

CHECKPOINT

1. What is the smallest biological unit that can evolve?

2. Define microevolution.

3. Which term in the Hardy-Weinberg formula ($p^2 + 2pq + q^2 = 1$) corresponds to the frequency of individuals who have no alleles for the disease PKU?

4. Which of the following variations in a human population is the best example of polymorphism: height, ABO blood group, number of fingers, or math proficiency?

5. Which process, mutation or sexual recombination, results in most of the generation-to-generation variability in human populations?

Answers: 1. A population **2.** Microevolution is a change in a population's frequencies of alleles. **3.** p^2 **4.** Blood group **5.** Sexual recombination

---

Mechanisms of Microevolution

What mechanisms can change a gene pool? Four causes of microevolution are genetic drift, gene flow, mutations, and, of course, natural selection. We'll see that each of these evolutionary mechanisms represents a departure from conditions required for Hardy-Weinberg equilibrium.

Genetic Drift

Flip a coin a thousand times, and a result of 700 heads and 300 tails would make you very suspicious about that coin. But flip a coin ten times, and an outcome of seven heads and three tails would seem within reason. The smaller the sample, the greater the chance of deviation from an idealized result—an equal number of heads and tails, in the case of a sample of coin tosses.

Let's apply coin toss logic to a population's gene pool. If a new generation draws its alleles at random from the previous generation, then the larger the population (the sample size), the better the new generation will represent the gene pool of the previous generation. Thus, one requirement for a gene pool to maintain the status quo—Hardy-Weinberg equilibrium—is a large population size. The gene pool of a small population may not be accurately represented in the next generation because of sampling error. It is analogous to the erratic outcome from a small sample of coin tosses.

Figure 13.20 applies this concept of sampling error to a small population of wildflowers. Chance causes the frequencies of the alleles for red (R) and white (r) flowers to change over the generations. And that fits our definition of microevolution. This evolutionary mechanism, a change in the gene pool of a small population due to chance, is called **genetic drift.** Two situations that can shrink populations down to a size where genetic drift occurs are known as the bottleneck effect and the founder effect.

The Bottleneck Effect Disasters such as earthquakes, floods, droughts, and fires may reduce the size of a population drastically. The small surviving

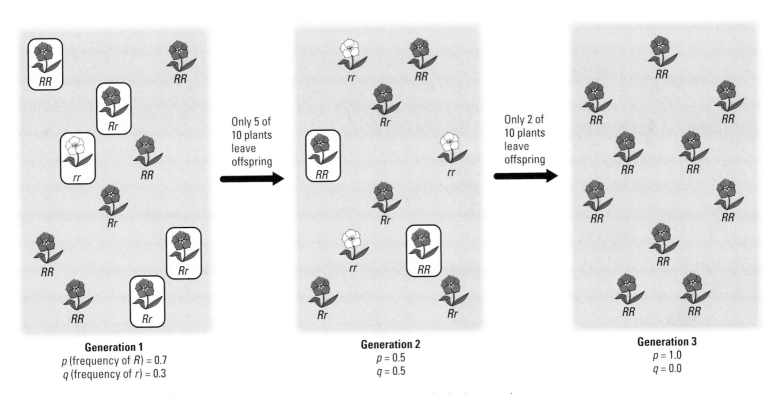

Generation 1
p (frequency of R) = 0.7
q (frequency of r) = 0.3

Only 5 of 10 plants leave offspring

Generation 2
p = 0.5
q = 0.5

Only 2 of 10 plants leave offspring

Generation 3
p = 1.0
q = 0.0

Figure 13.20 Genetic drift.   This small wildflower population has a stable size of only about ten plants. For generation 1, only the five boxed plants produce fertile offspring. Only two plants of generation 2 manage to leave fertile offspring. Over the generations, genetic drift can completely eliminate some alleles, as is the case for the r allele in generation 3 of this imaginary population.

Figure 13.21
The bottleneck effect.

The colored marbles in this analogy represent three morphs in an imaginary population. Shaking just a few of the marbles through the bottleneck is like drastically reducing the size of a population due to some environmental disaster. By chance, blue marbles are overrepresented in the new population, and gold marbles are absent. Similarly, bottlenecking a population of organisms tends to reduce variability.

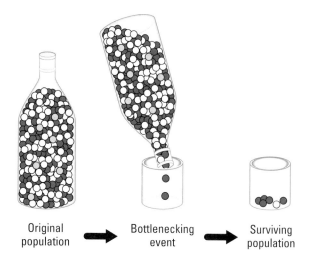

Original population → Bottlenecking event → Surviving population

population may not be representative of the original population's gene pool. By chance, certain alleles will be overrepresented among the survivors. Other alleles will be underrepresented. And some alleles may be eliminated altogether. Chance may continue to change the gene pool for many generations until the population is again large enough for sampling errors to be insignificant. The analogy in **Figure 13.21** illustrates why genetic drift due to a drastic reduction in population size is called the **bottleneck effect.**

Bottlenecking usually reduces the overall genetic variability in a population because at least some alleles are likely to be lost from the gene pool. An important application of this concept is the potential loss of individual variation, and hence adaptability, in bottlenecked populations of endangered species, such as the cheetah (**Figure 13.22**). The fastest of all running animals, cheetahs are magnificent cats that were once widespread in Africa and Asia. Like many African mammals, the number of cheetahs fell drastically during the last ice age some 10,000 years ago. At that time, the species may have suffered a severe bottleneck, possibly as a result of disease, human hunting, and periodic droughts. Some researchers think that the South African cheetah population suffered a second bottleneck during the nineteenth century when South African farmers hunted the animals to near extinction. Today, only three small populations of cheetahs exist in the wild. Genetic variability in these populations is very low compared to populations of other mammals. In fact, genetic uniformity in cheetahs rivals that of highly inbred varieties of laboratory mice! This lack of variability, coupled with an increasing loss of habitat, makes the cheetah's future precarious. The cheetahs remaining in Africa are being crowded into nature preserves and parks as human demands on the land increase. Along with crowding comes an increased potential for the spread of disease. With so little variability, the cheetah may have a reduced capacity to adapt to such environmental challenges. Captive breeding programs are already under way and may be required for the cheetah's long-term survival.

The Founder Effect Genetic drift is also likely when a few individuals colonize an isolated island, lake, or some other new habitat. The smaller the colony, the less its genetic makeup will represent the gene pool of the larger population from which the colonists emigrated. If the colony succeeds, random drift will continue to affect the frequency of alleles until the population is large enough for sampling error to be minimal.

Genetic drift in a new colony is known as the **founder effect.** The effect undoubtedly contributed to the evolutionary divergence of the finches and other South American organisms that arrived as strays on the remote Galápagos Islands that Darwin visited.

Genetic Drift and Hereditary Disorders in Human Populations

The founder effect explains the relatively high frequency of certain inherited disorders among some human populations established by small numbers of colonists. In 1814, 15 people founded a British colony on Tristan da Cunha, a group of small islands in the middle of the Atlantic

Figure 13.22 Implications of the bottleneck effect in conservation biology. Some endangered species, such as this cheetah, may have low genetic variability. As a result, they may be less adaptable to environmental changes, such as new diseases, than are species with a greater resource of genetic variation.

Ocean (**Figure 13.23**). Apparently, one of the colonists carried a recessive allele for retinitis pigmentosa, a progressive form of blindness. Of the 240 descendants who still lived on the islands in the 1960s, 4 had retinitis pigmentosa. At least 9 others were known to be carriers (heterozygous, with one copy of the recessive allele). That frequency of the retinitis pigmentosa allele is much higher than in Great Britain, the source of the colonists.

Gene Flow

Hardy-Weinberg equilibrium—the stagnant gene pool of a nonevolving population—requires genetic isolation from other populations. Most populations, however, are *not* completely isolated. A population may gain or lose alleles by **gene flow,** which is genetic exchange with another population. Gene flow occurs when fertile individuals or gametes migrate between populations. Perhaps, for example, a population neighboring our hypothetical wildflowers consists entirely of white-flowered individuals. A windstorm may blow pollen to our wildflowers from the neighboring population. Interbreeding would increase the frequency of the white-flower allele in our imaginary population. That qualifies as microevolution.

Gene flow tends to reduce genetic differences between populations. If it is extensive enough, gene flow can eventually amalgamate neighboring populations into a single population with a common gene pool. As humans began to move about the world more freely, gene flow became an important agent of microevolution in populations that were previously isolated (**Figure 13.24**).

Mutations

A **mutation,** remember, is a change in an organism's DNA. Hardy-Weinberg equilibrium requires that the gene pool is unaffected by mutations. A new mutation that is transmitted in gametes can immediately change the gene pool of a population by substituting one allele for another. For example, a mutation that causes a white-flowered plant *(rr)* in our hypothetical wildflower population to produce gametes bearing the dominant allele for red flowers *(R)* would decrease the frequency of the *r* allele in the population and increase the frequency of the *R* allele.

For any one gene locus, however, mutation alone does not have much quantitative effect on a large population in a single generation. This is because a mutation at any given gene locus is a very rare event. If some new allele produced by mutation increases its frequency by a significant amount in a population, it is not because mutation is generating the allele in abundance, but because individuals carrying the mutant allele are producing a disproportionate number of offspring as a result of natural selection or genetic drift.

Although mutations at a particular gene locus are rare, the cumulative impact of mutations at *all* loci can be significant. This is because each individual has thousands of genes, and many populations have thousands or millions of individuals. Certainly over the long term, mutation is, in itself, very important to evolution because it is the original source of the genetic variation that serves as raw material for natural selection.

Natural Selection: A Closer Look

Genetic drift, gene flow, and mutation can cause microevolution, but they do not necessarily lead to adaptation. In fact, only blind luck could result in random drift, migrant alleles, or shot-in-the-dark mutations improving a population's fit to its environment. Of all causes of microevolution, only natural

Figure 13.23 Residents of Tristan da Cunha in the early 1900s.

Figure 13.24 Gene flow and human evolution.
The migration of people throughout the world is transferring alleles between populations that were once isolated. This magazine cover celebrates our changing gene pools and culture with a computer-generated image blending facial features from several races.

Activity 13E on the Web & CD
Review the causes of microevolution in a hypothetical bug population.

selection is generally adaptive. And this adaptive evolution, remember, is a blend of chance and sorting—chance in the random generation of genetic variability, and sorting in the differential reproductive success among the varying individuals. Darwin explained the basics of natural selection. But it took the population genetics of the modern synthesis to fill in the details.

The Hardy-Weinberg equilibrium, which defines a nonevolving population, demands that all individuals in a population be equal in their ability to survive and reproduce. This condition is probably never completely met. On average, those individuals that work best in the environment leave the most offspring and therefore have a disproportionate impact on the gene pool. When farmers began spraying their fields with pesticides, resistant pests started outreproducing other members of the insect populations. This increased the frequency of alleles for pesticide resistance in gene pools. Microevolution occurred. And it was adaptive. It was evolution by natural selection.

Darwinian Fitness The phrases "struggle for existence" and "survival of the fittest" are misleading if we take them to mean direct competitive contests between individuals. There *are* animal species in which individuals, usually the males, lock horns or otherwise do combat to determine mating privilege. But reproductive success is generally more subtle and passive. In a varying population of moths, certain individuals may average more offspring than others because their wing colors hide them from predators better. Plants in a wildflower population may differ in reproductive success because some are better able to attract pollinators, owing to slight differences in flower color, shape, or fragrance (**Figure 13.25**). A frog may produce more eggs than her neighbors because she is better at catching insects for food. These examples point to a biological definition of fitness: **Darwinian fitness** *is the contribution an individual makes to the gene pool of the next generation relative to the contributions of other individuals.*

Survival to sexual maturity, of course, is prerequisite to reproductive success. But the biggest, fastest, toughest frog in the pond has a Darwinian fitness of zero if it is sterile. Production of fertile offspring is the only score that counts in natural selection.

Figure 13.25 Darwinian fitness of some flowering plants depends in part on competition in attracting pollinators.

Three General Outcomes of Natural Selection Imagine a population of deer mice. Individuals range in fur color from very light to very dark brown. If we graph the number of mice in each color category, we get a bell-shaped curve, like the grading curve some of your instructors draw after an exam. If natural selection favors certain fur color phenotypes over others, the population of mice will change over the generations. Three general outcomes are possible, depending on which phenotypes are favored. These three modes of natural selection are called directional selection, diversifying selection, and stabilizing selection.

Directional selection (**Figure 13.26a**) shifts the phenotypic curve of a population by selecting in favor of some extreme phenotype—the darkest mice, for example. Directional selection is most common when the local environment changes or when organisms migrate to a new environment. An example is the shift of insect populations toward a greater frequency of pesticide-resistant individuals.

Diversifying selection (**Figure 13.26b**) can lead to a balance between two or more contrasting morphs (phenotypic forms) in a population. A patchy environment, which favors different phenotypes in different patches, is one situation associated with diversifying selection. The polymorphic snake population in Figure 13.17 is an example of diversifying selection.

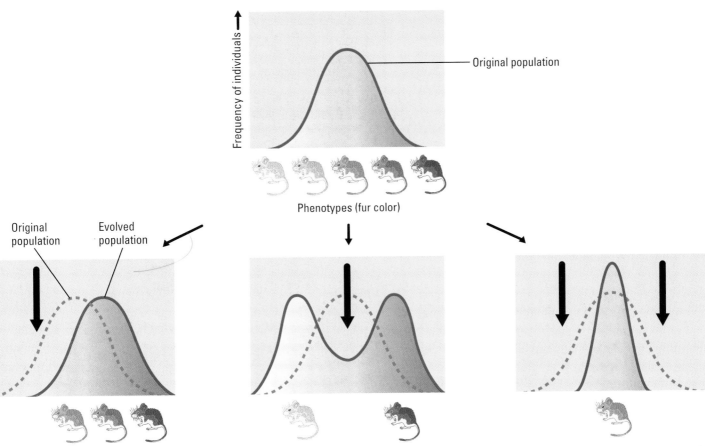

Frequency of individuals

Phenotypes (fur color)

Original population

Original population

Evolved population

(a) Directional selection shifts the overall makeup of the population by favoring variants of one extreme. In this case, the trend is toward darker color, perhaps because the landscape has been shaded by the growth of trees.

(b) Diversifying selection favors variants of opposite extremes over intermediate individuals. Here, the relative frequencies of very light and very dark mice have increased. Perhaps the mice have colonized a patchy habitat where a background of light soil is studded with dark rocks.

(c) Stabilizing selection culls extreme variants from the population, in this case eliminating individuals that are unusually light or dark. The trend is toward reduced phenotypic variation and maintenance of the status quo.

Stabilizing selection maintains variation for a particular trait within a narrow range (**Figure 13.26c**). It typically occurs in a relatively stable environment to which populations are already well adapted. This evolutionary conservatism works by selecting against the more extreme phenotypes. For example, stabilizing selection keeps the majority of human birth weights between 3 and 4 kg (approximately 6.5–9 pounds). For babies much smaller or larger than this, infant mortality is greater.

Of the three selection modes, stabilizing selection probably prevails most of the time, resisting change in well-adapted populations. Evolutionary spurts occur when a population is stressed by a change in the environment or migration to a new place. When challenged with a new set of environmental problems, a population either adapts through natural selection or becomes extinct in that locale. The fossil record tells us that extinction is the most common result. Those populations that do survive crises often change enough to be designated new species, as we will see in Chapter 14.

Figure 13.26 Three general effects of natural selection on a phenotypic character. Here are three possible outcomes for selection working on fur color in imaginary populations of deer mice. The large downward arrows symbolize the pressure of natural selection working against certain phenotypes.

CHECKPOINT

1. Compare and contrast the bottleneck effect and the founder effect as causes of genetic drift.

2. Why might new diseases pose a greater threat to cheetah populations than to mammalian populations having more genetic variation?

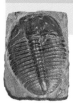

3. Which mechanism of microevolution has been most affected by the ease of human travel resulting from new modes of transportation?
4. What is the best measure of Darwinian fitness?
5. The thickness of fur in a bear population increases over several generations as the climate in the region becomes colder. This is an example of which type of selection: directional, diversifying, or stabilizing?

Answers: 1. Both processes result in populations being small enough for significant sampling error in the gene pool. A bottleneck reduces the size of an existing population. The founder effect is a new, small population consisting of individuals from a larger population. **2.** Because cheetah populations have so little variation, there is the potential for some new disease against which no individuals are resistant. **3.** Gene flow **4.** The number of fertile offspring an individual leaves **5.** Directional

Evolution Connection

Population Genetics of the Sickle-Cell Allele

As the capstone of every chapter in this textbook, the Evolution Connection section reinforces biology's unifying theme. But *this* unit of chapters is all about evolution itself. Here we'll use this section to relate evolutionary biology to society. And for this chapter, our connection is **Darwinian medicine,** the study of health problems in an evolutionary context.

About one out of every 500 African Americans has sickle-cell disease. The disease is named for the abnormal shape of red blood cells in individuals who inherit the disorder (**Figure 13.27** inset). Sickle-cell disease is caused by a recessive allele. Only homozygous individuals, who inherit the recessive alleles from both parents, have the disorder. About one in 11 African Americans has a single copy of the sickle-cell allele. These heterozygous individuals do not have sickle-cell disease, but can pass the allele for the disorder on to their children. (You can review the biology of sickle-cell disease in Chapters 3 and 9.)

Evolution Connection on the Web
Learn more about sickle-cell disease.

Why is the sickle-cell allele so much more common in African Americans than in the general U.S. population? And how can we explain such a high frequency among African Americans for an allele with the potential to shorten life (and hence reproductive success)? Evolutionary biology holds the answers.

In the African tropics, the sickle-cell allele is both boon and bane. It is true that when inherited in double dosage, the allele causes sickle-cell disease. But heterozygous individuals, who have just one copy of the sickle-cell allele, are relatively resistant to malaria. This is an important advantage in tropical regions where malaria, caused by a parasitic microorganism, is a major cause of death. The frequency of the sickle-cell allele in Africa is generally highest in areas where the malaria parasite is most common (see Figure 13.27).

How do we weigh the health impact of an allele that is both beneficial (in heterozygotes) and harmful (in homozygotes)? The Hardy-Weinberg formula is the tool we need. In some African populations, the sickle-cell allele has a frequency of 0.2, or 20%. This is q, the frequency of the recessive allele in the gene pool. And q^2, the frequency of homozygous recessives, is 0.2×0.2, or 0.04. So about 4% of individuals in these populations have sickle-cell disease. Now let's calculate the frequency of heterozygous individuals who

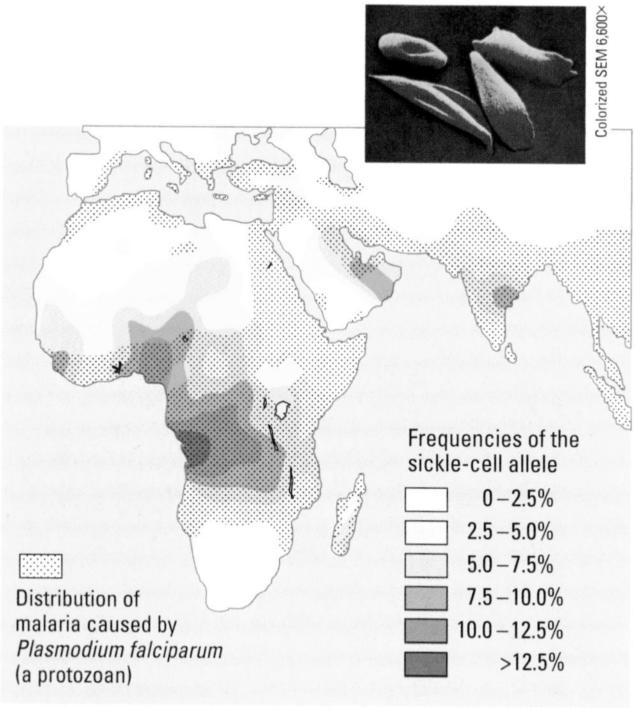

Colorized SEM 6,600×
Distribution of malaria caused by *Plasmodium falciparum* (a protozoan)
Frequencies of the sickle-cell allele
0–2.5%
2.5–5.0%
5.0–7.5%
7.5–10.0%
10.0–12.5%
>12.5%

Figure 13.27
Mapping malaria and the sickle-cell allele.

have enhanced resistance to malaria. That would be $2pq$ in the Hardy-Weinberg formula: $2 \times 0.8 \times 0.2$, or 0.32. So in this case, the sickle-cell allele benefits about 32% of the population, while it causes sickle-cell disease in only about 4% of the population. That explains the high frequency of the sickle-cell allele compared to other alleles that cause debilitating diseases. And the representation of the allele among black Americans is a vestige of African roots.

This example of Darwinian medicine is a reminder that biology is the foundation of all medicine. And evolution is the foundation of all biology.

Chapter Review

Summary of Key Concepts

For study help, go to the Essential Biology Website (www.essentialbiology.com) or CD-ROM to explore the Activities and Case Studies in the Process of Science.

Charles Darwin and *The Origin of Species*

- Charles Darwin established the ideas of evolution and natural selection in his 1859 publication *On the Origin of Species by Means of Natural Selection.*

- **Darwin's Cultural and Scientific Context** During his around-the-world voyage on the *Beagle,* Darwin observed adaptations of organisms that inhabited diverse environments. In particular, Darwin was struck by the geographic distribution of organisms on the Galápagos Islands off the South American coast. When Darwin considered his observations in light of new evidence for a very old Earth that changed slowly, he formulated ideas that were at odds with the long-held notion of a young Earth populated by unrelated and unchanging species.

Activity 13A *The Voyage of the* Beagle*: Darwin's Trip Around the World*

Activity 13B *Darwin and the Galápagos Islands*

- **Descent with Modification** Darwin made two proposals in *The Origin of Species:* (1) Modern species descended from ancestral species, and (2) natural selection is the mechanism of evolution.

Evidence of Evolution

- **The Fossil Record** The fossil record shows that organisms have appeared in a historical sequence, and many fossils link ancestral species with those living today.

- **Biogeography** Biogeography, the study of the geographic distribution of species, suggests that species evolved from ancestors that inhabited the same region.

- **Comparative Anatomy** Homologous structures among species bear witness to evolutionary history.

Activity 13C *Reconstructing Forelimbs*

- **Comparative Embryology** Closely related species often have similar stages in their embryonic development.

- **Molecular Biology** All species share a common genetic code, suggesting that all forms of life are related through branching evolution from the earliest organisms. Also, species that seem to be closely related have similar DNA and protein sequences.

Natural Selection and Adaptive Evolution

- **Darwin's Theory of Natural Selection** Individuals best suited for a particular environment are more likely to survive and reproduce than less fit individuals.

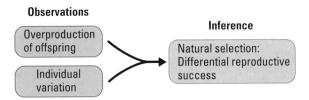

- **Natural Selection in Action** Natural selection can be observed in the development of pesticide-resistant insects and drug-resistant microbes.

Case Study in the Process of Science *What Are the Patterns of Antibiotic Resistance?*

Case Study in the Process of Science *How Do Environmental Changes Affect a Population of Leafhoppers?*

The Modern Synthesis: Darwinism Meets Genetics

- The modern synthesis fused genetics (Mendelism) and evolutionary biology (Darwinism) in the mid-1900s.

- **Populations as the Units of Evolution** A population is the smallest biological unit that can evolve. Population genetics emphasizes the extensive genetic variation within populations and tracks the genetic makeup of populations over time.

- **Genetic Variation in Populations** Polygenic ("many gene") inheritance produces traits that vary continuously, whereas traits that are determined by one genetic locus may be polymorphic, producing two or more distinct forms. Mutations and sexual recombination produce genetic variation.

Activity 13D *Genetic Variation from Sexual Recombination*

- **Analyzing Gene Pools** The gene pool consists of all alleles in all the individuals making up a population. The Hardy-Weinberg formula can be used to calculate the frequencies of genotypes in a gene pool from the frequencies of alleles, and vice versa.

Case Study in the Process of Science *How Can Frequency of Alleles Be Calculated?*

- **Population Genetics and Health Science** The Hardy-Weinberg formula can be used to estimate the frequency of a harmful allele, which is useful information for public health programs dealing with genetic diseases.

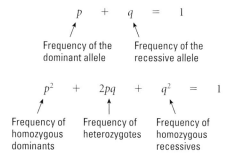

$$p + q = 1$$

Frequency of the dominant allele   Frequency of the recessive allele

$$p^2 + 2pq + q^2 = 1$$

Frequency of homozygous dominants   Frequency of heterozygotes   Frequency of homozygous recessives

- **Microevolution as Change in a Gene Pool** Microevolution is a generation-to-generation change in a population's frequencies of alleles.

Mechanisms of Microevolution

- **Genetic Drift** Genetic drift is a change in the gene pool of a small population due to chance. Bottlenecking and the founder effect are two situations leading to genetic drift.

- **Gene Flow** A population may gain or lose alleles by gene flow, which is genetic exchange with another population.

- **Mutations** Individual mutations have relatively little short-term effect on a large gene pool. Longer-term, mutation is the cumulative source of genetic variation.

- **Natural Selection: A Closer Look** Of all causes of microevolution, only natural selection is generally adaptive. Darwinian fitness is the contribution an individual makes to the gene pool of the next generation relative to the contributions of other individuals. The outcome of natural selection may be directional, diversifying, or stabilizing.

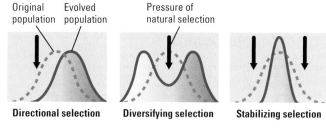

Original population   Evolved population   Pressure of natural selection

Directional selection   **Diversifying selection**   **Stabilizing selection**

Activity 13E *Causes of Microevolution*

Self-Quiz

1. Which of the following is *not* an observation or inference on which Darwin's theory of natural selection is based?
    a. There is heritable variation among individuals.
    b. Poorly adapted individuals never produce offspring.
    c. Because only a fraction of offspring survive, there is a struggle for limited resources.
    d. Individuals whose inherited characteristics best fit them to the environment will generally produce more offspring.

2. Which of the following is a true statement about Charles Darwin?
    a. He was the first to discover that living things can change, or evolve.
    b. He based his theory on the inheritance of acquired characteristics.
    c. He proposed natural selection as the mechanism of evolution.
    d. He was the first to realize that Earth is more than 6,000 years old.

3. In a population with two alleles for a particular genetic locus, *B* and *b*, the allele frequency of *B* is 0.7. If this population is in Hardy-Weinberg equilibrium, the frequency of heterozygotes is _____, the frequency of homozygous dominants is _____, and the frequency of homozygous recessives is _____.

4. In a population that is in Hardy-Weinberg equilibrium, 16% of the individuals show the recessive trait. What is the frequency of the dominant allele in the population?

5. Which of the following is an example of a polymorphism in humans?
    a. variation in height
    b. variation in intelligence
    c. the presence or absence of a widow's peak (see Figure 9.14)
    d. variation in the number of fingers

6. As a mechanism of microevolution, natural selection can be most closely equated with
    a. random mating
    b. genetic drift.
    c. differential reproductive success.
    d. gene flow.

7. Why does a founder event favor microevolution in the founding population?

8. In a particular bird species, individuals with average-sized wings survive severe storms more successfully than other birds in the same population with longer or shorter wings. Of the three general outcomes of natural selection, this example illustrates _____.

9. Which of the following statements is true about a population that is in Hardy-Weinberg equilibrium? (More than one may be true.)
    a. The population is quite small.
    b. The population is not evolving.
    c. Gene flow between the population and surrounding populations does not occur.
    d. Natural selection is not occurring.

10. What environmental factor accounts for the relatively high frequency of the sickle-cell allele in tropical Africa?

Answers to the Self-Quiz questions can be found in Appendix B.

Go to the website or CD-ROM for more Self-Quiz questions.

The Process of Science

1. A population of snails has recently become established in a new region. The snails are preyed on by birds that break the snails open on rocks, eat the soft bodies, and leave the shells. The snails occur in both striped and unstriped forms. In one area, researchers counted both live snails and broken shells. Their data are summarized here:

	Striped	Unstriped
Living	264	296
Broken	486	377
Total	750	673

Based on these data, which snail form is more subject to predation by birds? Predict how the frequencies of striped and unstriped individuals might change over the generations.

2. Imagine that the presence or absence of stripes on the snails from the previous question is determined by a single gene locus, with the dominant allele (*S*) producing striped snails and the recessive allele (*s*) producing unstriped snails. Combining the data from both the living and broken snails, calculate the following: the frequency of the dominant allele, the frequency of the recessive allele, and the number of heterozygotes in the observed groups.

Case Study in the Process of Science on the Web & CD *Investigate case studies of antibiotic resistance.*

Case Study in the Process of Science on the Web & CD *Explore how different environments affect a population of leafhoppers.*

Case Study in the Process of Science on the Web & CD *Practice using the Hardy-Weinberg formula.*

Biology and Society

To what extent are humans in a technological society exempt from natural selection? Explain your answer.

Biology and Society on the Web *Learn about the evolution of bacteria that are resistant to antibiotics.*

CHAPTER 14

How Biological Diversity Evolves

**Biology and Society:
The Impact of Asteroids   271**

**Macroevolution and the
Diversity of Life   271**

The Origin of Species   272

What Is a Species?

*Reproductive Barriers Between
Species*

Mechanisms of Speciation

What Is the Tempo of Speciation?

**The Evolution of Biological
Novelty   280**

*Adaptation of Old Structures for
New Functions*

*"Evo-Devo": Development and
Evolutionary Novelty*

**Earth History and
Macroevolution   282**

Geologic Time and the Fossil Record

*Continental Drift and
Macroevolution*

*Mass Extinctions and Explosive
Diversifications of Life*

**Classifying the Diversity
of Life   287**

Some Basics of Taxonomy

Classification and Phylogeny

*Arranging Life into Kingdoms:
A Work in Progress*

**Evolution Connection:
Just a Theory?   292**

Biologists estimate that there are 5–30 million species of organisms currently on Earth.

Our own scientific name, *Homo sapiens,* means "wise man."

North America and Europe are drifting apart at a rate of about 2 cm per year.

Globally, the rate of species loss may be 50 times higher now than at any time in the past 100,000 years.

The Impact of Asteroids

For 150 million years, dinosaurs dominated Earth's land and air, while mammals were few and small, resembling modern-day rodents. Then, about 65 million years ago, all the dinosaurs (as well as about half the other species living on Earth at the time) became extinct in less than 10 million years—a brief period in geologic time. For decades, scientists have been debating the cause of this rapid dinosaur die-off.

What do we know about this unusual time? The fossil record shows that the climate had cooled and that shallow seas were receding from continental lowlands. It also shows that many plants required by plant-eating dinosaurs died out first. Perhaps most telling of all, the rock deposited at the time contains a thin layer of clay rich in iridium, an element very rare on Earth but common in meteorites. Many paleontologists conclude that the iridium layer is the result of fallout from a huge cloud of dust that billowed into the atmosphere when a large meteorite or asteroid hit Earth. The cloud would have blocked light and disturbed climate severely for months, perhaps killing off many plant species.

Science is based on the testable predictions of hypotheses, and the asteroid hypothesis fits the bill: It predicts that we should find a huge impact crater of the right age. In fact, in 1981, two petroleum geologists discovered the Chicxulub crater near the Yucatán Peninsula in the Caribbean Sea (**Figure 14.1**). This impact site, 150–300 km wide and dating from the predicted time, was created when a meteorite or asteroid the size of San Francisco slammed into Earth, releasing thousands of times more energy than the world's current combined stockpile of nuclear weapons.

Biology and Society on the Web
Learn about the evidence for an asteroid impact 250 million years ago that may have contributed to mass extinctions at that time.

Debate continues about whether this impact alone caused the dinosaurs to die out or whether other factors also contributed, but most scientists believe that the collision that created the Chicxulub crater could indeed have caused global climatic changes and mass extinctions. Could such an event happen again? Some astronomers have called for increased funding to catalog near-Earth objects that could potentially collide with our planet.

The history of life on Earth includes many similar stories of species dying out while others multiply and diversify. In this chapter, we study the birth of species and the methods biologists use to trace the evolution of biological diversity.

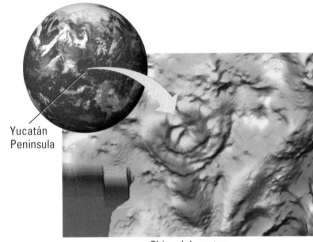

Yucatán Peninsula

Chicxulub crater

Figure 14.1 Crater of doom? The impact that created the horseshoe-shaped Chicxulub crater (shown in blue) may have caused widespread devastation, contributing to the demise of the dinosaurs and other species.

Macroevolution and the Diversity of Life

When Darwin traveled to the Galápagos Islands, he realized that he was visiting a place of genesis. Though the volcanic islands are geologically young, they are already home to many plants and animals known nowhere else in the world. Among the islands' unique inhabitants are its giant tortoises, for which the Galápagos are named (*galápago* is the Spanish word for "tortoise"). After visiting the Galápagos, Darwin wrote in his diary: "Both in space and time, we seem to be brought somewhat near to that great fact—that mystery of mysteries—the first appearance of new beings on this Earth."

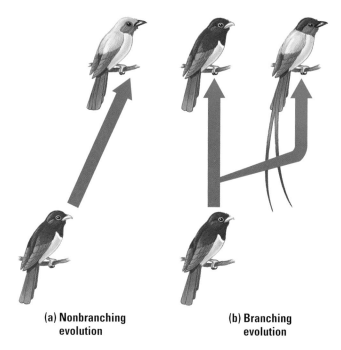

(a) Nonbranching evolution

(b) Branching evolution

Figure 14.2 Two patterns of speciation.
(a) Nonbranching evolution can transform a population enough for it to be designated a new species, but this linear pattern of evolution cannot multiply species. **(b)** Branching evolution splits a lineage into two or more species.

Figure 14.3 Ernst Mayr (on right) in New Guinea, 1927. Mayr was one of the architects of the modern synthesis, which you learned about in Chapter 13. One of his many contributions to the modern synthesis was the biological species concept. Still an active scientist 75 years after his first New Guinea expedition, Mayr continues to influence modern biology by publishing new books and articles.

To understand the "appearance of new beings," as Darwin put it, it is not enough to explain microevolution and the adaptation of populations, which you learned about in Chapter 13. If that were all that ever happened, then Earth would be populated only by a highly adapted version of the first form of life. Evolutionary theory must also explain macroevolution, which encompasses the major biological changes evident in the fossil record. **Macroevolution** includes the multiplication of species, which has generated biological diversity; the origin of evolutionary novelty, such as the wings and feathers of birds and the big brains of humans; the explosive diversification that follows some evolutionary breakthrough, such as the origin of thousands of plant species after the flower evolved; and mass extinctions, which cleared the way for new adaptive explosions, such as the diversification of mammals that followed the disappearance of dinosaurs.

Activity 14A on the Web & CD See animations of various aspects of macroevolution.

The beginning of new forms of life—the origin of species—is at the focal point of our study of macroevolution, for it is in new species that biological diversity arises. **Figure 14.2** contrasts two patterns of **speciation**, the origin of new species. In the case of nonbranching evolution, a population may change so much through adaptation to a changing environment that it becomes a new species (**Figure 14.2a**). Such linear evolution cannot multiply species. The second major pattern is branching evolution (**Figure 14.2b**). In this case, one or more new species branch from a parent species that may continue to exist. Branching evolution is the more important pattern in the history of life; it is probably more common than the nonbranching pattern, and only branching evolution can generate biological diversity by increasing the number of species. How did evolution produce over 250,000 species of flowering plants? Or about 35,000 species of fishes? Or over a million insect species? To explain such diversification, we must understand how the origin of new species makes branching evolution possible.

CHECKPOINT

1. Contrast microevolution with macroevolution.

2. Explain why nonbranching evolution can't increase the number of species.

Answers: 1. Microevolution is a change in the gene pool of a population, often associated with adaptation; macroevolution is a change in a life-form, such as the origin of a new species, that is noticeable enough to be evident in the fossil record. **2.** In nonbranching evolution, an ancestral species changes gradually to form a new species. Thus, there is no effect on species number.

The Origin of Species

Species is a Latin word meaning "kind" or "appearance." Indeed, we learn to distinguish between the kinds of plants and animals—between dogs and cats, for example—from differences in their appearance. Although the basic idea of species as distinct life-forms seems intuitive, devising a more formal definition is not so easy.

What Is a Species?

In 1927, a young biologist named Ernst Mayr led an expedition into the remote Arafak Mountains of New Guinea to study the wildlife (**Figure 14.3**). He found a great diversity of birds, identifying 138 species on the basis of differences in their appearance. Mayr was surprised to learn that the local Papuan

(a) Similarity between different species

(b) Diversity within one species

people had given names of their own to 137 birds (two birds assigned by Mayr to separate species are very similar, and the Papuans did not distinguish between them). Mayr and the indigenous people agreed almost exactly in their inventory of the local birdlife. The experience sharpened Mayr's view that species are demarcated from one another as distinct forms of life. Put another way, biological diversity is not a continuum of form, but is instead divided into the separate forms that we identify as species. Mayr distilled this observation into the biological species concept.

The **biological species concept** defines species as "groups of interbreeding natural populations that are reproductively isolated from other such groups" (Mayr's words). Put another way, a species is a population or group of populations whose members have the potential to interbreed with one another in nature to produce fertile offspring, but who cannot successfully interbreed with members of other species (**Figure 14.4**). Geography and culture may conspire to keep a Manhattan businesswoman and a Mongolian dairyman apart. But if the two *did* get together, they could have viable babies that develop into fertile adults. All humans belong to the same species. In contrast, humans and chimpanzees, similar and closely related in their evolution as they are, remain distinct species even where they share territory. The two species do not interbreed. Such reproductive isolation blocks exchange of genes between species and keeps their gene pools separate.

We cannot apply the biological species concept to all situations. For example, basing the definition of species on reproductive compatibility excludes organisms that only reproduce asexually (producing offspring from a single parent rather than from a mating pair). Fossils aren't doing much sexual reproduction either, so they cannot be evaluated by the biological species concept. Biologists distinguish fossil species mainly by their differences in appearance. But in spite of its limitations, and even though there are alternative species concepts that some biologists now favor, the biological species concept still proves useful in its premise of reproductive barriers that keep species separate.

Figure 14.4 The biological species concept is based on reproductive compatibility rather than physical similarity. **(a)** The eastern meadowlark (*Sturnella magna*, left) and the western meadowlark (*Sturnella neglecta*) are very similar in appearance, but they are separate species that cannot interbreed. Distinct songs help these birds choose mates of the same species. **(b)** In contrast, humans, as diverse in appearance as we are, belong to a single species (*Homo sapiens*), defined by our capacity to interbreed.

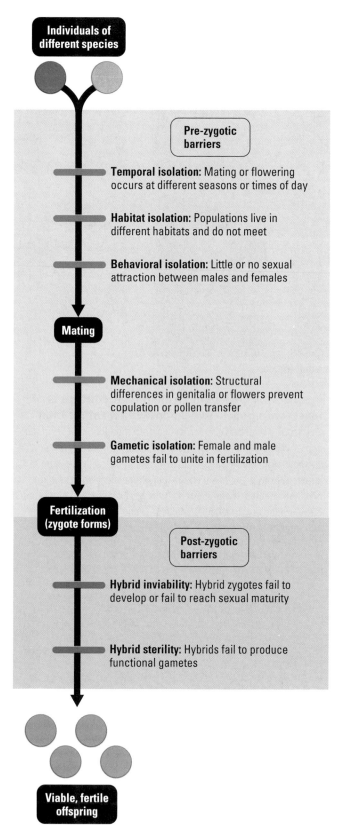

Individuals of different species

Pre-zygotic barriers

Temporal isolation: Mating or flowering occurs at different seasons or times of day

Habitat isolation: Populations live in different habitats and do not meet

Behavioral isolation: Little or no sexual attraction between males and females

Mating

Mechanical isolation: Structural differences in genitalia or flowers prevent copulation or pollen transfer

Gametic isolation: Female and male gametes fail to unite in fertilization

Fertilization (zygote forms)

Post-zygotic barriers

Hybrid inviability: Hybrid zygotes fail to develop or fail to reach sexual maturity

Hybrid sterility: Hybrids fail to produce functional gametes

Viable, fertile offspring

Figure 14.5 Reproductive barriers between closely related species.

Reproductive Barriers Between Species

Clearly, a fly will not mate with a frog or a fern. But what prevents biological species that are very similar—that is, closely related—from interbreeding? What, for example, maintains the species boundary between the western spotted skunk and the eastern spotted skunk? Their geographic ranges overlap in the Great Plains region, and they are so similar that only expert zoologists can tell them apart. And yet, these two skunk species do not interbreed. Let's examine the reproductive barriers that isolate the gene pools of species. We can classify reproductive barriers as either pre-zygotic or post-zygotic, depending on whether they block interbreeding before or after the formation of zygotes, or fertilized eggs.

Pre-zygotic barriers impede mating between species or hinder fertilization of eggs if members of different species should attempt to mate (**Figure 14.5**). The barrier may be time based (*temporal isolation*). For example, western spotted skunks breed in the fall, but the eastern species breeds in late winter. Temporal isolation keeps the species separate even where they coexist on the Great Plains. In other cases, species living in the same region are spatially segregated (*habitat isolation*). For example, one species of garter snake lives mainly in water, while a closely related species lives on land. Courtship rituals or other identification requirements for mates can also function as reproductive barriers (*behavioral isolation*). In many bird species, for example, courtship behavior is so elaborate that individuals are unlikely to mistake a bird of a different species as one of their kind (**Figure 14.6**). In still other cases, the male and female sex organs of different species are anatomically incompatible (*mechanical isolation*). For example, even if insects of closely related species attempt to mate, the male and female copulatory organs may not fit together correctly and no sperm is transferred. In still other cases, individuals of different species may actually copulate, but their gametes are incompatible and fertilization does not occur (*gametic isolation*). In many mammals, for example, sperm may not survive in females of a different species.

Figure 14.6 Courtship ritual as a behavioral barrier between species. These blue-footed boobies, inhabitants of the Galápagos Islands, will mate only after a specific ritual of courtship displays. Part of the "script" calls for the male to high-step, a dance that advertises the bright blue feet characteristic of the species.

Horse

Mule (hybrid)

Donkey

Figure 14.7 Hybrid sterility, a post-zygotic barrier. Horses and donkeys remain separate species because their hybrid offspring, mules, are sterile.

Post-zygotic barriers are mechanisms that operate should interspecies mating actually occur and form hybrid zygotes (hybrid here meaning that the egg comes from one species and the sperm from another species). In some cases, hybrid offspring die before reaching reproductive maturity (*hybrid inviability*). For example, although certain closely related frog species will hybridize, the offspring fail to develop normally because of genetic incompatibilities between the two species. In other cases of hybridization, offspring may become vigorous adults, but are infertile (*hybrid sterility*). A mule, for example, is the hybrid offspring of a female horse and a male donkey (**Figure 14.7**). Because mules are sterile, there is no avenue for gene transfer between the two parental species. The horse and donkey remain separate species because their gene pools remain isolated.

In most cases, it is not a single reproductive barrier but some combination of two or more that reinforces boundaries between species. If it is reproductive isolation that keeps species separate, then the evolution of these barriers is the key to the origin of new species.

Mechanisms of Speciation

A key event in the potential origin of a species occurs when the gene pool of a population is somehow severed from other populations of the parent species. With its gene pool isolated, the splinter population can follow its own evolutionary course. Changes in its allele frequencies caused by genetic drift and natural selection are undiluted by gene flow from other populations. Such reproductive isolation can result from two general scenarios: allopatric speciation and sympatric speciation. In **allopatric speciation,** the initial block to gene flow is a geographic barrier that physically isolates the splinter population (the term *allopatric* is from the Greek, meaning "other country"). In contrast, **sympatric speciation** is the origin of a new species without geographic isolation (*sympatric* means "together country"). The splinter population becomes reproductively isolated right in the midst of the parent population (**Figure 14.8**).

Allopatric Speciation Several geologic processes can fragment a population into two or more isolated populations. A mountain range may emerge and gradually split a population of organisms that can inhabit only lowlands. A creeping glacier may gradually divide a population. A land bridge, such as the Isthmus of Panama, may form and separate the marine life on either side. A large lake may subside until

Activity 14B on the Web & CD
Explore speciation on islands.

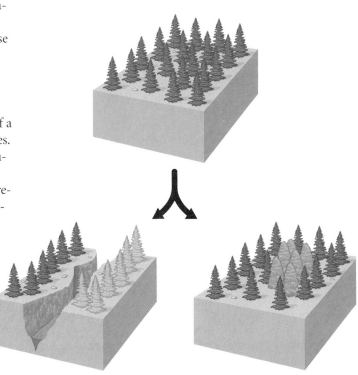

(a) Allopatric speciation

(b) Sympatric speciation

Figure 14.8 Two modes of speciation. These drawings simplify the geographic relationships of new species to their parent species. **(a)** In allopatric speciation, a population forms a new species while geographically isolated from its parent population. **(b)** In sympatric speciation, a small population becomes a new species in the midst of its parent population.

Figure 14.9 Allopatric speciation of squirrels in the Grand Canyon. Two species of antelope squirrels inhabit opposite rims of the Grand Canyon. On the south rim is Harris's antelope squirrel (*Ammospermophilus harrisi*). Just a few miles away on the north rim is the closely related white-tailed antelope squirrel (*Ammospermophilus leucurus*). Birds and other organisms that can disperse easily across the canyon have not diverged into different species on opposite rims.

Populations become sympatric again but do not interbreed

Populations become sympatric again and interbreed

Geographic barrier

Population becomes allopatric

(a) Speciation has not occurred

(b) Speciation has occurred

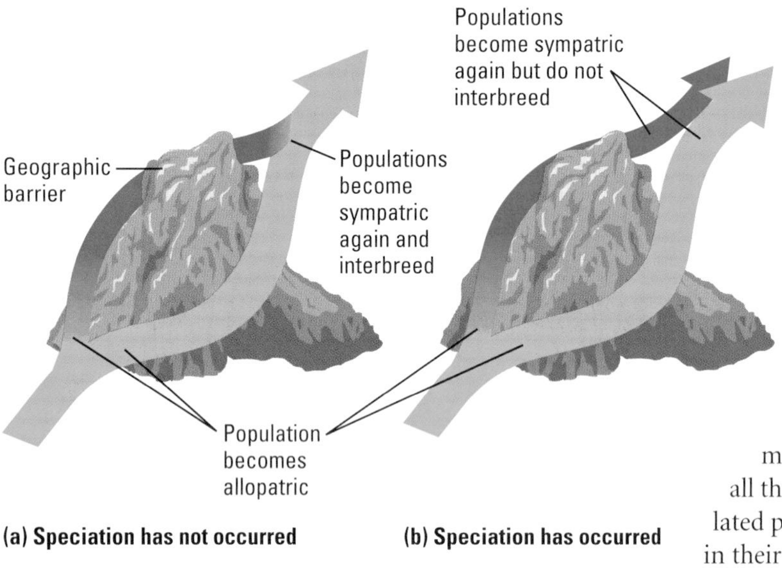

Figure 14.10 Has speciation occurred during geographic isolation? In this analogy, the arrows track populations over time. The mountain symbolizes a period of geographic isolation. What happens if the two populations come back into contact after a long period of separation? **(a)** If the two populations interbreed freely when they become sympatric again, their gene pools merge. Speciation has not occurred. **(b)** If the evolutionary divergence of the two populations results in reproductive isolation, then they will not interbreed, even if they come back into contact. Speciation has occurred.

there are several smaller lakes, with their populations now isolated. Even without such geologic remodeling, geographic isolation and allopatric speciation can occur if individuals colonize a new, geographically remote area and become isolated from the parent population. An example is the speciation that occurred on the Galápagos Islands following colonization by mainland organisms.

How formidable must a geographic barrier be to keep allopatric populations apart? The answer depends on the ability of the organisms to move about. Birds, mountain lions, and coyotes can cross mountain ranges, rivers, and canyons. Nor do such barriers hinder the windblown pollen of pine trees, and the seeds of many plants may be carried back and forth on animals. In contrast, small rodents may find a deep canyon or a wide river a formidable barrier (**Figure 14.9**).

The likelihood of allopatric speciation increases when a population is both small and isolated. A small, isolated population is more likely than a large population to have its gene pool changed substantially by factors such as genetic drift and natural selection. For example, in less than 2 million years, small populations of stray animals and plants from the South American mainland that managed to colonize the Galápagos Islands gave rise to all the new species that now inhabit the islands. But for each small, isolated population that becomes a new species, many more simply perish in their new environment. Life on the frontier is harsh, and most pioneer populations probably become extinct.

Even if a small, isolated population survives, it does not necessarily evolve into a new species. The population may adapt to its local environment and begin to look very different from the ancestral population, but that doesn't make it a new species. Speciation occurs only with the evolution of reproductive barriers between the isolated population and its parent population. In other words, if speciation occurs during geographic separation, the new species will not breed with its ancestral population, even if the two populations should come back into contact (**Figure 14.10**).

Sympatric Speciation How can a subpopulation become reproductively isolated while in the midst of its parent population? This can occur in a single generation if a genetic change produces a reproductive barrier between mutants and the parent population. Sympatric speciation does not seem to be widespread among animals, but it may account for over 25% of all plant species.

Many plant species have originated from accidents during cell division that resulted in extra sets of chromosomes (see Evolution Connection in Chapter 8). A new species that evolves this way has polyploid cells, meaning that each cell has more than two complete sets of chromosomes. The new species cannot produce fertile hybrids with its parent species, and reproductive isolation and speciation have occurred in a single generation without geographic isolation. This mechanism of sympatric speciation was first discovered in the early 1900s by Dutch botanist Hugo de Vries. His experiments produced a new species of primrose (**Figure 14.11**).

Polyploids do not always come from a single parent species. In fact, most polyploid species arise from the hybridization of two parent species (**Figure 14.12**). This mechanism of sympatric speciation accounts for many of the plant species we grow for food, including oats, potatoes, bananas, peanuts, barley, plums, apples, sugarcane, coffee, and wheat. Wheat makes a good case study. The most widely cultivated plant in the world, what we call wheat is actually represented by 20 different species. Humans began domesticating wheat from

Activity 14C on the Web & CD Learn how polyploid plants arise.

Case Study in the Process of Science on the Web & CD Investigate how new species arise by genetic isolation.

O. lamarckiana

O. gigas

Figure 14.11 Botanist Hugo de Vries and his new primrose species. Working in the early 1900s, de Vries studied variation in evening primroses. *Oenothera lamarckiana* (upper left) is a diploid species of primrose with 14 chromosomes. During his breeding experiments, de Vries noted a new primrose variety with 28 chromosomes. It was a tetraploid, which could not interbreed with its parent species. De Vries named the new primrose species *Oenothera gigas,* for its large size (lower right).

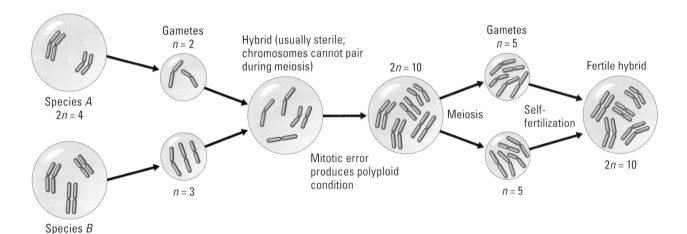

Gametes $n = 2$

Hybrid (usually sterile; chromosomes cannot pair during meiosis)

$2n = 10$

Gametes $n = 5$

Fertile hybrid

Species *A* $2n = 4$

Meiosis

Self-fertilization

Species *B* $2n = 6$

$n = 3$

Mitotic error produces polyploid condition

$n = 5$

$2n = 10$

Figure 14.12 Polyploid formation by hybridization of two parent species. A hybrid between two plant species is normally sterile because its chromosomes are not homologous and cannot pair during meiosis (see Chapter 8). However, the hybrids may be able to reproduce asexually. This diagram shows one mechanism that can produce new species from such hybrids. At some instant in the history of the hybrid clone, a mitotic error affecting the reproductive tissue of an individual may double chromosome number. The hybrid will then be able to make gametes because each chromosome has a homologue with which to pair during meiosis. Union of gametes from this hybrid may give rise to a new species of interbreeding plants, reproductively isolated from both parent species (because chromosome sets don't match). The new species has a chromosome number equal to the sum of the chromosomes in the two parent species. Such combinations of hybridization and cell division accidents make it possible for a new species to arise without geographic isolation from parent species—that is, for sympatric speciation to occur.

Figure 14.13 The evolution of wheat.

The uppercase letters in this diagram represent not alleles but sets of chromosomes that we are tracing in the evolution of wheat, or *Triticum*. ❶ The evolutionary descent of wheat began with hybridization between two diploid wheats. One was the first domesticated species, *T. monococcum* (its diploid chromosome set is symbolized by AA in the diagram). The other species was a wild relative of wheat that probably grew as a weed at the edges of cultivated fields (we're using BB for its diploid chromosome set). Both of these species had a diploid number of 14 chromosomes, producing a hybrid with 14 chromosomes. ❷ Chromosome sets A and B of the two species would not have been able to pair at meiosis, making the AB hybrid sterile. However, an error in cell division in this sterile hybrid and self-fertilization among the resulting gametes produced a new species (AABB) with 28 chromosomes. Today, this species, known as emmer wheat (*T. turgidum*), is still grown widely in Eurasia and western North America. It is used mainly for making macaroni and other noodle products because its proteins have a tendency to hold their shape better than bread wheat proteins. ❸ The next step in the evolution of bread wheat is believed to have occurred in early farming villages on the shores of European lakes over 8,000 years ago. Cultivated emmer wheat, with its 28 chromosomes, hybridized spontaneously with a closely related wild species that has 14 chromosomes. ❹ The hybrid (ABD), with 21 chromosomes, was sterile. A cell division error in this hybrid and self-fertilization doubled the chromosome number to 42. The result was bread wheat (*T. aestivum*), with two each of the three ancestral sets of chromosomes (AABBDD).

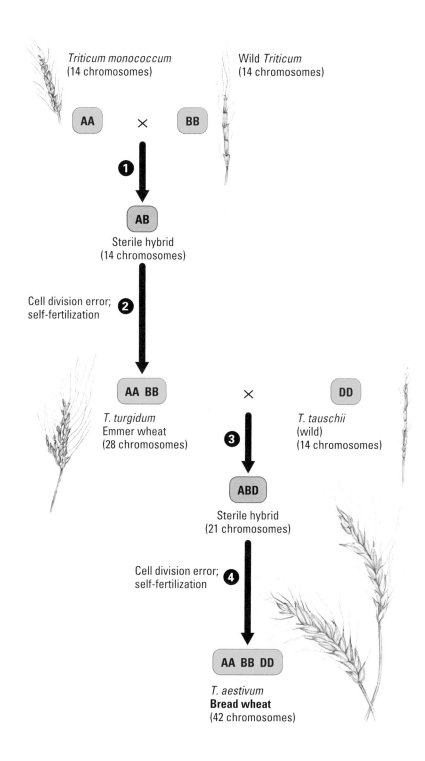

wild grasses at least 11,000 years ago in the Middle East. **Figure 14.13** traces the evolutionary path from that first cultivated wheat species to bread wheat, the most important wheat species today.

What Is the Tempo of Speciation?

Traditional evolutionary trees that diagram the descent of species sprout branches that diverge gradually (**Figure 14.14a**). Such trees are based on the idea that the changes big enough to distinguish a new species are an accumulation of many smaller changes occurring over vast spans of time. This concept applies the processes of microevolution you learned about in Chapter 13 to the multiplication of species. To Darwin, the origin of species was an extension of adaptation by natural selection, with isolated populations from common ancestral stock evolving differences gradually as they adapted to their local environments.

On the time scale of the fossil record, new species often appear more abruptly than the traditional model would predict. Paleontologists do find gradual transitions of fossil forms in some cases. However, the more common observation is of species appearing as new forms rather suddenly (in geologic terms) in a layer of rocks, persisting essentially unchanged for their tenure on Earth, and then disappearing from the fossil record as suddenly as they appeared. Darwin himself was bewildered by the dearth of connecting fossils and wrote: "Although each species must have passed through numerous transitional stages, it is probable that the periods during which each underwent modification, though many and long as measured by years, have been short in comparison with the periods during which each remained in an unchanged condition."

Over the past 30 years, some evolutionary biologists have addressed the nongradual appearance of species in the fossil record by developing a model now known as **punctuated equilibrium.** According to this model, species most often diverge in spurts of relatively rapid change, instead of slowly and gradually (**Figure 14.14b**). In other words, species undergo most of their modification in appearance as they first bud from parent species, and then they change little. The term *punctuated equilibrium* is derived from the idea of long periods of stasis (equilibrium) punctuated by episodes of speciation.

Is it likely that most species evolve abruptly and then remain essentially unchanged for most of their existence? One cause of sudden speciation is the polyploidy you learned about as a mechanism of sympatric speciation in plants. Proponents of punctuated equilibrium point out that allopatric speciation can also be quite rapid. In just a few hundred to a few thousand generations, genetic drift and natural selection can cause significant change in the gene pool of a small population cloistered in a challenging new environment.

How can speciation in a few thousand generations, which may require several thousand years, be called an abrupt episode? The fossil record indicates that successful species last for a few million years, on average. Suppose that a particular species survives for 5 million years, but most of its evolutionary changes in anatomy occur during the first 50,000 years of its existence. In this case, the evolution of the species-defining characteristics is compressed into just 1% of the lifetime of the species. In the fossil record, the species will appear suddenly in rocks of a certain age and then linger with little or no change before becoming extinct. During its formative millennia, the species may accumulate its modifications gradually, but relative to the overall history of the species, its inception is abrupt. This scenario of an evolutionary spurt preceding a much longer period of morphological stasis would help explain why paleontologists find relatively few smooth transitions in the fossil record of species. Also, since the best candidates for speciation are *small* populations, we are less likely to have fossils of transitional stages than if they occurred in a large population.

Once it is acknowledged that "sudden" may be many thousands of years on the vast scale of geologic time, the debate over the tempo of speciation is

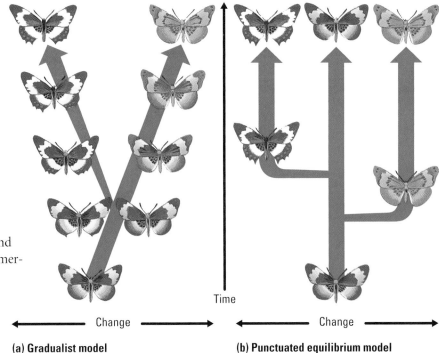

Time

Change

Change

(a) Gradualist model

(b) Punctuated equilibrium model

Figure 14.14 Two models for the tempo of evolution. **(a)** In the gradualist model, species descended from a common ancestor diverge more and more in form as they acquire unique adaptations. **(b)** According to the punctuated equilibrium model, a new species changes most as it buds from a parent species. After this speciation episode, there is little change for the rest of the species' existence.

muted somewhat. The degree to which a species changes after its origin is another issue. If the species is adapted to an environment that stays the same, then natural selection would not favor changes in the gene pool. In this view, stabilizing selection tends to hold a population in a long period of stasis. But some evolutionary biologists argue that stasis is an illusion. They propose that species may continue to change after they come into existence, but in non-anatomical ways that cannot be detected from fossils.

The Evolution of Biological Novelty

The two squirrels in Figure 14.9 are different species, but they are, after all, very similar animals that live very much the same way. When most people think of evolution, they envision much more dramatic transformation. How can we account for such evolutionary products as flight in birds and braininess in humans?

Adaptation of Old Structures for New Functions

Birds are derived from a lineage of earthbound reptiles (**Figure 14.15**). How could flying vertebrates evolve from flightless ancestors? More generally, how do major novelties of biological structure and function evolve? One mechanism is the gradual refinement of existing structures that take on new functions.

Most biological structures have an evolutionary plasticity that makes alternative functions possible. Biologists use the term **exaptation** for a structure that evolves in one context and becomes adapted for other functions. This concept does not imply that a structure somehow evolves in anticipation of future use. Natural selection cannot predict the future and can only refine a structure in the context of its current utility. Birds have lightweight skeletons with honeycombed bones, which are homologous to the bones of the earthbound reptilian ancestors of birds. However, honeycombed bones could not have evolved in the ancestors as an adaptation for upcoming

Figure 14.15 An artist's reconstruction of an extinct bird. Called *Archaeopteryx* ("ancient wing"), this animal lived near tropical lagoons in central Europe about 150 million years ago. Like modern birds, it had flight feathers, but otherwise it was more like some small bipedal dinosaurs of its era; for instance, like those dinosaurs, *Archaeopteryx* had teeth, wing claws, and a tail with many vertebrae. *Archaeopteryx*'s flight feathers could have enabled it to fly by flapping its wings. However, it had a fairly heavy body and probably relied mainly on gliding from trees. Despite its feathers, *Archaeopteryx* is not considered an ancestor of modern birds. Instead, it probably represents an extinct side branch of the bird lineage.

Wing claw (like reptile)

Teeth (like reptile)

Feathers

Long tail with many vertebrae (like reptile)

flights. If light bones predated flight, as is clearly indicated by the fossil record, then they must have had some function on the ground. The probable ancestors of birds were agile, bipedal reptiles that also would have benefited from a light frame. Winglike forelimbs and feathers would have increased the surface area of the forelimbs. These enlarged forelimbs could have been adapted for flight after functioning in some other capacity, such as courtship displays. The first flights may have only been glides or extended hops in an effort to pursue prey or escape from a predator. Once flight itself became an advantage, natural selection would have remodeled feathers and wings to better fit their additional function.

Exaptation is a mechanism for novel features to arise gradually through a series of intermediate stages, each of which has some function in the organism's current context. Harvard zoologist Karel Liem puts it this way: "Evolution is like modifying a machine while it's running." The concept that biological novelties can evolve by the remodeling of old structures for new functions is in the Darwinian tradition of large changes being an accumulation of many small changes crafted by natural selection.

"Evo-Devo": Development and Evolutionary Novelty

Gradual evolutionary remodeling, such as the accumulation of flight adaptations in birds, probably involves a large number of genetic changes in populations. In other cases, relatively few genetic changes can cause major structural modifications. How can slight genetic divergence become magnified into major differences between organisms? Scientists working at the interface of evolutionary biology and developmental biology—the research field abbreviated "evo-devo"—are finding some answers to this question.

Genes that program development control the rate, timing, and spatial pattern of changes in an organism's form as it is transfigured from a zygote into an adult. (Chapter 11 provides a closer look at how genes control development.) A subtle change in a species' developmental program can have profound effects. The animal in **Figure 14.16** is a salamander called an axolotl. It illustrates a phenomenon called **paedomorphosis,** which is the retention of juvenile body features in the adult (the term comes from the Greek *paedos,* child, and *morphosis,* shaping). The axolotl grows to full size and reproduces without losing its external gills, a juvenile feature in most species of salamanders.

Paedomorphosis has also been important in human evolution. Humans and chimpanzees are even more alike in body form as fetuses than they are as adults. In the fetuses of both species, the skulls are rounded and the jaws are small, making the face rather flat (**Figure 14.17**). As development proceeds, uneven bone growth makes the chimpanzee skull sharply angular, with heavy browridges and massive jaws. The adult chimpanzee has much greater jaw strength than we have, and its teeth are proportionately larger. In contrast, the adult human has a skull with decidedly rounded, more fetus-like contours. Put another way, our skull is paedomorphic; it retains fetal features even after we are mature. Our large, paedomorphic skull is one of our most distinctive features. Our large, complex brain, which fills that bulbous skull, is another. The human

Activity 14D on the Web & CD
Morph baby chimps and humans into adults.

Figure 14.16 Paedomorphosis. Some species retain as adults features that were juvenile in ancestors. The axolotl, a salamander, becomes an adult and reproduces while retaining certain tadpole characteristics, including gills. It is an example of how changes in the genes controlling development can produce a very different organism.

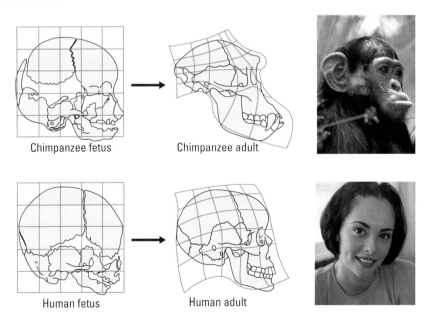

Chimpanzee fetus   Chimpanzee adult

Human fetus   Human adult

Figure 14.17 Comparison of human and chimpanzee skull development. Starting with fetal skulls that are very similar (left), the differential growth rates of different bones making up the skulls produce adult heads with very different proportions. The grid lines will help you relate the fetal skulls to the adult

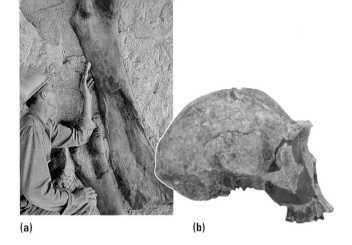

(a)

(b)

(c)

(d)

(e)

(f)

(g)

(h)

brain is proportionately larger than the chimpanzee brain because growth of the organ is switched off much later in human development. Compared to the brain of chimpanzees, our brain continues to grow for several more years, which can be interpreted as the prolonging of a juvenile process. We owe our culture to this evolutionary novelty, coupled with an extended childhood during which parents and teachers can influence what the young, growing brain stores.

CHECKPOINT

1. Explain why the concept of exaptation does not imply that a structure evolves *in anticipation of* some future environmental change.

2. How is the developmental cause of our large, roundish skulls similar to the developmental cause of the gills of axolotls?

Answers: 1. Although an exaptation is adapted for new or additional functions in a new environment, it existed because it worked as an adaptation to the old environment. **2.** Both structures are paedomorphic (juvenile anatomies retained into adulthood).

Earth History and Macroevolution

Having examined mechanisms of speciation and the origin of evolutionary novelty, we are ready to turn our attention to the history of biological diversity. Macroevolution is closely tied to the history of Earth.

Geologic Time and the Fossil Record

The fossil record is an archive of macroevolution. **Figure 14.18** surveys the diverse ways that organisms can fossilize. Sedimentary rocks are the richest sources of fossils and provide a record of life on Earth in their layers, or strata (see Chapter 13). The trapping of dead organisms in sediments freezes fossils in time. Thus, the fossils in each stratum of sedimentary rock are a local sample of the organisms that existed at the time the sediment was deposited. Because younger sediments are superimposed on older ones, the layers of sediment tell the relative ages of fossils.

By studying many different sites, geologists have established a **geologic time scale,** reflecting a consistent sequence of geologic periods (**Table 14.1**).

Figure 14.18 A gallery of fossils. **(a)** Sedimentary rocks are the richest hunting grounds for paleontologists, scientists who study the fossil record. This researcher is excavating a fossilized dinosaur skeleton from sandstone in Dinosaur National Monument, located in Utah and Colorado. **(b)** The hard parts of organisms are the most common fossils. This is a skull of *Homo erectus,* an ancestor of humans that lived about 1.5 million years ago. **(c)** Fossils may become even harder if minerals replace their organic matter. These petrified (stone) trees in the Petrified Forest National Park in Arizona are about 190 million years old. **(d)** Some sedimentary fossils, such as this 40-million-year-old leaf, retain organic material, including DNA, which scientists can analyze. **(e)** Buried organisms that decay may leave molds that are filled by minerals dissolved in water. The casts that form when the minerals harden are replicas of the organisms, as in the case of these casts of shelled marine animals called ammonites. **(f)** Trace fossils are footprints, burrows, and other remnants of an ancient organism's behavior. A dinosaur left these footprints in a creek bed in what is now Oklahoma. **(g)** This 30-million-year-old scorpion is embedded in amber (hardened resin from a tree). **(h)** These tusks belong to a whole 23,000-year-old mammoth, which scientists discovered in Siberian ice in 1999.

| Table 14.1 | The Geologic Time Scale |

Relative time span of eras	Era	Period	Epoch	Age (millions of years ago)	Some important events in the history of life
Cenozoic	Cenozoic	Quaternary	Recent		Historical time
Mesozoic				0.01	
			Pleistocene		Ice ages; humans appear
Paleozoic				1.8	
		Tertiary	Pliocene		Apelike ancestors of humans appear
				5	
			Miocene		Continued radiation of mammals and angiosperms
				23	
			Oligocene		Origins of many primate groups, including apes
				35	
			Eocene		Angiosperm dominance increases; origins of most modern mammalian orders
				57	
			Paleocene		Major radiation of mammals, birds, and pollinating insects
				65	
	Mesozoic	Cretaceous			Flowering plants (angiosperms) appear; many groups of organisms, including most dinosaur lineages, become extinct at end of period (Cretaceous extinctions)
				144	
		Jurassic			Gymnosperms continue as dominant plants; dinosaurs dominant
				206	
		Triassic			Cone-bearing plants (gymnosperms) dominate landscape; radiation of dinosaurs, early mammals, and birds
				245	
Pre-cambrian	Paleozoic	Permian			Extinction of many marine and terrestrial organisms (Permian extinctions); radiation of reptiles; origins of mammal-like reptiles and most modern orders of insects
				290	
		Carboniferous			Extensive forests of vascular plants; first seed plants; origin of reptiles; amphibians dominant
				363	
		Devonian			Diversification of bony fishes; first amphibians and insects
				409	
		Silurian			Diversity of jawless fishes; first jawed fishes; colonization of land by vascular plants and arthropods
				439	
		Ordovician			Origin of plants; marine algae abundant
				510	
		Cambrian			Origin of most modern animal phyla (Cambrian explosion)
				543	
	Precambrian			600	Diverse soft-bodied invertebrate animals; diverse algae
				700	Oldest animal fossils
				2,200	Oldest eukaryotic fossils
				2,700	Oxygen begins accumulating in atmosphere
				3,500	Oldest fossils known (prokaryotes)
				4,600	Approximate time of origin of Earth

These periods are grouped into four eras: the Precambrian, Paleozoic, Mesozoic, and Cenozoic eras. Each era represents a distinct age in the history of Earth and its life. The boundaries are marked in the fossil record by explosive diversification of many new forms of life. Mass extinctions also mark many of the boundaries between periods and between eras. For example, the beginning of the Cambrian period is delineated by a great diversity of fossilized animals that are absent in rocks of the late Precambrian. And most of the animals that lived during the late Precambrian became extinct at the end of that era.

Fossils are reliable historical documents only if we can determine their ages. The record of the rocks chronicles the *relative* ages of fossils. It tells us the order in which groups of species evolved. However, the series of sedimentary rocks does not tell the absolute ages of the embedded fossils. The difference is analogous to peeling the layers of wallpaper from the walls of a very old house that has been inhabited by many owners. You could determine the sequence in which the wallpapers had been applied, but not the year that each layer was added. Paleontologists use a variety of methods to determine the ages of fossils in years. The most common method is **radiometric dating,** which is based on the decay of radioactive isotopes (**Figure 14.19**). The dates you see on the geologic time scale in Table 14.1 were established by the radiometric dating of rocks and the fossils they contain.

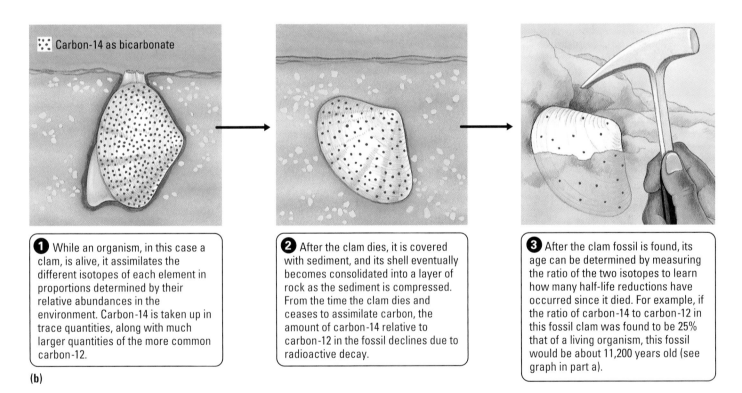

(a)

(b)

❶ While an organism, in this case a clam, is alive, it assimilates the different isotopes of each element in proportions determined by their relative abundances in the environment. Carbon-14 is taken up in trace quantities, along with much larger quantities of the more common carbon-12.

❷ After the clam dies, it is covered with sediment, and its shell eventually becomes consolidated into a layer of rock as the sediment is compressed. From the time the clam dies and ceases to assimilate carbon, the amount of carbon-14 relative to carbon-12 in the fossil declines due to radioactive decay.

❸ After the clam fossil is found, its age can be determined by measuring the ratio of the two isotopes to learn how many half-life reductions have occurred since it died. For example, if the ratio of carbon-14 to carbon-12 in this fossil clam was found to be 25% that of a living organism, this fossil would be about 11,200 years old (see graph in part a).

Figure 14.19 Radiometric dating.
Amounts of radioactive isotopes can be measured by the radiation they emit as they decompose to more stable atoms (see Chapter 2). The decay rate is unaffected by temperature, pressure, and other environmental variables. Paleontologists use this clock-like decay to date fossils. **(a)** Carbon-14 is a radioactive isotope with a half-life of 5,600 years. From the time an organism dies, it takes 5,600 years for half of the radioactive carbon-14 to decay; half of the remainder is present in another 5,600 years; and so on. Because the half-life of carbon-14 is relatively short, this isotope is only useful for dating fossils less than about 50,000 years old. To date older fossils, paleontologists use radioactive isotopes with longer half-lives. Radiometric dating has an error factor of less than 10%. Other methods for dating fossils generally confirm the ages determined by radiometric dating. **(b)** Here we use carbon-14 dating to determine the vintage of a fossilized clam shell.

Continental Drift and Macroevolution

The continents are not locked in place. They drift about Earth's surface like passengers on great plates of crust floating on the hot, underlying mantle (see Evolution Connection in Chapter 2). Unless two landmasses are embedded in the same plate, their positions relative to each other change. For example, North America and Europe are presently drifting apart at a rate of about 2 cm per year. Many important geologic processes, including mountain building, volcanic activity, and earthquakes, occur at plate boundaries. California's infamous San Andreas Fault is at the border where two plates slide past each other.

Plate movements rearrange geography constantly, but two chapters in the continuing saga of continental drift had an especially strong influence on life. About 250 million years ago, near the end of the Paleozoic era, plate movements brought all the landmasses together into a supercontinent that has been named Pangaea ("all land") (**Figure 14.20**). Imagine some of the possible effects on life. Species that had been evolving in isolation came together and competed. When the landmasses coalesced, the total amount of shoreline was reduced. There is also evidence that the ocean basins increased in depth, which lowered sea level and drained the shallow coastal seas. Then, as now, most marine species inhabited shallow waters, and the formation of Pangaea destroyed a considerable amount of that habitat. It was probably a long, traumatic period for terrestrial life as well. The continental interior, which has a drier and more erratic climate than coastal regions, increased in area substantially when the land came together. Changing ocean currents also would have affected land life as well as sea life. The formation of Pangaea had a tremendous environmental impact that reshaped biological diversity by causing extinctions and providing new opportunities for the survivors, which diversified through branching evolution.

The second dramatic chapter in the history of continental drift was written about 180 million years ago, during the Mesozoic era. Pangaea began to break up, causing geographic isolation of colossal proportions. As the continents drifted apart, each became a separate evolutionary arena, and the organisms of the different biogeographic realms diverged.

The pattern of continental separations is the solution to many biogeographic puzzles. For example, paleontologists have discovered matching fossils of Mesozoic reptiles in Ghana (West Africa) and Brazil. These two parts of the world, now separated by 3,000 km of ocean, were contiguous during the early Mesozoic era. Continental drift also explains much about the current distribution of organisms, such as why the Australian fauna and flora contrast so sharply with that of the rest of the world (see Figure 13.9).

Mass Extinctions and Explosive Diversifications of Life

The evolutionary road from ancient to modern life has not been smooth. The fossil record reveals an episodic history, with long, relatively stable periods punctuated by briefer intervals when the turnover in species composition was much more extensive. These biological makeovers include mass extinctions as well as explosive diversifications of certain forms of life.

As discussed at the start of the chapter, the world lost an enormous number of species—more than half of its marine animals and many groups of terrestrial plants and animals, including dinosaurs—at the end of the Cretaceous period, about 65 million years ago.

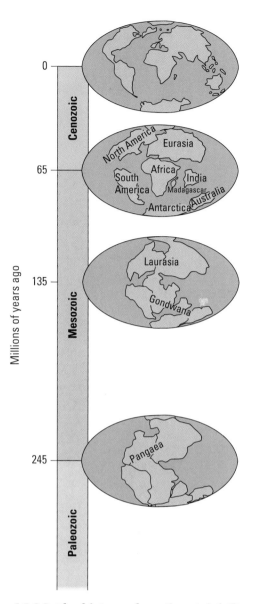

Figure 14.20 The history of continental drift.
Pangaea formed about 250 million years ago. About 180 million years ago, Pangaea began to split into northern (Laurasia) and southern (Gondwana) landmasses, which later separated into the modern continents. India collided with Eurasia just 10 million years ago, forming the Himalayas, the tallest and youngest of Earth's mountain ranges. The continents continue to drift, though not at a rate that's likely to cause any motion sickness for their passengers.

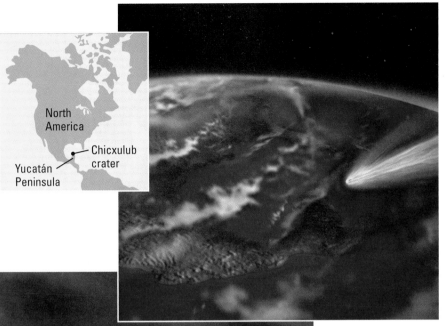

Figure 14.21 Trauma for planet Earth and its Cretaceous life. The 65-million-year-old Chicxulub impact crater is located in the Caribbean Sea near the Yucatán Peninsula of Mexico. The horseshoe shape of the crater (see Figure 14.1) and the pattern of debris in sedimentary rocks indicate that an asteroid or comet struck at a low angle from the southeast. This artist's interpretation represents the impact and its immediate effect—a cloud of hot vapor and debris that could have killed most of the plants and animals in North America within minutes. That would explain the higher extinction rates of land animals and plants in North America than elsewhere around the globe. This proposed impact scenario may have been only one of a series of events, including a global cooling trend, that contributed to the Cretaceous crisis.

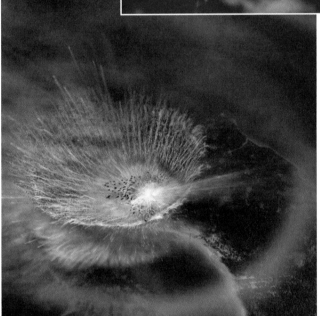

Many scientists believe that the impact that produced the Chicxulub crater could have caused global climatic changes and mass extinctions (**Figure 14.21**). Other researchers propose that climatic changes due to continental drift could have caused the extinctions, whether an asteroid collided with Earth or not. Still others point to evidence in the fossil record in India indicating that during the late Cretaceous, massive volcanic activity released particles into the atmosphere, blocking sunlight and thereby contributing to climatic cooling. The various hypotheses are not mutually exclusive, and researchers continue to debate the extent to which each environmental change contributed to the extinctions.

Extinction is inevitable in a changing world. A species may become extinct because its habitat has been destroyed or because of unfavorable climatic changes. Extinctions occur all the time, but extinction rates have not been steady. There have been six distinct periods of mass extinction over the last 600 million years. During these times, losses escalated to nearly six times the average rate. Of all the mass extinctions, the ones marking the ends of the Cretaceous and Permian periods have been the most intensively studied. The Permian extinctions, at about the time the continents merged to form Pangaea, claimed over 90% of the species of marine animals and took a tremendous toll on terrestrial life as well. Whatever their causes, mass extinctions affect biological diversity profoundly.

But there is a creative side to the destruction. Each massive dip in species diversity has been followed by explosive diversification of certain survivors. Extinctions seem to have provided the surviving organisms with new environmental opportunities. For example, mammals existed for at least 75 million years before undergoing an explosive increase in diversity just after the Cretaceous. Their rise to prominence was undoubtedly associated with the void left by the extinction of dinosaurs. The world would be a very different place today if many of the dinosaur lineages had escaped the Cretaceous extinctions or if none of the mammals that lived in the Cretaceous had survived. In addition to new diversifications in the wake of mass extinctions, explosive diversification can also result from some major new evolutionary innovation, such as flight.

Activity 14E on the Web & CD
Scroll through the history of life on Earth.

Classifying the Diversity of Life

Reconstructing evolutionary history is part of the science of **systematics,** the study of biological diversity, past and present. Systematics includes **taxonomy,** which is the identification, naming, and classification of species.

Some Basics of Taxonomy

Assigning scientific names to species is an essential part of systematics. Common names, such as monkey, fruit fly, crayfish, and garden pea, may work well in everyday communication, but they can be ambiguous because there are many species of each of these organisms. Biology's more formal taxonomic system dates back to Carolus Linnaeus (1707–1778), a Swedish physician and botanist (plant specialist). The system has two main characteristics: a two-part name for each species and a hierarchical classification of species into broader and broader groups of organisms.

Naming Species Linnaeus's system assigns to each species a two-part latinized name, or **binomial.** The first part of a binomial is the **genus** (plural, *genera*) to which the species belongs. The second part of a binomial refers to one **species** within the genus. An example of a binomial is *Panthera pardus,* the scientific name of the large cat we commonly call the leopard. Notice that the first letter of the genus is capitalized and that the whole binomial is italicized and latinized (you can name a bug you discover after a friend, but you must add the appropriate Latin ending). Our own scientific name, *Homo sapiens,* which Linnaeus made up in a show of optimism, means "wise man."

Hierarchical Classification In addition to defining and naming species, a major objective of systematics is to group species into broader taxonomic categories. The first step of such a hierarchical classification is built into the binomial for a species. We group species that are closely related into the same genus. For example the leopard, *Panthera pardus,* belongs to a genus that also includes the African lion (*Panthera leo*) and the tiger (*Panthera tigris*). Grouping species is natural for us, at least in concept. We lump together several trees we know as oaks and distinguish them from several other species of trees we call maples. Indeed, oaks and maples belong to separate genera. Biology's taxonomic scheme formalizes our tendency to group related objects as a way of structuring our view of the world.

Beyond the grouping of species within genera, taxonomy extends to progressively broader categories of classification. It places similar genera in

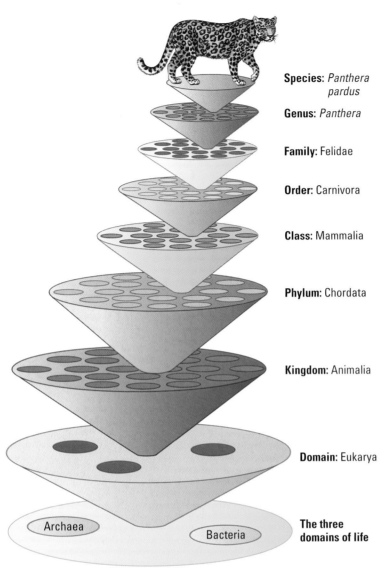

Species: *Panthera pardus*

Genus: *Panthera*

Family: Felidae

Order: Carnivora

Class: Mammalia

Phylum: Chordata

Kingdom: Animalia

Domain: Eukarya

Archaea

Bacteria

The three domains of life

Figure 14.22 Hierarchical classification.
The taxonomic scheme classifies species into groups belonging to more comprehensive groups. The leopard (*Panthera pardus*) belongs to the genus *Panthera,* which also includes the African lion and tiger. These wild felines belong to the cat family (Felidae), along with the genus *Felis,* which includes the domestic cat and several closely related wild cats, such as the lynx. Family Felidae belongs to the order Carnivora, which also includes the dog family, Canidae, and several other families. Order Carnivora, the carnivores, is grouped with many other orders in the class Mammalia, the mammals. Mammalia is one of several classes belonging to the phylum Chordata in the kingdom Animalia, which is part of the domain Eukarya.

the same **family,** puts families into **orders,** orders into **classes,** classes into **phyla** (singular, *phylum*), phyla into **kingdoms,** and kingdoms into **domains.** (see Chapter 1 to review the three domains of life.) **Figure 14.22** places the leopard in this taxonomic scheme of groups within groups. Classifying a species by kingdom, phylum, and so on, is analogous to sorting mail, first by zip code and then by street, house number, and specific member of the household.

Classification and Phylogeny

Ever since Darwin, systematics has had a goal beyond simple organization: to have classification reflect **phylogeny,** which is the evolutionary history of a species. As a systematist classifies species into groups subordinate to other groups in the taxonomic hierarchy, the final product takes on the branching pattern of a **phylogenetic tree.** The principle of common descent is reflected in the branches of the tree (**Figure 14.23**).

Sorting Homology from Analogy Homologous structures are one of the best sources of information about phylogenetic relationships. Recall from Chapter 13 that homologous structures may look different and function very differently in different species, but they exhibit fundamental similarities because they evolved from the same structure that existed in a common ancestor. Among the vertebrates, for instance, the whale limb is adapted for steering in the water; the bat wing is adapted for flight. Nonetheless, there is a basic similarity in the bones supporting these two structures (see Figure 13.10). The greater the number of homologous structures between two species, the more closely the species are related.

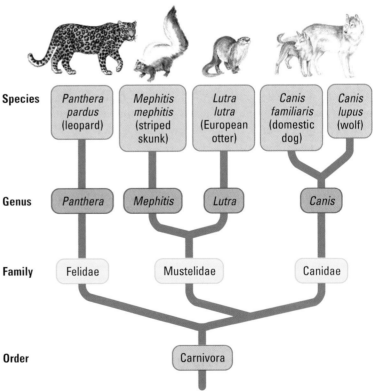

Species	*Panthera pardus* (leopard)	*Mephitis mephitis* (striped skunk)	*Lutra lutra* (European otter)	*Canis familiaris* (domestic dog)	*Canis lupus* (wolf)
Genus	*Panthera*	*Mephitis*	*Lutra*	*Canis*	
Family	Felidae	Mustelidae		Canidae	
Order			Carnivora		

Figure 14.23 The relationship of classification and phylogeny for some members of the order Carnivora.
The hierarchical classification is reflected in the finer and finer branching of the phylogenetic tree. Each branch point in the tree represents an ancestor common to species above that branch point.

There are pitfalls in the search for homology: Not all likeness is inherited from a common ancestor. Species from different evolutionary branches may have certain structures that are superficially similar if natural selection has shaped analogous adaptations. This is called **convergent evolution.** Similarity due to convergence is termed **analogy,** not homology. For example, the wings of insects and those of birds are analogous flight equipment: They evolved independently and are built from entirely different structures.

To develop phylogenetic trees and classify organisms according to evolutionary history, we must use only homologous similarities. This guideline is simpler in principle than it is in practice. Adaptation can obscure homologies, and convergence can create misleading analogies. As we saw in Chapter 13, comparing the embryonic development of the species in question can often expose homology that is not apparent in the mature structures.

There is another clue to identifying homology and sorting it from analogy: The more complex two similar structures are, the less likely it is they evolved independently. Compare the skulls of a human and a chimpanzee, for example (see Figure 14.17). The skulls are not single bones, but a fusion of many, and the two skulls match almost perfectly, bone for bone. It is highly improbable that such complex structures matching in so many details could have separate origins. Most likely, the genes required to build these skulls were inherited from a common ancestor.

Molecular Biology as a Tool in Systematics If homology is about common ancestry, then comparing the genes and gene products (proteins) of organisms gets right to the heart of their evolutionary relationships. Sequences of nucleotides in DNA are inherited, and they program corresponding sequences of amino acids in proteins (see the Evolution Connection in Chapter 3). At the molecular level, the evolutionary divergence of species parallels the accumulation of differences in their genomes. The more recently two species have branched from a common ancestor, the more similar their DNA and amino acid sequences should be.

Today, both the amino acid sequences for many proteins and the nucleotide sequences for a rapidly increasing number of genomes are in databases that are available via the Internet (see Chapter 12). This has catalyzed a boom in systematics as researchers use the databases to compare the hereditary information of different species in search of homology at its most basic level.

> **Case Study in the Process of Science on the Web & CD**
> Analyze proteins to build phylogenetic trees.

Molecular systematics provides a new way to test hypotheses about the phylogeny of species. The strongest support for any such hypothesis is agreement between molecular data and other means of tracing phylogeny, such as evaluating anatomical homology and analyzing the fossil record. And speaking of fossils, some are preserved in such a way that it is possible to extract DNA fragments for comparison with modern organisms (**Figure 14.24**).

The Cladistic Revolution Systematics entered a vigorous new era in the 1960s. Just as molecular methods became readily available for comparing species, computer technology helped usher in a new approach called **cladistic analysis.**

Cladistic analysis is the scientific search for clades (from the Greek word for "branch"). A clade consists of an ancestor and all its descendants—a distinctive branch in the tree of life. The items in the clade may be species or

Figure 14.24 Tyrolean Ice Man, a source of prehistoric human DNA. Hikers discovered the frozen 5,300-year-old Stone Age man in a melting glacier in the Alps in 1991. Some skepticism that the man was a fraudulently transplanted South American mummy persisted until 1994. Then researchers analyzed DNA fragments recovered from the fossil. The DNA closely matches that of central northern Europeans, not Native Americans. As researchers continue to analyze the Stone Age man's DNA, they may uncover more clues about his place in human evolution.

Figure 14.25 A simplified example of cladistic analysis.

Cladistic analysis requires a comparison between a so-called in-group and an out-group. In our example, the three mammals make up the in-group, while the turtle, a reptile, is the out-group. This provides a reference point for distinguishing primitive characters from derived characters. A primitive character is a homology present in all the organisms being compared and so must also have been present in the common ancestor. All four animals in our example have vertebral columns (backbones), which is a primitive character. The derived characters, such as hair and gestation, are the evolutionary innovations that define the sequence of branch points in the phylogeny of the in-group. ❶ Hair and mammary glands are among the homologies that distinguish mammals from reptiles. ❷ Among the derived characters defining the next branch point is gestation, the carrying of offspring in the womb of the female parent. The duck-billed platypus, though a mammal, lays eggs, as do turtles and other reptiles. We can infer that gestation evolved in an ancestor common to the kangaroo and beaver that is more recent than the ancestor shared by all mammals, including the platypus. ❸ The last branch point in our cladistic analysis is defined by the much longer gestation of the beaver compared to the kangaroo (in which the embryo emerges from the womb very early and completes its development within the mother's pouch).

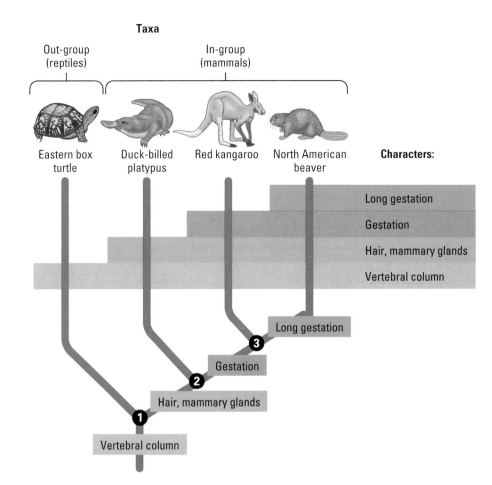

Figure 14.26 How cladistic analysis is shaking phylogenetic trees.

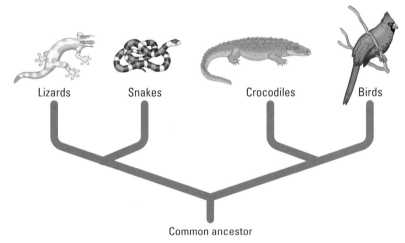

Strict application of cladistic analysis sometimes produces phylogenetic trees that conflict with classical taxonomy. Most biologists (and this textbook) will probably continue to place reptiles and birds in separate vertebrate classes, but the tree you see here is more consistent with cladistic analysis. Based on derived characters, crocodiles are more closely related to birds than they are to lizards and snakes. Birds and crocodiles make up one clade, and lizards and snakes make up another. Or if we go back as far as the ancestor that crocodiles share with lizards and snakes to make up a clade, then the class Reptilia must also include birds. The counterargument is that the numerous flight adaptations of birds justify separating them taxonomically from the reptiles. In other words, classical taxonomy factors in the degree of evolutionary divergence between birds and the organisms we commonly think of as reptiles.

higher taxonomic groups, such as classes or phyla. Identifying clades centers on the analysis of homologies unique to each group (**Figure 14.25**). In other words, cladistic analysis focuses on the evolutionary innovations that define the branch points in evolution. Identifying clades makes it possible to construct classification schemes that reflect phylogeny.

Cladistic analysis has become the most widely used method in systematics. Its strict application underscores some taxonomic dilemmas. For instance, biologists traditionally place birds and reptiles in separate classes of vertebrate animals (class Aves and class Reptilia, respectively). This classification, however, is inconsistent with a strict application of cladistic analysis. An inventory of homologies suggests that crocodiles are more closely related to birds than they are to lizards and snakes (**Figure 14.26**).

Arranging Life into Kingdoms: A Work in Progress

Phylogenetic trees are hypotheses about evolutionary history. Like all hypotheses, they are revised, or in some cases completely rejected, in accordance with new evidence. Molecular systematics and cladistic analysis are combining to remodel phylogenetic trees and challenge conventional classifications, even at the kingdom level.

Biologists have traditionally considered the kingdom to be the highest—the most inclusive—taxonomic category. Many of us grew up with the notion that there are only two kingdoms of life—plants and animals—because we live in a macroscopic, terrestrial realm where we are rarely aware of organisms that do not fit neatly into a plant-animal dichotomy. The two-kingdom system also had a long tradition in formal taxonomy; Linnaeus divided all known forms of life between the plant and animal kingdoms. The two-kingdom system prevailed in biology for over 200 years, but it was beset with problems. Where do prokaryotes fit in such a system? Can they be considered members of the plant kingdom? And what about fungi?

In 1969, American ecologist Robert H. Whittaker argued effectively for a **five-kingdom system.** It places prokaryotes in the kingdom Monera (**Figure 14.27a**). Organisms of the other four kingdoms all consist of eukaryotic cells (see Chapter 4). Kingdoms Plantae, Fungi, and Animalia consist of multicellular eukaryotes that differ in structure, development, and modes of nutrition. Plants make their own food by photosynthesis. Fungi live by decomposing the remains of other organisms and absorbing small organic molecules. Most animals live by ingesting food and digesting it within their bodies.

The kingdom Protista contains all eukaryotes that do not fit the definitions of plant, fungus, or animal. Protista is a taxonomic grab bag in the five-kingdom system. Most protists are unicellular. Amoebas and other so-called protozoa are examples. But kingdom Protista also includes certain large, multicellular organisms that are believed to be direct descendants of unicellular protists. For example, many biologists now classify the seaweeds as protists because they are more closely related to certain single-celled algae than they are to true plants.

In the last decade, molecular studies and cladistic analysis have led to some reevaluation of the five-kingdom system. **Figure 14.27b** shows a **three-domain system** as one alternative to the five-kingdom system. This newer scheme recognizes three basic groups: two domains of prokaryotes, Bacteria and Archaea, and one domain of eukaryotes, called Eukarya. The domains Bacteria

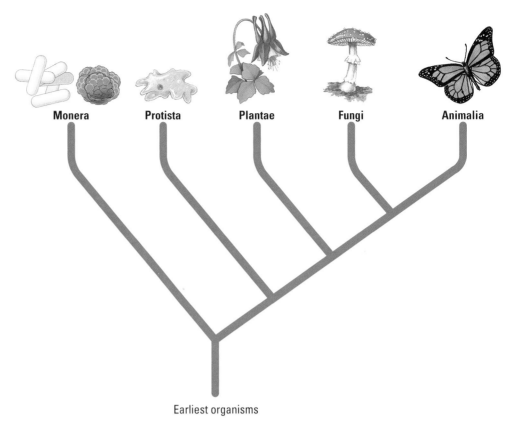

(a) The five-kingdom classification scheme

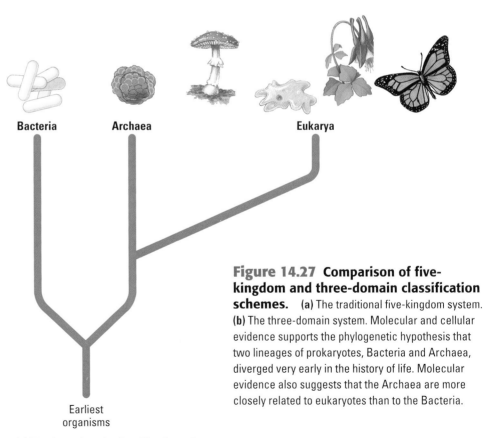

Figure 14.27 Comparison of five-kingdom and three-domain classification schemes. **(a)** The traditional five-kingdom system. **(b)** The three-domain system. Molecular and cellular evidence supports the phylogenetic hypothesis that two lineages of prokaryotes, Bacteria and Archaea, diverged very early in the history of life. Molecular evidence also suggests that the Archaea are more closely related to eukaryotes than to the Bacteria.

(b) The three-domain classification scheme

and Archaea differ in a number of important structural, biochemical, and functional features, which we will discuss in Chapter 15.

What is most important to understand here is that classifying Earth's diverse species of life will always be a work in progress as we learn more about organisms and their evolution. Charles Darwin envisioned the goals of modern systematics when he wrote in *The Origin of Species*, "Our classifications will come to be, as far as they can be so made, genealogies."

Activity 14F on the Web & CD
Review the history of classification schemes.

CHECKPOINT

1. How much of the classification in Figure 14.22 do we share with the leopard?

2. Our forearms and the wings of a bat are derived from the same ancestral prototype, and thus they are _____. In contrast, the wings of a bat and the wings of a bee are derived from totally unrelated structures, and thus they are _____.

3. To distinguish a particular clade of mammals within the larger clade that corresponds to class Mammalia, why is hair not a useful characteristic?

4. In comparing the five-kingdom system with the three-domain system, how many of the kingdoms correspond to domain Eukarya?

Answers: 1. We are classified the same down to the class level: both the cat and human are mammals. We do not belong to the same order. **2.** homologous; analogous **3.** Hair is a primitive character common to *all* mammals and cannot be helpful in distinguishing different mammalian subgroups. **4.** Four

Evolution Connection

Just a Theory?

So far in our study of evolution, we have come across several key debates. Classification schemes (five kingdoms versus three domains) and the tempo of evolution (gradualism versus punctuated equilibrium) are two examples. These and other controversies stimulate research and lead to refinements in evolutionary theory. But such debates among evolutionary biologists also seem to fuel the "just a theory" argument that certain groups use to pressure politicians to discredit or eliminate the study of evolution in the science curriculum in public schools.

The "just a theory" tactic for nullifying biology's unifying theme has two flaws. First, it fails to separate Darwin's two main points: the evidence that modern species evolved from ancestral forms and the theory that natural selection is the main mechanism for this evolution. To biologists, Darwin's theory of evolution is natural selection—the mechanism Darwin proposed to explain the historical record of evolution documented by fossils, biogeography, and other types of evidence.

This brings us to the second flaw in the "just a theory" argument. The term *theory* has a very different meaning in science than in general use. The colloquial use of "theory" comes close to what scientists mean by "hypothesis." In science, a theory is more comprehensive than a hypothesis (see Chapter 1). A theory, such as Newton's theory of gravity or Darwin's theory

**Evolution Connection
on the Web**
Explore a website that
explains common mis-
conceptions about
evolution.

of natural selection, accounts for many facts and attempts to explain a great variety of phenomena. Such
a unifying theory does not become widely accepted
in science unless its predictions stand up to thorough
and continual testing by experiments and observations. In other words, a comprehensive theory, to
be accepted, is held to a higher standard of evidence than a hypothesis. Even
then, science does not allow theories to become dogma. For example, many
evolutionary biologists now question whether natural selection alone accounts for the evolutionary history observed in the fossil record. Unpredictable events, such as asteroids crashing into Earth, may be as important
as adaptive evolution in the patterns of biological diversity we observe.

The study of evolution is more robust than ever as a branch of science,
and questions about how life evolves in no way imply that most biologists
consider evolution itself to be "just a theory." Debates about evolutionary
theory are like arguments over competing theories about gravity; we know
that objects keep right on falling while we debate the cause.

By attributing the diversity of life to natural causes rather than to supernatural forces, Darwin gave biology a sound, scientific basis. Nevertheless,
the diverse products of evolution are elegant and inspiring. As Darwin said
in the closing paragraph of *The Origin of Species,* "There is grandeur in this
view of life."

Chapter Review

Summary of Key Concepts

For study help, go to the Essential Biology Website (www.essentialbiology.com)
or CD-ROM to explore the Activities and Case Studies in the Process of Science.

Macroevolution and the Diversity of Life

• Microevolution is a change in the gene pool of a population; macroevolution is a change in life-forms noticeable enough to be evident in the fossil
record. Macroevolution includes the multiplication of species, the origin of
evolutionary novelty, the explosive diversification that follows some evolutionary breakthroughs, and mass extinctions.

Activity 14A Mechanisms of Macroevolution

The Origin of Species

• **What Is a Species?**   According to the biological species concept, a species
is a population or group of populations whose members have the potential to
interbreed with one another in nature to produce fertile offspring, but who
cannot successfully interbreed with members of other species.

• **Reproductive Barriers Between Species**

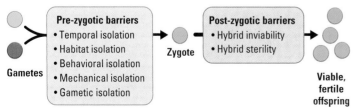

• **Mechanisms of Speciation**

When the gene pool of a population is sev-
ered from other populations of the parent species, the splinter population can
follow its own evolutionary course.

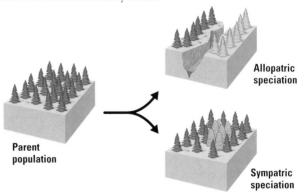

Hybridization leading to polyploids is a common mechanism of sympatric
speciation in plants.

Activity 14B Exploring Speciation on Islands

Activity 14C Polyploid Plants

Case Study in the Process of Science *How Do New Species Arise by Genetic
Isolation?*

- **What Is the Tempo of Speciation?** According to the punctuated equilibrium model, the time required for speciation in most cases is relatively short compared to the overall duration of the species' existence. This accounts for the relative rarity of transitional fossils linking newer species to older ones.

The Evolution of Biological Novelty

- **Adaptation of Old Structures for New Functions** An exaptation is a structure that evolves in one context and gradually becomes adapted for other functions.

- **"Evo-Devo": Development and Evolutionary Novelty** A subtle change in the genes that control a species' development can have profound effects. In paedomorphosis, for example, the adult retains juvenile body features.

Activity 14D Paedomorphosis: Morphing Chimps and Humans

Earth History and Macroevolution

- **Geologic Time and the Fossil Record** Geologists have established a geologic time scale divided into four eras: Precambrian, Paleozoic, Mesozoic, and Cenozoic. The most common method for determining the ages of fossils is radiometric dating.

- **Continental Drift and Macroevolution** About 250 million years ago, plate movements brought all the landmasses together into the supercontinent Pangaea, causing extinctions and providing new opportunities for the survivors to diversify. About 180 million years ago, Pangaea began to break up, causing geographic isolation.

- **Mass Extinctions and Explosive Diversifications of Life** The fossil record reveals long, relatively stable periods punctuated by mass extinctions followed by explosive diversification of certain survivors. For example, during the Cretaceous extinctions, about 65 million years ago, the world lost an enormous number of species, including dinosaurs. Mammals greatly increased in diversity after the Cretaceous.

Activity 14E The Geologic Time Scale

Classifying the Diversity of Life

- Systematics, the study of biological diversity, includes taxonomy, which is the identification, naming, and classification of species.

- **Some Basics of Taxonomy** Each species is assigned a two-part name consisting of the genus and the species. In the taxonomic hierarchy, domain > kingdom > phylum > class > order > family > genus > species.

- **Classification and Phylogeny** The goal of classification is to reflect phylogeny, which is the evolutionary history of a species. Classification is based on the fossil record, homologous structures, comparisons of DNA and amino acid sequences, and cladistic analysis.

Case Study in the Process of Science How Is Phylogeny Determined Using Protein Comparisons?

- **Arranging Life into Kingdoms: A Work in Progress** Biologists have reevaluated the five-kingdom system and currently classify life into a three-domain system.

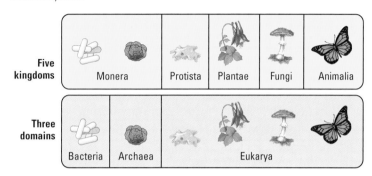

Activity 14F Classification Schemes

Self-Quiz

1. Bird guides once listed the myrtle warbler and Audubon's warbler as distinct species that lived side by side in parts of their ranges. However, recent books show them as eastern and western forms of a single species, the yellow-rumped warbler. Apparently, it has been found that the two kinds of warblers
   a. live in the same areas.
   b. successfully interbreed.
   c. are almost identical in appearance.
   d. are merging to form a single species.

2. Label each of the following reproductive barriers as pre-zygotic or post-zygotic:
   a. One lilac species lives on acid soil, another on basic soil.
   b. Mallard and pintail ducks mate at different times of year.
   c. Two species of leopard frogs have different mating calls.
   d. Hybrid offspring of two species of jimsonweed always die before reproducing.
   e. Pollen of one kind of tobacco cannot fertilize another kind.

3. Why is a small, isolated population more likely to undergo speciation than a large one?

4. Many species of plants and animals adapted to desert conditions probably did not arise there. Their success in living in deserts could be due to _____, the evolution of a structure in one context that becomes adapted for different functions.

5. Mass extinctions that occurred in the past
   a. cut the number of species to the few survivors left today.
   b. resulted mainly from the separation of the continents.
   c. occurred regularly, about every million years.
   d. were followed by diversification of the survivors.

6. The animals and plants of India are almost completely different from the species in nearby Southeast Asia. Why might this be true?
   a. They have become separated by convergent evolution.
   b. The climates of the two regions are completely different.
   c. India is in the process of separating from the rest of Asia.
   d. India was a separate continent until relatively recently.

7. Place these hierarchies of classification in order from least inclusive to most inclusive: class, domain, family, genus, kingdom, order, phylum, species.

8. A paleontologist estimates that when a particular rock formed, it contained 12 mg of the radioactive isotope potassium-40. The rock now contains 3 mg of potassium-40. The half-life of potassium-40 is 1.3 billion years. From this information, you can conclude that the rock is approximately _____ billion years old.

9. If you were using cladistic analysis to build a phylogenetic tree of cats, which of the following would be the best choice for an out-group? (*Hint:* See Figure 14.25 legend.)

    a. lion   c. wolf

    b. domestic cat   d. tiger

10. In the three-domain system, which two domains contain prokaryotic organisms?

Answers to the Self-Quiz questions can be found in Appendix B.

Go to the website or CD-ROM for more Self-Quiz questions.

The Process of Science

1. Distinguish "hypothesis" from "theory" in the vocabulary of science.

2. Imagine you are conducting fieldwork and discover two groups of mice living on opposite sides of a river. Assuming that you will not disturb the organisms, design a study to determine whether or not these two groups belong to the same species. If you could capture some of the mice and return them to the lab, would that affect your experimental design?

Case Study in the Process of Science on the Web & CD *Investigate how new species arise by genetic isolation.*

Case Study in the Process of Science on the Web & CD *Analyze proteins to build phylogenetic trees.*

Biology and Society

Experts estimate that human activities cause the extinction of hundreds of species every year. The natural rate of extinction is thought to be a few species per year. As we continue to alter the global environment, especially by cutting down tropical rain forests, the resulting extinction will probably rival that at the end of the Cretaceous period. Most biologists are alarmed at this prospect. What are some reasons for their concern? Consider that life has endured numerous mass extinctions and has always bounced back. How is the present mass extinction different from previous extinctions? What might be some of the consequences for the surviving species?

Biology and Society on the Web *Learn about the evidence for an asteroid impact 250 million years ago that may have contributed to mass extinctions at that time.*

CHAPTER 15

The Evolution of Microbial Life

**Biology and Society:
Bioterrorism   297**

**Major Episodes in the
History of Life   297**

The Origin of Life   300

Resolving the Biogenesis Paradox

*A Four-Stage Hypothesis for the
Origin of Life*

*From Chemical Evolution to
Darwinian Evolution*

Prokaryotes   303

They're Everywhere!

*The Two Main Branches of
Prokaryotic Evolution: Bacteria
and Archaea*

*The Structure, Function, and
Reproduction of Prokaryotes*

*The Nutritional Diversity of
Prokaryotes*

The Ecological Impact of Prokaryotes

Protists   311

The Origin of Eukaryotic Cells

The Diversity of Protists

**Evolution Connection: The Origin
of Multicellular Life   317**

The number of bacteria in one human's mouth is greater than the total number of people who ever lived.

Bacterial fermentation is used to produce cheese, yogurt, buttermilk, and many types of sausage.

Each year more than 200 million people become infected with malaria.

More than half of our antibiotics come from soil bacteria of the genus *Streptomyces*.

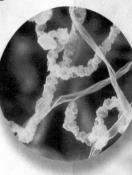

Bioterrorism

During the fall of 2001, five Americans died from the disease anthrax in a presumed terrorist attack (**Figure 15.1**). Unfortunately, this act of bioterrorism was hardly unprecedented; there is a long and ugly history of humans using organisms as weapons. Animals, plants, fungi, and viruses have all served this purpose, but the most frequently employed biowarfare agents have been bacteria.

During the Middle Ages, the bacterium *Yersinia pestis* (the cause of bubonic plague) played a role in battle when armies hurled the bodies of plague victims into enemy ranks. Early conquerors, settlers, and warring armies in South and North America gave native people items purposely contaminated with infectious bacteria. In the 1930s, the Japanese government instituted a biowarfare program that killed tens of thousands of Chinese soldiers and civilians using bacteria that cause plague, anthrax, and cholera. In 1984, members of an Oregon cult contaminated restaurant salad bars with the *Salmonella* bacteria (a cause of food poisoning); over 700 people became sick, and 45 were hospitalized. During the 1990s, another cult tried to start an anthrax epidemic in Tokyo, and the Iraqi army loaded missiles with harmful bacteria. Luckily, neither of these latter two attempts resulted in casualties.

Biology and Society on the Web
Learn more about the history of biowarfare.

What is the status of bioweapon research in the United States? The U.S. opened its first biological weapons research facility in 1943 at Fort Detrick, Maryland. There, the military studied and bred new strains of bacteria that cause such illnesses as anthrax, botulism, and tularemia. To "weaponize" naturally occurring pathogens, researchers selected highly virulent strains, made them resistant to antibiotic medications, and developed formulations for effective dispersion. But the practical difficulties of controlling such weapons—and a measure of ethical repugnance—led the U.S. to end this bioweapons program in 1969 and to order its products destroyed. In 1975, the U.S. signed the Biological Weapons Convention, pledging never to develop or store biological weapons. Eventually, 103 nations joined the ban, although not all signatories have honored it.

Of course, not all bacteria are harmful to humans. In fact, nearly all life on Earth depends on bacteria and other microbial life in one way or another. In this chapter, we'll examine the origins, structures, and functions of microbes. We'll begin by examining some of the key events in the history of life on Earth.

Figure 15.1 Cleaning up after a bioterrorist attack. This 2001 photo shows workers decontaminating a site where anthrax spores were released.

Major Episodes in the History of Life

Life began when Earth was young. The planet was born about 4.5 billion years ago, and its crust began to solidify about 4.0 billion years ago (see Evolution Connection in Chapter 2). A few hundred million years later, by 3.5 billion years ago, Earth was already inhabited by a diversity of organisms. Those earliest organisms were all prokaryotes, their cells lacking true nuclei (see Chapter 4). Within the next billion years, two distinct groups of prokaryotes—bacteria and archaea—diverged.

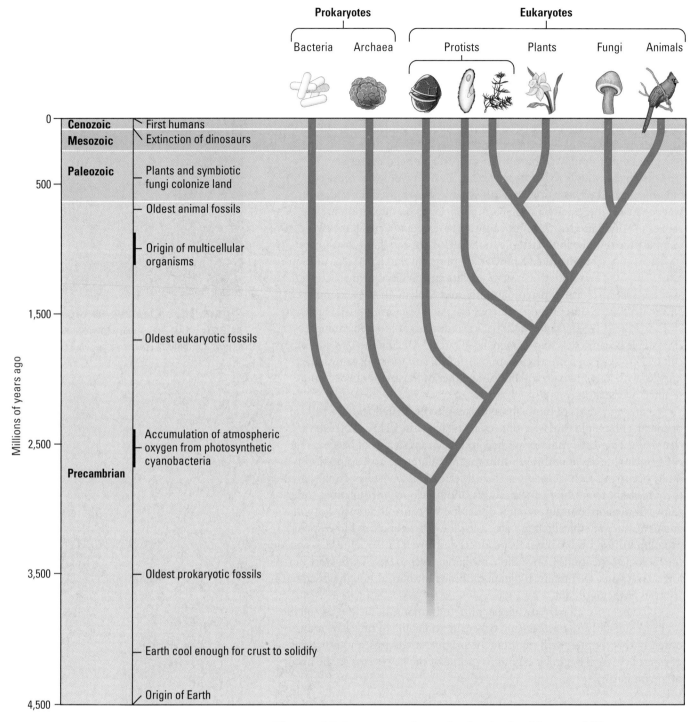

Prokaryotes
Bacteria   Archaea

Eukaryotes
Protists   Plants   Fungi   Animals

Millions of years ago

0	
Cenozoic	First humans
Mesozoic	Extinction of dinosaurs
Paleozoic	Plants and symbiotic fungi colonize land
500	Oldest animal fossils
	Origin of multicellular organisms
1,500	Oldest eukaryotic fossils
2,500	Accumulation of atmospheric oxygen from photosynthetic cyanobacteria
Precambrian	
3,500	Oldest prokaryotic fossils
	Earth cool enough for crust to solidify
4,500	Origin of Earth

Figure 15.2 Some major episodes in the history of life.
The timing of events is based on fossil evidence and molecular analysis.

An oxygen revolution began about 2.5 billion years ago (**Figure 15.2**). Photosynthetic prokaryotes that split water molecules released oxygen gas, changing Earth's atmosphere profoundly (see Evolution Connection in Chapter 7). The corrosive O_2 doomed many prokaryotic groups. Among the survivors, a diversity of metabolic modes evolved, including cellular respiration, which uses O_2 to extract energy from food (see Chapter 6). All of this metabolic evolution occurred during the almost 2 billion years that prokaryotes had Earth to themselves.

The oldest eukaryotic fossils are about 1.7 billion years old. Eukaryotic cells, remember, contain nuclei and many other organelles that are absent in prokaryotic cells (see Chapter 4). The eukaryotic cell evolved from a prokaryotic community, a host cell containing even smaller prokaryotes. The mitochondria of our cells and those of every other eukaryote are descendants of those smaller prokaryotes. And so are the chloroplasts of plants and algae.

The origin of more complex cells launched an explosive diversification of eukaryotic forms. They were the protists. Represented today by a great diversity of organisms, protists are mostly microscopic and unicellular. The organisms we call protozoans are protists. So are a great variety of single-celled algae, including diatoms.

The next great evolutionary "experiment" was multicellularity. The first multicellular eukaryotes evolved, perhaps a billion years ago, as colonies of single-celled ancestors. Their modern descendants include multicellular protists, such as seaweeds. Other evolutionary branches stemming from the ancient protists gave rise to animals, fungi, and plants.

The greatest diversification of animals was the so-called Cambrian explosion. The Cambrian was the first period of the Paleozoic era, which began about 570 million years ago. The earliest animals lived in late Precambrian seas, but they diversified extensively over a span of just 10 million years during the early Cambrian. In fact, all the major body plans (phyla) of animals had evolved by the end of that evolutionary eruption.

For over 85% of biological history—life's first 3 billion years—life was mostly confined to aquatic habitats. The colonization of land was a major milestone in the history of life. Plants, in the company of fungi, led the way about 475 million years ago. Even today, the roots of most plants are associated with fungi that aid in the absorption of water and minerals from soil. Plants transformed the landscape, creating new opportunities for all life-forms, especially herbivorous (plant-eating) animals and their predators.

The evolutionary venture onto land included vertebrate animals in the form of the first amphibians. These prototypes of today's frogs and sala-

Activity 15A on the Web & CD
Test your knowledge of the history of life.

manders descended from air-breathing fish with fleshy fins that could support the animal's weight on land. Reptiles evolved from amphibians, and birds and mammals evolved from reptiles. Among the mammals are the primates, the animal group that includes humans and their closest relatives, apes and monkeys. But trace our genealogy back far enough, and we count certain protists, and before them certain prokaryotes, as our ancestors. To understand life on Earth, we must go back to the origin and diversification of microbes.

CHECKPOINT

Put the following events in order, from the earliest to the most recent: diversification of animals (Cambrian explosion), evolution of eukaryotic cells, first humans, colonization of land by plants and fungi, origin of prokaryotes, evolution of land animals, evolution of multicellular organisms.

Answer: origin of prokaryotes, evolution of eukaryotic cells, evolution of multicellular organisms, diversification of animals (Cambrian explosion), colonization of land by plants and fungi, evolution of land animals, first humans

Figure 15.3 An artist's rendition of Earth about 3 billion years ago. The pad-like objects in the scene represent colonies of prokaryotes known from the fossil record.

Figure 15.4 Making organic molecules in a laboratory simulation of early-Earth chemistry.
A few decades after his famous experiments, Stanley Miller posed with a version of the contraption he built to simulate chemical dynamics on the early Earth. At the bottom of the photo, his hand cups a warmed flask containing the primordial "sea." Above, electrical discharge simulates lightning in an ancient "atmosphere" consisting of ammonia, hydrogen, methane, and water vapor. Glass tubes circulate the water between the "atmosphere" and the "sea." Products of chemical reactions in the "atmosphere" dissolve in the water and accumulate in the "sea" flask. And those products include amino acids and many other organic molecules that eventually color the "sea" a murky brown.

The Origin of Life

We will never know for sure, of course, how life on Earth began. But we must start with the assumption that science seeks *natural* causes for natural phenomena.

Resolving the Biogenesis Paradox

From the time of the ancient Greeks, it was common "knowledge" that life could arise from nonliving matter. This idea of life emerging from inanimate material is called **spontaneous generation,** and it persisted well into the nineteenth century as an explanation for the rapid growth of microorganisms in spoiled foods. Then in 1862, a series of experiments by Louis Pasteur confirmed what many others had suspected: All life today, including microbes, arises only by the reproduction of preexisting life. This "life-from-life" principle is called **biogenesis.**

But wait! What about the *first* organisms? If *they* arose by biogenesis, then they wouldn't be the first organisms. Although there is no evidence that spontaneous generation occurs today, conditions on the early Earth were very different. For instance, there was little atmospheric oxygen to tear apart complex molecules. And such energy sources as lightning, volcanic activity, and ultraviolet sunlight were all more intense than what we experience today (see Evolution Connection in Chapter 2). The resolution to the biogenesis paradox is that life did not begin on a planet anything like the modern Earth, but on a young Earth that was a very different world (**Figure 15.3**). Most biologists now think that it is at least a credible hypothesis that chemical and physical processes in Earth's primordial environment could eventually have produced very simple cells through a sequence of stages. Debate abounds about the nature of those steps.

A Four-Stage Hypothesis for the Origin of Life

According to one hypothetical scenario, the first organisms were products of chemical evolution in four stages: (1) the abiotic (nonliving) synthesis of small organic molecules, such as amino acids and nucleotides; (2) the joining of these small molecules (monomers) into polymers, including proteins and nucleic acids; (3) the origin of self-replicating molecules that eventually made inheritance possible; and (4) the packaging of all these molecules into pre-cells, droplets with membranes that maintained an internal chemistry different from the surroundings. This is all speculative, of course, but what makes it science is that the hypothesis leads to predictions that can be tested in the laboratory.

Stage 1: Abiotic Synthesis of Organic Monomers In 1953, Stanley Miller, a 23-year-old graduate student at the University of Chicago, made front-page news. Miller had built a contraption that simulated key conditions on the early Earth. His apparatus produced a variety of small organic molecules that are essential for life, including amino acids, the monomers of proteins (**Figure 15.4**). In the half century since Miller's seminal experiments, many scientists have repeated and extended the research, varying such conditions as the composition of the ancient "atmosphere" and "sea." These laboratory analogues of the primeval Earth have produced all 20 amino acids, several sugars, lipids, the nucleotides that are the monomers of DNA and RNA, and even ATP, the molecule that powers most biological work. The abiotic synthesis of organic molecules on the early Earth is certainly a plausible scenario.

It is also plausible that some organic compounds reached Earth from space. In 2000, Indian scientists reported that their computer models showed how molecules such as adenine, an ingredient of DNA, could form by reactions of cyanide in the clouds of gas between stars. These simulations would explain why some meteorites that have crashed to Earth contain organic molecules. But whether the primordial Earth was stocked with organic monomers made here or elsewhere, the key point is that the molecular ingredients of life were probably present very early.

Stage 2: Abiotic Synthesis of Polymers The hypothesis of an abiotic origin of life makes another prediction that can be tested in the laboratory. If the hypothesis is correct, then it should be possible to link organic monomers to form polymers such as proteins and nucleic acids without the help of enzymes and other cellular equipment. Researchers have observed such polymerization by dripping solutions of organic monomers onto hot sand, clay, or rock. The heat vaporizes the water in the solutions and concentrates the monomers on the underlying substance. Some of the monomers then spontaneously bond together to form polymers, including polypeptides, the chains of amino acids that make up proteins. On the early Earth, raindrops or waves may have splashed dilute solutions of organic monomers onto fresh lava or other hot rocks and then rinsed polypeptides and other polymers back into the sea. Alternatively, deep-sea vents, where gases and superheated water with dissolved minerals escape from Earth's interior, may have been locales for the abiotic synthesis of organic monomers and polymers. (You'll learn about these unique ecosystems in Chapter 19.)

Stage 3: Origin of Self-Replicating Molecules Life is defined partly by the process of inheritance, which is based on self-replicating molecules. Today's cells store their genetic information as DNA. They transcribe the information into RNA and then translate RNA messages into specific enzymes and other proteins (see Chapter 10). This mechanism of information flow probably emerged gradually through a series of refinements to much simpler processes. What were the first genes like?

One hypothesis is that the first genes were short strands of RNA that replicated without the assistance of enzymes (proteins). In laboratory experiments, short RNA molecules can assemble spontaneously from nucleotide monomers without the presence of cells or enzymes (**Figure 15.5**). The result is a population of RNA molecules, each with some random sequences of monomers. Some of the molecules self-replicate, but they vary in their success at this reproduction. What happens can only be described as molecular evolution in a test tube. The RNA varieties that replicate fastest increase their frequency in the population.

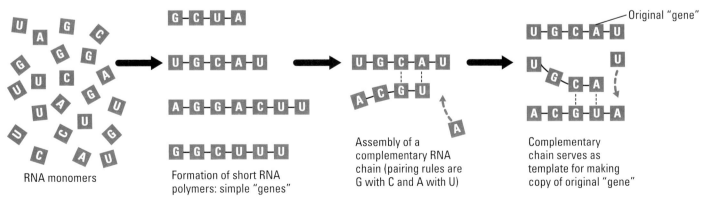

Figure 15.5 Abiotic replication of RNA "genes."

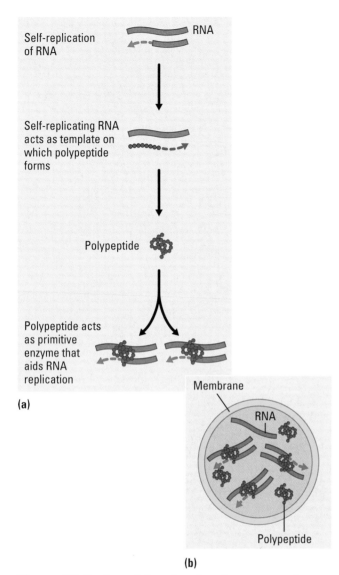

(a)

Self-replication of RNA

↓

Self-replicating RNA acts as template on which polypeptide forms

↓

Polypeptide

↓

Polypeptide acts as primitive enzyme that aids RNA replication

Membrane

RNA

Polypeptide

(b)

Figure 15.6 The origin of molecular cooperation in the prebiotic world. **(a)** Possible origin of cooperation between nucleic acids and polypeptides. **(b)** Molecular cooperation within a membrane-enclosed compartment. In such a pre-cell, RNA genes would benefit exclusively from their protein products rather than sharing them with other molecular complexes.

Figure 15.7 Laboratory versions of pre-cells.
(a) These tiny spheres are made by cooling solutions of polypeptides that are produced abiotically. The spheres grow by absorbing more polypeptides until they reach an unstable size, when they split to form "daughter" spheres. Of course, this division lacks the precision of cellular reproduction. **(b)** Lipids mixed with water self-assemble into these membrane-enclosed droplets (see Evolution Connection in Chapter 4). In some cases, large droplets bud to give rise to smaller "offspring."

In addition to the experimental evidence, there is another reason the idea of RNA genes in the primordial world is attractive. Cells actually have catalytic RNAs, which are called **ribozymes.** Perhaps early ribozymes catalyzed their own replication. That would help with the "chicken and egg" paradox of which came first, enzymes or genes. Maybe the "chicken and egg" came together in the same RNA molecules. The molecular biology of today may have been preceded by an "RNA world."

Stage 4: Formation of Pre-Cells The properties of life emerge from an interaction of molecules organized into higher levels of order. For example, **Figure 15.6a** diagrams one way that RNA and polypeptides could have cooperated in the prebiotic world. But such molecular teams would be much more efficient if they were packaged within membranes that kept the molecules close together in a solution that could be different from the surrounding environment. We'll call these molecular aggregates pre-cells—not really cells, but molecular packages with some of the properties of life (**Figure 15.5b**).

Laboratory experiments demonstrate that pre-cells could have formed spontaneously from abiotically produced organic compounds (**Figure 15.7**). Some of the pre-cells are coated by membranes that are selectively permeable; osmosis causes the droplets to swell or shrink when placed in solutions of different salt concentrations. Certain types of pre-cells store energy in the form of a voltage across their membranes. They can even discharge the voltage much as nerve cells do. If specific enzymes are included among the ingredients that assemble into pre-cells, the aggregates display a rudimentary metabolism; they absorb substrates from their surroundings and release the products of the reactions.

Case Study in the Process of Science on the Web & CD Perform virtual experiments on the four hypothetical stages of the origin of life.

From Chemical Evolution to Darwinian Evolution

If pre-cells with self-replicating RNA, and later DNA, did form on the young Earth, they would be refined by natural selection—Darwinian evolution. Mutations, errors in the copying of the "genes," would result in variation among the droplets. And the most successful of these droplets would grow, divide (reproduce), and continue to evolve. Of course, the gap between such pre-cells and even the simplest of true cells is enormous. But with millions of years of incremental refinement through natural selection, these molecular cooperatives could have become more and more cell-like. The point at which we stop calling membrane-enclosed units that metabolize and replicate genetic programs pre-cells and start calling them living cells is as fuzzy as our definition of life. But we do know that prokaryotes were already flourishing at least 3.5 billion years ago and that all branches of life arose from those ancient prokaryotes.

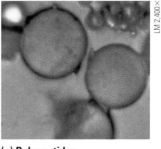

(a) Polypeptides

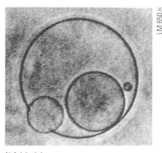

(b) Lipids

Prokaryotes

The history of prokaryotic life is a success story spanning billions of years (**Figure 15.8**). Prokaryotes lived and evolved all alone on Earth for 2 billion years. They have continued to adapt and flourish on an evolving Earth and in turn have helped to change Earth. In this section, you will become more familiar with prokaryotes by studying their diversity and ecological significance.

They're Everywhere!

Today, prokaryotes are found wherever there is life, and they outnumber all eukaryotes combined. More prokaryotes inhabit a handful of fertile soil or the mouth or skin of a human than the total number of people who have ever lived. Prokaryotes also thrive in habitats too cold, too hot, too salty, too acidic, or too alkaline for any eukaryote. In 1999, biologists even discovered prokaryotes growing on the walls of a gold mine 2 miles below Earth's surface.

Though individual prokaryotes are relatively small organisms, they are giants in their collective impact on Earth and its life. We hear most about a few species that cause serious illness. During the fourteenth century, Black Death—bubonic plague, a bacterial disease—spread across Europe, killing an estimated 25% of the human population. Tuberculosis, cholera, many sexually transmissible diseases, and certain types of food poisoning are some other human diseases caused by bacteria.

However, prokaryotic life is no rogues' gallery. Far more common than harmful bacteria are those that are benign or beneficial. Bacteria in our intestines provide us with important vitamins, and others living in our mouth prevent harmful fungi from growing there. Prokaryotes also recycle carbon and other vital chemical elements back and forth between organic matter and the soil and atmosphere. For example, there are prokaryotes that decompose dead organisms. Found in soil and at the bottom of lakes, rivers, and oceans, these decomposers return chemical elements to the environment in the form of inorganic compounds that can be used by plants, which in turn feed animals. If prokaryotic decomposers were to disappear, the chemical cycles that sustain life would come to a halt. All forms of eukaryotic life would also be doomed. In contrast, prokaryotic life would undoubtedly persist in the absence of eukaryotes, as it once did for 2 billion years.

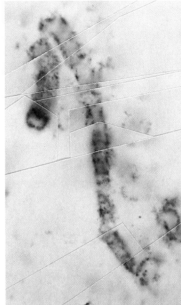

LM 1,300×

(a) Fossilized prokaryotes

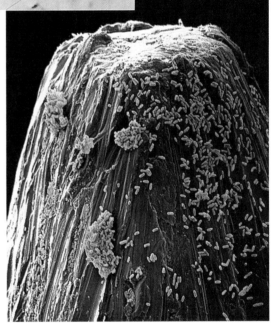

Colorized SEM 550×

(b) Modern-day prokaryotes

Figure 15.8 Over 3 billion years of prokaryotes.
(a) This microscopic fossil is a filamentous species consisting of a chain of prokaryotic cells. It is one of a diversity of prokaryotes found in western Australian rocks that are about 3.5 billion years old. **(b)** The orange rods are individual modern bacteria, each about 5 µm long, on the head of a pin. Most prokaryotic cells have diameters in the range of 1–10 µm, much smaller than most eukaryotic cells (typically 10–100 µm). This micrograph will help you understand why a pin prick can cause infection. And it will help you remember to flame the tip of a needle before using it to remover a splinter. The heat kills the bacteria.

The Two Main Branches of Prokaryotic Evolution: Bacteria and Archaea

Prokaryotes have a cellular organization fundamentally different from that of eukaryotes. Whereas eukaryotic cells have a membrane-enclosed nucleus and numerous other membrane-enclosed organelles, prokaryotic cells lack these structural features (see Chapter 4). The traditional five-kingdom classification scheme emphasizes this fundamental difference in cellular organization. Prokaryotes make up the kingdom Monera, separate from the four eukaryotic kingdoms (Protista, Plantae, Fungi, and Animalia). In the past two decades, however, researchers have learned that a single prokaryotic kingdom may not fit evolutionary history. By comparing genomes of diverse prokaryotes, biologists have identified two major branches of prokaryotic evolution: the **bacteria** and the **archaea.** Though they have prokaryotic cell organization in common, bacteria and archaea differ in many structural, biochemical, and physiological characteristics. And there is also evidence that archaea are more closely related to eukaryotes than they are to bacteria. It was these discoveries that prompted the three-domain classification—domains Bacteria, Archaea, and Eukarya—which you can review in Figure 14.27b. The majority of prokaryotes are bacteria, but the archaea are worth studying for their evolutionary and ecological significance.

The term *archaea* ("ancient") refers to the antiquity of the group's origin from the earliest cells. Even today, most species of archaea inhabit extreme environments, such as hot springs and salt ponds. Few other modern organisms (if any) can survive in these environments, which may resemble habitats on the early Earth.

Biologists refer to some archaea as "extremophiles," meaning "lovers of the extreme." There are extreme halophiles ("salt lovers"), archaea that thrive in such environments as Utah's Great Salt Lake and seawater-evaporating ponds used to produce salt (**Figure 15.9**). There are also extreme thermophiles ("heat lovers") that live in very hot water; some even populate the deep-ocean vents that gush superheated water hotter than 100°C, the boiling point of water at sea level. And then there are the methanogens, archaea that live in anaerobic environments and give off methane as a waste product. They are abundant in the anaerobic mud at the bottom of lakes and swamps. You may have seen methane, also called marsh gas, bubbling up from a swamp. Great numbers of methanogens also inhabit the digestive tracts of animals. In humans, intestinal gas is largely the result of their metabolism. More importantly, methanogens aid digestion in cattle, deer, and other animals that depend heavily on cellulose for their nutrition. Normally, bloating does not occur because these animals regularly belch out large volumes of gas produced by the methanogens and other microorganisms that enable them to utilize cellulose. And that may be more than you wanted to know about these gas-producing microbes.

The Structure, Function, and Reproduction of Prokaryotes

You can use Figure 4.5 to review the general structure of prokaryotic cells. Note again the absence of a true nucleus and the other membrane-enclosed organelles characteristic of the much more complex eukaryotic cells. Another feature to note in prokaryotes is that nearly all species have cell walls exterior to their plasma membranes. These walls are chemically different from the cellulose walls of plant cells. Some

Figure 15.9 Extreme halophiles ("salt-loving" archaea).   These are seawater-evaporating ponds at the edge of San Francisco Bay. The colors of the ponds result from dense growth of the prokaryotes that thrive when the salinity of the water reaches 15–20% (before evaporation, seawater has a salt concentration of about 3%). The ponds are used for commercial salt production; the halophilic archaea are harmless.

Activity 15B on the Web & CD
Review prokaryotic cell structure and function.

antibiotics, including the penicillins, kill certain bacteria by incapacitating an enzyme the microbes use to make their walls.

Determining cell shape by microscopic examination is an important step in identifying prokaryotes (**Figure 15.10**). Spherical species are called **cocci** (singular, *coccus*), from the Greek word for "berries." Cocci that occur in clusters are called staphylococci ("cluster of grapes"), or staph for short (as in "staph infections"). Other cocci occur in chains; they are called streptococci ("twisted grapes"). The bacterium that causes strep throat in humans is a streptococcus. Rod-shaped prokaryotes are called **bacilli** (singular, *bacillus*). A third group of prokaryotes are curved or spiral-shaped. The largest spiral-shaped prokaryotes are called **spirochetes.** The bacterium that causes syphilis, for example, is a spirochete. The culprit that causes Lyme disease is also a spirochete.

Most prokaryotes are unicellular and very small, but there are exceptions to both of these generalizations. Some species tend to aggregate transiently into groups of two or more cells, such as the streptococci already mentioned. Others form true colonies, which are permanent aggregates of identical cells (**Figure 15.11a**). And some species even exhibit a simple multicellular organization, with a division of labor between specialized types of cells (**Figure 15.11b**). Among unicellular species, there are some giants that actually dwarf most eukaryotic cells (**Figure 15.11c**).

(a) Cocci (spheres)

Colorized SEM 8,000×

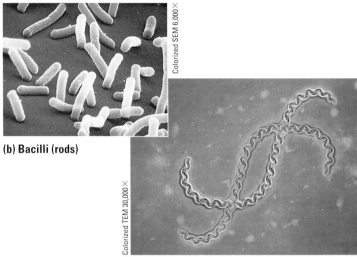

(b) Bacilli (rods)

Colorized SEM 6,000×

Colorized TEM 30,000×

(c) Spirochetes (spirals)

Figure 15.10 Three common shapes of prokaryotes.

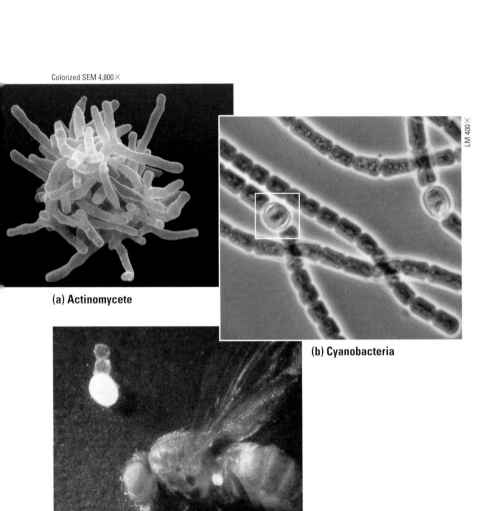

Colorized SEM 4,800×

(a) Actinomycete

LM 400×

(b) Cyanobacteria

20×

**(c) Giant bacterium
(and fruit fly for comparison)**

Figure 15.11 Some examples of bacterial diversity.
(a) This prokaryotic organism, called an actinomycete, is a mass of branching chains of rod-shaped cells. These bacteria are very common in soil, where they break down organic substances. The filaments enable the organism to bridge dry gaps between soil particles. Most species secrete antibiotics, which inhibit the growth of competing bacteria. Pharmaceutical companies use various species of actinomycetes to produce antibiotic drugs, including streptomycin. **(b)** This filamentous prokaryote belongs to a photosynthetic group called the cyanobacteria. Many species are truly multicellular in having a division of labor among specialized cells. The box on this micrograph highlights a cell that converts atmospheric nitrogen to ammonia, which can then be incorporated into amino acids and other organic compounds. **(c)** There are actually some prokaryotic cells that are gigantic, even by eukaryotic standards. The bright ball in this photo is a marine bacterium discovered in 1997 (the two smaller spheres above it are dead cells). This prokaryotic cell is over half a millimeter in diameter, about the size of a fruit fly's head.

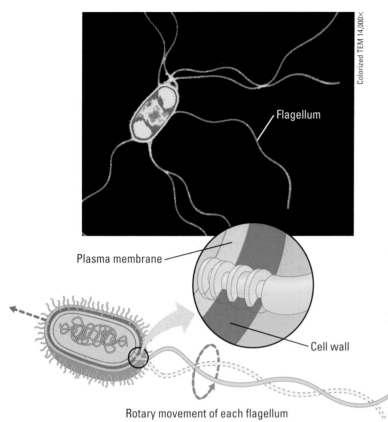

Plasma membrane

Cell wall

Rotary movement of each flagellum

Flagellum

Colorized TEM 14,000×

Figure 15.12 Prokaryotic flagella. These locomotor appendages are entirely different in structure and mechanics from the eukaryotic flagella discussed in Chapter 4. At the base of the prokaryotic version is a motor and set of rings embedded in the plasma membrane and cell wall. This machinery actually spins like a wheel, rotating the filament of the flagellum.

About half of all prokaryotic species are motile. Many of those travelers have one or more flagella that propel the cells away from unfavorable places or toward more favorable places, such as nutrient-rich locales (**Figure 15.12**).

Although few bacteria can thrive in the extreme environments favored by many archaea, some bacteria can survive extended periods of very harsh conditions by forming specialized "resting" cells, or **endospores** (**Figure 15.13**). Some endospores can remain dormant for centuries. Not even boiling water kills most of these resistant cells. To sterilize laboratory equipment, microbiologists use an appliance called an autoclave, a pressure cooker that kills endospores by heating to a temperature of 121°C (250°F) with high-pressure steam. The food-canning industry uses similar methods to kill endospores of dangerous soil bacteria such as *Clostridium botulinum*, which produces a toxin that causes the potentially fatal disease botulism.

Most prokaryotes can reproduce at a phenomenal rate if conditions are favorable. The cells copy their DNA almost continuously and divide again and again by the process called **binary fission.** To understand how this makes explosive population growth possible, flash back to that childhood numbers game: "Would you rather have a million dollars or start out with just a penny and have it doubled every day for a month?" If you opt for the penny, you feel like a loser at midmonth, when you have only a few hundred dollars. But by the end of the month, you've bagged about 10 million bucks. This is the exponential growth that repeated doublings make possible. Now apply that concept to bacterial reproduction, except double the number every 20 minutes. That's the rate at which some bacteria can divide if there is plenty of food and space. In just 24 hours, a single tiny cell could give rise to a bacterial colony equivalent in mass to about 15,000 humans! Fortunately, few bacterial populations can sustain exponential growth for long. Environments are usually limiting in resources

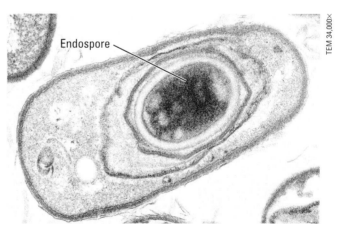

Endospore

TEM 34,000×

Figure 15.13 An anthrax endospore. This prokaryote is *Bacillus anthracis,* the notorious bacterium that produces the deadly disease called anthrax in cattle, sheep, and humans. There are actually two cells here, one inside the other. The outer cell produced the specialized inner cell, called an endospore. The endospore has a thick, protective coat. Its cytoplasm is dehydrated, and the cell does not metabolize. Under harsh conditions, the outer cell may disintegrate, but the endospore survives all sorts of trauma, including lack of water and nutrients, extreme heat or cold, and most poisons. When the environment becomes more hospitable, the endospore absorbs water and resumes growth. In late 2001, one or more bioterrorists disseminated anthrax spores through the U.S. postal system.

such as food and space. The bacteria also produce metabolic waste products that may eventually pollute the colony's environment. Still, you can understand why certain bacteria can make you sick so soon after just a few cells infect you—or why food can spoil so rapidly. Refrigeration retards food spoilage not because the cold kills the bacteria on food but because most microorganisms reproduce only very slowly at such low temperatures.

The Nutritional Diversity of Prokaryotes

Prokaryotic evolution "invented" every type of nutrition we observe throughout life, plus some nutritional modes unique to prokaryotes. Nutrition refers here to how an organism obtains two resources for synthesizing organic compounds: energy and a source of carbon. Species that use light energy are termed phototrophs. Chemotrophs obtain their energy from chemicals taken from the environment. If an organism needs only the inorganic compound carbon dioxide (CO_2) as a carbon source, it is called an autotroph. Heterotrophs require at least one organic nutrient—the sugar glucose, for instance—as a source of carbon for making other organic compounds (see Chapter 6). We can combine the phototroph-versus-chemotroph (energy source) and autotroph-versus-heterotroph (carbon source) criteria to group organisms according to four major modes of nutrition:

Case Study in the Process of Science on the Web & CD
See if you can provide the ideal energy source and carbon source for various prokaryotes.

1. **Photoautotrophs** are photosynthetic organisms that harness light energy to drive the synthesis of organic compounds from CO_2. Among the diverse groups of photosynthetic prokaryotes are the cyanobacteria, such as the species in Figure 15.11b. All photosynthetic eukaryotes—plants and algae—also fit into this nutritional category.

2. **Chemoautotrophs** need only CO_2 as a carbon source. However, instead of using light for energy, these prokaryotes extract energy from certain inorganic substances, such as hydrogen sulfide (H_2S) or ammonia (NH_3). This mode of nutrition is unique to certain prokaryotes. For example, prokaryotic species living around the hot-water vents deep in the seas are the main food producers in those bizarre ecosystems (see Chapter 19).

3. **Photoheterotrophs** can use light to generate ATP but must obtain their carbon in organic form. This mode of nutrition is restricted to certain prokaryotes.

4. **Chemoheterotrophs** must consume organic molecules for both energy and carbon. This nutritional mode is found widely among prokaryotes, certain protists, and even some plants. And all fungi and animals are chemoheterotrophs.

Table 15.1 reviews the four major modes of nutrition.

The Ecological Impact of Prokaryotes

Organisms as pervasive, abundant, and diverse as prokaryotes are guaranteed to have tremendous impact on Earth and all its inhabitants. Here we survey just a few examples of prokaryotic clout.

Bacteria That Cause Disease We are constantly exposed to bacteria, some of which are potentially harmful (**Figure 15.14**). Bacteria and other microorganisms that cause disease are called **pathogens.** Most of us are

Table 15.1	Nutritional Classification of Organisms	
Nutritional Type	**Energy Source**	**Carbon Source**
Photoautotroph (photosynthesizer)	Sunlight	CO_2
Chemoautotroph	Inorganic chemicals	CO_2
Photoheterotroph	Sunlight	Organic compounds
Chemoheterotroph	Organic compounds	Organic compounds

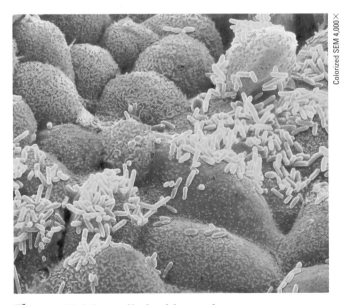

Colorized SEM 4,000×

Figure 15.14 Really bad bacteria. The yellow rods are *Haemophilus influenzae* bacteria on skin cells lining the interior of a human nose. These pathogens are transmitted through the air. *H. influenzae,* not to be confused with influenza (flu) viruses, causes pneumonia and other lung infections that kill about 4 million people worldwide per year. Most victims are children in developing countries, where malnutrition lowers resistance to all pathogens.

healthy most of the time only because our body defenses check the growth of pathogens. Occasionally, the balance shifts in favor of a pathogen, and we become ill. Even some of the bacteria that are normal residents of the human body can make us sick when our defenses have been weakened by poor nutrition or by a viral infection.

Most pathogenic bacteria cause disease by producing poisons. There are two classes of these poisons: exotoxins and endotoxins. **Exotoxins** are poisonous proteins secreted by bacterial cells. A single gram of the exotoxin that causes botulism could kill a million people. Another exotoxin producer is *Staphylococcus aureus* (abbreviated *S. aureus*). It is a common, usually harmless resident of our skin surface. However, if *S. aureus* enters the body through a cut or other wound or is swallowed in contaminated food, it can cause serious diseases. One type of *S. aureus* produces exotoxins that cause layers of skin to slough off; another can cause vomiting and severe diarrhea; yet another can produce the potentially deadly toxic shock syndrome.

In contrast to exotoxins, **endotoxins** are not cell secretions but are chemical components of the cell walls of certain bacteria. All endotoxins induce the same general symptoms: fever, aches, and sometimes a dangerous drop in blood pressure (shock). The severity of symptoms varies with the host's condition and with the bacterium. Different species of *Salmonella,* for example, produce endotoxins that cause food poisoning and typhoid fever.

During the last 100 years, following the nineteenth-century discovery that "germs" cause disease, the incidence of bacterial infections has declined, particularly in developed nations. Sanitation is generally the most effective way to prevent bacterial disease. The installation of water treatment and sewage systems continues to be a public health priority throughout the world. Antibiotics have been discovered that can cure most bacterial diseases. However, resistance to widely used antibiotics has evolved in many of these pathogens (see Chapter 1).

In addition to sanitation and antibiotics, a third defense against bacterial disease is education. A case in point is Lyme disease, currently the most widespread pest-carried disease in the United States. The disease is caused by a spirochete bacterium carried by ticks that live on deer and field mice (**Figure 15.15**). Lyme disease usually starts as a red rash shaped like a bull's-eye around a tick bite. Antibiotics can cure the disease if administered within about a month after exposure. If untreated, Lyme disease can cause debilitating arthritis, heart disease, and nervous disorders. A vaccine is now available, but it does not give full protection. The best defense is public education about avoiding tick bites and the importance of seeking treatment if a rash develops. When walking through brush, using insect repellent and wearing light-colored clothing can reduce contact with ticks.

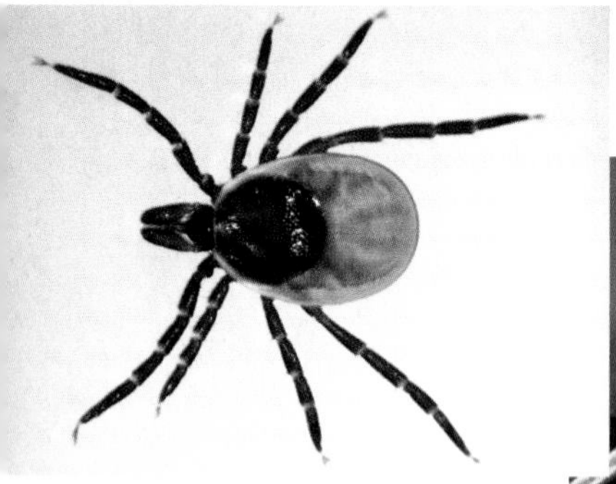

Pathogenic bacteria are in the minority among prokaryotes. Far more common are species that are essential to our well-being, either directly or indirectly. Let's turn our attention now to the vital role that prokaryotes play in sustaining the biosphere.

(a) Tick that carries the Lyme disease bacterium

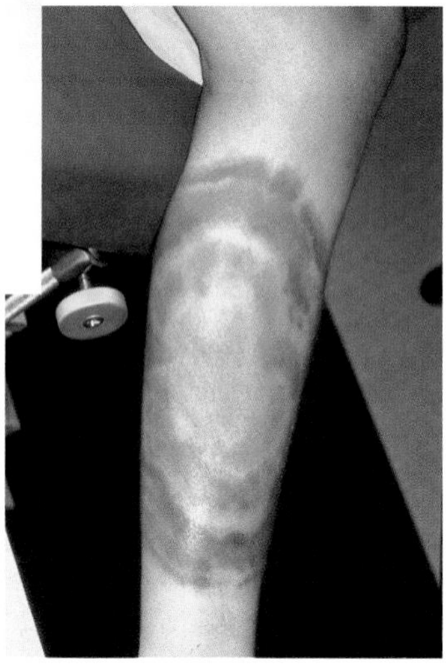

(b) "Bull's-eye" rash

Figure 15.15 Lyme disease, a bacterial disease transmitted by ticks.

Prokaryotes and Chemical Recycling Not too long ago, in geologic terms, the atoms of the organic molecules in your body were parts of the inorganic compounds of soil, air, and water, as they will be again. Life depends on the recycling of chemical elements between the biological and physical components of ecosystems. Prokaryotes play essential roles in these chemical cycles. For example, cyanobacteria not only restore oxygen to the atmosphere; some of them also convert nitrogen gas (N_2) in the atmosphere to nitrogen compounds that plants can absorb from

Activity 15C on the Web & CD
Explore the diversity of prokaryotes.

soil and water (see Figure 15.11b). Other prokaryotes, including bacteria living within the roots of bean plants and other legumes, also contribute large amounts of nitrogen compounds to soil. In fact, all the nitrogen that plants use to make proteins and nucleic acids comes from prokaryotic metabolism in the soil. In turn, animals get their nitrogen compounds from plants.

Another vital function of prokaryotes is the breakdown of organic wastes and dead organisms. Prokaryotes decompose organic matter and, in the process, return elements to the environment in inorganic forms that can be used by other organisms. If it were not for such decomposers, carbon, nitrogen, and other elements essential to life would become locked in the organic molecules of corpses and waste products. You'll learn more about the roles that prokaryotes play in chemical cycling in Chapter 19.

Prokaryotes and Bioremediation Humans have put the metabolically diverse prokaryotes to work in cleaning up the environment. The use of organisms to remove pollutants from water, air, and soil is called **bioremediation.** The most familiar example of bioremediation is the use of prokaryotic decomposers to treat our sewage. Raw sewage is first passed through a series of screens and shredders, and solid matter is allowed to settle out from the liquid waste. This solid matter, called sludge, is then gradually added to a culture of anaerobic prokaryotes, including both bacteria and archaea. The microbes decompose the organic matter in the sludge, converting it to material that can be used as landfill or fertilizer after chemical sterilization. Liquid wastes are treated separately from the sludge (**Figure 15.16**).

We are just beginning to explore the great potential that prokaryotes offer for bioremediation. Certain bacteria that occur naturally on ocean beaches can decompose petroleum and are useful in cleaning up oil spills (**Figure 15.17**). Genetically engineered bacteria may be able to degrade oil more rapidly than the naturally occurring oil-eaters. Bacteria may also help us clean up old mining sites. The water that drains from mines is highly acidic and is also laced with poisons—often compounds of arsenic, copper, zinc, and the heavy

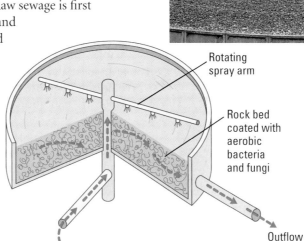

Rotating spray arm

Rock bed coated with aerobic bacteria and fungi

Liquid wastes

Outflow

Figure 15.16 Putting prokaryotes to work in sewage treatment facilities.   This is a trickling filter system, one type of mechanism for treating liquid wastes after sludge is removed. The long horizontal pipes rotate slowly, spraying liquid wastes through the air onto a thick bed of rocks. Bacteria and fungi growing on the rocks remove much of the organic material dissolved in the waste. Outflow from the rock bed is sterilized and then released, usually into a river or ocean.

Figure 15.17 Treatment of an oil spill in Alaska. The workers are spraying fertilizers onto an oil-soaked beach. The fertilizers stimulate growth of naturally occurring bacteria that initiate the breakdown of the oil. This technique is the fastest and least expensive way yet devised to clean up spills on beaches. Of course, it would be much better to keep oil off the beaches in the first place!

metals lead, mercury, and cadmium. Contamination of our soils and groundwater by these toxic substances poses a widespread threat, and cleaning up the mess is extremely expensive. Although there are no simple solutions to the problem, prokaryotes may be able to help. Bacteria called *Thiobacillus* thrive in the acidic waters that drain from mines. Some mining companies use these microbes to extract copper and other valuable metals from low-grade ores. While obtaining energy by oxidizing sulfur or sulfur-containing compounds, the bacteria also accumulate metals from the mine waters. Unfortunately, their use in cleaning up mine wastes is limited because their metabolism also adds sulfuric acid to the water. If this problem is solved, perhaps through genetic engineering, *Thiobacillus* and other prokaryotes may help us overcome some environmental dilemmas that seem intractable today. One current research focus is a bacterium that tolerates radiation doses thousands of times stronger than those that would kill people. This species may help clean up toxic dump sites that include radioactive wastes.

It is the nutritional diversity of prokaryotes that makes such benefits as chemical recycling and bioremediation possible. The various modes of nutrition and metabolic pathways we find throughout life are all variations on themes that prokaryotes "invented" during their long reign as Earth's exclusive inhabitants. Prokaryotes are at the foundation of life in both the ecological sense and the evolutionary sense. The subsequent breakthroughs in evolution were mostly structural, including the origin of the eukaryotic cell and the diversification of the organisms we call protists.

CHECKPOINT

1. As different as archaea and bacteria are, both groups are characterized by _____ cells, which lack nuclei and other membrane-bounded organelles.

2. Upon microscopic examination, how do you think you could distinguish the cocci that cause "staph" infections from those that cause "strep" throat?

3. A species of bacterium requires only the amino acid methionine as an organic nutrient and lives in very deep caves where no light penetrates. Based on its mode of nutrition, this species would be classified as a _____.

4. Why are some archaea referred to as "extremophiles"?

5. Why do microbiologists autoclave their laboratory instruments and glassware?

6. Contrast exotoxins with endotoxins.

7. How do bacteria help restore the atmospheric CO_2 required by plants for photosynthesis?

8. What is the pitfall in believing that genetically engineered bacteria will solve all our pollution problems?

Answers: 1. prokaryotic **2.** By the arrangement of the cell aggregates: grapelike clusters for staphylococcus and chains of cells for streptococcus **3.** chemoheterotroph **4.** Because they can thrive in extreme environments too hot, too salty, or too acidic for other organisms **5.** To kill bacterial endospores, which can survive boiling water **6.** Exotoxins are poisons secreted by pathogenic bacteria; endotoxins are components of the cell walls of pathogenic bacteria. **7.** By decomposing the organic molecules of dead organisms and organic refuse such as leaf litter, the metabolism of bacteria releases carbon from the organic matter in the form of CO_2. **8.** They won't, and falsely believing in such an elixir might increase our public tolerance for pollution.

Protists

"No more pleasant sight has met my eye than this of so many thousands of living creatures in one small drop of water," wrote Anton van Leeuwenhoek after his discovery of the microbial world more than three centuries ago. It is a world every biology student should have the opportunity to rediscover by peering through a microscope into a droplet of pond water filled with diverse creatures we call **protists.**

Protists are eukaryotic, and thus even the simplest are much more complex than the prokaryotes. The first eukaryotes to evolve from prokaryotic ancestors were protists. The very word implies great antiquity (from the Greek *protos,* first). The primal eukaryotes were not only the predecessors of the great variety of modern protists, but were also ancestral to all other eukaryotes—plants, fungi, and animals. Two of the most significant chapters in the history of life—the origin of the eukaryotic cell and the subsequent emergence of multicellular eukaryotes—unfolded during the evolution of protists.

The Origin of Eukaryotic Cells

The many differences between prokaryotic and eukaryotic cells far outnumber the differences between plant and animal cells. The fossil record indicates that eukaryotes evolved from prokaryotes more than 1.7 billion years ago. One of biology's most engaging questions is how this happened—in particular, how the membrane-enclosed organelles of eukaryotic cells arose. A widely accepted theory is that eukaryotic cells evolved through a combination of two processes. In one process, the eukaryotic cell's endomembrane system—all the membrane-enclosed organelles except mitochondria and chloroplasts (see Chapter 4)—evolved from inward folds of the plasma membrane of a prokaryotic cell (**Figure 15.18a**). A second, very different process, called **endosymbiosis,** generated mitochondria and chloroplasts.

Symbiosis is a close association between organisms of two or more species. The word *symbiosis* is from the Greek for "living together," and endosymbiosis refers to one species living within another, called the host. Chloroplasts and mitochondria evolved from small symbiotic prokaryotes that established residence within other, larger host prokaryotes (**Figure 15.18b**). The ancestors of mitochondria may have been aerobic bacteria that were able to use oxygen to release large amounts of energy from organic molecules by cellular respiration. At some point, such a prokaryote might have been an internal parasite of a larger heterotroph, or an ancestral host cell may have ingested some of these aerobic cells for food. If some of the smaller cells were indigestible, they might have remained alive and continued to perform respiration in the host cell. In a similar way, photosynthetic bacteria ancestral to chloroplasts may have come to live inside a larger host cell. Because almost all eukaryotes have mitochondria but only some have chloroplasts, it is likely that mitochondria evolved first.

By whatever means the relationships began, it is not hard to imagine the symbiosis eventually becoming mutually beneficial. In a world that was becoming increasingly aerobic, a cell that was itself an anaerobe would have benefited

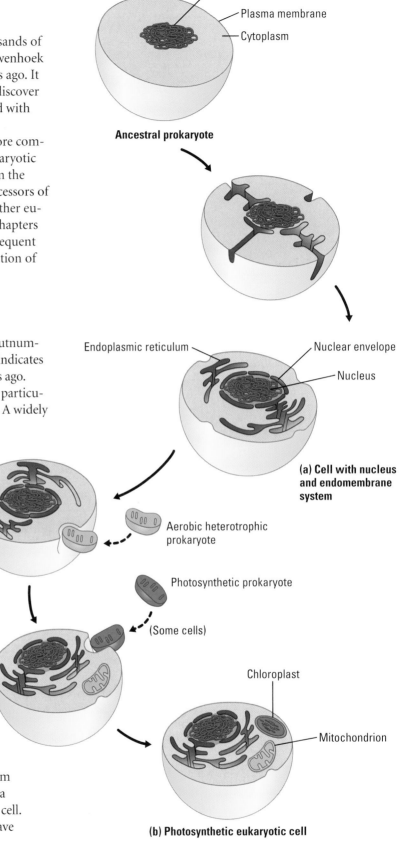

Ancestral prokaryote

- DNA
- Plasma membrane
- Cytoplasm

- Endoplasmic reticulum
- Nuclear envelope
- Nucleus

(a) Cell with nucleus and endomembrane system

Aerobic heterotrophic prokaryote

Photosynthetic prokaryote

(Some cells)

Chloroplast

Mitochondrion

(b) Photosynthetic eukaryotic cell

Figure 15.18 How did eukaryotic cells evolve?
(a) Origin of the endomembrane system. **(b)** Origin of mitochondria and chloroplasts.

from aerobic endosymbionts that turned the oxygen to advantage. And a heterotrophic host could derive nourishment from photosynthetic endosymbionts. In the process of becoming more interdependent, the host and endosymbionts would have become a single organism, its parts inseparable.

Developed most extensively by Lynn Margulis, the endosymbiosis theory is supported by extensive evidence. Present-day mitochondria and chloroplasts are similar to prokaryotic cells in a number of ways. For example, both types of organelles contain small amounts of DNA, RNA, and ribosomes that resemble prokaryotic versions more than eukaryotic ones. These components enable chloroplasts and mitochondria to exhibit some autonomy in their activities. The organelles transcribe and translate their DNA into polypeptides, contributing to some of their own enzymes. They also replicate their own DNA and reproduce within the cell by a process resembling the binary fission of prokaryotes.

The origin of the eukaryotic cell made more complex organisms possible, and a vast variety of protists evolved.

The Diversity of Protists

All protists are eukaryotes, but they are so diverse that few other general characteristics can be cited. In fact, protists vary in structure and function more than any other group of organisms. Most protists are unicellular, but there are some colonial and multicellular species. Because most protists are unicellular, they are justifiably considered the simplest eukaryotic organisms. But at the cellular level, many protists are exceedingly complex—the most elaborate of all cells. We should expect this of organisms that must carry out, within the boundaries of a single cell, all the basic functions performed by the collective of specialized cells that make up the bodies of plants and animals. Each unicellular protist is not at all analogous to a single cell from a human, but is itself an organism as complete as any whole animal or plant.

For our survey of these diverse organisms, we'll look at four major categories of protists, grouped more by lifestyle than by their evolutionary relationships: protozoans, slime molds, unicellular algae, and seaweeds.

Protozoans Protists that live primarily by ingesting food, a mode of nutrition that is animal-like, are called **protozoans** ("first animal"). Protozoans thrive in all types of aquatic environments, including wet soil and the watery environment inside animals. Most species eat bacteria or other protozoans, but some can absorb nutrients dissolved in the water. Protozoans that live as parasites in animals, though in the minority, cause some of the world's most harmful human diseases. We'll examine five groups of protozoans: flagellates, amoebas, forams, apicomplexans, and ciliates.

Flagellates are protozoans that move by means of one or more flagella. Most species are free-living (nonparasitic). However, there are also some nasty parasites that make humans sick. An example is *Giardia*, a flagellate that infects the human intestine and can cause abdominal cramps and severe diarrhea. People become infected mainly by drinking water contaminated with feces from infected animals. *Giardia* can ruin a camping trip. Another group of dangerous flagellates are the trypanosomes, including a species that causes sleeping sickness, a serious illness prevalent in tropical Africa and transmitted by the tsetse fly (**Figure 15.19a**).

Amoebas are characterized by great flexibility and the absence of permanent locomotor organelles. Most species move and feed by means of **pseudopodia** (singular, *pseudopodium*), temporary extensions of the cell (**Figure 15.19b**). Amoebas can assume virtually any shape as they creep

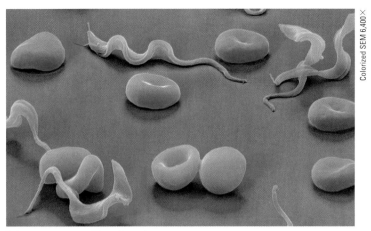

(a) Trypanosomes (flagellates)

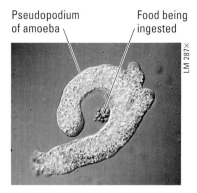

Pseudopodium of amoeba · Food being ingested

(b) An amoeba ingesting food

(c) A foram

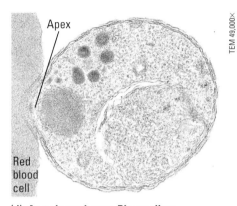

Apex

Red blood cell

(d) An apicomplexan: *Plasmodium*

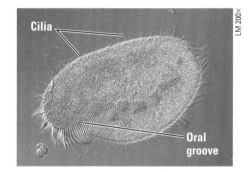

Cilia

Oral groove

(e) *Paramecium*

Figure 15.19 Examples of protozoans.
(a) Trypanosomes are flagellates that live as parasites in the bloodstream of vertebrate animals. The squiggles among these human red blood cells are trypanosomes that cause sleeping sickness, a debilitating disease common in parts of Africa. Trypanosomes escape being killed by their host's defenses by being quick-change artists. They alter the molecular structure of their coats frequently, thus preventing immunity from developing in the host. **(b)** This amoeba is ingesting a smaller protozoan as food. The amoeba's pseudopodia arch around the prey and engulf it into a food vacuole (also see Chapter 5). **(c)** Forams are almost all marine. The foram cell secretes a porous, multi-chambered shell made of organic material hardened with calcium carbonate, the same mineral that makes up limestone. Thin strands of cytoplasm (pseudopodia) extend through the pores, functioning in swimming, shell formation, and feeding. The shells of fossilized forams are major components of the limestone rocks that are now land formations. **(d)** *Plasmodium,* the apicomplexan that causes malaria, uses its apical complex to enter red blood cells of its human host. The parasite feeds on the host cell from within, eventually destroying it. **(e)** The ciliate *Paramecium* uses its cilia to move through pond water. Cilia also line an indentation called the oral groove, and their beating keeps a current of water containing bacteria and small protists moving toward the cell "mouth" at the base of the groove.

over rocks, sticks, or mud at the bottom of a pond or ocean. Other protozoans with pseudopodia include the **forams** (**Figure 15.19c**).

Apicomplexans are all parasitic, and some cause serious human diseases. They are named for an apparatus at their apex that is specialized for penetrating host cells and tissues. This protozoan group includes *Plasmodium,* the parasite that causes malaria (**Figure 15.19d**). Spread by mosquitoes, malaria is one of the most debilitating and widespread human diseases. Each year in the tropics, more than 200 million people become infected, and at least a million die in Africa alone. As part of the effort to combat malaria, scientists determined the complete sequence of the *Plasmodium* genome in 2002.

Ciliates are protozoans that use locomotor structures called cilia to move and feed. Nearly all ciliates are free-living (nonparasitic). The best-known example is the freshwater ciliate *Paramecium* (**Figure 15.19e**).

Figure 15.20 A plasmodial slime mold.
Pseudopodia of the huge cell engulf small food particles in mulch or moist soil. The weblike form is an adaptation that enlarges the organism's surface area, increasing its contact with food, water, and oxygen. Within the fine channels of the plasmodium, cytoplasm streams first one way and then the other, in pulses that are beautiful to watch with a microscope. The cytoplasmic streaming helps distribute nutrients and oxygen within the giant cell.

Slime Molds These protists are more attractive than their name. Slime molds resemble fungi in appearance and lifestyle, but the similarities are due to convergent evolution; slime molds and fungi are not at all closely related. The filamentous body of a slime mold, like that of a fungus, is an adaptation that increases exposure to the environment. This suits the role of these organisms as decomposers. The two main groups of these protists are plasmodial slime molds and cellular slime molds.

Plasmodial slime molds are named for the feeding stage in their life cycle, an amoeboid mass called a plasmodium (not to be confused with *Plasmodium*, the parasite that causes malaria). You can find plasmodial slime molds among the leaf litter and other decaying material on a forest floor, and you won't need a microscope to see them. A plasmodium can measure several centimeters across, with its network of fine filaments taking in bacteria and bits of dead organic matter amoeboid style. Large as it is, the plasmodium is actually a single cell with many nuclei (**Figure 15.20**).

Cellular slime molds pose a semantic question about what it means to be an individual organism. The feeding stage in the life cycle of a cellular slime mold consists of solitary amoeboid cells. They function individually, using their pseudopodia to feed on decaying organic matter. But when food is depleted, the cells aggregate to form a sluglike colony that moves and functions as a single unit (**Figure 15.21**).

Unicellular Algae Photosynthetic protists are called **algae** (singular, *alga*). Their chloroplasts support food chains in freshwater and marine ecosystems. Many unicellular algae are components of **plankton** (from the Greek *planktos,* wandering), the communities of organisms, mostly microscopic, that drift or swim weakly near the surfaces of ponds, lakes, and oceans. More specifically, planktonic algae are referred to as phytoplankton. We'll look at three groups of unicellular algae: dinoflagellates, diatoms, and green algae (a group that also includes colonial and truly multicellular species).

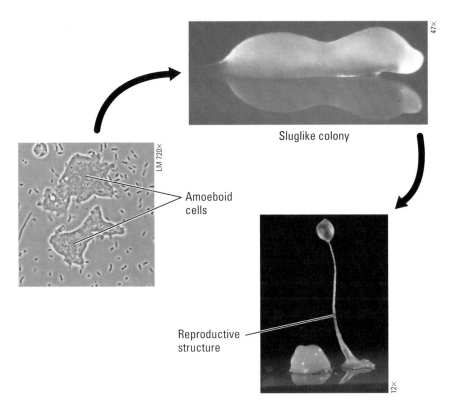

Sluglike colony

Amoeboid cells

Reproductive structure

Figure 15.21 Life cycle of a cellular slime mold.
Most of the time, cellular slime molds live as solitary amoeboid cells, using their pseudopodia to creep through compost and engulf bacteria. When food is in short supply, the amoeboid cells swarm together, forming a colony that looks and moves like a slug. After wandering around for a short time, the colony extends a stalk and develops into a multicellular reproductive structure.

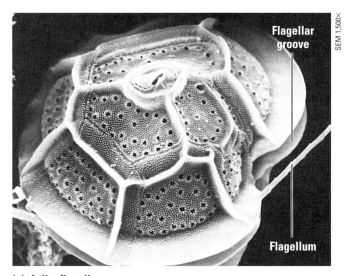

(a) A dinoflagellate

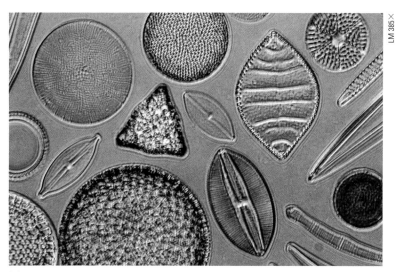

(b) Diatoms

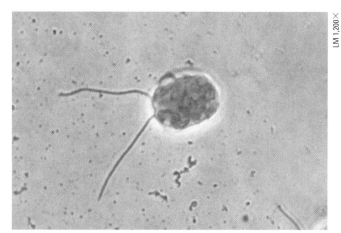

(c) *Chlamydomonas*

Dinoflagellates are abundant in the vast aquatic pastures of phyto-plankton. Each dinoflagellate species has a characteristic shape reinforced by external plates made of cellulose (**Figure 15.22a**). The beating of two flagella in perpendicular grooves produces the spinning movement for which these organisms are named (from the Greek *dinos,* whirling). Di-noflagellate blooms—population explosions—sometimes cause warm coastal waters to turn pinkish orange, a phenomenon known as a red tide. Toxins produced by some red-tide dinoflagellates have caused massive fish kills, especially in the tropics, and are poisonous to humans as well.

Diatoms have glassy cell walls containing silica, the mineral used to make glass (**Figure 15.22b**). The cell wall consists of two halves that fit together like the bottom and lid of a shoe box. Diatoms store their food reserves in the form of an oil that provides buoyancy, keeping diatoms floating as phytoplankton near the sunlit surface. Massive accumulations of fossilized diatoms make up thick sediments known as diatomaceous earth, which is mined for its use as both a filtering material and an abrasive.

Green algae are named for their grass-green chloroplasts. Unicellular green algae flourish in most freshwater lakes and ponds. Some species are flagellated (**Figure 15.22c**). The green algal group also includes colonial forms, such as the *Volvox* in Figure **15.22d**. Each *Volvox* colony is a ball of flagellated cells (the small green dots in the photo) that are very similar to certain unicellular green algae. The balls within the balls in Figure 15.22d are "daughter" colonies that will be released when the parent colonies rup-ture. Of all photosynthetic protists, green algae are the most closely related to true plants. (We'll examine the evidence of this evolutionary relationship in the next chapter.)

Seaweeds Defined as large, multicellular marine algae, **seaweeds** grow on rocky shores and just offshore beyond the zone of the pounding surf. Their cell walls have slimy and rubbery substances that cushion their bodies against the agitation of the waves. Some seaweeds are as large and complex as many plants. Even the word *seaweed* implies plantlike appearance, but the similarities between these algae and true plants are a consequence of

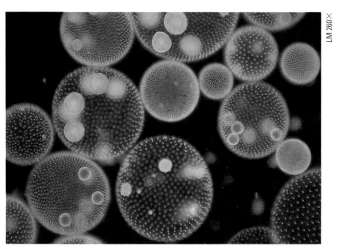

(d) *Volvox*

Figure 15.22 Unicellular and colonial algae.
(a) A dinoflagellate, with its wall of protective plates. **(b)** A sample of diverse diatoms, which have glassy walls. **(c)** *Chlamydomonas,* a unicellular green alga with a pair of flagella. **(d)** *Volvox,* a colonial green alga.

convergent evolution. In fact, the closest relatives of seaweeds are certain unicellular algae, which is why many biologists include seaweeds with the protists. Seaweeds are classified into three different groups, based partly on the types of pigments present in their chloroplasts: green algae, red algae, and brown algae (**Figure 15.23**).

Coastal people, particularly in Asia, harvest seaweeds for food. For example, in Japan and Korea, some seaweed species are ingredients in soups. Other seaweeds are used to wrap sushi. Marine algae are rich in iodine and other essential minerals. However, much of their organic material consists of unusual polysaccharides that humans cannot digest, which prevents seaweeds from becoming staple foods. They are ingested mostly for their rich tastes and unusual textures. The gel-forming substances in the cell walls of seaweeds are widely used as thickeners for such processed foods as puddings, ice cream, and salad dressing. And the seaweed extract called agar provides the gel-forming base for the media microbiologists use to culture bacteria in Petri dishes.

Case Study in the Process of Science on the Web & CD Identify protists collected from different habitats.

(a) Green algae

(b) Red algae

(c) Brown algae

Figure 15.23 The three major groups of seaweeds. **(a)** Green algae. This sea lettuce is an edible species that inhabits the intertidal zone. In addition to seaweeds, the green algal group includes unicellular and colonial species, such as those in Figures 15.22c and d. **(b)** Red algae. These seaweeds are most abundant in the warm coastal waters of the tropics. Of all the seaweeds, red algae can generally live in the deepest water. Their chloroplasts have special pigments that absorb the blue and green light that penetrates best through water. The species in this photo is an example of corraline algae, which contribute to the architecture of some coral reefs. The cell walls are hardened by a mineral. **(c)** Brown algae. This group includes the largest seaweeds, known as kelp, which grow as marine "forests" in relatively deep water beyond the intertidal zone. Some species grow to a length of over 60 m in a single season, the fastest linear growth of any organism. Kelp is a renewable resource reaped by special boats that cut and collect the tops of the algae. More importantly, kelp forests provide habitat for many animals, including a great diversity of fishes. If you have walked on a beach covered with kelp that has washed ashore after a storm, you may have noticed the organs called floats, which keep the photosynthetic blades of the kelp in the light near the water's surface. Maybe you even picked up and popped some of those floats, the way you do those irresistible packing-material bubbles.

Evolution Connection

The Origin of Multicellular Life

An orchestra can play a greater variety of musical compositions than a violin soloist can. Put simply, increased complexity makes more variations possible. Thus, the origin of the eukaryotic cell led to an evolutionary radiation of new forms of life. Unicellular protists, which are organized on the complex eukaryotic plan, are much more diverse in form than the simpler prokaryotes. The evolution of multicellular bodies broke through another threshold in structural organization.

Multicellular organisms are fundamentally different from unicellular ones. In a unicellular organism, all of life's activities occur within a single cell. In contrast, a multicellular organism has various specialized cells that perform different functions and are dependent on each other. For example, some cells procure food, while others transport materials or provide movement.

The evolutionary links between unicellular and multicellular life were probably colonial forms, in which unicellular protists stuck together as loose federations of independent cells (**Figure 15.24**). The gradual transition from colonies to truly multicellular organisms involved the cells becoming increasingly interdependent as a division of labor evolved. We can see one level of specialization and cooperation in the colonial green alga *Volvox* (see Figure 15.22d). *Volvox* produces gametes (sperm and ova), which depend on nonreproductive cells, or somatic cells, while developing. Cells in truly multicellular organisms are specialized for many more nonreproductive functions, including feeding, waste disposal, gas exchange, and protection, to name a few.

Multicellularity evolved many times among the ancestral stock of protists, leading to new waves of biological diversification. The diverse seaweeds are examples of the descendants, and so are plants, fungi, and animals. In the next chapter, we'll trace the long evolutionary movement of plants and fungi onto land. Then we'll follow the threads of animal evolution in Chapter 17.

Evolution Connection on the Web
Learn about what may be the oldest fossils of multicellular organisms.

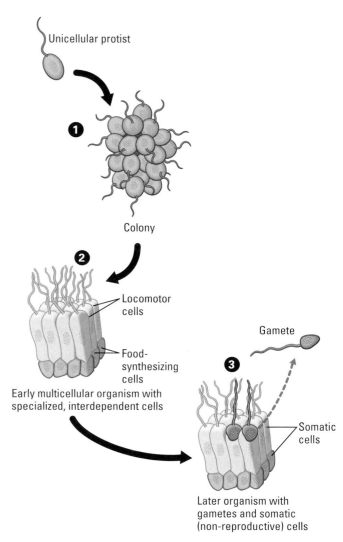

Figure 15.24 A model for the evolution of multicellular organisms from unicellular protists.
1 An ancestral colony may have formed, as colonial protists do today, when a cell divided and its offspring remained attached to one another. **2** The cells in the colony may have become somewhat specialized and interdependent, with different cell types becoming more and more efficient at performing specific, limited tasks. Cells that retained a flagellum may have become specialized for locomotion, while others that lost their flagellum could have assumed functions such as ingesting or synthesizing food. **3** Additional specialization among the cells in the colony may have led to distinctions between sex cells (gametes) and non-reproductive cells (somatic cells).

Chapter Review

Summary of Key Concepts

For study help, go to the Essential Biology Website (www.essentialbiology.com) or CD-ROM to explore the Activities and Case Studies in the Process of Science.

Major Episodes in the History of Life

Millions of years ago	Major episode
475	Plants and fungi colonize land
570	All major animal phyla established
1,000	First multicellular organisms
1,700	Oldest eukaryotic fossils
2,500	Accumulation of atmospheric O_2
3,500	Oldest prokaryotic fossils
4,500	Origin of Earth

Activity 15A *The History of Life*

The Origin of Life

- **Resolving the Biogenesis Paradox**   All life today arises only by the reproduction of preexisting life. However, most biologists now think it possible that chemical and physical processes in Earth's primordial environment produced very simple cells through a sequence of stages.

- **A Four-Stage Hypothesis for the Origin of Life**   One scenario suggests that the first organisms were products of chemical evolution in four stages:

Inorganic compounds

➊ Abiotic synthesis

Organic monomers

➋ Polymerization

Polymer

➌ Self-replication

Complementary chain

➍ Packaging

Membrane-enclosed compartment

Case Study in the Process of Science *How Did Life Begin on Early Earth?*

- **From Chemical Evolution to Darwinian Evolution**   Over millions of years, natural selection favored the most efficient pre-cells, which evolved into the first prokaryotic cells.

Prokaryotes

- **They're Everywhere!**   Prokaryotes are found wherever there is life. They outnumber all eukaryotes combined. Prokaryotes thrive in habitats where eukaryotes cannot live. A few prokaryotic species cause serious diseases, but most are either benign or beneficial.

- **The Two Main Branches of Prokaryotic Evolution: Bacteria and Archaea**

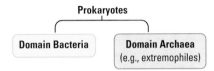

Prokaryotes

Domain Bacteria   Domain Archaea (e.g., extremophiles)

- **The Structure, Function, and Reproduction of Prokaryotes**   Prokaryotic cells lack nuclei and other membrane-enclosed organelles. Most have cell walls. Prokaryotes come in several shapes, most commonly spheres (cocci), rods (bacilli), and curves or spirals (the largest called spirochetes). About half of all prokaryotic species are motile, most of these using flagella to move. Some prokaryotes can survive extended periods of harsh conditions by forming endospores. Most prokaryotes can reproduce by binary fission at phenomenal rates if conditions are favorable, but growth is usually restricted by limited resources.

Activity 15B *Prokaryotic Cell Structure and Function*

- **The Nutritional Diversity of Prokaryotes**   Prokaryotes exhibit four major modes of nutrition.

Nutritional Mode	Energy Source	Carbon Source
Photoautotroph	Sunlight	CO_2
Chemoautotroph	Inorganic chemicals	
Photoheterotroph	Sunlight	Organic compounds
Chemoheterotroph	Organic compounds	

Case Study in the Process of Science *What Are the Modes of Nutrition in Prokaryotes?*

- **The Ecological Impact of Prokaryotes**   Most pathogenic bacteria cause disease by producing exotoxins or endotoxins. Sanitation, antibiotics, and education are the best defenses against bacterial disease. Prokaryotes help recycle chemical elements between the biological and physical components of ecosystems. Humans can use bacteria to remove pollutants from water, air, and soil in the process called bioremediation.

Activity 15C *Diversity of Prokaryotes*

Protists

- **The Origin of Eukaryotic Cells**   The nucleus and endomembrane system of eukaryotes probably evolved from infoldings of the plasma membrane of ancestral prokaryotes. Mitochondria and chloroplasts probably evolved from symbiotic prokaryotes that took up residence inside larger prokaryotic cells.

- **The Diversity of Protists** Protists are unicellular eukaryotes and their closest multicellular relatives. Protozoans primarily live by ingesting food. They include flagellates, amoebas, apicomplexans, and ciliates. Most protozoans thrive in all types of aquatic environments, but a few live in animals and cause some of the world's most harmful diseases. Slime molds resemble fungi in appearance and lifestyle as decomposers, but are not at all closely related. Plasmodial slime molds are multinuclear organisms named for the plasmodium feeding stage in their life cycle. Cellular slime molds have unicellular and multicellular life stages. Unicellular algae are photosynthetic protists that support food chains in freshwater and marine ecosystems. They include the dinoflagellates, diatoms, and unicellular green algae. Seaweeds are large, multicellular marine algae that grow on and near rocky shores. Green, red, and brown algae are classified partly by the types of pigments present in their chloroplasts.

Case Study in the Process of Science *What Kinds of Protists Are Found in Various Habitats?*

Self-Quiz

1. Place these events in the history of life on Earth in the order that they occurred.
   a. origin of multicellular organisms
   b. colonization of land by plants and fungi
   c. origin of eukaryotes
   d. origin of prokaryotes
   e. colonization of land by animals

2. Place the following steps in the origin of life in the order that they are hypothesized to have occurred.
   a. integration of self-replicating molecules into membrane-enclosed pre-cells
   b. origin of the first molecules capable of self-replication
   c. abiotic joining of organic monomers into polymers
   d. abiotic synthesis of organic monomers
   e. natural selection among pre-cells

3. Part of the explanation for the absence of spontaneous generation of life on Earth today is that
   a. there is too much lightning today.
   b. the abundance of O_2 in our modern atmosphere would destroy complex organic molecules that are not inside organisms.
   c. too much ultraviolet light is reaching Earth today.
   d. all habitable places are already filled.

4. The two main evolutionary branches of prokaryotic life are _____ and _____.

5. How do penicillins kill certain bacteria?

6. What is the difference between autotrophs and heterotrophs in terms of the source of their organic compounds?

7. The bacteria that cause tetanus can be killed only by prolonged heating at temperatures considerably above boiling. What does this suggest about tetanus bacteria?

8. Which of the following is an autotrophic protist?
   a. amoeba   c. cyanobacteria
   b. slime mold   d. diatom

9. Of the following, which describes protists most inclusively?
   a. multicellular eukaryotes
   b. protozoans
   c. eukaryotes that are not plants, fungi, or animals
   d. single-celled organisms closely related to bacteria

10. Which algal group is most closely related to plants?
    a. diatoms   c. dinoflagellates
    b. green algae   d. seaweeds

Answers to the Self-Quiz questions can be found in Appendix B.

Go to the website or CD-ROM for more Self-Quiz questions.

The Process of Science

1. Imagine you are on a team designing a moon base that will be self-contained and self-sustaining. Once supplied with building materials, equipment, and organisms from Earth, the base will be expected to function indefinitely. One of the members of your team has suggested that everything sent to the base be chemically treated or irradiated so that no bacteria of any kind are present. Do you think this is a good idea? Predict some of the consequences of eliminating all bacteria from an environment.

2. Your classmate says that organisms that require oxygen existed before photosynthetic organisms. Do you support this idea? Explain why or why not.

Case Study in the Process of Science on the Web & CD *Perform virtual experiments on the four hypothetical stages of the origin of life.*

Case Study in the Process of Science on the Web & CD *See if you can provide the ideal energy source and carbon source for various prokaryotes.*

Case Study in the Process of Science on the Web & CD *Identify protists collected from different habitats.*

Biology and Society

1. Many local newspapers publish a weekly list of restaurants that have been cited by inspectors for poor sanitation. Locate such a report and highlight the cases that are likely associated with potential food contamination by pathogenic prokaryotes.

2. What do you think should be done to prevent bioterrorism?

Biology and Society on the Web *Learn more about the history of biowarfare.*

CHAPTER 16

Plants, Fungi, and the Move onto Land

Biology and Society:
The Balancing Act of Forest
Conservation 321

Colonizing Land 321

Terrestrial Adaptations of Plants

The Origin of Plants from
Green Algae

Plant Diversity 324

Highlights of Plant Evolution

Bryophytes

Ferns

Gymnosperms

Angiosperms

Plant Diversity as a
Nonrenewable Resource

Fungi 334

Characteristics of Fungi

The Ecological Impact of Fungi

Evolution Connection:
Mutual Symbiosis 340

Some giant sequoia trees weigh more than a dozen space shuttles.

A mushroom is probably more closely related to humans than it is to any plant.

Flowering plants such as corn, rice, and wheat provide nearly all our food.

Every 2 seconds, humans destroy an area of tropical rain forest equal to the area of 3 football fields.

The Balancing Act of Forest Conservation

With a soft floor underfoot and the scent of pine needles in the air, few places are as pleasing to the senses as coniferous forests filled with cone-bearing plants such as pines, firs, spruces, and redwoods. Today, about 190 million acres of coniferous forests in the United States, mostly in the western states and Alaska, are designated national forests. Some of these areas are set aside as unspoiled wilderness and wildlife habitats. But most national forests are working forests, managed by the U.S. Forest Service for harvesting lumber, grazing, mining, and public recreation.

Coniferous forests are highly productive; you probably use products harvested there every day. For example, conifers provide much of our lumber for building and wood pulp for paper production. Currently, our demand for wood and paper is so great—the average U.S. citizen consumes about

Biology and Society on the Web
Learn more about the impact of deforestation.

50 times more paper than the average person in less developed nations—that clear-cut areas have become commonplace (**Figure 16.1**). In many areas, only about 10% of the original forest remains intact. Some forests have been replanted, but the rate of cutting often exceeds the rate at which new trees can grow. Moreover, many scientists predict that an increase in global temperatures, which now seems to be occurring, poses an additional threat to coniferous forests.

The loss of coniferous forests threatens more than just the trees themselves. The original forests of North America were more biologically diverse than the forests that are now regrowing. Balancing the uses of coniferous forests while simultaneously trying to sustain them for future generations is a formidable challenge. What can you do? Reducing paper waste, increasing recycling, and expanding the use of electronic media can all help.

The trees that fill coniferous forests are just one of several major types of vegetation that share the planet with us. In this chapter, we explore the diversity of plants and fungi, starting with a look at how aquatic life first adapted to living on land.

Colonizing Land

Plants are terrestrial (land-dwelling) organisms. True, some, such as water lilies, have returned to the water, but they evolved secondarily from terrestrial ancestors (as did several species of aquatic animals, such as porpoises).

What exactly is a plant? A plant is a multicellular eukaryote that makes organic molecules by photosynthesis. Photosynthesis distinguishes plants from the animal and fungal kingdoms. But what about large algae, including seaweeds, which we classified as protists in the preceding chapter? They, too, are multicellular, eukaryotic, and photosynthetic. It is a set of terrestrial adaptations that distinguishes plants from algae.

Terrestrial Adaptations of Plants

Structural Adaptations Living on land poses very different problems from living in water (**Figure 16.2**). In terrestrial habitats, the resources that

Figure 16.1 A clear-cut area in a coniferous forest.

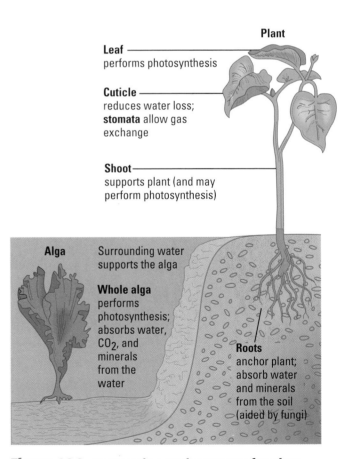

Figure 16.2 Contrasting environments for algae and plants.

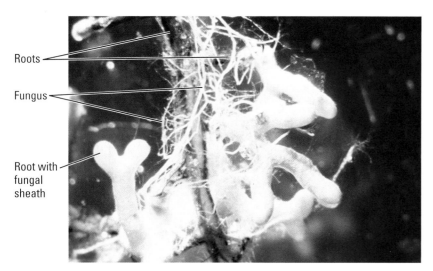

Figure 16.3 Mycorrhizae: symbiotic associations of fungi and roots. The finely branched filaments of the fungus provide an extensive surface area for absorption of water and minerals from the soil. The fungus provides some of those materials to the plant and benefits in turn by receiving sugars and other organic products of the plant's photosynthesis.

Roots

Fungus

Root with fungal sheath

Figure 16.4 Network of veins in a leaf. The vascular tissue of the veins delivers water and minerals absorbed by the roots and carries away the sugars produced in the leaves.

a photosynthetic organism needs are found in two very different places. Light and carbon dioxide are mainly available above-ground, while water and mineral nutrients are found mainly in the soil. Thus, the complex bodies of plants show varying degrees of structural specialization into subterranean and aerial organs—**roots** and leaf-bearing **shoots,** respectively.

Most plants have symbiotic fungi associated with their roots. These root-fungus combinations are called **mycorrhizae** ("fungus root"). For their part, the fungi absorb water and essential minerals from the soil and provide these materials to the plant. The sugars produced by the plant nourish the fungi. Mycorrhizae are evident on some of the oldest plant fossils. They are key adaptations that made it possible to live on land (**Figure 16.3**).

Leaves are the main photosynthetic organs of most plants. Exchange of carbon dioxide and oxygen between the atmosphere and the photosynthetic interior of a leaf occurs via **stomata,** the microscopic pores through the leaf's surface (see Figure 7.3). A waxy layer called the **cuticle** coats the leaves and other aerial parts of most plants, helping the plant body retain its water. (Think of the waxy surface of a cucumber or unpolished apple.)

Differentiation of the plant body into root and shoot systems solved one problem but created new ones. For the shoot system to stand up straight in the air, it must have support. This is not a problem in the water: Huge seaweeds need no skeletons because the surrounding water buoys them. An important terrestrial adaptation of plants is **lignin,** a chemical that hardens the cell walls. Imagine what would happen to you if your skeleton were to disappear or suddenly turn mushy. A tree would also collapse if it were not for its "skeleton," its framework of lignin-rich cell walls.

Specialization of the plant body into roots and shoots also introduced the problem of transporting vital materials between the distant organs. The terrestrial equipment of most plants includes **vascular tissue,** a system of tube-shaped cells that branch throughout the plant (**Figure 16.4**). The vascular tissue actually has two types of tissues specialized for transport: **xylem,** consisting of dead cells with tubular cavities for transporting water and minerals from roots to leaves; and **phloem,** consisting of living cells that distribute sugars from the leaves to the roots and other nonphotosynthetic parts of the plant.

Reproductive Adaptations Adapting to land also required a new mode of reproduction. For algae, the surrounding water ensures that gametes (sperm and eggs) and developing offspring stay moist. The aquatic environment also provides a means of dispersing the gametes and offspring. Plants, however, must keep their gametes and developing offspring from drying out in the air. Plants (and some algae) produce their gametes in protective structures called **gametangia** (singular, *gametangium*). A gametangium has a jacket of protective cells surrounding a moist chamber where gametes can develop without dehydrating.

Activity 16A on the Web & CD See animations of the terrestrial adaptations of plants.

In most plants, sperm reach the eggs by traveling within pollen, which is carried by wind or animals. The egg remains within tissues of the mother plant and is fertilized there. In plants, but not algae, the zygote (fertilized egg) develops into an embryo while still contained within the female parent, which protects the embryo and keeps it from dehydrating (**Figure 16.5**). Most plants rely on wind or animals, such as fruit-eating birds or mammals, to disperse their offspring, which are in the form of embryos contained in seeds.

The reproductive "strategy" of plants is analogous to how mammals manage to reproduce on land. As in plants, mammalian fertilization is internal (within the mother's body). And in most mammals, embryonic development also occurs within the mother's body, as it does in plants.

The Origin of Plants from Green Algae

The move onto land and the spread of plants to diverse terrestrial environments was incremental. It paralleled the gradual accumulation of terrestrial adaptations, beginning with populations that descended from algae. Green algae are the protists most closely related to plants. More specifically, molecular comparisons and other evidence place a group of multicellular green algae called **charophyceans** closest to plants (**Figure 16.6**).

The evolutionary "walk" onto land was more like adaptive baby steps. Many species of modern charophyceans are found in shallow water around the edges of ponds and lakes. Some of the ancient charophyceans that lived about the time that land was first colonized may have inhabited shallow-water habitats subject to occasional drying. Natural selection would have favored individual algae that could survive through periods when they were not submerged. The protection of developing gametes and embryos within jacketed organs (gametangia) on the parent is one adaptation to living in shallow water that would also prove essential on land. We know that by about 475 million years ago, the vintage of the oldest plant fossils, an accumulation of adaptations allowed permanent residency above water. The plants that color our world today diversified from those early descendants of green algae.

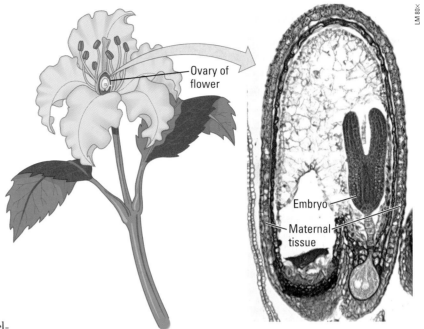

Figure 16.5 The protected embryo of a plant.
Internal fertilization, with sperm and egg combining within a moist chamber on the mother plant, is an adaptation for living on land. The female parent continues to nurture and protect the plant embryo, which develops from the zygote.

(a) Chara

(b) Coleochaete

Figure 16.6 Charophyceans, closest algal relatives of plants.
(a) *Chara* is a particularly elaborate green alga. **(b)** *Coleochaete,* though less plantlike than *Chara* in appearance, is actually more closely related to plants.

Plant Diversity

As we survey the diversity of modern plants, remember that the past is the key to the present. The history of the plant kingdom is a story of adaptation to diverse terrestrial habitats.

Highlights of Plant Evolution

The fossil record chronicles four major periods of plant evolution, which are also evident in the diversity of modern plants (**Figure 16.7**). Each stage is marked by the evolution of structures that opened new opportunities on land.

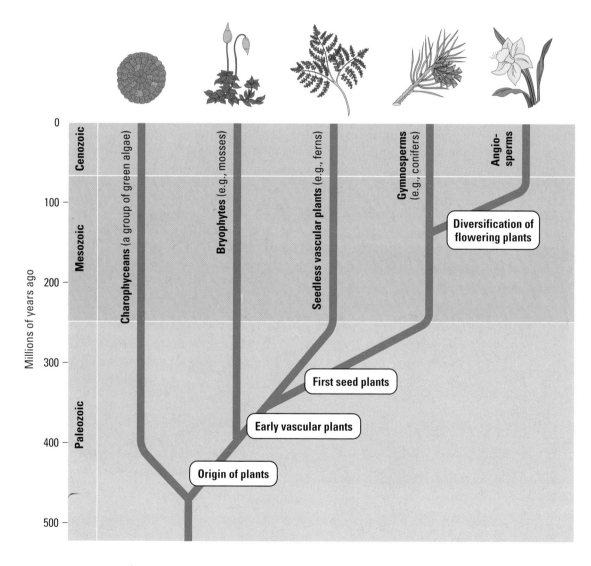

Figure 16.7 Highlights of plant evolution.
Modern representatives of the major evolutionary branches are illustrated at the top of this phylogenetic tree. As we survey the diversity of plants, miniature versions of this tree will help you place each plant group in its evolutionary context.

The first period of evolution was the origin of plants from their aquatic ancestors, the green algae called charophyceans. The first terrestrial adaptations included gametangia, which protected gametes and embryos. This made it possible for the plants known as **bryophytes,** including the mosses, to diversify from early plants. Vascular tissue also evolved relatively early in plant history. However, most bryophytes lack vascular tissue, which is why they are categorized as nonvascular plants.

The second period of plant evolution was the diversification of vascular plants (plants with vascular tissue that conducts water and nutrients). The earliest vascular plants lacked seeds. Today, this seedless condition is retained by **ferns** and a few other groups of vascular plants.

The third major period of plant evolution began with the origin of the seed. Seeds advanced the colonization of land by further protecting plant embryos from desiccation (drying) and other hazards. A seed consists of an embryo packaged along with a store of food within a protective covering. The seeds of early seed plants were not enclosed in any specialized chambers. These plants gave rise to many types of **gymnosperms** ("naked seed"). Today, the most widespread and diverse gymnosperms are the conifers, which are the pines and other plants with cones.

The fourth major episode in the evolutionary history of plants was the emergence of flowering plants, or **angiosperms** ("seed container").

Activity 16B on the Web & CD Review the major periods of plant evolution.

The flower is a complex reproductive structure that bears seeds within protective chambers (containers) called ovaries. This contrasts with the bearing of naked seeds by gymnosperms. The great majority of modern-day plants are angiosperms.

With these highlights as our framework, we are now ready to survey the four major groups of modern plants: bryophytes, ferns, gymnosperms, and angiosperms.

Bryophytes

The most familiar bryophytes are **mosses.** A mat of moss actually consists of many plants growing in a tight pack, helping to hold one another up (**Figure 16.8**). The mat has a spongy quality that enables it to absorb and retain water.

Mosses are not totally liberated from their ancestral aquatic habitat. They do display two of the key terrestrial adaptations that made the move onto land possible: a waxy cuticle that helps prevent dehydration; and the retention of developing embryos within the mother plant's gametangium. However, mosses need water to reproduce. Their sperm are flagellated, like those of most green algae. These sperm must swim through water to reach eggs. (A film of rainwater or dew is often enough moisture for the sperm to travel.) In addition, most mosses have no vascular tissue to carry water from soil to aerial parts of the plant. This explains why damp, shady places are the most common habitats of mosses. These plants also lack lignin, the wall-hardening material that enables other plants to stand tall. Mosses may sprawl as mats over acres, but they always have a low profile.

If you look closely at some moss growing on your campus, you may actually see two distinct versions of the plant. The greener, spongelike plant

Figure 16.8 A peat moss bog in Norway.
Although mosses are short in stature, their collective impact on Earth is huge. For example, peat mosses, or *Sphagnum,* carpet at least 3% of Earth's terrestrial surface, with greatest density in high northern latitudes. The accumulation of "peat," the thick mat of living and dead plants in wetlands, ties up an enormous amount of organic carbon because peat has an abundance of chemical materials that are not easily degraded by microbes. That explains why peat makes an excellent fuel as an alternative to coal and wood. More importantly, the carbon storage by peat bogs plays an important role in stabilizing Earth's atmospheric carbon dioxide concentrations, and hence climate, through the CO_2-related greenhouse effect (see Chapter 7).

Spore capsule

Sporophyte

Gametophytes

Figure 16.9 The two forms of a moss. The feathery plant we generally know as a moss is the gametophyte. The stalk with the capsule at its tip is the sporophyte. This photo shows the capsule releasing its spores, reproductive cells that can develop into new gametophytes.

that is the more obvious is called the **gametophyte.** You may see the other version of the moss, called a **sporophyte,** growing out of a gametophyte as a stalk with a capsule at its tip (**Figure 16.9**). The cells of the gametophyte are haploid (one set of chromosomes; see Chapter 8). In contrast, the sporophyte is made up of diploid cells (two chromosome sets). These two different stages of the plant life cycle are named for the types of reproductive cells they produce. Gametophytes produce gametes (sperm and eggs), while sporophytes produce spores. As reproductive cells, **spores** differ from gametes in two ways: A spore can develop into a new organism without fusing with another cell (two gametes must fuse to form a zygote); and spores usually have tough coats that enable them to resist harsh environments.

The gametophyte and sporophyte are alternating generations that take turns producing each other. Gametophytes produce gametes that unite to form zygotes, which develop into new sporophytes. And sporophytes produce spores that give rise to new gametophytes. This type of life cycle, called

Activity 16C on the Web & CD
Review the moss life cycle.

alternation of generations, occurs only in plants and certain algae (**Figure 16.10**). Among plants, mosses and other bryophytes are unique in having the gametophyte as the dominant generation—the larger, more obvious plant. As we continue our survey of plants, we'll see an increasing dominance of the sporophyte as the more highly developed generation.

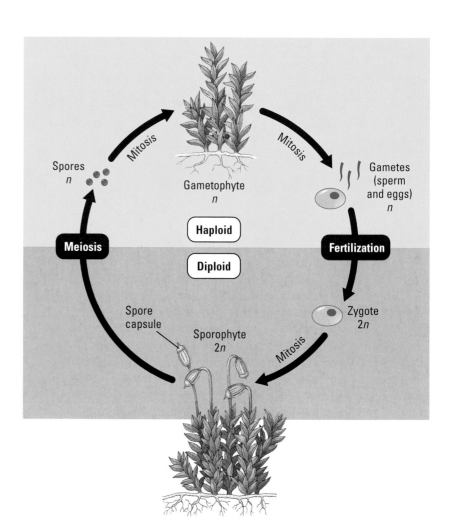

Figure 16.10 Alternation of generations. Plants have life cycles very different from ours. Each of us is a diploid individual; the only haploid stages in the human life cycle, as for nearly all animals, are sperm and eggs. By contrast, plants have alternating generations: Diploid (2*n*) individuals (sporophytes) and haploid (*n*) individuals (gametophytes) generate each other in the life cycle. In the case of mosses, the gametophyte is the dominant stage. In fact, the moss sporophyte remains attached to the gametophyte, depending on its parent for water and nutrients. In other plant groups, this balance is reversed, with the sporophyte being the more developed of the two generations.

Ferns

Charophyceans Bryophytes Ferns Gymnosperms Angiosperms

Ferns took terrestrial adaptation to the next level with the evolution of vascular tissue. However, the sperm of ferns, like those of mosses, are flagellated and must swim through a film of water to fertilize eggs. Ferns are also seedless, which helps explain why they do not dominate most modern terrestrial landscapes. However, of all seedless vascular plants, ferns are by far the most diverse today, with more than 12,000 species. Most of those species inhabit the tropics, although many species are found in temperate forests, such as most woodlands of the United States (**Figure 16.11**).

During the Carboniferous period, about 290–360 million years ago, ancient ferns were among a much greater diversity of seedless plants that formed vast, swampy forests that covered much of what is now Eurasia and North America (**Figure 16.12**). At that time, these continents were close to the equator and had tropical climates. The tropical swamp forests of the Carboniferous period generated great quantities of organic matter. As the plants died, they fell into stagnant wetlands and did not decay completely. Their remains formed thick deposits of organic rubble, or peat. Later, seawater flooded the swamps, marine sediments covered the peat, and pressure and heat gradually converted the peat to coal. Coal is black sedimentary rock made up of fossilized plant material. It formed during several geologic periods, but the most extensive coal beds are derived from Carboniferous deposits. (The name Carboniferous comes from the Latin *carbo*, coal, and *fer-*, bearing.) Coal, oil, and natural gas are **fossil fuels**—fuels formed from the remains of extinct organisms. Fossil fuels are burned to generate much of our electricity. As we deplete our oil and gas reserves, the use of coal is likely to increase.

Activity 16D on the Web & CD
Learn more about how ferns reproduce.

Case Study in the Process of Science on the Web & CD
See photos of the stages of the fern life cycle.

Figure 16.11 Ferns (seedless vascular plants).
This species grows on the forest floor in the eastern United States. The "fiddleheads" in the inset on the right are young fronds (leaves) ready to unfurl. The fern generation familiar to us is the sporophyte generation. The inset on the left is the underside of a sporophyte leaf specialized for reproduction. The yellow dots consist of spore capsules that can release numerous tiny spores. The spores develop into gametophytes. However, you would have to crawl on the forest floor and explore with careful hands and sharp eyes to find fern gametophytes, tiny plants growing on or just below the soil surface.

Figure 16.12
A "coal forest" of the Carboniferous period.
This painting, based on fossil evidence, reconstructs one of the great seedless forests. Most of the large trees with straight trunks are seedless plants called lycophytes. On the left, the tree with numerous feathery branches is another type of seedless plant called a horsetail. The plants near the base of the trees are ferns. Note the giant bird-sized dragonfly, which would have made quite a buzz.

Figure 16.13 A coniferous forest near Peyto Lake in the Canadian Rockies.

Gymnosperms

"Coal forests" dominated the North American and Eurasian landscapes until near the end of the Carboniferous period. At that time, global climate turned drier and colder, and the vast swamps began to disappear. This climatic change provided an opportunity for seed plants, which can complete their life cycles on dry land and withstand long, harsh winters. Of the earliest seed plants, the most successful were the gymnosperms, and several kinds grew along with the seedless plants in the Carboniferous swamps. Their descendants include the **conifers,** or cone-bearing plants.

Conifers Perhaps you have had the fun of hiking or skiing through a forest of conifers, the most common gymnosperms. Pines, firs, spruces, junipers, cedars, and redwoods are all conifers. A broad band of coniferous forests covers much of northern Eurasia and North America and extends southward in mountainous regions (**Figure 16.13**).

Conifers are among the tallest, largest, and oldest organisms on Earth. Redwoods, found only in a narrow coastal strip of northern California, grow to heights of more than 110 m; only certain eucalyptus trees in Australia are taller. The largest (most massive) organisms alive are the giant sequoias, relatives of redwoods that grow in the Sierra Nevada mountains of California. One, known as the General Sherman tree, has a trunk with a circumference of 26 m and weighs more than the combined weight of a dozen space shuttles. Bristlecone pines, another species of California conifer, are among the oldest organisms alive. One bristlecone, named Methuselah, is more than 4,600 years old; it was a young tree when humans invented writing.

Nearly all conifers are evergreens, meaning they retain leaves throughout the year. Even during winter, a limited amount of photosynthesis occurs on sunny days. And when spring comes, conifers already have fully developed leaves that can take advantage of the sunnier days. The needle-shaped leaves of pines and firs are also adapted to survive dry seasons. A thick cuticle covers the leaf, and the stomata are located in pits, further reducing water loss.

We get most of our lumber and paper pulp from the wood of conifers. What we call wood is actually an accumulation of vascular tissue with lignin, which gives the tree structural support.

Terrestrial Adaptations of Seed Plants Compared to ferns, conifers and most other gymnosperms have three additional adaptations that make survival in diverse terrestrial habitats possible: (1) further reduction of the gametophyte; (2) the evolution of pollen; and (3) the advent of the seed.

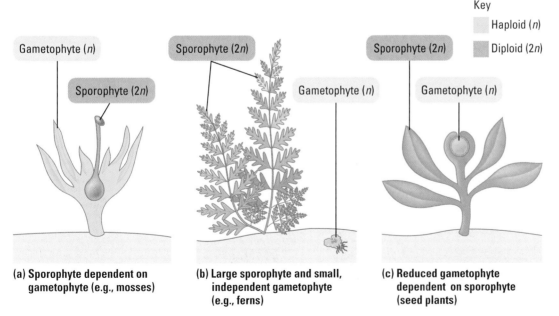

Key
Haploid (*n*)
Diploid (2*n*)

(a) **Sporophyte dependent on gametophyte (e.g., mosses)**

(b) **Large sporophyte and small, independent gametophyte (e.g., ferns)**

(c) **Reduced gametophyte dependent on sporophyte (seed plants)**

Gametophyte (*n*)   Sporophyte (2*n*)   Sporophyte (2*n*)   Gametophyte (*n*)   Sporophyte (2*n*)   Gametophyte (*n*)

Figure 16.14 Three variations on alternation of generations in plants.

The first adaptation is an even greater development of the diploid sporophyte compared to the haploid gametophyte generation (**Figure 16.14**).

A pine tree or other conifer is actually a sporophyte with tiny gametophytes living in cones (**Figure 16.15**). The gametophytes, though multicellular, are totally dependent on and protected by the tissues of the parent sporophyte. Some plant biologists speculate that the shift toward diploidy in land plants was related to the harmful impact of the sun's ionizing radiation, which causes mutations. This damaging radiation is more intense on land than in aquatic habitats, where organisms are somewhat protected by the light-filtering properties of water. Of the two generations of land plants, the diploid form (sporophyte) may cope better with mutagenic radiation. A diploid organism homozygous for a particular essential allele has a "spare tire" in the sense that one copy of the allele may be sufficient for survival if the other is damaged.

A second adaptation of seed plants to dry land was the evolution of **pollen.** A pollen grain is actually the much-reduced male gametophyte. It houses cells that will develop into sperm. In the case of conifers, wind carries the pollen from male to female cones, where eggs develop within female gametophytes (see Figure 16.15). This mechanism for sperm transfer contrasts with the swimming sperm of mosses and ferns. In seed plants, this use of resistant, airborne pollen to bring gametes together is a terrestrial adaptation that led to even greater success and diversity of plants on land.

The third important terrestrial adaptation of seed plants is, of course, the seed itself. A **seed,** remember, consists of a plant embryo packaged along with a food supply within a protective coat. Seeds develop from structures called ovules (**Figure 16.16**). In conifers, the ovules are located on the scales of female cones. Conifers and other gymnosperms, lacking ovaries, bear their seeds "naked" on the cone scales (though the seeds do have protective coats, of course). Once released from the parent plant, the resistant seed can remain dormant for days, months, or even years. Under favorable conditions, the seed can then **germinate,** its embryo emerging through the seed coat as a seedling. Some seeds drop close to their parents. Others are carried far by the wind or animals.

Activity 16E on the Web & CD
Follow the life cycle of a pine tree.

Figure 16.15 A pine tree, a conifer. The tree bears two types of cones. The hard, woody ones we usually notice are female cones. Each scale of the female cone (upper left inset) is actually a modified leaf bearing a pair of structures called ovules on its upper surface. An ovule contains the egg-producing female gametophyte. The smaller male cones (lower right inset) produce the male gametophytes, which are pollen grains. Mature male cones release clouds of millions of pollen grains. You may have seen yellowish conifer pollen covering car tops or floating on ponds in the spring. Some of the windblown pollen manages to land on female cones on trees of the same species. The female cones generally develop on the higher branches, where they are unlikely to be dusted with pollen from the same tree. Sperm released by pollen fertilizes eggs in the ovules of the female cones. The ovules eventually develop into seeds.

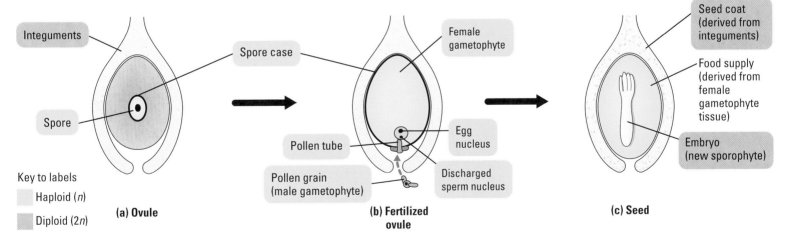

Key to labels
- Haploid (*n*)
- Diploid (2*n*)

(a) Ovule

(b) Fertilized ovule

(c) Seed

Figure 16.16 From ovule to seed. (a) The sporophyte produces spores within a tissue surrounded by a protective layer called integuments. **(b)** The spore develops into a female gametophyte, which produces one or more eggs. If a pollen grain enters the ovule through a special pore in the integuments, it discharges sperm cells that fertilize eggs. **(c)** Fertilization initiates the transformation of ovule to seed. The fertilized egg (zygote) develops into an embryo; the rest of the gametophyte forms a tissue that stockpiles food; and the integuments of the ovule harden to become the seed coat.

Angiosperms

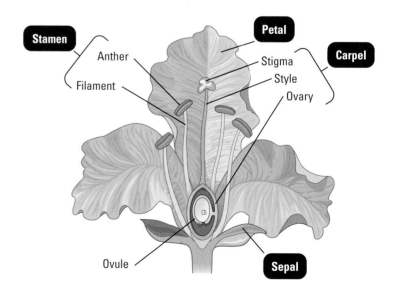

The photograph of the coniferous forest in Figure 16.13 could give us a somewhat distorted view of today's plant life. Conifers do cover much land in the northern parts of the globe, but it is the angiosperms, or flowering plants, that dominate most other regions. There are about 250,000 angiosperm species versus about 700 species of conifers and other gymnosperms. Whereas gymnosperms supply most of our lumber and paper, angiosperms supply nearly all our food and much of our fiber for textiles. Cereal grains, including wheat, corn, oats, and barley, are flowering plants, as are citrus and other fruit trees, garden vegetables, cotton, and flax. Fine hardwoods from flowering plants such as oak, cherry, and walnut trees supplement the lumber we get from conifers.

Case Study in the Process of Science on the Web & CD Learn to identify trees by their leaves.

Several unique adaptations account for the success of angiosperms. For example, refinements in vascular tissue make water transport even more efficient in angiosperms than in gymnosperms. Of all terrestrial adaptations, however, it is the flower that accounts for the unparalleled success of the angiosperms.

Flowers, Fruits, and the Angiosperm Life Cycle No organisms make a showier display of their sex lives than angiosperms. From roses to dandelions, flowers display a plant's male and female parts. For most angiosperms, insects and other animals transfer pollen from the male parts of one flower to the female sex organs of another flower. This targets the pollen rather than relying on the capricious winds to blow the pollen between plants of the same species.

A **flower** is actually a short stem with four whorls of modified leaves: sepals, petals, stamens, and carpels (**Figure 16.17**). At the bottom of the flower are the **sepals,** which are usually green. They enclose the flower before it opens (think of a rosebud). Above the sepals are the **petals,** which are usually the most striking part of the flower and are often important in attracting insects and other pollinators. The actual reproductive structures are multiple stamens and one or more carpels. Each **stamen** consists of a stalk bearing a sac called an **anther,** the male organ in which pollen grains develop. The **carpel** consists of a stalk, the **style,** with an ovary at the base and a sticky tip

Figure 16.17 Structure of a flower.

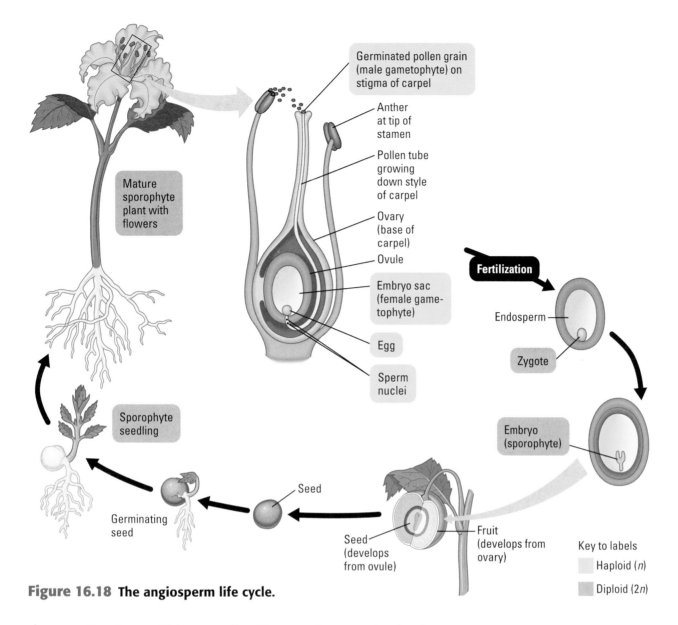

Figure 16.18 The angiosperm life cycle.

Labels in figure:
- Germinated pollen grain (male gametophyte) on stigma of carpel
- Anther at tip of stamen
- Pollen tube growing down style of carpel
- Ovary (base of carpel)
- Ovule
- Embryo sac (female gametophyte)
- Egg
- Sperm nuclei
- Mature sporophyte plant with flowers
- Fertilization
- Endosperm
- Zygote
- Embryo (sporophyte)
- Sporophyte seedling
- Seed
- Germinating seed
- Seed (develops from ovule)
- Fruit (develops from ovary)
- Key to labels: Haploid (*n*), Diploid (2*n*)

known as the **stigma,** which traps pollen. The **ovary** is a protective chamber containing one or more ovules, in which the eggs develop.

Figure 16.18 highlights key stages in the angiosperm life cycle. The plant familiar to us is the sporophyte. As in gymnosperms, the pollen grain is the male gametophyte of angiosperms. The female gametophyte is located within an ovule, which in turn resides within a chamber of the ovary. Pollen that lands on the sticky stigma of a carpel extends a tube down to an ovule and deposits two sperm nuclei within the female gametophyte. This **double fertilization** is an angiosperm characteristic. One sperm cell fertilizes an egg in the female gametophyte. This produces a zygote, which develops into an embryo. The second sperm cell fertilizes another female gametophyte cell, which then develops into a nutrient-storing tissue called **endosperm.** Double fertilization thus synchronizes the development of the embryo and food reserves within an ovule. The whole ovule develops into a seed. The seed's enclosure within an ovary is what distinguishes angiosperms from the naked-seed condition of gymnosperms.

A **fruit** is the ripened ovary of a flower. As seeds are developing from ovules, the ovary wall thickens, forming the fruit that encloses the seeds.

Activity 16F on the Web & CD
Quiz yourself on the angiosperm life cycle.

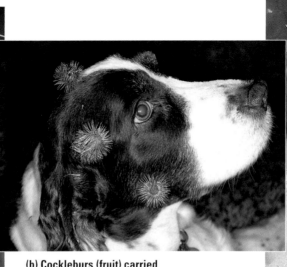

(b) Cockleburs (fruit) carried by animal fur

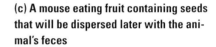

(c) A mouse eating fruit containing seeds that will be dispersed later with the animal's feces

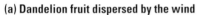

(a) Dandelion fruit dispersed by the wind

Figure 16.19 Fruits and seed dispersal.

(a) Some angiosperms depend on wind for seed dispersal. For example, the dandelion fruit acts like a kite, carrying a tiny seed far away from its parent plant. (b) Some fruits are adapted to hitch free rides on animals. The cockleburs attached to the fur of this dog are fruit that may be carried miles before opening and releasing seeds. (c) Many angiosperms produce fleshy, edible fruits that are attractive to animals as food. When a mouse eats a berry, it digests the fleshy part of the fruit, but most of the tough seeds pass unharmed through the mouse's digestive tract. The mouse later deposits the seeds, along with a fertilizer supply, some distance from where it ate the fruit. Many types of garden produce are the edible fruits from plants we have domesticated. Examples include tomatoes, squash, melons, strawberries, apples, oranges, and cherries.

A pea pod is an example of a fruit, with seeds (mature ovules, the peas) encased in the ripened ovary (the pod). Fruits protect and help disperse seeds. As **Figure 16.19** demonstrates, many angiosperms depend on animals to disperse seeds. Conversely, most land animals, including humans, rely on angiosperms as a food source.

Angiosperms and Agriculture Flowering plants provide nearly all our food. All of our fruit and vegetable crops are angiosperms. Corn, rice, wheat, and the other grains are grass fruits. Grains are also the main food source for domesticated animals, such as cows and chickens. We also grow angiosperms for fiber, medications, perfumes, and decoration.

Like other animals, early humans probably collected wild seeds and fruits. Agriculture was gradually invented as humans began sowing seeds and cultivating plants to have a more dependable food source. As they domesticated certain plants, humans began to intervene in plant evolution by selective breeding designed to improve the quantity and quality of the foods. Agriculture is a unique kind of evolutionary relationship between plants and animals.

Plant Diversity as a Nonrenewable Resource

The exploding human population, with its demand for space and natural resources, is extinguishing plant species at an unprecedented rate. The problem is especially critical in the tropics, where more than half the human population lives and population growth is fastest. Tropical rain forests are being destroyed at a frightening pace. The most common cause of this destruction is slash-and-burn clearing of the forest for agricultural use. Fifty million acres, an area about the size of the state of Washington, are cleared each year, a rate that would completely eliminate Earth's tropical forests within 25 years. As the forest disappears, so do thousands of plant species. Insects and other rain forest animals that depend on these plants are also vanishing. In all, researchers estimate that the destruction of habitat in the rain forest and other ecosystems is claiming

Activity 16G on the Web & CD
Learn about the biodiversity crisis in Madagascar.

hundreds of species each year. The toll is greatest in the tropics because that is where most species live; but environmental assault is a generically human tendency. Europeans eliminated most of their forests centuries ago, and habitat destruction is now endangering many species in North America. Extinction is irrevocable; plant diversity is a nonrenewable resource.

Many people have ethical concerns about contributing to the extinction of living forms. But there are also practical reasons to be concerned about the loss of plant diversity. We depend on plants for thousands of products, including food, building materials, and medicines (**Table 16.1**). So far, we have explored the potential uses of only a tiny fraction of the 300,000 known plant species. For example, almost all our food is based on the cultivation of only about two dozen species. More than 120 prescription drugs are extracted from plants. However, researchers have investigated fewer than 5,000 plant species as potential sources of medicine. And pharmaceutical companies were led to most of these species by local peoples who use the plants in preparing their traditional medicines.

The tropical rain forest may be a medicine chest of healing plants that could become extinct before we even know they exist. This is only one reason to value what is left of plant diversity and to search for ways to slow the loss. The solutions we propose must be economically realistic. If the goal is only profit for the short term, then we will continue to slash and burn until the forests are gone. If, however, we begin to see rain forests and other ecosystems as living treasures that can regenerate only slowly, we may learn to harvest their products at sustainable rates.

Table 16.1	A Sampling of Medicines Derived from Plants		
Compound	**Example of Source**		**Example of Use**
Atropine	Belladonna plant		Pupil dilator in eye exams
Digitalin	Foxglove		Heart medication
Menthol	Eucalyptus tree		Ingredient in cough medicines
Morphine	Opium poppy		Pain reliever
Quinine	Quinine tree		Malaria preventive
Taxol	Pacific yew		Ovarian cancer drug
Tubocurarine	Curare tree		Muscle relaxant during surgery
Vinblastine	Periwinkle		Leukemia drug

Source: Adapted from Randy Moore et al., *Botany,* 2nd ed. Dubuque, IA: Brown, 1998. Table 2.2, p. 37.

We have seen in our survey of plants, especially the angiosperms, how entangled the botanical world is with other terrestrial life. We switch our attention now to that other group of organisms that moved onto land with plants, the kingdom Fungi.

CHECKPOINT

1. Which of the following structures is common to all four major plant groups: vascular tissue, flowers, seeds, cuticle, pollen?

2. Gametophyte is to _____ as _____ is to diploid.

3. Why are coal, oil, and natural gas called "fossil" fuels?

4. How does the evergreen nature of pines and other conifers adapt the plants for living where the growing season is very short?

5. Contrast the mode of sperm delivery in ferns with sperm delivery in conifers.

6. (a) In what way are forests renewable resources? (b) Under what conditions does reality belie the "renewable" designation?

7. _____ are to conifers as flowers are to _____.

8. What are the four main organs of a flower?

9. What is a fruit?

Answers: **1.** Cuticle **2.** haploid; sporophyte **3.** Because they are derived from ancient organisms that did not decay completely after dying **4.** Because the plants do not lose their leaves during autumn and winter, the leaves are already fully developed for photosynthesis when the short growing season begins in spring. **5.** The flagellated sperm of ferns must swim through water to reach eggs. In contrast, the airborne pollen of conifers brings gametes together without need of an aqueous route, and the pollen releases sperm near the egg after pollination occurs. **6.** (a) Forests are renewable in the sense that new trees can grow where old growth has been removed by logging. (b) The current situation is the harvesting of wood at a much faster rate than new growth can replace, and thus "renewable" is an unrealistic designation for forest resources. **7.** Cones; angiosperms **8.** Sepals, petals, stamens, and carpels **9.** A ripened ovary of a flower that protects and aids in the dispersal of seeds contained in the fruit

Fungi

The word *fungus* often evokes some unpleasant images. Fungi rot timbers, spoil food, and afflict humans with athlete's foot and worse maladies. However, ecosystems would collapse without fungi to decompose dead organisms, fallen leaves, feces, and other organic materials, thus recycling vital chemical elements back to the environment in forms other organisms can assimilate. And you have already learned that nearly all plants have mycorrhizae, fungus-root associations that absorb minerals and water from the soil. In addition to these ecological roles, fungi have been used by humans in various ways for centuries. We eat some fungi (mushrooms and truffles, for instance), culture fungi to produce antibiotics and other drugs, add them to dough to make bread rise, culture them in milk to produce a variety of cheeses, and use them to ferment beer and wine.

Case Study in the Process of Science on the Web & CD Conduct experiments to find out how the fungus *Pilobolus* ends up in cow manure.

Fungi are eukaryotes, and most are multicellular. They were once grouped with plants. But in fact, molecular studies indicate that fungi and animals probably arose from a common ancestor. In other words, a mushroom is probably more closely related to you than it is to any plant! However, fungi are actually a form of life so distinctive that they are accorded their own kingdom, the kingdom Fungi (**Figure 16.20**).

(a) Fly agaric mushrooms

(b) A "fairy ring"

(c) *Pilobolus*

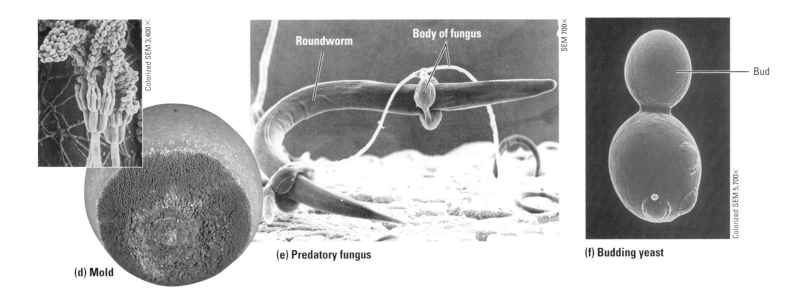

Colorized SEM 3,400×

(d) Mold

Roundworm

Body of fungus

SEM 700×

(e) Predatory fungus

Bud

Colorized SEM 5,700×

(f) Budding yeast

Figure 16.20 A gallery of diverse fungi. (a) These mushrooms are the reproductive structures of a fungus that absorbs nutrients as it decomposes compost on a forest floor. (b) Some mushroom-producing fungi poke up "fairy rings," which can appear on a lawn overnight. The legendary explanation of these circles is that mushrooms spring up where fairies have danced in a ring on moonlit nights. Afterward, the tired fairies sit down on some of the mushrooms, but toads use other mushrooms as stools; hence the name toadstools. Biology offers an alternative explanation. A ring develops at the edge of the main body of the fungus, which consists of an underground mass of tiny filaments within the ring. The filaments secrete enzymes that digest soil compost. As the underground fungal mass grows outward from its center, the diameter of the fairy rings produced at its expanding perimeter increases annually. (c) This fungus, *Pilobolus,* decomposes animal dung. The bulbs at the tips of the stalks are sacs of spores, which are reproductive cells. *Pilobolus* can actually aim these spore sacs. The stalks bend toward light, where grass is likely to be growing, and then shoot their spore sacs like cannonballs. Grazing animals eat the spore sacs and scatter the spores in feces, where the spores grow into new fungi. (d) The fungi we call molds grow rapidly on their food sources, often on *our* food sources. The mold on this orange reproduces asexually by producing chains of microscopic spores (inset) that are dispersed via air currents. (e) This predatory fungus traps and feeds on tiny roundworms in the soil. The fungus is equipped with hoops that can constrict around a worm in less than a second. (f) Yeasts are unicellular fungi. This yeast cell is reproducing asexually by a process called budding. For centuries, humans have domesticated yeasts and put their metabolism to work in breweries and bakeries.

Characteristics of Fungi

In this section, we'll examine the structure and function of fungi, beginning with an overview of how fungi obtain nutrients.

Fungal Nutrition Fungi are heterotrophs that acquire their nutrients by **absorption.** In this mode of nutrition, small organic molecules are absorbed from the surrounding medium. A fungus digests food outside its body by secreting powerful hydrolytic enzymes into the food. The enzymes decompose complex molecules to the simpler compounds that the fungus can absorb. For example, fungi that are decomposers absorb nutrients from nonliving organic material, such as fallen logs, animal corpses, or the wastes of live organisms. Parasitic fungi absorb nutrients from the cells or body fluids of living hosts. Some of these fungi, such as certain species infecting the lungs of humans, are pathogenic. In other cases, such as mycorrhizae, the relationships between fungi and their hosts are mutually beneficial.

Fungal Structure Fungi are structurally adapted for their absorptive nutrition. The bodies of most fungi are constructed of structures called **hyphae** (singular, *hypha*). Hyphae are minute threads composed of tubular walls surrounding plasma membranes and cytoplasm. The hyphae form an interwoven mat called a **mycelium** (plural, *mycelia*), which is the feeding network of a fungus (**Figure 16.21**). Fungal mycelia can be huge, although they usually escape our notice because they are often subterranean. In 2000, scientists discovered the mycelium of one humongous fungus in Oregon that is 5.5 km (3.4 miles) in diameter and spreads through 2,200 acres of forest (equivalent to over 1,600 football fields). This fungus is at least 2,400 years old and hundreds of tons in weight, qualifying it among Earth's oldest and largest organisms.

Most fungi are multicellular, with hyphae divided into cells by cross-walls. The cross-walls generally have pores large enough to allow ribosomes, mitochondria, and even nuclei to flow from cell to cell. The cell walls of

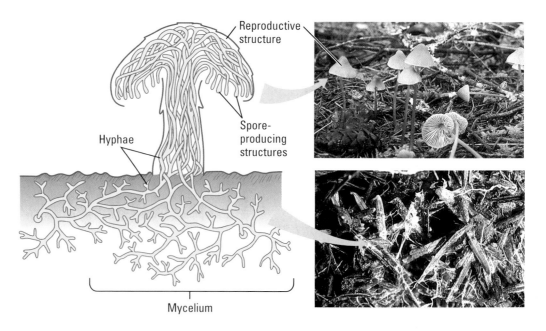

Figure 16.21 The fungal mycelium.   The mushroom we see is like the tip of an iceberg. It is a reproductive structure consisting of tightly packed hyphae that extend upward from a much more massive mycelium of hyphae growing underground. The photos show mushrooms and the mycelium of cottony threads that decompose organic litter.

Reproductive structure

Spore-producing structures

Hyphae

Mycelium

fungi differ from the cellulose walls of plants. Most fungi build their cell walls mainly of chitin, a strong but flexible polysaccharide similar to the chitin found in the external skeletons of insects.

Mingling with the organic matter it is decomposing and absorbing, a mycelium maximizes contact with its food source. Ten cubic centimeters of rich organic soil may contain as much as a kilometer of hyphae. And a fungal mycelium grows rapidly, adding as much as a kilometer of hyphae each day as it branches within its food. Fungi are nonmotile organisms; they cannot run, swim, or fly in search of food. But the mycelium makes up for the lack of mobility by swiftly extending the tips of its hyphae into new territory.

Fungal Reproduction Fungi reproduce by releasing spores that are produced either sexually or asexually. The output of spores is mind-boggling. For example, puffballs, which are the reproductive structures of certain

Activity 16H on the Web & CD Watch an animation of fungal reproduction and nutrition.

fungi, can puff out clouds containing trillions of spores (see Figure 13.14). Carried by wind or water, spores germinate to produce mycelia if they land in a moist place where there is food. Spores thus function in dispersal and account for the wide geographic distribution of many species of fungi. The airborne spores of fungi have been found more than 160 km (100 miles) above Earth. Closer to home, try leaving a slice of bread out for a week or two and you will observe the furry mycelia that grow from the invisible spores raining down from the surrounding air. It's a good thing those particular molds cannot grow in our lungs.

The Ecological Impact of Fungi

Fungi have been major players in terrestrial communities ever since they moved onto land in the company of plants. Let's examine a few examples of how fungi continue to have an enormous ecological impact.

Fungi as Decomposers Fungi and bacteria are the principal decomposers that keep ecosystems stocked with the inorganic nutrients essential for plant growth. Without decomposers, carbon, nitrogen, and other elements would accumulate in organic matter. Plants and the animals they feed would starve because elements taken from the soil would not be returned (see Chapter 19).

Fungi are well adapted as decomposers of organic refuse. Their invasive hyphae enter the tissues and cells of dead organic matter and hydrolyze polymers, including the cellulose of plant cell walls. A succession of fungi, in concert with bacteria and, in some environments, invertebrate animals, is responsible for the complete breakdown of organic litter. The air is so loaded with fungal spores that as soon as a leaf falls or an insect dies, it is covered with spores and is soon infiltrated by fungal hyphae.

We may applaud fungi that decompose forest litter or dung, but it is a different story when molds attack our fruit or our shower curtains. Between 10% and 50% of the world's fruit harvest is lost each year to fungal attack. And a wood-digesting fungus does not distinguish between a fallen oak limb and the oak planks of a boat. During the Revolutionary War, the British lost more ships to fungal rot than to enemy attack. What's more, soldiers stationed in the tropics during World War II watched as their tents, clothing, boots, and binoculars were destroyed by molds. Some fungi can even decompose certain plastics.

Figure 16.22 Parasitic fungi that cause plant disease. **(a)** This photo shows American elm trees after the arrival of the parasitic fungus that causes Dutch elm disease. The fungus evolved with European species of elm trees, and it is relatively harmless to them. But it is deadly to American elms. The fungus was accidentally introduced into the United States on logs sent from Europe to pay World War I debts. Insects called bark beetles carried the fungus from tree to tree. Since then, the disease has destroyed elm trees all across North America. **(b)** The seeds of some kinds of grain, including rye, wheat, and oats, are sometimes infected with fungal growths called ergots, the dark structures on this seed head of rye. Consumption of flour made from ergot-infested grain can cause gangrene, nervous spasms, burning sensations, hallucinations, temporary insanity, and death. One epidemic in Europe in the year A.D. 944 killed more than 40,000 people. During the Middle Ages, the disease (ergotism) became known as Saint Anthony's fire because many of its victims were cared for by a Catholic nursing order dedicated to Saint Anthony. Several kinds of toxins have been isolated from ergots. One called lysergic acid is the raw material from which the hallucinogenic drug LSD is made. Certain other chemical extracts are medicinal in small doses. One ergot compound is useful in treating high blood pressure, for example.

(a) American elm trees killed by a fungus

(b) Ergots on rye

Parasitic Fungi Of the 100,000 known species of fungi, about 30% make their living as parasites, mostly on or in plants. In some cases, fungi that infect plants have literally changed landscapes. One species, for example, has eliminated most American elm trees (**Figure 16.22a**). Fungi are also serious agricultural pests. Some species infect grain crops and cause tremendous economic losses each year (**Figure 16.22b**).

Animals are much less susceptible to parasitic fungi than are plants. Only about 50 species of fungi are known to be parasitic in humans and other animals. However, their effects are significant enough to make us take them seriously. Among the diseases that fungi cause in humans are yeast infections of the lungs, some of which can be fatal, and vaginal yeast infections. Other fungal parasites produce a skin disease called ringworm, so named because it appears as circular red areas on the skin. The ringworm fungi can infect virtually any skin surface. Most commonly, they attack the feet and cause intense itching and sometimes blisters. This condition, known as athlete's foot, is highly contagious but can be treated with various fungicidal preparations.

Commercial Uses of Fungi It would not be fair to fungi to end our discussion with an account of diseases. In addition to their positive global impact as decomposers, fungi also have a number of practical uses for humans.

Most of us have eaten mushrooms, although we may not have realized that we were ingesting the reproductive extensions of subterranean fungi. Mushrooms are often cultivated commercially in artificial caves in which cow manure is piled (be sure to wash your store-bought mushrooms thoroughly). Edible mushrooms also grow wild in fields, forests, and backyards,

(a) Truffles

(b) Blue cheese

Figure 16.23 Feeding on fungi. **(a)** Truffles (the fungal kind, not the chocolates) are the reproductive structures of fungi that grow with tree roots as mycorrhizae. Truffles release strong odors that attract mammals and insects that excavate the fungi and disperse their spores. In some cases, the odors mimic sex attractants of certain mammals. Truffle hunters traditionally used pigs to locate their prizes. However, dogs are now more commonly used because they have the nose for the scent without the fondness for the flavor. Gourmets describe the complex flavors of truffles as nutty, musky, cheesy, or some combination of those tastes. At about $400 per pound for truffles, you probably won't get a chance to do a taste test of your own in the campus cafeteria. **(b)** The turquoise streaks in blue cheese and Roquefort are the mycelia of a specific fungus.

but so do poisonous ones. There are no simple rules to help the novice distinguish edible from deadly mushrooms. Only experts in mushroom taxonomy should dare to collect the fungi for eating.

Mushrooms are not the only fungi we eat. The fungi called truffles are highly prized by gourmets (**Figure 16.23a**). And the distinctive flavors of certain kinds of cheeses come from the fungi used to ripen them (**Figure 16.23b**). Particularly important in food production are unicellular fungi, the yeasts. As discussed in Chapter 6, yeasts are used in baking, brewing, and winemaking.

Fungi are medically valuable as well. Some fungi produce antibiotics that are used to treat bacterial diseases. In fact, the first antibiotic discovered was penicillin, which is made by the common mold called *Penicillium* (**Figure 16.24**).

As sources of antibiotics and food, as decomposers, and as partners with plants in mycorrhizae, fungi play vital roles in life on Earth.

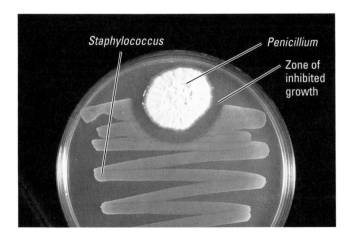

Figure 16.24 Fungal production of an antibiotic. The first antibiotic discovered was penicillin, which is made by the common mold called *Penicillium*. In this petri dish, the clear area between the mold and the bacterial colony is where the antibiotic produced by *Penicillium* inhibits the growth of the bacteria, a species of *Staphylococcus*.

1. What are mycorrhizae?

2. Contrast the heterotrophic nutrition of a fungus with your own heterotrophic nutrition.

3. Describe how a fungal mycelium adapts the organism for its absorptive nutrition.

4. What is "athlete's foot"?

5. What do you think is the function of the antibiotics that fungi produce in their natural environments?

Answers: 1. Root-fungus symbiotic associations that enhance the uptake of water and minerals by the plant and provide organic nutrients to the fungus **2.** A fungus digests its food externally by secreting digestive juices into the food and then absorbing the small nutrients that result from digestion. In contrast, humans and most other animals "eat" relatively large pieces of food and digest the food within their bodies. **3.** The extensive network of hyphae puts much surface area in touch with the food source, and rapid growth of the mycelium extends hyphae into new territory. **4.** Infection of the foot's skin with ringworm fungus **5.** The antibiotics block the growth of microorganisms, especially bacteria, that compete with the fungi for nutrients and other resources.

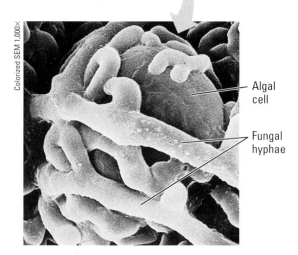

Colorized SEM 1,000×

Algal cell

Fungal hyphae

Figure 16.25 Lichens: symbiotic associations of fungi and algae. Lichens generally grow very slowly, sometimes in spurts of less than a millimeter per year. You can date the oldest lichens you see here by the engraving on the gravestone. Elsewhere, there are lichens that are thousands of years old, rivaling the oldest plants as Earth's elders. The close relationship between the fungal and algal partners is evident in the microscopic blowup of a lichen.

Evolution Connection

Mutual Symbiosis

Evolution is not just about the origin and adaptation of individual species. Relationships between species are also an evolutionary product. **Symbiosis** is the term used to describe ecological relationships between organisms of different species that are in direct contact. Parasitism is a symbiotic relationship in which one species, the parasite, benefits while harming its host in the process. Our focus here, however, is on **mutualism,** symbiosis that benefits both species.

We have seen many examples of mutualism over the past two chapters. Eukaryotic cells evolved from mutual symbiosis among prokaryotes. And today, bacteria living in the roots of certain plants provide nitrogen compounds to their host and receive food in exchange. We have our own mutually symbiotic bacteria that help keep our skin healthy and produce certain vitamins in our intestines. Particularly relevant to this chapter is the symbiotic association of fungi and plant roots—mycorrhizae—which made life's move onto land possible.

Lichens, symbiotic associations of fungi and algae, are striking examples of how two species can become so merged that the cooperative is essentially a new life-form. At a distance, it is easy to mistake lichens for mosses or other simple plants growing on rocks, rotting logs, trees, roofs, or gravestones (**Figure 16.25**). In fact, lichens are not mosses or any other kind of plant, nor are they even individual organisms. A lichen is a symbiotic association of millions of tiny algae embraced by a mesh of fungal hyphae. The photosynthetic algae feed the fungi. The fungal mycelium, in turn, provides a suitable habitat for the algae, helping to absorb and retain water and minerals. The mutualistic merger of partners is so complete that lichens are actually named as species, as though they are individual organisms. Mutualisms such as lichens and mycorrhizae showcase the web of life that has evolved on Earth.

Evolution Connection on the Web
Discover more about lichens.

Chapter Review

Summary of Key Concepts

For study help, go to the Essential Biology Website (www.essentialbiology.com) or CD-ROM to explore the Activities and Case Studies in the Process of Science.

Colonizing Land

- **Terrestrial Adaptations of Plants** Plants are multicellular photosynthetic eukaryotes with adaptations for living on land.

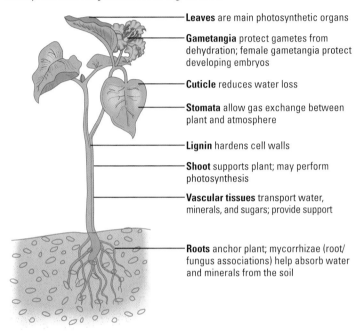

Leaves are main photosynthetic organs

Gametangia protect gametes from dehydration; female gametangia protect developing embryos

Cuticle reduces water loss

Stomata allow gas exchange between plant and atmosphere

Lignin hardens cell walls

Shoot supports plant; may perform photosynthesis

Vascular tissues transport water, minerals, and sugars; provide support

Roots anchor plant; mycorrhizae (root/fungus associations) help absorb water and minerals from the soil

Activity 16A Terrestrial Adaptations of Plants

- **The Origin of Plants from Green Algae** Plants probably evolved gradually from green algae. Molecular comparisons and other evidence place a group of multicellular green algae called charophyceans closest to plants.

Plant Diversity

- **Highlights of Plant Evolution** Four major periods of plant evolution are marked by terrestrial adaptations.

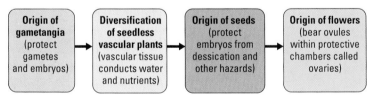

Origin of gametangia (protect gametes and embryos) → **Diversification of seedless vascular plants** (vascular tissue conducts water and nutrients) → **Origin of seeds** (protect embryos from dessication and other hazards) → **Origin of flowers** (bear ovules within protective chambers called ovaries)

Activity 16B Highlights of Plant Evolution

- **Bryophytes** The most familiar bryophytes are mosses. Mosses display two key terrestrial adaptations: a waxy cuticle that prevents dehydration and the retention of developing embryos within the mother plant's gametangium. Mosses are most common in moist environments because their sperm must swim to the eggs and because they lack lignin in their cell walls and thus cannot stand tall. Bryophytes are unique among plants in having the gametophyte as the dominant generation in the life cycle.

Activity 16C Moss Life Cycle

- **Ferns** Ferns are seedless plants that have vascular tissues but still use flagellated sperm to fertilize eggs. During the Carboniferous period, many giant ferns were among the plants that formed thick deposits of organic matter that were gradually converted into coal.

Activity 16D Fern Life Cycle

Case Study in the Process of Science What Are the Different Stages of a Fern Life Cycle?

- **Gymnosperms** A drier and colder global climate near the end of the Carboniferous favored the evolution of the first seed plants. The most successful were the gymnosperms, represented by conifers. Needle-shaped leaves with thick cuticles and sunken stomata are adaptations to dry conditions. Conifers and most other gymnosperms have three additional terrestrial adaptations: (1) further reduction of the gametophyte generation and greater development of the diploid sporophyte; (2) the evolution of pollen, which doesn't require water for transport; and (3) the advent of the seed, which consists of a plant embryo packaged along with a food supply within a protective coat.

Activity 16E Pine Life Cycle

- **Angiosperms** Angiosperms supply nearly all our food and much of our fiber for textiles. The evolution of the flower and more efficient water transport help account for the success of the angiosperms. The dominant stage is a sporophyte with gametophytes in its flowers. The female gametophyte is located within an ovule, which in turn resides within a chamber of the ovary. Fertilization of an egg in the female gametophyte produces a zygote, which develops into an embryo. The whole ovule develops into a seed. The seed's enclosure within an ovary is what distinguishes angiosperms from the naked-seed condition of gymnosperms. A fruit is the ripened ovary of a flower. Fruits protect and help disperse seeds. Angiosperms are a major food source for animals, while animals aid plants in pollination and seed dispersal. Agriculture constitutes a unique kind of evolutionary relationship among humans, plants, and animals.

Case Study in the Process of Science How Are Trees Identified by Their Leaves?

Activity 16F Angiosperm Life Cycle

- **Plant Diversity as a Nonrenewable Resource** The exploding human population, with its demand for space and natural resources, is causing the extinction of plant species at an unprecedented rate.

Activity 16G Madagascar and the Biodiversity Crisis

Fungi

- **Characteristics of Fungi** Fungi are unicellular or multicellular eukaryotes that are probably more closely related to animals than to plants. Fungi are heterotrophs that digest their food externally and absorb the nutrients. A fungus usually consists of a mass of threadlike hyphae forming a mycelium. The cell walls of fungi are mainly composed of chitin. Although most fungi are nonmotile, the mycelium can swiftly extend the tips of its hyphae into new territory. Fungi reproduce and disperse by releasing spores that are produced either sexually or asexually.

Case Study in the Process of Science How Does the Fungus Pilobolus Succeed as a Decomposer?

Activity 16H Fungal Reproduction and Nutrition

- **The Ecological Impact of Fungi** Fungi and bacteria are the principal decomposers of ecosystems. Many molds destroy fruit, wood, and human-made materials. About 50 species of fungi are known to be parasitic in humans and other animals. Fungi are also commercially important as food and in baking, beer and wine production, and the manufacture of antibiotics.

Self-Quiz

1. Angiosperms are different from all other plants because only angiosperms have _____.

2. Ovule is to seed as ovary is to _____.

3. Under a microscope, a piece of a mushroom would look most like
   a. jelly.
   c. grains of sand.
   b. a tangle of string.
   d. a sponge.

4. During the Carboniferous period, the dominant plants, which later formed the great coal beds, were mainly
   a. mosses and other bryophytes.
   b. ferns and other seedless vascular plants.
   c. charophyceans and other green algae.
   d. conifers and other gymnosperms.

5. You discover a new species of plant. Under the microscope, you find that it produces flagellated sperm. A genetic analysis shows that its dominant generation has diploid cells. What kind of plant do you have?

6. Which of the following terms includes all others in the list: angiosperm, fern, vascular plant, gymnosperm, seed plant.

7. Plant diversity is greatest in
   a. tropical forests.
   c. deserts.
   b. the temperate forests of Europe.
   d. the oceans.

8. Name five products you've used today that come from angiosperms.

9. Most lichens are symbionts of photosynthetic _____ with _____.

10. Fungi acquire nutrients by _____.

Answers to the Self-Quiz questions can be found in Appendix B.
Go to the website or CD-ROM for more Self-Quiz questions.

The Process of Science

1. In April 1986, an accident at a nuclear power plant in Chernobyl, Ukraine, scattered radioactive fallout for hundreds of miles. In assessing the biological effects of the radiation, researchers found mosses to be especially valuable as organisms for monitoring the damage. Radiation damages organisms by causing mutations. Explain why it is faster to observe the genetic effects of radiation on mosses than on other types of plants. Imagine that you are conducting tests shortly after a nuclear accident. Using potted moss plants as your experimental organisms, design an experiment to test the hypothesis that the frequency of mutations decreases with the organism's distance from the source of radiation.

2. You discover what you think may be one extremely large underground fungal mycelium living beneath your campus. How could you prove that it is, in fact, one individual organism spread across a very large area, as opposed to a group of separate organisms?

Case Study in the Process of Science on the Web & CD *See photos of the stages of the fern life cycle.*

Case Study in the Process of Science on the Web & CD *Learn to identify trees by their leaves.*

Case Study in the Process of Science on the Web & CD *Conduct experiments to find out how the fungus* Pilobolus *ends up in cow manure.*

Biology and Society

1. Why are tropical rain forests being destroyed at such an alarming rate? What kinds of social, technological, and economic factors are responsible? Most forests in developed Northern Hemisphere countries have already been cut. Do the developed nations have a right to pressure the developing nations in the Southern Hemisphere to slow or stop the destruction of their forests? Defend your answer. What kinds of benefits, incentives, or programs might slow the assault on the rain forests?

2. Imagine you were charged with the task of managing a coniferous forest. How would you balance the need for productive use of the forest (to provide lumber, for example) with preservation of its diversity? What activities would you allow or prohibit in the forest (for example, snowmobiling, logging, hiking, mushroom harvesting, mining, camping, grazing, making campfires)? How would you defend your choices to the public?

Biology and Society on the Web *Learn more about the impact of deforestation.*

CHAPTER 17

The Evolution of Animals

Biology and Society:
Invasion of the Killer Toads 344

The Origins of Animal Diversity 344

What Is an Animal?

Early Animals and the Cambrian Explosion

Animal Phylogeny

Major Invertebrate Phyla 349

Sponges

Cnidarians

Flatworms

Roundworms

Mollusks

Annelids

Arthropods

Echinoderms

The Vertebrate Genealogy 360

Characteristics of Chordates

Fishes

Amphibians

Reptiles

Birds

Mammals

The Human Ancestry 368

The Evolution of Primates

The Emergence of Humankind

Evolution Connection:
Earth's New Crisis 376

Zoologists estimate that about a **billion billion** (10^{18}) individual arthropods populate Earth.

The blue whale, an endangered species that grows to lengths of nearly 30 meters, is the **largest animal** that has ever existed.

Tapeworms can reach lengths of 20 meters in the human intestine.

A reptile can survive on less than 10% of the calories required by a mammal of equivalent size.

(a) A quoll

(b) A cane toad

Figure 17.1 Native versus invasive animals.
(a) Populations of the Australian quoll seriously declined after introduction of **(b)** the non-native cane toad.

Biology and Society on the Web
Learn about how invasive species arrive in the United States.

Biology and Society

Invasion of the Killer Toads

Of the 1.7 million species of organisms known to science, over two-thirds are animals. This incredible diversity arose through hundreds of millions of years of evolution as natural selection shaped animal adaptations to Earth's many environments. But this diversity can be quickly threatened by actions that throw ecosystems into disarray—as when a small catlike creature meets an unstoppable overseas toad.

That catlike creature is the Australian quoll (**Figure 17.1a**). Like the majority of Australian animals, quolls are native only to that continent. And when such local species encounter invasive non-native species, the results can be disastrous.

Quolls are predators, hunting smaller animals, including many types of frogs. Their diet didn't present a problem until 1935, when sugarcane growers decided to import a boxful of special toads to fight beetles that were damaging their sugar crops. The non-native amphibians from South America, known as cane toads (**Figure 17.1b**), didn't do much to stop the beetles. Instead, the 102 toads in that box quickly turned into one of Australia's biggest wildlife disasters.

The invaders bred and spread quickly. Adult cane toads have voracious appetites, eating everything from dog food to mice and devouring many native insects and small animals along the way. But cane toads themselves are not good prey: They are poisonous to almost all their predators, including quolls. Together with habitat loss and other environmental pressures, the introduction of cane toads to Australia has driven native quolls into serious decline.

This scenario of invasive non-native organisms versus native organisms has been repeated over and over in Australia and other parts of the world as humans have introduced species that threaten natural ecosystems. Australia's ongoing efforts to undo the damage from invasive species—at a cost of millions of dollars each year—have become a continent-wide example of how precarious diversity can be. Millions of years worth of evolutionary changes can be threatened by just a few years of human-caused changes to native habitats.

In this chapter on animal diversity, we'll look at 9 of the roughly 35 phyla (major groups) in the kingdom Animalia. These major phyla contain the greatest number of species and are the most abundant and widespread. Along the way, we'll give special attention to the major milestones in animal evolution.

The Origins of Animal Diversity

Animal life began in Precambrian seas with the evolution of multicellular creatures that ate other organisms. We are among their descendants.

What Is an Animal?

Animals are eukaryotic, multicellular, heterotrophic organisms that obtain nutrients by ingestion. *That's* a mouthful. And speaking of mouthfuls, **ingestion** means eating food. This mode of nutrition contrasts animals with fungi, which obtain nutrients by absorption after digesting the food outside the body (see Chapter 16). Animals digest their food within their bodies after ingesting other organisms, dead or alive, whole or by the piece (**Figure 17.2**).

A few key features of life history also distinguish animals. Most animals reproduce sexually. The zygote (fertilized egg) develops into an early embryonic stage called a **blastula,** which is usually a hollow ball of cells (**Figure 17.3**). The next embryonic stage in most animals is a gastrula, which has layers of cells that will eventually form the adult body parts. The gastrula also has a primitive gut, which will develop into the animal's digestive compartment. Continued development, growth, and maturation transform some animals directly from the embryo into an adult. However, the life histories of many animals include larval stages. A **larva** is a sexually immature form of an animal. It is anatomically distinct from the adult form, usually eats different foods, and may even have a different habitat. Think how different a frog is from its larval form, which we call a tadpole. A change of body form, called **metamorphosis,** eventually remodels the larva into the adult form.

Most animals have muscle cells, as well as nerve cells that control the muscles. The evolution of this equipment for coordinated movement enhanced feeding, even enabling some animals to search for or chase their food. The most complex animals, of course, can use their muscular and nervous systems for many functions other than eating. Some species even use massive networks of nerve cells called brains to think.

Figure 17.2 Nutrition by ingestion, the animal way of life. Most animals ingest relatively large pieces of food, though rarely as large as the prey in this case. In this amazing scene, a rock python is beginning to ingest a gazelle. The snake will spend two weeks or more in a quiet place digesting its meal.

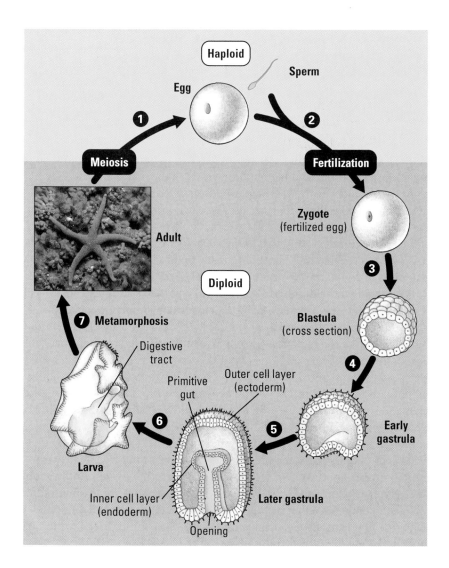

Figure 17.3 Life cycle of a sea star as an example of animal development. ❶ Male and female adult animals produce haploid gametes (eggs and sperm) by meiosis. ❷ An egg and a sperm fuse to produce a diploid zygote. ❸ Early mitotic divisions lead to an embryonic stage called a blastula, common to all animals. Typically, the blastula consists of a ball of cells surrounding a hollow cavity. ❹ Later, in the sea star and many other animals, one side of the blastula cups inward, forming an embryonic stage called a gastrula. ❺ The gastrula develops into a saclike embryo with a two-layered wall and an opening at one end. Eventually, the outer layer (ectoderm) develops into the animal's epidermis (skin) and nervous system. The inner layer (endoderm) forms the digestive tract. Still later in development, in most animals, a third layer (mesoderm) forms between the other two and develops into most of the other internal organs (not shown in the figure). ❻ Following the gastrula, many animals continue to develop and then mature directly into adults. But others, including the sea star, develop into one or more larval stages first. ❼ The larva undergoes a major change of body form, called metamorphosis, in becoming an adult—a mature animal capable of reproducing sexually.

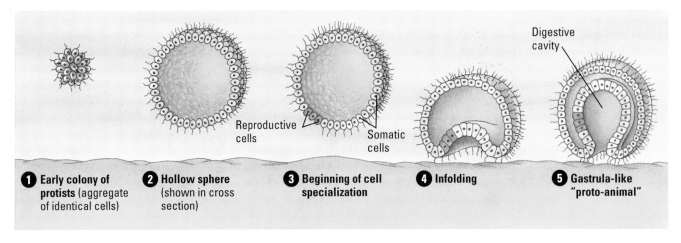

Reproductive cells

Somatic cells

Digestive cavity

❶ **Early colony of protists** (aggregate of identical cells)

❷ **Hollow sphere** (shown in cross section)

❸ **Beginning of cell specialization**

❹ **Infolding**

❺ **Gastrula-like "proto-animal"**

Figure 17.4 One hypothesis for a sequence of stages in the origin of animals from a colonial protist. ❶ The earliest colonies may have consisted of only a few cells, all of which were flagellated and basically identical. ❷ Some of the later colonies may have been hollow spheres—floating aggregates of heterotrophic cells—that ingested organic nutrients from the water.

❸ Eventually, cells in the colony may have specialized, with some cells adapted for reproduction and others for somatic (nonreproductive) functions, such as locomotion and feeding. ❹ A simple multicellular organism with cell layers may have evolved from a hollow colony, with cells on one side of the colony cupping inward, the way they do in the gastrula of an animal embryo (see Figure 17.3).

❺ A layered body plan would have enabled further division of labor among the cells. The outer flagellated cells would have provided locomotion and some protection, while the inner cells could have specialized in reproduction or feeding. With its specialized cells and a simple digestive compartment, the proto-animal shown here could have fed on organic matter on the seafloor.

Early Animals and the Cambrian Explosion

Animals probably evolved from a colonial, flagellated protist that lived in Precambrian seas (**Figure 17.4**). By the late Precambrian, about 600–700 million years ago, a diversity of animals had already evolved. Then came the Cambrian explosion. At the beginning of the Cambrian period, 545 million years ago, animals underwent a relatively rapid diversification. In fact, during a span of only about 10 million years, all the major animal body plans we see today evolved. It is an evolutionary episode so boldly marked in the fossil record that geologists use the dawn of the Cambrian period as the beginning of the Paleozoic era (see Table 14.1). Many of the Cambrian animals seem bizarre compared to the versions we see today, but most zoologists now agree that the Cambrian fossils can be classified as ancient representatives of contemporary animal phyla (**Figure 17.5**).

What ignited the Cambrian explosion? Hypotheses abound. Most researchers now believe that the Cambrian explosion simply extended animal diversification that was already well under way during the late Precambrian. But what caused the radiation of animal forms to accelerate so dramatically during the early Cambrian? One hypothesis emphasizes increasingly complex predator-prey relationships that led to diverse adaptations for feeding, motility, and protection. This would help explain why most Cambrian animals had shells or hard outer skeletons, in contrast to Precambrian animals, which were mostly soft-bodied. Another hypothesis focuses on the evolution of genes that control the development of animal form, such as the placement of body parts in embryos. At least some of these genes are common to diverse animal phyla. However, variation in how, when, and where these genes are expressed in an embryo can produce some of the major differences in body form that distinguish the phyla. Perhaps this developmental plasticity was partly responsible for the relatively rapid diversification of animals during the early Cambrian.

Figure 17.5 A Cambrian seascape. This drawing is based on fossils collected at a site called the Burgess Shale in British Columbia, Canada.

Continuing research will help test hypotheses about the Cambrian explosion. But as the explosion becomes less mysterious, it will seem no less wonderful. In the last half billion years, animal evolution has mainly generated new variations of old "designs" that originated in the Cambrian seas.

Animal Phylogeny

Because animals diversified so rapidly on the scale of geologic time, it is difficult, using only the fossil record, to sort out the sequence of branching in animal phylogeny. To reconstruct the evolutionary history of animal phyla, researchers must depend mainly on clues from comparative anatomy and embryology (see Chapter 13). Molecular methods are now providing additional tools for testing hypotheses about animal phylogeny. **Figure 17.6** represents one set of hypotheses about the evolutionary relationships

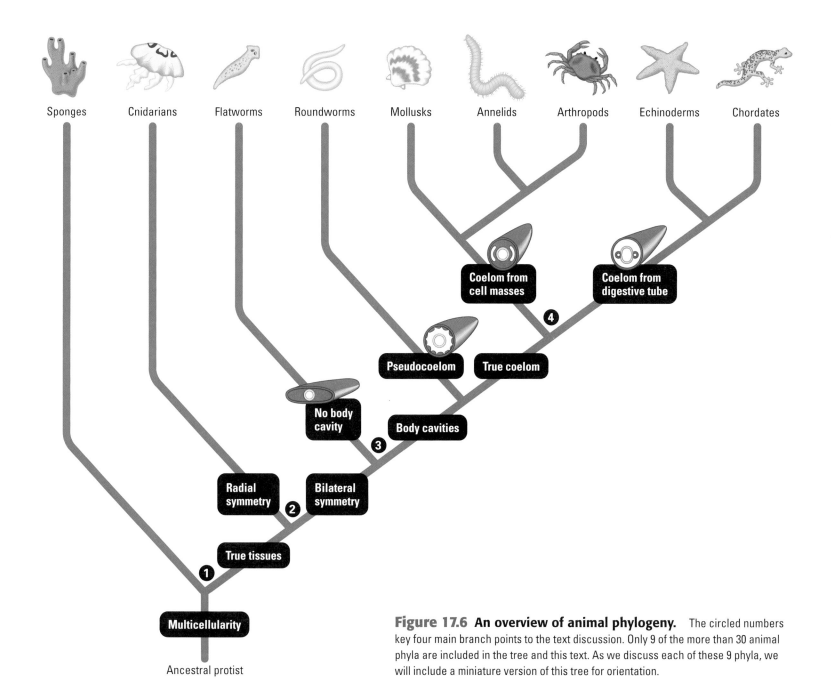

Figure 17.6 An overview of animal phylogeny. The circled numbers key four main branch points to the text discussion. Only 9 of the more than 30 animal phyla are included in the tree and this text. As we discuss each of these 9 phyla, we will include a miniature version of this tree for orientation.

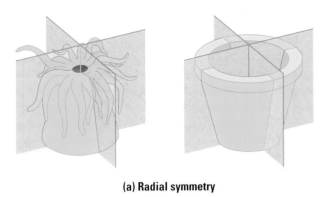

(a) Radial symmetry

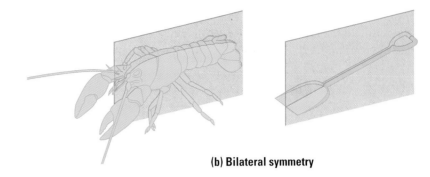

(b) Bilateral symmetry

Figure 17.7 Body symmetry. **(a)** The parts of a radial animal, such as this sea anemone, radiate from the center. Any imaginary slice through the central axis would divide the animal into mirror images. **(b)** A bilateral animal, such as this lobster, has a left and right side. Only one imaginary cut would divide the animal into mirror-image halves.

among nine major animal phyla. The circled numbers on the tree highlight four key evolutionary branch points, and these numbers are keyed to the following discussion.

❶ The first branch point distinguishes sponges from all other animals based on structural complexity. Sponges, though multicellular, lack the true tissues, such as epithelial (skin) tissue, that characterize more complex animals.

❷ The second major evolutionary split is based partly on body symmetry: radial versus bilateral. To understand this difference, imagine a pail and shovel. The pail has **radial symmetry,** identical all around a central axis. The shovel has **bilateral symmetry,** which means there's only one way to split it into two equal halves—right down the midline. **Figure 17.7** contrasts a radial animal with a bilateral one.

The symmetry of an animal generally fits its lifestyle. Many radial animals are sessile forms (attached to a substratum) or plankton (drifting or weakly swimming aquatic forms). Their symmetry equips them to meet the environment equally well from all sides. Most animals that move actively from place to place are bilateral. A bilateral animal has a definite "head end" that encounters food, danger, and other stimuli first when the animal is traveling. In most bilateral animals, a nerve center in the form of a brain is at the head end, near a concentration of sense organs such as eyes. Thus, bilateral symmetry is an adaptation for movement, such as crawling, burrowing, or swimming.

❸ The evolution of body cavities led to more complex animals. A **body cavity** is a fluid-filled space separating the digestive tract from the outer body wall. A body cavity has many functions. Its fluid cushions the suspended organs, helping to prevent internal injury. The cavity also enables the internal organs to grow and move independently of the outer body wall. If it were not for your body cavity, exercise would be very hard on your internal organs. And every beat of your heart or ripple of your intestine would deform your body surface. It would be a scary sight. In soft-bodied animals such as earthworms, the noncompressible fluid of the body cavity is under pressure and functions as a hydrostatic skeleton against which muscles can work. In fact, body cavities may have first evolved as adaptations for burrowing.

Activity 17A on the Web & CD
Classify animals using key characteristics.

Figure 17.8 Body plans of bilateral animals. The various organ systems of these animals develop from the three tissue layers that form in the embryo. **(a)** Flatworms are examples of animals that lack a body cavity. **(b)** Roundworms have a pseudocoelom, a body cavity only partially lined by mesoderm, the middle tissue layer. **(c)** Earthworms and other annelids are examples of animals with a true coelom. A coelom is a body cavity completely lined by mesoderm. Mesenteries, also derived from mesoderm, suspend the organs in the fluid-filled coelom.

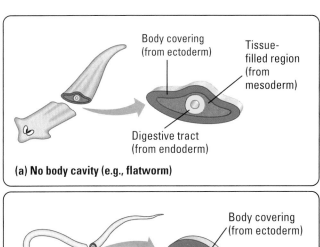

(a) No body cavity (e.g., flatworm)

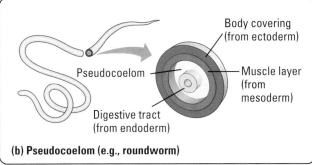

(b) Pseudocoelom (e.g., roundworm)

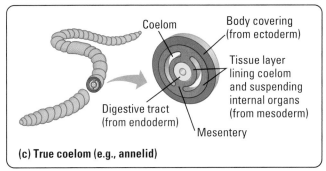

(c) True coelom (e.g., annelid)

Among animals with a body cavity, there are differences in how the cavity develops. In all cases, the cavity is at least partly lined by a middle layer of tissue, called mesoderm, which develops between the inner (endoderm) and outer (ectoderm) layers of the gastrula embryo. If the body cavity is not completely lined by tissue derived from mesoderm, it is termed a **pseudocoelom** (**Figure 17.8**). A true **coelom,** the type of body cavity humans and many other animals have, is completely lined by tissue derived from mesoderm.

❹ Among animals with a true coelom, there are two main evolutionary branches. They differ in several details of embryonic development, including the mechanism of coelom formation. One branch includes mollusks (such as clams, snails, and squids), annelids (such as earthworms), and arthropods (such as crustaceans, spiders, and insects). The two major phyla of the other branch are echinoderms (such as sea stars and sea urchins) and chordates (including humans and other vertebrates).

With the overview of animal evolution in Figure 17.6 as our guide, we're ready to take a closer look at some animal phyla.

CHECKPOINT

1. What embryonic stage is common to the development of all animals?

2. What is metamorphosis?

3. In key features of embryonic development, humans and other chordates are most like which other animal group?

4. What mode of heterotrophic nutrition distinguishes animals from fungi, which are also heterotrophs?

5. Why is animal evolution during the early Cambrian referred to as an "explosion"?

6. A round pizza is to _____ symmetry as a fork is to _____ symmetry.

7. The fully lined cavity between your outer body wall and your digestive tract is an example of a true _____.

Answers: 1. Blastula **2.** The transformation from a larval form of an animal to an adult form **3.** Echinoderms **4.** Ingestion **5.** Because a great diversity of animals evolved in a relatively short time span **6.** radial; bilateral **7.** coelom

---

Major Invertebrate Phyla

Living as we do on land, our sense of animal diversity is biased in favor of vertebrates, which are animals with a backbone. Vertebrates are well represented on land in the form of such animals as reptiles, birds, and mammals. However, vertebrates make up only one subphylum within the phylum Chordata, or less than 5% of all animal species. If we were to sample the animals in an aquatic habitat, such as a pond, tide pool, or coral reef, we

Figure 17.9 **A sponge.**

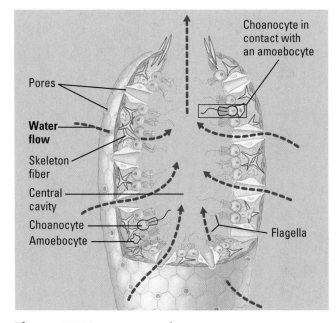

Figure 17.10 **Anatomy of a sponge.** Feeding cells called choanocytes have flagella that sweep water through the sponge's body. Choanocytes trap bacteria and other food particles, and amoebocytes distribute the food to other cells. To obtain enough food to grow by 100 g (about 3 ounces), a sponge must filter 1,000 kg (about 275 gallons) of seawater.

would find ourselves in the realm of **invertebrates.** These are the animals *without* backbones. It is traditional to divide the animal kingdom into vertebrates and invertebrates, but this makes about as much zoological sense as sorting animals into flatworms and nonflatworms. We give special attention to the vertebrates only because we humans are among the backboned ones. However, by exploring the other 95% of the animal kingdom—the invertebrates—we'll discover an astonishing diversity of beautiful creatures that too often escape our notice.

Sponges

Sponges (phylum **Porifera**) are sessile animals that appear so sedate to the human eye that the ancient Greeks believed them to be plants (**Figure 17.9**). The simplest of all animals, sponges probably evolved very early from colonial protists. Sponges range in height from about 1 cm to 2 m. Sponges have no nerves or muscles, but the individual cells can sense and react to changes in the environment. The cell layers of sponges are loose federations of cells, not really tissues, because the cells are relatively unspecialized. Of the 9,000 or so species of sponges, only about 100 live in fresh water; the rest are marine.

The body of a sponge resembles a sac perforated with holes (the name of the sponge phylum, Porifera, means "pore bearer"). Water is drawn through the pores into a central cavity, then flows out of the sponge through a larger opening (**Figure 17.10**). Most sponges feed by collecting bacteria from the water that streams through their porous bodies. Flagellated cells called **choanocytes** trap bacteria in mucus and then engulf the food by phagocytosis (see Chapter 4). Cells called **amoebocytes** pick up food from the choanocytes, digest it, and carry the nutrients to other cells. Amoebocytes are the "do-all" cells of sponges. Moving about by means of pseudopodia, they digest and distribute food, transport oxygen, and dispose of wastes. Amoebocytes also manufacture the fibers that make up a sponge's skeleton. In some sponges, these fibers are sharp and spur-like. Other sponges have softer, more flexible skeletons; we use these pliant, honeycombed skeletons as bathroom sponges.

Cnidarians

Cnidarians (phylum **Cnidaria**) are characterized by radial symmetry and tentacles with stinging cells. Jellies, sea anemones, hydras, and coral animals are all cnidarians. Most of the 10,000 cnidarian species are marine. The basic body plan of a cnidarian is a sac with a central digestive compartment, the **gastrovascular cavity.** A single opening to this cavity functions as both mouth and anus. This basic body plan has two variations: the sessile **polyp** and the floating **medusa** (**Figure 17.11**). Polyps adhere to the substratum and extend their tentacles, waiting for prey. Examples of the polyp form are hydras, sea anemones, and coral animals (**Figure 17.12**). A medusa is a flattened, mouth-down version of the polyp. It moves freely in the water by a combi-

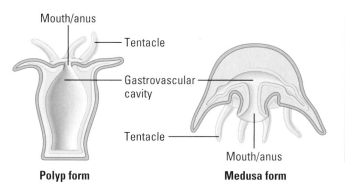

(a) Sea anemone: a polyp

(b) Jelly: a medusa

Figure 17.11 Polyp and medusa forms of cnidarians. Note that cnidarians have two tissue layers, distinguished in the diagrams by blue and yellow. The gastrovascular cavity is a digestive sac, meaning that it has only one opening, which functions as both mouth and anus. **(a)** Sea anemones are examples of the polyp form of the basic cnidarian body plan. **(b)** Jellies are examples of the medusa form.

nation of passive drifting and contractions of its bell-shaped body. The animals we generally call jellies are medusas (jellyfish is another common name, though these animals are not really fishes, which are vertebrates). Some cnidarians exist only as polyps, others only as medusas, and still others pass sequentially through both a medusa stage and a polyp stage in their life cycle.

Cnidarians are carnivores that use tentacles arranged in a ring around the mouth to capture prey and push the food into the gastrovascular cavity, where digestion begins. The undigested remains are eliminated through the mouth/anus. The tentacles are armed with batteries of **cnidocytes** ("stinging cells"), unique structures that function in defense and in the capture of prey (**Figure 17.13**). The phylum Cnidaria is named for these stinging cells.

Figure 17.12 Coral animals. Each polyp in this colony is about 3 mm in diameter. Coral animals secrete hard external skeletons of calcium carbonate (limestone). Each polyp builds on the skeletal remains of earlier generations to construct the "rocks" we call coral. Though individual coral animals are small, their collective construction accounts for such biological wonders as Australia's Great Barrier Reef, which Apollo astronauts were able to identify from the moon. Tropical coral reefs are home to an enormous variety of invertebrates and fishes.

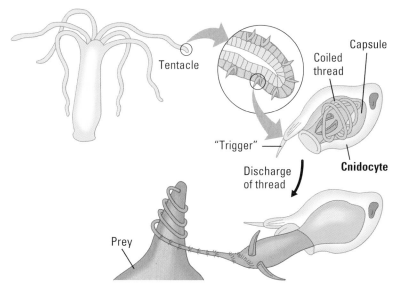

Figure 17.13 Cnidocyte action. Each cnidocyte contains a fine thread coiled within a capsule. When a trigger is stimulated by touch, the thread shoots out. Some cnidocyte threads entangle prey, while others puncture the prey and inject a poison.

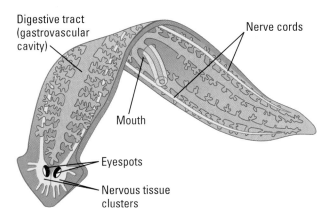

Figure 17.14 labels: Digestive tract (gastrovascular cavity), Nerve cords, Mouth, Eyespots, Nervous tissue clusters

Figure 17.14 Anatomy of a planarian.

This free-living flatworm has a head with two light-detecting eyespots and a flap at each side that detects certain chemicals in the water. Dense clusters of nervous tissue form a simple brain. The digestive tract is highly branched, providing an extensive surface area for the absorption of nutrients. When the animal feeds, a muscular tube projects through the mouth and sucks food in. The digestive tract, like that of cnidarians, is a gastrovascular cavity (a single opening functions as both mouth and anus). Planarians live on the undersurfaces of rocks in freshwater ponds and streams.

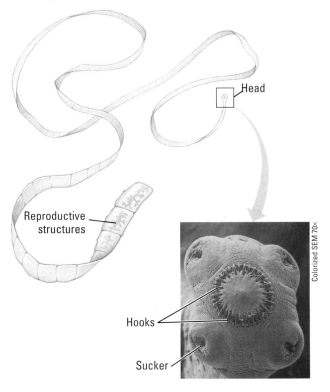

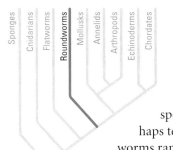

Labels: Head, Reproductive structures, Hooks, Sucker, Colorized SEM 70×

Figure 17.15 Anatomy of a tapeworm.

Humans acquire larvae of these parasites by eating undercooked meat that is infected. The head of a tapeworm is armed with suckers and menacing hooks that lock the worm to the intestinal lining of the host. Behind the head is a long ribbon of units that are little more than sacs of sex organs. At the back of the worm, mature units containing thousands of eggs break off and leave the host's body with the feces.

Flatworms

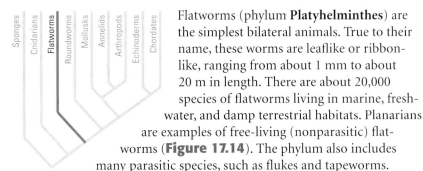

Cladogram labels: Sponges, Cnidarians, **Flatworms**, Roundworms, Mollusks, Annelids, Arthropods, Echinoderms, Chordates

Flatworms (phylum **Platyhelminthes**) are the simplest bilateral animals. True to their name, these worms are leaflike or ribbon-like, ranging from about 1 mm to about 20 m in length. There are about 20,000 species of flatworms living in marine, freshwater, and damp terrestrial habitats. Planarians are examples of free-living (nonparasitic) flatworms (**Figure 17.14**). The phylum also includes many parasitic species, such as flukes and tapeworms.

Parasitic flatworms called blood flukes are a huge health problem in the tropics. These worms have suckers that attach to the inside of the blood vessels near the human host's intestines. This causes a long-lasting disease with such symptoms as severe abdominal pain, anemia, and dysentery. About 250 million people in 70 countries suffer from blood fluke disease. Flukes generally have complex life cycles that require more than one host species. People are most commonly exposed to blood flukes while working in irrigated fields contaminated with human feces. Blood flukes living in a human host reproduce sexually, and fertilized eggs pass out in the host's feces. If an egg lands in a pond or stream, a motile larva hatches. This larva can enter a snail, the next host. Asexual reproduction in the snail eventually produces other larvae that can infect humans. A person becomes infected when these larvae penetrate the skin.

Tapeworms parasitize many vertebrates, including humans. In contrast to planarians and flukes, most tapeworms have a very long, ribbon-like body with repeated parts (**Figure 17.15**). They also differ from other flatworms in not having any digestive tract at all. Living in partially digested food in the intestines of their hosts, tapeworms simply absorb nutrients across their body surface. Like parasitic flukes, tapeworms have a complex life cycle, usually involving more than one host. Humans can become infected with tapeworms by eating rare beef containing the worm's larvae. The larvae are microscopic, but the adults can reach lengths of 20 m in the human intestine. Such large tapeworms can cause intestinal blockage and rob enough nutrients from the human host to cause nutritional deficiencies. An orally administered drug called niclosamide kills the adult worms.

Roundworms

Cladogram labels: Sponges, Cnidarians, Flatworms, **Roundworms**, Mollusks, Annelids, Arthropods, Echinoderms, Chordates

Roundworms (phylum **Nematoda**) get their common name from their cylindrical body, which is usually tapered at both ends (**Figure 17.16a**). Roundworms are among the most diverse (in species number) and widespread of all animals. About 90,000 species of roundworms are known, and perhaps ten times that number actually exist. Roundworms range in length from about a millimeter to a meter. They are found in most aquatic habitats, in wet soil, and as parasites in the body fluids and tissues of plants and animals. Free-living roundworms in the soil are important decomposers. Other species are major agricultural pests that attack the roots of plants. Humans host at least 50 parasitic roundworm species, including pinworms, hookworms, and the parasite that causes trichinosis (**Figure 17.16b**).

Roundworms exhibit two evolutionary innovations not found in flatworms. First, roundworms have a **complete digestive tract,** which is a digestive tube with two openings, a mouth and an anus. This anatomy contrasts with the digestive sac, or gastrovascular cavity, of cnidarians and flatworms, which uses a single opening as both mouth and anus. All the remaining animals in our survey of animal phyla have a complete digestive tract. A complete digestive tract can process food and absorb nutrients as a meal moves in one direction from one specialized digestive organ to the next. In humans, for example, the mouth, stomach, and intestines are examples of digestive organs. A second evolutionary innovation we see for the first time in roundworms is a body cavity, which in this case is a pseudocoelom (see Figure 17.8).

Mollusks

Snails and slugs, oysters and clams, and octopuses and squids are all mollusks (phylum **Mollusca**). Mollusks are soft-bodied animals, but most are protected by a hard shell. Slugs, squids, and octopuses have reduced shells, most of which are internal, or they have lost their shells completely during their evolution. Many mollusks feed by using a straplike rasping organ called a **radula** to scrape up food. Garden snails use their radulas like tiny saws to cut pieces out of leaves. Most of the 150,000 known species of mollusks are marine, though some inhabit fresh water, and there are land-dwelling mollusks in the form of snails and slugs.

Despite their apparent differences, all mollusks have a similar body plan (**Figure 17.17**). The body has three main parts: a muscular foot, usually used for movement; a visceral mass containing most of the internal organs; and a fold of tissue called the mantle. The **mantle** drapes over the visceral mass and secretes the shell (if one is present).

(a) A free-living roundworm

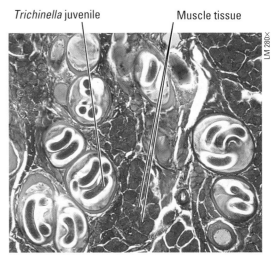

Trichinella juvenile   Muscle tissue

(b) Parasitic roundworms

Figure 17.16 Roundworms. **(a)** This species has the classic roundworm shape: cylindrical with tapered ends. You can see the mouth at the end that is more blunt. Not visible is the anus at the other end of a complete digestive tract. This worm looks like it's wearing a corduroy coat, but the ridges actually indicate muscles that run the length of the body. **(b)** The disease called trichinosis is caused by these roundworms, encysted here in human muscle tissue. Humans acquire the parasite by eating undercooked pork or other meat that is infected. The worms then burrow into the human intestine and eventually travel to other parts of the body, encysting in muscles and other organs.

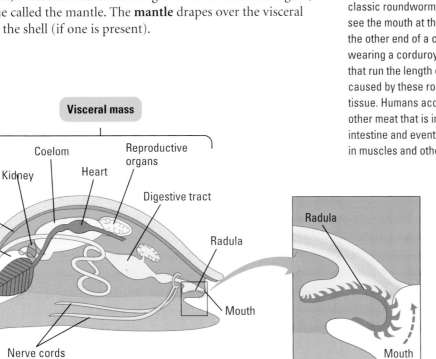

Figure 17.17 The general body plan of a mollusk. Note the body cavity (a true coelom, though a small one) and the complete digestive tract, with both mouth and anus.

(a) Gastropod shells

(b) A bivalve (a scallop)

(c) A cephalopod (an octopus)

Figure 17.18 Mollusks. **(a)** Shell collectors are delighted by the variety of gastropods. **(b)** This scallop, a bivalve, has many eyes peering out between the two halves of the hinged shell. **(c)** An octopus is a cephalopod without a shell. All cephalopods have large brains and sophisticated sense organs, which contribute to the success of these animals as mobile predators. This octopus lives on the seafloor, where it scurries about in search of crabs and other food. Its brain is larger and more complex, proportionate to body size, than that of any other invertebrate. Octopuses are very intelligent and have shown remarkable learning abilities in laboratory experiments.

The three major classes of mollusks are gastropods (including snails and slugs), bivalves (including clams and oysters), and cephalopods (including squids and octopuses). Most **gastropods** are protected by a single, spiraled shell into which the animal can retreat when threatened (**Figure 17.18a**). Many gastropods have a distinct head with eyes at the tips of tentacles (think of a garden snail). **Bivalves**, including numerous species of clams, oysters, mussels, and scallops, have shells divided into two halves hinged together (**Figure 17.18b**). Most bivalves are sedentary, living in sand or mud in marine and freshwater environments. They use their muscular foot for digging and anchoring. **Cephalopods** generally differ from gastropods and sedentary bivalves in being built for speed and agility. A few cephalopods have large, heavy shells, but in most the shell is small and internal (as in squids) or missing altogether (as in octopuses). Cephalopods are marine predators that use beaklike jaws and a radula to crush or rip prey apart. The cephalopod mouth is at the base of the foot, which is drawn out into several long tentacles for catching and holding prey (**Figure 17.18c**).

Annelids

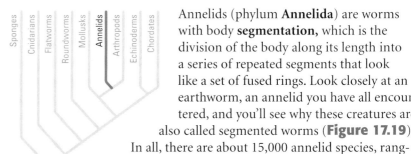

Annelids (phylum **Annelida**) are worms with body **segmentation,** which is the division of the body along its length into a series of repeated segments that look like a set of fused rings. Look closely at an earthworm, an annelid you have all encountered, and you'll see why these creatures are also called segmented worms (**Figure 17.19**).

In all, there are about 15,000 annelid species, ranging in length from less than 1 mm to the 3-m giant Australian earthworm. Annelids live in the sea, most freshwater habitats, and damp soil. The three main classes of annelids are the earthworms and their relatives, the polychaetes, and the leeches.

Earthworms eat their way through the soil, extracting nutrients as the soil passes through the digestive tube (**Figure 17.20a**). Undigested material, mixed with mucus secreted into the digestive tract, is eliminated as castings through the anus. Farmers value earthworms because the animals till the earth, and the castings improve the texture of the soil. Charles Darwin estimated that each acre of British farmland had about 50,000 earthworms that produced 18 tons of castings per year.

In contrast to earthworms, most polychaetes are marine, mainly crawling or burrowing in the seafloor. Segmental appendages with hard bristles help the worm wriggle about in search of small invertebrates to eat. The appendages also increase the animal's surface area for taking up oxygen and disposing of metabolic wastes, including carbon dioxide (**Figure 17.20b**).

The third group of annelids, leeches, are notorious for the bloodsucking habits of some species. However, most species are free-living carnivores that eat small invertebrates such as snails and insects. The majority of leeches live in fresh water, but a few terrestrial species inhabit moist vegetation in the tropics. Until the twentieth century, bloodsucking leeches were frequently used by physicians for bloodletting, the removal of what was considered "bad blood" from sick patients. Some leeches have razorlike jaws that cut through the skin, and they secrete saliva containing a strong anesthetic and an anticoagulant into the wound. The anesthetic makes the bite virtually painless, and the anticoagulant keeps the blood from clotting. Leech

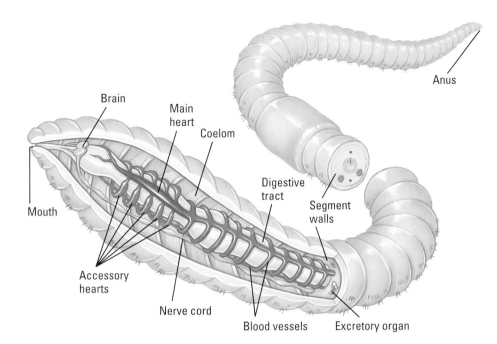

Brain

Main heart

Coelom

Anus

Mouth

Accessory hearts

Nerve cord

Digestive tract

Segment walls

Blood vessels

Excretory organ

Figure 17.19
Segmented anatomy of an earthworm.
Annelids are segmented both externally and internally. Many of the internal structures are repeated, segment by segment. The coelom (body cavity) is partitioned by walls (only two segment walls are fully shown here). The nervous system (yellow) includes a nerve cord with a cluster of nerve cells in each segment. Excretory organs (green), which dispose of fluid wastes, are also repeated in each segment. The digestive tract, however, is not segmented; it passes through the segment walls from the mouth to the anus. Segmental blood vessels connect continuous vessels that run along the top (dorsal location) and bottom (ventral location) of the worm. The segmental vessels include five pairs of accessory hearts. The main heart is simply an enlarged region of the dorsal blood vessel near the head end of the worm.

anticoagulant is now being produced commercially by genetic engineering and may find wide use in human medicine. Tests show that it prevents blood clots that can cause heart attacks. Leeches are also still occasionally used to remove blood from bruised tissues and to help relieve swelling in fingers or toes that have been sewn back on after accidents (**Figure 17.20c**). Blood tends to accumulate and cause swelling in a reattached finger or toe until small veins have a chance to grow back into it. Leeches are applied to remove the excess blood.

(a) A giant Australian earthworm

(b) Polychaetes

(c) A leech

Figure 17.20 Annelids.
(a) Giant Australian earthworms are bigger than most snakes. Perhaps you've slipped on slimy worms, but imagine actually *tripping* over one!
(b) Polychaetes have segmental appendages that function in movement and as gills. On the left is a sandworm. The beautiful polychaete on the right is an example of a fan worm, which lives in a tube it constructs by mixing mucus with bits of sand and broken shells. Fan worms use their feathery headdresses as gills and to extract food particles from the seawater. This species is called a Christmas tree worm. **(c)** A nurse applied this medicinal leech (*Hirudo medicinalis*) to a patient's sore thumb to drain blood from a hematoma (abnormal accumulation of blood around an internal injury).

Arthropods

Arthropods (phylum **Arthropoda**) are named for their jointed appendages. Crustaceans (such as crabs and lobsters), arachnids (such as spiders and scorpions), and insects are all examples of arthropods. Zoologists estimate that the total arthropod population of Earth numbers about a billion billion (10^{18}) individuals. Researchers have identified about a million arthropod species, mostly insects. In fact, two out of every three species of life that have been described are arthropods. And arthropods are represented in nearly all habitats of the biosphere. On the criteria of species diversity, distribution, and sheer numbers, arthropods must be regarded as the most successful of all animal phyla.

General Characteristics of Arthropods Arthropods are segmented animals. In contrast to the matching segments of annelids, however, arthropod segments and their appendages have become specialized for a great variety of functions. This evolutionary flexibility contributed to the great diversification of arthropods. Specialization of segments, or fused groups of segments, also provides for an efficient division of labor among body regions. For example, the appendages of different segments are variously modified for walking, feeding, sensory reception, copulation, and defense (**Figure 17.21**).

The body of an arthropod is completely covered by an **exoskeleton** (external skeleton). This coat is constructed from layers of protein and a polysaccharide called chitin. The exoskeleton can be a thick, hard armor over some parts of the body yet paper-thin and flexible in other locations, such as the joints. The exoskeleton protects the animal and provides points of attachment for the muscles that move the appendages. There are, of course, advantages to wearing hard parts on the outside. Our own skeleton is interior to most of our soft tissues, an arrangement that doesn't provide much protection from injury. But our skeleton does offer the advantage of being able to grow along with the rest of our body. In contrast, a growing arthropod must occasionally shed its old exoskeleton and secrete a larger one. This process, called **molting,** leaves the animal temporarily vulnerable to predators and other dangers.

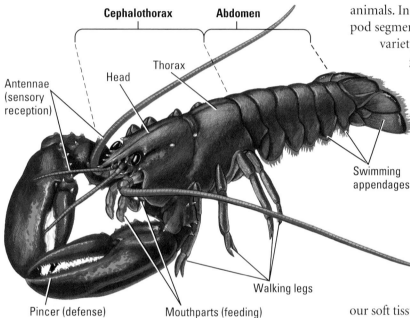

Figure 17.21 Arthropod characteristics of a lobster. The whole body, including the appendages, is covered by an exoskeleton. The two distinct regions of the body are the cephalothorax (consisting of the head and thorax) and the abdomen. The head bears a pair of eyes, each situated on a movable stalk. The body is segmented, but this characteristic is only obvious in the abdomen. The animal has a tool kit of specialized appendages, including pincers, walking legs, swimming appendages, and two pairs of sensory antennae. Even the multiple mouthparts are modified legs, which is why they work form side to side rather than up and down (as our jaws do).

Arthropod Diversity The four main groups of arthropods are the arachnids, the crustaceans, the millipedes and centipedes, and the insects.

Most **arachnids** live on land. Scorpions, spiders, ticks, and mites are examples (**Figure 17.22**). Arachnids usually have four pairs of walking legs and a specialized pair of feeding appendages. In spiders, these feeding appendages are fanglike and equipped with poison glands. As a spider uses these appendages to immobilize and chew its prey, it spills digestive juices onto the torn tissues and sucks up its liquid meal.

Crustaceans are nearly all aquatic. Crabs, lobsters, crayfish, shrimps, and barnacles are all crustaceans (**Figure 17.23**). They all exhibit the crustacean hallmark of multiple pairs of specialized appendages (see Figure 17.21). One group of crustaceans, the isopods, is represented on land by pill bugs, which you have probably found on the undersides of moist leaves and other organic debris.

(a) A scorpion (about 8 cm long)

(b) A black widow spider (about 3 cm long)

(c) A dust mite

Figure 17.22 Arachnids. **(a)** Scorpions are nocturnal hunters. Their ancestors were among the first terrestrial carnivores, preying on herbivorous arthropods that fed on the early land plants. Scorpions have a pair of appendages modified as large pincers that function in defense and food capture. The tip of the tail bears a poisonous stinger. Scorpions eat mainly insects and spiders. They will sting people only when prodded or stepped on. (If you camp in the desert and leave your shoes on the ground when you go to bed, make sure there are no scorpions in those shoes before putting them on in the morning.) **(b)** Spiders are usually most active during the daytime, hunting insects or trapping them in webs. Spiders spin their webs of liquid silk, which solidifies as it comes out of specialized glands. Each spider engineers a style of web that is characteristic of its species, getting the web right on the very first try. Besides building their webs of silk, spiders use the fibers in many other ways: as droplines for rapid escape; as cloth that covers eggs; and even as "gift wrapping" for food that certain male spiders offer to seduce females. **(c)** This magnified house dust mite is a ubiquitous scavenger in our homes. Each square inch of carpet and every one of those dust balls under a bed are like cities to thousands of dust mites. Unlike some mites that carry pathogenic bacteria, dust mites are harmless except to people who are allergic to the mites' feces.

(a) A shrimp

(b) Barnacles

Figure 17.23 Crustaceans. **(a)** A grass shrimp. **(b)** Easily confused with bivalve mollusks, barnacles are actually sessile crustaceans with exoskeletons hardened into shells by calcium carbonate (lime). The jointed appendages projecting from the shell capture small plankton.

Figure 17.24 A millipede.

Millipedes and **centipedes** have similar segments over most of the body and superficially resemble annelids, but their jointed legs give them away as arthropods. Millipedes are landlubbers that eat decaying plant matter (**Figure 17.24**). They have two pairs of short legs per body segment. Centipedes are terrestrial carnivores, with a pair of poison claws used in defense and to paralyze prey, such as cockroaches and flies. Each of their body segments bears a single pair of long legs.

In species diversity, **insects** outnumber all other forms of life combined. They live in almost every terrestrial habitat and in fresh water, and flying insects fill the air. Insects are rare, though not absent, in the seas, where crustaceans are the dominant arthropods. There is a whole big branch of biology, called **entomology,** that specializes in the study of insects.

The oldest insect fossils date back to about 400 million years ago, during the Paleozoic era (see Table 14.1). Later, the evolution of flight sparked an explosion in insect variety (**Figure 17.25**).

Case Study in the Process of Science on the Web & CD Learn to identify insects based on characteristics such as mouthparts and wings.

Like the grasshopper in Figure 17.25, most insects have a three-part body: head, thorax, and abdomen. The head usually bears a pair of sensory antennae and a pair of eyes. Several pairs of mouthparts are adapted for particular kinds of eating—for example, for biting and chewing plant material in grasshoppers; for lapping up fluids in houseflies; and for piercing skin and sucking blood in mosquitoes and other biting flies. Most adult insects have three pairs of legs and one or two pairs of wings, all borne on the thorax.

Flight is obviously one key to the great success of insects. An animal that can fly can escape many predators, find food and mates, and disperse to new habitats much faster than an animal that must crawl about on the ground. Because their wings are extensions of the exoskeleton and not true appendages, insects can fly without sacrificing any walking legs. By contrast, the flying vertebrates—birds and bats—have one of their two pairs of walking legs modified for wings, which explains why these vertebrates are generally not very swift on the ground.

Many insects undergo metamorphosis in their development. In the case of grasshoppers and some other insect groups, the young resemble adults but are smaller and have different body proportions. The animal goes

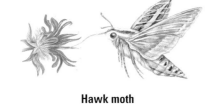

Hawk moth

Mosquito

Paper wasp

Damselfly

Water strider

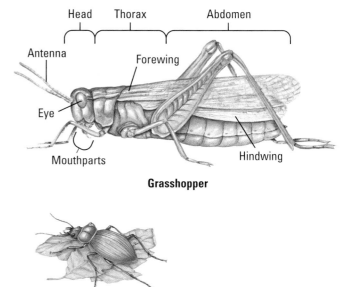

Head   Thorax   Abdomen

Antenna   Forewing

Eye

Mouthparts   Hindwing

Grasshopper

Ground beetle

Figure 17.25 A small sample of insect diversity.

through a series of molts, each time looking more like an adult, until it reaches full size. In other cases, insects have distinctive larval stages specialized for eating and growing that are known by such names as maggots, grubs, or caterpillars. The larval stage looks entirely different from the adult stage, which is specialized for dispersal and reproduction. Metamorphosis from the larva to the adult occurs during a pupal stage (**Figure 17.26**).

Animals so numerous, diverse, and widespread as insects are bound to affect the lives of all other terrestrial organisms, including humans. On one hand, we depend on bees, flies, and many other insects to pollinate our crops and orchards. On the other hand, insects are carriers of the microorganisms that cause many diseases, including malaria and African sleeping sickness. Insects also compete with humans for food. In parts of Africa, for instance, insects claim about 75% of the crops. Trying to minimize their losses, farmers in the United States spend billions of dollars each year on pesticides, spraying crops with massive doses of some of the deadliest poisons ever invented. Try as they may, not even humans have challenged the preeminence of insects and their arthropod kin. As Cornell University's Thomas Eisner puts it: "Bugs are not going to *inherit* the Earth. They own it now. So we might as well make peace with the landlord."

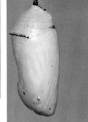

(a) Larva (caterpillar)

(b) Pupa

(c) Pupa

(d) Emerging adult

(e) Adult

Echinoderms

The echinoderms (phylum **Echinodermata**) are named for their spiny surfaces. Sea urchins, the porcupines of the invertebrates, are certainly echinoderms that live up to the phylum name. Among the other echinoderms are sea stars, sand dollars, and sea cucumbers (**Figure 17.27**).

Echinoderms are all marine. Most are sessile or slow moving. Echinoderms lack body segments, and most have radial symmetry as adults. Both the external and the internal parts of a sea star, for instance, radiate from the center like spokes of a

Figure 17.26
Metamorphosis of a monarch butterfly.

(a) The larva (caterpillar) spends its time eating and growing, molting as it grows. **(b)** After several molts, the larva encases itself in a cocoon and becomes a pupa. **(c)** Within the pupa, the larval organs break down and adult organs develop from cells that were dormant in the larva. **(d)** Finally, the adult emerges from the cocoon. **(e)** The butterfly flies off and reproduces, nourished mainly from the calories it stored when it was a caterpillar.

(a) A sea star

(b) A sea urchin

(c) A sea cucumber

Figure 17.27 Echinoderms.

(a) The mouth of a sea star, not visible here, is located in the center of the undersurface. The inset shows how the tube feet function in feeding. When a sea star encounters an oyster or clam, its favorite foods, it grips the mollusk's shell with its tube feet and positions its mouth next to the narrow opening between the two halves of the prey's shell. The sea star then pushes its stomach out through its mouth and through the crack in the mollusk's shell. The predator then digests the soft tissue of its prey. **(b)** In contrast to sea stars, sea urchins are spherical and have no arms. If you look closely, you can see the long tube feet projecting among the spines.

Unlike sea stars, which are mostly carnivorous, sea urchins mainly graze on seaweed and other algae. **(c)** On casual inspection, sea cucumbers do not look much like other echinoderms. Sea cucumbers lack spines, and the hard endoskeleton is much reduced. However, closer inspection reveals many echinoderm traits, including five rows of tube feet.

wheel. In contrast to the adult, the larval stage of echinoderms is bilaterally symmetrical. This supports other evidence that echinoderms are not closely related to cnidarians or other radial animals that never show bilateral symmetry. Most echinoderms have an **endoskeleton** (interior skeleton) constructed from hard plates just beneath the skin. Bumps and spines of this endoskeleton account for the animal's rough or prickly surface. Unique to echinoderms is the **water vascular system,** a network of water-filled canals that circulate water throughout the echinoderm's body, facilitating gas exchange and waste disposal. The water vascular system also branches into extensions called tube feet. A sea star or sea urchin pulls itself slowly over the seafloor using its suction-cup-like tube feet. Sea stars also use their tube feet to grip prey during feeding (see Figure 17.27a).

Looking at sea stars and other adult echinoderms, you may think they have little in common with humans and other vertebrates. But if you return

Activity 17B on the Web & CD
Construct a table to review invertebrates.

to Figure 17.6, you'll see that echinoderms share an evolutionary branch with chordates, the phylum that includes vertebrates. Analysis of embryonic development reveals this relationship. The mechanism of coelom formation and many other details of embryology differentiate the echinoderms and chordates from the evolutionary branch that includes mollusks, annelids, and arthropods. With this phylogenetic context, we're now ready to make the transition from invertebrates to vertebrates.

CHECKPOINT

1. In what main way does the digestive tract of a jelly differ from that of a roundworm?

2. In what fundamental way does the structure of a sponge differ from that of all other animals?

3. A sea anemone is to the phylum _____ as a blood fluke is to the phylum _____.

4. Why is it a mistake for someone who eats pork chops to order them "rare" in a restaurant?

5. As representatives of classes of mollusks, a garden snail is an example of a _____; a clam is an example of a _____; and a squid is an example of a _____.

6. The phylum Arthropoda is named for its members' _____.

7. Which major arthropod group is mainly aquatic?

8. Contrast the skeleton of an echinoderm with that of an arthropod.

Answers: 1. The digestive tract of a jelly is a sac with one opening; the roundworm tract is a tube with a separate mouth and anus. **2.** A sponge has no true tissues. **3.** Cnidaria; Platyhelminthes **4.** Incomplete cooking doesn't kill nematodes and other parasites that might be present in the meat. **5.** gastropod; bivalve; cephalopod **6.** jointed appendages **7.** Crustaceans **8.** An echinoderm has an endoskeleton; an arthropod has an exoskeleton.

The Vertebrate Genealogy

Most of us are curious about our genealogy. On the personal level, we wonder about our family ancestry. As biology students, we are interested in tracing human ancestry within the broader scope of the entire animal kingdom. In this quest, we ask three questions: What were our ancestors like? How are we related to other animals? and What are our closest relatives?

In this section, we trace the evolution of the vertebrates, the group that includes humans and their closest relatives. Mammals, birds, reptiles, amphibians, and the various classes of fishes are all classified as **vertebrates.** Among the unique vertebrate features are the cranium and backbone, a series of vertebrae for which the group is named (**Figure 17.28**). Our first step in tracing the vertebrate genealogy is to determine where vertebrates fit in the animal kingdom.

Characteristics of Chordates

Vertebrates make up one subphylum of the phylum **Chordata.** Our phylum also includes two subphyla of invertebrates, animals lacking a backbone: **lancelets** and **tunicates** (**Figure 17.29**). These invertebrate chordates and vertebrates all share four key features that appear in the embryo and sometimes in the adult. These four chordate hallmarks are (1) a **dorsal, hollow nerve cord** (the chordate brain and spinal cord); (2) a **notochord,** which is a flexible, longitudinal rod located between the digestive tract and the nerve cord; (3) **pharyngeal slits,** which are gill structures in the pharynx, the region of the digestive tube just behind the mouth; and (4) a **post-anal tail,** which is a tail to the rear of the anus (**Figure 17.30**). Though these chordate characteristics are often difficult to recognize in the adult animal, they are always present in chordate embryos. For example, the notochord, for which our phylum is named, persists in adult humans only in the form of the cartilage disks that function as cushions between the vertebrae. (Back injuries described as "ruptured disks" or "slipped disks" refer to these structures.)

Figure 17.28 Backbone, extra long. Vertebrates are named for their backbone, which consists of a series of vertebrae. The vertebrate hallmark is apparent in this snake skeleton. You can also see the skull, the bony case protecting the brain. The backbone and skull are parts of an endoskeleton, a skeleton inside the animal rather than covering it.

(a) A lancelet

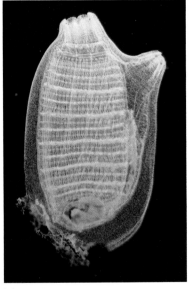

(b) A tunicate

Figure 17.29 Invertebrate chordates. **(a)** Lancelets owe their name to their bladelike shape. Marine animals only a few centimeters long, lancelets wiggle backward into sand, leaving only their head exposed. The animal filters tiny food particles from the seawater. **(b)** This adult tunicate, or sea squirt, is a sessile filter feeder that bears little resemblance to other chordates. However, a tunicate goes through a larval stage that is unmistakably chordate.

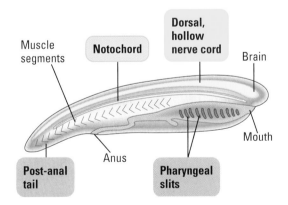

Figure 17.30 Chordate characteristics.

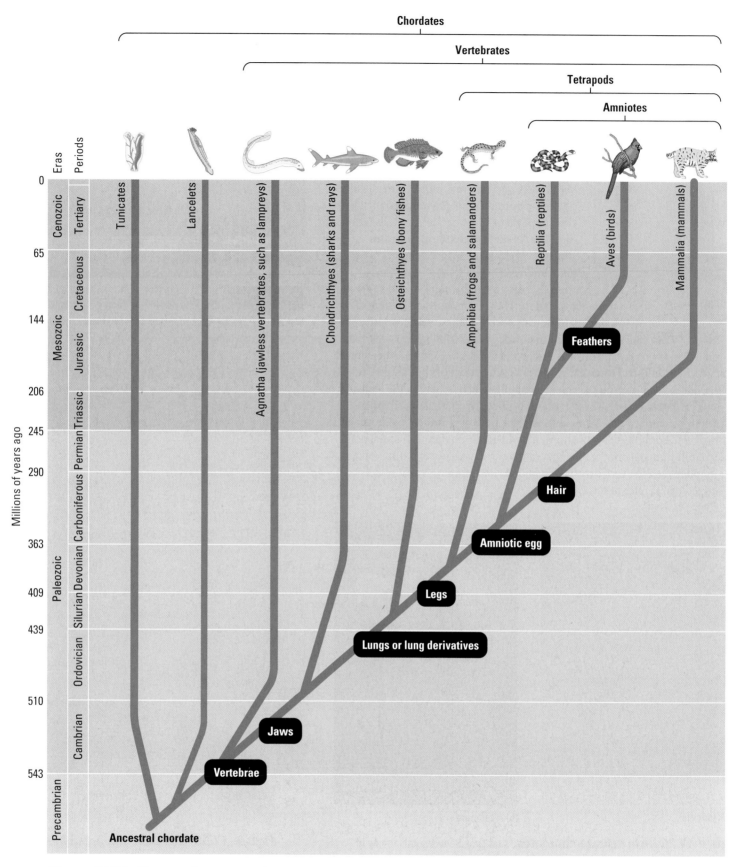

Figure 17.31 The vertebrate genealogy. "Tetrapods" refers to the four legs of terrestrial vertebrates. "Amniotes" refers to the evolution of the amniotic egg, a shelled egg that made it possible for vertebrates to reproduce on land (a chicken egg is one example).

Body segmentation is another chordate characteristic, though not a unique one. The chordate version of segmentation probably evolved independently of the segmentation we observe in annelids and arthropods. Chordate segmentation is apparent in the backbone of vertebrates (see Figure 17.28) and in the segmental muscles of all chordates (see the chevron-shaped— >>>> —muscles in the lancelet of Figure 17.29a). Segmental musculature is not so obvious in adult humans unless one is motivated enough to sculpture those washboard "abs of steel."

Vertebrates retain the basic chordate characteristics, but have additional features that are unique, including, of course, the backbone (see Figure 17.28). **Figure 17.31** is an overview of chordate and vertebrate evolution that will provide a context for our survey of the vertebrate classes.

(a) A cartilaginous fish (a sand bar shark)

Fishes

The first vertebrates probably evolved during the early Cambrian period about 540 million years ago. These early vertebrates, the **agnathans,** lacked jaws. Agnathans are represented today by vertebrates called lampreys (see Figure 17.31). Some lampreys are parasites that use their jawless mouths as suckers to attach to the sides of large fishes and draw blood. In contrast, most vertebrates have jaws, which are hinged skeletons that work the mouth. We know from the fossil record that the first jawed vertebrates were fishes that replaced most agnathans by about 400 million years ago. In addition to jaws, these fishes had two pairs of fins, which made them maneuverable swimmers. Some of those fishes were more than 10 m long. Sporting jaws and fins, some of the early fishes were active predators that could chase prey and bite off chunks of flesh. Even today, most fishes are carnivores. The two major groups of living fishes are the class **Chondrichthyes** (cartilaginous fishes—the sharks and rays) and the class **Osteichthyes** (the bony fishes, including such familiar groups as tuna, trout, and goldfish).

(b) A bony fish (a yellow perch)

Figure 17.32 Two classes of fishes.
(a) A member of the class Chondrichthyes, the cartilaginous fishes.
(b) A member of the class Osteichthyes, the bony fishes.

Cartilaginous fishes have a flexible skeleton made of cartilage. Most sharks are adept predators—fast swimmers with streamlined bodies, acute senses, and powerful jaws (**Figure 17.32a**). A shark does not have keen eyesight, but its sense of smell is very sharp. In addition, special electrosensors on the head can detect minute electrical fields produced by muscle contractions in nearby animals. Sharks also have a **lateral line system,** a row of sensory organs running along each side of the body. Sensitive to changes in water pressure, the lateral line system enables a shark to detect minor vibrations caused by animals swimming in its neighborhood. There are fewer than 1,000 living species of cartilaginous fishes, nearly all of them marine.

Bony fishes (**Figure 17.32b**) have a skeleton reinforced by hard calcium salts. They also have a lateral line system, a keen sense of smell, and excellent eyesight. On each side of the head, a protective flap called the **operculum** (plural, *opercula*) covers a chamber housing the gills. Movement of the operculum allows the fish to breathe without swimming. (By contrast, sharks lack opercula and must swim to pass water over their gills.) Bony fishes also have a specialized organ that helps keep them buoyant—the **swim bladder,** a gas-filled sac. Thus, many bony fishes can conserve energy by remaining almost motionless, in contrast to sharks, which sink to the bottom if they stop swimming. Some bony fishes have a connection between the swim bladder and the digestive tract that enables them to gulp air and extract oxygen from it when

(a) Tadpole

(b) Tadpole undergoing metamorphosis

(c) Adult frog

Figure 17.33 **The "dual life" of an amphibian.**
Though not all amphibians have aquatic larval stages, this class of vertebrates is named for the familiar tadpole-to-frog metamorphosis of many species.

the dissolved oxygen level in the water gets too low. In fact, swim bladders evolved from simple lungs that augmented gills in absorbing oxygen from the water of stagnant swamps, where the first bony fishes lived.

The largest class of vertebrates (about 30,000 species), bony fishes are common in the seas and in freshwater habitats. Most bony fishes, including trout, bass, perch, and tuna, are **ray-finned fishes.** Their fins are supported by thin, flexible skeletal rays (see Figure 17.32b). A second evolutionary branch of bony fishes includes the **lungfishes** and **lobe-finned fishes.** Lungfishes live today in the Southern Hemisphere. They inhabit stagnant ponds and swamps, surfacing to gulp air into their lungs. The lobe-fins are named for their muscular fins supported by stout bones. Lobe-fins are extinct except for one species, a deep-sea dweller that may use its fins to waddle along the seafloor. Ancient freshwater lobe-finned fishes with lungs played a key role in the evolution of amphibians, the first terrestrial vertebrates.

Amphibians

In Greek, the word *amphibios* means "living a double life." Most members of the class **Amphibia** exhibit a mixture of aquatic and terrestrial adaptations. Most species are tied to water because their eggs, lacking shells, dry out quickly in the air. The frog in **Figure 17.33** spends much of its time on land, but it lays its eggs in water. An egg develops into a larva called a tadpole, a legless, aquatic algae-eater with gills, a lateral line system resembling that of fishes, and a long finned tail. In changing into a frog, the tadpole undergoes a radical metamorphosis. When a young frog crawls onto shore and begins life as a terrestrial insect-eater, it has four legs, air-breathing lungs instead of gills, a pair of external eardrums, and no lateral line system. Because of metamorphosis, many amphibians truly live a double life. But even as adults, amphibians are most abundant in damp habitats, such as swamps and rain forests. This is partly because amphibians depend on their moist skin to supplement lung function in exchanging gases with the environment. Thus, even those frogs that are adapted to relatively dry habitats spend much of their time in humid burrows or under piles of moist leaves. The amphibians of today, including frogs and salamanders, account for only about 8% of all living vertebrates, or about 4,000 species.

Amphibians were the first vertebrates to colonize land. They descended from fishes that had lungs and fins with muscles and skeletal supports strong enough to enable some movement, however clumsy, on land (**Figure 17.34**).

Figure 17.34
The origin of tetrapods.
Fossils of some lobe-finned fishes have skeletal supports extending into their fins. Early amphibians left fossilized limb skeletons that probably functioned in movement on land.

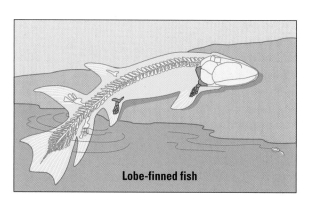

Lobe-finned fish

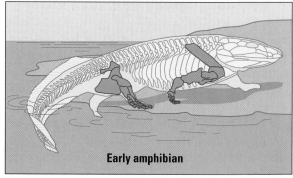

Early amphibian

The fossil record chronicles the evolution of four-limbed amphibians from fishlike ancestors. Terrestrial vertebrates—amphibians, reptiles, birds, and mammals—are collectively called **tetrapods,** which means "four legs." Had our amphibian ancestors had three pairs of legs on their undersides instead of just two, we might be hexapods. This image seems silly, but serves to reinforce the point that evolution, as descent with modification, is constrained by history.

Reptiles

Class **Reptilia** includes snakes, lizards, turtles, crocodiles, and alligators. The evolution of reptiles from an amphibian ancestor paralleled many additional adaptations for living on land. Scales containing a protein called keratin waterproof the skin of a reptile, helping to prevent dehydration in dry air. Reptiles cannot breathe through their dry skin and obtain most of their oxygen with their lungs. Another breakthrough for living on land that evolved in reptiles is the **amniotic egg,** a water-containing egg enclosed in a shell (**Figure 17.35**). The amniotic egg functions as a "self-contained pond" that enables reptiles to complete their life cycle on land. With adaptations such as waterproof skin and amniotic eggs, reptiles broke their ancestral ties to aquatic habitats. There are about 6,500 species of reptiles alive today.

Reptiles are sometimes labeled "cold-blooded" animals because they do not use their metabolism extensively to control body temperature. But reptiles do regulate body temperature through behavioral adaptations. For example, many lizards regulate their internal temperature by basking in the sun when the air is cool and seeking shade when the air is too warm. Because they absorb external heat rather than generating much of their own, reptiles are said to be **ectotherms,** a term more accurate than *cold-blooded.* By heating directly with solar energy rather than through the metabolic breakdown of food, a reptile can survive on less than 10% of the calories required by a mammal of equivalent size.

As successful as reptiles are today, they were far more widespread, numerous, and diverse during the Mesozoic era, which is sometimes called the "age of reptiles." Reptiles diversified extensively during that era, producing a dynasty that lasted until the end of the Mesozoic, about 65 million years ago. Dinosaurs, the most diverse group, included the largest animals ever to inhabit land. Some were gentle giants that lumbered about while browsing vegetation. Others were voracious carnivores that chased their larger prey on two legs (**Figure 17.36**).

Figure 17.35 Terrestrial equipment of reptiles.
This bull snake displays two reptilian adaptations to living on land: a waterproof skin with keratinized scales; and amniotic eggs, with shells that protect a watery, nutritious internal environment where the embryo can develop on land. Snakes evolved from lizards that adapted to a burrowing lifestyle.

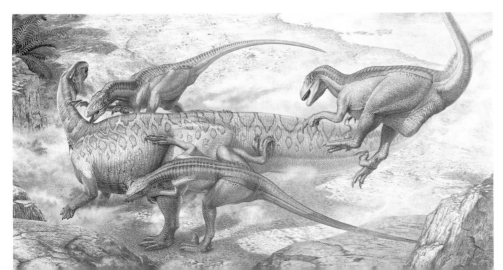

Figure 17.36 A Mesozoic feeding frenzy.
Hunting in packs, *Deinonychus* (meaning "terrible claw") probably used its sickle-shaped claws to slash at larger prey.

The age of reptiles began to wane about 70 million years ago. During the Cretaceous, the last period of the Mesozoic era, global climate became cooler and more variable (see Chapter 14). This was a period of mass extinctions that claimed all the dinosaurs by about 65 million years ago, except for one lineage. That lone surviving lineage is represented today by birds.

Birds

Birds (class **Aves**) evolved during the great reptilian radiation of the Mesozoic era. Amniotic eggs and scales on the legs are just two of the reptilian features we see in birds. But modern birds look quite different from modern reptiles because of their feathers and other distinctive flight equipment. Almost all of the 8,600 living bird species are airborne. The few flightless species, including the ostrich, evolved from flying ancestors. Appreciating the avian world is all about understanding flight.

Almost every element of bird anatomy is modified in some way that enhances flight. The bones have a honeycombed structure that makes them strong but light (the wings of airplanes have the same basic construction). For example, a huge seagoing species called the frigate bird has a wingspan of more than 2 m, but its whole skeleton weighs only about 113 g (4 ounces). Another adaptation that reduces the weight of birds is the absence of some internal organs found in other vertebrates. Female birds, for instance, have only one ovary instead of a pair. Also, modern birds are toothless, an adaptation that trims the weight of the head (no uncontrolled nosedives). Birds do not chew food in the mouth but grind it in the gizzard, a chamber of the digestive tract near the stomach.

Flying requires a great expenditure of energy and an active metabolism. In contrast to the ectothermic reptiles, birds are **endotherms.** That means they use their own metabolic heat to maintain a warm, constant body temperature.

A bird's most obvious flight equipment is its wings. Bird wings are airfoils that illustrate the same principles of aerodynamics as the wings of an airplane (**Figure 17.37**). A bird's flight motors are its powerful breast muscles, which are anchored to a keel-like breastbone. It is mainly these flight muscles that we call "white meat" on a turkey or chicken. Some birds, such as eagles and hawks, have wings adapted for soaring on air currents and flap their wings only occasionally. Other birds, including hummingbirds, excel at maneuvering but must flap continuously to stay aloft. In either case, it is the shape and arrangement of the feathers that form the wing into an airfoil. Feathers are made of keratin, the same protein that forms our hair and fingernails as well as the scales of reptiles. Feathers may have functioned first as insulation, helping birds retain body heat, only later being co-opted as flight gear.

In tracing the ancestry of birds back to the Mesozoic era, we must search for the oldest fossils with feathered wings that could have functioned in flight. Fossils of an ancient bird named *Archaeopteryx* have been found in Bavarian limestone in Germany and date back some 150 million years into the Jurassic period (see Figure 14.15). *Archaeopteryx* is not considered the ancestor of modern birds, and paleontologists place it on a side branch of the avian lineage. Nonetheless, *Archaeopteryx* probably was derived from

Figure 17.37 A bald eagle in flight. Bird wings are airfoils, which have shapes that create lift by altering air currents. Air passing over a wing must travel farther in the same amount of time than air passing under the wing. This expands the air above the wing relative to the air below the wing. And this makes the air pressure pushing upward against the lower wing surface greater than the pressure of the expanded air pushing downward on the wing. The wings of birds and airplanes owe their "lift" to this pressure differential.

Case Study in the Process of Science on the Web & CD Analyze bones from dinosaurs, birds, and mammals to investigate the origin of birds.

ancestral forms that also gave rise to modern birds. Its skeletal anatomy indicates that it was a weak flyer, perhaps mainly a tree-dwelling glider. A combination of gliding downward and jumping into the air from the ground may have been the earliest mode of flying in the bird lineage.

Mammals

Agnathans
Cartilaginous fishes
Bony fishes
Amphibians
Reptiles
Birds
Mammals

Mammals (class **Mammalia**) evolved from reptiles about 225 million years ago, long before there were any dinosaurs. During the peak of the age of reptiles, there were a number of mouse-sized, nocturnal mammals that lived on a diet of insects. Mammals became much more diverse after the downfall of the dinosaurs. Most mammals are terrestrial. However, there are nearly 1,000 species of winged mammals, the bats. And about 80 species of dolphins, porpoises, and whales are totally aquatic. The blue whale, an endangered species that grows to lengths of nearly 30 m, is the largest animal that has ever existed.

Two features—hair and mammary glands that produce milk that nourishes the young—are mammalian hallmarks. The main function of hair is to insulate the body and help maintain a warm, constant internal temperature; mammals, like birds, are endotherms.

There are three major groups of mammals: the monotremes, the marsupials, and the eutherians. The duck-billed platypus is one of only three existing species of **monotremes,** the egg-laying mammals. The platypus lives along rivers in eastern Australia and on the nearby island of Tasmania. It eats mainly small shrimps and aquatic insects. The female usually lays two eggs and incubates them in a leaf nest. After hatching, the young nurse by licking up milk secreted onto the mother's fur. The animals called echidnas are also monotremes (**Figure 17.38a**).

Most mammals are born rather than hatched. During gestation in marsupials and eutherians, the embryos are nurtured inside the mother by an organ called the **placenta.** Consisting of both embryonic and maternal tissues, the placenta joins the embryo to the mother within the uterus. The embryo is nurtured by maternal blood that flows close to the embryonic blood system in the placenta. The embryo is bathed in fluid contained by an amniotic sac, which is homologous to the fluid compartment within the amniotic eggs of reptiles.

Marsupials are the so-called pouched mammals, including kangaroos and koalas. These mammals have a brief gestation and give birth to tiny, embryonic offspring that complete development while attached to the mother's nipples. The nursing young are usually housed in an external pouch, called the marsupium, on the mother's abdomen (**Figure 17.38b**). Nearly all marsupials live in Australia, New Zealand, and Central and South America. Australia has been a marsupial sanctuary for much of the past 60 million years. Australian marsupials have diversified extensively, filling terrestrial habitats that on other continents are occupied by eutherian mammals.

Eutherians are also called placental mammals because their placentas provide more intimate and long-lasting association between the mother and her developing young than do marsupial placentas (**Figure 17.38c**). Eutherians

(a) A monotreme (an echidna)

(b) A young marsupial in its mother's pouch (a brushtail opossum)

(c) Eutherians (placental mammals) (zebras)

Figure 17.38 The three major groups of mammals.
(a) Monotremes, such as this echidna, are the only mammals that lay eggs (inset). **(b)** The young of marsupials, such as this brushtail opossum, are born very early in their development. They finish their growth while nursing from a nipple in their mother's pouch. **(c)** In eutherians (placentals), such as these zebras, young develop within the uterus of the mother. There they are nurtured by the flow of blood though the dense network of vessels in the placenta. The reddish portion of the afterbirth clinging to the newborn zebra in this photograph is the placenta.

Activity 17C on the Web & CD
Test your knowledge of chordates.

make up almost 95% of the 4,500 species of living mammals. Dogs, cats, cows, rodents, rabbits, bats, and whales are all examples of eutherian mammals. One of the eutherian groups is the order **Primates,** which includes monkeys, apes, and humans.

The Human Ancestry

We have now traced the animal genealogy to the mammalian group that includes *Homo sapiens* and its closest kin. We are primates. To understand what that means, we must trace our ancestry back to the trees, where some of our most treasured traits originated as arboreal adaptations.

The Evolution of Primates

Primate evolution provides a context for understanding human origins. The fossil record supports the hypothesis that primates evolved from insect-eating mammals during the late Cretaceous period, about 65 million years ago. Those early primates were small, arboreal (tree-dwelling) mammals. Thus, the order Primates was first distinguished by characteristics that were shaped, through natural selection, by the demands of living in the trees. For example, primates have limber shoulder joints, which make it possible to brachiate (swing from one branch to another). The dexterous hands of primates can hang on to branches and manipulate food. Nails have replaced claws in many primate species, and the fingers are very sensitive. The eyes of primates are close together on the front of the face. The overlapping fields of vision of the two eyes enhance depth perception, an obvious advantage when brachiating. Excellent eye-hand coordination is also important for arboreal maneuvering (**Figure 17.39**). Parental care is essential for young animals in the trees. Mammals devote more energy to caring for their young than most other vertebrates, and primates are among the most attentive parents of all mammals. Most primates have single births and nurture their offspring for a long time. Though humans do not live in trees, we retain in modified form many of the traits that originated there.

Taxonomists divide the primates into two main groups: prosimians and anthropoids. The oldest primate fossils are **prosimians.** Modern prosimians include the lemurs of Madagascar and the lorises, pottos, and tarsiers that

Figure 17.39 The arboreal athleticism of primates.
This orangutan displays primate adaptations for living in the trees: limber shoulder joints, manual dexterity, and stereo vision due to eyes on the front of the face.

live in tropical Africa and southern Asia (**Figure 17.40a**). **Anthropoids** include monkeys, apes, and humans. All monkeys in the New World (the Americas) are arboreal and are distinguished by prehensile tails that function as an extra appendage for brachiating (**Figure 17.40b**). (If you see a monkey in a zoo swinging by its tail, you know it's from the New World.) Although some Old World monkeys are also arboreal, their tails are not prehensile. And many Old World monkeys, including baboons, macaques, and mandrills, are mainly ground-dwellers (**Figure 17.40c**).

Our closest anthropoid relatives are the apes: gibbons, orangutans, gorillas, and chimpanzees (**Figure 17.40d–g**). Modern apes live only in tropical regions of the Old World. With the exception of some gibbons, apes are larger than monkeys, with relatively long arms and short legs and no tail. Although all the apes are capable of brachiation, only gibbons and orangutans are primarily arboreal. Gorillas and chimpanzees are highly social. Apes have larger brains proportionate to body size than monkeys, and their behavior is consequently more adaptable.

Activity 17D on the Web & CD Sort out the primates.

(a) A prosimian (a slender loris)

(b) A New World monkey (a woolly spider monkey with young)

(c) Old World monkeys (pig-tail macaque with young)

(d) A white-handed gibbon

(e) An orangutan

(f) A gorilla with young

(g) A chimpanzee

(h) A human with young

Figure 17.40 Primate diversity.
(a) A prosimian. (b)–(c) Monkeys. (d)–(g) Apes. (h) A human.

The Emergence of Humankind

Humanity is one very young twig on the vertebrate branch, just one of many branches on the tree of life. In the continuum of life spanning 3.5 billion years, humans and apes have shared a common ancestry for all but the last 5–7 million years (**Figure 17.41**). Put another way, if we compressed the history of life to a year, humans and chimpanzees diverged from a common ancestor less than 18 hours ago. The fossil record and molecular systematics concur in that vintage for the human lineage. **Paleoanthropology,** the study of human evolution, focuses on this very thin slice of biological history.

Some Common Misconceptions Certain misconceptions about human evolution that were generated during the early part of the twentieth century still persist today in the minds of many, long after these myths have been debunked by the fossil evidence.

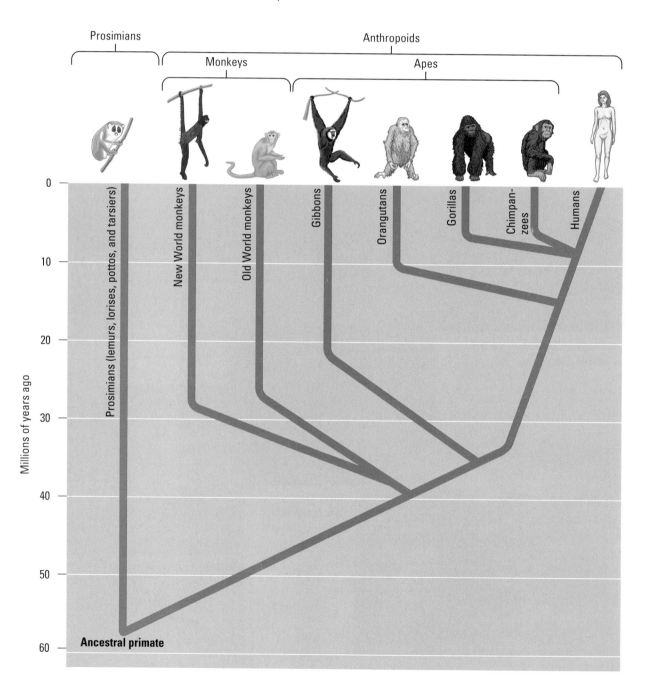

Figure 17.41
Primate phylogeny.

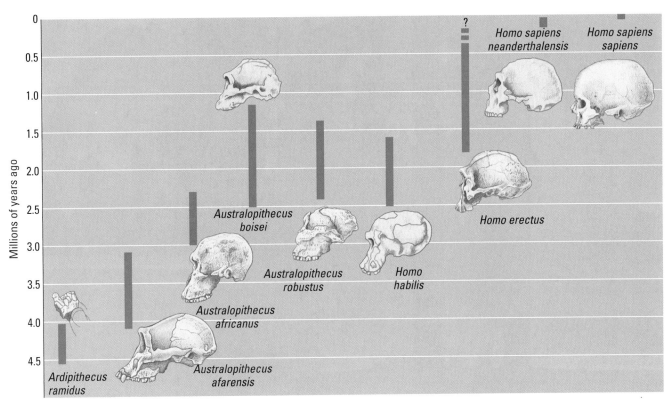

Figure 17.42 A timeline of human evolution.
Notice that there have been times when two or more hominid species coexisted. The skulls are all drawn to the same scale to enable you to compare the sizes of craniums and hence brains.

Let's first dispose of the myth that our ancestors were chimpanzees or any other modern apes. Chimpanzees and humans represent two divergent branches of the anthropoid tree that evolved from a common, less specialized ancestor. Chimps are not our parent species, but more like our phylogenetic siblings or cousins.

Another misconception envisions human evolution as a ladder with a series of steps leading directly from an ancestral anthropoid to *Homo sapiens*. This is often illustrated as a parade of fossil **hominids** (members of the human family) becoming progressively more modern as they march across the page. If human evolution is a parade, then it is a disorderly one, with many splinter groups having traveled down dead ends. At times in hominid history, several different human species coexisted (**Figure 17.42**). Human phylogeny is more like a multibranched bush than a ladder, with our species being the tip of the only twig that still lives.

One more myth we must bury is the notion that various human characteristics, such as upright posture and an enlarged brain, evolved in unison. A popular image is of early humans as half-stooped, half-witted cave-dwellers. In fact, we know from the fossil record that different human features evolved at different rates, with erect posture, or bipedalism, leading the way. Our pedigree includes ancestors who walked upright but had ape-sized brains.

After dismissing some of the folklore on human evolution, however, we must admit that many questions about our ancestry have not yet been resolved.

***Australopithecus* and the Antiquity of Bipedalism** Before there was *Homo*, several hominid species of the genus *Australopithecus* walked the African savanna (grasslands with clumps of trees). Paleoanthropologists have focused much of their attention on *A. afarensis*, an early species of *Australopithecus*. Fossil evidence now pushes bipedalism in *A. afarensis* back

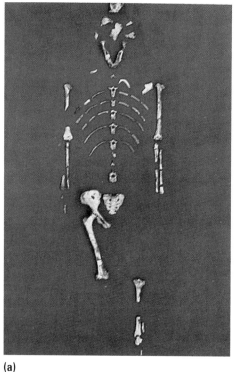

(a)

(b)

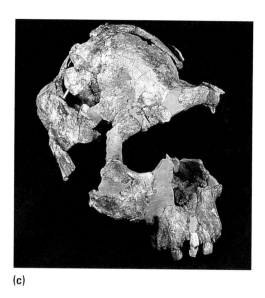

(c)

Figure 17.43 The antiquity of upright posture. (a) Lucy, a 3.18-million-year-old skeleton, represents the hominid species *Australopithecus afarensis*. Fragments of the pelvis and skull put *A. afarensis* on two feet. (b) Some 3.7 million years ago, several bipedal (upright-walking) hominids left footprints in damp volcanic ash in what is now Tanzania in East Africa. The prints fossilized and were discovered by British anthropologist Mary Leakey in 1978. The footprints are part of the strong evidence that bipedalism is a very old human trait. (c) This *A. afarensis* skull, 3.9 million years old, articulated with a vertical backbone. The upright posture of humans is at least that old.

to at least 4 million years ago (**Figure 17.43**). Doubt remains whether an even older hominid, *Ardipithecus ramidus,* which dates back at least 4.4 million years, was bipedal.

One of the most complete fossil skeletons of *A. afarensis* dates to about 3.2 million years ago in East Africa. Nicknamed Lucy by her discoverers, the individual was a female, only about 3 feet tall, with a head about the size of a softball (see Figure 17.43a). Lucy and her kind lived in savanna areas and may have subsisted on nuts and seeds, bird eggs, and whatever animals they could catch or scavenge from kills made by more efficient predators such as large cats and dogs.

All *Australopithecus* species were extinct by about 1.4 million years ago. Some of the later species overlapped in time with early species of our own genus, *Homo* (see Figure 17.42). Much debate centers on the evolutionary relationships of the *Australopithecus* species to each other and to *Homo.* Were the *Australopithecus* species all evolutionary side branches? Or were some of them ancestors to later humans? Either way, these early hominids show us that the fundamental human trait of bipedalism evolved millions of years before the other major human trait—an enlarged brain. As evolutionary biologist Stephen Jay Gould put it, "Mankind stood up first and got smart later."

***Homo habilis* and the Evolution of Inventive Minds** Enlargement of the human brain is first evident in fossils from East Africa dating to the latter part of the era of *Australopithecus,* about 2.5 million years ago.

Anthropologists have found skulls with brain capacities intermediate in size between those of the latest *Australopithecus* species and those of *Homo sapiens*. Simple handmade stone tools are sometimes found with the larger-brained fossils, which have been dubbed *Homo habilis* ("handy man"). After walking upright for about 2 million years, humans were finally beginning to use their manual dexterity and big brains to invent tools that enhanced their hunting, gathering, and scavenging on the African savanna.

Homo erectus and the Global Dispersal of Humanity The first species to extend humanity's range from its birthplace in Africa to other continents was *Homo erectus,* perhaps a descendant of *H. habilis.* The global dispersal began about 1.8 million years ago. But don't picture this migration as a mad dash for new territory or even as a casual stroll. If *H. erectus* simply expanded its range from Africa by about a mile per year, it would take only about 12,000 years to populate many regions of Asia and Europe. The gradual spread of humanity may have been associated with a change in diet to include a larger proportion of meat. In general, animals that hunt require more geographic territory than animals that feed mainly on vegetation.

Homo erectus was taller than *H. habilis* and had a larger brain capacity. During the 1.5 million years the species existed, the *H. erectus* brain increased to as large as 1,200 cubic centimeters (cm^3), a brain capacity that overlaps the normal range for modern humans. Intelligence enabled humans to continue succeeding in Africa and also to survive in the colder climates of the north. *Homo erectus* resided in huts or caves, built fires, made clothes from animal skins, and designed stone tools that were more refined than the tools of *H. habilis*. In anatomical and physiological adaptations, *H. erectus* was poorly equipped for life outside the tropics but made up for the deficiencies with cleverness and social cooperation.

Some African, Asian, European, and Australasian (from Indonesia, New Guinea, and Australia) populations of *H. erectus* gave rise to regionally diverse descendants that had even larger brains. Among these descendants of *H. erectus* were the Neanderthals, who lived in Europe, the Middle East, and parts of Asia from about 130,000 years ago to about 35,000 years ago. (They are named Neanderthals because their fossils were first found in the Neander Valley of Germany.) Compared with us, Neanderthals had slightly heavier browridges and less pronounced chins, but their brains, on average, were slightly larger than ours. Neanderthals were skilled toolmakers, and they participated in burials and other rituals that required abstract thought. Much current research on the Neanderthal skull addresses an intriguing question: Did Neanderthals have the anatomical equipment necessary for speech?

The Origin of _Homo sapiens_ The oldest known post–*H. erectus* fossils, dating back over 300,000 years, are found in Africa. Many paleoanthropologists group these African fossils along with Neanderthals and various Asian and Australasian fossils as the earliest forms of our species, *Homo sapiens.*

Activity 17E on the Web & CD Quiz yourself on human evolution.

These regionally diverse descendants of *H. erectus* are sometimes referred to as "archaic *Homo sapiens.*" The oldest fully modern fossils of *H. sapiens*—skulls and other bones that look essentially like those of today's humans—are about 100,000 years old and are located in Africa. Similar fossils almost as ancient have also been discovered in caves in Israel. The famous fossils of modern humans from the Cro-Magnon caves of France date back about 35,000 years.

The relationship of the Cro-Magnons and other modern humans to archaic *Homo sapiens* is a question that will continue to engage paleoanthropologists

Figure 17.44 "Out of Africa"—but when? Two hypotheses for the origin of modern *Homo sapiens*. There is no question that Africa is the cradle of humanity; we can all trace our ancestry back to that continent. But how recently was the ancestor common to all the world's modern humans still in Africa? **(a)** According to the multiregional hypothesis, modern humans throughout the world evolved from the *Homo erectus* populations that spread to different regions beginning almost 2 million years ago. Interbreeding between populations (dashed line in the diagram) kept humans very similar genetically. **(b)** In contrast, the "Out of Africa" hypothesis postulates that all regional descendants of *H. erectus* became extinct except in Africa. And according to this model, it was a *second* dispersal out of Africa, just 100,000 years ago, that populated the world with modern humans, which had evolved from *H. erectus* in Africa.

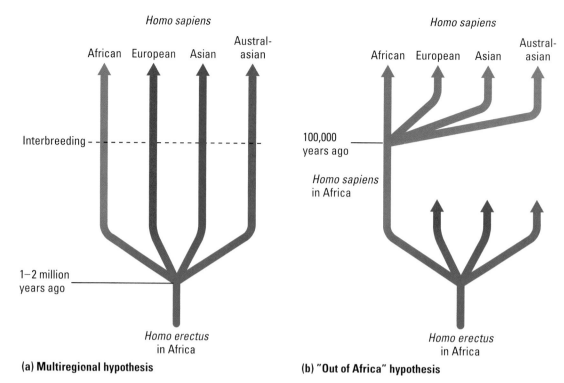

(a) **Multiregional hypothesis**

(b) **"Out of Africa" hypothesis**

for decades. What was the fate of the various descendants of *Homo erectus* who populated different parts of the world? In the view of some anthropologists, these archaic *H. sapiens* populations gave rise to modern humans. According to this hypothesis, called the **multiregional hypothesis,** modern humans evolved simultaneously in different parts of the world (**Figure 17.44a**). If this view is correct, then the geographic diversity of humans originated relatively early, when *H. erectus* spread from Africa into the other continents between 1 and 2 million years ago. This hypothesis accounts for the great genetic similarity of all modern people by pointing out that interbreeding among neighboring populations has always provided corridors for gene flow throughout the entire geographic range of humanity.

In sharp contrast to the multiregional hypothesis is a hypothesis that modern *H. sapiens* arose from a single archaic group in Africa. According to this **"Out of Africa" hypothesis** (also called the **replacement hypothesis**), the Neanderthals and other archaic peoples outside Africa were evolutionary dead ends (**Figure 17.44b**). Proponents of this hypothesis argue that modern humans spread out of Africa about 100,000 years ago and completely replaced the archaic *H. sapiens* in other regions. So far, genetic evidence mostly supports the replacement hypothesis. Based on comparisons of mitochondrial DNA (mtDNA) and of Y chromosomes between samples from various human populations, today's global human population seems to be very genetically uniform. Supporters of the replacement hypothesis argue that such uniformity could only stem from a single ancestral stock, not from diverse, geographically isolated populations. Proponents of the multiregional model counter that interbreeding between populations can explain our great genetic similarity. Debate about which hypothesis is correct—a multiregional or an "Out of Africa" origin of modern humans—centers mainly on differences in interpreting the fossil record and molecular data. As research continues, perhaps anthropologists will come closer to a consensus on our origins.

Cultural Evolution An erect stance was the most radical anatomical change in our evolution; it required major remodeling of the foot, pelvis, and vertebral column. Enlargement of the brain was a secondary alteration made possible by prolonging the growth period of the skull and its contents (see Figure 14.17). The primate brain continues to grow after birth, and the period of growth is longer for a human than for any other primate. The extended period of human development also lengthens the time parents care for their offspring, which contributes to the child's ability to benefit from the experiences of earlier generations. This is the basis of **culture**—the transmission of accumulated knowledge over generations. The major means of this transmission is language, written and spoken.

Cultural evolution is continuous, but there have been three major stages. The first stage began with nomads who hunted and gathered food on the African grasslands 2 million years ago. They made tools, organized communal activities, and divided labor. Beautiful ancient art is just one example of our cultural roots in these early societies (**Figure 17.45**). The second main stage of cultural evolution came with the development of agriculture in Africa, Eurasia, and the Americas about 10,000 to 15,000 years ago. Along with agriculture came permanent settlements and the first cities. The third major stage in our cultural evolution was the Industrial Revolution, which began in the eighteenth century. Since then, new technology has escalated exponentially; a single generation spanned the flight of the Wright brothers and Neil Armstrong's walk on the moon. It took less than a decade for the World-Wide Web to transform commerce, communication, and education. Through all this cultural evolution, from simple hunter-gatherers to high-tech societies, we have not changed biologically in any significant way. We are probably no more intelligent than our cave-dwelling ancestors. The same toolmaker who chipped away at stones now designs microchips and software. The know-how to build skyscrapers, computers, and spaceships is stored not in our genes but in the cumulative product of hundreds of generations of human experience, passed along by parents, teachers, books, and electronic media.

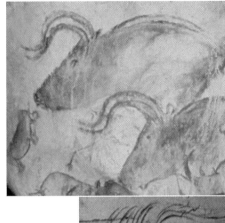

European Bison

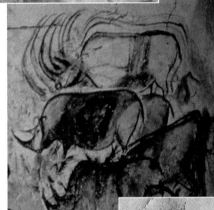

Rhinoceroses

Human hand

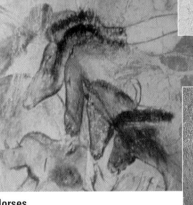

Horses

Owl

CHECKPOINT

1. To which mammalian order do we belong? What are the two main subgroups of this order?
2. Based on the fossil evidence in Figure 17.42, how many hominid species existed 1.7 million years ago?
3. Humans first evolved on which continent?
4. What role does a long period of parental care play in culture?
5. Why is *Homo habilis* known as "handy man"?
6. Lucy is an example of an early hominid belonging to the species _____ _____.
7. Which hypothesis—multiregional or "Out of Africa"—postulates that all *Homo erectus* populations outside of Africa became extinct?

Answers: **1.** Primates; prosimians and anthropoids **2.** Four **3.** Africa **4.** Extends opportunity for parents to transmit the lessons of the past to offspring **5.** Because it was the earliest hominid that definitely created and used stone tools **6.** *Australopithecus afarensis* **7.** "Out of Africa"

Figure 17.45 Art history goes way back—and so does our fascination with and dependence on animal diversity. Cro-Magnon wildlife artists created these remarkable paintings beginning about 30,000 years ago. Three cave explorers found this prehistoric art gallery on Christmas eve, 1994, when they ventured into a cavern near Vallon-Pont d'Arc in southern France.

Evolution Connection

Earth's New Crisis

Evolution of the human brain may have been anatomically simpler than acquiring an upright stance, but the global consequences of cerebral expansion have been enormous. Cultural evolution made *Homo sapiens* a new force in the history of life—a species that could defy its physical limitations and shortcut biological evolution. We do not have to wait to adapt to an environment through natural selection; we simply change the environment to meet our needs. We are the most numerous and widespread of all large animals, and everywhere we go, we bring environmental change. There is nothing new about environmental change. The history of life is the story of biological evolution on a changing planet. But it is unlikely that change has ever been as rapid as in the age of humans. Cultural evolution outpaces biological evolution by orders of magnitude. We are changing the world faster than many species can adapt; the rate of extinctions in the twentieth century was 50 times greater than the average for the past 100,000 years.

This rapid rate of extinction is mainly a result of habitat destruction, which is a function of human cultural changes and overpopulation. Feeding, clothing, and housing 6 billion people imposes an enormous strain on Earth's capacity to sustain life. If all these people suddenly assumed the high standard of living enjoyed by many people in developed nations, it is likely that Earth's support systems would be overwhelmed. Already, for example, current rates of fossil-fuel consumption, mainly by developed nations, are so great that waste carbon dioxide may be causing the temperature of the atmosphere to increase enough to alter world climates. Today, it is not just individual species that are endangered, but entire ecosystems, the global atmosphere, and the oceans. Tropical rain forests, which play a vital role in moderating global weather, are being cut down at a startling rate. Scientists have hardly begun to study these ecosystems, and many species in them may become extinct before they are even discovered.

Of the many crises in the history of life, the impact of one species, *Homo sapiens*, is the latest and potentially the most devastating.

Evolution Connection on the Web
Learn about what is being done to combat the effects of habitat destruction.

Chapter Review

Summary of Key Concepts

For study help, go to the Essential Biology Website (www.essentialbiology.com) or CD-ROM to explore the Activities and Case Studies in the Process of Science.

The Origins of Animal Diversity

• **What Is an Animal?** Animals are eukaryotic, multicellular, heterotrophic organisms that obtain nutrients by ingestion. Most animals reproduce sexually and develop from a zygote to a blastula, gastrula, and perhaps a larval stage before becoming an adult.

• **Early Animals and the Cambrian Explosion** Animals probably evolved from a colonial, flagellated protist more than 700 million years ago. At the beginning of the Cambrian period, 545 million years ago, animal diversity exploded.

• **Animal Phylogeny** Major branches of animal evolution are defined by four key evolutionary differences: the presence or absence of true tissues; radial versus bilateral body symmetry; presence or absence of a body cavity at least partly lined by mesoderm; and details of embryonic development.

Activity 17A Overview of Animal Phylogeny

Major Invertebrate Phyla

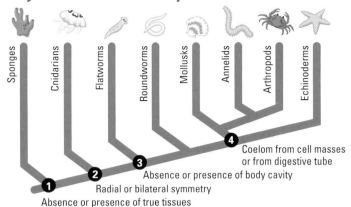

Coelom from cell masses
or from digestive tube

Absence or presence of body cavity

Radial or bilateral symmetry

Absence or presence of true tissues

- **Sponges**   Sponges (phylum Porifera) are sessile animals with porous bodies and choanocytes but no true tissues. They filter-feed by drawing water through pores in the sides of the body.

- **Cnidarians**   Cnidarians (phylum Cnidaria) have radial symmetry, a gastrovascular cavity, and tentacles with cnidocytes. The body is either a sessile polyp or floating medusa.

- **Flatworms**   Flatworms (phylum Platyhelminthes) are the simplest bilateral animals. They may be free-living or parasitic in or on plants and animals.

- **Roundworms**   Roundworms (phylum Nematoda) are unsegmented and cylindrical with tapered ends. They have a complete digestive tract and a pseudocoelom. They may be free-living or parasitic in plants and animals.

- **Mollusks**   Mollusks (phylum Mollusca) are soft-bodied animals often protected by a hard shell. The body has three main parts: a muscular foot, a visceral mass, and a fold of tissue called the mantle.

- **Annelids**   Annelids (phylum Annelida) are segmented worms. They may be free-living or parasitic on other animals.

- **Arthropods**   Arthropods (phylum Arthropoda) are segmented animals with an exoskeleton and specialized, jointed appendages. Arthropods consist of four main groups: Arachnids, crustaceans, millipedes and centipedes, and insects. In species diversity, insects outnumber all other forms of life combined.

Case Study in the Process of Science How Are Insect Species Identified?

- **Echinoderms**   Echinoderms (phylum Echinodermata) are sessile or slow-moving marine animals that lack body segments and possess a unique water vascular system. Bilaterally symmetrical larvae usually change to radially symmetrical adults.

Activity 17B Characteristics of Invertebrates

The Vertebrate Genealogy

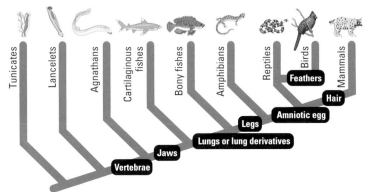

Feathers

Hair

Amniotic egg

Legs

Lungs or lung derivatives

Jaws

Vertebrae

- **Characteristics of Chordates**   Chordates (phylum Chordata) are defined by a dorsal, hollow nerve cord; a flexible notochord; pharyngeal slits; and a post-anal tail. Tunicates and lancelets are invertebrate chordates. All other chordates are vertebrates, possessing a cranium and backbone.

- **Fishes**   Agnathans are jawless vertebrates. Cartilaginous fishes (class Chondrichthyes), such as sharks, are mostly predators with powerful jaws and a flexible skeleton made of cartilage. Bony fishes (class Osteichthyes) have a stiff skeleton reinforced by hard calcium salts. Bony fishes are further classified into ray-finned fishes, lungfishes, and lobe-finned fishes.

- **Amphibians**   Amphibians (class Amphibia) are tetrapod vertebrates that usually deposit their eggs (lacking shells) in water. Aquatic larvae typically undergo a radical metamorphosis into the adult stage. Their moist skin requires that amphibians spend much of their adult life in moist environments.

- **Reptiles**   Reptiles (class Reptilia) are terrestrial ectotherms with lungs and waterproof skin covered by scales. Their amniotic eggs enhanced reproduction on land.

- **Birds**   Birds (class Aves) are endothermic vertebrates with amniotic eggs, wings, feathers, and other adaptations for flight.

Case Study in the Process of Science How Does Bone Structure Shed Light on the Origin of Birds?

- **Mammals**   Mammals (class Mammalia) are endothermic vertebrates with hair and mammary glands. There are three major groups: Monotremes lay eggs; marsupials use a placenta but give birth to tiny offspring that usually complete development while attached to nipples inside the mother's pouch; and eutherians, or placental mammals, use their placenta in a longer-lasting association between the mother and her developing young.

Activity 17C Characteristics of Chordates

The Human Ancestry

- **The Evolution of Primates**   The first primates were small, arboreal mammals that evolved from insect-eating mammals about 65 million years ago. Two subgroups of primates are the prosimians and anthropoids. Anthropoids consist of New World monkeys (with prehensile tails), Old World monkeys (without prehensile tails), apes, and humans. Modern apes are confined to tropical regions of the Old World.

Activity 17D Primate Diversity

- **The Emergence of Humankind**   Chimpanzees and humans evolved from a common, less specialized ancestor about 5–7 million years ago. Upright posture evolved in several hominid species of the genus *Australopithecus* at least 4 million years ago. Enlargement of the human brain came later, about 2.5 million years ago, when *Homo habilis* coexisted with smaller-brained hominids of the genus *Australopithecus*. *Homo erectus* was the first species to extend humanity's range from its birthplace in Africa to other continents. *Homo erectus* gave rise to regionally diverse descendants, such as the Neanderthals. According to the multiregional hypothesis, modern humans evolved in several locations from archaic *Homo sapiens*. In contrast, the Out-of-Africa hypothesis views all but the African archaic *Homo sapiens* as evolutionary dead ends. According to this model, a relatively recent dispersal (100,000 years ago) of modern Africans gave rise to today's human diversity. Human cultural evolution began in Africa with wandering hunter-gatherers, then progressed to the development of agriculture and eventually to the Industrial Revolution, which today continues with accelerating technological change.

Activity 17E Human Evolution

1. Bilateral symmetry in the animal kingdom is best correlated with
   a. an ability to sense equally in all directions.
   b. the presence of a skeleton.
   c. motility and active predation and escape.
   d. development of a true coelom.

2. A small group of worms classified in the phylum Gnathostomulida (not discussed in this chapter) live between sand grains on ocean beaches. These animals are unsegmented and lack a body cavity, but some have a complete digestive tract. These characteristics suggest that gnathostomulid worms might branch from the main trunk of the animal phylogenetic tree between which two phyla?

3. Which of the following categories includes all others in the list: arthropod, arachnid, insect, butterfly, crustacean, millipede?

4. Reptiles are much more extensively adapted to life on land than amphibians in that reptiles
   a. have a complete digestive tract.
   b. lay eggs that are enclosed in shells.
   c. are endothermic.
   d. go through a larval stage.

5. What is the name of the phylum to which humans belong? For what anatomical structure is the phylum named? Where in your body is this anatomical structure found?

6. The two major groups of primates are _____ and _____. Name an example of each.

7. Fossils suggest that the first major trait distinguishing human primates from other primates was _____.

8. Which of the following types of animals is not included in the human ancestry? (*Hint:* See Figure 17.31.)
   a. a bird   c. an amphibian
   b. a bony fish   d. a primate

9. Put the following list of species in order, from the oldest to the most recent: *Homo erectus, Australopithecus* species, *Homo habilis, Homo sapiens.*

10. Match each of the following animals to their phylum:
   a. Human   1. Porifera
   b. Leech   2. Arthropoda
   c. Sponge   3. Cnidaria
   d. Lobster   4. Chordata
   e. Sea anemone   5. Annelida

Answers to the Self-Quiz questions can be found in Appendix B.

Go to the website or CD-ROM for more Self-Quiz questions.

The Process of Science

1. Imagine that you are a marine biologist. As part of your exploration, you dredge up an unknown animal from the seafloor. Describe some of the characteristics you should look at to determine the animal phylum to which the creature should be assigned.

2. Many people describe themselves as vegetarians. Strictly speaking, vegetarians eat only plant products. Most vegetarians are not, in fact, that strict. Interview acquaintances who describe themselves as vegetarians and determine which taxonomic groups they avoid eating (see Figure 17.6 and 17.31). Try to generalize about their diet. For example, do they avoid eating vertebrates but eat some invertebrates? Do they avoid only birds and mammals?

3. Some researchers think that drying and cooling of the climate caused expansion of the African savanna, and that this environment favored upright walking in early humans. Why might bipedalism be advantageous in the savanna? How might an erect posture relate to the evolution of a larger brain?

Case Study in the Process of Science on the Web & CD *Learn to identify insects based on characteristics such as mouthparts and wings.*

Case Study in the Process of Science on the Web & CD *Analyze bones from dinosaurs, birds, and mammals to investigate the origin of birds.*

Biology and Society

1. Coral reefs harbor a greater diversity of animals than any other environment in the sea. Australia's Great Barrier Reef has been protected as a marine reserve and is a mecca for scientists and nature enthusiasts. Elsewhere, such as Indonesia and the Philippines, coral reefs are in danger. Many reefs have been depleted of fish, and runoff from the shore has covered coral heads with sediment. Nearly all the changes in the reefs can be traced back to human activities. What kinds of activities do you think might be contributing to the decline of the reefs? What are some reasons to be concerned about this decline? Do you think the situation is likely to improve or worsen in the future? Why? What might the local people do to halt the decline? Should the developed countries help? Why or why not?

2. In the opening section, the disastrous consequences of the introduction of cane toads to Australia were described. This is just one of many examples of invasive non-native species displacing native species. Can you name other examples? Can you name some that have occurred in your home area? What species was introduced? What species was displaced? What were the consequences? What could have been done to reduce the consequences?

3. The human body has not changed much in the last 100,000 years, but human culture has changed a great deal. As a result of our culture, we change the environment at a rate far greater than many species, including our own, can evolve. What evidence of rapid environmental change do you see regularly? What aspects of human culture are responsible for these changes? Do you see any evidence of a decrease in the rate of human-caused environmental changes?

Biology and Society on the Web *Learn about how invasive species arrive in the United States.*

UNIT FOUR

Ecology

CHAPTER 18

The Ecology of Organisms and Populations

Biology and Society: The Human Population Explosion 381

An Overview of Ecology 381
Ecology as Scientific Study
A Hierarchy of Interactions
Ecology and Environmentalism
Abiotic Factors of the Biosphere

The Evolutionary Adaptations of Organisms 387
Physiological Responses
Anatomical Responses
Behavioral Responses

What Is Population Ecology? 388
Population Density
Patterns of Dispersion
Population Growth Models
Regulation of Population Growth
Human Population Growth

Life Histories and Their Evolution 398
Life Tables and Survivorship Curves
Life History Traits as Evolutionary Adaptations

Evolution Connection: Testing a Darwinian Hypothesis 401

Every four years, the world's population increases by the population equivalent of the United States.

The biosphere is currently undergoing a mass extinction.

Some ecological communities depend on periodic fires.

After several days at high altitude, your body will begin to produce more red blood cells.

The Human Population Explosion

Sometime during the summer of 1999, Earth's human population passed the 6 billion mark—and kept right on exploding into the new millennium. The population had doubled from 3 billion in just 40 years. At the present growth rate of 73 million people per year, it takes only four days to add the population equivalent of a San Francisco, or just under four years to add the population equivalent of the United States. We are by far the most abundant large animals, and given our technological prowess, we have a disproportionately high impact on the environment (**Figure 18.1**).

Biology and Society on the Web
This U.S. census website will provide you with some impressive numbers.

Every day, we hear about local and global problems that threaten our well-being or provoke disputes between individuals or nations—global warming, toxic waste, conflicts over oil in the Middle East, the declining health of the oceans, civil strife aggravated by depressed economics. Contributing to all these apparently unrelated problems is a common factor: the continued increase of the human population in the face of limited resources. The human population explosion is now Earth's most significant biological phenomenon. Our species requires vast amounts of materials and space, including places to live, land to grow our food, and places to dump our waste. Incessantly expanding our presence on Earth, we have devastated the environment for many other species and now threaten to make it unfit for ourselves.

Every 20 minutes, our population increases by nearly 3,000 individuals. In the same 20 minutes, on average, one species of plant or animal becomes extinct. Most ecologists believe that the two trends are connected, at least over the long term—a connection between the exploding human population and the mass extinction the biosphere is now experiencing. The most important factor is our encroachment on and destruction of habitat as we spread out and "tame" more land to satisfy our growing demand for space, food, shelter, fuel, water, and other resources.

To understand the problem of human population growth on more than a superficial level, we must understand the principles of ecology that apply to all species. This chapter therefore begins with an overview of ecology as a scientific discipline. Next, we'll consider how organisms adapt to Earth's varied ecological arenas. We will then focus on the study of populations of individual species, eventually returning to our discussion of the human population.

Figure 18.1 **"Subduing the Earth."** This photo taken on Lyell Island in British Columbia, Canada, shows clear-cutting of an old-growth forest. The lumber is harvested to support a growing demand for housing.

An Overview of Ecology

The scientific study of the interactions between organisms and their environments is called **ecology.** All organisms, including humans, interact continuously with their environments. We breathe, exchanging gases with the atmosphere. We eat other organisms, taking in secondhand energy that entered plants as sunlight. We add urea and other wastes to water and soil. We absorb and radiate heat. We are inextricably connected to the outside world, as are all of Earth's creatures.

This outside world, the environment, can be divided into two major components. The **abiotic component** consists of nonliving chemical and

(a)

(b)

Figure 18.2 Ecological research in a rain forest canopy. **(a)** This hot-air dirigible is placing a giant rubber raft on the treetops of a tropical rain forest in French Guiana, in northeastern South America. **(b)** Living and working on this field station 30 m above the forest floor, an international research team studies the canopy environment. French biologist Pierre Grard and two other scientists are cataloging plants and the insects that feed on those plants. Among Earth's diverse environments, the canopy (upper tier) of tropical rain forests is especially rich in its diversity of insects, birds, and other animals.

Figure 18.3 Experimental ecology on a grand scale. In this classic experiment, David Schindler (University of Alberta) and his colleagues tested the hypothesis that the availability of mineral nutrients such as nitrogen and phosphorus can promote the growth of algae in lakes. The researchers used a plastic curtain to partition this lake into an "experimental lake" (top) and a "control lake." The team added certain mineral nutrients to the experimental lake, but not to the control, which served as a basis of comparison. Within two months after the scientists added phosphorus, a bloom (population explosion) of algae gave the experimental lake the cloudy, whitish appearance you see in the photograph. The experiment helped ecologists understand how the pollution of lakes with phosphorus from sewage and runoff from fertilized farmland can cause algal blooms. In such "overfertilized" lakes, bacteria and other decomposers of the algal mass sometimes consume all the oxygen, leading to the death of fish and other aerobic organisms. Scientific evidence has resulted in better controls on water quality, including the banning of phosphate-containing detergents in many states. (Reprinted with permission from D. W. Schindler, "Eutrophication and Recovery in Experimental Lakes: Implications for Lake Management," *Science* vol. 184, page 897, Figure 1.49 (1974). Copyright © 1974 American Association for the Advancement of Science.)

physical factors, such as temperature, light, water, minerals, and air. The **biotic component** includes the living factors—all the other organisms that are part of an individual's environment. Other organisms may compete with an individual for food and other resources, prey upon it, or change its physical and chemical environment.

In this section, we'll take a closer look at three key words in our definition of ecology: the *scientific* study of the *interactions* between organisms and their *environments*. This straightforward definition masks an enormously complex and exciting area of biology that is also of crucial practical importance. The science of ecology provides a basic understanding of how natural processes and organisms interact, giving us the tools we need to manage the planet's limited resources over the long term.

Ecology as Scientific Study

Humans have always had an interest in other organisms and their environments. As hunters and gatherers, prehistoric people had to learn where game and edible plants could be found in greatest abundance. And naturalists, from Aristotle to Darwin, made the process of observing and describing organisms in their natural habitats an end in itself rather than simply a means of survival. Extraordinary insight can still be gained from this descriptive approach of watching nature and recording its structure and processes. (As Yogi Berra once put it, "You can observe a lot just by watching.") Thus, natural history as a "discovery science" (see Chapter 1) remains fundamental to ecology (**Figure 18.2**). But in the past few decades, ecology has become increasingly experimental. In spite of the difficulty of conducting experiments that often involve large amounts of time and space, many ecologists are testing hypotheses in the laboratory and in the field. Ecologists also complement descriptive and experimental studies with computer simulations of large-scale experiments that might be impossible to conduct in the field. Whatever the approach, ecology employs that most basic of scientific processes, the posing of hypotheses and the use of observations and experiments to test those hypotheses. In the experiment depicted in **Figure 18.3**, for example, scientists tested the hypothesis that mineral nutrients can promote the growth of algae in lakes.

A Hierarchy of Interactions

When we study the interactions between organisms and their environments, it is convenient to divide ecology into four increasingly comprehensive levels: organismal ecology, population ecology, community ecology, and ecosystem ecology (see Figure 2.2).

Organismal ecology is concerned with the evolutionary adaptations that enable individual organisms to meet the challenges posed by their abiotic environments. For example, an organismal ecologist might be interested in the special equipment that enables diving whales to stay under water for so long (**Figure 18.4a**). The distribution of organisms is limited by the abiotic conditions they can tolerate. Earthworms, for instance, are restricted to moist environments because their skin does not prevent dehydration, as is apparent the day after a rain, when many desiccated earthworms can be found on recently rain-soaked pavement.

The next level of organization in ecology is the **population,** a group of individuals of the same species living in a particular geographic area. **Population ecology** concentrates mainly on factors that affect population density and growth. What factors, for example, limit the number of striped mice that can inhabit a particular area (**Figure 18.4b**)? The latter part of this chapter focuses on population ecology.

A **community** consists of all the organisms that inhabit a particular area; it is an assemblage of populations of different species. Questions in **community ecology** focus on how interactions between species, such as predation, competition, and symbiosis, affect community structure and organization. For example, what factors influence the diversity of tree species that make up a particular forest (**Figure 18.4c**)?

An **ecosystem** includes all the abiotic factors in addition to the community of species in a certain area. For example, a forest ecosystem includes not only the organisms, such as diverse plants and animals, but also the soil, water sources, sunlight, and other abiotic factors of the environment. In **ecosystem ecology,** questions concern energy flow and the cycling of chemicals among the various biotic and abiotic factors. What processes, for instance, recycle vital chemical elements such as nitrogen within a savanna ecosystem (**Figure 18.4d**)? (Chapter 19 focuses on community and ecosystem ecology.)

The **biosphere** is the global ecosystem—the sum of all the planet's ecosystems, or all of life and where it lives. The most complex level in ecology, the biosphere includes the atmosphere to an altitude of several kilometers, the land down to water-bearing rocks about 1,500 m deep, lakes and streams, caves, and the oceans to a depth of several kilometers. Isolated in space, the biosphere is self-contained, or closed, except that its photosynthetic producers derive energy from sunlight, and it loses heat to space.

Ecology and Environmentalism

The science of ecology should be distinguished from the informal use of the word *ecology* to refer to environmental concerns. And yet, we need to understand the complicated and delicate relationships between organisms and their environments to address environmental problems.

Our current awareness of the biosphere's limits stems mainly from the 1960s, a time of growing disillusionment with environmental practices of the past. In the 1950s, technology seemed poised to free humankind from

(a) Organismal ecology: How do diving whales stay under water for so long?

(b) Population ecology: What factors limit the number of striped mice that can inhabit a particular area?

(c) Community ecology: What factors influence the diversity of tree species that make up a particular forest?

(d) Ecosystem ecology: What processes recycle vital chemical elements such as nitrogen within a savanna ecosystem?

Figure 18.4 Examples of questions at different levels of ecology.

Figure 18.5 Rachel Carson and *Silent Spring*.
Although her book, seminal to the modern environmental movement, focused on the biosphere's hangover from the pesticide DDT, Carson's message was much broader: "The 'control of nature' is a phrase conceived in arrogance, born of the Neanderthal age of biology and philosophy, when it was supposed that nature exists for the convenience of man."

Figure 18.6 Environmental activism. Students in Vancouver, British Columbia, continue the tradition of environmental activism by advocating for preservation of an old growth forest.

several age-old bonds. New chemical fertilizers and pesticides, for example, showed great promise for increasing agricultural productivity and eliminating insect-borne diseases. Fertilizers were applied extensively, and pests were attacked with massive aerial spraying of pesticides, including a poison called DDT. The immediate results were astonishing: Increases in farm productivity enabled developed nations such as the United States to grow surplus food and market it overseas, and the worldwide incidence of malaria and several other insect-borne diseases was markedly reduced. The pesticide DDT was hailed as a miracle weapon to wield anywhere insects caused problems.

Our enthusiasm for chemical fertilizers and pesticides began to wane as some of the side effects of DDT and other widely used poisons began appearing in the late 1950s. One of the first to perceive the global dangers of pesticide abuse was Rachel Carson (**Figure 18.5**). We can trace our current environmental awareness to Carson's 1962 book, *Silent Spring*. Her warnings were underscored when scientists began reporting that DDT was threatening the survival of predatory birds and was showing up in human milk.

> **Activity 18A on the Web & CD**
> Analyze the pros and cons of DDT.

Another serious problem was the genetic resistance to pesticides that evolved in an increasing number of pest populations (see Chapter 13). At the same time, some of the ill effects of other technological developments began to be widely publicized. By the early 1970s, disillusionment with the overuse of chemicals and a realization that our finite biosphere could not tolerate unlimited exploitation had developed into widespread concern about the environment. The environmental movement that Carson helped catalyze continues, with students of your generation sustaining a tradition of activism on behalf of the biosphere's future health (**Figure 18.6**).

Today, it's clear that no part of the biosphere is untouched by the abusive impact of human populations and their technology. Depletion of natural resources, localized famine aggravated by land misuse and expanding population, the growing list of species extinguished or endangered by loss of habitat, the poisoning of soil and streams with toxic wastes, and global warming caused by deforestation and combustion of fossil fuels—these are just a few of the problems that we have created and now must solve. Analyzing environmental issues and planning for better practices start with a basic understanding of ecology, which should be part of every student's education.

Abiotic Factors of the Biosphere

The biosphere is patchy. We can see this environmental patchwork on several levels. On a global scale, ecologists have long recognized striking regional patterns in the distribution of terrestrial and aquatic life (**Figure 18.7**). These patterns mainly reflect regional differences in climate and other abiotic factors. **Figure 18.8** shows patchiness on a local scale; we can see a mixture of forest, small lakes, a meandering river, and open meadows. If we moved even closer, into any one of these different environments, we would find patchiness on yet a smaller scale. For example, we would find that each lake has several different **habitats**—environmental situations in which organisms live. And each habitat has a characteristic community of organisms. As with the global patchiness of the biosphere, smaller-scale environmental variation is based mainly on differences in abiotic factors. We'll now consider some of the most important abiotic factors and how they might affect the organisms that interact with them.

Sunlight Solar energy powers nearly all ecosystems. In aquatic environments, the availability of sunlight has a significant effect on the growth and

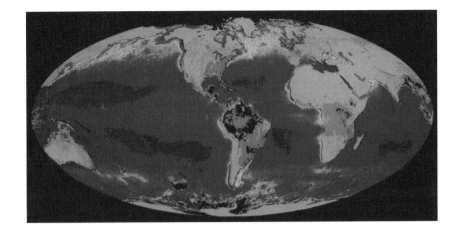

Figure 18.7 Regional distribution of life in the biosphere. In this image of Earth, based on data sent from satellites, colors are keyed to the relative abundance of chlorophyll, which correlates with the regional densities of life. Green areas on land are dense forests, including the tropical forests of South America and Africa. Orange areas on land are relatively barren regions, such as the Sahara of Africa. The patchy productivity of the oceans is also evident, with green regions having a greater abundance of phytoplankton (algae) than darker regions. The patchiness of the biosphere is mainly due to regional variations in abiotic factors such as temperature and the availability of water and mineral nutrients.

distribution of algae. Because water itself and the microorganisms in it absorb light and keep it from penetrating very far, most photosynthesis occurs near the surface of the water. In terrestrial environments, light is often not the most important factor limiting plant growth. In many forests, however, shading by trees creates intense competition for light at ground level.

Water Aquatic organisms have a seemingly unlimited supply of water, but they face problems of water balance if their own solute concentration does not match that of their surroundings. (Figure 4.7 shows water balance in cells.) For a terrestrial organism, the main water problem is the threat of drying out. Many land species have watertight coverings that reduce water loss. For example, a waxy coating on the leaves and other aerial parts of most plants helps prevent dehydration. (Rubbing a cucumber is a good way to sample this wax.) And humans and other mammals have a layer of dead outer skin containing a waterproofing protein. Moreover, the ability of our kidneys to excrete very concentrated urine is an evolutionary adaptation that enables us to rid our body of urea, a waste product, with minimal water loss.

> **Case Study in the Process of Science on the Web & CD**
> Devise an experiment to determine whether pillbugs prefer a wet or dry environment.

Temperature Environmental temperature is an important abiotic factor because of its effect on metabolism. Few organisms can maintain a sufficiently active metabolism at temperatures close to 0°C (32°F), and temperatures above 50°C (122°F) destroy the enzymes of most organisms. Still, extraordinary adaptations enable some species to live outside this temperature range. For example, some North American frogs and turtles can freeze during the winter months and still survive, and prokaryotes living in hot springs have enzymes that function optimally at extremely high temperatures (**Figure 18.9**). Mammals and birds can remain considerably warmer than their surroundings and can be active in a fairly wide range of temperatures, but even these animals function best at certain temperatures.

> **Case Study in the Process of Science on the Web & CD**
> Make changes in abiotic factors and see the effects on various plants and animals.

Wind Some organisms—for example, bacteria, protists, and many insects that live on snow-covered mountain peaks—depend on nutrients blown to them by winds. Many plants depend on wind to disperse their pollen and seeds. Local wind damage often creates openings in forests, contributing to patchiness in ecosystems. Wind also increases an organism's rate of water loss by

Figure 18.8 Patchiness of the environment in an Alaskan wilderness.

Figure 18.9 Home of hot prokaryotes. "Heat-loving" prokaryotes thrive in this pool at Yellowstone National Park, where temperatures may exceed 80°C (176°F).

Figure 18.10 Wind as an abiotic factor that shapes trees. The prevailing winds gradually caused the "flagging" of these fir trees. The mechanical disturbance of the wind inhibits limb growth on the windward side of the trees, while limbs on the leeward side grow normally. This growth response is an evolutionary adaptation that reduces the number of limbs that are broken during strong winds.

evaporation. The consequent increase in evaporative cooling can be advantageous on a hot summer day, but can cause dangerous wind chill in the winter. Wind can also affect the pattern of a plant's growth (**Figure 18.10**).

Rocks and Soil The physical structure and chemical composition of rocks and soil limit the distribution of plants and of the animals that feed on the vegetation. Soil variation contributes to the patchiness we see in terrestrial landscapes such as in Figure 18.8. In streams and rivers, the composition of the substrate can affect water chemistry, which in turn influences the resident plants and animals. In marine environments, the structure of underlying substrates determines the types of organisms that can attach or burrow in those habitats.

Periodic Disturbances Catastrophic disturbances, such as fires, hurricanes, tornadoes, and volcanic eruptions, can devastate biological communities. After the disturbance, the area is recolonized by organisms or repopulated by survivors, but the structure of the community undergoes a succession of changes during the rebound (**Figure 18.11**). Some disturbances, such as volcanic eruptions, are so infrequent and irregular over space and time that organisms have not acquired evolutionary adaptations to them. Fire, on the other hand, although unpredictable over the short term, recurs frequently in some communities, and many plants have adapted to this periodic disturbance. In fact, several communities actually depend on periodic fire to maintain them.

Figure 18.11 Recovery after a forest fire. Just a few months after a forest fire swept through Yellowstone National Park, wildflowers and other small plants were already colonizing the area. The increased sunlight and soil nutrients released from the trees that burned were among the abiotic factors contributing to the regreening of the scorched land.

CHECKPOINT

1. A (an) _____ consists of a biological _____, or all the biotic factors in the area, along with the nonliving, or _____, factors.

2. Why is it more accurate to define the biosphere as the global *ecosystem* rather than the global *community*?

3. Rachel Carson's *Silent Spring* focused on the destructive consequences of toxic pollutants, especially _____.

4. _____ energy is such an important abiotic factor because _____ provides most of the organic fuel and building material for the organisms of most ecosystems.

Answers: 1. ecosystem; community; abiotic **2.** Because the biosphere includes both abiotic and biotic factors **3.** the pesticide DDT **4.** Solar (or light); photosynthesis

The Evolutionary Adaptations of Organisms

Having surveyed the abiotic factors of the biosphere, we can now turn to the organisms themselves. In this section, we'll look at some of the evolutionary adaptations that allow species to meet the challenges of their environments.

The fields of ecology and evolutionary biology are tightly linked. Charles Darwin was an ecologist (although he predated the word *ecology*). It was the geographic distribution of organisms and their exquisite adaptations to specific environments that provided Darwin with evidence for evolution. Evolutionary adaptation via natural selection results from the interaction of organisms with their environments, which brings us back to our definition of ecology. Thus, events that occur in the time frame of what is sometimes called ecological time translate into effects over the longer scale of evolutionary time. For instance, hawks feeding on field mice have an impact on the gene pool of the prey population by curtailing the reproductive success of certain individuals. One long-term effect of such a predator-prey interaction may be the prevalence in the mouse population of fur coloration that camouflages the animals.

In our brief survey of organismal ecology, we'll focus on three types of adaptations—physiological, anatomical, and behavioral—that enable plants and animals to adjust to changes in their environments. Note that these changes occur during the lifetime of an individual, so they do not qualify as evolution, which is change in a population over time (see Chapter 13). But an individual organism's abilities to adjust to environmental change during ecological time are themselves adaptations refined by natural selection during evolutionary time.

Physiological Responses

You may have seen a cat's fur fluff up on a cold day, a response that helps insulate the animal's body. The mechanism for this response is the contraction of tiny muscles attached to the hairs. (Our own muscles do this, too, but we just get "goose bumps" instead of a furry insulation.) The blood vessels in the cat's skin also constrict, which slows the loss of body heat. (This works for humans, too.) These are examples of physiological responses to environmental change. In these mechanisms of temperature regulation (thermoregulation), the response occurs in just seconds.

Physiological response that is longer term, though still reversible, is called **acclimation.** For example, suppose you moved from Boston, which is essentially at sea level, to the mile-high city of Denver. One physiological response to the lower oxygen supply in your new environment would be a gradual increase in the number of your red blood cells, which transport O_2 from your lungs to other parts of your body. Acclimation can take days or weeks. This is why high-altitude climbers (such as those attempting to scale Mount Everest) need extended stays at a base camp before proceeding to the summit.

The ability to acclimate is generally related to the range of environmental conditions the species naturally experiences. Species that live in very warm climates, for example, usually do not acclimate to extreme cold. Among vertebrates, birds and mammals can generally tolerate the greatest temperature extremes because, as endotherms, they use their metabolism to regulate internal temperature (see Chapter 17). In contrast, reptiles, which are ectotherms, are more limited in the climates they can tolerate (**Figure 18.12**). Contrary to popular legend, it wasn't Saint Patrick who

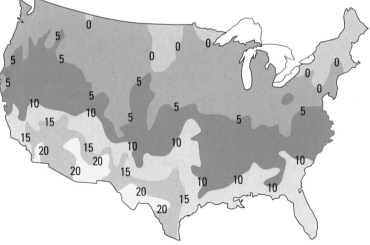

Figure 18.12 The number of lizard species in different regions of the contiguous United States. The northward decrease in the diversity of reptiles reflects their ectothermic physiology, which depends on environmental heat for keeping the body warm enough for the animal to be active.

chased the snakes out of Ireland; the ice ages did. Even now, there are no snakes in Ireland. They have a hard time getting there across the sea, and Ireland is apparently too cool for snake populations to become established even if they manage to reach the island from warmer parts of the world.

Anatomical Responses

Many organisms respond to environmental challenge with some type of change in body shape or anatomy (structure). In some cases, these responses are examples of acclimation, since they are reversible. Many mammals and birds, for example, grow a heavier coat of fur or feathers in winter; sometimes coat color changes seasonally as well, camouflaging the animal against winter snow and summer vegetation.

Other anatomical changes are irreversible over the lifetime of an individual. Environmental variation can affect growth and development so much that there may be remarkable differences in body shape within a population. You can see an example in Figure 18.10, which shows the "flagging" that wind causes in certain trees. In general, plants are more anatomically plastic than animals. Rooted and unable to move to a better location, plants rely entirely on their anatomical and physiological responses to survive environmental fluctuations.

Behavioral Responses

In contrast to plants, most animals can respond to an unfavorable change in the environment by moving to a new location. Such movement may be fairly localized. For example, many desert ectotherms, including reptiles, maintain a reasonably constant body temperature by shuttling between sun and shade. Some animals, however, are capable of migrating great distances in response to such environmental cues as the changing seasons. Many migratory birds overwinter in Central and South America, returning to northern latitudes to breed during summer. And humans, with their large brains and available technology, have an especially rich range of behavioral responses available to them (**Figure 18.13**).

Activity 18B on the Web & CD
Explore evolutionary adaptations of desert and forest organisms.

Figure 18.13 Behavioral responses have expanded the geographic range of humans. Dressing for the weather is a thermoregulatory behavior unique to humans.

CHECKPOINT

1. Why does the set of evolutionary adaptations characterizing a species tend to limit the geographic distribution of that species?

2. What is acclimation?

3. Contrast ecological time with evolutionary time.

Answers: 1. To the extent that the species is adapted to particular environmental conditions, it is not so well equipped to survive and reproduce where there are different conditions to which the species is not adapted. **2.** A gradual, reversible change in anatomy or physiology in response to an environmental change **3.** Ecological time is the temporal scale for the present interactions between organisms and their environments. Evolutionary time is the longer-term consequence of those interactions in the adaptations that evolve via natural selection.

What Is Population Ecology?

Now that you have a basic understanding of the factors that make up the environment and how organisms respond to them, we will turn our attention to one of the four hierarchical levels of ecology: population ecology.

No population can grow indefinitely. Species other than humans sometimes exhibit population explosions, but their populations inevitably crash. In contrast to these radical booms and busts, many populations are relatively stable over time, with only minor increases or decreases in population size. **Population ecology** focuses on the factors that influence a population's size (number of individuals), growth rate (rate of change in population size), density (number of individuals per unit area or volume), and population structure (including relative numbers of individuals of different ages).

Before we go on, it's important to understand what biologists mean by a population. A **population** is a group of individuals of the same species living in a given area at a given time. A population's geographic boundaries may be natural, as with certain species of trout in an isolated lake. But ecologists often define a population's boundaries in more arbitrary ways that fit their research questions. For example, an ecologist studying the contribution of asexual reproduction to the population growth of sea anemones might define a population as all the anemones of one species in a tide pool. Another researcher studying the effects of hunting on deer might define a population as all the deer within a particular state. Yet another researcher, attempting to determine which segment of the human population will be most affected by the AIDS epidemic, might study the HIV infection rate of the human population in one nation or throughout the world. Whatever the scale of the population we're studying, there are some common principles of population structure and growth that will guide our analysis.

Population Density

Population density is the number of individuals of a species per unit area or volume—the number of oak trees per square kilometer (km^2) in a forest, for example, or the number of earthworms per cubic meter (m^3) in the forest's soil.

How do we measure population density? In rare cases, it is possible to actually count all individuals within the boundaries of the population. For example, we could count the total number of oak trees (say, 200) in a forest covering 50 km^2. The population density would be the total number of trees divided by the area, or 4/km^2 (or 4 trees per square kilometer).

In most cases, it is impractical or impossible to count all individuals in a population. Instead, ecologists use a variety of sampling techniques to estimate population densities. For example, they might estimate the density of alligators in the Florida Everglades based on a count of individuals in a few sample plots of 1 km^2 each. The larger the number and size of sample plots, the more accurate the estimates. In some cases, population densities are estimated not by counts of organisms but by indirect indicators, such as number of bird nests or rodent burrows (**Figure 18.14**).

Another sampling technique commonly used to estimate wildlife populations is the **mark-recapture method.** The researcher places traps within the boundaries of the population under study. The researcher then marks the captured animals with tags, collars, bands, or spots of dye. The marked animals are released. After a few days or weeks—enough time for the marked individuals to mix randomly with unmarked members of the population—traps are set again. This second capture will yield both marked and unmarked individuals. From these data, the total number of individuals in the population can be estimated.

Figure 18.15 shows this method of estimating the population size for a bird population. The mark-recapture method assumes that each marked individual

Activity 18C on the Web & CD
Imagine you're a wildlife biologist as you estimate population densities of mice and the plants they use for food.

Figure 18.14 An indirect census of a prairie dog population.   We could estimate the number of prairie dogs in this colony in South Dakota by counting the number of mounds constructed by the rodents. The estimate is rough because the animals are social, and the number of individuals that cohabit a system of tunnels under each mound varies.

(a) Capturing sanderlings in a mist net

(b) Marking birds with leg bands so they can be identified among birds trapped during a second capture

Figure 18.15 A mark-recapture estimate of population size.   This biologist is using the mark-recapture method to estimate the number of individuals in a population of birds called sanderlings. **(a)** Let's say that he has captured 50 sanderlings in a harmless trap called a mist net. **(b)** He marks the birds with leg bands and then releases them back to the whole population. A second capture two weeks later yields a total of 100 sanderlings, of which 10 are marked birds that have been recaptured. We can estimate that 10% of the total sanderling population is marked. Since the biologist marked 50 birds, our estimate for the entire population is about 500 birds.

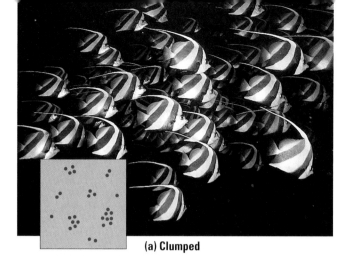

(a) Clumped

(b) Uniform

(c) Random

Figure 18.16 Patterns of dispersion within a population's geographic range. **(a)** Butterfly fish, like many fishes, often clump into what are called schools. Schooling may enhance the hydrodynamic efficiency of swimming, reduce predation ("safety in numbers"), and increase feeding efficiency. **(b)** Birds nesting on small islands, such as these king penguins on South Georgia Island in the South Atlantic, often exhibit uniform spacing. Territorial behavior helps stabilize the distance between individuals. **(c)** Trees of the same species are often randomly distributed among trees of other species in tropical rain forests.

has the same probability of being trapped as each unmarked individual. This is not always a safe assumption, because an animal that has been trapped once may be wary of traps in the future.

Patterns of Dispersion

The **dispersion pattern** of a population is the way individuals are spaced within the population's geographic range. A **clumped** pattern, in which individuals are aggregated in patches, is the most common in nature. Clumping often results from an unequal distribution of resources in the environment. For instance, cottonwood trees are usually clumped along a streamside in patches of moist and sandy soil. Clumping of animals is often associated with uneven food distribution or with mating or other social behavior. For instance, mosquitoes often swarm in great numbers, which increases their chances for mating. Schooling fishes are another example (**Figure 18.16a**).

A **uniform** pattern of dispersion often results from interactions among the individuals of a population. For instance, creosote bushes in the desert tend to be uniformly spaced because their roots compete for water and dissolved nutrients. Animals often exhibit uniform dispersion as a result of social interactions. Examples are birds nesting in large numbers on small islands (**Figure 18.16b**).

In a third type of dispersion, a **random** dispersion, individuals in a population are spaced in a patternless, unpredictable way. This only occurs in the absence of strong attractions or repulsions among individuals in a population. Clams living in a coastal mudflat, for instance, might be randomly dispersed at times of the year when they are not breeding and when resources are plentiful and do not affect their distribution. Forest trees are also randomly distributed in some cases (**Figure 18.16c**). However, environmental conditions and social interactions make random dispersion rare.

Some populations exhibit both clumped and uniform dispersion patterns, but on different scales. For instance, if you studied dispersion patterns of the human population in the northeastern United States (**Figure 18.17**), you would find most of the population clumped in metropolitan areas, such as New York City and Boston. Within each clump, however, individuals or family groups might be more or less uniformly dispersed—in housing tracts of fixed lot size, for example.

Population Growth Models

To appreciate the explosive potential for population increase, consider a single bacterium that can reproduce by fission every 20 minutes under ideal laboratory conditions. After 20 minutes, there would be two bacteria, four after 40 minutes, and so on. If this continued for only a day and a half—a mere 36 hours—there would be bacteria enough to form a layer a foot deep over the entire Earth. At the other extreme, elephants may produce only six young in a 100-year life span. Still, Darwin calculated that it would take only 750 years for a single mating pair of elephants to give rise to a population of 19 million. Obviously, indefinite increase does not occur, either in the laboratory or in nature. A population that begins at a low level in a favorable environment may increase rapidly for a while, but eventually the numbers must, as a result of limited resources and other factors, stop growing. We'll take a look at two models that will help us understand the raw potential and limitations for population growth.

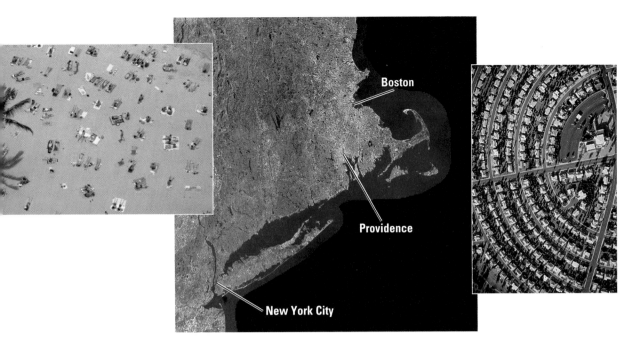

The Exponential Growth Model: The Ideal of an Unlimited Environment The rate of expansion of a population under ideal, unregulated conditions is described by the **exponential growth model,** in which the whole population multiplies by a constant factor during constant time intervals (the generation time). For example, the constant factor for the bacterial population represented in **Figure 18.18** is 2, because each parent cell splits to produce two daughter cells. The generation time for these bacteria is 20 minutes. The progression for bacterial growth—2, 4, 8, 16, and so on—is the number 2 raised to a successively higher power (exponent) each generation (that is, 2^1, 2^2, 2^3, 2^4, and so on).

Here's a key feature of the exponential growth model: The rate at which a population grows depends on the number of individuals already in the population. It's like compound interest on a long-term savings account at a fixed annual yield. At 7% per annum, you make only $70 the first year on a $1,000 deposit. But leave it in the bank for 50 years, and you'll be earning about $2,200 per year in interest on an account that has grown to over $30,000. Similarly, in our model for exponential population growth, the bigger the population size, the faster the population increases. As time goes by, the population gets bigger faster and faster.

The exponential growth model gives an idealized picture of the unregulated growth of a population. For bacteria, unregulated growth means there is no restriction on the ability of the cells to live, grow, and reproduce. Given a few days of unregulated growth, bacteria would smother every other living thing. Obviously, long periods of exponential increases are not common in the real world, or life could not continue on Earth. Where we do observe exponential growth in nature, it is generally a short-lived consequence of organisms being introduced to a new or underexploited environment.

The Logistic Growth Model: The Reality of a Limited Environment
In nature, a population may grow exponentially for a while, but eventually, one or more environmental factors will limit its growth. Population size then stops increasing or may even crash. Environmental factors that restrict population growth are called **population-limiting factors.** The graph for

Time	Number of Cells	
0 minutes	1	$= 2^0$
20	2	$= 2^1$
40	4	$= 2^2$
60	8	$= 2^3$
80	16	$= 2^4$
100	32	$= 2^5$
120 (= 2 hours)	64	$= 2^6$
3 hours	512	$= 2^9$
4 hours	4096	$= 2^{12}$
8 hours	16,777,216	$= 2^{24}$
12 hours	68,719,476,736	$= 2^{36}$

(a)

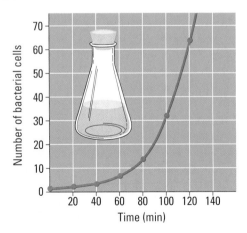

(b)

Figure 18.18 Exponential growth of a bacterial colony. (a) A table of data for the colony. (b) Plotting the data as a graph.

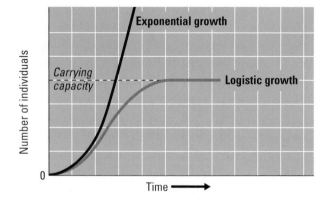

the seal population in **Figure 18.19** resembles what is called the **logistic growth model,** a description of idealized population growth that is slowed by limiting factors. **Figure 18.20** contrasts the logistic growth model with the exponential growth model.

Carrying capacity is the number of individuals in a population that the environment can just maintain ("carry") with no net increase or decrease. For the fur seal population in Figure 18.19, for instance, the carrying capacity is about 10,000 mated males. This value varies, depending on species and habitat. Carrying capacity might be considerably less than 10,000 for a fur seal population on a smaller island with fewer breeding sites, for example. At carrying capacity, the population is as big as it can theoretically get in its environment, and the population growth rate is zero.

The logistic model predicts that a population's growth rate will be low when the population size is either small or large, and highest when the population is at an intermediate level relative to the carrying capacity. At a low population level, resources are abundant, and the population is able to grow nearly exponentially. At this point, however, the increase is small because the population is small. In contrast, at a high population level, population-limiting factors strongly oppose the population's potential to increase. In nature, there might be less food available per individual or fewer breeding sites, nest sites, or shelters. The limiting factors make the birth rate decrease, the death rate increase, or both. Eventually, the population stabilizes at the carrying capacity, when the birth rate equals the death rate.

Both the logistic growth model and the exponential growth model are theoretical ideals. No natural population fits either one perfectly. However, these models are useful starting points for studying population growth. Ecologists use them to predict how populations will grow in certain environments and as a basis for constructing more complex models. The models have stimulated research leading to a greater understanding of populations in nature.

Regulation of Population Growth

Let's take a closer look at the population-limiting factors that contribute to carrying capacity. Ecologists classify these factors into two categories: density-dependent factors and density-independent factors.

Density-Dependent Factors The major biological implication of the logistic model is that increasing population density reduces the resources available for individual organisms, ultimately limiting population growth. The logistic model is actually a description of **intraspecific competition**—competition between individuals of the same species for the same limited resources. As population size increases, competition becomes more intense, and the growth rate declines in proportion to the intensity of competition. Thus, population growth rate is density dependent. A **density-dependent factor** is a population-limiting factor whose effects intensify as the population increases in size. Put another way, density-dependent factors affect a greater percentage of individuals in a population as the number of individuals increases. Limited food supply and the buildup of poisonous wastes are examples of density-dependent factors. Such factors depress a population's growth rate by increasing the death rate, decreasing the birth rate, or both (**Figure 18.21**).

We often see density-dependent regulation in laboratory populations. For example, when a pair of fruit flies is placed in a jar with a limited amount of food added each day, population growth fits the logistic model.

Figure 18.19 Effect of population-limiting factors on the growth of an animal population. The graph profiles the growth of a population of fur seals on Saint Paul Island, off the coast of Alaska. (For simplicity, only the mated bulls were counted. Each has a harem of females, as shown in the photograph.) Before 1925, the seal population on the island remained low because of uncontrolled hunting, although it changed from year to year. After hunting was controlled, the population increased rapidly until about 1935, when it leveled off and began fluctuating around a population size of about 10,000 bull seals. At this point, a number of population-limiting factors, including some hunting and the amount of space suitable for breeding, restricted population growth.

Figure 18.20 Logistic growth and exponential growth compared.

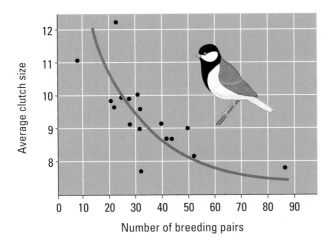

(a) Decreasing birth rate with increasing density

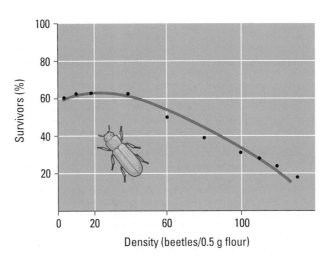

(b) Increasing death rate with increasing density

Figure 18.21 Density-dependent regulation of population growth. **(a)** This graph relates the clutch size (a "litter" of eggs a female bird lays) to population density for a forest population of a species called the great tit. Note that clutch size decreases (birth rate decreases) as population density increases.

(b) In this laboratory culture of an insect called the flour beetle, the percentage of individuals that survive from the egg stage to reproductive maturity decreases (death rate increases) as population density increases.

After a rapid increase, population growth levels off as the flies become so numerous that they outstrip their limited food supply. Each individual in a large population has a smaller share of the limited food than it would in a small population. Also, the more flies, the more concentrated the poisonous wastes become in the jar.

Laboratory populations are one thing; natural populations are another. We do not often see clear-cut cases of density-dependent factors regulating populations in nature. To test whether such factors are operating, it is necessary to change the density of individuals in the population while keeping other factors constant. This is sometimes done in managing game populations. For instance, state agencies often allow hunters to reduce populations of white-tailed deer to levels that keep the animals from permanently damaging the plants they use for food. White-tailed deer are browsers, preferring the highly nutritious parts of woody shrubs—young stems, leaves, and buds. When deer populations are kept low and high-quality food is therefore abundant, a high percentage of females become pregnant and bear offspring; in fact, many of them produce twins (**Figure 18.22**). On the other hand, when populations are high and food quality is poor, many females fail to reproduce at all. These observations support the hypothesis that food supply is a density-dependent factor regulating white-tailed deer populations.

Density-Independent Factors A population-limiting factor whose intensity is unrelated to population density is called a **density-independent factor.** Examples are such abiotic factors as unfavorable changes in the weather. A freeze in the fall, for example, may kill a certain percentage of insects in a population. The date and severity of the first freeze obviously are not affected by the density of the insect population. Density-independent factors such as a killing frost affect the same percentage of individuals regardless of population size. (In larger populations, of course, greater numbers will die.)

In many natural populations, density-independent factors limit population size well before resources or other density-dependent factors become important. In such cases, the population may decline suddenly. If we look at the

Figure 18.22 Increased birth rates in times of plenty. In deer populations, the birth of twin fawns is much more common when population densities are low.

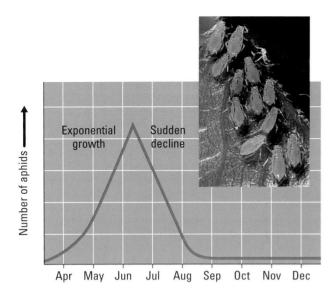

Figure 18.23 Weather change as a density-independent factor limiting growth of an aphid population.

Figure 18.24 Population cycles of the snowshoe hare and the lynx.

growth curve of such a population, we see something like exponential growth followed by a rapid decline rather than a leveling off. Ecologists have observed such history, for example, in certain populations of aphids, insects that feed on the sap of plants (**Figure 18.23**). Aphids and many other insects often show virtually exponential growth in the spring and then rapid die-offs when it becomes hot and dry in the summer. A few individuals may remain, and these may allow population growth to resume again if favorable conditions return. Some insect populations—many mosquitoes and grasshoppers, for instance—will die off entirely, leaving only eggs, which will initiate population growth the following year. In addition to seasonal changes in the weather, abrupt environmental trauma, such as fire, floods, storms, and habitat disruption by human activity, can affect populations in a density-independent manner.

Over the long term, most populations are probably regulated by a mixture of density-independent and density-dependent factors. Many populations remain fairly stable in size and are presumably close to a carrying capacity that is determined by density-dependent factors. In addition, however, many show short-term fluctuations due to density-independent factors. In some cases, the distinction between density-dependent and density-independent factors is not clear. In the case of white-tailed deer, for example, many individuals may starve to death in very cold, snowy areas. The severity of this effect is related to the harshness of the winter; cold temperatures increase energy requirements (and therefore the need for food), while deeper snow makes it harder to find food. But the severity of the effect is also density dependent, because the larger the population, the less food available per individual.

Population Cycles Some populations of insects, birds, and mammals have regular boom-and-bust cycles. Perhaps the most striking is that of the periodical cicadas, grasshopper-like insects that complete their life cycle every 13 or 17 years, emerging from the ground at phenomenal densities (as high as 600 individuals per square meter). This long life cycle may be an adaptation that reduces predation; few predators can wait 13 or 17 years for their prey to appear.

As a case study of population cycles, let's examine the boom-and-bust cycles of the snowshoe hare and one of its predators, the lynx. Both animals inhabit the northern coniferous forests of North America (**Figure 18.24**). About every ten years, both hare and lynx populations have a rapid increase (a "boom") followed by a sharp decline (a "bust"). What causes these boom-and-bust cycles? The ups and downs in the two populations seem to almost match each other on the graph. Does this mean that changes in one directly affect the other? In other words, does predation by the lynx make the hare population fluctuate, and do the ups and downs of the hare population cause the changes in the lynx population? For the lynx and many other predators that depend heavily on a single species of prey, the availability of prey can influence population changes. Thus, the ten-year cycle in the lynx population probably results at least in part from the ten-year cycle in the hare population. But we cannot conclude from the graph that lynx predation alone causes the cyclical changes in the hare population. In fact, researchers have discovered that hare populations will cycle about every ten years, whether or not lynx are present. Another hypothesis is that fluctuations in hare populations are tied to fluctuations in the populations of plants eaten by the hares. There is evidence that when certain plants are damaged by herbivores, the nutrient content of the plants decreases. Experimental studies performed in the field support the hypothesis that the ten-year cycle of snowshoe hares results from the combined effects of predation and fluctuations in the hare's food sources.

Populations of many rodents, including lemmings, also exhibit boom-and-bust changes, often cycling every three to five years. For such short-term cycles, some researchers postulate that either predation or food supply alone may be the underlying cause. Another hypothesis, based on laboratory studies of mice and other small rodents, is that stress from crowding may alter hormonal balance and reduce fertility. The causes of cycles probably vary among species and maybe even among populations of the same species.

Human Population Growth

Now that we have examined some general concepts of population dynamics, let's return to the specific case of the human population.

The History of Global Population Growth The human population has been growing almost exponentially for centuries. In fact, if we compare the history of human population growth in **Figure 18.25** with the exponential growth model in Figure 18.18b, it almost looks as if we've been multiplying like bacteria. Of course, our generation span is about 20 years instead of the mere 20 minutes for bacteria, so our population explosion has been stretched out in time. Still, when we compare the two graphs, it seems as though we've been proliferating into the space and resources of the biosphere as though it were an enormous petri dish. And like bacterial growth in a real petri dish, exponential growth cannot continue forever.

You can see in Figure 18.25 that the human population increased relatively slowly until about 1650, when approximately 500 million people inhabited Earth. The population doubled to 1 billion within the next two centuries, doubled again to 2 billion between 1850 and 1930, and doubled still again by 1975 to more than 4 billion. If the present growth rate persists, the population will reach 8 billion people by the year 2017. Recall that this hockey-stick-like arch, from a relatively horizontal to an almost vertical curve, is characteristic of exponential growth.

Human population growth is based on the same two general parameters that affect other animal and plant populations: birth rates and death rates. Birth rates increased and death rates decreased when agricultural societies replaced a lifestyle of hunting and gathering about 10,000 years ago.

Since the Industrial Revolution, virtually exponential growth has resulted mainly from a drop in death rates, especially infant mortality, even in the least developed countries. Improved nutrition, better medical care, and sanitation have all contributed to an increased percentage of newborns that survive long enough to leave offspring of their own. A decrease in mortality coupled with birth rates that are still relatively high in most developing countries results in an actual increase in population growth rates. In Sri Lanka, for example, the birth rate has been decreasing over the past 60 years, but it has never declined enough to offset the drop in the death rate. Thus, the Sri Lankan population continues to grow (**Figure 18.26**).

Activity 18D on the Web & CD
Learn more about the history of human population growth.

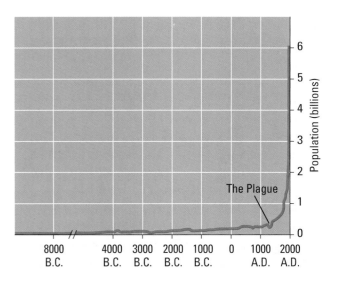

Figure 18.25 The history of human population growth.

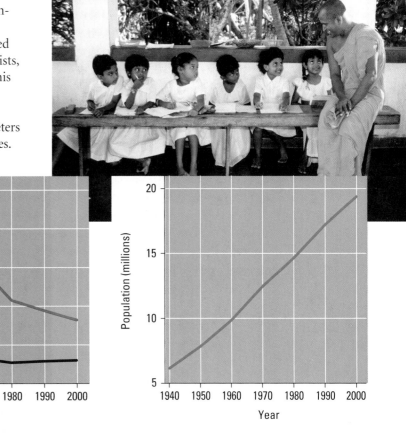

Changes in birth rate and death rate in Sri Lanka

Population growth in Sri Lanka

Figure 18.26 A case study of population growth in developing countries: Sri Lanka. Although family-planning education and better medical care reduced both birth rate and death rate in Sri Lanka, the population continued to grow because of the large difference between birth rate and death rate.

Age Structure and Population Growth Worldwide, human population growth is a mosaic of various rates of growth in different countries. Some developed countries, such as Sweden, have stable populations because birth rates and death rates balance. In sharp contrast to Sweden, most developing nations have burgeoning populations in which birth rates greatly exceed death rates. Partly as a result of such unchecked growth, many people in such countries face serious housing, water, and food shortages, as well as severe pollution problems.

A population characteristic called age structure can help us predict the future growth of populations in different countries. The **age structure** of a population is the proportion of individuals in different age-groups. The relatively uniform age distribution in Italy, for instance, contributes to that country's stable population size; individuals of reproductive age or younger are not disproportionately represented in the population (**Figure 18.27**). In contrast, Kenya has an age structure that is bottom-heavy, skewed toward young individuals who will grow up and sustain the explosive growth with their own reproduction.

Notice in Figure 18.27 that the age structure for the United States is relatively even except for a bulge that corresponds to the "baby boom" that lasted for about two decades after the end of World War II. Even though couples born during those years had an average of fewer than two children, population continued to increase during the past few decades because there were so many "boomers" of reproductive age. Because the boomers now range in age from 40 to 58, they are having less impact on population growth. The U.S. population is still growing, however, because the combination of birth rate and immigration exceeds the death rate. Immigration (legal and illegal combined) now contributes about 40% of the current growth of the U.S. population. And

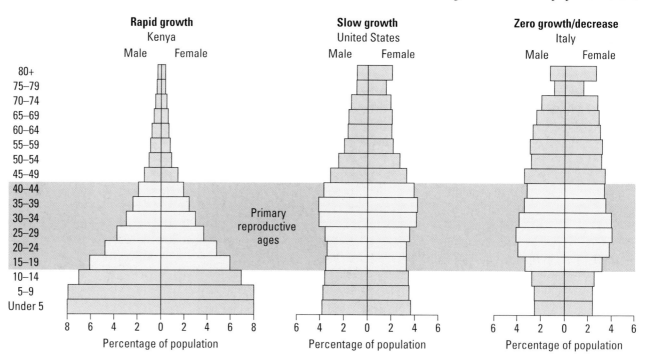

Figure 18.27 Age structures of three nations. The proportion of individuals in different age-groups has a significant impact on the potential for future population growth. Kenya, like many developing countries, has a disproportion- ate number of young people, and their future reproduction will sustain a steep population increase for many years. In contrast, Italy's population is distributed more evenly over all age classes, with a high proportion of individuals past their prime reproductive years. The United States has a fairly even age distribution except for the bulge corresponding to the post-World War II "baby boom." (The graphs are based on data from 1995.)

(a)

while the average woman in the United States tends to have only two children, the total number of women having children remains high enough for the birth rate to exceed the death rate. Population researchers predict that the U.S. population will continue to grow well into the twenty-first century, perhaps increasing from about 280 million today to about 390 million by the year 2050.

The Sociology, Economics, and Politics of Population Growth

Age-structure diagrams not only reveal a population's growth trends; they also relate to social conditions. Based on the diagrams in Figure 18.27, we can predict, for instance, that employment for an increasing number of working-age people will continue to be a significant problem for Kenya in the foreseeable future. For Italy and the United States, a decreasing proportion of working-age people—mostly those of college age today—will be supporting an increasing proportion of retired people. Programs such as the U.S. Social Security system and Medicare, which are crucial to many older citizens, will become severely strained as the proportion of senior citizens swells.

Activity 18E on the Web & CD
Analyze age-structure diagrams of different countries and make predictions about future conditions.

Predictions of future trends in global human population growth vary widely. The gloomiest models predict a continuing high rate of growth through the twenty-first century, with a doubling to about 12 billion as early as 2050. Perhaps more realistic is a computer model, based on the almost-global trend toward smaller families, that predicts that by about 2080, the human population will peak at about 10.6 billion and then begin a slight decline to about 10.4 billion by the end of the twenty-first century. Either way, there will be a lot more people consuming resources and dumping pollutants, bruising a biosphere that already ails (**Figure 18.28**).

A unique feature of human population growth is our ability to control it with voluntary contraception and government-sponsored family planning. Social change and the rising educational and career aspirations of women in many cultures encourage them to delay marriage and postpone reproduction. Delayed reproduction dramatically decreases population growth rates. You can get a sense of this phenomenon by imagining two populations in which women each produce three children but begin reproduction at different ages. In one population, females first give birth at age 15, and in the other, at age 30. If we start with a group of newborn girls, then after 30 years the women in the first population will already begin to have grandchildren, whereas women in the second population will be giving birth to their first children. After 60 years, women in the first population will have a large number of great-great-grandchildren (who will themselves begin to reproduce 15 years later), but women in the second population will just begin to see their grandchildren being born.

There is a great deal of heated disagreement among leaders in almost every country as to how much support should be provided for family planning. The issue is certainly socially charged, but there are also political and economic threads to the debate. For example, the United States and many other developed countries, as well as some developing ones, currently face labor shortages, a problem cited by some opponents as stemming from policies that encourage family planning. But the debate is also related in part to the difficulty of answering the question, How many people are too many? The problem of defining carrying capacity for humans is confounded by the observation that carrying capacity has changed with human cultural evolution. The advent of agricultural and industrial technology has significantly increased carrying capacity at least twice during human history, and opponents of population control are counting on some new, as yet unidentified

(b)

Figure 18.28 Two very crowded places.

(a) This scene captures the stifling, unsanitary conditions in a slum that has arisen within a garbage dump in Manila, in the Philippines. One of the world's largest cities, Manila has well over 10 million people. (b) Commuters who crawl along on the Los Angeles freeways to go to work each day take more than their share of our growing population's toll on the biosphere. Although developing countries are generally growing in population faster than developed ones, it is unfair to assess the problem of human population size in terms of numbers alone. On a per capita basis, people in developed countries have much more impact on the environment than people in developing countries with lower standards of living. In the United States, for example, each person, on average, consumes ten times as much energy (mostly as fossil fuels) as each person in a developing country.

technological breakthrough that will allow our population to grow and plateau at some higher level.

Technology has undoubtedly increased Earth's carrying capacity for humans, but no population can continue to grow indefinitely. Ideally, human populations will reach carrying capacity smoothly and then level off. This will occur when birth rates and death rates are equal, and a decrease in birth rate is more desirable than an increase in death rate. If, however, the population fluctuates about carrying capacity, we can expect periods of increase followed by mass death, as has occurred during plagues, localized famines, and international military conflicts. In any case, the human population must eventually stop growing. Unlike other organisms, we can decide whether zero population growth will be attained through social changes involving individual choice or government intervention or through increased mortality due to resource limitation and environmental degradation. For better or worse, we have the unique responsibility to decide the fate of our species and that of the rest of the biosphere.

CHECKPOINT

1. What is the relationship between a population and a species?

2. What is the approximate size of Earth's human population?

3. An aquarium population of guppies has reached a stable population size. We decide to add twice as much guppy food per day to the aquarium, but this turns out to have no effect on population size. What is the most likely explanation for this observation?

4. Of the following factors with the potential to limit the growth of a human population, which one is most density independent? (a) lowering of fertility due to hormonal changes in very crowded conditions; (b) a famine; (c) mass drowning caused by hurricane floods; (d) epidemic of a highly contagious disease; (e) freezing deaths due to a shortage of housing

5. What new research question arises from evidence that the population cycle of snowshoe hares is caused at least partly by a cycle in the availability or nutritional value of the plants eaten by the hares?

6. What causes a population's growth to level off if its behavior approximates the logistic model?

7. (a) How does the age structure of the U.S. population explain the current surplus in the Social Security fund? (b) If the system is not changed, why will the surplus give way to a deficit sometime in the next few decades?

8. This question will tell you how well you interpret a graph. From the graph in Figure 18.26, what was the approximate percentage increase in the Sri Lankan population over the 20-year span from 1960 to 1980?

Answers: 1. A population is a localized subset of a species. **2.** 6 billion **3.** The population was already at carrying capacity before we increased food supply, and the key limiting factor was something other than food availability. **4.** c **5.** What causes the cyclical changes in the plants? **6.** The population size reaches the environment's carrying capacity. **7.** (a) The largest population segment, the boomers, are currently in the work force in their peak earning years, paying into the system. (b) The boomers will retire over the next few decades and begin drawing from the system at a time when there will be fewer employees paying into Social Security. **8.** 50%, an increase from about 10 million to 15 million people

---

Life Histories and Their Evolution

Earlier in this chapter, we looked at how certain physiological, anatomical, and behavioral responses to environmental variation can increase an organism's chances of survival. However, natural selection does not act only on

Table 18.1   Life Table for the U.S. Population in 1999

Age Interval	Number Living at Start of Age Interval (N)	Number Dying During Interval (D)	Mortality (Death Rate) During Interval (D/N)	Chance of Surviving Interval (1 − D/N)
0–10	100,000	930	0.009	0.991
10–20	99,070	448	0.005	0.995
20–30	98,621	937	0.010	0.990
30–40	97,684	1,362	0.014	0.986
40–50	96,321	2,800	0.029	0.971
50–60	93,521	6,068	0.065	0.935
60–70	87,453	13,169	0.151	0.849
70–80	74,284	23,757	0.320	0.680
80–90	50,526	32,082	0.635	0.365
90+	18,445	18,445	1.000	0.000

traits that increase survival; organisms that survive but do not reproduce are not at all "fit" in the Darwinian sense (see Chapter 13). Clearly, an organism can pass along its genes only if it survives long enough to reproduce. But how long is long enough? In many cases, there are trade-offs between survival traits and traits that enhance reproductive output—traits such as frequency of reproduction, investment in parental care, and the number of offspring per reproductive episode (usually called seed crop for seed plants and litter size or clutch size for animals).

The traits that affect an organism's schedule of reproduction and death make up its **life history.** Of course, a particular life history, like most characteristics of an organism, is the result of natural selection operating over evolutionary time. In this section, we'll see how life history traits affect population growth.

Life Tables and Survivorship Curves

When the life insurance industry was established about a century ago, insurance companies based their business plans on some of the early scientific studies of human populations. Needing to determine how long, on average, an individual of a given age could be expected to live, the insurance companies began using what are called life tables.

A **life table** tracks survivorship and mortality (death) in a population. For example, **Table 18.1** arranges the survivorship/mortality data for a sample of 100,000 U.S. citizens over a ten-year period, ending in 1999. Using this table, an insurance agent could predict that a 21-year-old has about a 0.99 (99%) chance of surviving to age 30. Borrowing the basic idea from the insurance industry, population ecologists construct life tables for plants and nonhuman animals.

A graphic way of representing some of the data in a life table is to draw a **survivorship curve,** a plot of the number of people still alive at each age (**Figure 18.29**). We can classify survivorship curves for diverse organisms into three general types. A Type I curve is relatively flat at the start, reflecting low death rates during early and middle life, and dropping steeply as death rates increase among older age-groups. Humans and many other large mammals that produce relatively few offspring but provide them with good care often exhibit this kind of curve. In contrast, a Type III curve indicates high

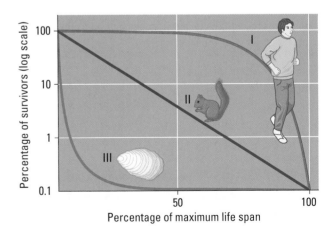

Figure 18.29
Three idealized types of survivorship curves.

death rates for the very young and then a period when death rates are much lower for those few individuals who survive to a certain age. Species with this type of survivorship curve usually produce very large numbers of offspring but provide little or no care for them. An oyster, for instance, may release millions of eggs, but most offspring die as larvae from predation or other causes. A Type II curve is intermediate, with mortality more constant over the life span. This type of survivorship has been observed in several invertebrates, including hydras, and in certain rodents, such as the gray squirrel.

A population's pattern of mortality is certainly a key feature of life history. But we have defined life history as the set of traits that affect an organism's schedule of *reproduction* and death. Let's take a closer look now at how natural selection affects the reproductive strategies that evolve in populations.

Life History Traits as Evolutionary Adaptations

Some key life history traits affecting population growth are the age at which reproduction first occurs, the number of offspring produced for each reproductive episode, the amount of parental care committed to offspring, and the overall energy cost of reproduction. For a given population living in a particular environmental context, natural selection will favor the combination of life history traits that maximizes an individual's output of viable, fertile offspring. In other words, life history traits, like anatomical features, are shaped by adaptive evolution (**Figure 18.30**).

Life history strategies vary with species and may change even for a single population as the environmental context changes. However, it is useful to contrast two extreme types of life history: opportunistic life history and equilibrial life history.

We can find examples of **opportunistic life history** among certain populations of small-bodied species. Individuals reproduce when young, and they produce many offspring. The population tends to grow exponentially when conditions are favorable (hence the term *opportunistic*). Such a population typically lives in an unpredictable environment and is controlled by density-independent factors, such as the weather. For example, dandelions and many other annual weeds grow quickly in open, disturbed areas, producing a large number of seeds in a brief time when the weather is favorable. Although most of the seeds will not produce mature plants, their large number and ability to disperse to new habitats ensure that at least some will grow and eventually produce seeds themselves. Many insects, including locusts, also exhibit an opportunistic life history, maximizing reproductive output whenever environmental opportunity knocks. For such species, natural selection has reinforced quantity of reproduction more than individual survivorship. In general, populations with an opportunistic life history exhibit a Type III survivorship curve (see Figure 18.29).

In contrast, some populations, mostly larger-bodied species, exhibit an **equilibrial life history,** which generally results in a Type I survivorship curve. Individuals usually mature later and produce few offspring but care for their young. The population size may be quite stable, held near the carrying capacity by density-dependent factors (stabilization around carrying capacity accounts for the term *equilibrial*). Natural selection has resulted in the production of better-endowed offspring that can become established in the well-adapted population into which they are born. The life histories of many large terrestrial vertebrates fit this model. Among polar bears, for instance, a female has only one or two offspring every three years, but the cubs remain in her protective custody for over two years. In the plant kingdom, the coconut palm tree also fits the equilibrial model. Compared with most other trees, it

Figure 18.30 The "big bang" reproduction of century plants. A century plant, a type of agave, grows for many years (a century in some cases) without flowering or reproducing. Then one spring it grows a floral stalk that may be as tall as a telephone pole. That season, the plant produces many seeds and then withers and dies, its food reserves and water spent in the formation of its massive bloom. The century plant's life history is an example of what ecologists call the "big bang" strategy of reproduction. It's an evolutionary adaptation to the organism's environmental situation. Century plants grow in arid climates with sparse and unpredictable rainfall. Their shallow roots catch water after rain showers but are dry during droughts. This unpredictable water supply may prevent seed production for several years at a time. By growing and storing nutrients until an unusually wet year and then putting all its resources into one grand burst of seed production, the plant maximizes its reproductive success.

Table 18.2

Characteristic	Opportunistic Populations (e.g., many wildflowers)	Equilibrial Populations (e.g., many large mammals)
Climate	Relatively unpredictable	Relatively predictable
Maturation time	Short	Long
Life span	Short	Long
Death rate	Often high	Usually low
Number of offspring produced per reproductive episode	Many	Few
Number of reproductions per lifetime	Usually one	Often several
Timing of first reproduction	Early in life	Later in life
Size of offspring or eggs	Small	Large
Parental care	None	Often extensive

Source: Adapted from E. R. Pianka, *Evolutionary Ecology,* 6th ed. (San Francisco, CA: Benjamin/Cummings, 2000), p. 186.

produces relatively few, very large seeds, which provide nutrients (including the coconut "milk") for the embryo; this is a plant's version of parental care.

Table 18.2 contrasts some life history traits between opportunistic and equilibrial strategies. Most populations probably fall between the opportunistic and equilibrial extremes. Thus, these two life history strategies are only hypothetical models, starting points for studying the complex interplay of the forces of natural selection on reproductive characteristics.

Activity 18F on the Web & CD
Investigate the life histories of oysters, elephants, robins, and deer.

CHECKPOINT

1. How do the terms *opportunistic* and *equilibrial* contrast the key characteristics of these life history strategies?

2. What is the key feature of the mortality column in the life table for a population exhibiting a Type II survivorship curve?

Answers: 1. An opportunistic life history is characterized by an ability to produce a large number of offspring very rapidly when the environment affords a temporary opportunity for exponential growth; an equilibrial life history is characterized by a population size that fluctuates only slightly from carrying capacity. **2.** The mortality is about the same for every age interval.

Evolution Connection

Testing a Darwinian Hypothesis

A body of evidence supports the Darwinian view that a population's life history traits are evolutionary adaptations shaped by natural selection. As a case study of the process of science, let's examine a classic set of experiments on the evolution of life history.

For many years, David Reznick of the University of California, Riverside, and John Endler of the University of California, Santa Barbara, have been investigating the life histories of guppy populations in Trinidad, a Caribbean island. Guppies are small freshwater fish you probably recognize as popular aquarium pets. In the Aripo River system of Trinidad, guppies live in small pools as populations that are relatively isolated from one another. In some cases, two populations inhabiting the same stream live less than 100 m apart, but they are separated by a waterfall that impedes the migration of guppies between the two ponds.

Early in their research, Reznick and Endler recognized that certain life history traits among guppy populations correlated with the main type of predator in a stream pool. Certain guppy populations live in pools where the predator is the killifish, which eats mainly small, immature guppies. Other guppy populations live where larger fish, called pike-cichlids, eat mostly large, mature guppies. Where preyed on by pike-cichlids, guppies tend to be smaller, mature earlier, and produce more offspring each time they give birth than those in areas without pike-cichlids. If the differences between the populations result from natural selection, the life history traits should be heritable. To test for heritability, the scientists raised guppies from both types of populations in the laboratory without predators. The two populations retained their life history differences when followed through several generations, indicating that the differences were indeed inherited. A reasonable hypothesis is that the selective predation of larger versus smaller guppies results in the life history adaptations that the researchers observed. Apparently, when predators such as pike-cichlids prey mainly on reproductively mature adults, the chance that a guppy will survive to reproduce several times is relatively low. The guppies with the greatest reproductive success should then be the individuals that mature at a young age and small size and reproduce at least once before growing to a size preferred by the local predator.

Reznick and Endler tested their hypothesis with field experiments (**Figure 18.31**). Here is the scientific reasoning, based on the "*If . . . then*" logic you learned about in Chapter 1:

Hypothesis: *If* the feeding preferences of different predators cause contrasting life histories in different guppy populations by natural selection,

Experiment: and guppies are transplanted from locations with pike-cichlids (predators of mature guppies) to guppy-free sites inhabited by killifish (predators of juvenile guppies),

Predicted Result: *then* the transplanted guppy populations should show a generation-to-generation trend toward later maturation and larger size—life history traits typical of natural populations that coexist with killifish.

Reznick and Endler performed these guppy transplant experiments and studied the populations for 11 years, measuring age and size at maturity and other life history traits (**Figure 18.32**). Over the 11 years, or 30 to 60 guppy generations, the average weight at maturity for guppies in the transplanted (experimental) populations increased by about 14% compared with control populations. Other life history traits also changed in the direction predicted by the hypothesis of evolutionary adaptation due to selective predation.

Without a control group for comparison, there would be no way to tell whether it was the killifish or some other factor that caused the transplanted guppy populations to change. But because control sites and experimental sites were usually nearby pools of the same stream, the main variable was probably

Figure 18.31 David Reznick conducting field experiments on guppy evolution in Trinidad.

the presence of different predators. And these careful scientists observed similar results when guppy populations were reared in artificial streams that were identical except for the type of predator. Reznick and Endler had in fact tested a Darwinian hypothesis and documented evolution in a natural setting over a relatively short time.

Evolutionary biology qualifies as science because its questions stimulate scientists to articulate hypotheses they can test by further observation and experiments.

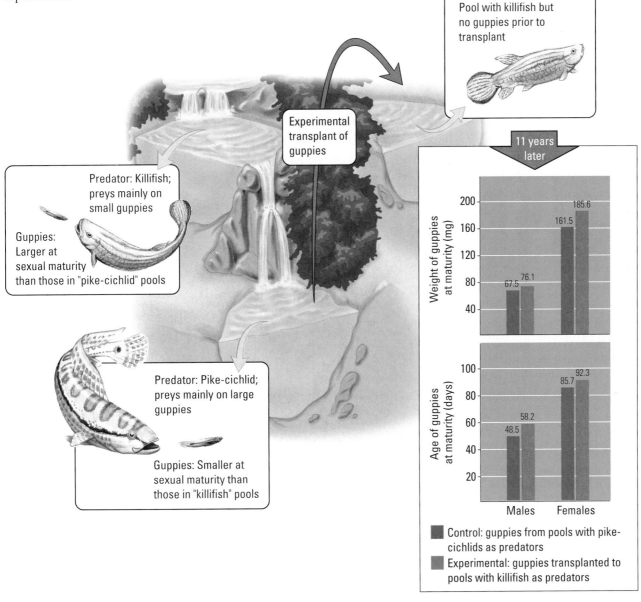

Pool with killifish but no guppies prior to transplant

Experimental transplant of guppies

Predator: Killifish; preys mainly on small guppies

Guppies: Larger at sexual maturity than those in "pike-cichlid" pools

Predator: Pike-cichlid; preys mainly on large guppies

Guppies: Smaller at sexual maturity than those in "killifish" pools

11 years later

Weight of guppies at maturity (mg)
- Males: Control 67.5, Experimental 76.1
- Females: Control 161.5, Experimental 185.6

Age of guppies at maturity (days)
- Males: Control 48.5, Experimental 58.2
- Females: Control 85.7, Experimental 92.3

■ Control: guppies from pools with pike-cichlids as predators
■ Experimental: guppies transplanted to pools with killifish as predators

Figure 18.32 Testing a Darwinian hypothesis. This illustration represents three pools along streams of Trinidad's Aripo River system. In one pool (upper left), the predators are killifish, which prey on relatively small, immature guppies. In another pool (lower left), the predators are pike-cichlids, which eat primarily large, mature guppies. Guppy populations living in pools where the predator is the pike-cichlid tend to be smaller at sexual maturity than those in the "killifish pools." Researchers tested the hypothesis that selective predation accounts for the life history differences between guppy populations. The scientists transplanted guppies from pike-cichlid pools to pools that contained killifish but had no natural guppy populations (upper right). The biologists then tracked the evolution of weight and age at sexual maturity in the experimental guppy populations for 11 years, comparing their measurements to guppies from control pools inhabited by pike-cichlids. As shown in the graphs, the researchers observed that the average weight and age at sexual maturity of the transplanted populations increased significantly as compared to the control populations.

Chapter Review

Summary of Key Concepts

For study help, go to the Essential Biology Website (www.essentialbiology.com) or CD-ROM to explore the Activities and Case Studies in the Process of Science.

An Overview of Ecology

- Ecology is the scientific study of interactions between organisms and their environments. The environment includes abiotic (nonliving) and biotic (living) components.

- **Ecology as Scientific Study**   Ecologists use observation, experiments, and computer models to test hypothetical explanations of these interactions.

- **A Hierarchy of Interactions**   Ecologists study interactions at four increasingly complex levels.

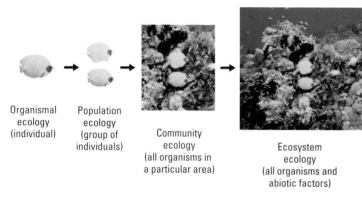

Organismal ecology (individual) → Population ecology (group of individuals) → Community ecology (all organisms in a particular area) → Ecosystem ecology (all organisms and abiotic factors)

- **Ecology and Environmentalism**   Human activities have had an impact on all parts of the biosphere. Ecology provides the basis for understanding and addressing these environmental problems.

Activity 18A   DDT and the Environment

- **Abiotic Factors of the Biosphere**   The biosphere is an environmental patchwork in which several abiotic factors affect the distribution and abundance of organisms. These include the availability of sunlight and water, temperature, wind, rock and soil characteristics, and catastrophic disturbances such as fires, hurricanes, tornadoes, and volcanic eruptions.

Case Study in the Process of Science   Do Pillbugs Prefer Wet or Dry Environments?

Case Study in the Process of Science   How Do Abiotic Factors Affect the Distribution of Organisms?

The Evolutionary Adaptations of Organisms

- **Physiological Responses**   Most organisms adjust their physiological conditions in response to changes in the environment. Acclimation is a longer-term response that can take days or weeks. The ability to acclimate is generally related to the range of environmental conditions that a species naturally experiences.

- **Anatomical Responses**   Many organisms respond to environmental change with reversible or irreversible changes in anatomy. Unable to move to a better location, plants are generally more anatomically plastic than animals.

- **Behavioral Responses**   Able to travel about, animals frequently adjust to poor environmental conditions by moving to a new location. These may be small adjustments such as shuttling between sun and shade or seasonal migrations to new regions.

Activity 18B   Evolutionary Adaptations

What Is Population Ecology?

- Population ecology focuses on the factors that influence a population's size, growth rate, density, and structure. A population consists of members of a species living in the same place at the same time.

- **Population Density**   Population density, the number of individuals of a species per unit area or volume, is usually estimated by a variety of sampling techniques. These include counting the number of individuals in a sample plot and the mark-recapture method.

Activity 18C   Techniques for Estimating Population Density and Size

- **Patterns of Dispersion**   Dispersion patterns of a population are determined by various environmental or social factors.

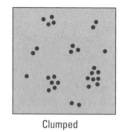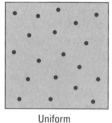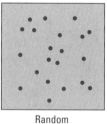

Clumped   Uniform   Random

- **Population Growth Models**

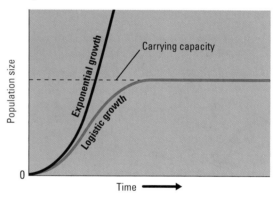

The exponential model of population growth describes an idealized population in an unlimited environment. This model predicts that the larger a population becomes, the faster it grows. Exponential growth in nature is generally a short-lived consequence of organisms being introduced to a new or underexploited environment. The logistic model of population growth describes an idealized population that is slowed by limiting factors. This model predicts that a population's growth rate will be small when the population size is either small or large, and highest when the population is at an intermediate level relative to the carrying capacity.

- **Regulation of Population Growth**   Over the long term, most population growth is limited by a mixture of density-independent and density-dependent factors. Density-dependent factors intensify as a population increases in size, increasing the death rate, decreasing the birth rate, or both. Density-independent factors affect the same percentage of individuals regardless of population size. Some populations have regular boom-and-bust cycles.

- **Human Population Growth**   The human population has been growing almost exponentially for centuries. Human population growth is based on the same two general parameters that affect other animal and plant populations: birth rates and death rates. Birth rates increased and death rates decreased when agricultural societies replaced a lifestyle of hunting and gathering. The age structure of the population is a major factor in the different growth rates of different countries. Population researchers predict that the U.S. population will continue to grow well into the twenty-first century. Age-structure diagrams also predict social predicaments. Predictions of future trends in global human population growth vary widely. The human species is unique in having the ability to consciously control its own population growth, the fate of our species, and the fate of the rest of the biosphere.

Activity 18D *Human Population Growth*

Activity 18E *Analyzing Age-Structure Diagrams*

Life Histories and Their Evolution

- **Life Tables and Survivorship Curves** A population's pattern of mortality is a key feature of life history. A life table tracks survivorship and mortality in a population. Survivorship curves can be classified into three general types, depending on the rate of mortality over the entire life span.

- **Life History Traits as Evolutionary Adaptations** Life history traits are shaped by adaptive evolution; they may vary within a species and may change as the environmental context changes. Most populations probably fall between the extreme opportunistic strategies of many insects and equilibrial strategies of many larger-bodied species.

Activity 18F *Investigating Life Histories*

Self-Quiz

1. Place these levels of ecological study in order, from the least to the most comprehensive: community ecology, ecosystem ecology, organismal ecology, population ecology.

2. Name several abiotic factors that might affect the community of organisms living inside a home fish tank.

3. The formation of goose bumps on your skin in cold weather is an example of a (an) _____ response, while seasonal migration is an example of a (an) _____ response.

4. What two values would you need to know to figure out the human population density of your community?

5. Pine trees in a forest tend to shade and kill pine seedlings that sprout nearby. What pattern of growth will this produce?

6. A uniform dispersion pattern for a population may indicate that
   a. the population is spreading out and increasing its range.
   b. resources are heterogeneously distributed.
   c. individuals of the population are competing for some resource, such as water and minerals for plants or nesting sites for animals.
   d. there is an absence of strong attractions or repulsions among individuals.

7. With regard to its percent increase, a population that is growing logistically grows fastest when its density is _____ compared to the carrying capacity.
   a. low   b. intermediate   c. high

8. Which of the following shows the effects of a density-dependent limiting factor?
   a. A forest fire kills all the pine trees in a patch of forest.
   b. Early rainfall triggers the explosion of a locust population.
   c. Drought decimates a wheat crop.
   d. Rabbits multiply, and their food supply begins to dwindle.

9. Skyrocketing growth of the human population since the beginning of the Industrial Revolution appears to be mainly a result of
   a. migration to thinly settled regions of the globe.
   b. better nutrition boosting the birth rate.
   c. a drop in the death rate due to better nutrition and health care.
   d. the concentration of humans in cities.

10. If members of a species produce a large number of offspring but provide minimal parent care, then a Type _____ survivorship curve is expected. In contrast, if members of a species produce few offspring and provide them with long-standing care, then a Type _____ survivorship curve is expected. (*Hint:* Refer to Figure 18.29.)

Answers to the Self-Quiz questions can be found in Appendix B.

Go to the website or CD-ROM for more Self-Quiz questions.

The Process of Science

1. Design a laboratory procedure to measure the effect of water temperature on the population growth of a certain phytoplankton species from a pond.

2. We estimate the size of a population of small mice in a particular field by the mark-recapture method. Our estimate is 350. Later, we learn from experiments on the behavior of these mice that they can locate a baited trap faster if they have already been rewarded with food by visiting that trap once before. Does this mean that our original estimate of 350 individuals was too low or too high? Explain your answer in terms of the principles that underlie the mark-recapture method.

Case Study in the Process of Science on the Web & CD *Devise an experiment to determine whether pillbugs prefer a wet or dry environment.*

Case Study in the Process of Science on the Web & CD *Make changes in abiotic factors and see the effects on various plants and animals.*

Biology and Society

1. During the summer of 1988, lightning ignited huge forest fires that burned a large portion of Yellowstone National Park (see Figure 18.11). The National Park Service has a natural-burn policy: Fires that start naturally are allowed to burn unless they endanger human settlements. The fires were allowed to spread and burn while firefighters primarily protected people. The public accused the Park Service of letting a national treasure go up in flames. Park Service scientists stuck with the natural-burn policy. Do you think this was the best decision? Support your position. More recently, in the spring of 2000, park officials authorized a controlled burn of a forested area near Los Alamos, New Mexico, with the intention of clearing away brush and dead wood in order to reduce the severity of future fires. Unfortunately, a weather warning was ignored and the fire escaped control, claiming over 200 homes and thousands of acres of forest. What impact do you think the Los Alamos fire has had on public opinion about controlled burning? Assuming that the basic concept of controlled burning is scientifically sound, how would you justify future burns to a public that remembers the Los Alamos fiasco?

2. Many people regard the rapid population growth of developing countries as our most serious environmental problem. Others think that the population growth in developed countries, though smaller, is actually a greater threat to the environment. What kinds of problems result from population growth in developing countries and industrialized countries? Which do you think is the greater threat, and why?

Biology and Society on the Web *This U.S. census website will provide you with some impressive numbers.*

CHAPTER 19

Communities and Ecosystems

Biology and Society:
Reefs: Coral and Artificial   407

Key Properties of
Communities   407
Diversity
Prevalent Form of Vegetation
Stability
Trophic Structure

Interspecific Interactions
in Communities   409
Competition Between Species
Predation
Symbiotic Relationships
The Complexity of Community
Networks

Disturbance of
Communities   415
Ecological Succession
A Dynamic View of Community
Structure

An Overview of
Ecosystem Dynamics   417
Trophic Levels and Food Chains
Food Webs

Energy Flow in Ecosystems   421
Productivity and the Energy Budgets
of Ecosystems
Energy Pyramids
Ecosystem Energetics and Human
Nutrition

Chemical Cycling
in Ecosystems   424
The General Scheme of Chemical
Cycling
Examples of Biogeochemical Cycles

Biomes   428
How Climate Affects Biome
Distribution
Terrestrial Biomes
Freshwater Biomes
Marine Biomes

Evolution Connection:
Coevolution in Biological
Communities   438

Some plants **recruit** **parasitic wasps** that lay their eggs in caterpillars that eat the plants.

The oceans **cover about** **75%** of Earth's surface.

Only a **tiny** **fraction** of the sunlight that shines on Earth is converted to chemical energy.

Sunken ships, army tanks, and subway cars are being used as **artificial reefs.**

Biology and Society

Reefs: Coral and Artificial

Found in warm tropical waters, coral reefs are distinctive and complex ecosystems. Coral reefs are dominated by the structure of the coral itself, formed mainly by cnidarians that secrete hard external skeletons made of calcium carbonate (see Figure 17.12). These skeletons vary in shape, forming a substrate on which other corals, sponges, and algae grow. The coral animals themselves feed on microscopic organisms and particles of organic debris. They also obtain organic molecules from the photosynthesis of symbiotic algae that live in their tissues. Reefs also support a diversity of invertebrates and fishes, making them a popular destination for people who enjoy diving and fishing (**Figure 19.1**).

Some coral reefs cover enormous expanses of shallow ocean, but these delicate ecosystems are easily degraded by pollution and development. While no part of the biosphere is safe from human intrusion, coral reef ecosystems are particularly vulnerable because they are very old and grow very slowly. Corals are also subject to damage from both native and introduced predators, such as the crown-of-thorns sea star, which has undergone a population explosion in many regions and actually destroyed coral reefs in parts of the western Pacific Ocean.

But there is a flip side to human intrusion: the possibility of creating new, artificial reefs through human intervention. Artificial reefs are formed by sinking large, stable objects to the ocean bottom. Some reefs, for example, have been constructed from steel and concrete, while others consist of sunken ships, army tanks, and even subway cars. Artificial reefs contain no coral, but they can serve as a base for new reef ecosystems. Within just a few years, artificial reefs teem with marine organisms that live in or on nearly every square inch. One large artificial reef program is run by the state of South Carolina. Since 1973, the state has established 44 offshore reefs. Two of them serve as research stations, while the rest are open to the public for fishing and diving, generating revenue for the state while also providing new marine habitats.

Biology and Society on the Web
Find out how old subway cars can be made into reefs.

In this chapter, we'll examine the diverse interactions among organisms—in reefs and all other parts of the biosphere—and how those relationships determine the species composition and other features of communities. On a larger scale, we'll explore the dynamics of ecosystems, such as forests, ponds, and oceanic zones. Ultimately, we'll widen our scope to view the biosphere as the global ecosystem, the sum of all ecosystems. Throughout the chapter, biology's core theme of evolution will be apparent in the adaptations that fit organisms to their ecological roles in communities.

Figure 19.1 A coral reef community. This underwater scene shows the stunning diversity of life in one coral reef community near the Fiji Islands.

Key Properties of Communities

On your next walk through a field or woodland, or even across campus or through a park or your own backyard, try to observe some of the interactions among the species present. You may see birds using trees as nesting sites, bees pollinating flowers, caterpillars feeding on leaves, spiders trapping insects in their webs, cats stalking small rodents, ferns growing in shade provided by trees—a sample of the many interactions that occur in any ecological theater.

Figure 19.2 Diverse species interacting in a Kenyan savanna community.

In addition to the physical and chemical factors (abiotic factors) we discussed in Chapter 18, an organism's environment includes other individuals in its population and populations of other species living in the same area. Such an assemblage of species living close enough together for potential interaction is called a **community**. In **Figure 19.2**, the lion, the zebra, the hyena, the vultures, and the grasses and other plants are all members of a community in Kenya. In this section, we'll examine four key properties of a community: its diversity, its prevalent form of vegetation, its stability, and its trophic structure.

Diversity

The diversity of a community—the variety of different kinds of organisms that make up the community—has two components. One is species richness, or the total number of different species in the community. The other is the relative abundance of the different species. For example, imagine two forest communities, each with 100 individuals distributed among four different tree species (A, B, C, and D) as follows:

Community 1: 25A, 25B, 25C, 25D

Community 2: 80A, 10B, 5C, 5D

The species richness is the same for both communities because they both contain four species, but the relative abundance is very different (**Figure 19.3**). You would easily notice the four different types of trees in community 1, but without looking carefully, you might see only the abundant species A in the second forest. Most observers would intuitively describe community 1 as the more diverse of the two communities. Indeed, the term **species diversity**, as used by ecologists, considers *both* diversity factors: richness and relative abundance.

> **Case Study in the Process of Science on the Web & CD**
> Try your hand at using species diversity to measure the effects of pollution on an environment.

Prevalent Form of Vegetation

The second property of a community, its prevalent form of vegetation, applies mainly to terrestrial situations. For example, deciduous trees are the prevalent components of the community in a temperate deciduous forest. When we look more closely at such a community, we see not only which plants are dominant, but also how the plants are arranged, or "structured." For instance, a deciduous forest has a pronounced vertical structure: The treetops form a top layer, or canopy, under which there is a subcanopy of lower branches, and small shrubs and herbs carpet the forest floor. The types and structural features of plants largely determine the kinds of animals that live in a community.

Stability

The third property of a community, **stability**, refers to the community's ability to resist change and return to its original species composition after being disturbed. Stability depends on both the type of community and the nature of disturbances. For example, a forest dominated by cedar and hemlock trees is a highly stable community in that it may last for thousands of years with little change in species composition. Large cedars and hemlocks even withstand most lightning-caused fires, which kill small trees and shrubs growing in forest openings or in the shade of the dominant trees.

Community 1

Community 2

Figure 19.3 Which forest is more diverse?
With the same four tree species in both forests, the two communities are equal in their species richness of trees. But if we factor in the relative abundance of species, then community 1 is the more diverse because of the more equitable representation of the different tree species.

However, when a fire does kill the dominant trees, a cedar/hemlock forest might seem less stable than, say, a grassland, because it will take much longer for the forest to return to its original species composition.

Trophic Structure

The fourth property of a community is its **trophic structure,** the feeding relationships among the various species making up the community. A community's trophic structure determines the passage of energy and nutrients from plants and other photosynthetic organisms to herbivores and then to carnivores. Trophic relationships are certainly implied in the scene in **Figure 19.4**, and we'll see many more examples as we take a closer look at community interactions.

Figure 19.4
Trophic structure: feeding relationships.
After spending four to six years feeding in the open ocean, this salmon is attempting to return to the stream of its birth. Along the way, however, it may become a meal for a grizzly bear.

CHECKPOINT

1. How could a community appear to have relatively little diversity even though it is rich in species?

2. A community's feeding relationships of producers and consumers is referred to as the community's _____ structure.

Answers: 1. If one or a few of the diverse species accounted for almost all the organisms in the community, with the other species being rare **2.** trophic

---

Interspecific Interactions in Communities

With the four main properties of a community in mind, we turn next to the various kinds of interactions between species—what ecologists call **interspecific interactions.** The three main types of interspecific interactions we'll explore are competition, predation, and symbiosis. In each case, we'll see how these community relationships function as environmental factors in the adaptive evolution of organisms through natural selection.

Competition Between Species

When populations of two or more species in a community rely on similar limiting resources, they may be subject to **interspecific competition.** You already learned in Chapter 18 that as a population's density increases and nears carrying capacity, every individual has access to a smaller share of some limiting resource, such as food. As a result, mortality rates increase, birth rates decrease, and population growth is curtailed. In interspecific competition, however, the population growth of a species may be limited by the density of competing species as well as by the density of its own population. For example, if several bird species in a forest feed on a limited population of insects, the density of each species may have a negative impact on population growth in the other species. Similarly, species may compete for nesting sites, shelters, or any other resource that is in short supply.

Competitive Exclusion Principle In 1934, Russian ecologist G. F. Gause studied the effects of interspecific competition in laboratory experiments with two closely related species of protists, *Paramecium aurelia* and *Paramecium caudatum* (**Figure 19.5**). Gause cultured the protists under stable

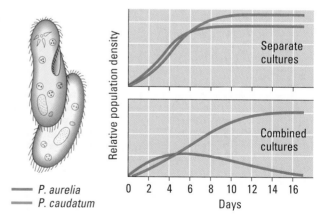

P. aurelia
P. caudatum

Figure 19.5 Competitive exclusion in laboratory populations of *Paramecium*. Cultured separately (upper graph) with constant amounts of food (bacteria) added daily, populations of each of the two *Paramecium* species grow to carrying capacity. But when the two species are cultured together (lower graph), *P. aurelia* has a competitive edge in obtaining food, an advantage that drives *P. caudatum* to extinction in the microcosm of the culture jar.

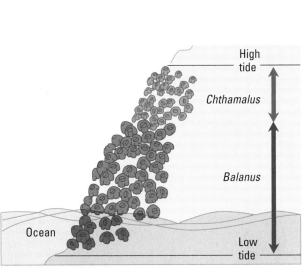

Figure 19.6 Testing the hypothesis of competitive exclusion in the field. These two types of barnacles, *Balanus* and *Chthamalus,* grow on rocks that are exposed during low tide. Both types are attached as adults, but have free-swimming larvae that may settle and begin to develop on virtually any rock surface. *Chthamalus* occupies the upper parts of the rocks, which are out of the water longer during low tides. *Balanus* fails to survive as high on the rocks as *Chthamalus,* apparently because *Balanus* dries out in the air. When ecologists experimented by removing *Balanus* from the lower rocks, *Chthamalus* spread lower, colonizing the unoccupied rocks. However, when both species colonize the same rock, *Balanus* eventually displaces *Chthamalus* on the lower part of the rock. The researchers concluded that the upper limit of *Balanus*'s distribution is set mainly by the availability of water, whereas the lower limit of *Chthamalus*'s distribution is set by competition.

conditions with a constant amount of food added every day. When he grew the two species in separate cultures, each population grew rapidly and then leveled off at what was apparently the carrying capacity of the culture. But when Gause cultured the two species together, *P. aurelia* apparently had a competitive edge in obtaining food, and *P. caudatum* was driven to extinction in the culture. Gause concluded that two species so similar that they compete for the same limiting resources cannot coexist in the same place. One will use the resources more efficiently and thus reproduce more rapidly. Even a slight reproductive advantage will eventually lead to local elimination of the inferior competitor. Ecologists called Gause's concept the **competitive exclusion principle.** Numerous tests of the hypothesis include classic field experiments with two species of barnacles that attach to intertidal rocks on the North Atlantic coast (**Figure 19.6**).

The Ecological Niche The sum total of a species' use of the biotic and abiotic resources in its environment is called the species' **ecological niche.** One way to grasp the concept is through an analogy made by ecologist Eugene Odum: If an organism's habitat is its address, the niche is that habitat plus the organism's occupation. Put another way, an organism's niche is its ecological role—how it "fits into" an ecosystem. The niche of a population of tropical tree lizards, for example, consists of, among many other components, the temperature range it tolerates, the size of trees on which it perches, the time of day in which it is active, and the size and type of insects it eats.

We can now restate the competitive exclusion principle to say that two species cannot coexist in a community if their niches are identical. However, ecologically similar species can coexist in a community if there are one or more significant differences in their niches.

Resource Partitioning There are two possible outcomes of competition between species having identical niches: Either the less competitive species will be driven to local extinction, or one of the species may evolve enough to use a different set of resources. This differentiation of niches that enables similar species to coexist in a community is called **resource partitioning. Figure 19.7** illustrates an example. We can think of resource partitioning within a community as "the ghost of competition past"—circumstantial evidence of earlier interspecific competition resolved by the evolution of niche differences.

Predation

In everyday usage, the term *community* is benign, maybe even implying the warmness of cooperation, as in *community spirit*. In contrast, an ecological community exhibits the Darwinian realities of competition and predation, where organisms eat other organisms. In the interspecific interaction called **predation,** the consumer is the **predator** and the food species is the **prey.** We include herbivory, the eating of plants by animals, as a form of predation, even in cases such as grazing, where the animal does not kill the whole plant.

It won't surprise you that predation is a potent factor in adaptive evolution. Eating and avoiding being eaten are prerequisite to reproductive success. Natural selection refines the adaptations of both predators and prey.

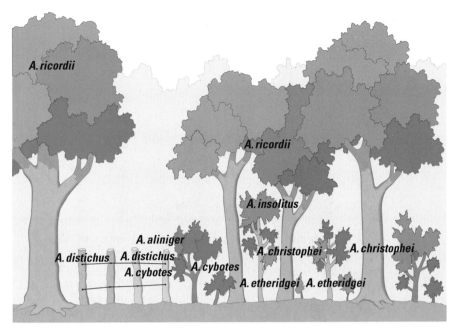

(a)

(b) *A. distichus*

(c) *A. insolitus*

Figure 19.7 Resource partitioning in a group of lizards. **(a)** Seven species of *Anolis* lizards live in close proximity at La Palma, in the Dominican Republic. The lizards all feed on insects and other small arthropods. However, competition for food is minimized because each lizard species perches in a certain microhabitat. **(b)** *Anolis distichus,* for example, perches on fence posts and other sunny surfaces (such as this leaf), whereas **(c)** *A. insolitus* usually perches on shady branches.

Predator Adaptations Many important feeding adaptations of predators are both obvious and familiar. Most predators have acute senses that enable them to locate and identify potential prey. In addition, many predators have adaptations such as claws, teeth, fangs, stingers, or poison that help catch and subdue the organisms on which they feed. Rattlesnakes and other pit vipers, for example, locate their prey with special heat-sensing organs located between each eye and nostril, and they kill small birds and mammals by injecting them with toxins through their fangs. Similarly, many herbivorous insects locate appropriate food plants by using chemical sensors on their feet, and their mouthparts are adapted for shredding tough vegetation. Predators that pursue their prey are generally fast and agile, whereas those that lie in ambush are often camouflaged in their environments. In our own case, perhaps it is valid to view our capacity for learning and teaching to be adaptations contributing to our success as predators; the invention and refinement of agriculture over the centuries has certainly increased Earth's carrying capacity for humans (see Chapter 18).

Plant Defenses Against Herbivores Plants cannot run away from herbivores. Chemical toxins, often in combination with various kinds of antipredator spines and thorns, are plants' main arsenals against being eaten to extinction. Among such chemical weapons are the poison strychnine, produced by a tropical vine called *Strychnos toxifera;* morphine, from the opium poppy; nicotine, produced by the tobacco plant; mescaline, from peyote cactus; and tannins (such as those found in tea and wine), from a variety of plant species. Other defensive compounds that are not toxic to humans but may be distasteful to other herbivores are responsible for the familiar flavors of cinnamon, cloves, and peppermint. Some plants even produce chemicals that imitate insect hormones and cause abnormal development in some insects that eat them.

Animal Defenses Against Predators Animals can avoid being eaten by using passive defenses, such as hiding, or active defenses, such as escaping or defending themselves against predators. Fleeing is a common antipredator response, though it can be very costly in terms of energy. Many animals

Figure 19.8 Mobbing, a behavioral defense against predators. Many prey species turn the tables and attack their predators. Here, two crows mob a barn owl, a predator that often kills and eats crow eggs and nestlings.

Figure 19.9 Protecting offspring by faking an injury. This killdeer uses deception to defend her nest against predators and human disturbance. When danger threatens, she leaves the nest and fakes a broken wing. This behavior distracts and draws a potential predator away from the nest, and then the mother just flies away. The trickery often saves the lives of a killdeer's offspring.

Figure 19.10 Camouflage. A dark brown Asian leaf frog blends in among the dead leaves of a forest floor.

flee into a shelter and avoid being caught without expending the energy required for a prolonged flight. Active self-defense is less common, though some large grazing mammals will vigorously defend their young from predators such as lions. Other behavioral defenses include alarm calls, which often bring in many individuals of the prey species that mob the predator (**Figure 19.8**). Distraction displays direct the attention of the predator away from a vulnerable prey, such as a bird chick, to another potential prey that is more likely to escape, such as the chick's parent (**Figure 19.9**).

Many other defenses rely on adaptive coloration, which has evolved repeatedly among animals. Camouflage, called **cryptic coloration,** is passive defense that makes potential prey difficult to spot against its background (**Figure 19.10**).

Some animals have mechanical or chemical defenses against would-be predators. Most predators are strongly discouraged by the familiar defenses of skunks and porcupines. Some animals, such as poisonous toads and frogs, can synthesize toxins. Others acquire chemical defense passively by accumulating toxins from the plants they eat. For example, monarch butterflies store poisons from the milkweed plants they eat as larvae, making the butterflies distasteful to some potential predators. Animals with chemical defenses are often brightly colored, a caution to predators known as **warning coloration** (**Figure 19.11**).

A species of prey may gain significant protection through mimicry, a "copycat" adaptation in which one species mimics the appearance of another (see figure 1.20). In **Batesian mimicry,** a palatable or harmless species mimics an unpalatable or harmful model. In one intriguing example, the larva of the hawk moth puffs up its head and thorax when disturbed, looking like the head of a small poisonous snake, complete with eyes (**Figure 19.12**). The mimicry even involves behavior; the larva weaves its head back and forth and hisses like a snake. In **Müllerian mimicry,** two or more unpalatable species resemble each other. Presumably, each species gains an additional

Figure 19.11 Warning coloration of the poison-arrow frog. The vivid markings of this pair of tree frogs, inhabitants of a rain forest in Costa Rica, warn of noxious chemicals in the frogs' skin; predators learn about this as soon as they touch the frog. In some parts of South America, human hunters in the rain forest tip their arrows with poisons from similar frogs to bring down large mammals.

(a) Hawk moth larva **(b) Snake**

Figure 19.12 Batesian mimicry. When disturbed, **(a)** the hawk moth larva resembles **(b)** a snake.

advantage, because the pooling of numbers causes predators to learn more quickly to avoid any prey with a particular appearance (**Figure 19.13**).

Predators also use mimicry in a variety of ways. For example, some snapping turtles have tongues that resemble a wriggling worm, thus luring small fish; any fish that tries to eat the "bait" is itself quickly consumed as the turtle's strong jaws snap closed.

Predation and Species Diversity in Communities You might think that organisms eating other organisms would always reduce species diversity, but field experiments reveal that predator-prey relationships can actually preserve diversity. Experiments by American ecologist Robert Paine in the 1960s were among the first to provide such evidence. Paine removed the dominant predator, a sea star of the genus *Pisaster*, from experimental areas within the intertidal zone of the Washington coast (**Figure 19.14**). The result

(a) Cuckoo bee

(b) Yellow jacket

Figure 19.13 Müllerian mimicry.
Both **(a)** the cuckoo bee and **(b)** the yellow jacket wasp have stingers that release toxins. The cross-mimicry in appearance presumably benefits both species because predators learn more quickly to avoid any prey with these distinctive markings.

Figure 19.14 Effect of a keystone predator on species diversity.
Ecologist Robert Paine removed the sea star *Pisaster* (inset, eating its favorite food, a mussel) from a tide pool in 1963. Mussels then gradually took over the rock space and eliminated other invertebrates and algae.

Figure 19.15 A case study in the evolution of host-parasite relations. In the 1940s, Australia was overrun by hundreds of millions of rabbits, all descended from just 12 pairs imported a century earlier onto an estate for sport. The rabbits destroyed vegetation in huge expanses of Australia and threatened the sheep and cattle industries. In 1950, in an effort to control the exploding rabbit population, biologists deliberately introduced a myxoma virus that specifically parasitizes rabbits. The virus spread rapidly and killed almost all (99.8%) the infected rabbits. However, a second exposure to the virus killed only 90% of the rabbit population derived from survivors of the first application. And a third exposure eliminated only about 50% of the rabbit population. Apparently, viral infection selected for host genotypes that were better able to resist the parasite. Over the generations, the myxoma virus had less and less effect, and the rabbit population rebounded. The Australian government is now having more success with a different virus, which was introduced into the rabbit population in 1995.

Figure 19.16 Mutualism between acacia trees and ants. Certain species of Central and South American acacia trees have hollow thorns that house stinging ants. The ants feed on sugar and proteins from specialized glands on the trees (the orange swellings at the tips of the leaflets). The acacia also benefits from housing and feeding this population of ants, for the ants will attack anything that touches the tree. The pugnacious ants sting other insects, remove fungal spores, and clip surrounding vegetation that happens to grow close to the foliage of the acacia.

was that *Pisaster*'s main prey, a mussel of the genus *Mytilus*, outcompeted many of the other shoreline organisms (barnacles and snails, for instance) for the important resource of space on the rocks. The number of species dropped.

The experiments of Paine and others led to the concept of keystone predators. A **keystone predator** is a species, such as *Pisaster*, that reduces the density of the strongest competitors in a community. In so doing, the predator helps maintain species diversity by preventing competitive exclusion of weaker competitors. By checking the population growth of mussels, the most successful (and most abundant) species, the sea star *Pisaster* helped maintain populations of several less competitive species in Paine's experimental tide pools. Predation has its constructive side.

Symbiotic Relationships

A **symbiotic relationship** is an interspecific interaction in which one species, the **symbiont,** lives in or on another species, the **host.** The two types of symbiotic relationships most important in structuring ecological communities are parasitism and mutualism.

Parasitism A symbiotic relationship is called **parasitism** if one organism benefits while the other is harmed. The **parasite,** usually the smaller of the two organisms, obtains its nutrients by living on or in its host organism. We can actually think of parasitism as a specialized form of predation. Tapeworms and the protozoa that cause malaria are examples of internal parasites (see Figures 17.15 and 15.19d). External parasites include mosquitoes, which suck blood from animals, and aphids, which tap into the sap of plants.

Natural selection has refined the relationships between parasites and their hosts. Many parasites, particularly microorganisms, have adapted to specific hosts, often a single species. In any parasite population, reproductive success is greatest for individuals that are best at locating and feeding on their hosts. For example, some aquatic leeches first locate a host by detecting movement in the water and then confirm its identity based on temperature and chemical cues on the host's skin. Natural selection has also favored the evolution of host defenses. In humans and other vertebrates, an elaborate immune system helps defend the body against specific internal parasites. With natural selection working on both host and parasite, the eventual outcome is usually a relatively stable relationship that does not kill the host quickly, an excess that would eliminate the parasite as well. An example of how rapidly natural selection can temper a host-parasite relationship is the evolution of resistance to a viral parasite in Australian rabbit populations (**Figure 19.15**).

Mutualism In contrast to the one-sidedness of parasitism, **mutualism** is a symbiosis that benefits both partners. Just two of the examples we encountered in earlier chapters are the root-fungus associations called mycorrhizae and the symbiotic association of fungi and algae in lichens (see Figures 16.3 and 16.25). **Figure 19.16** illustrates another case of mutual symbiosis, the relationship between acacia trees and the ants that protect these trees from herbivorous insects and competing plants.

Many mutualistic relationships may have evolved from predator-prey or host-parasite interactions. Certain angiosperm plants, for example, have adaptations that attract animals that function in pollination or seed dispersal; these adaptations may represent an evolutionary response to a history of the herbivores' feeding on pollen and seeds. In many cases, pollen is spared when the pollinator is able to consume nectar instead. Indigestible seeds within fruits are dispersed by animals, which are nourished by the fruits. Any plants that could derive some benefit by sacrificing organic materials other than pollen or seeds would increase their reproductive success, and the adaptations for mutualistic interactions would spread through the plant population over the generations.

The Complexity of Community Networks

So far, we have reduced the networks known as biological communities to interactions between species: competition, predator-prey interactions, and symbiosis. It is the branching of these interactions that makes communities so complex. For example, some plants recruit parasitic wasps that lay their eggs in caterpillars that eat the plants (**Figure 19.17**). Both the wasps and caterpillars are also eaten by predators such as spiders and birds, and all of these organisms are hosts to specific parasites. Ecologists are only beginning to sort out the complex networks of just a few biological communities. Contributing to the challenges of community ecology is the fact that the structure of a community may change, sometimes over relatively short periods of time, due to a variety of disturbances. The next section examines some of these changes.

> **Activity 19A on the Web & CD**
> Categorize various interactions between species.

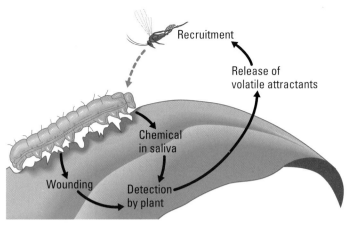

Figure 19.17 A three-way community interaction. Some plants recruit wasps that lay their eggs in caterpillars feeding on the plants. A leaf releases a wasp-attracting vapor in response to saliva from the caterpillar and injury of the plant's cells. The wasp stings the leaf-munching caterpillar and then lays its eggs within the caterpillar. The wasp larvae that hatch from these eggs will eat their way out of the caterpillar. (Adapted with permission from Edward Farmer, "Plant Biology: New Fatty Acid–Based Signals: A Lesson from the Plant World," *Science* vol. 276, page 912 (May 9, 1997). Copyright © 1997 American Association for the Advancement of Science.)

CHECKPOINT

1. Reexamine the distribution of the two barnacle species in Figure 19.6. What experiment could you perform to test the hypothesis that it is mainly susceptibility to drying, not competitive exclusion, that keeps *Balanus* from populating the upper parts of the rock?

2. What is the advantage to a keystone predator of being specialized to feed mainly on those prey species that are otherwise the most successful among potential prey species?

3. Which type of symbiotic relationship is represented by bacteria in the human colon (large intestine) that produce vitamin K, a nutrient that can be utilized by the human host?

Answers: **1.** You could remove *Chthamalus* from the upper parts of the rock to see whether *Balanus* still failed to spread upward, even in the absence of a competitor. **2.** The most competitive prey species probably represent the most abundant and dependable food source for the predator. **3.** Mutualism

---

Disturbance of Communities

Disturbances are episodes that damage biological communities, at least temporarily, by destroying organisms and altering the availability of resources such as mineral nutrients and water. Examples of disturbances are storms, fires, floods and the severe erosion they can cause, droughts, and human activities such as deforestation and the introduction of domesticated animals that graze.

Disturbances affect all communities, which rarely have time to settle into a constant state before another disturbance occurs. Put another way, most communities spend much of their time in various stages of recovery from various disturbances. Change seems to characterize most communities more than constancy and balance.

Ecological Succession

Communities may change drastically after a flood, fire, glacial advance and retreat, or volcanic eruption strips away their vegetation. A variety of species may colonize the disturbed area. Later these may be replaced as yet other species colonize the area. This process of community change is called **ecological succession.**

When a community arises in a virtually lifeless area with no soil, the change is called **primary succession.** Examples of such areas are new volcanic islands or the rubble left by a retreating glacier. Often the only life-forms initially present are autotrophic microorganisms. Lichens and mosses, which grow from windblown spores, are commonly the first large photosynthetic organisms to colonize the barren ground. Soil develops gradually as organic matter accumulates from the decomposed remains of the early colonizers. Once soil is present, the lichens and mosses are overgrown by grasses, shrubs, and trees that sprout from seeds blown in from nearby areas or carried in by animals. Eventually, the area may be colonized by plants that will become the community's prevalent form of vegetation. Primary succession from barren soil to a community such as a deciduous forest can take hundreds or thousands of years (**Figure 19.18**).

> **Activity 19B on the Web & CD**
> Practice putting the steps of primary succession in order.

❶ Retreating glacier

❷ Barren landscape after glacial retreat

❸ Moss and lichen stage

❹ Alders and cottonwoods covering the hillsides

❺ Spruce coming into the alder and cottonwood forest

❻ Spruce and hemlock forest

Figure 19.18 Succession after retreat of glaciers. Ecologists sometimes deduce the process of succession by studying several locations that are at different successional stages. For example, these photographs, taken at different sites where glaciers are retreating (melting), represent about 200 years of succession.

Secondary succession occurs where a disturbance has destroyed an existing community but left the soil intact. For instance, forested areas in the eastern United States that are cleared for farming will, if abandoned, undergo secondary succession and may eventually return to forest. The earliest plants to recolonize an area are often species with opportunistic life histories (see Chapter 18). Many are herbaceous (nonwoody) species that grow from windblown or animal-borne seeds. These plants thrive where there is little competition from other plants. Woody shrubs may eventually replace most of the herbaceous species. Later yet, trees may replace most of the shrubs.

A Dynamic View of Community Structure

Disturbances keep communities in a continual flux, making them mosaics of patches at various successional stages and preventing them from ever reaching a state of equilibrium or complete balance. We tend to think of disturbances as having negative impacts, but this is only part of the story, at least in the case of natural disturbances. Small-scale disturbances often have positive effects, such as creating new opportunities for species. For example, when a tree falls, it disturbs the immediate surroundings. However, the fallen log fosters new habitats, and the depression left by its roots may fill with water and be used as egg-laying sites by frogs, salamanders, and numerous insects (**Figure 19.19**).

If most communities never really stabilize, what is the effect of this constant change on species diversity? When disturbance is severe and relatively frequent, the community may include only good colonizers typical of early stages of succession. Temperate grasslands maintained by periodic fires are an example. At the other end of the disturbance scale, when disruptions are mild and rare in a particular location, then the late-successional species that are most competitive will make up the community to the exclusion of other species. Between these two extremes, in an area where disturbance is moderate in both severity and frequency, species diversity may be greatest. Studies of species diversity in tropical rain forests support this **intermediate disturbance hypothesis.** Scattered throughout these forests are gaps where trees have fallen. In these disturbed areas, species of various successional stages coexist within a relatively small space.

Figure 19.19 A small-scale disturbance. When this tree fell during a windstorm in a forest in Michigan, its root system and the surrounding soil uplifted, resulting in a depression that filled with water. The dead tree, the root mound, and the water-filled depression created new habitats.

CHECKPOINT

1. What is the main abiotic factor that distinguishes primary from secondary succession?

2. Why are trees unlikely to be among the first colonizers of a landscape cleared by the advance and retreat of a glacier?

Answers: 1. Absence of soil (primary succession) versus presence of soil (secondary succession) at the onset of succession **2.** Trees can't grow until earlier colonizers form soil.

An Overview of Ecosystem Dynamics

To this point in the chapter, we have focused on the study of communities. We will now broaden our scope to ecological systems, or **ecosystems,** the highest level of ecological complexity. An ecosystem consists of all the organisms in a given area plus the physical environment, including soil, water, and air. In other words, an ecosystem is a biological community and the abiotic factors with which the community interacts.

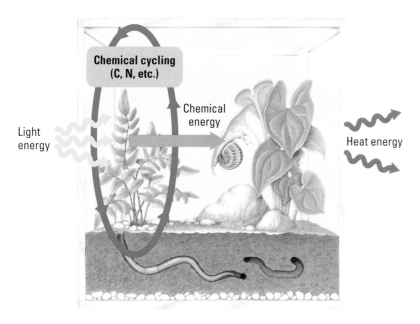

Figure 19.20 A terrarium ecosystem. Small and artificial as it is, this microcosm illustrates the two major ecosystem processes: energy flow and chemical cycling.

Perhaps you have seen or even made a terrarium such as the one in **Figure 19.20**. Such a microcosm qualifies as an ecosystem, because it exhibits the two major processes that sustain all ecosystems, from terrariums to forests and oceans all the way to the entire biosphere. Those two key processes of ecosystem dynamics are energy flow and chemical cycling. **Energy flow** is the passage of energy through the components of the ecosystem. **Chemical cycling** is the use and reuse of chemical elements such as carbon and nitrogen within the ecosystem.

Energy reaches most ecosystems in the form of sunlight. Plants and other photosynthetic producers convert the light energy to the chemical energy of food. Energy continues its flow through an ecosystem when animals and other consumers acquire organic matter by feeding on plants and other producers. Especially important consumers are the soil bacteria and fungi that obtain their energy by decomposing the dead remains of plants, animals, and other organisms. In using their chemical energy for work, all organisms dissipate heat energy to their surroundings. Thus, energy cannot be recycled within an ecosystem, but must flow continuously through the ecosystem, entering as light and exiting as heat (see Figure 19.20).

In contrast to the flow of energy through an ecosystem, chemical elements can be recycled between an ecosystem's living community and the abiotic environment. The plants and other producers acquire their carbon, nitrogen, and other chemical elements in inorganic form from the air and soil. Photosynthesis then enables plants to incorporate these elements into organic compounds such as carbohydrates and proteins. Animals acquire these elements in organic form by eating. The metabolism of all organisms returns some of the chemical elements to the abiotic environment in inorganic form. Cellular respiration, for example, breaks organic molecules down to carbon dioxide and water. This job of recycling is finished by microorganisms that decompose dead organisms and their organic wastes, such as feces and leaf litter. These decomposers restock the soil, water, and air with chemical elements in the inorganic form that plants and other producers can again build into organic matter. And the chemical cycles continue.

Activity 19C on the Web & CD
Increase your understanding of energy flow and chemical cycling.

Note again the key distinction between energy *flow* and chemical *cycling*. Because energy, unlike matter, cannot be recycled, an ecosystem must be powered by a continuous influx of energy from an external source, usually the sun. Notice also that an ecosystem's energy flow and chemical cycling are closely related. Both depend on transfer of substances in the feeding relationships, or trophic structure, of the ecosystem.

Trophic relationships determine an ecosystem's routes of energy flow and chemical cycling. In analyzing these feeding relationships, ecologists divide the species of an ecosystem into different **trophic levels** based on their main sources of nutrition. Let's see how these trophic levels connect as the living components of an ecosystem's energy flow and chemical cycling.

Trophic Levels and Food Chains

The sequence of food transfer from trophic level to trophic level is called a **food chain. Figure 19.21** compares a terrestrial food chain and a marine food chain. Starting at the bottom of such diagrams, the trophic level that supports all others consists of the **producers,** the autotrophic organisms (see Chapter 6). As photosynthetic organisms, producers use light energy to power the synthesis of organic compounds. Plants are the main producers on land. In aquatic ecosystems, the producers are mainly photosynthetic protists and bacteria, collectively known as phytoplankton. Multicellular algae and aquatic plants are also important producers in shallow waters.

All organisms in trophic levels above the producers are consumers, the heterotrophs directly or indirectly dependent on the output of producers. **Herbivores,** which eat plants, algae, or autotrophic bacteria, are the **primary consumers** of an ecosystem. Primary consumers on land include grasshoppers and many other insects, snails, and certain vertebrates, such as grazing mammals and birds that eat seeds and fruit. In aquatic ecosystems, primary consumers include a variety of zooplankton (mainly protists and microscopic animals such as small shrimps) that prey on the phytoplankton.

Above the primary consumers, the trophic levels are made up of **carnivores,** which eat the consumers from the levels below. On land, **secondary consumers** include many small mammals, such as rodents that eat herbivorous insects, and a great variety of small birds, frogs, and spiders, as well as lions and other large carnivores that eat grazers. In aquatic ecosystems, secondary consumers are mainly small fishes that eat zooplankton and bottom-dwelling invertebrates. Higher trophic levels include **tertiary consumers,** such as snakes that eat mice and other secondary consumers. Some ecosystems even have **quaternary consumers,** such as hawks in terrestrial ecosystems and killer whales in marine environments (see Figure 19.21).

Most diagrams of food chains are incomplete by not showing a critical trophic level of consumers called **detritivores,** another name for **decomposers.** Detritivores derive their energy from **detritus,** the dead material left by all trophic levels. Detritus includes animal wastes, plant litter, and all sorts of dead organisms. Most organic matter eventually becomes detritus and is consumed by detritivores. A great variety of animals, often called scavengers, eat detritus. For instance, earthworms, many rodents, and insects eat fallen leaves and other detritus. Other scavengers include crayfish, catfish, and vultures. But an ecosystem's main detritivores are the prokaryotes and fungi (**Figure 19.22**). Enormous numbers of microorganisms in the soil and in mud at the bottoms of lakes and oceans recycle most of the ecosystem's organic materials to inorganic compounds that plants and other producers can reuse.

Food Webs

Actually, few ecosystems are so simple that they are characterized by a single unbranched food chain. Several types of primary consumers usually feed on the same plant species, and one species of primary consumer may eat several different plants. Such branching of food chains occurs at the other trophic levels as well. For example, adult frogs, which are secondary consumers, eat several insect species that may also be eaten by various birds. In addition, some consumers feed at several different

Activity 19D on the Web & CD Practice interpreting food webs.

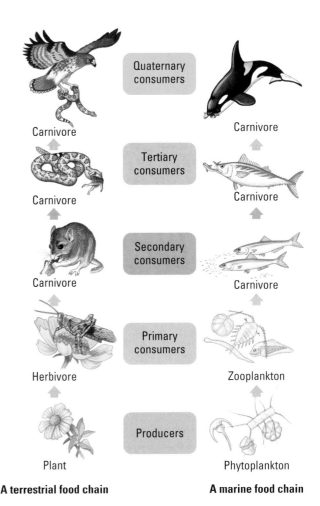

A terrestrial food chain   **A marine food chain**

Figure 19.21 Examples of food chains. The arrows trace the transfer of food from producers through various levels of consumers. Detritivores (decomposers), important consumers in all ecosystems, are not included in these simplified diagrams of terrestrial and marine ecosystems.

Figure 19.22 Fungi decomposing a dead log.

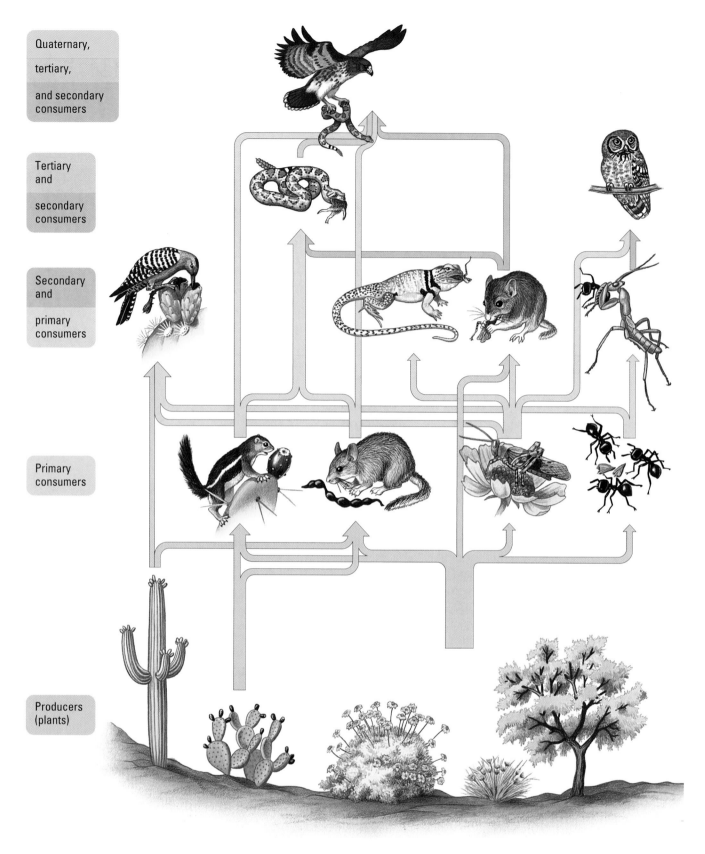

Quaternary, tertiary, and secondary consumers

Tertiary and secondary consumers

Secondary and primary consumers

Primary consumers

Producers (plants)

Figure 19.23
A simplified food web. As in the food chains of Figure 19.21, the arrows in this web indicate "who eats whom," the direction of nutrient transfers. We also continue the color coding for the trophic levels and food transfers introduced in Figure 19.21. Note that in a food web, such as this one for a Sonoran desert ecosystem, some species feed at more than one trophic level. The other branching factor that turns food chains into webs is the consumption of the same species by more than one consumer.

trophic levels. An owl, for instance, may eat mice, which are mainly primary consumers that may also eat some invertebrates; but an owl may also feed on snakes, which are strictly carnivorous. **Omnivores,** including humans, eat producers as well as consumers of different levels. Thus, the feeding relationships in an ecosystem are usually woven into elaborate **food webs.**

Though more realistic than a food chain, a food web diagram, such as the one in **Figure 19.23**, is still a highly simplified model of the feeding relationships in an ecosystem. An actual food web would involve many more organisms at each trophic level, and most of the animals would have a more diverse diet than shown in the figure.

CHECKPOINT

1. Why is the transfer of energy in an ecosystem referred to as energy *flow*, not energy *cycling?*

2. I'm eating a cheese pizza. At which trophic level(s) am I feeding?

3. In the "who eats whom" dynamics of a food web, even consumers of the highest level in the ecosystem eventually become food for

_____ .

Answers: 1. Because energy passes through an ecosystem, entering as sunlight and leaving as lheat. It is not recycled within the ecosystem. **2.** Primary consumer (flour and tomato sauce) and secondary consumer (cheese, a product from cows, which are primary consumers) **3.** detritivores, or decomposers

---

Energy Flow in Ecosystems

All organisms require energy for growth, maintenance, reproduction, and, in some species, locomotion (see Chapter 6). In this section, we take a closer look at energy flow through ecosystems. Along the way, we'll answer two key questions: What limits the length of food chains? and How do lessons about energy flow apply to human nutrition?

Productivity and the Energy Budgets of Ecosystems

Each day, Earth is bombarded by about 10^{19} kilocalories of solar radiation. This is the energy equivalent of about 100 million atomic bombs the size of the one that devastated Hiroshima in 1945. Most of this solar energy is absorbed, scattered, or reflected by the atmosphere or by Earth's surface. Of the visible light that reaches plants and other producers, only about 1% is converted to chemical energy by photosynthesis. But on a global scale, this is enough to produce about 170 billion tons of organic material per year.

Ecologists call the amount, or mass, of organic material in an ecosys-

Case Study in the Process of Science on the Web & CD
Conduct a virtual experiment to determine how light affects primary productivity in an aquatic habitat.

tem the **biomass.** And the rate at which plants and other producers build biomass, or organic matter, is called the ecosystem's **primary productivity.** Thus, the primary productivity of the entire biosphere is 170 billion tons of organic material per year. Different ecosystems vary considerably in their productivity as well as in their contribution to the total

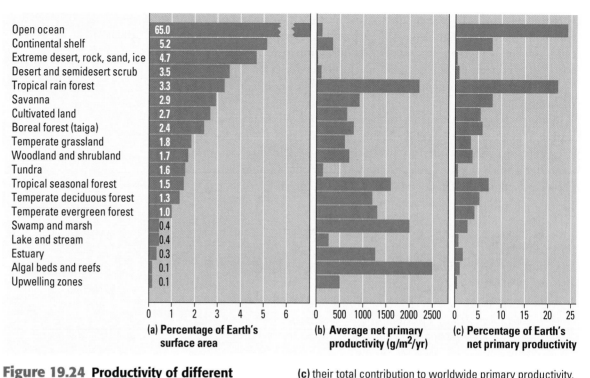

Open ocean	65.0
Continental shelf	5.2
Extreme desert, rock, sand, ice	4.7
Desert and semidesert scrub	3.5
Tropical rain forest	3.3
Savanna	2.9
Cultivated land	2.7
Boreal forest (taiga)	2.4
Temperate grassland	1.8
Woodland and shrubland	1.7
Tundra	1.6
Tropical seasonal forest	1.5
Temperate deciduous forest	1.3
Temperate evergreen forest	1.0
Swamp and marsh	0.4
Lake and stream	0.4
Estuary	0.3
Algal beds and reefs	0.1
Upwelling zones	0.1

(a) Percentage of Earth's surface area

(b) Average net primary productivity (g/m^2/yr)

(c) Percentage of Earth's net primary productivity

Figure 19.24 Productivity of different ecosystems. Primary productivity is the rate at which plants and other producers store chemical energy in biomass over a unit of time, in this case a year. Aquatic ecosystems are color-coded blue in these histograms; terrestrial ecosystems are brown. **(a)** The geographic extent and **(b)** the productivity per unit area of different ecosystems determine **(c)** their total contribution to worldwide primary productivity. Open ocean, for example, contributes a lot to the planet's productivity despite its low productivity per unit area because of its large size. In contrast, tropical rain forests contribute a lot to the biosphere's energy budget mainly because of their high productivity per unit area.

productivity of the biosphere (**Figure 19.24**). Whatever the ecosystem, primary productivity sets the spending limit for the energy budget of the entire ecosystem because consumers must acquire their organic fuels from producers. Now let's see how this energy budget is divided among the different trophic levels in an ecosystem's food web.

Energy Pyramids

When energy flows as organic matter through the trophic levels of an ecosystem, much of it is lost at each link in a food chain. Consider the transfer of organic matter from plants (producers) to herbivores (primary consumers). In most ecosystems, herbivores manage to eat only a fraction of the plant material produced, and they cannot digest all the organic compounds they do ingest. Of the organic compounds an herbivore *can* use, much of it is consumed as fuel for cellular respiration. Only the remainder is available as raw material for growth of the herbivore. For example, in the case of a caterpillar consuming leaves, only about 15% of the biomass of the food is transformed into caterpillar biomass (**Figure 19.25**). In general, an average of only about 10% of energy in the form of organic matter at each trophic level is stored as biomass in the next level of the food chain. The rest of the energy is accounted for by the heat released by all the working organisms in the ecosystem, including the detritivores decomposing feces and other organic wastes.

The cumulative loss of energy from a food chain can be represented in a diagram called an **energy pyramid.** The trophic levels are stacked in blocks, with the block representing producers forming the foundation of the pyra-

Plant material eaten by caterpillar

100 kilocalories (kcal)

Feces
50 kcal

35 kcal — Cellular respiration

15 kcal

Growth

Figure 19.25 What becomes of a caterpillar's food? Only about 15% of the calories of plant material this herbivore consumes becomes stored as biomass available to the next link in the food chain.

mid (**Figure 19.26**). The size of each block is proportional to the productivity of each trophic level, the rate of energy storage as biomass at that level. The pyramid owes its steep slopes to the loss of 90% or so of the energy with each food transfer in the chain.

An important implication of this stepwise decline of energy along a food chain is that the amount of energy available to top-level consumers is small compared with that available to lower-level consumers. Only a tiny fraction of the energy stored by photosynthesis flows through a food chain to a tertiary consumer, such as a snake feeding on a mouse (see Figure 19.26). This explains why top-level consumers such as lions and hawks require so much geographic territory; it takes a lot of vegetation to support trophic levels so many steps removed from photosynthetic production. You can also understand now why most food chains are limited to three to five levels; there is simply not enough energy at the very top of an energy pyramid to support another trophic level. There are, for example, no nonhuman predators of lions, eagles, and killer whales; the biomass in populations of these top-level consumers is insufficient to support yet another trophic level with a reliable source of nutrition.

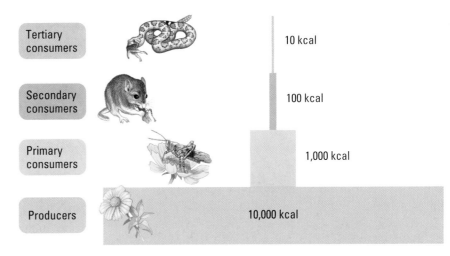

Figure 19.26 An idealized energy pyramid.
In this case, 10% of the chemical energy available at each trophic level is converted to new biomass in the trophic level above it.

Ecosystem Energetics and Human Nutrition

The dynamics of energy flow apply to the human population as much as to other organisms. Like other consumers, we depend entirely on productivity by plants for our food. As omnivores, most of us eat both plant material and meat. When we eat grain or fruit, we are primary consumers; when we eat beef or other meat cut from herbivores, we are secondary consumers. When we eat fish like salmon (which eat insects and other small animals) and tuna (which eat smaller fish), we are tertiary or quaternary consumers.

Figure 19.27 applies the concept of energy pyramids to show why consuming a high proportion of our calories in the form of meat is a relatively inefficient way of tapping photosynthetic productivity. About 90%

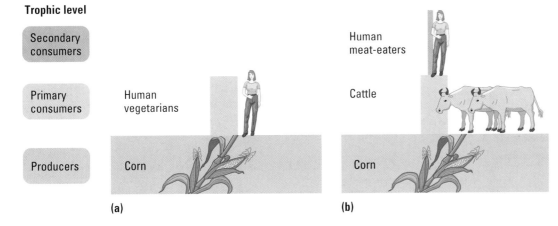

Figure 19.27 Food energy available to the human population at different trophic levels. Most humans have a diet between these two extremes. The point here is that a greater proportion of photosynthetic energy reaches us when we feed as primary consumers directly on plants than when we are nourished by photosynthesis indirectly by feeding as secondary consumers on animals.

of the food energy in a crop such as corn is lost to us if that corn is first "processed" by primary consumers such as hogs or cattle. Put still another way, it takes at least 10 pounds of feed corn to produce 1 pound of bacon or beef steak.

This analysis of energy flow to human consumers is not meant as a plea for vegetarianism. The main point is that eating meat of any kind is an expensive luxury, both economically and environmentally. In many developing countries, people are mainly vegetarian by necessity because they cannot afford to buy much meat or their country cannot afford to produce it. Wherever meat is on the menu, producing it requires that more land be cultivated, more water be used for irrigation, and more chemical fertilizers and pesticides be applied to croplands used for growing grain. It is likely that as the human population continues to expand, meat consumption will become even more of a luxury than it is today.

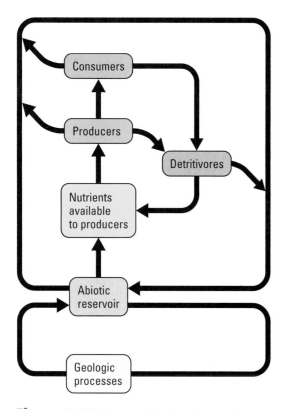

Figure 19.28 Generalized scheme for biogeochemical cycles.

Chemical Cycling in Ecosystems

The sun supplies ecosystems with energy, but there are no extraterrestrial sources of the chemical elements life requires. Ecosystems depend on a recycling of these chemical elements. Even while an individual organism is alive, much of its chemical stock is rotated continuously as nutrients are acquired and waste products are released. Atoms present in the complex molecules of an organism at the time of its death are returned in simpler compounds to the atmosphere, water, or soil by the action of bacterial and fungal detritivores. This decomposition replenishes the pools of inorganic nutrients that plants and other producers use to build new organic matter. In a sense, we only borrow an ecosystem's chemical elements, returning what is left in our bodies after we die. "Ashes to ashes, dust to dust" is one metaphor for this fact of life. Let's take a closer look at how chemicals cycle between organisms and the abiotic (nonliving) components of ecosystems.

The General Scheme of Chemical Cycling

Because chemical cycles in an ecosystem involve both biotic and abiotic components, they are also called **biogeochemical cycles. Figure 19.28** is a general scheme for these cycles. Here are three key points to note:

- Each circuit, whether for carbon, nitrogen, or some other chemical material required for life, has an abiotic reservoir through which the chemical cycles. Carbon's main **abiotic reservoir,** for example, is the atmosphere, where carbon atoms are stocked mainly in the form of CO_2 gas (carbon dioxide). Carbon makes the abiotic → biotic transition when it is incorporated from CO_2 into food by plants and other producers. And the carbon is returned to the abiotic reservoir by the cellular respiration of the ecosystem's organisms, including detritivores.

• A portion of chemical cycling can bypass the biotic components and rely completely on geologic processes. Water, for example, can exit lakes and oceans by evaporation and recycle to those abiotic reservoirs by precipitation, such as rain.

• Some chemical elements require "processing" by certain microorganisms before they are available to plants as inorganic nutrients. The element nitrogen is an example. Its main abiotic reservoir is the atmosphere, which is almost 80% nitrogen in the form of N_2 gas. However, plants can utilize nitrogen only in the forms of ammonium (NH_4^+) and nitrate (NO_3^-), which roots absorb from the soil. Certain soil bacteria convert atmospheric N_2 to ammonium and nitrate in the soil, making nitrogen available to plants (see Chapter 15). In fact, microorganisms, mainly prokaryotes, play a critical role in all biogeochemical cycles.

Examples of Biogeochemical Cycles

Like most generalizations, the basic scheme of Figure 19.28 can go only so far in helping us understand biogeochemical cycles. A chemical's specific route through an ecosystem varies with the particular element and the trophic structure of the ecosystem. **Figure 19.29**, on the next two pages, will help you apply the basic principles of chemical cycling to four key materials: carbon, nitrogen, phosphorus, and water (the only one that's a compound, not an element).

The carbon, nitrogen, phosphorus, and water cycles (and others) can be classified into two main categories based on how mobile the chemicals are within environments. Some chemicals, including phosphorus, are not very mobile and thus cycle almost entirely locally, at least over the short term. Soil is the main reservoir for these nutrients, which are absorbed by plant roots and eventually returned to the soil by detritivores living in the same vicinity. In contrast, for those materials that spend part of their time in gaseous form—carbon, nitrogen, and water are examples—the cycling is essentially global. For instance, some of the carbon atoms a plant acquires from the air as CO_2 may have been released to the atmosphere by the respiration of some animal living far away.

Activities 19F and 19G on the Web & CD
Review the carbon and nitrogen cycles.

CHECKPOINT

1. What is the main abiotic reservoir for carbon?

2. What would happen to the carbon cycle if all the detritivores suddenly went on "strike" and stopped working?

3. Over the short term, why does phosphorus cycling tend to be more localized than carbon, nitrogen, or water cycling?

Answers: 1. The atmospheric stock of CO_2 **2.** Carbon would accumulate in organic mass, the atmospheric reservoir of carbon would decline, and plants would eventually be starved for CO_2.
3. Because phosphorus is cycled almost entirely within the soil rather than being transferred over long distances via the atmosphere

Figure 19.29 Examples of biogeochemical cycles.

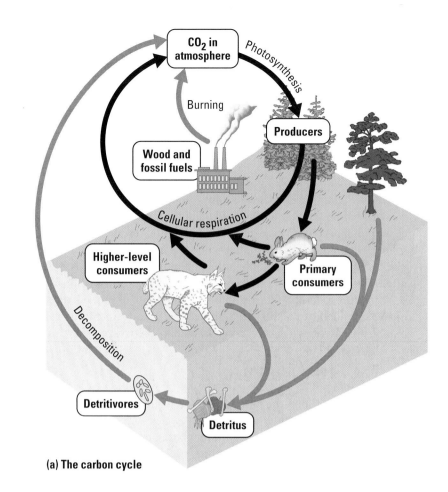

(a) The carbon cycle. Carbon is a major ingredient of all organic molecules. With the atmosphere as its abiotic reservoir, carbon cycles globally. The reciprocal metabolic processes of photosynthesis and respiration are mainly responsible for the cycling of carbon between the biotic and abiotic world (black arrows in diagram). On a global scale, the return of CO_2 to the atmosphere by respiration closely balances its removal by photosynthesis. However, the increased burning of wood (during slash-and-burn deforestation) and fossil fuels (coal and petroleum) is steadily raising the level of CO_2 in the atmosphere, causing significant environmental problems, including global warming.

(a) The carbon cycle

(b) The nitrogen cycle. As an ingredient of amino acids and other organic compounds essential to life, nitrogen is a key chemical element in all ecosystems. Earth's atmosphere is almost 80% N_2 gas, but that form of nitrogen is unavailable to plants. Soil bacteria called nitrogen fixers convert the N_2 to ammonium, and nitrifying bacteria convert the ammonium to nitrates. Ammonium and nitrates are the soil minerals the roots of plants absorb as a nitrogen source for synthesis of amino acids and other organic molecules. Notice in the diagram that there are two groups of nitrogen-fixing bacteria: free-living bacteria and those that live as symbiotic organisms in nodules of beans and other legumes (see Chapter 15). Note that other types of bacteria play key roles at various junctures in the nitrogen cycle. For example, denitrifying bacteria complete the nitrogen cycle by converting soil nitrates to atmospheric N_2. Much of the nitrogen cycling by bacteria in many ecosystems involves the inner cycle in the diagram (black arrows). The outer cycle (gray arrows) often moves only a tiny fraction of nitrogen into and out of natural ecosystems.

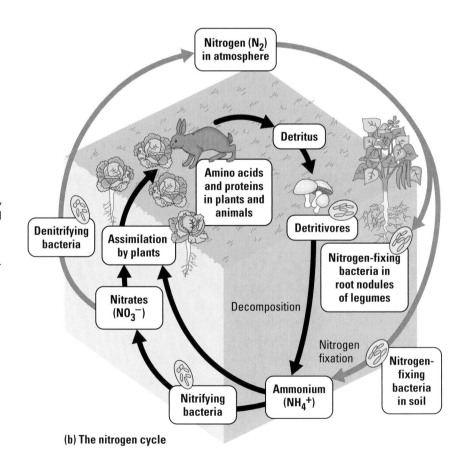

(b) The nitrogen cycle

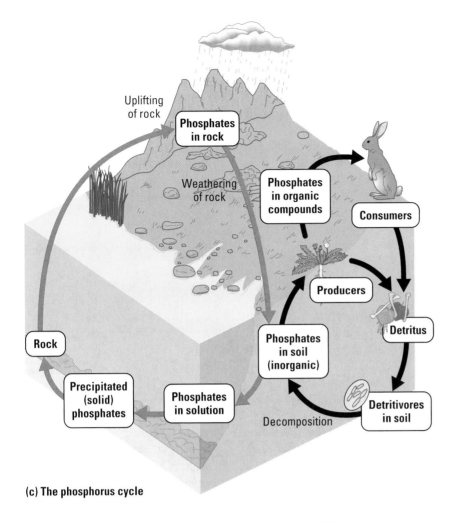

(c) The phosphorus cycle

(c) The phosphorus cycle. Organisms require phosphorus as an ingredient of ATP and nucleic acids (such as DNA) and (in vertebrates) as a major component of bones and teeth. In contrast to carbon and nitrogen, phosphorus does not have an atmospheric presence and tends to recycle only locally. The main abiotic reservoir is rock, which, upon weathering, releases phosphorus mainly in the form of the mineral phosphate (PO_4^{3-}). Plants absorb the dissolved phosphate ions in the soil and build them into organic compounds. Consumers obtain phosphorus in organic form from plants. Detritivores return phosphates to the soil. Some phosphates also precipitate out of solution at the bottom of deep lakes and oceans. The phosphates in this form may eventually become part of new rocks and will not cycle back into living organisms until geologic processes uplift the rocks and expose them to weathering. Thus, phosphorus actually cycles on two time scales: through a local ecosystem on the scale of ecological time (black arrows in diagram) and through strictly geologic processes on the much longer scale of geologic time (gray arrows).

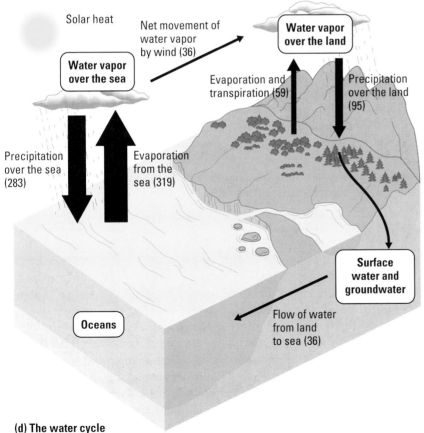

(d) The water cycle

(d) The water cycle. Water is essential to life, both because organisms are made mostly of water and because water's unique properties make environments on Earth habitable (see Chapter 2). In this diagram of the global water cycle, the numbers indicate water movements in billion billion (10^{18}) grams per year. That's a lot of water, but it's the relative comparisons that are important here. Those relative quantities are also indicated by the widths of the arrows. Three major processes driven by solar energy—precipitation, evaporation, and transpiration from plants (evaporation from leaves)—continuously move water between the land, the oceans, and the atmosphere. Note that over the oceans, evaporation exceeds precipitation. The result is a net movement of this excess amount of water vapor in clouds that are carried by winds from the oceans across the land. On land, precipitation exceeds evaporation and transpiration. The excess precipitation forms systems of surface water (such as lakes and rivers) and groundwater, all of which flow back to the sea, completing the water cycle. The water cycle has a global character because there is a large reservoir of water in the atmosphere. Thus, water molecules that have evaporated from the Pacific Ocean, for instance, may appear in a lake or in an animal's body far inland in North America.

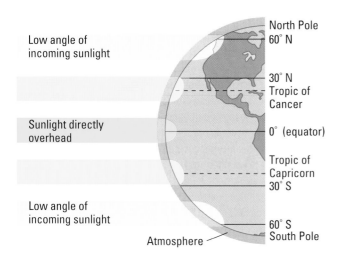

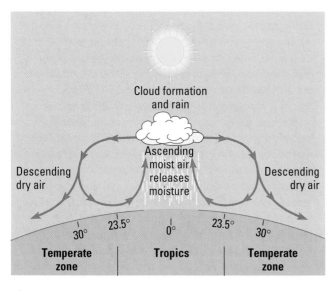

Figure 19.30 Uneven heating of Earth's surface.

Figure 19.31 How uneven heating of Earth causes rain and wind.

Biomes

Now that we've examined how ecosystems are organized, we can turn our attention to the ecosystems found on Earth. Major types of ecosystems that cover large geographic regions are called **biomes.** You can think of them as types of landscapes. Tropical forests and deserts are examples. The distribution of the biomes largely depends on climate, with temperature and rainfall often the key factors determining the kind of biome that exists in a particular region. Let's look at how climate is determined and how it helps explain the locations of various types of biomes.

How Climate Affects Biome Distribution

Because of its curvature, Earth receives an uneven distribution of solar energy (**Figure 19.30**). The equator receives the greatest intensity of solar radiation. Heated by the direct rays of the sun, air at the equator rises. It then cools, forms clouds, and drops rain (**Figure 19.31**). This largely explains why rain forests are concentrated in the **tropics**—the region from the Tropic of Cancer (23.5° N) to the Tropic of Capricorn (23.5° S).

After losing moisture over equatorial zones, dry high-altitude air masses spread away from the equator until they cool and descend again at latitudes of about 30° north and south. Many of the world's great deserts—the Sahara in North Africa and the Arabian on the Arabian Peninsula, for example—are centered at these latitudes because of the dry air they receive.

Latitudes between the tropics and the Arctic Circle in the north and the Antarctic Circle in the south are called **temperate zones**. Generally, these regions have milder climates than the tropics or the polar regions. Notice in Figure 19.31 that some of the descending dry air heads into the latitudes above 30°. At first these air masses pick up moisture, but they tend to drop it as they cool at higher latitudes. This is why the north and south temperate zones tend to be relatively wet. Coniferous forests dominate the landscape at the wet but cool latitudes around 60° N.

Proximity to large bodies of water and the presence of landforms such as mountain ranges also affect climate. Oceans and large lakes moderate climate by absorbing heat when the air is warm and releasing heat to cold air (see Chapter 2). Mountains affect climate in two major ways. First, air temperature declines as elevation increases. Second, mountains can block the flow of cool, moist air from a coast, causing radically different climates on opposite sides of a mountain range (**Figure 19.32**). Now let's survey the Earth's major ecosystems, noting the influence of climate.

Terrestrial Biomes

Figure 19.33 shows a map of Earth's major terrestrial biomes. Because there are latitudinal patterns of climate over Earth's surface, there are also latitudinal patterns of biome distribution. If the climate in two geographically separate areas is similar, the same type of biome may occur in both places. Coniferous forests, for instance, extend in a broad band across North America, Europe, and Asia. Note, however, that a biome is characterized by a type of biological community, not a collection of certain species of organisms. For example, the groups of species living in the Sahara Desert of Africa and the Gobi Desert of eastern Asia are different, but the species in both regions are adapted to desert conditions. Organisms in widely separated biomes may look alike because of convergence, the evolution of similar adaptations in independently evolved species living in similar environments (see Chapter 14).

Most biomes are named for major physical or climatic features and for their predominant vegetation. For example, temperate grasslands are dominated by various grass species and are generally found in the temperate zones. Each biome is also characterized by microorganisms, fungi, and animals adapted to that particular environment. Temperate grasslands, for example, are more likely than forests to be populated by large grazing mammals.

Activity 19H on the Web & CD Identify terrestrial biome photos and locate biomes on a map.

In most biomes today, extensive human activities have radically altered the natural patterns. Most of the eastern United States, for example, is classified as temperate deciduous forest, but human activity has eliminated all but a tiny percentage of the original forest. In fact, humans have altered much of Earth's surface, replacing original biomes with urban and agricultural ones.

Figure 19.34, spread over the next four pages, surveys the major terrestrial biomes, beginning near the equator and generally approaching the poles.

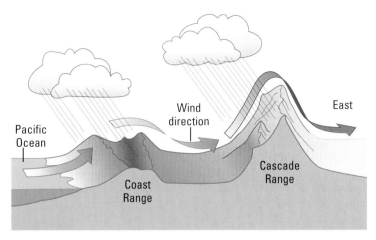

Figure 19.32 How mountains affect rainfall. In this example, moist air moves in off the Pacific Ocean and encounters the Coast Range in the state of Washington. Air flows upward, cools at higher altitudes, and drops a large amount of water. The biological community in this wet region is a temperate rain forest. Farther inland, precipitation increases again as the air moves up and over the Cascade Range. On the eastern side of the Cascades, there is little precipitation. As a result of this rain shadow, much of central Washington is very arid, almost qualifying as desert.

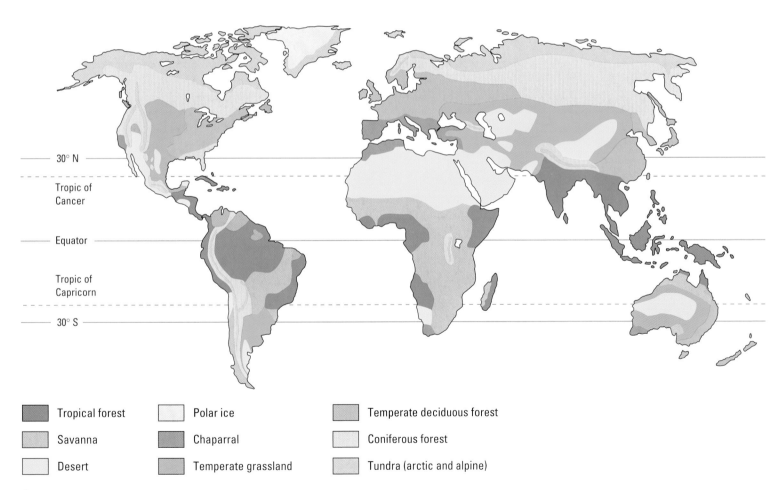

Tropical forest

Savanna

Desert

Polar ice

Chaparral

Temperate grassland

Temperate deciduous forest

Coniferous forest

Tundra (arctic and alpine)

Figure 19.33 Map of the major terrestrial biomes. Although this map has sharp boundaries, biomes actually grade into one another. We'll use smaller versions of this map, highlighted by color coding, during our closer look at the terrestrial biomes in Figure 19.34.

Figure 19.34 Major terrestrial biomes.

(a) Tropical forest. The photograph shows a tropical rain forest in Costa Rica. Tropical rain forests have pronounced vertical stratification, or layering. Trees in the canopy make up the topmost stratum. The canopy is often closed, so that little light reaches the ground below. When an opening does occur, perhaps because of a fallen tree, other trees and large woody vines grow rapidly, competing for light and space as they fill the gap. Many of the trees are covered with epiphytes (plants that grow on other plants rather than in soil), such as orchids and bromeliads. Rainfall, which is quite variable in the tropics, is the prime determinant of the vegetation growing in an area. In low-land areas that have a prolonged dry season or scarce rainfall at any time, tropical dry forests predominate. The plants found there are a mixture of thorny shrubs and trees and succulents. In regions with distinct wet and dry seasons, tropical deciduous trees are common.

(b) Savanna. Savannas are grasslands with scattered trees. This Kenyan savanna is a showcase of large herbi-vores and their predators. Actually, the dominant herbi-vores here and in other savannas are insects, especially ants and termites. Fire is an important abiotic component, and the dominant plant species are fire adapted. The luxu-riant growth of grasses and forbs (small broadleaf plants) during the rainy season provides a rich food source for animals. However, large grazing mammals must migrate to greener pastures and scattered watering holes during regular periods of seasonal drought.

(c) Desert. Sparse rainfall (less than 30 cm per year) largely determines that an area will be a desert. Some deserts have soil surface temperatures above 60°C (140°F) during the day. Other deserts, such as those west of the Rocky Mountains and in central Asia, are relatively cold. The Sonoran Desert of southern Arizona (shown here) is characterized by giant saguaro cacti and deeply rooted shrubs. Evolutionary adaptations of desert plants and animals include a remarkable array of mechanisms that store water. The "pleated" structure of saguaro cacti enables the plants to expand when they absorb water during wet periods. Some desert mice never drink, deriving all their water from the metabolic breakdown of the seeds they eat. Protective adaptations that deter feeding by mammals and insects, such as spines on cacti and poisons in the leaves of shrubs, are common in desert plants.

(d) Chaparral. This biome occurs in midlatitude coastal areas with mild, rainy winters and long, hot, dry summers. Dense, spiny, evergreen shrubs dominate chaparral. Plants of the chaparral, such as those in this California scrubland, are adapted to and dependent on periodic fires. The dry, woody shrubs are frequently ignited by lightning and by careless human activities, creating summer and autumn brushfires in the densely populated canyons of Southern California and elsewhere. Some of the shrubs produce seeds that will germinate only after a hot fire. Food reserves stored in their fire-resistant roots enable them to resprout quickly and use mineral nutrients released by fires.

Figure 19.34 Major terrestrial biomes, continued.

(e) Temperate grassland. The puszta of Hungary, the pampas of Argentina and Uruguay, the steppes of Russia, and the plains and prairies of central North America are all temperate grasslands. The key to the persistence of grasslands is seasonal drought, occasional fires, and grazing by large mammals, all of which prevent establishment of woody shrubs and trees. Temperate grasslands such as the tallgrass prairie in South Dakota (shown here) once covered much of central North America. Because grassland soil is both deep and rich in nutrients, these habitats provide fertile land for agriculture. Most grassland in the United States has been converted to farmland, and very little natural prairie exists today.

(f) Temperate deciduous forest. Dense stands of deciduous trees are trademarks of temperate deciduous forests, such as this one in Great Smoky Mountains National Park in North Carolina. Temperate deciduous forests occur throughout midlatitudes where there is sufficient moisture to support the growth of large trees. Deciduous forest trees drop their leaves before winter, when temperatures are too low for effective photosynthesis and water lost by evaporation is not easily replaced from frozen soil. Many temperate deciduous forest mammals also enter a dormant winter state called hibernation, and some bird species migrate to warmer climates. Virtually all the original deciduous forests in North America were destroyed by logging and land clearing for agriculture and urban development. In contrast to drier biomes, these forests tend to recover after disturbance, and today we see deciduous trees dominating undeveloped areas over much of their former range.

(g) Coniferous forest. Cone-bearing evergreen trees, such as pine, spruce, fir, and hemlock, dominate coniferous forests. Coastal coniferous forests of the U.S. Pacific Northwest, such as the one in Olympic National Park in western Washington shown here, are actually temperate rain forests. Warm, moist air from the Pacific Ocean supports these unique communities, which, like most coniferous forests, are dominated by one or a few tree species. Extending in a broad band across northern North America and Eurasia to the southern border of the arctic tundra, the northern coniferous forest, or taiga, is the largest terrestrial biome on Earth (see Figure 19.33). Taiga receives heavy snowfall during winter. The conical shape of many conifers prevents too much snow from accumulating on and breaking their branches. Coniferous forests are being logged at an alarming rate, and the old-growth stands of these trees may soon disappear.

(h) Tundra. Permafrost (permanently frozen subsoil), bitterly cold temperatures, and high winds are responsible for the absence of trees and other tall plants in this arctic tundra in central Alaska (photographed in autumn). Although the arctic tundra receives very little annual rainfall, water cannot penetrate the underlying permafrost and accumulates in pools on the shallow topsoil during the short summer. Tundra covers expansive areas of the Arctic, amounting to 20% of Earth's land surface. High winds and cold temperatures create similar plant communities, called alpine tundra, on very high mountaintops at all latitudes, including the tropics.

(a) Satellite view of the Great Lakes

(b) A stream flowing into the Snake River, Idaho

(c) A freshwater wetland in Georgia

Figure 19.35 Freshwater biomes.

Freshwater Biomes

Aquatic biomes, consisting of freshwater and marine ecosystems, occupy the largest part of the biosphere.

Lakes and Ponds Standing bodies of water range from small ponds only a few square meters in area to large lakes, such as North America's Great Lakes, that are thousands of square kilometers (**Figure 19.35a**).

In lakes and large ponds, the communities of plants, algae, and animals are distributed according to the depth of the water and its distance from shore. Shallow water near shore and the upper stratum of water away from shore make up the **photic zone,** so named because light is available for photosynthesis. **Phytoplankton** (microscopic algae and cyanobacteria) grow in the photic zone, joined by rooted plants and floating plants such as water lilies in photic regions near shore. If a lake or pond is deep enough or murky enough, it has an **aphotic zone,** where light levels are too low to support photosynthesis.

At the bottom of all aquatic biomes, the substrate is called the **benthic zone.** Made up of sand and organic and inorganic sediments ("ooze"), the benthic zone is occupied by communities of organisms, including a diversity of bacteria, collectively called benthos. (If you have ever waded barefoot into a pond or lake, you have felt the benthos squish between your toes.) A major source of food for the benthos is detritus that "rains" down from the productive surface waters of the photic zone.

Nitrogen and phosphorus are the mineral nutrients that usually limit the amount of phytoplankton growth in a lake or pond. Many lakes and ponds are affected by large inputs of nitrogen and phosphorus from sewage and from runoff from fertilized lawns and agricultural fields. These nutrients often produce blooms, or population explosions, of algae. Heavy algal growth reduces light penetration into the water, and when the algae die and decompose, a pond or lake can suffer serious oxygen depletion. Fish and other aerobic organisms may then die (see Figure 18.3).

Rivers and Streams Bodies of water flowing in one direction, rivers and streams generally support quite different communities of organisms than lakes and ponds (**Figure 19.35b**). A river or stream changes greatly between its source (perhaps a spring or snowmelt) and the point at which it empties into a lake or the ocean. Near a source, the water is usually cold, low in nutrients, and clear. The channel is often narrow, with a swift current that does not allow much silt to accumulate on the bottom. The current also inhibits the growth of phytoplankton; most of the organisms found here are supported by the photosynthesis of algae attached to rocks or organic material (such as leaves) carried into the stream from the surrounding land. The most abundant benthic animals are usually insects that eat algae, leaves, or one another. Trout are often the predominant fishes, locating their food, including insects, mainly by sight in the clear water.

Downstream, a river or stream generally widens and slows. Marshes, ponds, and other wetlands are common in downstream areas (**Figure 19.35c**). There the water is usually warmer and may be murkier because of sediments and phytoplankton suspended in it. Worms and insects that burrow into mud are often abundant, as are waterfowl, frogs, and catfish and other fishes that find food more by scent and taste than by sight.

Figure 19.36 Damming the Columbia River Basin.

If Lewis and Clark lived today, they would have a hard time navigating the Columbia. This map shows only the largest of the 250 dams that have altered freshwater ecosystems throughout the Pacific Northwest. The great concrete obstacles make it difficult for salmon to swim upriver to their breeding streams, though many dams now have "fish ladders" that provide detours.

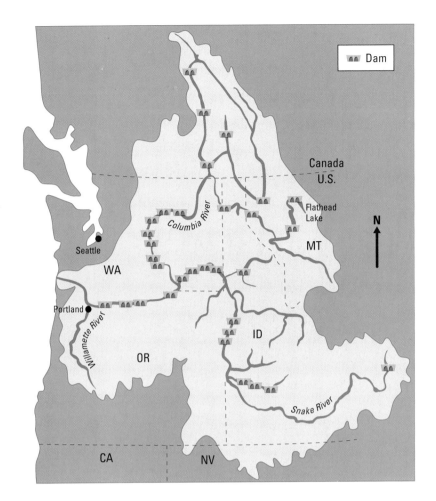

Many streams and rivers have been affected by pollution from human activities, by stream channelization to speed water flow, and by dams that hold water. For centuries, humans used streams and rivers as depositories of waste, thinking that these materials would be diluted and carried downstream. While some pollutants are carried far from their source, many settle to the bottom, where they can be taken up by aquatic organisms. Even the pollutants that are carried away contribute to ocean and lake pollution. In many cases, dams have completely changed the downstream ecosystems, altering the intensity and volume of water flow and affecting fish and invertebrate populations (**Figure 19.36**).

Marine Biomes

Life originated in the sea and evolved there for almost 3 billion years before plants and animals began colonizing land. Covering about 75% of the planet's surface, oceans have always had an enormous impact on the biosphere. Their evaporation provides most of Earth's rainfall, and photosynthesis by marine algae supplies a substantial portion of the biosphere's oxygen.

Estuaries The area where a freshwater stream or river merges with the ocean is called an **estuary.** Most estuaries are bordered by extensive coastal wetlands called mudflats and saltmarshes (**Figure 19.37**). Salinity (salt concentration) varies spatially within estuaries, from nearly that of fresh water to that of the ocean; it also varies over the course of a day with the rise and fall of the tides. Nutrients from the river enrich estuarine waters, making estuaries one of the most biologically productive environments on Earth.

Saltmarsh grasses and algae are the major producers in estuaries. This environment also supports a variety of worms, oysters, crabs, and many of the fish species that humans consume. A diversity of marine invertebrates and fishes use estuaries as a breeding ground or migrate through them to freshwater habitats upstream. Estuaries are also crucial feeding areas for a great diversity of birds.

Although estuaries support a wide variety of commercially valuable species, areas around estuaries are also prime locations for commercial and residential developments. In addition, estuaries are unfortunately at the receiving end for pollutants dumped upstream. Very little undisturbed estuarine habitat remains, and a large percentage has been totally eliminated by landfill and development. Many states have now—rather belatedly—taken steps to preserve their remaining estuaries.

Figure 19.37 An estuary on the edge of Chesapeake Bay, Maryland.

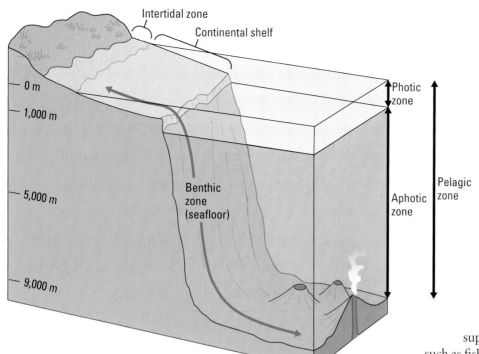

Figure 19.38 **Oceanic zones.**

Major Oceanic Zones Marine life is distributed according to depth of the water, degree of light penetration, distance from shore, and open water versus bottom (**Figure 19.38**). The area where land meets sea is called the **intertidal zone.** This environment is alternately submerged and exposed by the twice-daily cycle of tides. Intertidal organisms are therefore subject to huge daily variations in the availability of seawater (and the nutrients it carries) and in temperature. Also, intertidal organisms are subject to the mechanical forces of wave action, which can dislodge them from their habitats. The organisms that are adapted to this agitation attach to rocks or vegetation or burrow into mud or sand. Examples are certain seaweeds (large algae), sessile crustaceans such as barnacles, and echinoderms such as sea stars and sea urchins (**Figure 19.39**).

The open ocean itself, called the **pelagic zone,** supports communities dominated by motile animals such as fishes, squids, and marine mammals, including whales and dolphins. Microscopic algae and cyanobacteria—the phytoplankton—drift passively in the pelagic zone. Phytoplankton are the ocean's main photosynthetic producers, making most of the organic food molecules on which other ocean-dwellers depend (see Chapter 15). **Zooplankton** are animals that drift in the pelagic zone either because they are too small to resist ocean currents or because they don't swim. Zooplankton eat phytoplankton and in turn are consumed by other animals, including fishes.

The seafloor, like the lake bottom, is called the benthic zone. Depending on depth and light penetration, the benthic community consists of attached algae, fungi, bacteria, sponges, burrowing worms, sea anemones, clams, crabs, and fishes.

Note in Figure 19.38 that marine biologists often group the illuminated regions of the benthic and pelagic communities together, calling them the photic zone, as we did for lakes. Underlying the photic zone is the vast, dark

Figure 19.39
A sampling of organisms in a tide pool.

Figure 19.40
Exploring hydrothermal vent communities.

Research vessels such as *Alvin* have enabled scientists to explore the seafloor to depths of 2,500 m, more than a mile deeper than sunlight penetrates. In the late 1970s, researchers discovered seafloor life that does not depend on photosynthesis. It was on these expeditions that humans first encountered a diversity of organisms living around hydrothermal vents, where molten rock and hot gases escape from Earth's interior. Nearby lives a community of sea anemones, giant clams, shrimps, worms, and a few fish species. The food chain is supported by sulfur bacteria, which use the oxidation of hydrogen sulfide (H_2S) in the vent's effluent to power the synthesis of organic molecules from CO_2. Some of the animals eat the bacteria. Others, such as the tube worms in the inset, harbor the food-producing bacteria as symbionts. Unlike all the other ecosystems we know, life at the hydrothermal vents on the seafloor is driven by energy from Earth itself rather than from the sun.

aphotic zone. This is the most extensive part of the biosphere. Without light, there are no photosynthetic organisms, but life is still diverse in the aphotic zone. Many kinds of invertebrates, such as sea urchins and polychaete worms, and some fishes scavenge organic matter that sinks from the lighted waters above. The seafloor also includes **hydrothermal vent communities,** which are powered by chemical energy from Earth's interior rather than by sunlight (**Figure 19.40**).

Activity 19I on the Web & CD
Visit freshwater and marine biomes.

CHECKPOINT

1. Why is homeowner's insurance relatively expensive for people who live in the chaparral?

2. How do humans now use most of the North American land that was once temperate grasslands?

3. How does the loss of leaves function as an adaptation of deciduous trees to cold winters?

4. What two abiotic factors account for the rarity of trees in arctic tundra?

5. The _____, small photosynthetic organisms inhabiting the _____ zone of the pelagic zone, provide most of the food for oceanic life.

6. Why does sewage cause algal blooms in lakes?

7. What is the energy source for hydrothermal vent communities?

Answers: 1. High fire risk **2.** For farming **3.** By reducing loss of water from the trees when that water cannot be replaced because of frozen soil **4.** Long, very cold winters (short growing season) and permafrost **5.** phytoplankton; photic **6.** The sewage adds minerals that stimulate growth of the algae. **7.** Chemicals that spew up from Earth's interior

(a) *Heliconius* caterpillar

(b) *Heliconius* eggs

Sugar
deposits

(c) Passionflower leaves with sugar-secreting nectar glands

Figure 19.41 Coevolution of a plant and insect.
(a) Passionflower vines produce toxic chemicals that help protect leaves from most herbivorous insects. A counteradaptation has evolved in *Heliconius* butterflies; their larva can feed on passionflower leaves because they have digestive enzymes that break down the plant's toxins. (b) The female butterflies avoid laying eggs on passionflower leaves with egg clusters deposited by other females. This behavior ensures an adequate food supply when eggs hatch and the larvae begin feeding on the leaves. (c) Some passionflower species have yellow nectaries (sugar-secreting glands) that mimic *Heliconius* eggs. The butterflies avoid laying eggs on leaves with these yellow spots. The nectaries also attract ants that prey on the butterfly eggs.

Evolution Connection

Coevolution in Biological Communities

We have seen many examples in this chapter of adaptations that evolve in populations as a result of interactions with other species in a community. Defenses of plants against herbivores, animal defenses against predators,

> **Evolution Connection on the Web**
> Learn more about coevolution.

mutual symbiosis, and parasite-host relationships are all examples. Ecologists use the term **coevolution** when the adaptations of two species are closely connected—that is, when an adaptation in one species leads to a counter-adaptation in a second species. Let's examine one such case of coevolution, the reciprocal adaptations between certain butterfly species and passionflowers, plant species on which the caterpillars of the butterflies feed.

Passionflower vines of the genus *Passiflora* are protected against most herbivorous insects by their production of toxic compounds in young leaves and shoots. However, the larvae (caterpillars) of butterflies of the genus *Heliconius* can tolerate these defensive chemicals. This counteradaptation has enabled *Heliconius* larvae to become specialized feeders on plants that few other insects can eat (**Figure 19.41**). Survival of the larvae is further enhanced by a behavioral adaptation of the butterflies. The eggs that female *Heliconius* butterflies lay on the leaves of passionflower vines are bright yellow, and other females generally avoid laying eggs on leaves marked by these yellow dots. This behavior presumably reduces competition among the larvae for food.

An infestation of *Heliconius* larvae can devastate a passionflower vine, and these poison-resistant insects are likely to be a strong selection force favoring the evolution of more defenses in the plants. In some species of *Passiflora*, the leaves have conspicuous yellow spots that mimic *Heliconius* eggs, an adaptation that may divert the butterflies to other plants in their search for egg-laying sites.

On closer inspection, such seemingly clear-cut examples of coevolution between two species usually turn out to be more complicated. The yellow "spots" on the passionflower vine are actually nectar-secreting glands, which attract ants and wasps that prey on *Heliconius* eggs or larvae. And there is evidence that the mere presence of ants on a leaf will discourage a *Heliconius* butterfly from laying eggs there. (We saw a similar example in the mutualism of ants and acacia trees.) The relationship of the *Passiflora* and *Heliconius* is another example of the complexity—shaped by evolution—that characterizes Earth's communities and ecosystems.

Chapter Review

Summary of Key Concepts

For study help, go to the Essential Biology Website (www.essentialbiology.com) or CD-ROM to explore the Activities and Case Studies in the Process of Science.

Key Properties of Communities

- **Diversity** Community diversity includes the species richness and relative abundance of different species.

Case Study in the Process of Science How Are Impacts on Community Diversity Measured?

- **Prevalent Form of Vegetation** The types and structural features of plants largely determine the kinds of animals that live in a community.

- **Stability** A community's stability is its ability to resist change and return to its original species composition after being disturbed. Stability depends on the type of community and the nature of the disturbances.

- **Trophic Structure**   Trophic structure consists of the feeding relationships among the various species making up a community.

Interspecific Interactions in Communities

- **Competition Between Species**   When populations of two or more species in a community rely on similar limiting resources, they may be subject to interspecific competition. Two species cannot coexist in a community if their niches are identical. Resource partitioning is the differentiation of niches that enables similar species to coexist in a local community.

- **Predation**   Natural selection refines the adaptations of both predators and prey. Most predators have acute senses that enable them to locate and identify potential prey. Plants mainly use chemical toxins, spines, and thorns to defend against predators. Animals may defend themselves by using passive defenses, such as cryptic coloration (camouflage), or active defenses, such as escaping, alarm calls, mobbing, or distraction displays. Animals with chemical defenses are often brightly colored, a warning to potential predators. A species of prey may also gain protection through mimicry. In Batesian mimicry, a palatable species mimics an unpalatable model. In Müllerian mimicry, two or more unpalatable species resemble each other. Predator-prey relationships can preserve diversity. A keystone predator is a species that reduces the density of the strongest competitors in a community. This predator helps maintain species diversity by preventing competitive exclusion of weaker competitors.

- **Symbiotic Relationships**   A symbiotic relationship is an interspecific interaction in which a symbiont species lives in or on a host species. Many mutualistic relationships may have evolved from predator-prey or host-parasite interactions. Parasitism is a one-sided relationship in which one organism benefits at the expense of the other. Natural selection has refined the relationships between parasites and their hosts. Mutualism is a symbiosis that benefits both partners.

- **The Complexity of Community Networks**   The branching of interactions between species makes communities complex.

Activity 19A Interspecific Interactions

Disturbance of Communities

- Disturbances are episodes that damage biological communities, at least temporarily, by destroying organisms and altering the availability of resources such as mineral nutrients and water. Most communities spend much of their time in various stages of recovery from disturbances.

- **Ecological Succession**   The sequence of changes in a community after a disturbance is called ecological succession. Primary succession occurs where a community arises in a virtually lifeless area with no soil. Secondary succession occurs where a disturbance has destroyed an existing community but left the soil intact.

Activity 19B Primary Succession

- **A Dynamic View of Community Structure**   In general, disturbances keep communities in a continual flux, making them mosaics of patches at various successional stages and preventing them from reaching a completely stable state. Small-scale disturbances often have positive effects, such as creating new opportunities for species. Species diversity may be greatest in places where disturbances are moderate in severity and frequency.

An Overview of Ecosystem Dynamics

- An ecosystem is a biological community and the abiotic factors with which the community interacts. Energy must flow continuously through an ecosystem, from producers to consumers and decomposers. Chemical elements can be recycled between an ecosystem's living community and the abiotic environment.

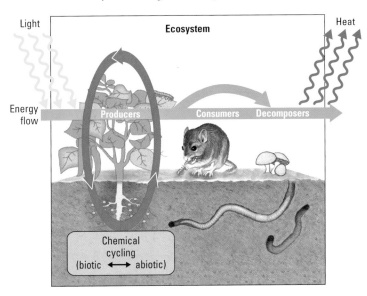

Activity 19C Energy Flow and Chemical Cycling

- **Trophic Levels and Food Chains**   Trophic relationships determine an ecosystem's routes of energy flow and chemical cycling.

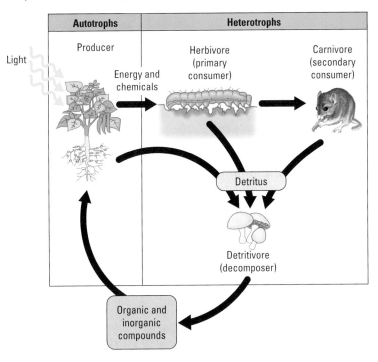

- **Food Webs**   The feeding relationships in an ecosystem are usually woven into elaborate food webs.

Activity 19D Food Webs

Energy Flow in Ecosystems

- **Productivity and the Energy Budgets of Ecosystems** The rate at which plants and other producers build biomass is the ecosystem's primary productivity. Ecosystems vary considerably in their productivity. Primary productivity sets the spending limit for the energy budget of the entire ecosystem because consumers must acquire their organic fuels from producers.

Case Study in the Process of Science How Does Light Affect Primary Productivity?

- **Energy Pyramids**

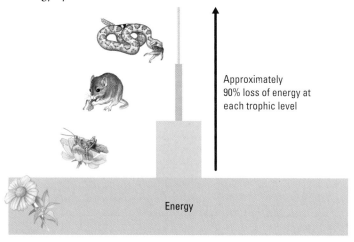

Approximately 90% loss of energy at each trophic level

Energy

Activity 19E Energy Pyramids

- **Ecosystem Energetics and Human Nutrition** Eating producers instead of consumers requires less photosynthetic productivity and reduces the impact to the environment.

Chemical Cycling in Ecosystems

- **The General Scheme of Chemical Cycling** Biogeochemical cycles involve biotic and abiotic components. Each circuit has an abiotic reservoir through which the chemical cycles. Some chemical elements require "processing" by certain microorganisms before they are available to plants as inorganic nutrients.

- **Examples of Biogeochemical Cycles** A chemical's specific route through an ecosystem varies according to the trophic structure. Phosphorus is not very mobile and is cycled locally. Carbon, nitrogen, and water spend part of their time in gaseous form and are cycled globally.

Activity 19F The Carbon Cycle

Activity 19G The Nitrogen Cycle

Biomes

- **How Climate Affects Biome Distribution** The geographic distribution of terrestrial biomes is based mainly on regional variations in climate. Climate is largely determined by the uneven distribution of solar energy on Earth. Proximity to large bodies of water and the presence of landforms such as mountains also affect climate.

- **Terrestrial Biomes** If the climate in two geographically separate areas is similar, the same type of biome may occur in them. Most biomes are named for major physical or climatic features and for their predominant vegetation. The major terrestrial biomes include the tropical forest, savanna, desert, chaparral, temperate grassland, temperate deciduous forest, coniferous forest, and tundra.

Activity 19H Terrestrial Biomes

- **Freshwater Biomes** Freshwater biomes include lakes, ponds, rivers, streams, and wetlands. Lakes are often stratified vertically with regard to light penetration, temperature, nutrients, oxygen levels, and community structure. Rivers change greatly from their source to the point at which they empty into a lake or an ocean.

- **Marine Biomes** Estuaries, located where a freshwater river or stream merges with the ocean, are one of the most biologically productive environments on Earth. As in freshwater environments, marine life is distributed into distinct zones (intertidal, pelagic, and benthic) according to the depth of the water, degree of light penetration, distance from shore, and open water versus bottom. Hydrothermal vent communities are a marine deepwater biome powered by chemical energy from Earth's interior instead of by sunlight.

Activity 19I Aquatic Biomes

Self-Quiz

1. The concept of trophic structure of a community emphasizes the
   a. prevalent form of vegetation.
   b. keystone predator.
   c. feeding relationships within a community.
   d. species richness of the community.

2. Match each organism with its trophic level:
   a. alga   1. detritivore
   b. grasshopper   2. producer
   c. zooplankton   3. tertiary consumer
   d. eagle   4. secondary consumer
   e. fungi   5. primary consumer

3. According to the concept of competitive exclusion,
   a. two species cannot coexist in the same habitat.
   b. extinction or emigration is the only possible result of competitive interactions.
   c. intraspecific competition results in the success of the best-adapted individuals.
   d. two species cannot share the same niche in a community.

4. How can a keystone predator help maintain species diversity within a community?

5. Match the defense mechanism with the term that describes it.
   a. a harmless beetle that resembles a scorpion
   b. the bright markings of a poisonous tropical frog
   c. the mottled coloring of moths that rest on lichens
   d. two poisonous frogs that resemble each other in coloration

   1. camouflage coloration
   2. warning coloration
   3. Batesian mimicry
   4. Müllerian mimicry

6. Over a period of many years, grass grows on a sand dune, then shrubs grow, and then eventually trees grow. This is an example of ecological _____.

7. According to the energy pyramid model, why is eating grain-fed beef a relatively inefficient means of obtaining the energy trapped by photosynthesis?

8. Local conditions, such as heavy rainfall or the removal of plants, may limit the amount of nitrogen, phosphorus, or calcium available to a particular ecosystem, but the amount of carbon available to the system is seldom a problem. Why?

9. We are on a coastal hillside on a hot, dry summer day among evergreen shrubs that are adapted to fire. We are most likely standing in a _____ biome.

10. In *volume,* which biome is largest?

    a. deserts

    b. the intertidal zone

    c. the photic zone of the oceans

    d. the pelagic zone of the oceans

Answers to the Self-Quiz questions can be found in Appendix B.

Go to the website or CD-ROM for more Self-Quiz questions.

The Process of Science

1. An ecologist studying desert plants performed the following experiment. She staked out two identical plots that included a few sagebrush plants and numerous small annual wildflowers. She found the same five wildflower species in similar numbers in both plots. Then she enclosed one of the plots with a fence to keep out kangaroo rats, the most common herbivores in the area. After two years, four species of wildflowers were no longer present in the fenced plot, but one wildflower species had increased dramatically. The control plot had not changed significantly in species composition. Using the concepts discussed in the chapter, what do you think happened?

2. Imagine that you have been chosen as the biologist for the design team for a self-contained space station to be assembled in orbit. It will be stocked with organisms you choose to create, an ecosystem that will support you and five other people for two years. Describe the main functions you expect the organisms to perform. List the types of organisms you would select, and explain why you chose them.

Case Study in the Process of Science on the Web & CD *Try your hand at using species diversity to measure the effects of pollution on an environment.*

Case Study in the Process of Science on the Web & CD *Conduct a virtual experiment to determine how light affects primary productivity in an aquatic habitat.*

Biology and Society

1. Sometime in 1986, near Detroit, a freighter pumped out water ballast containing larvae of European zebra mussels. The mollusks multiplied wildly, spreading through Lake Erie and entering Lake Ontario. In some places, they have become so numerous that they have blocked the intake pipes of power plants and water treatment plants, fouled boat hulls, and sunk buoys. What makes this type of population explosion occur? What might happen to native organisms that suddenly must share the Great Lakes ecosystem with zebra mussels? Why is this a problem? How would you suggest trying to solve the mussel population problem?

2. Near Lawrence, Kansas, there was a rare patch of the original North American temperate grassland that had never been plowed. It was home to numerous native grasses, annual plants, and grassland animals. Among the species present were two endangered plants. Environmental activists thought the area should be set aside as a nature preserve, and they started to raise money to save the patch of land. In 1990, the owner of the land plowed it, stating that there are no federal laws protecting endangered plants on private grasslands and that he did not want to be told what he could do with his property. What issues and values are in conflict in this situation? How could this story have had a more satisfactory ending for all concerned? What would you have done if you were an environmental activist? If you were the farmer?

Biology and Society on the Web *Find out how old subway cars can be made into reefs.*

CHAPTER 20

Human Impact on the Environment

Biology and Society: Aquarium Menaces   443

Human Impact on Biological Communities   443

Human Disturbance of Communities

Introduced Species

Human Impact on Ecosystems   446

Impact on Chemical Cycles

Deforestation and Chemical Cycles: A Case Study

The Release of Toxic Chemicals to Ecosystems

Human Impact on the Atmosphere and Climate

The Biodiversity Crisis   452

The Three Levels of Biodiversity

The Loss of Species

The Three Main Causes of the Biodiversity Crisis

Why Biodiversity Matters

Conservation Biology   456

Biodiversity "Hot Spots"

Conservation at the Population and Species Levels

Conservation at the Ecosystem Level

The Goal of Sustainable Development

Evolution Connection: Biophilia and an Environmental Ethic   462

One-third of all plant and vertebrate species live on just 1.5% of Earth's land.

Every year, humans destroy an area of tropical rain forest equal to the size of West Virginia.

About 11% of all known bird species are endangered.

Purposefully introduced invasive species have cost the U.S. economy $130 billion.

Biology and Society

Aquarium Menaces

In May 2002, in a pond near Crofton, Maryland, a fisherman caught, photographed, and released an 18-inch exotic-looking fish. A month later, another fisherman caught and froze a 26-inch specimen, which was identified as *Channa argus,* the northern snakehead (**Figure 20.1a**). State officials found hundreds of juvenile snakeheads living in the pond.

The fish soon caught the attention of the press, which dubbed it "Frankenfish" because of its odd anatomy and physiology. It has a snake-like head with large sharp teeth and can grow up to 3 feet in length. In addition, it is able to gulp air and wiggle across land for several days (if it can stay moist). Native to eastern Asia, where it is prized as a delicacy, the snakehead poses a significant risk to ecosystems outside its native habitat because it voraciously eats smaller fishes, crustaceans, frogs, and insects, while having no predators to keep its numbers in check. By the end of the summer, the state successfully eradicated the snakehead—but only by killing all the fish in the pond (and in two neighboring ponds) with huge doses of poison.

How did an Asian fish get into a Maryland pond? Investigators discovered that a local man bought two snakeheads in New York, raised them as pets, and then released them after they outgrew his home aquarium. This situation is hardly unique. In 2000, an alga called *Caulerpa*—a Caribbean native bred as an aquarium "plant" that is particularly hearty and resistant to disease, predators, and herbicides—was found growing in a California lagoon, probably after someone dumped a home saltwater aquarium (**Figure 20.1b**). An earlier invasion of the Mediterranean Sea by this super seaweed is displacing many of the native algae there, and the same thing could happen now all along the Pacific coast of North America.

These stories of human-introduced species disrupting native ecosystems make two important points. First, humans have a disproportionately high impact on the environment. We therefore begin the chapter with an exploration of how human activities are putting communities, ecosystems, and biological diversity at risk. Second, these stories illustrate the continuing human fascination with the natural world (that is, after all, why we keep aquariums). Thus, we end the chapter (and the unit) with a discussion of conservation biology and the challenge facing us as individuals and a society to preserve biodiversity.

Biology and Society on the Web
Learn what the U.S. government has to say about introduced species.

(a)

(b)

Figure 20.1 Non-native species in your aquarium can cause significant damage to local ecosystems.
(a) The northern snakehead. (b) An aquarium-bred, hypervigorous variety of the alga *Caulerpa*. In this underwater photo, you can see the *Caulerpa* (foreground) crowding out the native eelgrass.

Human Impact on Biological Communities

In Chapter 19, you learned that natural disturbances that are intermediate in scale can be constructive in enhancing species diversity in a community. Unfortunately, human disturbance of biological communities is almost always destructive, reducing species diversity.

Figure 20.2 A large-scale human disturbance.
This half-mile-deep open-pit mine in Butte, Montana, is known as the Berkeley Pit. The mine has not operated since 1983, when its high-quality copper ore was used up. Water in the pit is highly acidic and loaded with heavy-metal ions. It threatens groundwater, soils, adjacent rivers, and communities for hundreds of miles downstream. The Berkeley Pit is one of the most troublesome cleanup sites in the United States, and the federal government has spent millions to initiate waste removal at this site.

Human Disturbance of Communities

Of all animals, humans have the greatest impact on communities worldwide (**Figure 20.2**). Logging and clearing for urban development, mining, and farming have reduced large tracts of forests to small patches of disconnected woodlots in many parts of the United States and throughout Europe. Similarly, agricultural development has disrupted what were once the vast grasslands of the North American prairie.

Much of the United States is now a hodgepodge of early successional growth where more mature communities once prevailed. After forests are clear-cut and abandoned, weedy and shrubby vegetation often colonizes the area and dominates it for many years. This type of vegetation is also found extensively in agricultural fields that are no longer under cultivation and in vacant lots and construction sites that are periodically cleared.

Human disturbance of communities is by no means limited to the United States and Europe; nor is it a recent problem. Tropical rain forests are quickly disappearing as a result of clear-cutting for lumber and pastureland. Centuries of overgrazing and agricultural disturbance have contributed to the current famine in parts of Africa by turning seasonal grasslands into great barren areas.

Human disturbance usually reduces species diversity in communities. We currently use about 60% of Earth's land in one way or another, mostly as cropland and rangeland. Most crops are grown in monocultures, intensive cultivations of a single plant variety over large areas. Even forests that are used to produce pulpwood and lumber are often replanted in single-species stands. And the effects of intensive grazing on rangelands often include the removal of several native plant species and replacement with only a few introduced species. Let's take a closer look at the problem of introduced species.

Introduced Species

Sometimes called exotic species, **introduced species** are those that humans move from the species' native locations to new geographic regions (as with the fish and alga discussed in the chapter-opening story). In some cases, the introductions are intentional. Australia's rabbit fiasco is an example (see Figure 19.15). In other cases, humans transplant species accidentally. For instance, the fungus that causes Dutch elm disease stowed away with logs shipped from Europe to the United States after World War I (see Figure 16.22). Whether purposeful or unintentional, introduced species that gain a foothold usually disrupt their adopted community, often by preying on native organisms or outcompeting native species that use some of the same resources.

Throughout human history, moving plants and animals from place to place was considered "improving on nature." This was especially true of domesticated species. As Europeans began colonizing the Americas in the fifteenth and sixteenth centuries, plants and animals that provide food were transported back and forth across the Atlantic in what we now call the "Columbian Exchange." Corn, potatoes, tomatoes, beans, and certain kinds of squash are just a few examples of New World species that became European staples.

Humans have also introduced many nonfood species with the best of intentions. For example, the U.S. Department of Agriculture encouraged the import of a Japanese plant called kudzu to the American South in the 1930s to help control erosion, especially along irrigation canals. At first, the government paid farmers to plant kudzu vines. The enthusiasm for the new vines led to kudzu festivals in southern towns, complete with the crowning of kudzu queens. But kudzu celebrations ended decades ago as the invasive

(a) Kudzu

(b) European starlings

plant took over vast expanses of the southern landscape (**Figure 20.3a**). Another introduced plant called purple loosestrife is claiming over 200,000 acres of wetlands per year, crowding out native plants and the animals that feed on the native flora. The story is similar for the introduction to the United States of a bird called the European starling. A citizens group intent on introducing all plants and animals mentioned in Shakespeare's plays imported 120 starlings to New York's Central Park in 1890 (the starling is mentioned in just one line of Shakespeare's *Henry IV*). From that foothold, starlings spread rapidly across North America. In less than a century, the population increased to about 100 million, displacing many of the native songbird species in the United States and Canada (**Figure 20.3b**).

The ease of travel by ships and airplanes has accelerated the transplant of species, especially unintentional introductions. For example, fire ants, which can inflict very painful beelike stings, reached the southeastern United States in the early 1900s from South America, probably in the hold of a produce ship. Fire ants have been extending their range northward and westward ever since. In Texas, for example, fire ants have apparently managed to eliminate about two-thirds of the native ant species. Another accidentally introduced ant species, the Argentine ant, is decimating populations of native ants in California (**Figure 20.3c**). Still another recent troublesome invader in the United States is the zebra mussel, a mollusk that entered the St. Lawrence Seaway in the mid-1980s in ballast water released by a cargo ship that had traveled from the mussels' native Caspian Sea (**Figure 20.3d**). By the summer of 1993, zebra mussels were found in the Mississippi River as far south as New Orleans. Of the many problems caused by the mussels, those that have provoked the most uproar have resulted from their competition with humans—by clogging reservoir intake pipes, for example. However, zebra mussels also compete with native shellfish for space and with fish for the plankton used for nourishment. The full extent to which this new competitor is altering community structure has yet to be determined.

All told, the United States has at least 50,000 introduced species, with a cost to the economy of over $130 billion in damage and control efforts. And that does not include the priceless loss of native species. Fortunately, most introduced species fail to thrive in their new homes. But for those exotic species that succeed, elimination of native species is a common consequence. In fact, introduced species rank second only to habitat destruction as a cause of extinctions and loss of Earth's biodiversity. Why should some introduced species be able to outcompete native organisms? In many cases,

Activity 20A on the Web & CD
Learn more about fire ants, an introduced species.

(c) Argentine ants

(d) Zebra mussels

Figure 20.3 A few of America's 50,000 introduced species. (a) Kudzu. (b) European starlings. (c) Argentine ants ganging up on a red ant native to California. (d) Zebra mussels. (The shopping cart, covered with the mussels, was retrieved from a lake.)

the answer is that there are relatively few pathogens, parasites, and predators to hold population explosions of the introduced species in check once the globetrotters arrive. In contrast, the native species of a biological community fit into a network of interactions shaped by the adaptive evolution of multiple species.

Human Impact on Ecosystems

Human population growth plus technology add up to a badly bruised biosphere. We have managed to intrude in one way or another into the dynamics of most ecosystems. Even where we have not completely destroyed a natural system, our actions have disrupted the trophic structure, energy flow, and chemical cycling of ecosystems in most areas of the world. The effects are sometimes local or regional, but the ecological impact of humans can be far-reaching or even global. For example, acid precipitation may be carried by prevailing winds and fall as rain hundreds or thousands of miles from the smokestacks emitting the chemicals that produce it (see Chapter 2). And the carbon dioxide (CO_2) exhaust of our machinery is probably causing a global warming that will affect all life on Earth, including humans (**Figure 20.4**). In this section, we'll examine a few of our ecological footprints on both local ecosystems and the biosphere, the global ecosystem.

Impact on Chemical Cycles

Human activities often intrude in biogeochemical cycles by removing nutrients from one location and adding them to another. This may result in the depletion of key nutrients in one area, excesses in another place, and the disruption of the natural chemical cycling in both locations. For example, nutrients in the soil of croplands soon appear in the wastes of humans and livestock and then appear in streams and lakes through runoff from stockyards and discharge as sewage. Someone eating a salad in Washington, D.C., is consuming nutrients that only days before might have been in the soil in California. And a short time later, some of these nutrients will be in the Potomac River on their way to the sea, having passed through an individual's digestive system and the local sewage facilities.

Humans have altered chemical cycles to such an extent that it is no longer possible to understand any cycle without taking the human impact into account. Let's consider just a few examples.

Impact on the Carbon Cycle The increased burning of fossil fuels (coal and petroleum) as well as wood from deforested areas is steadily raising the level of CO_2 in the atmosphere. This is leading to significant environmental problems, such as global warming (discussed later).

Figure 20.4 Carbon dioxide producers.
Burning fossil fuels, such as gasoline, natural gas, and coal, produces CO_2. The rising level of carbon dioxide in the atmosphere is probably contributing to an increase in the average global temperature.

Impact on the Nitrogen Cycle Sewage treatment facilities typically empty large amounts of dissolved inorganic nitrogen compounds into rivers and streams. Farmers routinely apply large amounts of inorganic nitrogen fertilizers, mainly ammonium and nitrates, to croplands. Lawns and golf courses also receive sizable doses of fertilizer. Crop and lawn plants take up some of the nitrogen compounds, and soil bacteria convert some into atmospheric nitrogen (N_2) (see Figure 19.29b). However, chemical fertilizers usually exceed the soil's natural recycling capacity. The excess nitrogen compounds often enter streams, lakes, and groundwater. In lakes and streams, these nitrogen compounds continue to fertilize, causing heavy growth of algae. Groundwater pollution by nitrogen fertilizers is also a serious problem in many agricultural areas. In the human digestive tract, nitrates in drinking water are converted to nitrites, which can be toxic.

Activity 20B on the Web & CD Explore the problem of water pollution from nitrates.

Impact on the Phosphorus Cycle Like nitrogen compounds, phosphates are a major component of sewage outflow. They are also used extensively in agricultural fertilizers and are a common ingredient in pesticides. Phosphate pollution of lakes and rivers, like nitrate pollution, stimulates a heavy algal growth. This leads to population explosions of detritivores that can eventually suffocate all the aerobic life, including the fish, in a lake or pond. You can see the results of such "overfertilization," or **eutrophication,** of lakes in Figure 18.3.

Impact on the Water Cycle One of the main sources of atmospheric water is transpiration (evaporation) from the dense vegetation of tropical rain forests. The destruction of these forests, which is occurring rapidly today, will change the amount of water vapor in the air (**Figure 20.5**). This, in turn, will probably alter local, and perhaps global, weather patterns. Another change in the water cycle caused by humans results from pumping large amounts of groundwater to the surface to use for crop irrigation. This practice can increase the rate of evaporation from soil, and unless this loss is balanced by increased rainfall over land, groundwater supplies are depleted. Large areas in the midwestern United States, the southwestern American desert, parts of California, and areas bordering the Gulf of Mexico currently face this problem.

Figure 20.5 Deforestation and the water cycle. Destruction of tropical forests, such as this one in Belize, reduces return of water to the atmosphere via transpiration.

Deforestation and Chemical Cycles: A Case Study

Since 1963, a team of scientists has been conducting a long-term study of chemical cycling in a forest ecosystem both under natural conditions and after vegetation is removed. The study site is the Hubbard Brook Experimental Forest in the White Mountains of New Hampshire. It is a deciduous forest with several valleys, each drained by a small creek that is a tributary of Hubbard Brook. Bedrock impenetrable to water is close to the surface of the soil, and each valley constitutes a watershed that can drain only through its creek.

The research team first determined the mineral budget for each of six valleys by measuring the inflow and outflow of several key nutrients. They collected rainfall at several sites to measure the amount of water and dissolved minerals added to the ecosystem. To monitor the loss of water and minerals, the scientists constructed small concrete dams, each with a V-shaped spillway,

(a) A dam at the Hubbard Brook study site

(b) Logged watersheds in the Hubbard Brook Forest

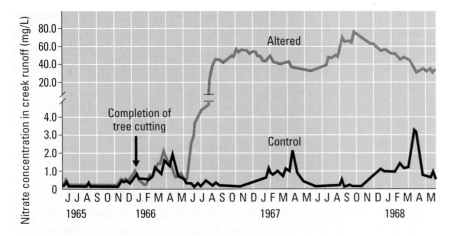

(c) The loss of nitrate from a deforested watershed

Figure 20.6 Chemical cycling in an experimental forest: the Hubbard Brook study. (a) A dam at a water-sampling station in the Hubbard Brook Forest. (b) Logged ("experimental group") and unlogged ("control group") watersheds in the Hubbard Brook Forest. (c) The loss of nitrate from a deforested watershed. Nitrate levels in the runoff from the deforested area began to rise markedly about five months after the plants were killed. Within eight months, the nitrate loss was about 60 times greater in the altered watershed than in the control area. Without plants to take up and hold nitrate, this mineral nutrient drained out of the ecosystem.

across the creek at the bottom of each valley (**Figure 20.6a**). About 60% of the water added to the ecosystem as rainfall and snow exits through the streams, and the remaining 40% is lost by transpiration from plants and evaporation from the soil.

Measurements confirmed that local cycling within a terrestrial ecosystem conserves most of the mineral nutrients. Mineral inflow and outflow balanced and were relatively small compared with the quantity of minerals being recycled within the forest ecosystem.

In 1966, one of the valleys, with an area of 15.6 hectares, was completely logged and then sprayed with herbicides for three years to prevent regrowth of plants (**Figure 20.6b**). All the original plant material was left in place to decompose. The inflow and outflow of water and minerals in the experimentally altered watershed were compared with those in a control watershed for three years. Water runoff from the altered watershed increased by 30–40%, apparently because there were no plants to absorb and transpire water from the soil. Net losses of minerals from the altered watershed were huge. Most remarkable was the loss of nitrate, which increased in concentration in the creek 60-fold (**Figure 20.6c**). Not only was this vital mineral nutrient drained from the ecosystem, but nitrate in the creek reached a level considered unsafe for drinking water.

The Hubbard Brook research demonstrated that the amount of nutrients leaving an intact forest ecosystem is controlled mainly by plants. Nutrients were lost from the system when plants were not present to retain them. These effects were almost immediate, occurring within a few months of logging, and continuing as long as plants were absent. However, as new plants grew in the treatment area, transpiration increased and runoff decreased. While scientists designed the Hubbard Brook experiments to assess natural ecosystem dynamics, the results also provided important insights into the mechanisms by which human activities such as deforestation affect these processes.

The Release of Toxic Chemicals to Ecosystems

In addition to transporting vital chemical elements from one location to another, we have added entirely new materials, many of them toxic, to ecosystems. Humans produce an immense variety of these toxic chemicals, including thousands of synthetics previously unknown in nature. Many of these poisons cannot be degraded by microorganisms and consequently persist in the environment for years or even decades. In other cases, chemicals released into the environment may be relatively harmless but are converted to more toxic products by reaction with other substances or by the metabolism of microorganisms. For example, mercury, a by-product of plastic production, was once routinely expelled into rivers and the sea in an insoluble form. Bacteria in the bottom mud converted the waste to methyl mercury, an extremely toxic soluble compound that then accumulated in the tissues of organisms, including humans who consumed fish from the contaminated waters.

Organisms acquire toxic substances from the environment along with nutrients and water. Some of the poisons are metabolized or excreted, but others accumulate in specific tissues, especially fat. Examples of industrially synthesized compounds that act in this manner are the chlorinated hydrocarbons (which include many pesticides, such as DDT) as well as the industrial chemicals called PCBs (polychlorinated biphenols).

One of the reasons the toxins we add to ecosystems are such ecological disasters is that they become more concentrated in successive trophic levels of a food web, a process called **biological magnification.** Magnification occurs because the biomass at any given trophic level is accumulated from a much larger toxin-containing biomass ingested from the level below (see Figure 19.26 to review energy pyramids). Thus, top-level carnivores are usually the organisms most severely damaged by toxic compounds that have been released into the environment.

A classic example of biological magnification involves DDT, the poisonous pollutant that Rachel Carson warned about more than 40 years ago (see Chapter 18). DDT was used to control insects such as mosquitoes and agricultural pests. But DDT persisted in the environment and was transported by water to areas far from where it was sprayed, rapidly becoming a global menace. Because the compound is soluble in lipids, it collects in the fatty tissues of animals, and its concentration is magnified in higher trophic levels (**Figure 20.7**). Traces of DDT have been found in nearly every organism tested; it has even been detected in human breast milk throughout the world. One of the first signs that DDT was a serious environmental problem was a decline in the populations of pelicans, ospreys, and eagles, birds that feed at the tops of food chains. DDT was banned in the United States in 1971, and a dramatic recovery in populations of the affected bird species followed. The pesticide is still used in many developing countries, however. And in the United States, we have replaced DDT with other pesticides that may also be biologically magnified in the food chains of ecosystems. For example, in 1999, several areas of New York City were sprayed with insecticides called pyrethroids. The spraying was a precaution against a pathogen called the West Nile virus, which is carried by mosquitoes. Within months, there was a massive die-off of lobsters in Long Island Sound, and there is evidence that the insecticides became magnified in these commercially important animals. Scientists hypothesize that the poisons reached the ocean in rainstorm runoff from the city.

Activity 20C on the Web & CD
Analyze the pros and cons of DDT.

Human Impact on the Atmosphere and Climate

We are causing radical changes in the composition of the atmosphere and, consequently, in the global climate. Our activities release a variety of gaseous waste products. We once thought that the vastness of the atmosphere could absorb these materials without significant consequences, but an astronaut's view of our little planet squashes such naive notions (**Figure 20.8**). One pressing problem that relates directly to one of the chemical cycles we examined is the rising level of carbon dioxide in the atmosphere.

Carbon Dioxide Emissions, the Greenhouse Effect, and Global Warming Since the Industrial Revolution, the concentration of CO_2 in the atmosphere has been increasing as a result of the combustion of fossil fuels and the burning of enormous quantities of wood removed by deforestation. Various methods have estimated that the average CO_2 concentration in the atmosphere before 1850 was about 274 parts per million (ppm). When a monitoring station on Hawaii's Mauna Loa peak began making

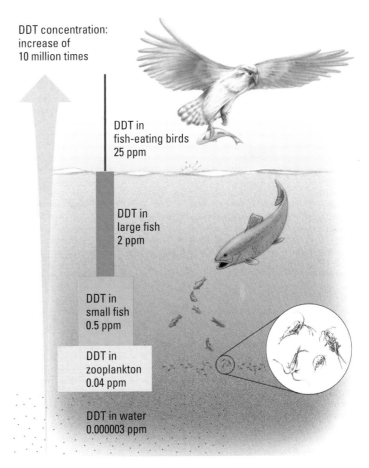

DDT concentration: increase of 10 million times

DDT in fish-eating birds 25 ppm

DDT in large fish 2 ppm

DDT in small fish 0.5 ppm

DDT in zooplankton 0.04 ppm

DDT in water 0.000003 ppm

Figure 20.7 Biological magnification of DDT in a food chain. DDT concentration in a Long Island Sound food chain was magnified by a factor of 10 million, from 0.000003 ppm in the water to 25 ppm in fish-eating birds. (The abbreviation "ppm" stands for "parts per million," a unit of measurement commonly used for toxins.)

Figure 20.8 An earthrise photographed from the moon. Apollo astronaut Rusty Schweickart, reflecting on such a view, once remarked: "On that small blue-and-white planet below is everything that means anything to you. National boundaries and human artifacts no longer seem real. Only the biosphere, whole and home of life."

Figure 20.9 **Increase in atmospheric CO₂ in the past four decades.**

Activity 20D on the Web & CD Conduct virtual experiments to change carbon dioxide levels and see the effects on global temperatures.

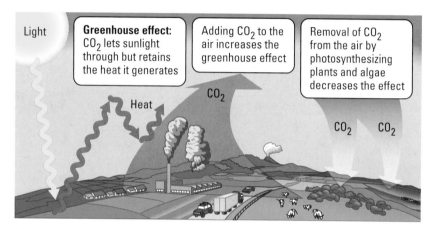

Greenhouse effect: CO₂ lets sunlight through but retains the heat it generates

Adding CO₂ to the air increases the greenhouse effect

Removal of CO₂ from the air by photosynthesizing plants and algae decreases the effect

Light

Heat

CO₂

CO₂   CO₂

Figure 20.10 **Factors influencing the greenhouse effect.**

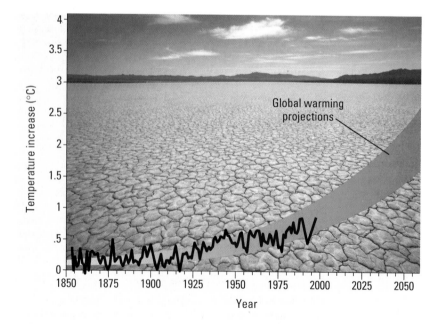

Global warming projections

Figure 20.11 **Average atmospheric temperatures and projections for global warming.**

very accurate measurements in 1958, the CO_2 concentration was 316 ppm (**Figure 20.9**). Today, the concentration of CO_2 in the atmosphere exceeds 360 ppm, an increase of about 14% since the measurements began just over 40 years ago. If CO_2 emissions continue to increase at the present rate, by the year 2075 the atmospheric concentration of this gas will be double what it was at the start of the Industrial Revolution. It is difficult to predict the multiple ways this intrusion in the carbon cycle will affect the biosphere and its various ecosystems.

One factor that complicates predictions about the long-term effects of rising atmospheric CO_2 concentration is its possible influence on Earth's heat budget. Much of the solar radiation that strikes the planet is reflected back into space. Although CO_2 and water vapor in the atmosphere are transparent to visible light, they intercept and absorb much of the reflected heat radiation, bouncing it back toward Earth. This process, called the **greenhouse effect,** retains some of the solar heat. **Figure 20.10** adds some details to the illustration of the greenhouse effect you saw in Chapter 7.

The marked increase in atmospheric CO_2 concentrations during the last 150 years concerns ecologists because of its potential effect on global temperature through the greenhouse effect. A number of studies predict that a doubling of CO_2 concentration by the end of the twenty-first century will cause an average global temperature increase of about 2°C (**Figure 20.11**). An increase of only 1.3°C would make the world warmer than at any time in the past 100,000 years. A worst-case scenario suggests that the warming would be greatest near the poles. Melting of polar ice might raise sea level by an estimated 100 m, gradually flooding coastal areas 150 km (or more) inland from the current coastline. New York, Miami, Los Angeles, and many other cities would then be underwater. A warming trend would also alter the geographic distribution of precipitation, making major agricultural areas of the central United States much drier. Most of Earth's natural ecosystems would also be affected, with boundaries between systems such as forests and grasslands shifting. However, the various mathematical models disagree about the details of how warming on a global level will change the climate in each region.

By studying how prehistoric periods of global warming and cooling affected plant communities, ecologists are using another strategy to help predict the consequences of future temperature changes. Records from fossilized pollen provide evidence that plant communities are altered dramatically by climate change. However, past climate changes occurred gradually, and plant and animal populations could spread into areas where conditions allowed them to survive. A major concern about the global warming under way now is that it is so rapid that many species may not be able to survive.

Relatively few scientists still doubt the evidence that human activity is heating up our planet. But what can be done to lessen the chances of greenhouse disaster? The burning of trees after deforestation to clear land in the tropics accounts for about 20% of the excess CO_2 released into the atmosphere. The burning of fossil fuels is the cause of the other 80%. International cooperation and national and individual action are needed to decrease fossil-fuel consumption and to reduce the destruction of forests. Many nations have called for efforts to cap the level of CO_2 output from fossil-fuel combustion. Because fossil fuels currently power much of our industry and economic growth, meeting the terms of any international agreement will require strong individual commitment and acceptance of some major lifestyle changes.

Those of us in developed countries have the greatest responsibility to reduce energy consumption. People in the United States consume more energy than do the total populations of Central and South America, Africa, India, and China combined—about 3.7 billion people compared to 270 million in the United States. Thus, the moderation of global warming depends mainly on the richest countries reducing their use of fossil fuels by conserving energy and by developing alternative energy sources, such as wind, solar, and geothermal energy. We can help individually by becoming more energy efficient at home and reducing our reliance on the automobile.

Depletion of Atmospheric Ozone Life on Earth is protected from the damaging effects of ultraviolet (UV) radiation by a very thin protective layer of ozone molecules (O_3) located in the atmosphere between 17 and 25 km above Earth's surface. This **ozone layer** absorbs UV radiation, preventing much of it from contacting organisms in the biosphere. Measurements by atmospheric scientists document that the ozone layer has been gradually thinning since 1975.

The destruction of atmospheric ozone probably results mainly from the accumulation of chlorofluorocarbons, chemicals used in refrigeration, as propellants in aerosol cans, and in certain manufacturing processes. When the breakdown products from these chemicals rise in the atmosphere, the chlorine they contain reacts with ozone, converting it to O_2. Subsequent chemical reactions liberate the chlorine, allowing it to react with other ozone molecules in a catalytic chain reaction. The effect is most apparent over Antarctica, where cold winter temperatures facilitate these atmospheric reactions (**Figure 20.12**). Scientists first described the "ozone hole" over Antarctica in 1985. Since then, the size of the ozone hole has increased, sometimes extending as far as the southernmost portions of Australia, New Zealand, and South America. And at the more heavily populated middle latitudes, ozone levels have decreased 2–10% during the past 20 years.

The consequences of ozone depletion may be quite severe for all life on Earth, including humans. Some scientists expect the growing intensity of UV radiation to increase the incidence of skin cancer and cataracts among humans. It is likely that there will also be damaging effects on crops and natural communities, especially the phytoplankton that are responsible for a large proportion of the biosphere's primary productivity. The danger posed by ozone depletion is so great that many nations agreed in 1987 to end the production of chlorofluorocarbons by the year 2010. (The United States and other developed nations have already substituted safer compounds for chlorofluorocarbons, but a grace period was allowed for developing countries.) Unfortunately, even if all chlorofluorocarbons were banned today, the chlorine molecules already in the atmosphere will continue to influence atmospheric ozone levels for at least a century. It is just one more example of how far our technological tentacles reach in disrupting the dynamics of ecosystems and the entire biosphere.

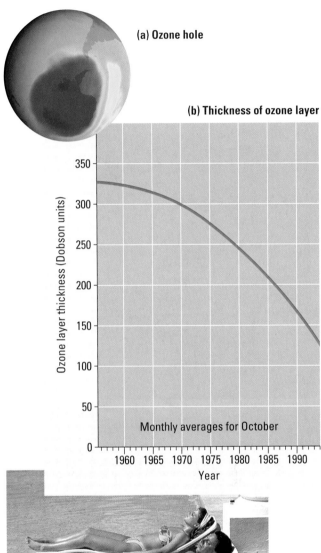

(a) Ozone hole

(b) Thickness of ozone layer

Monthly averages for October

y-axis: Ozone layer thickness (Dobson units) — 0, 50, 100, 150, 200, 250, 300, 350

x-axis: Year — 1960, 1965, 1970, 1975, 1980, 1985, 1990

(c) Exposure to UV radiation

Figure 20.12 Erosion of Earth's ozone shield.
(a) The ozone hole over the Antarctic is visible as the blue patch in this image based on atmospheric data. **(b)** This graph tracks the thickness of the ozone layer in units called Dobsons. **(c)** Australia, home to these sun lovers, already has the world's highest rate of skin cancer. Increased UV radiation through an eroded ozone shield won't help.

CHECKPOINT

1. How can clear-cutting a forest (removing all trees) damage the water quality of nearby lakes?
2. How does excessive addition of mineral nutrients to a lake eventually result in the loss of most fish in the lake?
3. How is biological magnification relevant to the health of most humans in developed countries?
4. Here's a question that requires interpreting data and graphs. From the data in Figure 20.9, what was the approximate percentage increase in atmospheric CO_2 between 1960 and 1990?

Answers: 1. Without the growing trees to assimilate minerals from the soil, more of the minerals run off and end up polluting water resources. **2.** The eutrophication (overfertilization) initially causes population explosions of algae and the organisms that feed on them. The respiration of so much life, including the detritivores working on all the organic refuse, consumes most of the lake's oxygen, which the fish require. **3.** People in developed countries generally eat more meat than do people in developing countries. As secondary or tertiary consumers in a food chain, meat-eaters acquire a greater dose of some toxic chemicals than if they fed exclusively on plants as primary consumers. **4.** About 12%, from 315 ppm to about 352 ppm; $352 - 315 = 37$, and $37/315 \times 100 = 11.7\%$

The Biodiversity Crisis

Now that we've surveyed the ways that humans can affect communities and ecosystems, we can understand one of the primary consequences: an alarming **biodiversity crisis,** a precipitous decline in Earth's great variety of life.

The Three Levels of Biodiversity

Biodiversity, short for biological diversity, has three main components. The first is the diversity of ecosystems. Each ecosystem, be it a rain forest, desert, or coral reef, has a unique biological community and characteristic patterns of energy flow and chemical cycling. And each ecosystem has a unique impact on the entire biosphere. For example, the productive "pastures" of phytoplankton in the oceans help moderate the greenhouse effect by consuming massive quantities of atmospheric CO_2 for photosynthesis and shell building (many microscopic protists in plankton secrete shells of bicarbonate, a derivative of CO_2). Some ecosystems are being erased from the biosphere at an astonishing rate. For example, the cumulative area of all tropical rain forests is only about the size of the 48 contiguous United States, and we lose an area equal to the state of West Virginia each year.

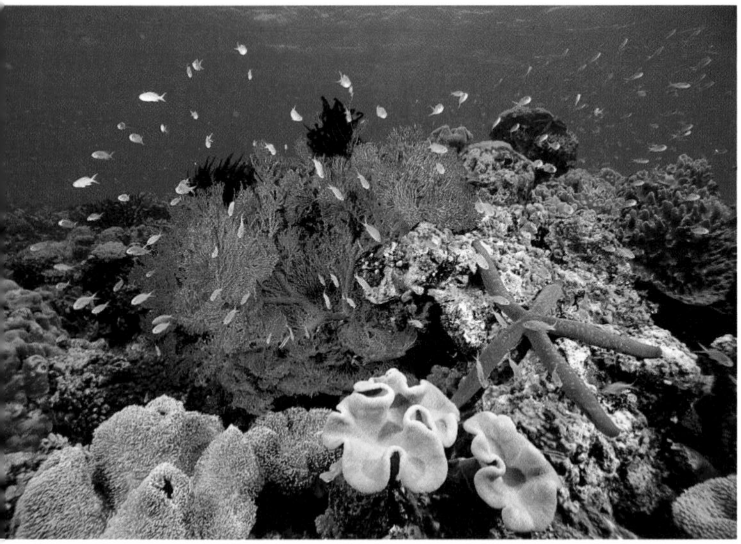

Figure 20.13 A coral reef is a showcase of biodiversity.

The second component of biodiversity is the variety of species that make up the biological community of any ecosystem (**Figure 20.13**). And the third component is the genetic variation within each species. You learned in Chapter 13 that the loss of genetic diversity—by a severe reduction in population size, for example—can hasten the demise of a dwindling species.

Though human impact reaches all three levels of biodiversity, most of the research focus so far has been on species extinction.

The Loss of Species

The seventh mass extinction in the history of life is well under way. Previous episodes, including the Cretaceous crunch that claimed the dinosaurs and many other groups, pale by comparison. The current mass extinction is both broader and faster, extinguishing species at a rate at least 50 times faster than just a few centuries ago. And unlike past poundings of biodiversity, which were triggered mainly by physical processes, such as climate change caused by volcanism or asteroid crashes, this latest mass extinction is due to the evolution of a single species—a big-brained, manually dexterous, environment-manipulating toolmaker that has named itself *Homo sapiens*.

We do not know the full scale of the biodiversity crisis in terms of a species "body count," for we are undoubtedly losing species that we didn't even know existed; the 1.7 million species that have been identified probably represent less than 10% of the true number of species. However, there are already enough signs to know that the biosphere is in deep trouble:

- About 11% of the 9,040 known bird species in the world are endangered. In the past 40 years, population densities of migratory songbirds in the mid-Atlantic United States dropped 50%.

- Of the approximately 20,000 known plant species in the United States, over 600 are very close to extinction.

- Throughout the world, 970 tree species have been classified as critically endangered. At least 5 of those species are down to fewer than a half dozen surviving individuals.

- About 20% of the known freshwater fishes in the world have either become extinct during historical times or are seriously threatened. The toll on amphibians and reptiles has been almost as great.

- Harvard biologist Edward O. Wilson, a renowned scholar of biodiversity, has compiled what he grimly calls the Hundred Heartbeat Club. The species that belong are those animals that number fewer than 100 individuals and so are only that many heartbeats away from extinction (**Figure 20.14**).

- Several researchers estimate that at the current rate of destruction, over half of all plant and animal species will be gone by the end of this new century.

The modern mass extinction is different from earlier biodiversity shakeouts in still another important way. The prehistoric crashes were all followed by rebounds in diversity as the survivors radiated and adapted to ecological niches left vacant by the extinctions. But as long as we humans are around to destroy habitats and degrade biodiversity at the ecosystem level, there can be no rebound in the evolutionary diversification of life. In fact, the trend is toward increased geographic range and prevalence of "disaster species," those life-forms such as house mice, kudzu and other weeds, cockroaches, and fire ants that seem to thrive in environments disrupted by human activities. Unless we can reverse the current trend of increasing loss of biodiversity, we will leave our children and grandchildren a biosphere that is much less interesting and much more biologically impoverished.

(a) Philippine eagle

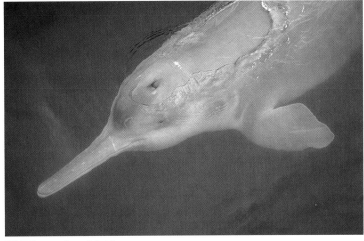

(b) Chinese river dolphin

(c) Javan rhinoceros

Figure 20.14 A hundred heartbeats from extinction.
These are just three of the many members of what E. O. Wilson calls the Hundred Heartbeat Club, species with fewer than 100 individuals remaining on Earth.

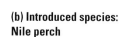

The Three Main Causes of the Biodiversity Crisis

Habitat Destruction Human alteration of habitats poses the single greatest threat to biodiversity throughout the biosphere (**Figure 20.15a**). Assaults on diversity at the ecosystem level result from the expansion of agriculture to feed the burgeoning human population, urban development, forestry, mining, and environmental pollution. The amount of human-altered land surface is approaching 50%, and we use over half of all accessible surface fresh water. Some of the most productive aquatic habitats in estuaries and intertidal wetlands are also prime locations for commercial and residential developments. The loss of marine habitats is also severe, especially in coastal areas and coral reefs.

(a) Habitat destruction: clearing a rain forest

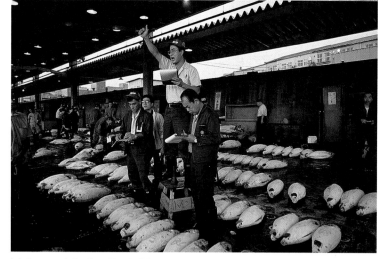

(b) Introduced species: Nile perch

Introduced Species Ranking second behind habitat loss as a cause of the biodiversity crisis is human introduction of exotic (non-native) species that eliminate native species through predation or competition. For example, if your campus is in an urban setting, there is a good chance that the birds you see most often as you walk between classes are starlings, rock doves (often called "pigeons"), and house sparrows—all introduced species that have replaced native birds in many areas of North America. One of the largest rapid-extinction events yet recorded is the loss of freshwater fishes in Lake Victoria in East Africa. About 200 of the 300 species of native fishes, found nowhere else but in this lake, have become extinct since Europeans introduced a non-native predator, the Nile perch, in the 1960s (**Figure 20.15b**).

Overexploitation As a third major threat to biodiversity, overexploitation of wildlife often compounds problems of shrinking habitat and introduced species. Animal species whose numbers have been drastically reduced by excessive commercial harvest or sport hunting include whales, the American bison, Galápagos tortoises, and numerous fishes. Many fish stocks in the ocean have been overfished to levels that cannot sustain further human exploitation (**Figure 20.15c**). In addition to the commercially important species, members of many other species are often killed by harvesting methods; for example, dolphins, marine turtles, and seabirds are caught in fishing nets, and countless numbers of invertebrates are killed by marine trawls (big nets). An expanding, often illegal world trade in wildlife products, including rhinocerous horns, elephant tusks, and grizzly bear gallbladders, also threatens many species.

(c) Overexploitation: North Atlantic bluefin tuna being auctioned in a Japanese fish market

Figure 20.15 The three main causes of the biodiversity crisis.
(a) Habitat destruction. This is an all-too-common scene in the tropics: the clearing of a rain forest for lumber, agriculture, housing projects, or roads.
(b) Introduced species. One of the largest freshwater fishes (up to 2 m long and weighing up to 450 kg), the Nile perch was introduced to Lake Victoria in East Africa to provide high-protein food for the growing human population. Unfortunately, the perch's main effect has been to wipe out about 200 smaller native species, reducing its own food supply to a critical level. **(c) Overexploitation.** Until the past few decades, the North Atlantic bluefin tuna was considered a sport fish of little commercial value—just a few cents per pound as cat food. Then, beginning in the 1980s, wholesalers began airfreighting fresh, iced bluefin to Japan for sushi and sashimi. In that market, the fish now brings up to $100 per pound! With that kind of demand, the results are predictable. It took just ten years to reduce the North Atlantic bluefin population to less than 20% of its 1980 size. In spite of quotas, the high price that bluefin tuna brings probably dooms the species to extinction.

Why Biodiversity Matters

Why should we care about the loss of biodiversity? First of all, we depend on many other species for food, clothing, shelter, oxygen, soil fertility—the list goes on and on. In the United States, 25% of all prescriptions dispensed from pharmacies contain substances derived from plants. For instance, two drugs effective against Hodgkin disease and certain other forms of cancer come from the rosy periwinkle, a flowering plant native to the island of Madagascar (**Figure 20.16**). Madagascar alone harbors some 8,000 species of flowering plants, 80% of which occur only there. Among these unique plants are several species of wild coffee trees, some of which yield beans lacking caffeine (naturally "decaffeinated"). With an estimated 200,000 species of plants and animals, Madagascar is among the top five most biologically diverse countries in the world. Unfortunately, most of Madagascar's species are in serious trouble. People have lived on the island for only about 2,000 years, but in that time, Madagascar has lost 80% of its forests and about 50% of its native species. Madagascar's dilemma represents that of much of the developing world. The island is home to over 10 million people, most of whom are desperately poor and hardly in a position to be concerned with environmental conservation. Yet the people of Madagascar as well as others around the globe could derive vital benefits from the biodiversity that is being destroyed.

> **Activity 20E on the Web & CD**
> Learn more about the biodiversity crisis in Madagascar.

Who knows which of the species disappearing from the biosphere could provide new sources of food or medicine? And who understands the dynamics of ecosystems well enough to know which species are least essential to biological communities? The great American naturalist Aldo Leopold explained it this way: "If the biota, in the course of aeons, has built something we like but do not understand, then who but a fool would discard seemingly useless parts? To keep every cog and wheel is the first precaution of intelligent tinkering."

Another reason to be concerned about the changes that underlie the biodiversity crisis is that the human population itself is threatened by large-scale alterations in the biosphere. Like all other species, we evolved in Earth's ecosystems, and we are dependent on the living and nonliving components of these systems. By allowing the extinction of species and the degradation of habitats to continue, we are taking a risk with our own species' survival.

In an attempt to counter what they see as a tendency of policymakers and governments to undervalue life-sustaining features of the biosphere, a team of ecologists and economists recently estimated the cost of losing ecosystem "services." For example, they estimated part of the value of a wetland from the cost of flood damage that occurred because of the loss of the wetland's ability to hold floodwater. For a single year in the late 1990s, the scientists estimated the average annual value of ecosystem dynamics in the biosphere at 33 trillion U.S. dollars. In contrast, the global gross national product for the same year was 18 trillion U.S. dollars. Although rough, these estimates help make the important point that we cannot afford to continue to take ecosystems for granted.

Figure 20.16 Madagascar's rosy periwinkle, a source of anticancer drugs.

CHECKPOINT

1. What are the three main levels of biodiversity?

2. What are the three main causes of the biodiversity crisis?

Answers: 1. Ecosystem diversity, species diversity, and the genetic diversity of populations
2. Habitat destruction, introduced species, and overexploitation

---

Conservation Biology

Conservation biology is a goal-oriented science that seeks to counter the loss of biodiversity. It began to take form in the late 1970s as an interdisciplinary collaboration of scientists working toward the long-term maintenance of functional ecosystems and a reduction of the rate of species extinction. Promoting research on biodiversity and the means to save it, an international group of scientists and educators founded the Society for Conservation Biology in 1985. Today, with about 5,000 members, the society helps unite the conservation efforts of biologists, anthropologists, sociologists, economists, government and industry officials, and private citizens. Conservation biologists recognize that biodiversity can be sustained only if the evolutionary mechanisms that have given rise to species and communities of organisms continue to operate. Thus, the goal is not simply to preserve individual species but to sustain ecosystems, where natural selection can continue to function, and to maintain the genetic variability on which natural selection acts. The front lines for conservation biology are geographic areas that are especially rich in endangered biodiversity.

Biodiversity "Hot Spots"

A **biodiversity hot spot** is a relatively small area with an exceptional concentration of species (**Figure 20.17**). Many of the organisms in biodiversity hot spots are **endemic species,** meaning they are found nowhere else. For example, nearly 30% of all bird species are confined to only about 2% of Earth's land area. And about 50,000 plant species, or 20% of all known plant species, inhabit 18 hot spots making up only about 0.5% of the global land surface. Overall, the "hottest" of the biodiversity hot spots, including rain

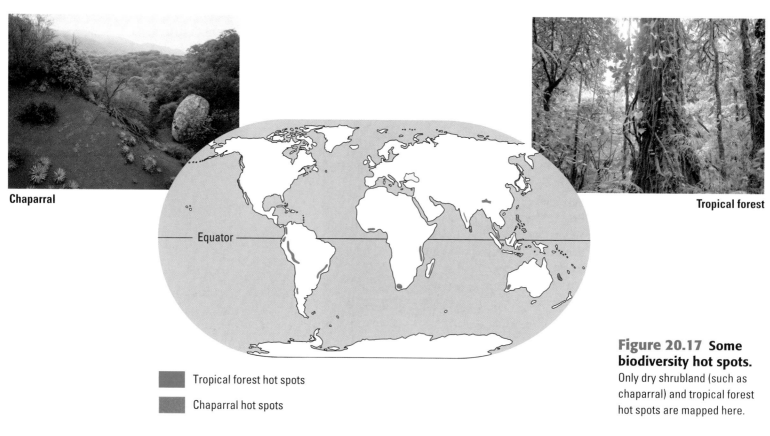

Chaparral

Tropical forest

Equator

■ Tropical forest hot spots

■ Chaparral hot spots

Figure 20.17 Some biodiversity hot spots. Only dry shrubland (such as chaparral) and tropical forest hot spots are mapped here.

forests and dry shrublands (such as California's chaparral), total less than 1.5% of Earth's land but are home to a third of all species of plants and vertebrates. Conservation biologists have also identified aquatic ecosystems, including certain river systems and coral reefs, that are biodiversity hot spots.

Because endemic species are limited to specific areas, they are highly sensitive to habitat degradation. To date, 6 of the 18 hot spots shown in Figure 20.17 have lost nearly 90% of their original habitats to human development. At the current rate of habitat destruction, the rest could lose similar amounts in the next two decades. Conservation biologists estimate that this loss of habitat will cause the extinction of about half the species in the hot spots. Thus, biodiversity hot spots are also hot spots of extinction and rank high on the list of areas demanding strong global conservation efforts. In the United States, the greatest numbers of endangered species (terrestrial and freshwater) occur in Hawaii, southern California, the southern Appalachians, and the southeastern coastal states (especially Florida), areas with the highest numbers of endemic species.

With so much biodiversity concentrated in a relatively small portion of the biosphere, there is reason for optimism that conservation biologists can accomplish much by focusing on these hot spots. The goal is to keep as many of these hot spots as wild as possible. On the other hand, the biodiversity crisis is a global problem, and focusing on hot spots should not detract from efforts to conserve habitats and species diversity in other areas.

Conservation at the Population and Species Levels

Much of the popular and political discussion of the biodiversity crisis centers on species. The U.S. Endangered Species Act (ESA) defines an **endangered species** as one that is "in danger of extinction throughout all or a significant portion of its range." Also defined for protection by the ESA, **threatened species** are those that are likely to become endangered in the foreseeable future throughout all or a significant portion of their geographic range.

Habitat Fragmentation and Subdivided Populations The focus of scientific studies concerned with sustaining species is on the dynamics of populations that have been reduced in numbers and fragmented by human activities. Because habitats are patchy, the populations of many species were subdivided into groups with varying degrees of isolation from one another before humans began altering habitats significantly. Gene flow among population subgroups varied according to the degree of isolation (see Chapter 13).

Today, severe **population fragmentation,** the splitting and consequent isolation of portions of populations by habitat degradation, is one of the most harmful effects of habitat loss due to human activities (**Figure 20.18**). We'll refer to these fragments of a population as subpopulations. A decrease in the overall size of populations and a reduction in gene flow among subpopulations usually accompany fragmentation.

In most cases, subpopulations are separated into habitat patches that vary in quality. For instance, in a fragmented forest, the presence of some large, dead trees can make the difference between a high-quality patch and a low-quality one for squirrels, owls, and many birds that require rot cavities for their nests. Patches with abundant high-quality resources tend to have stable, persistent subpopulations. Reproductive individuals in a high-quality patch tend to produce more offspring than the patch can sustain. Such an area of habitat where a subpopulation's reproductive success exceeds its death rate is

Northern spotted owl

Figure 20.18 Fragmentation of a forest ecosystem. This aerial photograph illustrates fragmentation of a coniferous forest in the Mount Hood National Forest in northwestern Oregon. The forest was originally contiguous. The open areas in the photo were logged, creating forest fragments, some of which are islands within clear-cut areas. A common result of human activities, this kind of habitat alteration has reduced and fragmented the populations of many species. An example is the northern spotted owl (inset), which inhabits coniferous forests of the U.S. Pacific Northwest. Owl populations declined markedly and were fragmented after these forests were logged.

called a **source habitat.** Source habitats produce enough new individuals that some disperse to other areas, often in search of food or a place to reproduce. In contrast, a habitat where a subpopulation's death rate exceeds its reproductive success is called a **sink habitat.** The persistence of many subpopulations in sink habitats depends on individuals dispersing from source habitats.

Today, because of habitat loss, more sink habitats exist than were present for most species historically, and the dispersal of individuals to sinks can sometimes threaten the survival of subpopulations in source habitats. For example, the remaining source habitats of the northern spotted owl are relatively small fragments of old-growth rain forests. The owls tend to disperse from small fragments of source habitats into larger, neighboring areas where trees are regrowing after logging. Such areas are sink habitats for the owls, and sustaining the owl populations depends in part on keeping enough reproductive individuals as robust breeding stock in the source habitats. Some researchers have suggested surrounding old-growth fragments with distinct boundaries—such as clear-cut areas—that the owls will not enter. The spotted owl situation illustrates the importance of identifying source and sink habitats and of protecting source habitats.

What Makes a Good Habitat? What factors make some habitats sources and others sinks? Identifying the specific combination of habitat factors that is critical for a species is fundamental in conservation biology.

As a case study in identifying critical habitat factors, we'll consider the red-cockaded woodpecker (*Picoides borealis*), an endangered, endemic species originally found throughout the southeastern United States. This species requires mature pine forests, preferably ones dominated by the longleaf pine. Most woodpeckers nest in dead trees, but the red-cockaded woodpecker drills its nest holes in mature, living pine trees (**Figure 20.19a**). The heartwood (deep wood) of mature longleaf pines is usually rotted and softened by fungi, allowing the woodpeckers adequate space for nesting once they excavate into the heartwood. Red-cockaded woodpeckers also drill small holes around the entrance to their nest cavity, which causes resin from the tree to ooze down the trunk. The resin seems to repel certain predators, such as corn snakes, that eat bird eggs and nestlings. Another critical habitat factor for this woodpecker is a low growth of plants among the mature pine trees (**Figure 20.19b**). Biologists have found that a habitat becomes a sink, with breeding birds tending to abandon nests, when vegetation among the pines is thick and higher than about 15 feet (**Figure 20.19c**). Apparently, the birds require a clear flight path between their home trees and the neighboring feeding grounds. Historically, periodic fires swept through longleaf pine forests, keeping the undergrowth low.

The recent recovery of the red-cockaded woodpecker from near-extinction to sustainable populations is largely due to recognizing the key habitat factors and protecting some longleaf pine forests that support viable numbers of the birds. The use of controlled fires to reduce forest undergrowth helps maintain mature pine trees as well as the woodpeckers.

Conserving Species amid Conflicting Demands Determining habitat requirements is only one aspect of the effort to save species. It is usually necessary to weigh a species' biological and ecological needs against the conflicting demands of our complex culture. Thus, conservation biology often highlights the relationships between biology and society. For example, an ongoing, sometimes bitter debate in the U.S. Pacific Northwest pits saving habitats for populations of the northern spotted owl, timber wolf, grizzly bear, and bull trout against demands for jobs in the timber, mining, and

(a) Red-cockaded woodpecker

(b) Forest that can sustain red-cockaded woodpeckers

(c) Forest that cannot sustain red-cockaded woodpeckers

Figure 20.19 Habitat requirements of the red-cockaded woodpecker. **(a)** A red-cockaded woodpecker at the entrance to its nest site in a longleaf pine tree. **(b)** Forest habitat with low undergrowth that sustains red-cockaded woodpeckers. **(c)** High, dense undergrowth that impedes the woodpeckers' access to feeding grounds.

other resource extraction industries. Programs to restock wolves and to bolster the populations of grizzly bears and other large carnivores are opposed by some recreationists concerned about safety and by many ranchers concerned with potential losses of livestock.

Large, high-profile vertebrates are not always the focal point in conflicts involving conservation biology. It is habitat use that is almost always at issue. Should work proceed on a new highway bridge if it destroys the only remaining habitat of a species of freshwater mussel? If you were the owner of a coffee plantation growing varieties that thrive in bright sunlight, do you think you would be willing to change to shade-tolerant coffee varieties that are less productive and less profitable but support large numbers of songbirds?

Conservation at the Ecosystem Level

Most conservation efforts in the past have focused on saving individual species, and this work continues. More and more, however, conservation biology aims at sustaining the biodiversity of entire communities and ecosystems. On an even broader scale, conservation biology considers the biodiversity of whole landscapes. Ecologically, a landscape is a regional assemblage of interacting ecosystems, such as an area with forest, adjacent fields, wetlands, streams, and streamside habitats. **Landscape ecology** is the application of ecological principles to the study of land-use patterns. Its goal is to make ecosystem conservation a functional part of the planning for land use.

Case Study in the Process of Science on the Web & CD Participate in a virtual project to restore a tall-grass prairie.

Edges and Corridors Edges between ecosystems are prominent features of landscapes, whether natural or altered by humans (**Figure 20.20**). Such edges have their own sets of physical conditions, such as soil type and surface features that differ from either side. Edges also may have their own type and amount of disturbance. For instance, the edge of a forest often has more blown-down trees than a forest interior because the edge is less protected from strong winds. Because of their specific physical features, edges also have their own communities of organisms. Some organisms depend on edges because they require resources of the two adjacent areas. For instance, white-tailed deer thrive in edge habitats where they can browse on woody shrubs, and their populations often expand when forests are logged.

Edges can have both positive and negative effects on biodiversity. A recent study in a tropical rain forest in western Africa indicated that natural edge communities are important sites of speciation (the origin of new species). On the other hand, landscapes where human activities have produced edges often have fewer species and are dominated by species that are adapted to edges. An example is the brown-headed cowbird, an edge-adapted species that is currently expanding its populations in many areas of North America. Cowbirds forage in open fields on insects disturbed by or attracted to cattle and other large herbivores. But the cowbirds also need forests, where they lay their eggs in the nests of other bird species. The "host" bird feeds the cowbird young as though they were the host's own offspring. This is a form of parasitism. Cowbird numbers are burgeoning where forests are being heavily cut and fragmented, creating more forest-edge habitats and open land for cattle, horses, and sheep. Increasing cowbird populations and loss of habitats are correlated with declining populations of several songbirds, such as warblers, that the cowbirds parasitize.

Another important landscape feature, especially where habitats have been severely fragmented, is the **movement corridor,** a narrow strip or

(a) Natural edges between ecosystems

(b) Edges created by human activities

Figure 20.20
Landscape edges between ecosystems.
(a) This relatively natural landscape in Australia includes a dry forest, a rocky area with grassy islands, and a flat, grass-covered lakeshore. **(b)** Human activities, such as logging and road building, often create edges that are more abrupt than those delineating natural landscapes. Sharp edges surround clear-cuts in this photograph of a heavily logged rain forest in Malaysia.

Figure 20.21 An artificial corridor. This highway underpass allows movement between protected areas for the few remaining Florida panthers. High fences along the highway reduce road kills of panthers and other species.

Figure 20.22 Zoned reserves in Costa Rica.
(a) The green areas on the map are national park lands, core areas where human disruption is minimized. Surrounding these conservation cores, gold areas map buffer zones. These are transition areas, mainly privately owned, where most of the human population live and work. Ideally, the most destructive practices—industries such as mining, large-scale monoculture (growth of a single type of crop over a large area), and urban development—are confined to the outermost fringes of the buffer zones. Within the buffer zones, the trend is toward sustainable agriculture and forestry, activities that can provide comfortable economic support for local residents without drastically altering habitats. **(b)** Local students marvel at the diversity of life in one of Costa Rica's reserves.

series of small clumps of quality habitat connecting otherwise isolated patches. Streamside habitats often serve as natural corridors, and government policy in some nations prohibits destruction of these areas. In places where there is extremely heavy human impact, government agencies sometimes construct artificial corridors (**Figure 20.21**). Corridors can promote dispersal and help sustain populations, and they are especially important to species that migrate between different habitats seasonally. On the other hand, a corridor can be harmful—as, for example, in the spread of diseases, especially among small subpopulations in closely situated habitat patches. A movement corridor that connects a source habitat to a sink habitat may also reduce population numbers in the source. The effects of movement corridors have not been thoroughly studied, and researchers tend to evaluate the potential effects of corridors on a case-by-case basis.

Zoned Reserves In an attempt to slow the disruption of ecosystems, a number of countries are setting up what they call zoned reserves. A **zoned reserve** is an extensive region of land that includes one or more areas undisturbed by humans. The undisturbed areas are surrounded by lands that have been changed by human activity and are used for economic gain. The key factor of the zoned reserve concept is the development of a social and economic climate in the surrounding lands that is compatible with ecosystem conservation. These surrounding areas continue to be used to support the human population, but they are protected from extensive alteration. As a result, they serve as buffer zones against further intrusion into the undisturbed areas.

The small Central American nation of Costa Rica has become a world leader in establishing zoned reserves. In exchange for reducing its international debt, the Costa Rican government established eight zoned reserves, called "conservation areas" (**Figure 20.22**). Costa Rica is making progress toward managing its zoned reserves so that the buffer zones provide a steady, lasting supply of forest products, water, and hydroelectric power and also support sustainable agriculture and tourism. An important goal is providing a stable economic base for the people living there. Destructive

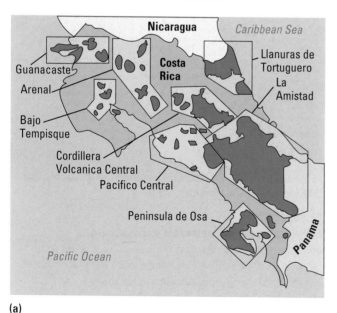

(a)

(b)

practices that are not compatible with long-term ecosystem conservation, and from which there is often little local profit, are gradually being discouraged. Such destructive practices include massive logging, large-scale single-crop agriculture, and extensive mining. Costa Rica looks to its zoned reserve system to maintain at least 80% of its native species.

The Goal of Sustainable Development

With conservation progress in countries such as Costa Rica as a model, many nations, scientific organizations, and private foundations are embracing the concept of **sustainable development.** Balancing human needs with the health of the biosphere, the goal of sustainable development is the long-term prosperity of human societies and the ecosystems that support them. The significance of that responsibility was nicely phrased by former Norwegian Prime Minister G. H. Brundtland: "We must consider our planet to be on loan from our children, rather than being a gift from our ancestors."

Activity 20F on the Web & CD
Review the concepts of conservation biology.

Sustainable development will depend on the continued research and applications of basic ecology and conservation biology. It will also require a cultural commitment to conserve ecosystem processes and biodiversity. That is a priority that, so far, relatively few nations have placed at the top of their political, economic, and social agendas. Those of us living in affluent developed nations are responsible for the greatest amount of environmental degradation. Reality demands that we rearrange some of our priorities, learn to revere the natural processes that sustain us, and temper our orientation toward short-term personal gain.

The current state of the biosphere demonstrates that we are treading precariously on uncharted ecological ground. But despite the uncertainties, now is not the time for gloom and doom, but a time to meet the challenges to pursue more knowledge about life and to work as individuals and a society toward long-term sustainability. Along the way, we will reap the bonus of appreciating our connections to the biosphere and its diversity of life.

CHECKPOINT

1. What is a biodiversity hot spot?
2. Why is the fragmentation of populations of endangered species increasing?
3. Are critical habitat factors generally more favorable in a source habitat or a sink habitat?
4. How is a landscape different from an ecosystem?
5. How can "living on the edge" be a good thing for some species, such as white-tailed deer and cowbirds?
6. Why is a concern for the well-being of future generations essential for progress toward sustainable development?

Answers: 1. A relatively small area with a disproportionate number of species, including endangered species **2.** By destroying habitat, human activities are fragmenting the populations of many species into subpopulations that live in the separate patches of habitat that remain. **3.** Source habitat **4.** A landscape is more inclusive in that it consists of several interacting ecosystems in the same region. **5.** They use a combination of resources from the two ecosystems on either side of the edge. **6.** Sustainable development is a long-term goal—longer than a human lifetime. Concern only with personal gain in the here and now is an obstacle to sustainable development because it discourages behavior that benefits future generations.

Figure 20.23 Doing what comes naturally.
One of modern biology's greatest naturalists, Edward O. Wilson has helped teach scientists and the general public a greater respect for Earth's biodiversity (two of his books have won Pulitzer Prizes). He coined the term *biophilia* to denote humans' inherent passion for nature. This photograph finds biophiliac Wilson in the woods near Massachusetts's Walden Pond, a landscape immortalized by another great naturalist and writer, Henry David Thoreau.

Evolution Connection

Biophilia and an Environmental Ethic

Not many people today live in truly wild environments or even visit such places often. Our modern lives are very different from those of early humans, who hunted and gathered and painted wildlife murals on cave walls (see Figure 17.45). But our behavior reflects remnants of our ancestral attachment to nature and the diversity of life. People keep pets, nurture houseplants, invite avian visitors with backyard birdhouses, and visit zoos, gardens, and nature parks. These pleasures are examples of what Harvard biologist Edward O. Wilson calls *biophilia*, a human desire to affiliate with other life in its many forms. (You met Wilson earlier in the context of the Hundred Heartbeat Club.) Wilson extends biophilia to include our attraction to pristine landscapes with clean water and lush vegetation (**Figure 20.23**). We evolved in natural environments rich in biodiversity, and we still have an affinity for such settings. Wilson makes the case that our biophilia is innate, an evolutionary product of natural selection acting on a brainy species whose survival depended on a close connection to the environment and a practical appreciation of plants and animals.

It will come as no surprise that most biologists have embraced the concept of biophilia (**Figure 20.24**). After all, these are people who have turned their passion for nature into careers. But biophilia strikes a harmonic chord with biologists for another reason. If biophilia is evolutionarily embedded in our genomes, then there is hope that we can become better custodians of the biosphere. If we all pay more attention to our biophilia, a new environmental ethic could catch on among individuals and societies. And that ethic is a resolve never to knowingly allow a single species to become extinct or any ecosystem to be destroyed as long as there are reasonable ways to prevent such ecological violence. It is an environmental ethic that balances out another human trait—our tendency to "subdue Earth." Yes, we should be motivated to preserve biodiversity because we depend on it for food, medicine, building materials, fertile soil, flood control, habitable climate, drinkable water, and breathable air. But maybe we can also work harder to prevent the extinction of other forms of life just because it is the ethical thing for us to do as the most thoughtful species in the biosphere. Again, Wilson sounds the call: "Right now, we're pushing the species of the world through a bottleneck. We've got to make it a major moral principle to get as many of them through this as possible. It's the challenge now and for the next century. And there's one good thing about our species: We like a challenge!"

Biophilia is a fitting capstone for this unit. Modern biology is the scientific extension of our human tendency to feel connected to and curious about all forms of life. We hope that our discussion of ecology has deepened your biophilia and broadened your education.

Evolution Connection on the Web
Read more about biophilia.

Figure 20.24 The face of biophilia.
Biologist Carlos Rivera Gonzales, who is participating in a biodiversity survey in a remote region of Peru, could not resist a closer look at a tiny tree frog. You can also see biophilia in the faces of the children in Figure 20.22b.

Chapter Review

Summary of Key Concepts

For study help, go to the Essential Biology Website (www.essentialbiology.com) or CD-ROM to explore the Activities and Case Studies in the Process of Science.

Human Impact on Biological Communities

- **Human Disturbance of Communities** Human disturbance usually reduces species diversity in communities. Much of the United States is now a hodgepodge of early successional growth where more mature communities once prevailed.

- **Introduced Species** Introduced species are those that humans intentionally or accidentally move from the species' native locations to new geographic regions. Introduced species rank second only to habitat destruction as a cause of extinctions and loss of Earth's biodiversity.

Activity 20A **Fire Ants: An Introduced Species**

Human Impact on Ecosystems

- **Impact on Chemical Cycles** Human activities often intrude in biogeochemical cycles by removing nutrients from one location and adding them to another. For example, the increased burning of fossil fuels is steadily raising CO_2 in the atmosphere, and sewage treatment facilities and fertilizers add large amounts of nitrogen and phosphorus to aquatic systems. Deforestation and extensive removal of groundwater change the water cycle.

Activity 20B **Water Pollution from Nitrates**

- **Deforestation and Chemical Cycles: A Case Study** Research at the Hubbard Brook Experimental Forest is demonstrating that the amount of nutrients leaving an intact forest ecosystem is controlled by plants.

- **The Release of Toxic Chemicals to Ecosystems** Humans have added entirely new materials, many of them toxic, to ecosystems. These toxins often become more concentrated through biological magnification.

Increasing DDT concentration

Activity 20C **DDT and the Environment**

- **Human Impact on the Atmosphere and Climate** Deforestation and the burning of fossil fuels have increased concentrations of CO_2 in the atmosphere, contributing to the greenhouse effect and global warming, with potentially disastrous consequences. Developed countries, which have the greatest energy consumption, have the greatest responsibility to reduce their use of fossil fuels by conserving energy and developing alternative energy sources. In addition, the protective ozone layer has been gradually thinning since 1975 because of the accumulation of chlorofluorocarbons. Scientists expect that a thinning ozone layer will result in an increase in skin cancer and cataracts among humans.

Activity 20D **The Greenhouse Effect**

The Biodiversity Crisis

- **The Three Levels of Biodiversity**

  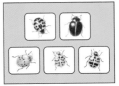

Diversity of ecosystems   Diversity of species within communities   Diversity within species

- **The Loss of Species** The current mass extinction, caused by human activities, is broader and faster than the Cretaceous extinction that claimed dinosaurs and many other groups. Species now go extinct at a rate at least 50 times faster than just a few centuries ago.

- **The Three Main Causes of the Biodiversity Crisis**

Habitat destruction   Introduction of non-native species   Overexploitation of wildlife

- **Why Biodiversity Matters** Humans have relied on biodiversity for food, clothing, shelter, oxygen, soil fertility, and medicinal substances. The loss of diversity limits the potential for new discoveries and reflects large-scale changes in the biosphere that could have catastrophic consequences.

Activity 20E **Madagascar and the Biodiversity Crisis**

Conservation Biology

- Conservation biology is a goal-oriented science that seeks to counter the loss of biodiversity.

- **Biodiversity "Hot Spots"** The front lines for conservation biology are relatively small geographic areas that are especially rich in endangered biodiversity.

- **Conservation at the Population and Species Levels** Severe population fragmentation is one of the most harmful effects of habitat loss due to human activities. Fragmentation usually results in a decrease in the overall size of populations and a reduction in gene flow among subpopulations. In addition, because of habitat loss, the survival of subpopulations in source habitats is threatened by the dispersal of individuals to an increasing number of sink habitats. Identifying the specific combination of habitat factors that is critical for a species is fundamental in conservation biology. Conservation biology often highlights the relationships between biology and society. Competing demands for habitat use is almost always at issue.

- **Conservation at the Ecosystem Level** Increasingly, conservation biology aims at sustaining the biodiversity of entire communities, ecosystems, and landscapes. Edges between ecosystems are prominent features of landscapes, with positive and negative effects on biodiversity. Natural edge communities are important sites of speciation. But human-produced edges often have fewer species and are dominated by species adapted to edges, such as cowbirds. Corridors can promote dispersal and help sustain populations. But corridors can also promote the spread of diseases and connect source and sink habitats. Zoned reserves are now used to slow the disruptions of ecosystems.

Case Study in the Process of Science *How Are Potential Prairie Restoration Sites Analyzed?*

- **The Goal of Sustainable Development** Balancing human needs with the health of the biosphere, the goal of sustainable development is the long-term prosperity of human societies and the ecosystems that support them.

Activity 20F *Conservation Biology Review*

Self-Quiz

1. What is an introduced species? Why are introduced species often able to outcompete native organisms?

2. The recent increase in atmospheric CO_2 concentration is mainly a result of an increase in
    a. primary productivity.
    b. the absorption of infrared radiation escaping from Earth.
    c. the burning of fossil fuels and wood.
    d. cellular respiration by the exploding human population.

3. The Hubbard Brook Experimental Forest study demonstrated all of the following *except:*
    a. Most minerals were recycled within the unaltered forest ecosystem.
    b. Mineral inflow and outflow within a natural watershed were nearly balanced.
    c. Deforestation resulted in an increase in water runoff.
    d. The nitrate concentration in waters draining the deforested area decreased.

4. Why are falcons and other top predators in food chains most severely affected by pesticides such as DDT?

5. What is the greenhouse effect? How could the greenhouse effect be related to global warming?

6. The people in which region consume the most energy each day?
    a. the United States
    b. South America
    c. China
    d. India
    e. Africa

7. The ozone layer
    a. affects the nitrogen cycle.
    b. contributes to global warming.
    c. absorbs UV radiation, possibly protecting organisms from UV damage.
    d. causes cancer.

8. Which of the following statements most comprehensively addresses what conservation biologists mean by the "biodiversity crisis"?
    a. Worldwide extinction rates are currently 50 times greater than at any time during the past 100,000 years.
    b. Introduced species, such as house sparrows and starlings, have rapidly expanded their ranges.
    c. Harvests of marine fish, such as cod and bluefin tuna, are declining.
    d. Many pest species have developed resistance and are no longer effectively controlled by insecticide applications.

9. Currently, the number one cause of biodiversity loss is _____.

10. A _____ is a local grouping of interacting ecosystems with several adjacent habitats.

Answers to the Self-Quiz questions can be found in Appendix B.

Go to the website or CD-ROM for more Self-Quiz questions.

The Process of Science

Biologists in the United States are concerned that populations of many migratory songbirds, such as warblers, are declining. Evidence suggests that some of these birds might be victims of pesticides. Most of the pesticides implicated in songbird mortality have not been used in the United States since the 1970s. Suggest a hypothesis to explain the current decline in songbird numbers. Design an experiment that could test your hypothesis.

Case Study in the Process of Science on the Web & CD *Participate in a virtual project to restore a tall-grass prairie.*

Biology and Society

1. By 1935, hunting and trapping had eliminated wolves from the United States outside Alaska. Since wolves have been protected as an endangered species, they have moved south from Canada and have become reestablished in the Rocky Mountains and northern Great Lakes. Conservationists who would like to speed up this process have reintroduced wolves into Yellowstone National Park. Local ranchers are opposed to bringing back the wolves because they fear predation on their cattle and sheep. What are some reasons for reestablishing wolves in Yellowstone Park? What effects might the reintroduction of wolves have on the ecological communities in the region? What might be done to mitigate the conflicts between ranchers and wolves?

2. Some organizations are starting to envision a sustainable society—one in which each generation inherits sufficient natural and economic resources and a relatively stable environment. The Worldwatch Institute, an environmental policy organization, estimates that we must reach sustainability by the year 2030 to avoid economic and environmental collapse. To get there, we must begin shaping a sustainable society during the next ten years or so. In what ways is our current system not sustainable? What might we do to work toward sustainability, and what are the major roadblocks to achieving it? How would your life be different in a sustainable society?

Biology and Society on the Web *Learn what the U.S. government has to say about introduced species.*

Measurement	Unit and Abbreviation	Metric Equivalent	Approximate Metric-to-English Conversion Factor	Approximate English-to-Metric Conversion Factor
Length	1 kilometer (km) 1 meter (m)	= 1000 (10^3) meters = 100 (10^2) centimeters = 1000 millimeters	1 km = 0.6 mile 1 m = 1.1 yards 1 m = 3.3 feet 1 m = 39.4 inches	1 mile = 1.6 km 1 yard = 0.9 m 1 foot = 0.3 m
	1 centimeter (cm)	= 0.01 (10^{-2}) meter	1 cm = 0.4 inch	1 foot = 30.5 cm 1 inch = 2.5 cm
	1 millimeter (mm) 1 micrometer (μm) 1 nanometer (nm) 1 angstrom (Å)	= 0.001 (10^{-3}) meter = 10^{-6} meter (10^{-3} mm) = 10^{-9} meter (10^{-3} μm) = 10^{-10} meter (10^{-4} μm)	1 mm = 0.04 inch	
Area	1 hectare (ha) 1 square meter (m^2)	= 10,000 square meters = 10,000 square centimeters	1 ha = 2.5 acres 1 m^2 = 1.2 square yards 1 m^2 = 10.8 square feet	1 acre = 0.4 ha 1 square yard = 0.8 m^2 1 square foot = 0.09 m^2
	1 square centimeter (cm^2)	= 100 square millimeters	1 cm^2 = 0.16 square inch	1 square inch = 6.5 cm^2
Mass	1 metric ton (t) 1 kilogram (kg) 1 gram (g)	= 1000 kilograms = 1000 grams = 1000 milligrams	1 t = 1.1 tons 1 kg = 2.2 pounds 1 g = 0.04 ounce 1 g = 15.4 grains	1 ton = 0.91 t 1 pound = 0.45 kg 1 ounce = 28.35 g
	1 milligram (mg) 1 microgram (μg)	= 10^{-3} gram = 10^{-6} gram	1 mg = 0.02 grain	
Volume (Solids)	1 cubic meter (m^3) 1 cubic centimeter (cm^3 or cc) 1 cubic millimeter (mm^3)	= 1,000,000 cubic centimeters = 10^{-6} cubic meter = 10^{-9} cubic meter (10^{-3} cubic centimeter)	1 m^3 = 1.3 cubic yards 1 m^3 = 35.3 cubic feet 1 cm^3 = 0.06 cubic inch	1 cubic yard = 0.8 m^3 1 cubic foot = 0.03 m^3 1 cubic inch = 16.4 cm^3
Volume (Liquids and Gases)	1 kiloliter (kL or kl) 1 liter (L)	= 1000 liters = 1000 milliliters	1 kL = 264.2 gallons 1 L = 0.26 gallon 1 L = 1.06 quarts	1 gallon = 3.79 L 1 quart = 0.95 L
	1 milliliter (mL or ml)	= 10^{-3} liter = 1 cubic centimeter	1 mL = 0.03 fluid ounce 1 mL = approx. $\frac{1}{4}$ teaspoon 1 mL = approx. 15–16 drops	1 quart = 946 mL 1 pint = 473 mL 1 fluid ounce = 29.6 mL 1 teaspoon = approx. 5 mL
Volume (Liquids and Gases)	1 microliter (μl or μL)	= 10^{-6} liter (10^{-3} milliliters)		
Time	1 second (s) 1 millisecond (ms)	= $\frac{1}{60}$ minute = 10^{-3} second		
Temperature	Degrees Celsius (°C)		°F = $\frac{9}{5}$°C + 32	°C = $\frac{5}{9}$(°F − 32)

Chapter 2

1. d
2. $MgCL_2$ and C_2H_6
3. element
4. electrons; neutrons; protons
5. 2
6. An ionic bond involves the transfer of electrons from one atom to another, while a covalent bond involves sharing electrons.
7. d
8. a
9. The positive and negative poles of water molecules make them bond together. The properties of water all arise from this atomic stickiness.
10. The cola is an aqueous solution, with water as the solvent, sugar as the solute, and the CO_2 making the solution acidic.

Chapter 3

1. dehydration synthesis; hydrolysis
2. b
3. fatty acid; glycerol
4. b
5. nitrogen
6. d
7. Hydrophobic amino acids are most likely to be found within the interior of a protein, far from the watery environment.
8. a
9. starch (or glycogen and cellulose); nucleotide
10. Both DNA and RNA are polynucleotides, both have the same phosphate group along the backbone, and both use A, C, and G bases. But DNA uses T while RNA uses U as a base, the sugar differs between them, and DNA is usually double-stranded while RNA is usually single-stranded.

Chapter 4

1. d
2. b
3. bacteria; archaea
4. b
5. smooth ER; rough ER
6. rough ER; Golgi; plasma membrane
7. Both organelles use membranes to organize enzymes, and both provide energy to the cell. But chloroplasts use pigments to capture energy from sunlight in photosynthesis, whereas mitochondria release energy from glucose using oxygen in cellular respiration. Chloroplasts are only in photosynthetic plants and protists whereas mitochondria are in almost all eukaryotic cells.
8. a3, b1, c5, d2, e4
9. nucleus, nuclear pores, ribosomes, rough ER, Golgi
10. a4, b2, c1, d3

Chapter 5

1. It is converted to potential energy and heat.
2. Energy; entropy
3. 10,000 g (or 10 kg) —remember that 1 calorie on a food label equals 1,000 calories of heat energy.
4. Negatively charged phosphate groups crowded together in the triphosphate tail repel each other. The release of a phosphate group makes some of this potential energy available to cells to perform work.
5. Hydrolases are enzymes that participate in hydrolysis reactions, breaking down polymers into the monomers that make them up.
6. An inhibitor can bind to another site that causes the enzyme's active site to change shape.
7. b
8. a
9. b
10. signal transduction pathway

Chapter 6

1. d
2. The electron transport chain
3. glucose; NAD^+
4. O_2
5. The majority of the energy provided by cellular respiration is generated during the electron transport chain. Shutting down that pathway will therefore deprive cells of energy very quickly.
6. b
7. d
8. Glycolysis
9. b
10. Because fermentation supplies only 2 ATPs per glucose molecule compared to 38 from cellular respiration, the yeast will have to consume 19 times as much glucose to produce the same amount of ATP.

Chapter 7

1. thylakoids; stroma
2. inputs: a, d, e; outputs: b, c
3. Green light is reflected by leaves, not absorbed, and therefore cannot drive photosynthesis.
4. H_2O
5. c
6. The reactions of the Calvin cycle require the outputs of the light reactions (ATP and NADPH) to proceed.
7. In hot, dry environments, most plants close their stomata, which saves water but decreases the amount of available CO_2.
8. C_4 and CAM plants can close their stomata and save water without shutting down photosynthesis.
9. c
10. CO_2

Chapter 8

1. c
2. They are in the form of very long, thin strands.
3. b
4. Prophase and telophase
5. c
6. 39
7. Prophase II or metaphase II. It cannot be during meiosis I because then you would see an even number of chromosomes. It cannot be during a later stage in meiosis II because then you would see the sister chromatids separated.
8. benign; malignant
9. 16 ($2n = 8$, so $n = 4$, and $2^n = 2^4 = 16$)
10. Nondisjunction would create just as many gametes with an extra copy of chromosome 3 or 16, but extra copies of chromosomes 3 or 16 are probably fatal.

Chapter 9

1. alleles; homozygous; heterozygous
2. genotype; phenotype
3. c
4. c
5. d
6. d
7. d

More Genetics Problems

1. The parental-type gametes are *WS* and *ws*. Recombinant gametes are *Ws* and *wS*, produced by crossing over.
2. Height appears to result from polygenic inheritance, like human skin color. See Figure 9.22.
3. The brown allele appears to be dominant, the white allele recessive. The brown parent appears to be homozygous dominant, *BB*, and the white mouse is homozygous recessive, *bb*. The F_1 mice are all heterozygous, *Bb*. If two of the F_1 mice are mated, 3/4 of the F_2 mice will be brown.
4. The best way to find out whether a brown F_2 mouse is homozygous dominant or heterozygous is to do a testcross: Mate the brown mouse with a white mouse. If the brown mouse is homozygous, all the offspring will be brown. If the brown mouse is heterozygous, you would expect half the offspring to be brown and half to be white.
5. Freckles is dominant, so Tim and Jan must both be heterozygous. There is a 3/4 chance that they will produce a child with freckles, a 1/4 chance they will produce a child without freckles. The probability that the next two children will have freckles is 3/4 × 3/4 = 9/16.
6. Half their children will be heterozygous and have elevated cholesterol levels. There is a 1/4 chance that their next child will be homozygous, *hh*, and have an extremely high cholesterol level, like Zoe.
7. The bristle-shape alleles are sex-linked, carried on the X chromosome. Normal bristles is dominant (F) and forked is recessive (f). The genotype of the female parent is $X^f X^f$. The genotype of the male parent is $X^F Y$. Their female offspring are $X^F X^f$; their male offspring $X^f Y$.
8. The mother is a heterozygous carrier, and the father is normal. See Figure 9.31 for a pedigree. 1/4 of their children will be boys suffering from hemophilia; 1/4 will be female carriers.
9. In order for a woman to be color-blind, she must inherit X chromosomes bearing the color blindness allele from both parents. Her father has only one X chromosome, which he passes on to all his daughters, so he must be color-blind. A male only needs to inherit the color blindness allele from a carrier mother; both his parents are usually phenotypically normal.

Chapter 10

1. polynucleotides; nucleotides; sugar (deoxyribose); phosphate; nitrogenous base
2. b
3. Each daughter DNA molecule will have half the radioactivity of the parent molecule, since one polynucleotide from the original parental DNA molecule winds up in each daughter DNA molecule.
4. CAU; GUA; histidine (His)
5. A gene is the polynucleotide sequence with information for making one polypeptide. Each codon—a triplet of bases in DNA or RNA—codes for one amino acid. Transcription occurs when RNA polymerase produces mRNA using one strand of DNA as a template. A ribosome is the site of translation, or polypeptide synthesis, and tRNA molecules serve as interpreters of the genetic code. Each tRNA molecule has an amino acid attached at one end, and a three-base anticodon at the other end. Beginning at the start codon, mRNA moves relative to the ribosome a codon at a time. A tRNA with a complementary anticodon pairs with each codon, adding its amino acid to the polypeptide chain. The amino acids are linked by peptide bonds. Translation stops at a stop codon, and the finished polypeptide is released. The polypeptide folds to form a functional protein, sometimes in combination with other poly-peptides.
6. a3, b3, c1, d2, e2 and 3
7. d
8. d
9. reverse transcriptase
10. The process of reverse transcription occurs only in RNA-containing retroviruses like HIV. Human cells never undergo reverse transcription, so reverse transcriptase can be knocked out without harming the human host.

Chapter 11

1. c
2. The ability of these cells to produce entire organisms through cloning
3. nuclear transplantation
4. Which genes are active in a particular sample of cells
5. b
6. operon
7. b
8. a
9. Proto-oncogenes are normal genes involved in the control of the cell cycle. Mutation or viruses can cause them to be converted to oncogenes, or cancer-causing genes. Proto-oncogenes are necessary for normal control of cell division.
10. Homeotic genes are master control genes that regulate many other genes during development.

Chapter 12

1. c, d, b, a
2. Because it does not contain introns
3. vector
4. Such an enzyme creates DNA fragments with "sticky ends," single-stranded regions whose unpaired bases can hydrogen-bond to the complementary sticky ends of other fragments created by the same enzyme.
5. PCR
6. It depends. If the nucleotide differences are not part of any recognition sequences, the two pieces of DNA will produce the same RFLP pattern. If the nucleotide differences affect one or more restriction sequences, the RFLP patterns will be different.
7. a
8. b
9. the whole genome shotgun method
10. c, b, a, d

Chapter 13

1. b
2. c
3. *Bb*: 0.42; *BB*: 0.49; *bb*: 0.09
4. $q^2 = 0.16$, so $q = 0.4$; $p = 1 - q$, so $p = 0.6$
5. c
6. c
7. A small founding population is subject to extensive sampling error in the composition of its gene pool.
8. stabilizing selection
9. b, c, d
10. The prevalence of the disease malaria

Chapter 14

1. b
2. pre-zygotic: a, b, c, e; post-zygotic: d
3. Because a small gene pool is more likely to be changed substantially by genetic drift and natural selection
4. exaptation
5. d
6. d
7. species, genus, family, order, class, phylum, kingdom, domain
8. 2.6
9. c
10. Archaea and Bacteria

Chapter 15

1. d, c, a, b, e
2. d, c, b, a, e
3. b
4. bacteria; archaea
5. By preventing the bacteria from making cell walls
6. Autotrophs make their own organic compounds from CO_2, while heterotrophs must obtain at least one type of organic compound from another organism.
7. They can form endospores.
8. d
9. c
10. b

Chapter 16

1. flowers
2. fruit
3. b
4. b
5. a fern
6. vascular plant
7. a
8. Answers might include nearly any food, such as grains (wheat, corn, oats, barley, and so on), fruits from trees, or garden vegetables; textile fibers, such as cotton; or products made from hardwoods such as oak, cherry, or walnut.

9. algae; fungi
10. absorption

Chapter 17

1. c
2. flatworms (Platyhelminthes) and round-worms (Nematoda)
3. arthropod
4. b
5. Chordata; notochord; cartilage disks between your vertebrae
6. prosimians; anthropoids. Examples of prosimians: lemurs, lorises, pottos, tarsiers. Examples of anthropoids: New World monkeys, Old World monkeys (baboons, macaques, mandrills), apes (gibbons, orangutans, gorillas, chimpanzees), humans
7. bipedalism
8. a
9. *Australopithecus* species, *Homo habilis*, *Homo erectus*, *Homo sapiens*
10. a4, b5, c1, d2, e3

Chapter 18

1. organismal ecology, population ecology, community ecology, ecosystem ecology
2. light, water, temperature, chemicals added
3. physiological; behavioral
4. The number of people and the land area in which they live
5. uniform
6. d
7. intermediate
8. d
9. c
10. III; I

Chapter 19

1. c
2. a2, b5, c5, d3 or 4, e1
3. d
4. By preying on the dominant competitor
5. a3, b2, c1, d4

6. succession
7. Because only 10% of the energy trapped by photosynthesis is turned into biomass by the plant, and only 10% of that energy is turned into meat of a grazing animal, so eating grain-fed beef obtains only about 1% of the energy of photosynthesis.
8. Many nutrients come from the soil, but carbon comes from the air.
9. chaparral
10. d

Chapter 20

1. An introduced species is a species that has been accidentally or intentionally transferred by humans from one location to another location where it does not occur naturally. Introduced species usually have relatively few pathogens, parasites, and predators to slow population growth and so may outcompete native organisms for resources.
2. c
3. d
4. Because the pesticides become concentrated in their prey.
5. Carbon dioxide and other gases in the atmosphere absorb infrared radiation and thus slow the escape of heat from Earth. This is called the greenhouse effect. As the carbon dioxide concentration in the atmosphere increases, more heat is retained, which could account for increased global temperatures (global warming).
6. a
7. c
8. a
9. habitat destruction
10. landscape

APPENDIX C: CREDITS

Photo Credits

Detailed Contents: p. xvii top NYU/Strongin; **bottom** Dorling Kindersley **p. xviii top** PhotoDisc; **bottom** David M. Phillips/Visuals Unlimited **p. xix** Eyewire **p. xx** CORBIS **p. xxi top** Craig Tuttle/CORBIS; **bottom** Dennis Kunkel/Phototake **p. xxii** Hans Reinhard/Bruce Coleman Photography **p. xxiii** Alfred Pasieka/Photo Researchers, Inc. **p. xxiv top** Jose Cibelli/Advanced Cell Technology; **bottom** Hank Morgan/Photo Researchers, Inc. **p. xxv** Michael Fogden/DRK Photo **p. xxvi** A. Witte/Stone **p. xxvii top** Manfred Kage/Peter Arnold; **bottom** E.R. Degginger/Science Source/Photo Researchers, Inc. **p. xxviii** CORBIS **p. xxix** CORBIS **p. xxx** Lester Lefkowitz/CORBIS **p. xxxi** CORBIS

Unit openers: Unit I David Becker/Stone **Unit II** Stone **Unit III** Tom & Pat Leeson/PhotoResearchers, Inc. **Unit IV** Kennan Ward/CORBIS **Unit V** Digital Vision **Unit VI** Michael Pohuski/FoodPix

Chapter 1: Chapter opening photos top left H. Reinhard/The National Audubon Society Collection/Photo Researchers, Inc.; **top right** Dave King/Dorling Kindersley; **bottom left** N.L. Max, University of California/BPS; **bottom right** Brian Eyden/Science Photo Library/Photo Researchers, Inc. **1.1** top to bottom U.S. News & World Report Inc.; The New York Times Co. Reprinted by permission; Knight Ridder; The New York Times Co.; TimePix; National Georgaphic Society; U.S. News & World Report Inc. **1.2 counterclockwise from top** Tom Van Sant/Geosphere Project, Santa Monica/Science Source/Photo Researchers, Inc.; CNES/Spot Image Corporation/Science Source/Photo Researchers, Inc.; Roger Wilmshurst; Frank Lane Picgure Agency/CORBIS; Science Pictures Limited/CORBIS; N.L. Max, University of California/BPS **1.6** Hank Morgan/Photo Researchers, Inc. **1.7** Charles H. Phillips **1.8 top left** Oliver Meckes/Nicole Ottawa/Photo Researchers, Inc.; **top right** K. O. Stetter, R. Huber, and R. Rachel, University of Regensburg; **middle left** D. P. Wilson/Photo Researchers, Inc.; **middle right and bottom left and right** CORBIS **1.9 left** Manfred Kage/Peter Arnold, Inc.; **middle** W. L. Dentler, University of Kansas/BPS; **right** David M. Phillips/Visuals Unlimited **1.10** Mike Hettwer **1.12 left** Richard Milner; **right** Dept. of Library Services, American Museum of Natural History **1.15a left** Anne Dowie; **right** Inga Spence/Tom Stack & Assoc. **1.15b** Chris Colling/CORBIS **1.16 left** S. Lowry/Univ. Ulster/Stone; **right** Simon Fraser/Photo Researchers, Inc. **1.17** Mary DeChirico **1.20a** Breck P. Kent **1.20b** E. R. Degginger/Photo Researhers, Inc. **1.22** NYU/Strongin **1.23** Stone

Chapter 2: Chapter opening photos top left Anne Dowie; **top right** Dorling Kindersley; **bottom left** Michael Newman /PhotoEdit; **bottom right** PLG Photo Studio **2.1 top** PhotoDisc; **middle** CORBIS; **bottom** Benjamin Cummings **2.4a** Alison Wright/CORBIS **2.4b** Anne Dowie **2.6a** CTI, Inc. **2.6b** M.E. Raichle, Washington University School of Medicine **2.10** NASA **2.12 top** Alan Pappe/PhotoDisc; **bottom** R. Kessel-Shih/Visuals Unlimited **2.13** Dorling Kindersley **2.14** Harry How/Allsport (USA), Inc. **2.18** Oliver Strewe/Stone **2.19** Jagga Gudlaugnd

Chapter 3: Chapter opening photos top left Cary Groner; **top right** Alan Pappe/PhotoDisc; **bottom left** Benjamin Cummings; **bottom right** Juhn Giustina/PhotoDisc **3.1 left** PLG Photo Studio; **right** Donna Day/Stone **3.4** Anne Dowie **3.8** Scott Camazine/Photo Researchers, Inc. **3.12** Anne Dowie **3.13a and c** Biophoto Associates/Photo Researchers, Inc. **3.13b** L. M. Beidler, Florida State University **3.14 left** Jeremy Woodhouse/PhotoDisc; **right** T. J. Beveridge/Visuals Unlimited **3.17** Mike Neveux **3.18a** PhotoDisc **3.18b** Ian O'Leary/Dorling Kindersley **3.18c** William Sallaz/Duomo/CORBIS **3.18d** R. M. Motta & S. Correr/Science Source/Photo Researchers, Inc. **3.22** Stanley Flegler/Visuals Unlimited

Chapter 4: Chapter opening photos top left Anne Dowie; **top right** Tony Brain & David Parker/Science Photo Library/Photo Researchers, Inc.; **bottom left** R. M. Motta & S. Correr/Science Source/Photo Researchers, Inc.; **bottom right** Fred Felleman/Stone **4.1 left** Michael Newman/PhotoEdit; **right** Ron Boardman; Frank Lane picture Agency/CORBIS **4.2a** Victor Eroschenko **4.2b** Ars Natura **4.2c** David M. Phillips/Visuals Unlimited **4.4** CNRI/Science Photo Library/Photo Researchers, Inc. **4.11** Barry King/Biological Photo Service **4.13** Garry Cole/Biological Photo Service **4.15a** Roland Birke/Peter Arnold, Inc. **4.15b** Dr. Henry C. Aldrich/Visuals Unlimited **4.17** W. P. Wergin and E. H. Newcomb, University of Wisconsin/Biological Photo Service **4.18** Daniel S. Friend, Harvard Medical School **4.19a** M. Schliwa/Visuals Unlimited **4.19b** M. Abbey/Visuals Unlimited **4.20a** Dennis Kunkel/Phototake **4.20b** Karl Aufderheide/Visuals Unlimited **4.20c** Science Photo Library/Photo Researchers, Inc. **4.23** Peter B. Armstrong, University of California, Davis

Chapter 5: Chapter opening photos top left Susanna Price/Dorling Kindersley; **top right** Paul Barton/CORBIS; **bottom left** William Sallaz/Duomo/CORBIS; **bottom right** C Squared Studios **5.1** Greg Kuchik/Getty Images, Inc. **5.2** Russell Chun and Maureen Kennedy **5.14** Nigel Cattlin/Holt Studios International/Photo Researchers, Inc. **5.17** Mike Abbey/Visuals Unlimited **5.19** Eyewire

Chapter 6: Chapter opening photos top left CORBIS; **top right** Eyewire; **bottom left** Steve Allen/The Image Bank; **bottom right** Clive Brunskill/Getty Images **6.1** Ted Spiegel/CORBIS **6.2** Gail Shumway/Taxi **6.4** CORBIS **6.16a** Steve Welsh/Liaison Agency, Inc. **6.16b** Ian O'Leary/Dorling Kindersley

Chapter 7: Chapter opening photos top left Craig Tuttle/CORBIS; **top right** Digital Vision; **bottom left** Marge Lawson **7.1** Tim Volk **7.2a** Renee Lynn/Photo Researchers, Inc. **7.2b** Bob Evans/Peter Arnold, Inc. **7.2c** Dwight Kuhn **7.2d** Sue Barns **7.3 top** M.Eichelberger/Visuals Unlimited; **bottom** W.P. Wergin and E.H. Newcomb, University of Wisconsin/Biological Photo Service **7.6** Adam Smith/Taxi **7.7** Siegfried Layda/Stone **7.8b** Christine L. Case **7.8c** Tony Freeman/PhotoEdit **7.14a** C.F. Miescke/Biological Photo Service **7.14b** Stone **7.16** Tom and Pat Leeson

Chapter 8: Chapter opening photos top left Luke Dodd/Science Source/Photo Researchers, Inc.; **top right** David M. Phillips/Science Source/Photo Researchers, Inc.; **bottom left** David M. Phillips/Science Source/Photo Researchers, Inc.; **bottom right** Dennis Kunkel/Phototake **8.1** Phototake NYC **8.2a** Biophoto Associates/Photo Researchers, Inc. **8.2b** Brian Parker/Tom Stack & Associates, Inc. **8.3** Andrew Bajer/University of Oregon **8.4 top** A.L. Olins, Univ. of Tennessee/ Biological Photo Service; **bottom** Biophoto Associates/Photo Researchers, Inc. **8.7 all** Conly Rieder **8.8a** David M. Phillips/Visuals Unlimited **8.8b** Carolina Biological Supply/Phototake NYC **8.10** James King-Holmes/Science Source/Photo Researchers, Inc. **8.11** Gary Buss/Taxi

8.12 CNRI/SPL/Photo Researchers, Inc. **8.19 left** CNRI/Science Photo Library/ Photo Researchers, Inc.; **right** Greenlar/The Image Works **8.24** Dr. Martin Gallardo, Universidad Austral de Chile **Page 140** Carolina Biological Supply/ Phototake NYC

Chapter 9: Chapter opening photos top left CORBIS Bettmann; **top right** PhotoDisc; **bottom left** Sinclair Stammers/Science Photo Library/Photo Researchers, Inc.; **bottom right** AKG London Ltd **9.1** Yoav Levy/Phototake NYC **9.2** Hans Reinhard/Bruce Coleman Photography **9.4** CORBIS Bettmann **9.14 top left** CORBIS; **top right** Eyewire; **middle left and right** PhotoDisc; **bottom left and right** Anthony Loveday **9.17** Dick Zimmerman/Shooting Star International Photo Agency **9.21** Lawrence Berkeley National Laboratory **9.28a** Jean Claude Revy/Phototake NYC **9.28b** Carolina Biological Supply/Phototake NYC **9.31** Taxi **9.32** Dr. Tudor Parfitt, University of London **Page 169** Norma Jubinville

Chapter 10: Chapter opening photos top left Juda Ngwenya/REUTERS/Getty Images; **top right** Keith V. Wood, University of California, San Diego; **bottom left** Richard Wagner, UCSF Graphics; **bottom right** Alfred Pasieka/Photo Researchers, Inc. **10.3a** Barrington Brown/Photo Researchers Inc **10.3b** Cold Spring Harbor Laboratory Archives **10.5** Michael Freeman/Phototake **10.7** B. Daemmrich/The Image Works **10.12** Keith V. Wood, University of California, San Diego **10.23** A. Witte/Stone **10.24** Francis Leroy, Biocosmos/ Science Source/Photo Researchers, Inc. **10.25** Oliver Meckes/Photo Researchers **10.27** Holt Studios/Jurgen Dielenschneider/The National Audubon Society Collection/Photo Researchers, Inc. **10.30** NIBSC/Science Photo Library/Photo Researchers, Inc. **10.31a** CDC/Phototake, NYC **10.31b** Keith V. Wood/Photo Researchers, Inc. **10.32** Christian Keenan/Getty Images

Chapter 11: Chapter opening photos top left Anne Dowie; **top right** PhotoDisc; **bottom left** Geoff Brightling/Dorling Kinersley; **bottom right** Lee Snider/The Image Works **11.1 main** Craig Hammell/Stock Market/CORBIS; **inset** Courtesy of Cord Blood Registry **11.2 all** Ed Reschke **11.4b** Incyte Pharmaceuticals, Inc., Palo Alto, CA, from R. F. Service, Science (1998) 282:396-399, with permission from Science **11.6** Roslin Institute, Edinburgh **11.7a** Jim Curley/University of Missouri **11.7b** Advanced Cell Technology, Inc. **11.9** Jose Cibelli/Advanced Cell Technology **11.12** Dave King/Dorling Kindersley **11.21** Edward B.Lewis/California Institute of Technology

Chapter 12: Chapter opening photos top left both PhotoDisc; **top right** Andrew Brookes/CORBIS; **bottom left** T. J. Berveridge and S. Schultze/Biological Photo Service; **bottom right** James P. Blair/CORBIS **12.1** Douglas Graham/Roll Call/Corbis Sygma **12.3** SIU/Visuals Unlimited **12.4** Hank Morgan/Photo Researchers, Inc. **12.5a** CORBIS **12.5b** Ken Ostlie **12.6** Peter Berger, Institut fur Biologie, Freiburg **12.7** PPL Therapeutics **12.8 top** S. Cohen/Science Source/Photo Researchers, Inc.; **bottom** Huntington Potter, University of South Florida and David Dressler, Oxford University **12.14** Steve Miller/AP/Wide World Photos **12.19** Cellmark Diagnostics Inc., Germantown, Maryland **12.20** Koji Sasahara/AP Photo **12.22** Department of Energy, Joint Genome Institute. Photograph by Michael Anthony. **12.24** Alfred Wolf/Science Source/Photo Researchers, Inc. **12.25** AFP/CORBIS **12.26** University of California, San Francisco

Chapter 13: Chapter opening photos top left Lara Jo Regan/Getty Images; **top right** Leonard Lessin/Peter Arnold, Inc.; **bottom right** Larry Burrows, TimePix **13.1** CORBIS **13.2a** E. S. Ross, California Academy of Sciences **13.2b** Ken G. Preston-Mafham/Animals Animals **13.2c** P. and W. Ward/ Animals Animals **13.3 left** Larry Burrows, Life Magazine ©Time Inc.; **right** National Maritime Museum, London **13.4** William Paton/NHPA **13.7** CORBIS **13.8** Philip Gingerich 1991. Reprinted with permission of

Discover Magazine **13.9 left** Dorling Kindersley; **middle** Worldsat International and J. Knighton/Science Source/Photo Researchers, Inc.; **right** Ken Findlay/Dorling Kindersley **13.11a** Dwight Kuhn Photography **13.11b** Lennart Nilsson/Albert Bonniers Forlag AB, A Child Is Born, Dell Publishing. 1990 **13.13a** Tui de Roy /Bruce Coleman, Inc. **13.13b** Mike Putland/Ardea London Ltd. **13.13c** Tui de Roy /Bruce Coleman, Inc. **13.14** Michael Fogden/DRK Photo **13.16a** David Cavagnaro **13.16b** USAF, NOAA/NESDIS at Univ. of CO, CIRES/National Snow and Ice Data Center **13.17** Edmund Brodie, Indiana University **13.19** Anne Dowie **13.22** EyeWire **13.23** Hulton-Deutsch Collection/CORBIS **13.24** 1993 Time magazine **13.25** Michael Fogden/DRK Photo **13.27** Bill Longcore/Photo Researchers, Inc.

Chapter 14: chapter opening photos top left A. Witte/Stone; **top right** Auguste Rodin, Le Penseur, private collection/The BridgemanArt Library; **bottom right** George Bernard/The National Audubon Society Collection/Photo Researchers, Inc. **14.1** Mark Pilkington/Geological Survey of Canada/SPL/Photo Researchers, Inc. **14.3** Ernst Mayr, Museum of comparative Zoology, Harvard University **14.4a left** John Shaw/Tom Stack and Associates; **right** Don & Pat Valenti/Tom Stack & Associates **14.4b all** PhotoDisc **14.6** Wolfgang Kaehler/CORBIS **14.7 left** CORBIS; **middle** Karl Shone/ Dorling Kindersley; **right** EyeWire **14.9 main photo** CORBIS; **left inset** John Shaw/Bruce Coleman, Inc.; **right inset** Michael Fogden/Bruce Coleman **14.11 all** University of Amsterdam **14.15** Chip Clark **14.16** Stephen Dalton/ Photo Researchers, Inc. **14.17 both** PhotoDisc **14.18a** Georg Gerster/Science Source/Photo Researchers, Inc. **14.18b** John Reader/Science Photo Library/ Photo Researchers, Inc. **14.18c** Tom Bean/ CORBIS **14.18d** Manfred Kage/ Peter Arnold, Inc. **14.18e** Chip Clark **14.18f** Tom Bean/CORBIS **14.18g** David A. Grimald. Photo by Jacklyn Beckett/The American Museum of Natural History, N.Y. **14.18h** F. Latreille/Cerpolex/Cercles Polaires Expeditions **14.24** Hanny Paul/Liaison Agency, Inc.

Chapter 15 Chapter opening photos top left P. Motta/Science Source/Photo Researchers, Inc.; **top right** David M. Frazier/Photo Researchers, Inc.; **bottom left** USDA/The National Audubon Society Collection/Photo Researchers, Inc.; **bottom right** Frederick P. Mertz/Visuals Unlimited **15.1** Alex Wong/Getty Images **15.3** Artist: Peter Sawyer ©NMNH Smithsonian Inst. **15.4** Roger Ressmeyer/Corbis **15.7a** Sidney Fox, University of Miami/BPS **15.7b** F. M. Menger and Kurt Gabrielson, Emory University **15.8a** Stanley Awramik/ Biological Photo Service **15.8b** Dr. Tony Brain and David Parker/Science Photo Library/Photo Researchers inc. **15.9** Helen E. Carr/BPS **15.10a** David M. Phillips/Visuals Unlimited **15.10b** David M. Phillips/Visuals Unlimited **15.10c** CNRI/SPL/Photo Researchers Inc **15.11a** David M. Phillips/Science Source/Photo Researchers, Inc. **15.11b** Sue Barns **15.11c** Heide Schulz, Max Planck Institute for Marine Microbiology **15.12** Lee D. Simon/Science Source/ Photo Researchers, Inc. **15.13** H.S. Pankratz, T.C. Beaman/Biological Photo Service **15.14** Dr. Tony Brain/Science Source/Photo Researchers, Inc. **15.15a** R. Calentine/Visuals Unlimited **15.15b** Centers for Disease Control **15.16** Martin Bond/The National Audubon Society Collection/Photo Researchers, Inc. **15.17** Exxon Corporation **15.19a** Oliver Meckes/Science Source/Photo Researchers, Inc. **15.19b** M. Abbey/Visuals Unlimited **15.19c** Manfred Kage/Peter Arnold, Inc. **15.19d** Dr. Masamichi Aikawa **15.19e** M. Abbey/Science Source/Photo Researchers, Inc. **15.20** George Barron **15.21 left** Matt Springer, Stanford University; **right top and bottom** Robert Kay, MRC Cambridge **15.22a** Biophoto Associates/Photo Researchers, Inc. **15.22b** Kent Wood/Photo Researchers, Inc. **15.22c** Herb Charles Ohlmeyer, Fran Heyl Associates **15.22d** Manfred Kage/Peter Arnold **15.23a** A. Flowers and L. Newman/Photo Researchers, Inc. **15.23b** Gary Robinson/Visuals Unlimited **15.23c** David Hall/Photo Researchers, Inc.

Chapter 16: Chapter opening photos top left Sequoia National Park Service; **top right** Anne Dowie; **bottom left** CORBIS; **bottom right** G. Prance/Visuals Unlimited **16.1** Lara Hartley **16.3** Dana Richter/Visuals Unlimited **16.4** David

Middleton/NHPA **16.5** Graham Kent **16.6a** E.R. Degginger/Science Source/ Photo Researchers, Inc. **16.6b** Linda.E. Graham **16.8** J. Shaw, Biology Department, Duke University **16.9** Dwight Kuhn/Dwight Kuhn Photography **16.11 left inset** Milton Rand/Tom Stack and Associates; **main photo** John Shaw/Tom Stack & Associates; **bottom inset** Glenn Oliver/Visuals Unlimited **16.12** TheField Museum, Chicago **16.13** Ron Watts/CORBIS **16.15 left inset** Derrick Ditchburn/Visuals Unlimited; **main photo** Doug Sokell/Visuals Unlimited; **right inset** Gerald & Buff Corsi/Visuals Unlimited **16.19a** Taxi **16.19b** Scott Camazine/Photo Researchers, Inc. **16.19c** Dwight R. Kuhn **16.20a** H. Reinhard/ The National Audubon Society Collection/Photo Researchers, Inc. **16.20b** Rob Simpson/Visuals Unlimited **16.20c** G. L.Barron, University of Guelph/ Biological Photo Service **16.20d left** M. F. Brown/Visuals Unlimited; **right** Jack Bostrack/Visuals Unlimited **16.20e** N. Allin and G. L. Barron, University of Guelph/Biological Photo Service **16.20f** J. Forsdyke/Gene Cox/Science Source/Photo Researchers, Inc. **16.21** Fred Rhoades/Mycena Consulting **16.22a** Stuart Bebb/Oxford Scientific Films/Animals Animals/ Earth Scenes **16.22b** David Cavagnaro/Visuals Unlimited **16.23a** Robb Walsh **16.23b** PhotoDisc **16.24** Christine Case **16.25 top** Dr. Jeremy Burgess/ The National Audubon Society Collection/Photo Researchers, Inc.; **bottom** V. Ahmadijian/Visuals Unlimited

Chapter 17: Chapter opening photos top left Steve P. Hopkin/Taxi; **top right** Francois Gohier/The National Audubon Society Collection/Photo Researchers, Inc.; **bottom left** R. Calentine/Visuals Unlimited; **bottom right** Jeremy Woodhouse/Getty Images, Inc. **17.1a and b** David Hosking/ CORBIS **17.2** Gunter Ziesler/Peter Arnold, Inc. **17.3** Anthony Mercieca/Photo Researchers, Inc. **17.9** Photodisc Green **17.11a** CORBIS **17.11b** Claudia Mills/Friday Harbor Labs **17.12** Mike Bacon/Tom Stack & Associates, Inc. **17.15** Stanley Fleger/Visuals Unlimited **17.16a** Reprinted with permission from A. Eizinger and R. Sommer, Max Planck Institut fur Entwicklungs-biologie, Tubingen. Copyright 2000 American Association for the Advancement of Science **17.16b** Andrew Syred/Science Photo Library/Photo Researchers **17.18a** Tony Craddock/The National Audubon Society Collection/ Photo Researchers, Inc. **17.18b** H.W. Pratt/BPS **17.18c** Charles R. Wyttenbach/ Biological Photo Service **17.20a** A.N.T./NHPA **17.20b left** R. DeGoursey/ Visuals Unlimited; **right** CORBIS **17.20c** Astrid & Hanns-Frieder Michler/ Science Source/Photo Researchers, Inc. **17.22a** William Dow/CORBIS **17.22b** D. Suzio/Photo Researchers, Inc. **17.22c** Oliver Meckes/Photo Researchers, Inc. **17.23a** CORBIS **17.23b** A. Kerstitch/Visuals Unlimited **17.24** Carolina Biological Supply Company/Phototake NYC **17.26a-e** John Shaw/Tom Stack and Associates **17.27a left** Gerald Corsi/Visuals Unlimited; **right** Gary Milburn/Tom Stack **17.27b** David Wrobel **17.27c** Fred Bavendam/ Peter Arnold, Inc. **17.28** Biophoto Associates/Photo Researchers Inc. **17.29a** Runk/ Schoenberger/Grant Heilman Photography, Inc. **17.29b** Robert Brons/Biological Photo Service **17.32a** George Grall/National Geographic Image Collection **17.32b** J. M. Labat/Jacana/Photo Researchers Inc. **17.33a** Geoff Brightling/Dorling Kindersley **17.33b** Hans Pfletschinger/Peter Arnold Inc. **17.33c** Dr. Eckart Pott/Bruce ColemanLtd. **17.35** Robert and Linda Mitchell **17.36** The Natural History Museum, London **17.37** Stephen J. Kraseman/DRK Photo **17.38a** Mervyn Griffiths/CSIRO **17.38b** Dan Hadden/ Ardea London Ltd **17.38c** Mitch Reardon/The National Audubon Society Collection/Photo Researchers, Inc. **17.39** Digital Vision **17.40a** E. H. Rao/ Photo Researchers, Inc. **17.40b** Kevin Schafer/Photo Researchers, Inc. **17.40c** Digital Vision **17.40d** Digital Vision **17.40e** Digital Vision **17.40f** Nancy Adams/Tom Stack and Assoc. **17.40g** Digital Vision **17.40h** Karl Weatherly/ Getty Images, Inc. **17.43a** Cleveland Museum of Natural History **17.43b** John Reader/SPL/Photo Researchers, Inc. **17.43c** Institute of Human Origins, photo by Donald Johanson **17.45 all** Jean Clottes/Corbis/Sygma

Chapter 18: Chapter opening photos top left Pete Seaward/Stone; **top right** John D. Cunningham/Visuals Unlimited; **bottom left** Raymond Gehman/CORBIS; **bottom right** S. P. Gillette/CORBIS **18.1** Joel W. Rogers/

CORBIS **18.2a and b** Raphael Gaillarde/Liaison Agency, Inc. **18.3** Reprinted with permission from D.W. Schindler, Science 184 (1974): 897, Figure 1.49. 1974 American Association for the Advancement of Science **18.4a** Francois Gohier/The National Audubon Society Collection/Photo Researchers, Inc. **18.4b** Ingrid Van Den Berg/Animals Animals/Earth Scenes **18.4c** David Lazenby/Planet Earth Pictures **18.4d** Nigel J. Dennis/NHPA **18.5** Erich Hartmann/Magnum Photos, Inc. **18.6** Joel W. Rogers/CORBIS **18.7** Los Angeles Times. Reprinted with permission **18.8** Stephen Krasemann/Photo Researchers, Inc. **18.9** CORBIS **18.10** Brian Parker/Tom Stack & Associates, Inc. **18.11 left** Raymond Gehman/CORBIS; **right** Scott T. Smith/CORBIS **18.13** Robert Brenner/PhotoEdit **18.14** Jim Brandenburg/ Minden Pictures **18.15a and b** Frans Lanting/Minden Pictures **18.16a** Sophie de Wilde/Jacana/ Photo Researchers, Inc. **18.16b** Art Wolfe/The Image Bank **18.16c** Will & Deni McIntyre/Photo Researchers, Inc. **18.17** Earth Satellite Corporation **18.19** Roy Corral/CORBIS **18.22** Runk/Schoenberger/Grant Heilman Photography, Inc. **18.23** Nigel Cattlin/The National Audubon Society Collection/Photo Researchers, Inc. **18.24** Alan Carey/Photo Researchers, Inc. **18.26** Mahaux Photography/The Image Bank **18.28a** Alain Evrard/Photo Researchers, Inc. **18.28b** Pete Seaward/Stone **18.30** Tom Bean/CORBIS **18.31** David Reznick **Table 18.2 left** CORBIS; **right** Ernest Manewa/Visuals Unlimited

Chapter 19: Chapter opening photos top left Gary W. Carter/CORBIS; **top right** Darryl Torckler/Stone; **bottom left** Lester Lefkowitz/CORBIS; **bottom right** Jeffrey L. Rotman/CORBIS **19.1** A. Witte/Stone **19.2** Richard D. Estes/ Photo Researchers Inc. **19.4** Galen Rowell/Photo Researchers, Inc. **19.6 both** Heather Angel/Biofotos **19.7b** Joseph T. Collins/Photo Researchers, Inc. **19.7c** Kevin de Queiroz, National Museum of Natural History **19.8** Arthur Morris/©VIREO **19.9** Jeff Lepore/Photo Researchers, Inc. **19.10** Jerry Young/ Dorling Kindersley **19.11** Gail Shumway/Taxi **19.12a** Lincoln Brower, Sweet Briar College **19.12b** Peter J. Mayne **19.13a** Dr. Edward S. Ross **19.13b** Runk/ Schoenberger/Grant Heilman Photography, Inc. **19.14** William E. Townsend/ Photo Researchers, Inc. **19.15** Austrailian Embassy Photo Library **19.16** Michael Fogden/DRK Photo **19.18.1** David Muench/CORBIS **19.18.2** Tom Bean/Tom & Susan Bean, Inc. **19.18.3** Andrew Brown; Ecoscene/ CORBIS **19.18.4** Tom Bean/Tom & Susan Bean, Inc. **19.18.5** Tom Bean/Tom & Susan Bean, Inc. **19.18.6** Tom Bean/Tom & Susan Bean, Inc. **19.19** John Sohlden/ Visuals Unlimited **19.22** Gregory G. Dimijian/Photo Researchers, Inc. **19.31a** David Samuel Robbins/CORBIS **19.31b and c** Joe McDonald/CORBIS **19.31d** Charles Mauzy/CORBIS **19.31e** Jake Rajs/Stone **19.31f** Kennan Ward/ CORBIS **19.31g** Richard Hamilton Smith/CORBIS **19.31h** Darrell Gulin/ CORBIS **19.32a** WorldSat International/Science Source/Photo Researchers, Inc. **19.32b** Williams H. Mullins/Photo Researchers, Inc. **19.32c** David Muench/CORBIS **19.34** M.E. Warren /Photo Researchers, Inc. **19.36** Frans Lanting/Photo Researchers, Inc. **19.37 main photo** Emory Kristof/National Geographic Society Image Collection; **inset** D. Foster/Visuals Unlimited **19.38 all** Lawrence E. Gilbert/Biological Photo Service

Chapter 20: Chapter opening photos top left James P. Blair/National Geographic Image Collection; **top right** David Lazenby/Planet Earth Pictures; **bottom left** AFP/CORBIS; **bottom right** Lynda Richardson/ CORBIS **20.1a** 2002 Getty Images **20.1b** Rachel Woodfield, Merkel + Associates **20.2** Susan Barnett **20.3a** Buddy Mays/CORBIS **20.3b** Lynda Richardson/CORBIS **20.3c** Marc Dantzker **20.3d main** J. Lubner/Wisconsin Sea Grant; **inset** Scott Camazine/The National Audubon Society Collection/ Photo Researchers, Inc. **20.4** CORBIS **20.5** Nigel Tucker/Planet Earth Pictures **20.6a** John D. Cunningham/Visuals Unilimited. **20.6b** Northeastern Forest Experiment Station, Forest Service, United States Department of Agriculture **20.8** NASA Earth Observing System **20.9** CORBIS **20.11** Craig Aurness/ CORBIS **20.12a** NASA/Goddard Space Flight Center **20.12c** Bill Bachmann/ Photo Edit **20.13** Fred Bavendam/Minden Pictures **20.14a** AFP/CORBIS **20.14b** Mark Carwardine/Still Pictures/Peter Arnold, Inc. **20.14c** Dieter & Mary Plage/Bruce Coleman Inc. **20.15a** Wayne Lawler/CORBIS **20.15b** Gary

Illustration and Text Credits

GLOSSARY

A

abiotic component (ā´-bī-ot´-ik) The nonliving factors of an ecosystem, such as air, water, and temperature.

abiotic reservoir The part of an ecosystem where a chemical, such as carbon or nitrogen, accumulates or is stockpiled outside of living organisms.

ABO blood groups Genetically determined classes of human blood that are based on the presence or absence of carbohydrates A and B on the surface of red blood cells. The ABO blood group phenotypes, also called blood types, are A, B, AB, and O.

absorption The uptake of small nutrient molecules by an organism's own body; the third main stage of food processing, following digestion.

acclimation (ak´-li-mā´-shun) Long-term but reversible physiological response to an environmental change.

accommodation The automatic changes made by the eye as it focuses on near objects.

acetyl CoA (acetyl coenzyme A) The entry compound for the Krebs cycle in cellular respiration; formed from a fragment of pyruvate attached to a coenzyme.

acetylcholine (as´-uh-til-kō´-lēn) A nitrogen-containing neurotransmitter; among other effects, it slows the heart rate and makes skeletal muscles contract.

achondroplasia (uh-kon´-druh-plā´-zhuh) A form of human dwarfism caused by a single dominant allele. The homozygous condition is lethal.

acid A substance that increases the hydrogen ion (H$^+$) concentration in a solution.

acid precipitation Rain, snow, sleet, hail, drizzle, etc., with a pH below 5.6; can damage or destroy organisms by acidifying lakes, streams, and possibly land habitats.

actinomycete (ak-tin´-ō-mī´-sēt) One of a group of bacteria characterized by a mass of branching cell chains called filaments.

activation energy The amount of energy that reactants must absorb before a chemical reaction will start.

activator A protein that switches on a gene or group of genes.

active site The part of an enzyme molecule where a substrate molecule attaches (by means of weak chemical bonds); typically, a pocket or groove on the enzyme's surface.

active transport The movement of a substance across a biological membrane against its concentration gradient, aided by specific transport proteins and requiring input of energy (often as ATP).

adaptation *See* evolutionary adaptation.

adaptive radiation The emergence of numerous species from a common ancestor introduced to new and diverse environments.

adenine (A) (ad´-uh-nēn) A double-ring nitrogenous base found in DNA and RNA.

ADP Adenosine diphosphate (a-den´-ō-sēn dī-fos´-fāt) A molecule composed of an adenine (a ribose sugar) and two phosphate groups. The high-energy molecule ATP is made by combining a molecule of ADP with a third phosphate in an energy-consuming reaction.

adrenocorticotropic hormone (ACTH) (uh-drē´-nō-cōr´-ti-kō-trop´-ik) A protein hormone secreted by the anterior pituitary that stimulates the adrenal cortex to secrete corticosteroids.

adult stem cell A cell present in adult tissues that generates replacements for nondividing differentiated cells.

aerobic (ār-ō´-bik) Containing or requiring molecular oxygen (O$_2$).

age structure The relative number of individuals of each age in a population.

aggregate fruit A fruit such as a blackberry that develops from a single flower with many carpels.

Agnatha (ag-nā´-thuh) A class of vertebrate animals that are superficially fishlike but lack jaws and paired fins.

agnathan (ag-nā´-thun) A vertebrate animal that is superficially fishlike but lacks jaws and paired fins.

agonistic behavior (a´-gō-nis´-tik) Confrontational behavior involving a contest waged by threats, displays, or actual combat, which settles disputes over limited resources, such as food or mates.

AIDS Acquired immune deficiency syndrome; the late stages of HIV infection, characterized by a reduced number of T cells; usually results in death caused by opportunistic infections.

alcohol fermentation The conversion of the acid produced by glycolysis to carbon dioxide and ethyl alcohol.

alga (al´-guh) (plural, **algae**) One of a great variety of protists, most of which are unicellular or colonial photosynthetic autotrophs with chloroplasts containing the pigment chlorophyll *a*. Heterotrophic and multicellular protists closely related to unicellular autotrophs are also regarded as algae.

allele (uh-lē´-ul) An alternative form of a gene.

allopatric speciation The formation of a new species as a result of an ancestral population's becoming isolated by a geographic barrier. *See also* sympatric speciation.

alpha helix (al´-fuh hē´-liks) The spiral shape resulting from the coiling of a polypeptide in a protein's secondary structure.

alternation of generations A life cycle in which there is both a multicellular diploid form, the sporophyte, and a multicellular haploid form, the gametophyte; a characteristic of plants and multicellular green algae.

alternative RNA splicing A type of regulation at the RNA-processing level in which different mRNA molecules are produced from the same primary transcript, depending on which RNA segments are treated as exons and which as introns.

altruism (al´-trū-iz-um) Behavior that reduces an individual's fitness while increasing the fitness of another individual.

amine (uh-mēn´) An organic compound with one or more amino groups.

amino acid (uh-mēn´-ō) An organic molecule containing a carboxyl group and an amino group; serves as the monomer of proteins.

amino acid sequencing Determining the sequence of amino acids in a polypeptide.

amino group In an organic molecule, a functional group consisting of a nitrogen atom bonded to two hydrogen atoms.

ammonia A small and very toxic nitrogenous waste produced by metabolism.

amniocentesis (am´-nē-ō-sen-tē´-sis) A technique for diagnosing genetic defects while a fetus is in the uterus; a sample of amiotic fluid, obtained via a needle inserted into the amnion, is analyzed for telltale chemicals and defective fetal cells.

amniotic egg (am´-nē-ot´-ik) A shelled egg in which an embryo develops within a fluid-filled amniotic sac and is nourished by yolk. Produced by reptiles, birds, and egg-laying mammals, it enables them to complete their life cycles on dry land.

amoeba (uh-mē´-buh) A type of protist characterized by great flexibility and the presence of pseudopodia.

amoebocyte (uh-mē´-buh-sīt) An amoeba-like cell that moves by pseudopodia, found in most animals; depending on the species, may digest and distribute food, dispose of wastes, form skeletal fibers, fight infections, and change into other cell types.

Amphibia A class of vertebrate animals that consists of the amphibians, such as frogs, toads, and salamanders.

amygdala (uh-mig´-duh-la) An integrative center of the cerebrum; functionally the part of the limbic system that seems to label information to be remembered.

anabolic steroid (an´-uh-bol´-ik stār´-oyd) A synthetic variant of the male hormone testosterone that mimics some of its effects.

anaerobic (an´-ār-ō´-bik) Lacking or not requiring molecular oxygen (O$_2$).

analogy The similarity of structure between two species that are not closely related, attributable to convergent evolution.

anaphase The third stage of mitosis, beginning when sister chromatids separate from each other and ending when a complete set of daughter chromosomes has arrived at each of the two poles of the cell.

anchorage dependence The requirement that to divide, a cell must be attached to the substratum.

anchoring junction A junction that connects tissue cells to each other (or to an extracellular matrix) and allows materials to pass from cell to cell.

angiosperm (an´-jē-ō-sperm) A flowering plant, which forms seeds inside a protective chamber called an ovary.

animal behavior Externally observable muscular activity triggered by some stimulus; what an animal does when interacting with its environment and how it does it.

Animalia (an-eh-mal´-ē-uh) The kingdom that contains the animals.

Annelida (uh-nel´-ih-duh) The phylum that contains the segmented worms, or annelids; characterized by uniform segmentation; includes earthworms, polychaetes, and leeches.

anterior Pertaining to the front, or head, of a bilaterally symmetrical animal.

anther A sac in which pollen grains develop, located at the tip of a flower's stamen.

anthropoid (an´-thruh-poyd) A member of a primate group made up of the apes (gibbons, orangutans, gorillas, chimpanzees, and bonobos), monkeys, and humans.

anthropomorphism (an´-thruh-puh-mōr´-fiz-um) Ascribing human motivation, conscious awareness, or feelings to other animals.

anticodon (an´-tī-kō´-don) On a tRNA molecule, a specific sequence of three nucleotides that is complementary to a codon triplet on mRNA.

antidiuretic hormone (ADH) (an´-tē-dī´-yū-ret´-ik) A hormone made by the hypothalamus and secreted by the posterior pituitary that promotes water retention by the kidneys.

antihistamine (an´-tē-his´-tuh-mēn) A drug that interferes with the action of histamine, providing temporary relief from an allergic reaction.

aorta (ā-or´-tuh) An artery that conveys blood directly from the heart to other arteries.

aphotic zone (ā-fō´-tik) The region of an aquatic ecosystem beneath the photic zone, where light does not penetrate enough for photosynthesis to take place.

apicomplexan (ap´-ē-kom-pleks´-un) One of a group of parasitic protozoans, some of which cause human diseases.

aqueous solution (ā´-kwē-us) A solution in which water is the solvent.

arachnid A member of a major arthropod group that includes spiders, scorpions, ticks, and mites.

Archaea (ar´-kē-uh) One of two prokaryotic domains of life, the other being Bacteria.

archaean (plural, **archaea**) An organism that is a member of the domain Archaea.

archenteron (ar-ken´-tuh-ron) In a developing animal, the endoderm-lined cavity formed during gastrulation; the digestive cavity of a gastrula.

Arthropoda (ar-throp´-uh-duh) The most diverse phylum in the animal kingdom; includes the horseshoe crab, arachnids (e.g., spiders, ticks, scorpions, and mites), crustaceans (e.g., crayfish, lobsters, crabs, barnacles), millipedes, centipedes, and insects. Arthropods are characterized by a chitinous exoskeleton, molting, jointed appendages, and a body formed of distinct groups of segments.

artificial selection Selective breeding of domesticated plants and animals to promote the occurrence of desirable inherited traits in offspring.

asexual reproduction The creation of offspring by a single parent, without the participation of sperm and egg.

associative learning Learning that a particular stimulus or response is linked to a reward or punishment; includes classical conditioning and trial-and-error learning.

atom The smallest unit of matter that retains the properties of an element.

atomic number The number of protons in each atom of a particular element.

atomic weight The approximate total mass of an atom; given as a whole number, the atomic weight approximately equals the mass number.

ATP Adenosine triphosphate (a-den´-ō-sēn trī-fos´-fāt). The main energy source for cells.

ATP synthase A complex (cluster) of several proteins found in a cellular membrane (including the inner membrane of mitochondria, the thylakoid membrane of chloroplasts, and the plasma membrane of prokary-

otes) that functions in chemiosmosis with adjacent electron transport chains, using the energy of a hydrogen ion concentration gradient to make ATP. An ATP synthase provides a port through which hydrogen ions (H^+) diffuse.

augmentation of ecosystem processes The process of determining what factors have been removed from an area and are limiting its rate of recovery.

australopithecines (os-trā-lō-pith´-uh-sēns) The first hominids; scavenger-gatherer-hunters who lived on African savannas between about 4.4 million years ago and 1.5 million years ago.

autosome A chromosome not directly involved in determining the sex of an organism; in mammals, for example, any chromosome other than X or Y.

autotroph (ot´-ō-trōf) An organism that makes its own food, thereby sustaining itself without eating other organisms or their molecules. Plants, algae, and photosynthetic bacteria are autotrophs.

Aves A class of vertebrate animals that consists of the birds.

B

bacillus (buh-sil´-us) (plural, **bacilli**) A rod-shaped prokaryotic cell.

backbone A series of segmental units called vertebrae, present in all vertebrates.

Bacteria One of two prokaryotic domains of life, the other being Archaea.

bacterial chromosome The single, circular DNA molecule found in bacteria.

bacteriophage (bak-tēr´-ē-ō-fāj) A virus that infects bacteria; also called a phage.

bacterium (plural, **bacteria**) An organism that is a member of the domain Bacteria.

Bartholin's glands (bar´-tō-linz) Glands near the vaginal opening in a human female that secrete lubricating fluid during sexual arousal.

basal body (bā´-sul) A eukaryotic cell organelle consisting of a 9 + 0 arrangement of microtubule triplets; may organize the microtubule assembly of a cilium or flagellum; structurally identical to a centriole.

basal metabolic rate (BMR) The number of kilocalories a resting animal requires to fuel its essential body processes for a given time.

base A substance that decreases the hydrogen ion (H^+) concentration in a solution.

basement membrane The extracellular matrix, consisting of a dense mat of proteins and sticky polysaccharides, that anchors an epithelium to underlying tissues.

Batesian mimicry (bāt´-zē-un mim´-uh-krē) A type of mimicry in which a species that a predator can eat looks like a different species that is poisonous or otherwise harmful to the predator.

behavioral biology The use of scientific methods in the study of behavior.

behavioral ecology The scientific search for evolutionary bases of behavior.

behavioral isolation A type of prezygotic barrier between species; two species remain isolated because individuals of neither species are sexually attracted to individuals of the other species.

benign tumor An abnormal mass of cells that remains at its original site in the body.

benthic zone A seafloor or the bottom of a freshwater lake, pond, river, or stream.

bilateral symmetry An arrangement of body parts such that an organism can be divided equally by a single cut

passing longitudinally through it. A bilaterally symmetrical organism has mirror-image right and left sides.

binary fission A means of asexual reproduction in which a parent organism, often a single cell, divides into two individuals of about equal size.

binomial A two-part latinized name of a species; for example, *Homo sapiens*.

biodiversity All of the variety of life; usually refers to the variety of species that make up a community; concerns both species richness (the total number of different species) and the relative abundance of the different species.

biodiversity crisis The current rapid decline in the variety of life on Earth, largely due to the effects of human culture.

biodiversity hot spot A small geographic area with an exceptional concentration of species, especially endemic species (those found nowhere else).

biogenesis The principle that all life arises by the reproduction of preexisting life.

biogenic amines Neurotransmitters derived from amino acids.

biogeochemical cycle Any of the various chemical circuits occurring in an ecosystem, involving both biotic and abiotic components of the ecosystem.

biogeography The geographic distribution of species.

biological community *See* community.

biological magnification The accumulation of persistent chemicals in the living tissues of consumers in food chains.

biological species concept The definition of a species as a population or group of populations whose members have the potential in nature to interbreed and produce fertile offspring.

biology The scientific study of life.

biomass The amount, or mass, of organic material in an ecosystem.

biome (bī´-ōm) A major type of ecosystem that covers a large geographic region and that is largely determined by climate, usually classified according to predominant vegetation, and characterized by organisms adapted to the particular environments.

bioremediation The use of living organisms to detoxify and restore polluted and degraded ecosystems.

biosphere The global ecosystem; that portion of Earth that is alive; all of life and where it lives.

biotechnology The use of living organisms (often microbes) to perform useful tasks; today, usually involves DNA technology.

biotic component (bī-ot´-ik) The living factors of a biological community; all the organisms that are part of an individual's environment.

bivalve A member of a group of mollusks that includes clams, mussels, scallops, and oysters.

blastocoel (blas´-tuh-sēl) In a developing animal, a central, fluid-filled cavity in a blastula.

blastopore (blas´-tō-por) A small indentation on one side of a blastula where cells that will form endoderm and mesoderm leave the surface and move inward.

blastula (blas´-tyū-luh) An embryonic stage that marks the end of cleavage during animal development; a hollow ball of cells in many species.

blood-brain barrier A system of capillaries in the brain that restricts passage of most substances into the brain, thereby preventing large fluctuations in the brain's environment.

body cavity A fluid-containing space between the digestive tract and the body wall.

bone marrow Blood-cell forming tissue (red bone marrow) or stored fat (yellow bone marrow) found in cavities within bones.

bony fish A fish that has a stiff skeleton reinforced by calcium salts.

bottleneck effect Genetic drift resulting from a drastic reduction in population size.

Bowman's capsule A cup-shaped swelling at the receiving end of a nephron in the vertebrate kidney; collects the filtrate from the blood.

brown alga One of a group of marine, multicellular, autotrophic protists, the most common and largest type of seaweed. Brown algae include the kelps.

bryophyte (brī´-uh-fīt) One of a group of plants that lack xylem and phloem; a nonvascular plant. Bryophytes include mosses and their close relatives.

buffer A chemical substance that resists changes in pH by accepting H^+ ions from or donating H^+ ions to solutions.

bulk feeder Animals that eat relatively large pieces of food.

C

C_3 plant A plant that uses the Calvin cycle for the initial steps that incorporate CO_2 into organic material, forming a three-carbon compound as the first stable intermediate.

C_4 plant A plant that prefaces the Calvin cycle with reactions that incorporate CO_2 into four-carbon compounds, the end product of which supplies CO_2 for the Calvin cycle.

calorie The amount of energy that raises the temperature of 1 g of water by 1°C.

Calvin cycle The second of two stages of photosynthesis; a cyclic series of chemical reactions that occur in the stroma of a chloroplast, using the carbon in CO_2 and the ATP and NADPH produced by the light reactions to make the energy-rich sugar molecule G3P.

CAM plant A plant that uses crassulacean acid metabolism, an adaptation for photosynthesis in arid conditions. Carbon dioxide entering open stomata during the night is converted into organic acids, which release CO_2 for the Calvin cycle during the day, when stomata are closed.

cancer cell A cell that is not subject to normal cell cycle control mechanisms and that will therefore divide continuously.

cap Extra nucleotides added to the beginning of an RNA transcript in the nucleus of a eukaryotic cell.

capsule A sticky layer that surrounds the bacterial cell wall, protects the cell surface, and sometimes helps glue the cell to surfaces.

carbohydrate (kar´-bō-hī´-drāt) A biological molecule consisting of simple single-monomer sugars (monosaccharides), two-monomer sugars (disaccharides), and other multi-unit sugars (polysaccharides).

carbon fixation The incorporation of carbon from atmospheric CO_2 into the carbon in organic compounds. During photosynthesis in a C_3 plant, carbon is fixed into a three-carbon sugar as it enters the Calvin cycle. In C_4 and CAM plants, carbon is fixed into a four-carbon sugar.

carbon skeleton The chain of carbon atoms that forms the structural backbone of an organic molecule.

carbonyl group (kar´-buh-nēl´) In an organic molecule, a functional group consisting of a carbon atom linked by a double bond to an oxygen atom.

carboxyl group (kar-bok´-sil) In an organic molecule, a functional group consisting of an oxygen atom double-bonded to a carbon atom that is also bonded to a hydroxyl group.

carboxylic acid An organic compound containing a carboxyl group.

carcinogen (kar-sin´-uh-jin) A cancer-causing agent, either high-energy radiation (such as X-rays or UV light) or a chemical.

carcinoma (kar´-sih-nō´-muh) Cancer that originates in the coverings of the body, such as the skin or the lining of the intestinal tract.

carnivore An animal that eats other animals. *See also* herbivore; omnivore.

carpel (kar´-pul) The female part of a flower, consisting of a stalk with an ovary at the base and a stigma, which traps pollen, at the tip.

carrier An individual who is heterozygous for a recessively inherited disorder and who therefore does not show symptoms of that disorder.

carrying capacity In a population, the number of individuals that an environment can sustain.

cartilaginous fish (kar-ti-laj´-uh-nus) A fish that has a flexible skeleton made of cartilage.

Casparian strip (kas-par´-ē-un) A waxy barrier in the walls of endodermal cells in a root that prevents water and ions from entering the xylem without crossing one or more cell membranes.

cation exchange A process in which positively charged minerals are made available to a plant when hydrogen ions in the soil displace mineral ions from the clay particles.

cecum (sē´-kum) (plural, **ceca**) A blind outpocket of a hollow organ such as an intestine.

cell A basic unit of living matter separated from its environment by a plasma membrane; the fundamental structural unit of life.

cell cycle An orderly sequence of events (including interphase and the mitotic phase) that extends from the time a eukaryotic cell divides to form two daughter cells to the time those daughter cells divide again.

cell cycle control system A cyclically operating set of proteins that triggers and coordinates events in the eukaryotic cell cycle.

cell division The reproduction of a cell.

cell junction A structure that connects tissue cells to one another.

cell plate A double membrane across the midline of a dividing plant cell, between which the new cell wall forms during cytokinesis.

cell theory The theory that all living things are composed of cells and that all cells come from other cells.

cell wall A protective layer external to the plasma membrane in plant cells, bacteria, fungi, and some protists; protects the cell and helps maintain its shape.

cellular differentiation The specialization in the structure and function of cells that occurs during the development of an organism; results from selective activation and deactivation of the cells' genes.

cellular metabolism (muh-tab´-uh-lizm) The chemical activities of cells.

cellular respiration The aerobic harvesting of energy from food molecules; the energy-releasing chemical breakdown of food molecules, such as glucose, and the storage of potential energy in a form that cells can use to perform work; involves glycolysis, the Krebs cycle, the electron transport chain, and chemiosmosis.

cellular slime mold A type of protist that has unicellular amoeboid cells and multicellular reproductive bodies in its life cycle.

cellulose (sel´-yū-lōs) A large polysaccharide composed of many glucose monomers linked into cable-like fibrils that provide structural support in plant cell walls.

centipede A carnivorous terrestrial arthropod that has one pair of long legs for each of its numerous body segments, with the front pair modified as poison claws.

central vacuole (vak´-yū-ōl) A membrane-enclosed sac occupying most of the interior of a mature plant cell, having diverse roles in reproduction, growth, and development.

centriole (sen´-trē-ōl) A structure in an animal cell, composed of microtubule triplets arranged in a 9 + 0 pattern. An animal cell usually has a pair of centrioles within each of its centrosomes (*see*).

centromere (sen´-trō-mēr) The region of a chromosome where two sister chromatids are joined and where spindle microtubules attach during mitosis and meiosis. The centromere divides at the onset of anaphase during mitosis and anaphase II of meiosis.

centrosome (sen´-trō-sōm) Material in the cytoplasm of a eukaryotic cell that gives rise to microtubules; important in mitosis and meiosis; also called microtubule-organizing center.

cephalopod A member of a group of mollusks that includes squids and octopuses.

chaparral (shap´-uh-ral´) A biome dominated by spiny evergreen shrubs adapted to periodic drought and fires; found where cold ocean currents circulate offshore, creating mild, rainy winters and long, hot, dry summers.

charophycean (kār´-uh-fī´-sē-unz) A member of the green algal group that shares two ultrastructural features with land plants. They are considered the closest relatives of land plants.

chemical bond An attraction between two atoms resulting from a sharing of outer-shell electrons or the presence of opposite charges on the atoms. The bonded atoms gain complete outer electron shells.

chemical cycling The use and reuse of chemical elements such as carbon within an ecosystem.

chemical energy Energy stored in the chemical bonds of molecules; a form of potential energy.

chemical reaction A process leading to chemical changes in matter; involves the making and/or breaking of chemical bonds.

chemiosmosis (kem´-ē-oz-mō´-sis) The production of ATP using the energy of hydrogen-ion (H^+) gradients across membranes to phosphorylate ADP; powers most ATP synthesis in cells.

chemoautotroph An organism that obtains both energy and carbon from inorganic chemicals, making its own organic compounds from CO_2 without using light energy.

chemoheterotroph (kē´-mō-het´-er-ō-trōf) An organism that obtains energy and carbon from organic molecules.

chemotherapy (kē´-mo-thār´-uh-pē) Treatment for cancer in which drugs are administered to disrupt cell division of the cancer cells.

chiasma (kī-az´-muh) (plural, **chiasmata**) The microscopically visible site where crossing over has occurred between chromatids of homologous chromosomes during prophase I of meiosis.

chlorophyll *a* (klor´-ō-fil ā) A green pigment in chloroplasts that participates directly in the light reactions.

chloroplast (klō′-rō-plast) An organelle found in plants and photosynthetic protists. Enclosed by two concentric membranes, a chloroplast absorbs sunlight and uses it to power the synthesis of organic food molecules (sugars).

choanocyte (kō-an′-uh-sīt) A flagellated feeding cell found in sponges. Also called a collar cell, it has a collarlike ring that traps food particles around the base of its flagellum.

choanoflagellate An ancestral colonial protist from which sponges, and possibly all animals, probably arose.

Chondrichthyes (kon-drik′-thēz) A class of cartilaginous fishes that includes sharks, rays, and skates.

Chordata (kōr-dā′-tuh) The phylum of the chordates; characterized by a dorsal hollow nerve cord, a notochord, gill structures, and a post-anal tail; includes lancelets, tunicates, and vertebrates.

chromatin (krō′-muh-tin) The combination of DNA and proteins that constitute chromosomes; often used to refer to the diffuse, very extended form taken by the chromosomes when a eukaryotic cell is not dividing.

chromosome (krō′-muh-sōm) A threadlike, gene-carrying structure found in the nucleus of a eukaryotic cell and most visible during mitosis and meiosis; also, the main gene-carrying structure of a prokaryotic cell. Chromosomes consist of chromatin.

chromosome theory of inheritance A basic principle in biology stating that genes are located on chromosomes and that the behavior of chromosomes during meiosis accounts for inheritance patterns.

ciliate (sil′-ē-it) A type of protozoan that moves by means of cilia.

cilium (sil′-ē-um) (plural, **cilia**) A short appendage that propels some protists through the water and moves fluids across the surface of many tissue cells in animals. In common with eukaryotic flagella, cilia have a 9 + 2 arrangement of microtubules covered by the cell's plasma membrane.

circadian rhythm (ser-kā′-dē-un) In an organism, a biological cycle of about 24 hours that is controlled by a biological clock, usually under the influence of environmental cues; a pattern of activity that is repeated daily. *See also* biological clock.

clades Evolutionary branches that consist of an ancestor and all its descendants.

cladistic analysis (kluh-dis′-tik) The study of evolutionary history; specifically, the scientific search for monophyletic taxa (clades), taxonomic groups composed of an ancestor and all its descendants.

cladogram A dichotomous phylogenetic tree that branches repeatedly, suggesting a classification of organisms based on the time sequence in which evolutionary branches arise.

class In classification, the taxonomic category above order.

cleavage furrow The first sign of cytokinesis during cell division in an animal cell; a shallow groove in the cell surface near the old metaphase plate.

cline A gradation in an inherited trait along a geographic continuum; variation in a population's phenotypic features that parallels an environmental gradient.

clone As a verb, to produce genetically identical copies of a cell, organism, or DNA molecule. As a noun, the collection of cells, organisms, or molecules resulting from cloning; also (colloquially), a single organism that is genetically identical to another because it arose from the cloning of a somatic cell.

clumped Describing a dispersion pattern in which individuals are aggregated in patches.

Cnidaria (nī-dār′-ē-uh) The phylum that contains the hydras, jellies, sea anemones, corals, and related animals characterized by cnidocytes, radial symmetry, a gastrovascular cavity, polyps, and medusae.

cnidocyte (nī′-duh-sīt) A specialized cell for which the phylum Cnidaria is named; consists of a capsule containing a fine coiled thread, which, when discharged, functions in defense and prey capture.

coccus (kok′-us) (plural, **cocci**) A spherical prokaryotic cell.

codominance The expression of two different alleles of a gene in a heterozygote.

codon (kō′-don) A three-nucleotide sequence in mRNA that specifies a particular amino acid or polypeptide termination signal; the basic unit of the genetic code.

coelom (sē′-lōm) A body cavity completely lined with mesoderm.

coenzyme (kō-en′-zīm) An organic molecule (usually a vitamin or a compound synthesized from a vitamin) that acts as a cofactor, helping an enzyme catalyze a metabolic reaction.

coevolution Evolutionary change in which adaptations in one species act as a selective force on a second species, inducing adaptations that in turn act as a selective force on the first species; mutual influence on the evolution of two different interacting species.

cofactor A nonprotein substance (such as a copper, iron, or zinc atom, or an organic molecule) that helps an enzyme catalyze a metabolic reaction. *See also* coenzyme.

cognition The ability of an animal's nervous system to perceive, store, process, and use information obtained by its sensory receptors.

cognitive ethology The scientific study of cognition; the study of the connection between data processing by nervous systems and animal behavior.

cognitive map A representation within the nervous system of spatial relations among objects in an animal's environment.

cohesion (kō-hē′-zhun) The attraction between molecules of the same kind.

collecting duct A tube in the vertebrate kidney that concentrates urine while conveying it to the renal pelvis.

commensalism (kuh-men′-suh-lizm) A symbiotic relationship in which one partner benefits without significantly affecting the other.

communicating junction A channel between adjacent tissue cells through which water and other small molecules pass freely.

community An assemblage of all the organisms living together and potentially interacting in a particular area.

community ecology The study of how interactions between species affect community structure and organization.

companion cell In a plant, a cell connected to a sieve-tube member whose nucleus and ribosomes provide proteins for the sieve-tube member.

comparative anatomy The study of body structures in different organisms.

comparative embryology The study of the formation, early growth, and development of different organisms.

competitive exclusion principle The concept that populations of two species cannot coexist in a community if their niches are nearly identical. Using resources more efficiently and having a reproductive advantage, one of the populations will eventually outcompete and eliminate the other.

competitive inhibitor A substance that reduces the activity of an enzyme by binding to the enzyme's active site in place of the substrate; a competitive inhibitor's structure mimics that of the enzyme's substrate.

complementary DNA (cDNA) A DNA molecule made in vitro using mRNA as a template and the enzyme reverse transcriptase. A cDNA molecule therefore corresponds to a gene but lacks the introns present in the DNA of the genome.

complete digestive tract A digestive tube with two openings, a mouth and an anus.

complete metamorphosis (met′-uh-mōr′-fuh-sis) The transformation of a larva into an adult that looks very different, and often functions very differently in its environment, than the larva.

compound A substance containing two or more elements in a fixed ratio; for example, table salt (NaCl) consists of one atom of the element sodium (Na) for every atom of chlorine (Cl).

compound eye The photoreceptor in many invertebrates; made up of many tiny light detectors, each of which detects light from a tiny portion of the field of view.

computed tomography (CT) A technology that uses a computer to create X-ray images of a series of sections through the body.

concentration gradient An increase or decrease in the density of a chemical substance in an area. Cells often maintain concentration gradients of H^+ ions across their membranes. When a gradient exists, the ions or other chemical substances involved tend to move from where they are more concentrated to where they are less concentrated.

conduction The direct transfer of thermal motion (heat) between molecules of objects in direct contact with each other.

conifer A gymnosperm, or naked-seed plant, that produces cones.

coniferous forest A biome characterized by conifers, cone-bearing evergreen trees.

conjugation The union (mating) of two bacterial cells or protist cells and the transfer of DNA between the two cells.

consciousness Awareness; a mental state characterized by conscious thinking and self-awareness.

conservation biology The science of species preservation; the scientific study of ways to slow the current high rate of species loss.

conservation of energy The principle that energy can neither be created nor destroyed.

consumer An organism that obtains its food by eating plants or by eating animals that have eaten plants.

continental drift A change in the position of continents resulting from the incessant slow movement (floating) of the plates of Earth's crust on the underlying molten mantle. It has caused continents to fuse and break apart periodically throughout geologic history.

continental shelves The submerged parts of continents.

controlled experiment A component of the process of science whereby a scientist carries out two parallel tests, an experimental test and a control test. The experimental test differs from the control by one factor, the variable.

convection The mass movement of warmed air or liquid to or from the surface of a body or object.

convergent evolution Adaptive change resulting in non-homologous (analogous) similarities among organisms. Species from different evolutionary lineages come to resemble each other (evolve analogous structures) as a result of living in very similar environments.

coral reefs Warm water, tropical, ecosystems dominated by the hard skeletal structures secreted primarily by the resident cnidarians.

countercurrent exchange The transfer of a substance from a fluid or volume of air moving in one direction to another fluid or volume of air moving in the opposite direction.

countercurrent heat exchanger Parallel blood vessels that convey warm and cold blood in opposite directions, maximizing heat transfer to the cold blood.

covalent bond (kō-vā´-lent) An attraction between atoms that share one or more pairs of outer-shell electrons; symbolized by a single line between the atoms.

cranial nerves Nerves that leave the brain and innervate organs of the head and upper body.

creatine phosphate (krē´-uh-tin) A compound in muscle cells that provides energy.

crista (kris´-tuh) (plural, **cristae**) A fold of the inner membrane of a mitochondrion. Enzyme molecules embedded in cristae make ATP.

crop A pouchlike organ in a digestive tract where food is softened and may be stored temporarily.

cross *See* hybridization.

cross-fertilization The fusion of sperm and egg derived from two different individuals.

crossing over The exchange of segments between chromatids of homologous chromosomes during synapsis in prophase I of meiosis; also, the exchange of segments between DNA molecules in prokaryotes.

crustacean A member of a major arthropod group that includes lobsters, crayfish, crabs, shrimps, and barnacles.

cryptic coloration A type of camouflage that makes potential prey difficult to spot against its background.

CT *See* computed tomography.

culture The accumulated knowledge, customs, beliefs, arts, and other human products that are socially transmitted over the generations.

cuticle (kyū´-tuh-kul) (1) In animals, a tough, nonliving outer layer of the skin. (2) In plants, a waxy coating on the surface of stems and leaves that helps retain water.

cyanobacteria (sī-an´-ō-bak-tēr´-ē-uh) Photosynthetic, oxygen-producing bacteria, formerly called blue-green algae.

cystic fibrosis (sis´-tik fī-brō´-sis) A genetic disease that occurs in people with two copies of a certain recessive allele; characterized by an excessive secretion of mucus and consequent vulnerability to infection; fatal if untreated.

cytokinesis (sī´-tō-kuh-nē´-sis) The division of the cytoplasm to form two separate daughter cells. Cytokinesis usually occurs during telophase of mitosis, and the two processes make up the mitotic (M) phase of the cell cycle.

cytoplasm (sī´-tō-plaz´-um) Everything inside a cell between the plasma membrane and the nucleus; consists of a semifluid medium and organelles.

cytosine (C) (sī´-tuh-sin) A single-ring nitrogenous base found in DNA and RNA.

cytoskeleton A meshwork of fine fibers in the cytoplasm of a eukaryotic cell; includes microfilaments, intermediate filaments, and microtubules.

cytosol (sī´-tuh-sol) The semifluid medium of a cell's cytoplasm.

D

Darwinian fitness The contribution an individual makes to the gene pool of the next generation, rela-tive to the contribution of other individuals in the population.

Darwinian medicine The study of health problems in an evolutionary context.

decomposer An organism that derives its energy from organic wastes and dead organisms; also called a detritivore.

decomposition The breakdown of organic materials into inorganic ones.

dehydration synthesis (dē-hī-drā´-shun sin´-thuh-sis) A chemical process in which a polymer forms as monomers are linked by the removal of water molecules. One molecule of water is removed for each pair of monomers linked. Also called condensation.

dehydrogenase (dē´-hī-droj´-uh-nās) An enzyme that catalyzes a chemical reaction during which one or more hydrogen atoms are removed from a molecule.

deletion The loss of one or more nucleotides from a gene by mutation; the loss of a fragment of a chromosome.

demographic transition A shift from zero population growth in which birth rates and death rates are high to zero population growth characterized instead by low birth and death rates.

denaturation (dē-nā´-chur-ā´-shun) A process in which a protein unravels, losing its specific conformation and hence function; can be caused by changes in pH or salt concentration or by high temperature; also refers to the separation of the two strands of the DNA double helix, caused by similar factors.

density-dependent factor A population-limiting factor whose effects depend on population density.

density-dependent inhibition The arrest of cell division that occurs when cells grown in a laboratory dish touch one another; generally due to an inadequate supply of growth factors.

density-independent factor A population-limiting factor whose occurrence and effects are not affected by population density.

derived characters Homologous features that have changed from a primitive (ancestral) condition and that are unique to an evolutionary lineage; features found in members of a lineage but not found in ancestors of the lineage.

descent with modification Darwin's initial phrase for the general process of evolution.

desert A biome characterized by organisms adapted to sparse rainfall (less than 30 cm per year) and rapid evaporation.

desertification The conversion of semi-arid regions to desert.

detritivore (duh-trī´-tuh-vor) An organism that derives its energy from organic wastes and dead organisms; also called a decomposer.

detritus (duh-trī´-tus) Nonliving matter.

deuterostome (dū-ter´-ō-stōm) An animal with a coelom that forms from hollow outgrowths of the digestive tube of the early embryo. The deuterostomes include the echinoderms and the chordates.

diatom (dī´-uh-tom) A unicellular photosynthetic alga with a unique, glassy cell wall containing silica.

differentiation *See* cellular differentiation.

diffusion The spontaneous movement of particles of any kind from where they are more concentrated to where they are less concentrated.

dihybrid cross (dī-hī´-brid) An experimental mating of individuals differing at two genetic loci.

dikaryotic phase (dī-kār´-ē-ot´-ik) A series of stages in the life cycle of many fungi in which cells contain two nuclei.

dinoflagellate (dī-nō-flaj´-uh-let) A unicellular photosynthetic alga with two flagella situated in perpendicular grooves in cellulose plates covering the cell.

diploid (dip´-loid) Containing two homologous sets of chromosomes in each cell, one set inherited from each parent; referring to a 2n cell.

diploid cell In an organism that reproduces sexually, a cell containing two homologous sets of chromosomes, one set inherited from each parent; a 2n cell.

directional selection Natural selection that acts in favor of the individuals at one end of a phenotypic range.

disaccharide (dī-sak´-uh-rīd) A sugar molecule consisting of two monosaccharides linked by dehydration synthesis.

dispersion pattern The manner in which individuals in a population are spaced within their area. Three types of dispersion patterns are: clumped (individuals are aggregated in patches), uniform (individuals are evenly distributed), and random (unpredictable distribution).

distal tubule In the vertebrate kidney, the portion of a nephron that helps refine filtrate and empties it into a collecting duct.

disturbance In an ecological sense, a force that changes a biological community and usually removes organisms from it. Disturbances, such as fire and storms, play pivotal roles in structuring many biological communities.

diversifying selection Natural selection that favors extreme over intermediate phenotypes.

DNA Deoxyribonucleic acid (dē-ok´-sē-rī´-bō-nū-klā´-ik). The genetic material that organisms inherit from their parents; a double-stranded helical macromolecule consisting of nucleotide monomers with deoxyribose sugar and the nitrogenous bases adenine (A), cytosine (C), guanine (G), and thymine (T). *See also* gene.

DNA fingerprinting A procedure that analyzes an individual's unique collection of DNA restriction fragments, detected by electrophoresis and nucleic acid probes. DNA fingerprinting can be used to determine whether two samples of genetic material are from the same individual.

DNA ligase (lī´-gās) An enzyme, essential for DNA replication, that catalyzes the covalent bonding of adjacent DNA nucleotides; used in genetic engineering to paste a specific piece of DNA containing a gene of interest into a bacterial plasmid or other vector.

DNA microarray A glass slide containing thousands of different kinds of single-stranded DNA fragments arranged in an array. Tiny amounts of DNA fragments, representing different genes, are fixed to the glass slide. These fragments are tested for hybridization with various samples of cDNA molecules, thereby measuring the expression of thousands of genes at one time.

DNA polymerase (puh-lim´-er-ās) An enzyme that assembles DNA nucleotides into polynucleotides using a preexisting strand of DNA as a template.

DNA technology Methods used to study and/or manipulate DNA, including recombinant DNA technology.

doldrums (dol´-drums) An area of calm or very light winds near the equator, caused by rising warm air.

domain A taxonomic category above the kingdom level; the three domains of life are Archaea, Bacteria, and Eukarya.

dominance hierarchy The ranking of individuals based on social interactions; usually maintained by agonistic behavior.

dominant allele In a heterozygote, the allele that determines the phenotype with respect to a particular gene.

dorsal Pertaining to the back of a bilaterally symmetrical animal.

dorsal, hollow nerve cord One of the four hallmarks of chordates; the chordate brain and spinal cord.

double bond A type of covalent bond in which two atoms share two pairs of electrons; symbolized by a pair of lines between the bonded atoms.

double fertilization In flowering plants, the formation of both a zygote and a cell with a triploid nucleus, which develops into the endosperm.

double helix The form of native DNA, referring to its two adjacent polynucleotide strands wound into a spiral shape.

Down syndrome A human genetic disorder resulting from the presence of an extra chromosome 21; characterized by heart and respiratory defects and varying degrees of mental retardation.

Duchenne muscular dystrophy (duh-shen´ dis´-truh-fē) A human genetic disease caused by a sex-linked recessive allele and characterized by progressive weakening and a loss of muscle tissue.

duplication Repetition of part of a chromosome resulting from fusion with a fragment from a homologous chromosome; can result from an error in meiosis or from mutagenesis.

dynein arm (dī´-nin) A protein extension from a microtubule doublet in a cilium or flagellum; involved in energy conversions that drive the bending of cilia and flagella.

E

earthworm One of the three large groups of annelids. *See* Annelida.

Echinodermata (uh-kī´-nō-der´-ma-tuh) The phylum of echinoderms, including sea stars, sea urchins, and sand dollars; characterized by a rough or spiny skin, a water vascular system, an endoskeleton, and radial symmetry in adults.

ecological footprint A method to use multiple constraints to estimate the human carrying capacity of Earth by calculating the aggregate land and water area in various ecosystem categories that is appropriated by a nation to produce all the resources it consumes and to absorb all the waste it generates.

ecological niche A population's role in its community; the sum total of a species' use of the biotic and abiotic resources of its habitat.

ecological species concept The idea that ecological roles (niches) define species.

ecological succession The process of biological community change resulting from disturbance; transition in the species composition of a biological community, often following a flood, fire, or volcanic eruption. *See also* primary succession; secondary succession.

ecology The scientific study of how organisms interact with their environments.

ecosystem (ē´-kō-sis-tem) All the organisms in a given area, along with the nonliving (abiotic) factors with which they interact; a biological community and its physical environment.

ecosystem ecology The study of energy flow and the cycling of chemicals among the various biotic and abiotic factors in an ecosystem.

ectopic pregnancy (ek-top´-ik) The implantation and development of an embryo outside the uterus.

ectotherm (ek´-tō-therm) An animal that warms itself mainly by absorbing heat from its surroundings.

electromagnetic energy Solar energy, or radiation, which travels in space as rhythmic waves and can be measured in photons.

electromagnetic spectrum The full range of radiation, from the very short wavelengths of gamma rays to the very long wavelengths of radio signals.

electron A subatomic particle with a single unit of negative electrical charge. One or more electrons move around the nucleus of an atom.

electron carrier A molecule that conveys electrons within a cell; one of several membrane molecules that make up electron transport chains. Electron carriers shuttle electrons during the redox reactions that release energy ultimately used for ATP synthesis.

electron microscope (EM) An instrument that focuses an electron beam through, or onto the surface of, a specimen. An electron microscope achieves a thousandfold greater resolving power than a light microscope; the most powerful EM can distinguish objects as small as 0.2 nm ($2 + 10^{-10}$ m).

electron shell An energy level representing the distance of an electron from the nucleus of an atom.

electron transport A redox (oxidation-reduction) reaction in which one or more electrons are transferred to carrier molecules. A series of such reactions, called an electron transport chain, can release the energy stored in high-energy molecules, such as glucose.

electron transport chain A series of electron carrier molecules that shuttle electrons during the redox reactions that release energy used to make ATP; located in the inner membrane of mitochondria, the thylakoid membranes of chloroplasts, and the plasma membranes of prokaryotes.

electronegativity The tendency for an atom to pull electrons toward itself.

electrophoresis *See* gel electrophoresis.

element A substance that cannot be broken down to other substances by chemical means. Scientists recognize 92 chemical elements occurring in nature.

embryonic stem cell (ES cell) Any of the cells in the early animal embryo that differentiate during development to give rise to all the different kinds of specialized cells in the body.

embryophyte Another name for land plants, recognizing that land plants share the common derived trait of multicellular, dependent embryos.

endangered species As defined in the U.S. Endangered Species Act, a species that is in danger of extinction throughout all or a significant portion of its range.

endemic species A species of organism whose distribution is limited to a specific geographic area.

endergonic reaction (en´-der-gon´-ik) An energy-requiring chemical reaction, which yields products with more potential energy than the reactants. The amount of energy stored in the products equals the difference between the potential energy in the reactants and that in the products.

endocytosis (en´-dō-sī-tō´-sis) The movement of materials into the cytoplasm of a cell via membranous vesicles or vacuoles.

endomembrane system A network of membranous organelles that partition the cytoplasm of eukaryotic cells into functional compartments. Some of the organelles are structurally connected to each other, whereas others are structurally separate but functionally connected by the traffic of membranous vesicles between them.

endoplasmic reticulum (ER) An extensive membranous network in a eukaryotic cell, continuous with the outer nuclear membrane and composed of ribosome-studded (rough) and ribosome-free (smooth) regions. *See also* rough ER; smooth ER.

endoskeleton A hard skeleton located within the soft tissues of an animal; includes spicules of sponges, the hard plates of echinoderms, and the cartilage and bony skeletons of many vertebrates.

endosperm In flowering plants, a nutrient-rich mass formed by the union of a sperm cell with two polar nuclei during double fertilization; provides nourishment to the developing embryo in the seed.

endospore A thick-coated, protective cell produced within a bacterial cell exposed to harsh conditions.

endosymbiosis (en´-dō-sim´-bē-ō-sis) A process by which the mitochondria and chloroplasts of eukaryotic cells probably evolved, from symbiotic associations between small prokaryotic cells living inside larger ones.

endotherm An animal that derives most of its body heat from its own metabolism.

endotoxin A poisonous component of the cell walls of certain bacteria.

energy The capacity to perform work, or to move matter in a direction it would not move if left alone.

energy coupling In cellular metabolism, the use of energy released from an exergonic reaction to drive an endergonic reaction.

energy flow The passage of energy through the components of an ecosystem.

energy of activation (E_A) The amount of energy that reactants must absorb before a chemical reaction will start.

energy pyramid A diagram depicting the cumulative loss of energy from a food chain.

enhancer A eukaryotic DNA sequence that helps stimulate the transcription of a gene at some distance from it. An enhancer functions by means of a transcription factor called an activator, which binds to it and then to the rest of the transcription apparatus. *See* silencer.

entomology The study of insects.

entropy (en´-truh-pē) A measure of disorder. One form of disorder is heat, which is random molecular motion.

enzyme (en´-zīm) A protein that serves as a biological catalyst, changing the rate of a chemical reaction without itself being changed in the process.

enzyme inhibitor A chemical that interferes with an enzyme's activity.

equilibrial life history (ē´-kwi-lib´-rē-ul) Often seen in larger-bodied species, the pattern of reproducing when mature and producing few offspring but caring for the young.

essential fatty acids Certain unsaturated fatty acids that animals cannot make.

estivation (es´-tuh-vā´-shun) An animal's state of reduced activity (torpor) during periods of high environmental temperatures and reduced food and water supplies.

estuary (es´-chū-âr-ē) An area where fresh water merges with seawater.

eugenics (yū-jen´-iks) A socially rejected practice, among humans, of attempting to eliminate genetic disorders and "undesirable" inherited traits by selective breeding.

Eukarya (yū-kār-ē-uh) The domain of eukaryotes, organisms made of eukaryotic cells; includes all of the protists, plants, fungi, and animals.

eukaryotic cell (yū-kār-ē-ot´-ik) A type of cell that has a membrane-enclosed nucleus and other membrane-enclosed organelles. All organisms except bacteria and archaea are composed of eukaryotic cells.

eutherian (yū-thēr´-ē-un) A placental mammal; a mammal whose young complete their embryonic development within the uterus, joined to the mother by the placenta.

eutrophication (yū-trō-fuh-kā´-shun) An increase in productivity of an aquatic ecosystem.

evaporative cooling The property of a liquid whereby the surface becomes cooler during evaporation, owing to a loss of highly kinetic molecules to the gaseous state.

evo-devo The research field that combines evolutionary biology with developmental biology.

evolution Genetic change in a population or species over generations; all the changes that transform life on Earth; the heritable changes that have produced Earth's diversity of organisms.

evolutionary adaptation An inherited characteristic that enhances an organism's ability to survive and reproduce in a particular environment.

exaptation (ek´-sap-tā´-shun) A structure that has evolved in one environmental context and later becomes adapted for a different function in a different environmental context.

exergonic reaction (ek´-ser-gon´-ik) An energy-releasing chemical reaction in which the reactants contain more potential energy than the products. The reaction releases an amount of energy equal to the difference in potential energy between the reactants and the products.

exocytosis (ek´-sō-sī-tō´-sis) The movement of materials out of the cytoplasm of a cell via membranous vesicles or vacuoles.

exon (ek´-son) In eukaryotes, a coding portion of a gene. *See* intron.

exoskeleton A hard, external skeleton that protects an animal and provides points of attachment for muscles.

exotoxin A poisonous protein secreted by certain bacteria.

exponential growth model A mathematical description of idealized, unregulated population growth.

external fertilization The fusion of gametes that parents have discharged into the environment.

extracellular matrix A substance in which the cells of an animal tissue are embedded; consists of protein and polysaccharides.

extraembryonic membranes Four membranes (the yolk sac, amnion, chorion, and allantois) that form a life-support system for the developing embryo of a reptile, bird, or mammal.

extreme halophiles Microorganisms that live in unusually highly saline environments such as the Great Salt Lake or the Dead Sea.

extreme thermophiles Microorganisms that thrive in hot environments (often 60–80°C).

eye cup The simplest type of photoreceptor, a cluster of photoreceptor cells shaded by a cuplike cluster of pigmented cells; detects light intensity and direction.

F

F factor A piece of DNA that can exist as a bacterial plasmid; carries genes for making sex pili and other structures needed for conjugation, as well as a site where DNA replication can start. F stands for fertility.

F_1 generation The offspring of two parental (P generation) individuals. F_1 stands for first filial.

F_2 generation The offspring of the F_1 generation. F_2 stands for second filial.

facilitated diffusion The passage of a substance across a biological membrane down its concentration gradient, aided by specific transport proteins.

facultative anaerobe (fak´-ul-tā´-tiv an´-uh-rōb) A microorganism that makes ATP by aerobic respiration if oxygen is present, but that switches to fermentation when oxygen is absent.

family In classification, the taxonomic category above genus.

fat A large lipid molecule made from an alcohol called glycerol and three fatty acids; a triglyceride. Most fats function as energy-storage molecules.

feedback regulation A method of metabolic control in which the end product of a metabolic pathway acts as an inhibitor of an enzyme within that pathway.

fermentation The anaerobic harvest of food by some cells.

fern One of a group of seedless vascular plants.

fertilization The union of the nucleus of a sperm cell with the nucleus of an egg cell, producing a zygote.

fever An abnormally high internal body temperature, usually the result of an infection.

fiber (1) In animals, an elongate, supportive thread in the matrix of connective tissue; an extension of a neuron; a muscle cell. (2) In plants, a long, slender sclerenchyma cell that usually occurs in a bundle.

first law of thermodynamics The natural law stating that the total amount of energy in the universe is constant and that energy can be transferred and transformed, but never destroyed; also called the principle of energy conservation.

five-kingdom system The system of taxonomic classification based on five basic groups: Monera, Protista, Plantae, Fungi, and Animalia.

fixed action pattern (FAP) A genetically programmed, virtually unchangeable behavioral sequence performed in response to a certain stimulus.

flagellate (flaj´-uh-lit) A protist (protozoan) that moves by means of one or more flagella.

flagellum (fluh-jel´-um) (plural, **flagella**) A long appendage that propels protists through the water and moves fluids across the surface of many tissue cells in animals. A cell may have one or more flagella. Like cilia, flagella have a 9 + 2 arrangement of microtubules covered by the cell's plasma membrane.

flatworm A member of the phylum Platyhelminthes.

flower In an angiosperm, a short stem with four sets of modified leaves, bearing structures that function in sexual reproduction.

fluid feeder An animal that lives by sucking nutrient-rich fluids from another living organism.

fluid mosaic A description of membrane structure, depicting a cellular membrane as a mosaic of diverse protein molecules embedded in a fluid bilayer made of phospholipid molecules.

fluke One of a group of parasitic flatworms.

food chain A sequence of food transfers from producers through several levels of consumers in an ecosystem.

food vacuole (vak´-ū-ōl) The simplest type of digestive cavity, found in single-celled organisms.

food web A network of interconnecting food chains.

foot In an invertebrate animal, a structure used for locomotion or attachment such as the muscular organ extending from the ventral side of a mollusk.

foram A marine protozoan that secretes a shell and extends pseudopodia through pores in its shell.

forebrain One of three ancestral and embryonic regions of the vertebrate brain; develops into the thalamus, hypothalamus, and cerebrum.

forensics The scientific analysis of evidence for crime scene investigations and other legal proceedings.

fossil A preserved remnant or impression of an organism that lived in the past.

fossil fuel An energy deposit formed from the remains of extinct organisms.

fossil record The chronicle of evolution over millions of years of geologic time engraved in the order in which fossils appear in rock strata.

founder effect Random change in the gene pool that occurs in a small colony of a population.

free-living flatworm One of a group of non-parasitic flatworms.

fruit A ripened, thickened ovary of a flower, which protects dormant seeds and aids in their dispersal.

fruiting body A stage in an organism's life cycle that functions only in reproduction; for example, a mushroom is a fruiting body of many fungi.

functional group The atoms that form the chemically reactive part of an organic molecule.

Fungi (fun´-jē) The kingdom that contains the fungi.

fungus (plural, **fungi**) A heterotrophic eukaryote that digests its food externally and absorbs the resulting small nutrient molecules. Most fungi consist of a netlike mass of filaments called hyphae. Molds, mushrooms, and yeasts are examples of fungi.

G

gametangium (gam´-uh-tan´-jē-um) (plural, **gametangia**) A reproductive organ that houses and protects the gametes of a plant.

gamete (gam´-ēt) A sex cell; a haploid egg or sperm. The union of two gametes of opposite sex (fertilization) produces a zygote.

gametic isolation (guh-mē´-tik) A type of prezygotic barrier between species; the species remain isolated because male and female gametes of the different species cannot fuse, or they die before they unite.

gametophyte (guh-mē´-tō-fīt) The multicellular haploid form in the life cycle of organisms undergoing alternation of generations; mitotically produces haploid gametes that unite and grow into the sporophyte generation.

ganglion (gang´-glē-un) (plural, **ganglia**) A cluster (functional group) of nerve cell bodies in a centralized nervous system.

gap analysis A method researchers use to study the distribution of organisms relative to landscape features and habitat types.

gas exchange *See* respiration.

gastric ulcer An open sore in the lining of the stomach, resulting when pepsin and hydrochloric acid destroy the lining tissues faster than they can regenerate.

gastrin A digestive hormone that stimulates the secretion of gastric juice.

gastropod A member of the largest group of mollusks, including snails and slugs.

gastrovascular cavity A digestive compartment with a single opening, the mouth; may function in circulation, body support, waste disposal, and gas exchange, as well as digestion.

gel electrophoresis (jel´ ē-lek´-trō-fōr-ē´-sis) A technique for separating and purifying macromolecules. A mixture of molecules is placed on a gel between a positively charged electrode and a negatively charged one; negative charges on the molecules are attracted to the positive electrode, and the molecules migrate toward that electrode; the molecules separate in the gel according to their rates of migration.

gene A discrete unit of hereditary information consisting of a specific nucleotide sequence in DNA (or RNA, in some viruses). Most of the genes of a eukaryote are located in its chromosomal DNA; a few are carried by the DNA of mitochondria and chloroplasts.

gene cloning The production of multiple copies of a gene.

gene expression The process whereby genetic information flows from genes to proteins; the flow of genetic information from the genotype to the phenotype.

gene flow The gain or loss of alleles from a population by the movement of individuals or gametes into or out of the population.

gene pool All the genes in a population at any one time.

gene therapy A treatment for a disease in which the patient is provided with a new gene.

genealogical species concept The idea that species are defined as a set of organisms with a unique genetic history.

genetic code The set of rules giving the correspondence between nucleotide triplets (codons) in mRNA and amino acids in protein.

genetic drift A change in the gene pool of a population due to chance.

genetic marker (1) An allele tracked in a genetic study. (2) A specific section of DNA that earmarks a particular allele; may contain specific restriction sites (points where restriction enzymes cut the DNA) that occur only in DNA that contains the allele.

genetic recombination The production, by crossing over and/or independent assortment of chromosomes during meiosis, of offspring with allele combinations different from those in the parents. The term may also be used more specifically to mean the production by crossing over of eukaryotic or prokaryotic chromosomes with gene combinations different from those in the original chromosomes.

genetically modified (GM) organism An organism that has acquired one or more genes by artificial means. If the gene is from another species, the organism is also known as a transgenic organism.

genetics The scientific study of heredity and hereditary variations.

genome (jē´-nōm) A complete (haploid) set of an organism's genes; an organism's genetic material.

genomic library (juh-nō´-mik) A set of DNA segments from an organism's genome. Each segment is usually carried by a plasmid or phage.

genomics The study of whole sets of genes and their interactions.

genotype (jē´-nō-tīp) The genetic makeup of an organism.

genus (jē´-nus) (plural, **genera**) In classification, the taxonomic category above species; the first part of a species' binomial; for example, *Homo*.

geologic time scale A time scale established by geologists that reflects a consistent sequence of historical periods, grouped into four eras: Precambrian, Paleozoic, Mesozoic, and Cenozoic.

gizzard A pouchlike organ in a digestive tract, where food is mechanically ground.

global warming A slow but steady rise in Earth's surface temperature, caused by increasing concentrations of greenhouse gases (such as CO_2 and CH_4) in the atmosphere.

glomerulus (glō-mār´-ū-lus) (plural, **glomeruli**) In the vertebrate kidney, the part of a nephron consisting of the capillaries that are surrounded by Bowman's capsule. Together, a glomerulus and Bowman's capsule produce the filtrate from the blood.

glycogen (glī´-kō-jen) A complex, extensively branched polysaccharide of many glucose monomers; serves as an energy-storage molecule in liver and muscle cells.

glycolysis (glī-kol´-uh-sis) The multistep chemical breakdown of a molecule of glucose into two molecules of pyruvic acid; the first stage of cellular respiration in all organisms; occurs in the cytoplasmic fluid.

glycoprotein (glī-kō-prō´-tēn) A macromolecule consisting of one or more polypeptides linked to short chains of sugars.

goiter An enlargement of the thyroid gland resulting from a dietary iodine deficiency.

Golgi apparatus (gol´-jē) An organelle in eukaryotic cells consisting of stacks of membranous sacs that modify, store, and ship products of the endoplasmic reticulum.

Gondwana (gon-dwa´-na) The southern landmass formed during the Mesozoic era when continental drift split Pangaea (all land masses fused). *See also* Laurasia.

gradualist model The view that evolution occurs as a result of populations becoming isolated from common ancestral stock and gradually becoming genetically unique as they are adapted by natural selection to their local environments; Darwin's view of the origin of species.

granum (gran´-um) (plural, **grana**) A stack of hollow disks formed of thylakoid membrane in a chloroplast. Grana are the sites where light energy is trapped by chlorophyll and converted to chemical energy during the light reactions of photosynthesis.

green alga One of a group of photosynthetic protists that includes unicellular, colonial, and multicellular species. Green algae are plantlike in having biflagellated cells (gametes in colonial and multicellular species), chloroplasts with chlorophyll *a*, cellulose cell walls, and starch.

greenhouse effect The warming of the atmosphere caused by CO_2, CH_4, and other gases that absorb infrared radiation and slow its escape from Earth's surface.

growth factor A protein secreted by certain body cells that stimulates other cells to divide.

guanine (G) (gwa´-nēn) A double-ring nitrogenous base found in DNA and RNA.

gymnosperm (jim´-nō-sperm) A naked-seed plant. Its seed is said to be naked because it is not enclosed in a fruit.

H

habitat A place where an organism lives; an environmental situation in which an organism lives.

habitat isolation A type of prezygotic barrier between species; the species remain isolated because they breed in different habitats.

habituation Learning not to respond to a repeated stimulus that conveys little or no information.

hair cell A type of mechanoreceptor that detects sound waves and other forms of movement in air or water.

haploid Containing a single set of chromosomes; referring to an *n* cell.

haploid cell In the life cycle of an organism that reproduces sexually, a cell containing a single set of chromosomes; an *n* cell.

Hardy-Weinberg equilibrium The principle that the shuffling of genes that occurs during sexual reproduction, by itself, cannot change the overall genetic makeup of a population.

Hardy-Weinberg formula A formula for calculating the frequencies of genotypes in a gene pool from the frequencies of alleles, and vice versa.

HDL *See* high-density lipoprotein.

heat The amount of energy associated with the movement of the atoms and molecules in a body of matter. Heat is energy in its most random form.

hemophilia (hē´-mō-fil´-ē-uh) A human genetic disease caused by a sex-linked recessive allele and characterized by excessive bleeding following injury.

hepatic portal vessel A blood vessel that conveys blood from capillaries surrounding the intestine directly to the liver.

herbivore An animal that eats mainly plants or algae. *See also* carnivore; omnivore.

heterotroph (het´-er-ō-trōf) An organism that cannot make its own organic food molecules and must obtain them by consuming other organisms or their organic products; a consumer or a decomposer in a food chain.

heterozygote advantage Greater reproductive success of heterozygous individuals compared to homozygotes; tends to preserve variation in gene pools.

heterozygous (het´-er-ō-zī´-gus) Having two different alleles for a given gene.

hibernation Long-term torpor during cold weather; an animal's metabolic rate is reduced, and it survives on energy stored in body fat.

high-density lipoprotein (HDL) A cholesterol-carrying particle in the blood, made up of cholesterol and other lipids surrounded by a single layer of phospholipids in which proteins are embedded. An HDL particle carries less cholesterol than a related lipoprotein, LDL, and may be correlated with a decreased risk of blood vessel blockage.

hindbrain One of three ancestral and embryonic regions of the vertebrate brain; develops into the medulla oblongata, pons, and cerebellum.

hippocampus (hip´-uh-kam´-pus) An integrative center of the cerebrum; functionally, part of the limbic system that plays central roles in memory and learning.

histone (his´-tōn) A small basic protein molecule associated with DNA and important in DNA packing in the eukaryotic chromosome.

HIV Human immunodeficiency virus, the retrovirus that attacks the human immune system and causes AIDS.

homeobox (hō´-mē-ō-boks´) A 180-nucleotide sequence within a homeotic gene encoding the part of the protein that binds to the DNA of the genes regulated by the protein.

homeotic gene (hō´-mē-ot´-ik) A master control gene that determines the identity of a body structure of a developing organism, presumably by controlling the developmental fate of groups of cells. (In plants, such genes are called organ identity genes.)

hominid (hah´-mi-nid) A species on the human branch of the evolutionary tree; a member of the family Hominidae, including *Homo sapiens* and our ancestors.

hominoid A term that refers to great apes and humans.

homologous chromosomes (hō-mol´-uh-gus) The two chromosomes that make up a matched pair in a diploid cell. Homologous chromosomes are of the same length, centromere position, and staining pattern and possess genes for the same characteristics at corresponding loci. One homologous chromosome is inherited from the organism's father, the other from the mother.

homologous structures Structures that are similar in different species of common ancestry.

homology (hō-mol´-uh-jē) Anatomical similarity due to common ancestry.

homozygous (hō´-mō-zī´-gus) Having two identical alleles for a given gene.

horseshoe crab A bottom-dwelling organism that belongs to the phylum Arthropoda.

host The larger participant in a symbiotic relationship, serving as home and feeding ground to the symbiont.

Human Genome Project An international collaborative effort to map and sequence the DNA of the entire human genome.

humus (hyū´-mus) Decomposing organic material found in topsoil.

Huntington disease A human genetic disease caused by a dominant allele; characterized by uncontrollable body movements and degeneration of the nervous system; usually fatal 10 to 20 years after the onset of symptoms.

hybrid The offspring of parents of two different species or of two different varieties of one species; the offspring of two parents that differ in one or more inherited traits; an individual that is heterozygous for one or more pair of genes.

hybrid breakdown A type of postzygotic barrier between species; the species remain isolated because the offspring of hybrids are weak or infertile.

hybrid inviability A type of postzygotic barrier between species; the species remain isolated because hybrid zygotes do not develop or hybrids do not become sexually mature.

hybrid sterility A type of postzygotic barrier between species; the species remain isolated because hybrids fail to produce functional gametes.

hybridization The cross-fertilization of two different varieties of an organism or of two different species; also called a cross.

hydrocarbon A chemical compound composed only of the elements carbon and hydrogen.

hydrogen bond A type of weak chemical bond formed when the partially positive hydrogen atom participating in a polar covalent bond in one molecule is attracted to the partially negative atom participating in a polar covalent bond in another molecule (or in another part of the same macromolecule).

hydrolysis (hī-drol´-uh-sis) A chemical process in which macromolecules are broken down by the chemical addition of water molecules to the bonds linking their monomers; an essential part of digestion.

hydrophilic (hī-drō-fil´-ik) "Water-loving"; pertaining to polar or charged molecules (or parts of molecules) that are soluble in water.

hydrophobic (hī-drō-fō´-bik) "Water-fearing"; pertaining to nonpolar molecules (or parts of molecules) that do not dissolve in water.

hydrostatic skeleton A skeletal system composed of fluid held under pressure in a closed body compartment; the main skeleton of most cnidarians, flatworms, nematodes, and annelids.

hydrothermal vent community A seafloor community powered by chemical energy from Earth's interior rather than by sunlight.

hydroxyl group (hī-drok´-sil) In an organic molecule, a functional group consisting of a hydrogen atom bonded to an oxygen atom.

hypercholesterolemia (hī´-per-kō-les´-tur-ah-lēm´-ē-uh) An inherited human disease characterized by an excessively high level of cholesterol in the blood.

hypertonic In comparing two solutions, referring to the one with the greater concentration of solutes.

hyperventilating Taking several deep breaths so rapidly that the CO_2 level in the blood is reduced, causing the breathing control centers to temporarily shut down breathing movements.

hypha (hī´-fuh) (plural, **hyphae**) One of many filaments making up the body of a fungus.

hypothesis (hī-poth´-uh-sis) (plural, **hypotheses**) A tentative explanation a scientist proposes for a specific phenomenon that has been observed.

hypotonic In comparing two solutions, referring to the one with the lower concentration of solutes.

I

imitation Learning by observing and mimicking the behavior of others.

imprinting Learning that is limited to a specific critical period in an animal's life and that is generally irreversible.

in-group In a cladistic study of evolutionary relationships among taxa of organisms, the group of taxa that is actually being analyzed. *See also* out-group.

inbreeding The mating of close relatives.

incomplete dominance A type of inheritance in which the phenotype of a heterozygote (*Aa*) is intermediate between the phenotypes of the two types of homozygotes (*AA* and *aa*).

incomplete metamorphosis A type of development in certain insects, such as grasshoppers, in which the larvae resemble adults but are smaller and have different body proportions. The animal goes through a series of molts, each time looking more like an adult, until it reaches full size.

induced fit The interaction between a substrate molecule and the active site of an enzyme, which changes shape slightly to embrace the substrate and catalyze the reaction.

induction During embryonic development, the influence of one group of cells on another group of cells.

inferior vena cava (vē´-nuh kā´-vuh) A large vein that returns O_2-poor blood to the heart from the lower, or posterior, part of the body. *See also* superior vena cava.

ingestion The act of eating; the first main stage of food processing.

inhibiting hormone A kind of hormone released from the hypothalamus that makes the anterior pituitary stop secreting hormone.

innate behavior Behavior that appears to be performed in virtually the same way by all members of a species.

insect An arthropod that usually has three body segments (head, thorax, and abdomen), three pairs of legs, and one or two pairs of wings.

integumentary system (in-teg´-yū-men´-ter-ē) The organ system consisting of the skin and its derivatives, such as hair and nails in mammals; helps protect the body from drying out, mechanical injury, and infection.

intermediate One of the compounds that form between the initial reactant in a metabolic pathway, such as glucose in glycolysis, and the final product, such as pyruvic acid in glycolysis.

intermediate disturbance hypothesis The idea that species diversity is greatest in places where disturbance is moderate in both severity and frequency.

intermediate filament An intermediate-sized protein fiber that is one of the three main kinds of fibers making up the cytoskeleton of eukaryotic cells; ropelike, made of fibrous proteins.

intermembrane space One of the two fluid-filled internal compartments of the mitochondrion, the narrow region between the inner and outer membranes.

interphase The period in the eukaryotic cell cycle when the cell is not actually dividing. *See* mitosis.

interspecific competition Competition between populations of two or more species that require similar limited resources. Interspecific competition may inhibit population growth and help structure communities.

interspecific interaction Any interaction between members of different species.

intertidal zone (in´-ter-tīd´-ul) A shallow zone where the waters of an estuary or ocean meet land.

intestine The region of a digestive tract between the gizzard or stomach and the anus, where chemical digestion and nutrient absorption usually occur.

intraspecific competition Competition between individuals of the same species for a limited resource.

intrinsic rate of increase An organism's inherent capacity to reproduce.

introduced species A species that humans move from the species' native location to a new geographic region; sometimes called an exotic species.

intron (in´-tron) In eukaryotes, a nonexpressed (noncoding) portion of a gene that is excised from the RNA transcript. *See* exon.

inversion A change in a chromosome resulting from reattachment in a reverse direction of a chromosome fragment to the original chromosome. Mutagens and errors during meiosis can cause inversions.

invertebrate An animal that lacks a backbone.

ion (ī´-on) An atom or molecule that has gained or lost one or more electrons, thus acquiring an electrical charge.

ionic bond (ī-on´-ik) An attraction between two ions with opposite electrical charges. The electrical attraction of the opposite charges holds the ions together.

isomers (ī´-sō-mers) Organic compounds with the same molecular formula but different structures and thus different properties.

isotonic (ī-sō-ton´-ik) Having the same solute concentration as another solution.

isotope (ī´-sō-tōp) A variant form of an atom. Isotopes of an element have the same number of protons but different numbers of neutrons.

K

K-selection The concept that in certain (*K*-selected) populations, life history is centered around producing relatively few offspring that have a good chance of survival.

karyotype (kār´-ē-ō-tīp) A display of micrographs of the metaphase chromosomes of a cell, arranged by size and centromere position.

kelp A giant brown alga, up to 100 m long, that forms extensive undersea forests.

keystone predator A predator species that reduces the density of the strongest competitors in a community, thereby helping maintain species diversity.

keystone species Species that are not usually abundant in a community yet exert strong control on community structure by the nature of their ecological roles or niches.

kin selection The concept that altruism evolves because it increases the number of copies of a gene common to a genetically related group of organisms; a hypothesis about the ultimate cause of altruism.

kinesis (kuh-nē´-sis) Random movement in response to a stimulus.

kinetochore (kuh-net´-ō-kor) A specialized protein structure at the centromere region on a sister chromatid. Spindle microtubules attach to the kinetochore during mitosis and meiosis.

kinetic energy (kuh-net´-ik) Energy that is actually doing work; the energy of a mass of matter that is moving. Moving matter performs work by transferring its motion to other matter, such as leg muscles pushing bicycle pedals.

kingdom In classification, the broad taxonomic category above phylum or division.

Koch's postulates A set of criteria used to establish that a particular infectious agent causes a disease.

Krebs cycle The metabolic cycle that is fueled by acetyl CoA formed after glycolysis in cellular respiration. Chemical reactions in the Krebs cycle complete the metabolic breakdown of glucose molecules to carbon dioxide. The Krebs cycle occurs in the matrix of mitochondria and supplies most of the NADH molecules that carry energy to the electron transport chains.

L

labor A series of strong, rhythmic contractions of the uterus that expel a baby out of the uterus and vagina during childbirth.

lactic acid fermentation The conversion of pyruvate to lactate with no release of carbon dioxide.

lancelet One of a group of invertebrate chordates.

landmark A point of reference for orientation during navigation.

landscape ecology The application of ecological principles to the study of land-use patterns; the scientific study of the biodiversity of interacting ecosystems.

larva (lar´-vuh) (plural, **larvae**) A free-living, sexually immature form in some animal life cycles that may differ from the adult in morphology, nutrition, and habitat.

larynx (lār´-inks) The voice box, containing the vocal cords.

lateral Pertaining to the side of a bilaterally symmetrical animal.

lateral line system A row of sensory organs along each side of a fish's body. Sensitive to changes in water pressure, it enables a fish to detect minor vibrations in the water.

Laurasia (lah-rā´-zhuh) The northern landmass formed when continental drift split Pangaea (all land masses fused) during the Mesozoic era. *See also* Gondwana.

LDL *See* low-density lipoprotein.

learning A behavioral change resulting from experience.

leeches One of the three large groups of annelids. *See* Annelida.

leukemia (lū-kī´-mē-ah) A type of cancer of the blood-forming tissues, characterized by an excessive production of white blood cells and an abnormally high number of them in the blood; cancer of the bone marrow cells that produce leukocytes.

lichen (lī´-ken) A mutualistic association between a fungus and an alga or between a fungus and a cyanobacterium.

life cycle The entire sequence of stages in the life of an organism, from the adults of one generation to the adults of the next.

life history The series of events from birth through reproduction to death.

life table A listing of survivals and deaths in a population in a particular time period and predictions of how long, on average, an individual of a given age will live.

light microscope (LM) An optical instrument with lenses that refract (bend) visible light to magnify images and project them into a viewer's eye or onto photographic film.

light reactions The first of two stages in photosynthesis, the steps in which solar energy is absorbed and converted to chemical energy in the form of ATP and NADPH. The light reactions power the sugar-producing Calvin cycle but produce no sugar themselves.

lignin (lig´-nin) A chemical that hardens the cell walls of plants.

linkage map A map of a chromosome showing the relative positions of genes.

linked genes Genes located close enough together on a chromosome to be usually inherited together.

lipid An organic compound consisting mainly of carbon and hydrogen atoms linked by nonpolar covalent bonds and therefore mostly hydrophobic and insoluble in water. Lipids include fats, waxes, phospholipids, and steroids.

lobe-finned fish A bony fish with strong, muscular fins supported by bones. Lobe-fins are extinct except for one species, the coelacanth.

local regulator A chemical messenger that is secreted into the interstitial fluid and causes changes in cells very near the point of secretion. Prostaglandins and neurotransmitters are local regulators.

locomotion Active movement from place to place.

locus (plural, **loci**) The particular site where a gene is found on a chromosome. Homologous chromosomes have corresponding gene loci.

logistic growth model A mathematical description of idealized population growth that is restricted by limiting factors.

long-day plant A plant that flowers in late spring or early summer when day length is increasing.

long-term depression (LTD) A reduced responsiveness to an action potential (nerve signal) by a receiving neuron.

long-term memory The ability to hold, associate, and recall information over one's life.

long-term potentiation (LTP) An enhanced responsiveness to an action potential (nerve signal) by a receiving neuron.

loop of Henle (hen´-lē) In the vertebrate kidney, the portion of a nephron that helps concentrate the filtrate while conveying it between a proximal tubule and a distal tubule.

low-density lipoprotein (LDL) A cholesterol-carrying particle in the blood, made up of cholesterol and other lipids surrounded by a single layer of phospholipids in which proteins are embedded. An LDL particle carries more cholesterol than a related lipoprotein, HDL, and high LDL levels in the blood correlate with a tendency to develop blocked blood vessels and heart disease.

lungfish A bony fish that generally inhabits stagnant waters and gulps air into lungs connected to a pharynx.

Lyme disease A debilitating human disease caused by the bacterium *Borrelia burgdorferi*; characterized at first by a red rash at the site of a tick bite and, if not treated, by heart disease, arthritis, and nervous disorders.

lymphoma (lim-fō´-muh) Cancer of the tissues that form white blood cells.

lysogenic cycle (lī-sō-jen´-ik) A bacteriophage replication cycle in which the viral genome is incorporated into the bacterial host chromosome as a prophage. New phages are not produced, and the host cell is not killed or lysed unless the viral genome leaves the host chromosome.

lysosomal storage disease A hereditary disorder associated with abnormal lysosomes, where the sufferer is missing one of the lysosomal digestive enzymes.

lysosome (lī´-sō-sōm) A digestive organelle in eukaryotic cells; contains hydrolytic enzymes that digest the cell's food and wastes.

lytic cycle (lit´-ik) A viral replication cycle resulting in the release of new viruses by lysis (breaking open) of the host cell.

M

macroevolution Evolutionary change on a grand scale, encompassing the origin of new taxonomic groups, evolutionary trends, adaptive radiation, and mass extinction.

macromolecule A giant molecule in a living organism: a protein, polysaccharide, or nucleic acid.

magnetic resonance imaging (MRI) Imaging technology that uses magnetism and radio waves to induce hydrogen nuclei in water molecules to emit faint radio signals. A computer creates images of the body from the radio signals.

magnification An increase in the apparent size of an object.

major histocompatibility complex (MHC) *See* self protein.

malignant tumor An abnormal tissue mass that can spread into neighboring tissue and to other parts of the body; a cancerous tumor.

Mammalia The class that includes endothermic vertebrates that possess mammary glands and hair.

mantle In mollusks, the outgrowth of the body surface that drapes over the animal. The mantle produces the shell and forms the mantle cavity.

mark-recapture method A sampling technique used to estimate wildlife populations.

marsupial (mar-sū´-pē-ul) A pouched mammal, such as a kangaroo, opossum, or koala. Marsupials give birth to embryonic offspring that complete development while housed in a pouch and attached to nipples on the mother's abdomen.

marsupium (mar-sū´-pē-um) (plural, **marsupia**) The external pouch on the abdomen of a female marsupial.

mass The amount of matter in an object.

mass number The sum of the number of protons and neutrons in an atom's nucleus.

mating type A group of sexually compatible individuals in a population. Certain species of prokaryotes, protists, and fungi have mating types.

matter Anything that occupies space and has mass.

maximum sustained yield The level of harvest that produces a consistent yield without forcing a population into decline.

mechanical isolation A type of prezygotic barrier between species; the species remain isolated because structural differences between them prevent fertilization.

medusa (med-ū´-suh) (plural, **medusae**) One of two types of cnidarian body forms; an umbrella-like body form; also called a jelly.

meiosis (mī-ō´-sis) In a sexually reproducing organism, the division of a single diploid nucleus into four haploid daughter nuclei. Meiosis and cytokinesis produce haploid gametes from diploid cells in the reproductive organs of the parents.

membrane infolding A process by which the eukaryotic cell's endo-membrane system evolved from inward folds of the plasma membrane of a prokaryotic cell.

Mendel's principle of independent assortment A general rule in inheritance that when gametes form during meiosis, each pair of alleles for a particular characteristic segregate independently; also known as Mendel's second law of inheritance.

Mendel's principle of segregation A general rule in inheritance that individuals have two alleles for each gene and that when gametes form by meiosis, the two alleles separate, and each resulting gamete ends up with only one allele of each gene; also known as Mendel's first law of inheritance.

mesophyll (mes´-ō-fil) The green tissue in the interior of a leaf; a leaf's ground tissue system, the main site of photosynthesis.

messenger RNA (mRNA) The type of ribonucleic acid that encodes genetic information from DNA and conveys it to ribosomes, where the information is translated into amino acid sequences.

metabolism (muh-tab´-uh-liz-um) The many chemical reactions that occur in organisms.

metamorphosis (met´-uh-mōr´-fuh-sis) The transformation of a larva into an adult.

metaphase (met´-eh-fāz) The second stage of mitosis. During metaphase, all the cell's duplicated chromosomes are lined up at an imaginary plane equidistant between the poles of the mitotic spindle.

metastasis (muh-tas´-tuh-sis) The spread of cancer cells beyond their original site.

methanogens Microorganisms that obtain energy by using carbon dioxide to oxidize hydrogen, producing methane as a waste product.

microevolution A change in a population's gene pool over a succession of generations; evolutionary changes in species over relatively brief periods of geologic time.

microfilament The thinnest of the three main kinds of protein fibers making up the cytoskeleton of a eukaryotic cell; a solid, helical rod composed of the globular protein actin.

micrograph A photograph taken through a microscope.

microtubule The thickest of the three main kinds of fibers making up the cytoskeleton of a eukaryotic cell; a straight, hollow tube made of globular proteins called tubulins. Microtubules form the basis of the structure and movement of cilia and flagella.

microvillus (plural, **microvilli**) A microscopic projection on the surface of a cell. Microvilli increase a cell's surface area.

midbrain One of three ancestral and embryonic regions of the vertebrate brain; develops into sensory integrating and relay centers that send sensory information to the cerebrum.

migration The regular back-and-forth movement of animals between two geographic areas at particular times of the year.

millipede A terrestrial arthropod that has two pairs of short legs for each of its numerous body segments and that eats decaying plant matter.

mineralocorticoid (min´-er-uh-lō-kort´-uh-koyd) A corticosteroid hormone secreted by the adrenal cortex that helps maintain salt and water homeostasis and may increase blood pressure in response to long-term stress.

mitochondrial matrix (mī´-tō-kon´-drē-ul mā´-triks) The fluid contained within the inner membrane of a mitochondrion.

mitochondrion (mī´-tō-kon´-drē-on) (plural, **mitochondria**) An organelle in eukaryotic cells where cellular respiration occurs. Enclosed by two concentric membranes, it is where most of the cell's ATP is made.

mitosis (mī´-tō-sis) The division of a single nucleus into two genetically identical daughter nuclei. Mitosis and cytokinesis make up the mitotic (M) phase of the cell cycle.

mitotic phase The part of the cell cycle when mitosis divides the nucleus and distributes its chromosomes to the daughter nuclei and cytokinesis divides the cytoplasm, producing two daughter cells.

mitotic spindle A spindle-shaped structure formed of microtubules and associated proteins that is involved in the movement of chromosomes during mitosis and meiosis. (A spindle is shaped roughly like a football.)

modern synthesis A comprehensive theory of evolution that incorporates genetics and includes most of Darwin's ideas, focusing on populations as the fundamental units of evolution.

molecular biology The study of the molecular basis of genes and gene expression; molecular genetics.

molecular clock Evolutionary timing method based on the observation that at least some regions of genomes evolve at constant rates.

molecule A group of two or more atoms held together by covalent bonds.

Mollusca (mol-lus´-kuh) The phylum that contains the mollusks; characterized by a muscular foot, mantle, mantle cavity, and radula; includes gastropods (snails and slugs), bivalves (clams, oysters, scallops), and cephalopods (squids and octopuses).

molting In arthropods, the process of shedding an old exoskeleton and secreting a new, larger one.

monoculture The cultivation of a single plant variety in a large land area.

monocyte (mon´-ō-sīt) A phagocytic white blood cell that can engulf bacteria and viruses in infected tissue; has a large oval or horseshoe-shaped nucleus.

monoecious (muh-nē´-shus) Having individuals that produce both sperm and eggs (usually refers to plants).

monogenesis model (mon´-ō-jen´-uh-sis) The hypothesis that humans arose from a single archaic group in Africa.

monohybrid cross An experimental mating of individuals differing at one genetic locus.

monomer (mon´-uh-mer) A chemical subunit that serves as a building block of a polymer.

monophyletic (mon´-ō-fī-let´-ik) Pertaining to a taxon derived from a single ancestral species that gave rise to no species in any other taxa.

monosaccharide (mon´-ō-sak´-uh-rīd) The smallest kind of sugar molecule; a single-unit sugar; also known as a simple sugar. Monosaccharides are the building blocks of more complex sugars and polysaccharides.

monotreme (mon´-uh-trēm) An egg-laying mammal, such as the duck-billed platypus.

morphological species concept The idea that species are defined by measurable anatomical criteria.

morphs Two or more different kinds of individuals or forms of a phenotypic characteristic in a population.

moss One of a group of seedless nonvascular plants.

movement corridor A series of small clumps or a narrow strip of quality habitat (usable by organisms) that connects otherwise isolated patches of quality habitat.

mRNA *See* messenger RNA.

mucous membrane (myū´-kus) Smooth, moist epithelium that lines the digestive tract and air tubes leading to the lungs.

Müllerian mimicry (myū-lār´-ē-un mim´-uh-krē) A mutual mimicry by two species, both of which are poisonous or otherwise harmful to a predator.

multicellular green alga *See* green alga.

multiple fruit A fruit such as pineapple that develops from a group of flowers tightly clustered together. When the walls of the many ovaries start to thicken, they fuse together and become incorporated into one fruit.

multiregional hypothesis The hypothesis that modern humans evolved in each region of the world from local populations of *Homo erectus*.

mutagen (myū´-tuh-jen) A chemical or physical agent that interacts with DNA and causes a mutation.

mutagenesis (myū´-tuh-jen´-uh-sis) The creation of a mutation.

mutation A change in the nucleotide sequence of DNA; the ultimate source of genetic diversity.

mutualism A symbiotic relationship in which both partners benefit.

mycelium (mī-sē´-lē-um) (plural, **mycelia**) The densely branched network of hyphae in a fungus.

mycorrhiza (mī´-kō-rī´-zuh) (plural, **mycorrhizae**) A mutualistic association of plant roots and fungi.

N

NAD⁺ Nicotinamide adenine dinucleotide; a coenzyme that assists enzymes by conveying electrons (from hydrogen atoms) during the redox reactions of cellular metabolism. The plus sign indicates that the molecule is oxidized and ready to pick up hydrogens; the reduced, hydrogen (electron)-carrying form is $NADH_2$.

NADH A molecule that carries electrons from glucose and other fuel molecules and deposits them at the top of an electron transport chain. NADH is generated during glycolysis and the Krebs cycle.

NADPH An electron carrier involved in photosynthesis. Light drives electrons from chlorophyll to $NADP^+$, forming NADPH, which provides the high-energy electrons for the reduction of carbon dioxide to sugar in the Calvin cycle.

natural selection Differential success in reproduction by different phenotypes resulting from interactions with the environment. Evolution occurs when natural selection produces changes in the relative frequencies of alleles in a population's gene pool.

Nematoda (nem´-uh-tōd´-uh) The phylum that contains the roundworms, or nematodes; characterized by a pseudocoelom, a cylindrical, wormlike body form, and a tough cuticle.

nerve cord An elongated bundle of axons and dendrites, usually extending longitudinally from the brain or anterior ganglia. One or more nerve cords and the brain make up the central nervous system in many animals.

nerve net A weblike system of neurons, characteristic of radially symmetrical animals such as Hydra.

neural tube (nyūr´-ul) An embryonic cylinder that develops from ectoderm after gastrulation; gives rise to the brain and spinal cord.

neuromuscular junction A synapse between an axon of a motor neuron and a muscle fiber.

neurosecretory cell A nerve cell that synthesizes hormones and secretes them into the blood, as well as conducting nerve signals.

neutral variation Genetic variation that provides no apparent selective advantage for some individuals over others.

neutron An electrically neutral particle (a particle having no electrical charge), found in the nucleus of an atom.

niche (nich) A population's role in its community; the sum total of a population's use of the biotic and abiotic resources of its habitat.

nitrogenous base (nī-troj´-en-us) An organic molecule that is a base and contains the element nitrogen.

noncompetitive inhibitor A substance that impedes the activity of an enzyme without entering an active site—and thus without competing directly with the normal substrate. By binding elsewhere on the enzyme, a noncompetitive inhibitor changes the shape of the enzyme so that the active site no longer functions.

nondisjunction An accident of meiosis or mitosis in which a pair of homologous chromosomes or a pair of sister chromatids fail to separate at anaphase.

nonpolar covalent bond An attraction between atoms that share one or more pairs of electrons equally because the atoms have similar electronegativity.

nonself molecule A foreign antigen; a protein or other macromolecule that is not part of an organism's body. *See also* self protein.

notochord (nō´-tuh-kord) A flexible, cartilage-like, longitudinal rod located between the digestive tract and nerve cord in chordate animals, present only in embryos in many species.

nuclear envelope A double membrane, perforated with pores, that encloses the nucleus and separates it from the rest of the eukaryotic cell.

nuclear transplantation A technique in which the nucleus of one cell is placed into another cell that already has a nucleus or in which the nucleus has been previously destroyed.

nucleic acid (nū-klā´-ik) A polymer consisting of many nucleotide monomers; serves as a blueprint for proteins and, through the actions of proteins, for all cellular structures and activities. The two types of nucleic acids are DNA and RNA.

nucleic acid probe (nū-klā´-ik) In DNA technology, a labeled single-stranded nucleic acid molecule used to find a specific gene or other nucleotide sequence within a mass of DNA. The probe hydrogen-bonds to the complementary sequence in the targeted DNA.

nucleoid region (nū´-klē-oyd) The region in a prokaryotic cell consisting of a concentrated mass of DNA.

nucleolus (nū-klē´-ō-lus) A structure within the nucleus of a eukaryotic cell where ribosomal RNA is made and assembled with proteins to make ribosomal subunits; consists of parts of the chromatin DNA, RNA transcribed from the DNA, and proteins imported from the cytoplasm.

nucleosome (nū-klē-ō-sōm) The beadlike unit of DNA packing in a eukaryotic cell; consists of DNA wound around a protein core made up of eight histone molecules.

nucleotide (nū´-klē-ō-tīd) An organic monomer consisting of a five-carbon sugar covalently bonded to a nitrogenous base and a phosphate group. Nucleotides are the building blocks of nucleic acids.

nucleus (plural, **nuclei**) (1) An atom's central core, containing protons and neutrons. (2) The genetic control center of a eukaryotic cell.

O

obligate aerobe (ob´-li-get ār´-ōb) An organism that cannot survive without oxygen (O_2).

obligate anaerobe (ob´-li-get an´-uh-rōb) An organism that cannot survive in the presence of oxygen (O_2).

ocean current One of the riverlike flow patterns in the oceans.

ommatidia (ōm´-uh-tid´-ē-uh) The functional units of a compound eye; each ommatidium includes a cornea, lens, and photoreceptor cells.

omnivore An animal that eats both plants and animals. *See also* carnivore; herbivore.

oncogene (on´-kō-jēn) A cancer-causing gene; usually contributes to malignancy by abnormally enhancing the amount or activity of a growth factor made by the cell.

operator In prokaryotic DNA, a sequence of nucleotides near the start of an operon to which an active repressor can attach. The binding of repressor prevents RNA polymerase from attaching to the promoter and transcribing the genes of the operon.

operculum (ō-per´-kyū-lum) (plural, **opercula**) A protective flap on each side of a fish's head that covers a chamber housing the gills.

operon (op´-er-on) A unit of genetic regulation common in prokaryotes; a cluster of genes with related functions, along with the promoter and operator that control their transcription.

opportunistic life history Often seen in small-bodied species, the pattern of reproducing when young and producing many offspring.

optimal foraging Feeding behavior that provides maximal energy gain with minimal energy expense and minimal time spent searching for, securing, and eating food.

order In classification, the taxonomic category above family.

organ A structure consisting of several tissues adapted as a group to perform specific functions.

organ system A group of organs that work together in performing vital body functions.

organelle (ōr-guh-nel´) A structure with a specialized function within a cell.

organic chemistry The study of carbon compounds.

organic compound A chemical compound containing the element carbon and usually synthesized by cells.

organism An individual living thing, such as a bacterium, fungus, protist, plant, or animal.

organismal ecology The study of the evolutionary adaptations that enable individual organisms to meet the challenges posed by their abiotic environments.

orgasm Rhythmic contractions of the reproductive structures, accompanied by extreme pleasure, at the peak of sexual excitement in both sexes; includes ejaculation by the male.

osmoregulation The control of the gain or loss of water and dissolved solutes in an organism.

osmosis (oz-mō´-sis) The passive transport of water across a selectively permeable membrane.

Osteichthyes (os-tē-ik´-thēz) The vertebrate class of bony fishes; for example, trout and goldfish.

"Out of Africa" hypothesis The idea that modern humans evolved from a second migration out of Africa that occurred about 100,000 years ago, replacing all the regional populations of hominids derived from the first migrations of Homo erectus out of Africa about 1.5 million years ago.

out-group In a cladistic study of evolutionary relationships among taxa of organisms, a taxon or group of taxa with a known relationship to, but not a member of, the taxa being studied. *See also* in-group.

oval window In the vertebrate ear, a membrane-covered gap in the skull bone, through which sound waves pass from the middle ear into the inner ear.

ovary (1) In animals, the female gonad, which produces egg cells and reproductive hormones. (2) In flowering plants, the basal portion of a carpel in which the egg-containing ovules develop.

oviduct (ō´-vuh-dukt) The tube that conveys egg cells away from an ovary; also called a Fallopian tube.

ovule (ō´-vyūl) A reproductive structure in a seed plant, containing the female gametophyte and the developing egg. An ovule develops into a seed.

oxidation The loss of electrons from a substance involved in a redox reaction; always accompanies reduction.

ozone layer The layer of O_3 in the upper atmosphere that protects life on Earth from the harmful ultraviolet rays in sunlight.

P

P generation The parent individuals from which offspring are derived in studies of inheritance. P stands for parental.

paedomorphosis (pē´-duh-mōr´-fuh-sis) The retention of juvenile body features in an adult.

paleoanthropology (pā´-lē-ō-an´-thruh-pol´-uh-jē) The study of human origins and evolution.

paleontologist (pā´-lē-on-tol´-uh-jist) A scientist who studies fossils.

Pangaea (pan-jē´-uh) The supercontinent consisting of all the major landmasses of Earth fused together. Continental drift formed Pangaea near the end of the Paleozoic era.

parasite An organism that benefits at the expense of another organism, which is harmed in the process.

parasitism (pār´-uh-sit-izm) A symbiotic relationship in which the parasite, a type of predator, lives within or on the surface of a host, from which it derives its food.

parsimony (par´-suh-mō´-nē) In scientific studies, the search for the least complex explanation for an observed phenomenon.

partial pressure A measure of the relative amount of gas in a mixture.

passive transport The diffusion of a substance across a biological membrane, without any input of energy.

pathogen A disease-causing organism.

pattern formation During embryonic development, the emergence of a spatial organization in which the tissues and organs of the organism are all in their correct places.

pedigree A family tree representing the occurrence of heritable traits in parents and offspring across a number of generations.

pelagic zone (puh-laj´-ik) The region of an ocean occupied by seawater; the open ocean.

peptide bond The covalent linkage between two amino acid units in a polypeptide; formed by dehydration synthesis.

peptidoglycan (pep´-tid-ō-glī´-kan) A polymer of complex sugars cross-linked by short polypeptides; a material unique to eubacterial cell walls.

perception The brain's meaningful interpretation, or conscious understanding, of sensory information.

permafrost Continuously frozen ground found in the tundra.

PET *See* positron-emission tomography.

petal A modified leaf of a flowering plant. Petals are the often colorful parts of a flower that advertise it to insects and other pollinators.

pH scale A measure of the relative acidity of a solution, ranging in value from 0 (most acidic) to 14 (most basic). pH stands for potential hydrogen and refers to the concentration of hydrogen ions (H^+).

phage (fāj) *See* bacteriophage.

phagocyte (fag´-ō-sīt) A white blood cell (e.g., a neutrophil or a monocyte) that engulfs bacteria, foreign proteins, and the remains of dead body cells.

phagocytosis (fag´-ō-sī-tō´-sis) Cellular "eating"; a type of endocytosis whereby a cell engulfs macromolecules, other cells, or particles into its cytoplasm.

pharyngeal slit (fā-rin´-jē-ul) A gill structure in the pharynx, found in chordate embryos and some adult chordates.

phenotype (fē´-nō-tīp) The expressed traits of an organism.

phenylketonuria (PKU) (fen´-ul-kē´-tuh-nūr´-ē-uh) A recessive genetic disorder characterized by an inability to properly break down the amino acid phenylalanine; if untreated results in mental retardation.

phloem (flō´-um) The portion of a plant's vascular system that conveys sugars, nutrients, and hormones throughout a plant; made up of food-conducting cells.

phosphate group (fos´-fāt) A functional group consisting of a phosphorus atom covalently bonded to four oxygen atoms.

phospholipid (fos´-fō-lip´-id) A molecule that is a constituent of the inner bilayer of biological membranes, having a polar, hydrophilic head and a nonpolar, hydrophobic tail.

phospholipid bilayer A double layer of phospholipid molecules (each molecule consisting of a phosphate group bonded to two fatty acids) that is the primary component of all cellular membranes.

phosphorylation (fos´-fōr-uh-lā´-shun) The transfer of a phosphate group, usually from ATP, to a molecule. Nearly all cellular work depends on ATP energizing other molecules by phosphorylation.

photic zone (fō´-tik) The region of an aquatic ecosystem into which light penetrates and where photosynthesis occurs.

photoautotroph An organism that obtains energy from sunlight and carbon from CO_2 by photosynthesis.

photoheterotroph An organism that obtains energy from sunlight and carbon from organic sources.

photon (fō´-ton) A fixed quantity of light energy. The shorter the wavelength of light, the greater the energy of a photon.

photophosphorylation (fō-tō-fos´-fōr-uh-lā´-shun) The production of ATP by chemiosmosis during the light reactions of photosynthesis.

photorespiration In a plant cell, the breakdown of a two-carbon compound produced by the Calvin cycle. The Calvin cycle produces the two-carbon compound, instead of its usual three-carbon product G3P, when leaf cells fix O_2, instead of CO_2. Photorespiration produces no sugar molecules or ATP.

photosynthesis (fō´-tō-sin´-thuh-sis) The process by which plants, autotrophic protists, and some bacteria use light energy to make sugars and other organic food molecules from carbon dioxide and water.

photosystem A light-harvesting unit of a chloroplast's thylakoid membrane; consists of several hundred antenna molecules, a reaction-center chlorophyll, and a primary electron acceptor.

phylogenetic tree (fī´-lō-juh-net´-ik) A branching diagram that represents a hypothesis about evolutionary relationships among organisms.

phylogeny (fī-loj´-uh-nē) The evolutionary history of a species or group of related species.

phylum (fī´-lum) (plural, **phyla**) In classification, the taxonomic category above class and below kingdom; members of a phylum all have a similar general body plan.

phytochrome (fī´-tuh-krōm) A colored protein in plants that contains a special set of atoms that absorbs light.

phytoplankton (fī´-tō-plank´-ton) Algae and photosynthetic bacteria that drift passively in aquatic environments.

pili (pī´-lī) (singular, **pilus**) Short projections on the surface of prokaryotic cells that help prokaryotes attach to other surfaces; specialized sex pili are used in conjugation to hold the mating cells together.

pineal gland (pin´-ē-ul) An outgrowth of the vertebrate brain that secretes the hormone melatonin, which coordinates daily and seasonal body activities, such as reproductive activity, with environmental light conditions.

pinocytosis (pī-nō-sī-tō´-sis) Cellular "drinking"; a type of endocytosis in which the cell takes fluid and dissolved solutes into small membranous vesicles.

placenta (pluh-sen´-tuh) In most mammals, the organ that provides nutrients and oxygen to the embryo and helps dispose of its metabolic wastes; formed of the embryo's chorion and the mother's endometrial blood vessels.

placentals (pluh-sen´-tuls) Mammals whose young complete their embryonic development in the uterus, nourished via the mother's blood vessels in the placenta; also called placental mammals or eutherians.

plankton Algae and other organisms, mostly microscopic, that drift passively in ponds, lakes, and oceans.

Plantae (plan´-tā) The kingdom that contains the plants.

plasma cell An antibody-secreting B cell.

plasma membrane The thin layer of lipids and proteins that sets a cell off from its surroundings and acts as a selective barrier to the passage of ions and molecules into and out of the cell; consists of a phospholipid bilayer in which are embedded molecules of protein and cholesterol.

plasmid A small ring of DNA separate from the chromosome(s). Plasmids are found in prokaryotes and yeasts.

plasmodesma (plaz´-mō-dez´-muh) (plural, **plasmodesmata**) An open channel in a plant cell wall, through which strands of cytoplasm connect from adjacent walls.

plasmodial slime mold (plaz-mō´-dē-ul) A type of protist that has amoeboid cells, flagellated cells, and an amoeboid plasmodial feeding stage in its life cycle.

plasmodium (1) A single mass of cytoplasm containing many nuclei. (2) The amoeboid feeding stage in the life cycle of a plasmodial slime mold.

plasmolysis (plaz-mol´-uh-sis) A phenomenon that occurs in plant cells in a hypertonic environment. The cell loses water and shrivels, and its plasma membrane pulls away from the cell wall, usually killing the cell.

plate tectonics Forces within planet Earth that cause movements of the crust, resulting in continental drift, volcanoes, and earthquakes.

Platyhelminthes (plat´-ē-hel-min´-thēz) The phylum that contains the flatworms, the bilateral animals with a thin, flat body form, gastrovascular cavity or no digestive system, and no body cavity; the free-living flatworms, flukes, and tapeworms.

pleated sheet The folded arrangement of a polypeptide in a protein's secondary structure.

pleiotropy (plī´-uh-trō-pē) The control of more than one phenotypic characteristic by a single gene.

polar covalent bond An attraction between atoms that share electrons unequally. The shared electrons are pulled closer to one atom, making it partially negative and the other atom partially positive.

polar molecule A molecule containing polar covalent bonds.

pollen See pollen grain.

polychaetes (pol´-ē-kēts) The largest group of annelids. See Annelida.

polygenic inheritance (pol´-ē-jen´-ik) The additive effect of two or more gene loci on a single phenotypic characteristic.

polymer (pol´-uh-mer) A large molecule consisting of many identical or similar molecular units, called monomers, covalently joined together in a chain.

polymerase chain reaction (**PCR**) (puh-lim´-uh-rās) A technique used to obtain many copies of a DNA molecule or part of a DNA molecule. A small amount of DNA mixed with the enzyme DNA polymerase, DNA nucleotides, and a few other ingredients replicates repeatedly in a test tube.

polymorphic (pol´-ē-mōr´-fik) Referring to a population in which two or more physical forms are present in readily noticeable frequencies.

polymorphism (pol´-ē-mōr´-fizm) The coexistence of two or more distinct forms of individuals (polymorphic characters) in the same population.

polynucleotide (pol´-ē-nū´-klē-ō-tīd) A polymer made up of many nucleotides covalently bonded together.

polyp (pol´-ip) One of two types of cnidarian body forms; a columnar, hydralike body.

polypeptide A chain of amino acids linked by peptide bonds.

polyploid (pol´-ē-ploid) Containing more than two complete sets of homologous chromosomes in each somatic cell.

polyploid cell (pol´-ē-ployd) A cell with more than two complete sets of chromosomes.

polysaccharide (pol´-ē-sak´-uh-rīd) A carbohydrate polymer consisting of hundreds to thousands of monosaccharides (sugars) linked by covalent bonds.

population A group of interacting individuals belonging to one species and living in the same geographic area.

population density the number of individuals of a species per unit area or volume.

population ecology The study of how members of a population interact with their environment, focusing on factors that influence population density and growth.

population fragmentation The splitting and consequent isolation of portions of a biological population, usually by human-caused habitat degradation.

population genetics The study of genetic changes in populations; the science of microevolutionary changes in populations.

population viability analysis (**PVA**) Scientific analysis of the current status of a population or species and predictions of its chances for long-term survival.

population-limiting factor An environmental factor that restricts population growth.

Porifera (por-if´-er-uh) The phylum that contains the sponges, characterized by choanocytes, a porous body wall, and no true tissues.

positron-emission tomography (**PET**) Imaging technology that uses radioactively labeled biological molecules, such as glucose, to obtain information about metabolic processes at specific locations in the body. The labeled molecules are injected into the bloodstream, and a PET scan for radioactive emissions determines which tissues have taken up the molecules.

post-anal tail A tail posterior to the anus, found in chordate embryos and most adult chordates.

posterior Pertaining to the rear, or tail, of a bilaterally symmetrical animal.

post-zygotic barrier (pōst´-zī-got´-ik) A reproductive barrier that operates should interspecies mating occur and form hybrid zygotes.

potential energy Stored energy; the capacity to perform work that matter possesses because of its location or arrangement. Water behind a dam or chemical bonds both possess potential energy.

predation An interaction between species in which one species, the predator, eats the other, the prey.

predator A consumer in a biological community.

prevailing winds Winds that result from the combined effects of Earth's rotation and the rising and falling of air masses.

prey An organism eaten by a predator.

pre-zygotic barrier (prē´-zī-got´-ik) A reproductive barrier that impedes mating between species or hinders fertilization of eggs if members of different species should attempt to mate.

primary consumer An organism that eats only autotrophs.

primary phloem *See* phloem.

primary production The amount of solar energy converted to chemical energy (organic compounds) by autotrophs in an ecosystem during a given time period.

primary productivity The rate at which an ecosystem's plants and other producers build biomass, or organic matter.

primary structure The first level of protein structure; the specific sequence of amino acids making up a polypeptide chain.

primary succession A type of ecological succession in which a biological community arises in an area without soil. *See also* secondary succession.

primary xylem *See* xylem.

primitive characters Homologous features found in members of a lineage and also in the ancestors of the lineage; ancestral features.

probe In DNA technology, a labeled single-stranded nucleic acid molecule used to find a specific gene, or other nucleotide sequence, within a mass of DNA. The probe hydrogen-bonds to the complementary sequence in the targeted DNA.

producer An organism that makes organic food molecules from CO_2, H_2O, and other inorganic raw materials: a plant, alga, or autotrophic bacterium.

product An ending material in a chemical reaction.

progesterone (prō-jes´-teh-rōn) A steroid hormone secreted by the corpus luteum of the ovary; maintains the uterine lining during pregnancy.

prokaryotic cell (prō-kār´-ē-ot´-ik) A type of cell lacking a membrane-enclosed nucleus and other membrane-enclosed organelles; found only in the domains Bacteria and Archaea.

prokaryotic cell wall A fairly rigid, chemically complex wall that protects the prokaryotic cell and helps maintain its shape.

prokaryotic flagellum (plural, **flagella**) A long surface projection that propels a prokaryotic cell through its liquid environment; totally different from the flagellum of a eukaryotic cell.

promoter A specific nucleotide sequence in DNA, located at the start of a gene, that is the binding site for RNA polymerase and the place where transcription begins.

prophage (prō´-fāj) Phage DNA that has inserted by genetic recombination into the DNA of a prokaryotic chromosome.

prophase The first stage of mitosis, during which duplicated chromosomes condense to form structures visible with a light microscope and the mitotic spindle forms and begins moving the chromosomes toward the center of the cell.

prosimian (pro-sim´-ē-un) A member of the primate group comprised of lorises, bush babies, tarsiers, and lemurs.

protein A biological polymer constructed from amino acid monomers.

protist (prō´-tist) A member of the kingdom Protista.

Protista (prō-tis´-tuh) In the five-kingdom classification system, the kingdom that contains the unicellular eukaryotes (and closely related multicellular organisms) called protists.

proton A subatomic particle with a single unit of positive electrical charge, found in the nucleus of an atom.

proto-oncogene (prō´-tō-on´-kō-jēn) A normal gene that can be converted to a cancer-causing gene.

protoplast fusion The fusing of two protoplasts from different plant species that would otherwise be reproductively incompatible.

protostome An animal with a coelom that develops from solid masses of cells that arise between the digestive tube and the body wall of the embryo. The protostomes include the mollusks, annelids, and arthropods.

protozoan (prō´-tō-zō´-un) (plural, **protozoa**) A protist that lives primarily by ingesting food; a heterotrophic, animal-like protist.

provirus Viral DNA that inserts into a host genome.

proximal tubule In the vertebrate kidney, the portion of a nephron immediately downstream from Bowman's capsule that conveys and helps refine filtrate.

proximate cause In behavioral biology, the immediate explanation for an organism's behavior; the interactions of an organism with the environment or the particular environmental stimuli that trigger a behavioral response in the organism.

pseudocoelom (sū-dō-sē´-lōm) A body cavity that is in direct contact with the wall of the digestive tract.

pseudopodium (sū-dō-pō´-dē-um) (plural, **pseudopodia**) A temporary extension of an amoeboid cell. Pseudopodia function in moving cells and engulfing food.

pulmonary artery A large blood vessel that conveys blood from the heart to a lung.

pulmonary vein A blood vessel that conveys blood from a lung to the heart.

punctuated equilibrium The idea that speciation occurs in spurts followed by long periods of little change.

Punnett square A diagram used in the study of inheritance to show the results of random fertilization.

purine (pyū´-rēn) One of two families of nitrogenous bases found in nucleotides. Adenine (A) and guanine (G) are purines.

pyloric sphincter (pī-lōr´-ik sfink´-ter) In the vertebrate digestive tract, a muscular ring that regulates the passage of food out of the stomach and into the small intestine.

pyrimidine (puh-rim´-uh-dēn) One of two families of nitrogenous bases found in nucleotides. Cytosine (C), thymine (T), and uracil (U) are pyrimidines.

Q

quaternary consumer (kwot´-er-nār-ē) An organism that eats tertiary consumers.

quaternary structure The fourth level of protein structure; the shape resulting from the association of two or more polypeptide subunits.

R

R plasmid A bacterial plasmid that carries genes for enzymes that destroy particular antibiotics, thus making the bacterium resistant to the antibiotics.

r-selection The concept that in certain (*r*-selected) populations, a high reproductive rate is the chief determinant of life history.

radial symmetry An arrangement of the body parts of an organism like pieces of a pie around an imaginary central axis. Any slice passing longitudinally through a radially symmetrical organism's central axis divides it into mirror-image halves.

radiation The emission of electromagnetic waves by all objects warmer than absolute zero.

radiation therapy Treatment for cancer in which parts of the body that have cancerous tumors are exposed to high-energy radiation to disrupt cell division of the cancer cells.

radioactive isotope An isotope whose nucleus decays spontaneously, giving off particles and energy.

radiometric dating A method for determining the age of fossils and rocks from the ratio of a radioactive isotope to the nonradioactive istope(s) of the same element in the sample.

radula (rad´-yū-luh) A toothed, rasping organ found in many mollusks, used to scrape up or shred food.

random Describing a dispersion pattern in which individuals are spaced in a patternless, unpredictable way.

ray-finned fish A bony fish having fins supported by thin, flexible skeletal rays. All but one living species of bony fishes are ray-fins. *See* lobe-finned fish.

reactant A starting material in a chemical reaction.

reaction center In a photosystem in a chloroplast, the chlorophyll *a* molecule and the primary electron acceptor that trigger the light reactions of photosynthesis. The chlorophyll donates an electron excited by light energy to the primary electron acceptor, which passes an electron to an electron transport chain.

reading frame The way a cell's mRNA-translating machinery groups the mRNA nucleotides into codons.

receptor On or in a cell, a specific protein molecule whose shape fits that of a specific molecular messenger, such as a hormone.

receptor-mediated endocytosis (en´-dō-sī-tō´-sis) The movement of specific molecules into a cell by the inward budding of membranous vesicles. The vesicles contain proteins with receptor sites specific to the molecules being taken in.

recessive allele In a heterozygous individual, the allele that has no noticeable effect on the phenotype.

reciprocal altruism (al´-trū-izm) In animal behavior, a selfless act repaid at a later time by the beneficiary or by another member of the beneficiary's social system.

recombinant DNA A DNA molecule carrying genes derived from two or more sources.

recombinant DNA technology A set of techniques for synthesizing recombinant DNA in vitro and transferring it into cells, where it can be replicated and may be expressed; also known as genetic engineering.

recombination frequency With respect to two given genes, the number of recombinant progeny from a mating divided by the total number of progeny. Recombinant progeny carry combinations of alleles different from those in either of the parents as a result of independent assortment of chromosomes and crossing over.

red alga One of a group of marine, mostly multicellular, autotrophic protists, which includes the reef-building coralline algae.

red-green color blindness A category of common sex-linked human disorders involving several genes on the X chromosome and characterized by a malfunction of light-sensitive cells in the eyes; affects mostly males but also homozygous females.

redox reaction Short for oxidation-reduction reaction; a chemical reaction in which electrons are lost from one substance (oxidation) and added to another (reduction). Oxidation and reduction always occur together.

reduction The gain of electrons by a substance involved in a redox reaction; always accompanies oxidation.

regeneration The regrowth of body parts from pieces of an organism.

regulatory gene A gene that codes for a protein, such as a repressor, that controls the transcription of another gene or group of genes.

releasing hormone A hormone, secreted by the hypothalamus, that makes the anterior pituitary secrete hormones.

REM sleep Rapid-eye-movement sleep; a period of sleep when the brain is highly active, brain waves are fairly rapid and regular, and the eyes move rapidly under the closed eyelids. Most dreams occur during REM sleep. *See also* slow-wave (SW) sleep.

renal cortex The outer portion of the vertebrate kidney.

renal medulla The inner portion of the vertebrate kidney, beneath the renal cortex.

renewable resource management Management of a renewable natural resource so as not to damage the resource.

repetitive DNA Nucleotide sequences that are present in many copies in the DNA of a genome. The repeated sequences may be long or short and may be located next to each other or dispersed in the DNA.

replacement hypothesis Another name for the "Out of Africa" hypothesis.

repressor A protein that blocks the transcription of a gene or operon.

reproductive barrier A biological feature of a species that prevents it from interbreeding with other species even when populations of the two species live together.

reproductive cloning Using a somatic cell from a multicellular organism to make one or more genetically identical individuals.

reproductive system The body organ system responsible for reproduction.

Reptilia A class of vertebrate animals that consists of the reptiles, including snakes, lizards, turtles, crocodiles, and alligators.

resolution phase The final phase of the human sexual response cycle, where the structures return to normal size, muscles relax, and passion subsides.

resolving power A measure of the clarity of an image; the ability of an optical instrument to show two objects as separate.

resource partitioning The division of environmental resources by coexisting species such that the niche of each species differs by one or more significant factors from the niches of all coexisting species.

respiration (1) Gas exchange, or breathing; the exchange of O_2 and CO_2 between an organism and its environment. An aerobic organism takes up O_2 and gives off CO_2. (2) Cellular respiration; the aerobic harvest of energy from food molecules by cells.

restoration ecology The use of ecological principles to develop ways to return degraded ecosystems to conditions as similar as possible to their natural, predegraded state.

restriction enzyme A bacterial enzyme that cuts up foreign DNA, thus protecting bacteria against intruding DNA from phages and other organisms. Restriction enzymes are used in DNA technology to cut DNA molecules in reproducible ways.

restriction fragments Molecules of DNA produced from a longer DNA molecule cut up by a restriction enzyme; used in genome mapping and other applications.

restriction site A specific sequence on a DNA strand that is recognized as a "cut site" by a restriction enzyme.

reticular formation A system of neurons, containing over 90 separate nuclei, that passes through the core of the brain stem.

retrovirus An RNA virus that reproduces by means of a DNA molecule. It reverse-transcribes its RNA into DNA, inserts the DNA into a cellular chromosome, and then transcribes more copies of the RNA from the viral DNA. HIV and a number of cancer-causing viruses are retroviruses.

reverse transcriptase (tran-skrip´-tās) An enzyme that catalyzes the synthesis of DNA on an RNA template.

RFLP analysis A common method of DNA fingerprinting, the comparison of the set of restriction fragments produced by DNA from different individuals.

RFLPs (rif´-lips) Restriction fragment length polymorphisms; the differences in homologous DNA sequences that are reflected in different lengths of restriction fragments produced when the DNA is cut up with restriction enzymes.

ribosomal RNA (rRNA) (rī´-buh-sōm´-ul) The type of ribonucleic acid that, together with proteins, makes up ribosomes; the most abundant type of RNA.

ribosomal RNA (rRNA) sequence analysis Determination of the nucleotide sequence of ribosomal RNA molecules.

ribosome (rī´-buh-sōm) A cell organelle consisting of RNA and protein organized into two subunits and functioning as the site of protein synthesis in the cytoplasm. The ribosomal subunits are constructed in the nucleolus.

ribozyme (rī´-bō-zīm) An enzymatic RNA molecule that catalyzes chemical reactions.

RNA Ribonucleic acid (rī´-bō-nū-klā´-ik). A type of nucleic acid consisting of nucleotide monomers with a ribose sugar and the nitrogenous bases adenine (A), cytosine (C), guanine (G), and uracil (U); usually single-stranded; functions in protein synthesis and as the genome of some viruses.

RNA polymerase (puh-lim´-uh-rās) An enzyme that links together the growing chain of RNA nucleotides during transcription, using a DNA strand as a template.

RNA splicing The removal of introns and joining of exons in eukaryotic RNA, forming an mRNA molecule with a continuous coding sequence; occurs before mRNA leaves the nucleus.

RNA world A hypothetical period in the evolution of life when RNA served as rudimentary genes and the sole catalytic molecules.

rod A photoreceptor cell in the vertebrate retina, enabling vision in dim light.

root A plant structure that anchors the plant in the soil, absorbs and transports minerals and water, and stores food.

root pressure The upward push of xylem sap in a vascular plant, caused by the active pumping of minerals into the xylem by root cells.

rough ER (rough endoplasmic reticulum) (reh-tik´-yuh-lum) A network of interconnected membranous sacs in a eukaryotic cell's cytoplasm. Rough ER membranes are studded with ribosomes that make membrane proteins and secretory proteins. The rough ER constructs membrane from phospholipids and proteins.

roundworm A member of the phylum Nematoda.

rRNA *See* ribosomal RNA.

rule of addition A rule stating that the probability of an event that can occur in two or more alternative ways is the sum of the separate probabilities of the different ways.

rule of multiplication A rule stating that the probability of a compound event is the product of the separate probabilities of the independent events.

ruminant mammal (rū´-min-ent) A mammal with a four-chambered stomach housing microorganisms that can digest cellulose; examples are cattle, deer, and sheep.

saccule (sak´-yūl) A fluid-filled inner ear chamber containing hair cells that detect the position of the head relative to gravity.

salt A compound resulting from the formation of ionic bonds, also called an ionic compound.

sarcoma (sar-kō´-muh) Cancer of the supportive tissues, such as bone, cartilage, and muscle.

saturated Pertaining to fats and fatty acids whose hydrocarbon chains contain the maximum number of hydrogens and therefore have no double covalent bonds. Saturated fats and fatty acids solidify at room temperature.

savanna A biome dominated by grasses and scattered trees.

scanning electron microscope (SEM) A microscope that uses an electron beam to study the surface architecture of a cell or other specimen.

sclereid (sklār´-ē-id) In plants, a very hard, dead sclerenchyma cell found in nutshells and seed coats; a stone cell.

search image The mechanism that enables an animal to find a particular kind of food efficiently.

seaweed A large, multicellular marine alga.

second law of thermodynamics The natural law stating that energy conversions reduce the order of the universe, increasing its entropy.

secondary consumer An organism that eats primary consumers.

secondary phloem *See* phloem.

secondary structure The second level of protein structure; the regular patterns of coils or folds of a polypeptide chain.

secondary succession A type of ecological succession that occurs where a disturbance has destroyed an existing biological community but left the soil intact. *See also* primary succession.

secondary xylem *See* xylem.

secretory protein A protein that is secreted by a cell, such as an antibody.

seed A plant embryo packaged with a food supply within a protective covering.

seed dormancy The temporary suspension of growth and development of a seed.

segmentation Subdivision along the length of an animal body into a series of repeated parts called segments.

selectively permeable (per´-mē-uh-bul) Allowing some substances to cross a biological membrane more easily than others and blocking the passage of other substances altogether.

self protein A protein on the surface of an antigen-presenting cell that can hold a foreign antigen and display it to helper T cells. Each individual has a unique set of self proteins that serve as molecular markers for the body. Lymphocytes do not attack self proteins unless the proteins are displaying foreign antigens; therefore, self proteins mark normal body cells as off-limits to the immune system. The technical name for self proteins is *major histocompatibility complex (MHC) proteins. See also* nonself molecule.

self-fertilization The fusion of sperm and egg that are produced by the same individual organism.

self-fertilize To form a zygote through the fusion of sperm and egg produced by the same individual organism.

sensation A feeling, or general awareness, of stimuli resulting from sensory information reaching the central nervous system.

sensitive period A limited phase in an individual animal's development when learning of particular behaviors can take place.

sepal (sē´-pul) A modified leaf of a flowering plant. A whorl of sepals encloses and protects the flower bud before it opens.

sex chromosome A chromosome that determines whether an individual is male or female.

sex-linked gene A gene located on a sex chromosome.

sexual dimorphism (dī-mōr´-fizm) A special case of polymorphism based on the distinction between the secondary sex characteristics of males and females.

sexual reproduction The creation of offspring by the fusion of two haploid sex cells (gametes), forming a diploid zygote.

shoot The stem and leaves of a plant.

short-day plant A plant that flowers in late summer, fall, or winter, when day length is shortening.

short-term memory The ability to hold information, anticipations, or goals for a time and then release them if they become irrelevant.

sickle-cell disease A genetic disorder in which the red blood cells have abnormal hemoglobin molecules and take on an abnormal shape.

sieve-tube member A food-conducting cell in a plant; chains of sieve-tube members make up phloem tissue.

sign stimulus In animal behavior, a stimulus that triggers a fixed action pattern.

signal A behavior that causes a change in behavior in another animal.

signal-transduction pathway A series of molecular changes that converts a signal on a target cell's surface into a specific response inside the cell.

silencer A eukaryotic DNA sequence that inhibits the start of gene transcription; may act analogously to an enhancer, binding a repressor.

simple fruit A fruit such as an apple that develops from a flower with a single carpel and ovary.

single-lens eye The cameralike eye found in some jellies, polychaetes, spiders, and many mollusks.

sink habitat An area of habitat where a species' death rate exceeds its reproductive success.

sister chromatid (krō´-muh-tid) One of the two identical parts of a duplicated chromosome in a eukaryotic cell.

skeletal system The organ system that provides body support and protects body organs such as the brain, heart, and lungs.

skull The bony framework of the head.

sliding-filament model The theory explaining how muscle contracts, based on change within a sarcomere, the basic unit of muscle organization, stating that thin (actin) filaments slide across thick (myosin) filaments, shortening the sarcomere; the shortening of all sarcomeres in a myofibril shortens the entire myofibril.

slime mold See cellular slime mold; plasmodial slime mold.

smooth ER (smooth endoplasmic reticulum) A network of interconnected membranous tubules in a eukaryotic cell's cytoplasm. Smooth ER lacks ribosomes. Enzymes embedded in the smooth ER membrane function in the synthesis of certain kinds of molecules, such as lipids.

social behavior Any kind of interaction between two or more animals, usually of the same species.

sociobiology The study of the evolutionary basis of social behavior.

sodium-potassium (Na^+-K^+) pump A membrane protein that transports sodium ions out of, and potassium ions into, a cell against their concentration gradients. The process is powered by ATP.

soil horizon A distinct layer of soil.

solar energy Energy obtained from the sun.

solute (sol´-yūt) A substance that is dissolved in a solution.

solution A liquid consisting of a homogeneous mixture of two or more substances: a dissolving agent, the solvent, and a substance that is dissolved, the solute.

solvent The dissolving agent in a solution. Water is the most versatile known solvent.

somatic cell (sō-mat´-ik) Any cell in a multicellular organism except a sperm or egg cell or a cell that develops into a sperm or egg.

somite (sō´-mīt) A block of mesoderm in a chordate embryo that gives rise to vertebrae and other segmental structures.

source habitat An area of habitat where a species' reproductive success exceeds its death rate and from which new individuals often disperse to other areas.

speciation (spē´-sē-ā´-shun) The evolution of new species.

species A group whose members possess similar anatomical characteristics and have the ability to interbreed. See biological species concept.

species diversity The number and relative abundance of species in a biological community.

spirochete (spī´-rō-kēt) A large spiral-shaped (curved) prokaryotic cell.

sponge An aquatic animal characterized by a highly porous body.

spontaneous generation The incorrect notion that life can emerge from inanimate material.

sporangium (plural, **sporangia**) (spuh-ranj´-ē-um´) A capsule in fungi and plants in which meiosis occurs and haploid spores develop.

spore (1) In plants and algae, a haploid cell that can develop into a multicellular individual without fusing with another cell. (2) In prokaryotes, protists, and fungi, any of a variety of thick-walled life cycle stages capable of surviving unfavorable environmental conditions.

sporophyte (spōr´-uh-fīt) The multicellular diploid form in the life cycle of organisms undergoing alternation of generations; results from a union of gametes and meiotically produces haploid spores that grow into the gametophyte generation.

stability In an ecological sense, the tendency of a biological community to remain in more or less constant balance due largely to interactions among organisms.

stabilizing selection Natural selection that favors intermediate variants by acting against extreme phenotypes.

stamen (stā´-men) A pollen-producing (male) reproductive part of a flower, consisting of a stalk and an anther.

starch A storage polysaccharide found in the roots of plants and certain other cells; a polymer of glucose.

start codon (kō´-don) On mRNA, the specific three-nucleotide sequence (AUG) to which an initiator tRNA molecule binds, starting translation of genetic information.

steroid (stār´-oyd) A type of lipid whose carbon skeleton is in the form of four fused rings: three 6-sided rings and one 5-sided ring; examples are cholesterol, testosterone, and estrogen.

stigma (stig´-muh) (plural, **stigmata**) The sticky tip of a flower's carpel, which traps pollen grains.

stoma (stō´-muh) (plural, **stomata**) A pore surrounded by guard cells in the epidermis of a leaf. When stomata are open, CO_2 enters a leaf, and water and O_2 exit. A plant conserves water when its stomata are closed.

stop codon In mRNA, one of three triplets (UAG, UAA, UGA) that signal gene translation to stop.

stretch receptor A type of mechanoreceptor sensitive to changes in muscle length; detects the position of body parts.

strict aerobe An organism that can survive only in an atmosphere of oxygen, which is used in aerobic respiration.

stroma (strō´-muh) A thick fluid enclosed by the inner membrane of a chloroplast. Sugars are made in the stroma by the enzymes of the Calvin cycle.

stromatolite (strō-mat´-uh-līt) Rock formed of layered, fossilized bacterial mats.

style The stalk of a flower's carpel, with the ovary at the base and the stigma at the top.

substrate (1) A specific substance (reactant) on which an enzyme acts. Each enzyme recognizes only the specific substrate of the reaction it catalyzes. (2) A surface in or on which an organism lives.

substrate feeders Organisms that live in or on their food source, eating their way through the food.

substrate-level phosphorylation The formation of ATP occurring when an enzyme transfers a phosphate group from an organic molecule (e.g., one of the intermediates in glycolysis or the Krebs cycle) to ADP.

succession See ecological succession; primary succession; secondary succession.

sugar-phosphate backbone The alternating chain of sugar and phosphate to which DNA and RNA nitrogenous bases are attached.

summation (suh-mā´-shun) The overall effect of all the information a neuron receives at a particular instant.

superior vena cava (vē´-nuh kā´-vuh) A large vein that returns O_2-poor blood to the heart from the upper body and head. See also inferior vena cava.

surface tension A measure of how difficult it is to stretch or break the surface of a liquid.

survivorship curve A plot of the number of members of a cohort that are still alive at each age; one way to represent age-specific mortality.

suspension feeder An animal that extracts food particles suspended in the surrounding water.

sustainable development The long-term prosperity of human societies and the ecosystems that support them.

swim bladder A gas-filled internal sac that helps bony fishes maintain buoyancy.

symbiont (sim´-bē-unt) The smaller participant in a symbiotic relationship, living in or on the host.

symbiosis (sim´-bē-ō´-sis) An interspecific interaction in which one species, the symbiont, lives in or on another species, the host.

symbiotic relationship An interspecific interaction in which one species, the symbiont, lives in or on another species, the host; a close association between organisms of two or more species.

sympatric speciation The formation of a new species as a result of a genetic change that produces a reproductive barrier between the changed population (mutants) and the parent population; sympatric speciation occurs without a geographic barrier. See also allopatric speciation.

synapsis (sin-ap´-sis) The pairing of duplicated homologous chromosomes, forming a tetrad, during prophase I of meiosis. During synapsis, chromatids of homologous chromosomes can exchange segments by crossing over.

systematics The scientific study of biological diversity and its classification.

systemic acquired resistance A defensive response in plants infected with a pathogenic microbe; helps protect healthy tissue from the microbe.

T

T₃ *See* triiodothyronine.

T₄ *See* thyroxine.

taiga (tī´-guh) The northern (boreal) coniferous forest, which extends across North America and Eurasia, to the southern border of the arctic tundra; also found just below alpine tundra on mountainsides in temperate zones.

tail Extra nucleotides added at the end of an RNA transcript in the nucleus of a eukaryotic cell.

tapeworm A parasitic flatworm characterized by the absence of a digestive tract.

taxis (tak´-sis) (plural, **taxes**) Virtually automatic orientation toward or away from a stimulus.

taxon (tak´-son) (plural, **taxa**) A proper name, such as Phylum Chordata, Class Mammalia, or *Homo sapiens*, in the taxonomic hierarchy used to classify organisms.

taxonomy The branch of biology concerned with identifying, naming, and classifying species.

technology The practical application of scientific knowledge.

telomere (tel´-uh-mēr) The repetitive DNA at each end of a eukaryotic chromosome.

telophase The fourth and final stage of mitosis, during which daughter nuclei form at the two poles of a cell. Telophase usually occurs together with cytokinesis.

temperate deciduous forest A biome located throughout midlatitude regions where there is sufficient moisture to support the growth of large, broadleaf deciduous trees.

temperate grasslands Grassland regions maintained by seasonal drought, occasional fires, and grazing by large mammals.

temperate zones Latitudes between the tropics and the Arctic Circle in the north and the Antarctic Circle in the south; regions with milder climates than the tropics or polar regions.

temperature A measure of the intensity of heat, reflecting the average kinetic energy or speed of molecules.

temporal isolation A type of prezygotic barrier between species; the species remain isolated because they breed at different times.

terminator A special sequence of nucleotides in DNA that marks the end of a gene. It signals RNA polymerase to release the newly made RNA molecule, which then departs from the gene.

territory An area that an individual or individuals defend and from which other members of the same species are usually excluded.

tertiary consumer (ter´-shē-ār-ē) An organism that eats secondary consumers.

tertiary structure The third level of protein structure; the overall, three-dimensional shape of a polypeptide in a protein.

testcross The mating between an individual of unknown genotype for a particular characteristic and an individual that is homozygous recessive for that same characteristic.

testosterone (tes-tos´-tuh-rōn) An androgen hormone that stimulates an embryo to develop into a male and promotes male body features.

tetrad A paired set of homologous chromosomes, each composed of two sister chromatids. Tetrads form during prophase I of meiosis.

theory A widely accepted explanatory idea that is broad in scope and supported by a large body of evidence.

therapeutic cloning The cloning of human cells by nuclear transplantation for therapeutic purposes, such as the replacement of body cells that have been irreversibly damaged by disease or injury. *See* nuclear transplantation; reproductive cloning.

thermodynamics (ther´-mō-dī-nam´-iks) The study of energy transformations that occur in a collection of matter. *See* first law of thermodynamics; second law of thermodynamics.

threatened species As defined in the U.S. Endangered Species Act, a species that is likely to become endangered in the foreseeable future throughout all or a significant portion of its range.

three-domain system A system of taxonomic classification based on three basic groups: Bacteria, Archaea, and Eukarya.

thylakoid (thī´-luh-koyd) One of a number of disk-shaped membranous sacs inside a chloroplast. Thylakoid membranes contain chlorophyll and the enzymes of the light reactions of photosynthesis. A stack of thylakoids is called a granum.

thymine (T) (thī´-min) A single-ring nitrogenous base found in DNA.

thymus gland (thī´-mus) An endocrine gland in the neck region of mammals that is active in establishing the immune system; secretes several hormones that promote the development and differentiation of T cells.

thyroid-stimulating hormone (TSH) A protein hormone secreted by the anterior pituitary that stimulates the thyroid gland to secrete its hormones.

thyroxine (T₄) (thī-rok´-sin) An amine hormone secreted by the thyroid gland that stimulates metabolism in virtually all body tissues.

Ti plasmid A bacterial plasmid that induces tumors in plant cells that it infects; often used as a vector to introduce new genes into plant cells. Ti stands for tumor-inducing.

tight junction A junction that binds tissue cells together in a leakproof sheet.

tissue system Organized collection of plant tissues. The organs of plants (such as roots, stems, and leaves) are formed from the epidermis, vascular, and ground tissue systems.

topsoil A mixture of particles derived from rock, living organisms, and humus.

torpor (tor´-per) A state of reduced activity by an endotherm. Torpor reduces energy consumption because the metabolic rate, body temperature, heart rate, and breathing rate decrease.

trace element An element that is essential for the survival of an organism but is needed in only minute quantities.

tracheole The narrowest tube in an insect's tracheal system. *See also* trachea.

trade winds The movement of air in the tropics (those regions that lie between 23.5° north latitude and 23.5° south latitude).

transcription The synthesis of RNA on a DNA template.

transcription factor In the eukaryotic cell, a protein that functions in initiating or regulating transcription. Transcription factors bind to DNA or to other proteins that bind to DNA.

transduction (1) The transfer of bacterial genes from one bacterial cell to another by a phage. (2) *See* sensory transduction. (3) *See* signal-transduction pathway.

transfer RNA (tRNA) A type of ribonucleic acid that functions as an interpreter in translation. Each tRNA molecule has a specific anticodon, picks up a specific amino acid, and conveys the amino acid to the appropriate codon on mRNA.

transformation The incorporation of new genes into a cell from DNA that the cell takes up from the fluid around it.

transgenic organism An organism that contains genes from another species.

translation The synthesis of a polypeptide using the genetic information encoded in an mRNA molecule. There is a change of "language" from nucleotides to amino acids.

translocation (1) During protein synthesis, the movement of a tRNA molecule carrying a growing polypeptide chain from the A site to the P site on a ribosome. (The mRNA travels with it.) (2) A change in a chromosome resulting from a chromosomal fragment attaching to a nonhomologous chromosome; can occur as a result of an error in meiosis or from mutagenesis.

transmission electron microscope (TEM) A microscope that uses an electron beam to study the internal structure of thinly sectioned specimens.

transport protein A membrane protein that helps move substances across a cell membrane.

transport vesicle A tiny membranous sac in a cell's cytoplasm carrying molecules produced by the cell. The vesicle buds from the endoplasmic reticulum or Golgi and eventually fuses with another membranous organelle or the plasma membranes, releasing its contents.

transposon (tranz-pō´-zon) A transposable genetic element, or "jumping gene"; a segment of DNA that can move from one site to another within a cell and serve as an agent of genetic change.

trial-and-error learning Learning to associate a particular behavioral act with a positive or negative effect.

triglyceride (trī-glis´-uh-rīd) A fat, which consists of a molecule of glycerol linked to three molecules of fatty acid.

triiodothyronine (T₃) (trī´-ī-ō-dō-thī´-rō-nēn) An amine hormone secreted by the thyroid gland that stimulates metabolism in virtually all body tissues.

triplet code A set of three-nucleotide-long words that specify the amino acids for polypeptide chains. *See* genetic code.

trisomy 21 *See* Down syndrome.

tRNA *See* transfer RNA.

trophic level (trō´-fik) A level in a food chain.

trophic structure (trō´-fik) The feeding relationships in an ecosystem. Trophic structure determines the route of energy flow and the pattern of chemical cycling in an ecosystem.

tropics Latitudes between 23.5° north and south.

true-breeding Referring to organisms for which sexual reproduction produces offspring with inherited trait(s) identical to those of the parents. The organisms are homozygous for the characteristic(s) under consideration.

tumor An abnormal mass of cells that forms within otherwise normal tissue.

tumor-suppressor gene A gene whose product inhibits cell division, thereby preventing uncontrolled cell growth.

tundra A biome at the northernmost limits of plant growth and at high altitudes, characterized by dwarf woody shrubs, grasses, mosses, and lichens.

tunicate One of a group of invertebrate chordates.

U

ultimate cause In behavioral biology, the evolutionary explanation for an organism's behavior.

ultrasound imaging A technique for examining a fetus in the uterus. High-frequency sound waves echoing off the fetus are used to produce an image of the fetus.

uniform Describing a dispersion pattern in which individuals are evenly distributed.

unsaturated Pertaining to fats and fatty acids whose hydrocarbon chains lack the maximum number of hydrogen atoms and therefore have one or more double covalent bonds. Unsaturated fats and fatty acids do not solidify at room temperature.

uracil (U) (yū´-ruh-sil) A single-ring nitrogenous base found in RNA.

urea (yū-rē´-ah) A soluble form of nitrogenous waste excreted by mammals and most adult amphibians.

uric acid (yū´-rik) An insoluble precipitate of nitrogenous waste excreted by land snails, insects, birds, and some reptiles.

utricle (yū´-truh-kul) A fluid-filled inner ear chamber containing hair cells that detect the position of the head relative to gravity.

V

vaccination (vak´-suh-nā´-shun) A procedure that presents the immune system with a harmless variant or derivative of a pathogen, thereby stimulating the immune system to mount a long-term defense against the pathogen.

vaccine (vak-sēn´) A harmless variant or derivative of a pathogen used to stimulate a host organism's immune system to mount a long-term defense against the pathogen.

vacuole (vak´-ū-ōl) A membrane-enclosed sac, part of the endomembrane system of a eukaryotic cell, having diverse functions.

vascular bundle (vas´-kyū-ler) A strand of vascular tissues (both xylem and phloem) in a plant stem.

vascular plant A plant with xylem and phloem.

vascular tissue Plant tissue consisting of cells joined into tubes that transport water and nutrients throughout the plant body.

vector In molecular biology, a piece of DNA, usually a plasmid or a viral genome, that is used to move genes from one cell to another.

vegetative reproduction Asexual reproduction by a plant.

ventilation A mechanism that provides contact between an animal's respiratory surface and the air or water to which it is exposed. Contact between a respiratory surface and air or water enables gas exchange to occur.

ventral Pertaining to the underside, or bottom, of a bilaterally symmetrical animal.

vertebra (ver´-tuh-bruh) (plural, **vertebrae**) One of a series of segmented units making up the backbone of a vertebrate animal.

vertebrate (ver´-tuh-brāt) A chordate animal with a backbone; includes agnathans, cartilaginous fishes, bony fishes, amphibians, reptiles, birds, and mammals.

villus (vil´-us) (plural, **villi**) (1) A fingerlike projection of the inner surface of the small intestine. (2) A fingerlike projection of the chorion of the mammalian placenta. Large numbers of villi increase the surface areas of these organs.

visceral mass (vis´-uh-rul) One of the three main parts of a mollusk, it contains most of the internal organs.

visual acuity The ability of the eyes to distinguish fine detail.

vital capacity The maximum volume of air that a respiratory system can inhale and exhale.

W

warning coloration The often brightly colored markings of animals possessing chemical defenses. The coloration provides a caution to their predators.

water vascular system In echinoderms, a radially arranged system of water-filled canals that branch into extensions called tube feet. The system provides movement and circulates water, facilitating gas exchange and waste disposal.

wavelength The distance between crests of adjacent waves, such as those of the electromagnetic spectrum.

wax A type of lipid molecule consisting of one fatty acid linked to an alcohol; functions as a waterproof coating on many biological surfaces, such as apples and other fruits.

westerlies Winds that blow from west to east.

wetland An ecosystem intermediate between an aquatic one and a terrestrial one. Wetland soil is saturated with water permanently or periodically.

wild type The phenotype most commonly found in nature.

X

X chromosome inactivation In female mammals, the inactivation of one X chromosome in each somatic cell.

X-Ray Diagnostic test in which an image is created using low doses of radiation.

xylem (zī´-lum) The nonliving portion of a plant's vascular system that provides support and conveys water and inorganic nutrients from the roots to the rest of the plant. Xylem is made up of vessel elements and/or tracheids, water-conducting cells.

Y

yolk plug A cluster of endodermal cells at the surface of an amphibian gastrula. The yolk plug marks the position of the blastopore and the site of the future anus.

Z

zoned reserve An extensive region of land that includes one or more areas that are undisturbed by humans. The undisturbed areas are surrounded by lands that have been altered by human activity.

zooplankton (zō´-ō-plank´-tun) Animals that drift in aquatic environments.

zygote (zī´-gōt) The fertilized egg, which is diploid, that results from the union of a sperm cell nucleus and an egg cell nucleus.

INDEX

Pages numbers with *f* indicate a figure, and those with *t* indicate a table

A

Abiotic component of biosphere, 381, 384–86
Abiotic reservoir, biogeochemical cycles and, 424–25
ABO blood groups, 156
Absorption, nutrition by, 336
Acacia trees, mutualism between ants and, 414*f*
Acclimation, 387
Acetyl CoA, conversation of pyruvic acid to, in cellular respiration, 95*f*
Achondroplasia, 153
Acid, 31
Acid precipitation, 31, 32*f*, 446
Actinomycete, 305*f*
Activation energy, 78*f*
Activator proteins, 205, 206
Active site, enzyme, 78, 79*f*
Active transport, 82–83
Activities, calories spent in select, 76*f*
Adaptation, evolutionary. *See* Evolutionary adaptation
Adenine (A), 47, 173
Adenovirus, 188*f*
Adipose cells, 42
ADP (adenosine diphosphate) 76*f*, 77
Adult stem cells, 202
Advanced Cell Technology (ACT), 202
Aerobic process, cellular respiration as, 90. *See also* Cellular respiration
Age structure and population growth, 396*f*, 397
Agnathans, 363
Agriculture, angiosperms and, 332
AIDS (acquired immune deficiency syndrome), 191, 193. *See also* HIV (human immunodeficiency virus)
Algae, 314–16
    contrasting environments of plants and, 321*f*
    growth of, in lakes, 382*f*
    lichen as symbiotic relationship between fungus and, 340*f*
    seaweeds, 315, 316*f*
    unicellular, 314, 315*f*
Ali, Muhammad, 217*f*
Allele(s), 144
    dominant, and recessive, 146, 152–53
    fertilization and segregation of, as chance events, 149*f*
    multiple, and human ABO blood groups, 156*f*
    relationship of homologous chromosomes to, 146*f*, 147

Allopatric speciation, 275*f*, 276
Alternation of generations in plants, 326*f*
    three variations on, 328*f*
Alternative RNA splicing, 206*f*
Alzheimer disease, 217
Amino acids, 44*f*
    evolutionary relationships of, in six vertebrates, 49*f*
    inherited DNA nucleotide sequences linked to, and systematics, 289
    peptide bonds between, 44, 45*f*
    protein structure and, 45, 46*f*
    translation of gene into, 177, 178–79, 182–86
Amniocentesis, 142
Amniotic egg, 365
Amoebas, 312, 313*f*
    asexual reproduction in, 120*f*
    movement in, 66*f*, 313*f*
Amoebocytes, 350
Amphibians (Amphibia), 364–65
    metamorphosis in, 364*f*
    origin of tetrapods and, 364*f*, 365
Anabolic steroids, 43
Analogy versus homology, 288–89
Anaphase
    meiosis, I and II, 130*f*, 131*f*
    mitosis, 124, 125*f*
Anatomical responses of organisms to ecological conditions, 388
Anaximander, 245
Anchoring junctions, 68*f*
Angiosperms, 325, 330–32
    agriculture and, 332
    flowers of, 330*f*
    life cycle, 331*f*
    seeds and fruit of, 331, 332*f*
Animal(s)
    body plans of, 348*f*, 349
    cloning of, 200, 201*f*, 202
    defenses of, against predators, 15–16, 411–13
    evolution of diversity in, 344–49
    genetically modified farm, 220
    homologous structures in, 252*f*
    human (*see* Human(s))
    invertebrate, 349–60
    nutrition, 344, 345*f*
    phylogeny of, 347–49
    polyploid, 137–38
    reproduction and development in, 345*f*
    ruminants, 41*f*
    vertebrates, 360–68
    viruses of, 190–91

Animal cells
    cleavage furrow in, 125, 126*f*
    cytoskeleton and microtubules of, 65, 66*f*
    idealized, 57*f*
    storages of polysaccharides as glycogen in, 40, 41*f*
    surfaces and junctions of, 68*f*
    water balance in, 81*f*
Animalia, Kingdom, 6, 7
Annelids (Annelida), 354, 355*f*
Anther, 330*f*
Anthrax
    bioterrorism and, 297
    comparative genomics and tracking of, 231–32
    endospore of, 306*f*
Anthropoids, 369
Antibiotics, 53*f*
    bacterial resistant to, 12–13, 255, 308
Anticoagulants, 354–55
Anticodon, 183
Ants
    as introduced, invasive species, 445*f*
    mutualism between acacia trees and, 414*f*
Aphotic zone, lakes and ponds, 434
Apicomplexans, 313*f*
Aquariums, non-native species released from, 443
Aqueous solution, 30
Arachnids, 356, 357*f*
Archaea, Domain, 6, 7*f*, 291*f*
    types of, 304
Archaeopteryx fossil, 280*f*, 366–67
Arid climates, plant adaptations to, 112
Aristotle, 245
Art, earliest examples of, 375*f*
Arthropods, 356–59
    characteristics of, 356*f*
Artificial corridor, 460*f*
Artificial reefs, 407
Artificial selection, 11, 12*f*
Asexual reproduction, 120–21
Asteroid impact, effect of theorized, on Earth's climate and biological diversity, 271, 286*f*
Atherosclerosis, 42–43, 155
Athletes, metabolism and extreme fatigue in, 88
Atmosphere
    carbon dioxide and global warming in, 449–51
    C. Darwin and, 9
    depletion of ozone in, 451*f*
    earth's earliest, 32, 300*f*
    evolution of oxygen in, 114, 298
    human impact on, 449–51
    life in ancient anaerobic, 99

Atomic number, 23
Atoms, 23–25
    chemical bonding between, 25–26
    electron arrangement and chemical properties
        of, 25
    isotopes, 24
    structure of, 23
ATP (adenosine triphosphate), 65, 76
    cycle, 77f
    function of, in driving cellular work, 77f
    structure of, 76–77
    synthesis of, in cellular respiration, 94f, 96f
    synthesis of, by photosynthetic light reactions,
        105–6, 109f, 110
    yield of, in cellular respiration, 97f
ATP synthase, 96
Australopithecus, 371–72
Autosomes, 129, 151
    human genetic disorders and, 151t
Autotrophs, 89
    photosynthetic, 104f
Aves, 366–67. *See also* Birds
Axolotl, 281f
AZT drug therapy for AIDS, 171

B

Bacilli (rods), bacteria, 305f
Bacteria, 4, 6, 291f, 304. *See also* Prokaryotes;
        Prokaryotic cells
    antibiotic actions on, 53
    binary fission in, 306
    cell walls of, 304–5
    disease causing, 307–8
    diversity of, 305f
    ecological functions of, 308–10, 337
    endospores of, 306
    exponential growth of, 391f
    gene regulation in, 203–4
    giant, 305f
    infection of, by phages, 188f
    nutrition in, 307
    resistance of, to antibiotics, 12–13, 255, 308
    shapes of, 305f
    toxins produced by, 189, 308
Bacteria, Domain, 6, 7f, 291f
Bacterial chromosome, 188
Bacterial plasmids, 220, 221f
Bacteriophages (phages), 188–89
Barnacles, 357f
    competitive exclusion principle applied to two
        types of, 410f
Base, 31
Bases, DNA. *See* Nitrogenous base(s)
Batesian mimicry, 412, 413f
Bateson, William, 160
Beadle, George, 177–78
Beagle, Charles Darwin's voyage of, 246–47
Bears, evolutionary tree of, 8, 9f
Behavioral isolation as reproductive barrier, 274
Behavioral responses of organisms to ecological
        conditions, 388

Benign tumor, 126
Benthic zone
    lakes and ponds, 434
    ocean, 436
Bilateral symmetry, 348f
Binary fission, 306
Binomial naming system, 287
Biodiversity, 452. *See also* Biological diversity
Biogenesis, principle of, 300
Biogeochemical cycles, 424f, 425
    carbon cycle, 426f
    nitrogen cycle, 426f
    phosphorus cycle, 427f
    water cycle, 427f
Biogeography, as evidence for evolution, 251
Biological species concept, 272f, 273
Biological diversity, 6–8, 270–95
    cell-division errors and evolution of, 137–38
    classifying and naming organisms of, 287–92
    in communities, 408, 413–14
    conservation biology and, 456–61
    current crisis and loss of, 452–55
    effect of asteroid impact on, 271, 286f
    evolution of biological novelty, 280–82
    genetic variations in populations, 257–58
    hot spots of, 456f, 457
    levels of, 452
    macroevolution, 271–72, 282–86
    meaning of, 455
    mutations as cause of, 187f
    origin of species and, 272–80 (*see also*
        Speciation; Species)
    in plants, as nonrenewable, 332–34
Biological magnification of chemicals, 449f
Biological molecules, 39–48
    carbohydrates, 39–41
    lipids, 42–43
    nucleic acids, 47–48
    proteins, 44–47
Biological organization, levels of, 21f
Biological Weapons Convention of 1975, 297
Biology, 1–18
    evolution as unifying theme of, 8–13
    news about, 2f
    process of science and, 13–18
    scope of, 2–8
Biology and society
    asteroid impact, effect on Earth's climate and
        biological diversity, 271, 286f
    bioterrorism, 297
    drugs designed to target specific cells, 53
    egg donation and infertility, 119
    extreme fatigue in athletes, metabolism and, 88
    fluoridated water, 21
    forest conservation, 321
    genetic testing of fetus, 142
    human-introduced invasive species, 344, 443
    human population explosion, 381
    hunting for genes, and DNA technology, 217
    lactose intolerance in humans, 36
    pesticide resistance in insects, 243, 255
    plants as energy source (biomass), 103

reefs, natural and artificial, 407
    sabotaging HIV with specific drugs, 171
    stem cells in umbilical cord blood, 197
    stonewashing jeans with cellulase, 73
Biomass, 421
    energy produced from, 103
Biome
    freshwater, 434–35
    marine, 435–37
    terrestrial, 428f–33
Biophilia, 462
Bioremediation, 309–10
Biosphere, 383
    abiotic components of, 381–82, 384–86
    biotic components of, 382
    ecosystems in, 4
    regional distribution of life in, 384, 385f
Biotechnology, 218
Bioterrorism, 231, 297
Biotic component of environment, 382
Bipedalism, evolution of, 371, 372f
Birds, 366–67
    chick embryo, 252f
    courtship rituals in, 274f
    density-dependent regulation of population
        growth in, 393f
    evolution of, 280–81
    flight in, 366f
    fossil of extinct, 280f, 366–67
    inheritance patterns in budgies, 142, 143f
    introduced species, 445f
    migrations of, 388
    northern spotted owl, 457f
    red cockaded woodpecker, 458f
Birth rates, 395
Bishop, J. Michael, 209
Bivalves, 354f
Black Death, 303
Blastula, 345
Blood
    ABO groups, 156
    cells, 197f
    hemoglobin (*see* Hemoglobin)
    stem cells in umbilical cord, 197f
Blood flukes, 352
Body cavity, animal, 348f, 349
Body plans, animal, 348
    of bilateral animals, 348f
    radial and bilateral symmetry, 348f
Body segmentation, 354, 355f, 363
Bony fishes, 363
Bottleneck effect, 261–62
Brain
    evolution of human, 372–73
    skull size and, 281f
BRCA1 and *BRCA2* genes, 210–22
Breast cancer, 127f
    genes linked to, 210–11
Breathing, relation of cellular respiration to, 90, 91f
Brown algae, 316f
Bryophytes, 325–26
Bubonic plague, 303

Buffers, 31
Buffon, Georges, 245

C

C₃ plants, photosynthesis in, 112f
C₄ plants, photosynthesis in, 112f
Calorie, 75
    select activities and expense of, 76f
    in select foods, 76f
Calvin cycle, 105f, 106, 111f
Cambrian period, explosion of biological diversity
    in, 299
    early animals and, 346–47
Camouflage as evolutionary adaptation, 244f,
    412f
CAM plants, photosynthesis in, 112
Cancer, 126–27, 208–12
    breast, 127f, 210–11
    cell cycle control system and uncontrolled
        growth of, 126
    development of typical colon, 210f
    effects of lifestyle on, 211–12
    genes as cause of, 208–11
    growth and metastasis of breast, 127f
    "inherited," 210–11
    prevention and survival, 127
    treatment of, 127
    types of, 126–27
    types and causes of, in U.S., 211t
Cap nucleotide sequences, 181, 206
Carbohydrates, 39–41
    disaccharides, 40
    monosaccharides, 39
    polysaccharides, 40–41
"Carbo-loading," 40–41
Carbon
    atom, 25f
    isotopes of, 24f
    variations in skeletons of, 36f
Carbon chemistry, 36–38
Carbon cycle, 426f
    human impact on, 446
Carbon dioxide (CO_2)
    Calvin cycle and use of, 111f, 112
    greenhouse effect and, 113, 449–51
    human production of excessive amounts of,
        446f, 449–51
    as waste product of cellular respiration, 90f
Carboniferous "coal forest," 327f
Carcinogens, 211
Carcinomas, 126
Cardiovascular diseases, link between saturated
    fats in diet and, 42–43
Carnivores (order Carnivora), 419
    phylogenetic tree for, 288f
Carotenoids, 107
Carpel, 330f
Carriers of genetic disorders, 151
Carrying capacity, 392
    humans and, 397–98
Carson, Rachel, 384f, 449

Cartilaginous fishes, 363
Cats, X chromosome inactivation and tortoiseshell
    fur in, 205f
Celera Genomics, 230, 233
Cell(s), 4–5, 52–86
    ATP and work in, 76–78
    basic energy concepts and, 73–76
    daughter, 120, 122
    defined, 22
    diploid, and haploid, 129
    drugs that target specific, 53
    enzymes and chemical reactions in, 78–79, 84
    eukaryotic cells, 4, 55, 56, 57, 58–69 (see also
        Eukaryotic cells)
    membrane transport in, 80–83
    microscopy and study of, 53–55
    pre-cells, and origins of life, 302f
    prokaryotic cells, 4, 55–56 (see also Bacteria;
        Prokaryotic cells)
    recycling of macromolecules in, 206
    reproduction of (see Cell reproduction)
    size range of, 55f
    water balance in, 81–82
Cell-cell recognition, membrane proteins and,
    59f
Cell cycle, 121–27
    cancer as out-of-control, 126–27
    eukaryotic chromosomes and, 121–22
    mitosis and cytokinesis in, 124–26
    phases of, in eukaryotic cell, 122, 123f
Cell cycle control system, 126
Cell division, 120. See also Cell reproduction
Cell junctions, 68
Cell plate, 126
Cell reproduction, 118–40
    cell cycle and mitosis, 121–27
    functions of, 119–21
    new species from errors in, 137–38
    sexual reproduction and meiosis, 121, 128–37
Cell signaling, 37f, 207–8
    effects of cancer genes on, 209–10
    pathway of, to turn on genes, 208f
    role of membranes in, 83f
Cell theory, 55
Cellular differentiation, 198
Cellular respiration, 75, 87–101
    ATP yield in, 97f
    energy flow, chemical cycling, and, 88–89, 90f
    fermentation, 98–99
    life on anaerobic earth, 99
    metabolic pathway of, 93–97
    mitochondria as site of, 65f
    overall equation for, 91
    relationship of breathing to, 90, 91f
    roadmap for, 93f
    role of oxygen in, 91–92
    versatility of, 97
Cellular slime mold, 314f
Cellulase, stonewashing jeans with, 73
Cellulose, 41
    bacterial digestion of, 41f
Cell walls, 304–5

Centipedes, 358f
Central vacuole, 63
Centrioles, 125
Centromere, 122
Centrosomes, 125
Century plant, opportunistic life history strategy
    of, 400f
Cephalopods, 354f
Chaparral biome, 431f
Charophyceans, 323f, 325
Cheetahs, low genetic variability in populations of,
    153, 262
Chemical bonds, 25–26
Chemical cycling in ecosystems, 3, 88–89, 90f, 418,
    424–27
    case study on forest ecosystem and, 447, 448f
    examples of biogeochemical cycles, 425–27
    generalized scheme for, 424f
    human impact on, 446–47
    role of prokaryotes in, 308–9
Chemical energy, 74–75
Chemical evolution, 300–302
    Darwinian natural selection and, 302
Chemical reactions, 27
Chemicals, release of toxic, into ecosystems, 448,
    449f
Chemical work, 77f
Chemistry, 20–34
    acids, bases, and pH, 31
    atoms, 23–25
    chemical bonding and molecules, 25–26
    chemical reactions, 27
    of earth before living organisms, 32
    elements and compounds, 22–23
    fluoridated drinking water, 21
    life at the level of, 21–22
    organic, 36–38
    water, life-supporting properties of, 29–31
    water, structure of, 28
Chemoautotrophs, 307
Chemoheterotrophs, 307
Chemotherapy, 127
Chernobyl, Ukraine, nuclear accident, 24
Chiasma, 134
Chicxulub crater, Caribbean, 271f, 286f
Chimpanzees, skull and brain size, compared to
    humans, 281f
Chitin, 337, 356
Chlamydomonas, 315f
Chloramphenicol, 53
Chlorofluorocarbons, 451
Chlorophyll pigments
    behavior of isolated, 108f
    chlorophyll a, 107
Chloroplasts
    endosymbiosis and origins of, 311f, 312
    pigments in, 107
    as site of photosynthesis, 64f–65, 103–5
Choanocytes, 350
Cholesterol, 43f
    uptake of, by human cells, 83f
Chondrichthyes, 363

Chordates (Chordata), 361–63
  characteristics of, 361f
  invertebrate, 361f
  vertebrate, 361–68
Chorionic villus sampling (CVS), 142
Chromatin, 60, 121
  DNA packing and, 204–5
Chromosome(s), 60, 120
  DNA packing in, 122f, 204–5 (see also DNA (deoxyribonucleic acid))
  duplication of eukaryotic, 121–22, 123f
  errors in meiosis and altered numbers of, 134–37
  halving number of, during meiosis, 129f
  homologous, 128–29 (see also Homologous chromosomes)
  inactivation of, 205
  independent assortment of, 133–34
  karyotype of pairs of, 128f
  movement of, during mitosis and cytokinesis, 124–26
  sex, 129 (see also Sex chromosomes)
Chromosome theory of inheritance, 159–62
Cicadas, population cycles of, 394
Cilia, 66, 67f
  universal architecture of eukaryotic, 8f
Ciliates, 313
Ciprofloxacin, 53
Cities, population growth in, 397f
Clades, 289
Cladistic analysis, 289–90
  simplified example of, 290f
Classes, classification into, 288
Classification, taxonomic, 287, 288
  kingdoms and domains, 290–92
  phylogeny and, 288–90
Cleavage furrow, 125, 126f
Climate
  global warming of, 4, 451f, 452
  theorized asteroid impact and Earth's, 271, 286f
Clinton, Bill, 225
Clones, 200
Cloning
  animals, 200, 201f, 202
  gene, 221
  plants, 200f
  stem cells and therapeutic, 201–2
Clumped population dispersion pattern, 390f
Cnidarians (Cnidaria), 350–51
Cnidocytes, 351f
Coal, 327
Coal forest, 327f, 328
Cocci (spheres), bacteria, 305f
Codominance (allele), 156
Codon, 179
  genetic code listed by RNA, 179t
  stop, 184
  transcription and translation of, 178f
Coelom, 348f, 349
Coevolution, 438
Cohesion, 29
Colon cancer, development of, 210f
Color blindness, 165

Columbian Exchange, 444
Columbia River Basin, damming of, 435f
Communicating junctions, 68f
Community(ies), 407–17. See also Ecosystem(s)
  coevolution in biological, 438
  defined, 21–22, 383, 408
  disturbances of, 386, 415–17
  human impact on biological, 443–46
  interspecific interactions in, 408–15
  key properties of, 407–9
  ocean hydrothermal vent communities, 437f
Community ecology, 383
Comparative anatomy, as evidence for evolution, 251–52
Comparative embryology, as evidence for evolution, 252
Competition between species, 409–10
  competitive exclusion principle and, 409–10
  ecological niche, 410
  resource partitioning, 410, 411f
Competitive exclusion principle, 410–11
Complete digestive tract, evolution of, 353
Compounds, 23
Coniferous forests, 321, 328f
  biomes, 433f
Conifers, 328, 329f
Conservation biology, 456–61
  biodiversity "hot spots" and, 456f–57
  at ecosystem level, 459–61
  goals of sustainable development and, 461
  implications of bottleneck effect in, 262f
  at population and species levels, 457–59
Conservation of energy, 73–74
Consumers in ecosystems, 89, 419
Continental drift, 285
Contractile proteins, 44f
Contractile vacuole, 63f
Convergent evolution, 289
Coral reefs, 407
Corals, 351f
Corn syrup, 40
Corridors, movement, 459–50
Costa Rica, zoned reserved in, 460f
Courtship rituals, 274f
Covalent bonds, 26
Crassulacean acid metabolism (CAM) plants, 112
Crick, Francis, 173–74
Criminal cases. See Forensics and DNA fingerprinting
Cristae, 65
Cro-Magnons, 373–74
Cross (hybridization), 144
  dihybrid, 147–48
  monohybrid, 144–46
  polyploid formation by, 277f
Cross-fertilization, 144
Crossing over (genetic material), 130, 134f
  linked genes and, 161f
Crustaceans, 356, 357f
Cryptic coloration, 412
Culture, evolution of human, 375

Cuticle, plant, 321f, 322, 325
Cyanobacteria, 305f
  evolution of oxygen in atmosphere and role of, 114
Cystic fibrosis, 152
Cytokinesis, 124–26
  cell cycle, and, 123f
  meiosis and, 131f
  mitosis and, 124–26
  telophase and, 125f
Cytoplasm, 57
  gene regulation in, 206–7
Cytosine (C), 47, 173
Cytoskeleton, 65–67
  attachment of membrane proteins to, 59f
  cilia, flagella, and, 66–67
  maintaining cell shape as function of, 65–66
Cytosol, 57

D

Darwin, Charles, 9, 244–49
  descent with modification, 248, 249f
  on natural selection, 10–12, 244
  sea voyage of, and cultural and scientific context of, 245–48
Darwinian fitness, 264
Darwinian medicine, 266
Daughter cells, 120, 122, 126f
DDT pesticide, 384
  biological magnification of, in food chain, 449f
Deafness, inheritance of, 151f, 152
Death rates, 395
Decomposers, fungi and bacteria as, 337, 419f
Deductive reasoning, 14, 15
Deforestation, 321f, 381f, 444, 447f, 451
  case study of chemical cycling and, 447, 448f
  connection to global warming, 114
  loss of plant biological diversity due to, 332–34
  water cycle and, 447f
Dehydration synthesis, 38
  amino acids linked with peptide bonds by, 44, 45f
  fatty acid attachment to glycerol by, 42f
Denaturation of proteins, 47
Density-dependent regulation of population growth, 392, 393f
Density-independent regulation of population growth, 393, 394f
Descent with modification, 9, 238, 249f
Desert
  biome, 431f
  food web in, 420f
Detritus, 419
Detrivores (decomposers), 337, 419
Development
  animal, 345f
  evolutionary novelty and ("evo-devo"), 281–82
  plant, 326f, 328f, 329f
De Vries, Hugo, 277f
Dextrose, 39
Diabetes, 5

Diatoms, 315*f*
Diffusion, 80
Diffusion facilitated, 80
Digestive tract, complete, 353
Dihybrid cross, 147*f*, 148
Dinoflagellates, 315*f*
Dinosaurs, 8*f*, 271, 365*f*
Diploid cells, 129
  in animals, 345*f*
  in plants, 326*f*, 328*f*, 329
Directional selection, 264, 265*f*
Disaccharides, 40
Discovery science, 13–14
  ecology and natural history as, 382
Disease, human. *See also* Genetic disorders
  AIDS, 191, 193
  anthrax, 231–32, 297, 306*f*
  bacterial, 303, 307–8
  cancer (*see* Cancer)
  cardiovascular, 42–43
  caused by bacterial toxins, 189, 308
  emerging viruses, 192
  lifestyle linked to, 211–12
  Lyme disease, 308*f*
  malaria, 157–58, 313*f*
  protozoan, 312, 313*f*
  ringworm, 338
  sleeping sickness, 312, 313*f*, 359
  viral, 190–91
  as weapon, 297
Dispersion patterns of populations, 390
Disturbances of communities, 415–17
  dynamic view of structure of, 417
  ecological succession and, 416–17
  human-caused, 444
  periodic, as abiotic factor, 386
Diversifying selection, 264, 265*f*
DNA (deoxyribonucleic acid), 4, 47, 171–87
  amplification of, using PCR technique, 226*f*
  analysis of prehistoric, 289*f*
  chemical building blocks of, 5*f*
  cutting and pasting, with restriction enzymes, 221, 222*f*
  discovery of double helix structure of, 173–75
  gel electrophoresis of, 228*f*
  genetic code and, 179–80
  genetic information flow from genotype to phenotype, 177–78, 182–86, 198
  mutations in, 186–88
  nitrogenous bases of, 47*f*, 48*f*, 173 (*see also* Nitrogenous base(s))
  nucleotides of, 47*f*, 48*f*, 172–73
  packing of, in eukaryotic chromosomes, 122*f*
  recombinant, 218, 220–23
  repetitive, 230
  replication of, 175–76
  sequencing of, 232–33
  sticky ends of, 222
  structure of, 18, 48*f*, 171, 172*f*, 173, 174*f*
  three representations of, 174*f*
  transcription of, to RNA, 180–81 (*see also* Transcription)

translation of, to amino acid sequence, 178–79, 182–86 (*see also* Translation)
  ultraviolet damage to, 176*f*
DNA fingerprinting, 2, 224
  forensic applications of, 18*f*, 225–26, 229*f*
  overview, 224*f*
  techniques of, 226–29
DNA ligase, 222
DNA microarray, 198, 199*f*
DNA packing, 122*f*
  gene regulation and, 204–5
DNA polymerase, 176, 226
  multiple bubbles in, 176*f*
DNA sequences
  amino acid sequences linked to, and systematics, 289
  cap, and tail, 181, 206
  enhancers, 206, 230
  homeobox, 213
  operator, 203
  promoter, 180, 203, 230
  repressor, 204, 205
  terminator, 180–81
  silencers, 206
DNA technology, 216–40
  in forensic science, 18*f*, 224–29, 231–32
  genetically modified plants and farm animals, 219–20
  genomics, 229–33
  genomics and evolution, 237
  human gene therapy, 234
  Human Genome Project and, 217, 230, 237
  hunting for genes using, 217
  in pharmaceutical industry, 5, 218–19
  production of Humulin (human insulin), 218–19, 223
  recombinant DNA and, 217, 218*f*
  recombinant DNA techniques, 220–23
  safety and ethical issues related to, 235–37
Dogs
  artificial selection in, 12*f*
  independent assortment of genes in, 148*f*
Dolly, cloned sheep, 201*f*, 202
Domains, classification into three, 6, 7*f*, 288, 291*f*
Dominance, incomplete, 155
Dominant allele, 146
  genetic disorders linked to, 153
Dominant traits, human, 150*f*
Dorsal, hollow nerve cord, 361
Double fertilization, plant, 331
Double helix, DNA, 48*f*, 173, 174*f*, 176
Down syndrome, 134–35*f*
Drugs, addiction to, 62. *See also* Pharmaceutical drugs
Duchenne muscular dystrophy, 165
Dutch elm disease, 338*f*, 444

E

Earth
  atmosphere (*see* Atmosphere)
  biomes of, 428–37
  biosphere (*see* Biosphere)

earthrise from moon, 449*f*
  global warming, 114, 450*f*, 451
  original atmosphere of, 32
  water on, 28*f*
Earthworms, 354, 355*f*
Ebola virus, 192*f*
Echidna, 367*f*
Echinoderms, 359–60
Ecological niche, 410
Ecological succession, 416*f*, 417
Ecology, 3, 381
  evolutionary adaptations of organisms and, 387–88
  human population explosion and, 381
  landscape, 459–61
  levels of, 383
  life histories of organisms and, 398–403
  overview of, 381–86
  of populations, 388–98
Ecosystem(s), 3–4, 417–21
  chemical cycling in, 3, 88–89, 90*f*, 418, 424–27
  communities in (*see* Community(ies))
  conservation biology at level of, 459–61
  consumers and producers in, 89, 419
  defined, 21, 383, 417
  energy flow in, 3, 4*f*, 88–89, 90*f*, 418, 421–24
  food chains in, 419*f*
  food webs in, 419, 420f, 421
  fragmention of forest, 457*f*
  human impact on, 446–51 (*see also* Human impact on environment)
  survey of biomes, 428–37
  trophic levels in, 418, 419, 423*f*
Ecosystem ecology, 383
Ectotherms, 365, 387
Edges, landscape, 459*f*–60
Education, control of bacterial disease through, 308
Egg
  amniotic, 365
  donation of human, 119
  fertilization of, 129 (*see also* Fertilization)
  plant, 322
Electromagnetic spectrum, 106*f*
Electron, 23
  "fall" of, in redox reactions, 92*f*
  ionic bonding and transfer of, 25*f*
Electron microscope (EM), 54*f*, 55
Electron shells, 25
Electron transport chain, 92, 95–96
  ATP generation and, 96*f*
Element(s), 22
  periodic table of, 22*f*
  trace, 23
Elephants, family evolutionary tree of, 249*f*
Embryo
  bird, 252*f*
  human, 119*f*, 252*f*
  plant, 323*f*
Embryology, comparative, 252
Embryonic development
  lysosomes and, 63
  role of homeotic genes in, 212–13

Embryonic stem cells (ES cells), 202
Endangered species, 457
   conserving, 457–59
   DNA fingerprinting of, 226
Endangered Species Act, (ESA), U.S., 457
Endocytosis, 82f–83
Endomembrane system, 61–64, 311
Endoplasmic reticulum (ER), 61f, 62
Endoskeleton, echinoderm, 360
Endosperm, 331
Endospores, 306
Endosymbiosis, 311f–12
Endotherms, 366
Endotoxins, 308
Energy, 73–76
   chemical, 74–75
   conservation of, 73–74
   defined, 73
   electromagnetic, 106–7
   entropy, 74
   food calories and, 75–76
   kinetic, and potential, 74
   plants as source of, 103
Energy budgets of ecosystems, 421–22
Energy conversions
   in cars and in cells, 75f
   snowboarding, 74f
Energy coupling, 77
Energy flow in ecosystems, 3, 4f, 88–89, 90f, 418,
      421–24
   energetics and human nutrition, 423–24
   energy pyramids, 422–23
   productivity and ecosystem energy budgets,
      421–22
Energy pyramids, 422, 423f
Enhancer nucleotide sequences, 206, 230
Entropy, 74
Enveloped virus, life cycle of, 190f
Environment
   abiotic component of, 381–82, 384–86
   biotic component of, 382
   human genetics and role of, 158–59
   human impact on (see Human impact on
      environment)
Environmentalism, ecology and, 383-84
Enzyme(s), 78–79
   activation energy and, 78f
   induced fit and working of, 78–79f
   inhibitors of, 78, 79f
   membrane proteins as, 59f
   restriction, 222
Enzyme inhibitors, 78, 79f
Equilibrial life history, 400, 401t
Ergot, 338f
Erythromycin, 53
Erythropoietin (EPO), 219
Escherichia coli (E. coli) bacteria
   genetic studies of, 217–18
   lactose metabolism in, 203, 204f
Estrogen, 43f
Ethics
   biophilia and environmental, 462

DNA technology and issues of, 236–37
   human egg donation, 119
Eukarya, Domain, 6, 7f, 291f
Eukaryotes
   cilia of, universal architecture of, 8f
   evolution of, 299, 311f–12
Eukaryotic cells, 4, 55. See also Eukaryotes
   animal, 57f
   cell cycle and mitosis in, 121–27
   cell surfaces, 67–68
   cytoplasm, 57
   cytoskeleton, 65–67
   cytosol, 57
   endomembrane system, 61–64
   energy conversion in chloroplasts and
      mitochondria, 64–65
   flow of genetic information in, 177–78, 182–86,
      198
   gene expression in, 203f, 204–7
   membrane structure and function, 58–59
   nucleus and ribosomes of, 4f, 56f, 57, 59–60
   origin of (endosymbiosis), 311f–12
   origin of membranes in, 69
   overview of, 57
   plant, 57
   plasma membrane of, 57, 58–59
   prokaryotes compared to, 56f
   size and complexity, compared to prokaryotic
      cells, 56f
European starlings, 445f
Eutherians, 367f, 368
Evaporative cooling, 29, 30f
"Evo-Devo" (evolution and development), 281–82
Evolution, 8–13
   of animals, 344–49
   of atmospheric oxygen, 114
   biological diversity and (see Biological diversity)
   of biological novelty, 280–82
   branching and nonbranching, 272f
   convergent, 289
   current rate of mass extinctions and, 376
   C. Darwin and, 9, 244–49
   descent with modification idea and, 248, 249f
   DNA and proteins as measures of, 49
   of earth before life, and earliest atmosphere, 32
   of enzymes, 84
   evidence for, 249–53
   life histories and, 398–401
   macroevolution, and earth history, 282–86
   of membranes, 69
   microevolution, 260–65
   modern synthesis of theories on, 256–60, 272f
   of multicellular life, 299, 317f
   mutual symbiosis and, 340
   natural selection and, 10–13, 254–56
   origin of eukaryotic cells, 311–12
   origin of life, 99, 297–302
   origin of new species in cell division errors,
      137–38
   of plants, 321–23, 324f, 325
   of prokaryotes, 303–10
   of species (see Speciation; Species)

theory, meaning of, applied to, 292–93
   as unifying theme of biology, 8–9
Evolutionary adaptation, 244
   beaks adaptations of Galápagos finches, 254f
   camouflage as example of, 244f
   J B Lamarck and, 245
   life history traits as, 400, 401t, 401–3
   of organisms, ecology and, 387–88
   terrestrial adaptations of plants, 321–23, 328–29
   testing hypothesis of life history traits as, in
      guppy evolution, 401–3
Exaptation, 280–81
Exocytosis, 82f–83
Exons, 182
Exoskeleton, 356
Exotoxins, 308
Exponential model of population growth, 391
   logistic growth compared to, 392f
Extinctions, mass, 284, 285–86, 376, 381, 452–53
Extinct species
   animals close to extinction, 453f
   bird (Archaeopteryx), 280f, 366–67
Extracellular matrix, 68
   attachment of membrane proteins to, 59f
Eye color, sex-linked genes and, in fruit flies, 163,
   164f

F
F₁ generation, 144
F₂ generation, 144
Facilitated diffusion, 80
Facultative anaerobes, 99
Fallopian tubes, cilia of, 8f
Family, classification into, 288
Family pedigree, inheritance patterns and, 150–51
Fat(s), 42–43
   saturated versus unsaturated, 42
Feedback regulation, 79
Females, maternal age of, and Down syndrome,
   135f
Fermentation, 98–99
Ferns, 325, 327
Fertilization, 129
   as chance event, 149f
   cross-, 144
   after nondisjunction, 136f
   random, 134
   self-, 144
Fetus, genetic testing of, 142
Fire ants, 445f
Fire as abiotic component in biosphere, 386
Fishes, 363–64
   guppy evolution, study of, 401–3
   non-native, invasive fish released from
      aquariums, 443
Five-kingdom system, 291f
Flagella, 66, 67f
Flagellates, 312, 313f
Flatworms, 352
Flight
   in birds, 366f
   in insects, 358

Flower, 330f, 331
  Darwinian fitness in attracting pollinators to, 264f
Fluid mosaic model of membranes, 58f
Flukes, 352
Fluoride in drinking water, 21
Food
  calories in, 75–76f
  cancer prevention and, 212
  fungi as, 339f
  human nutrition, trophic levels, and, 423f, 424
  genetically modified, 219–20, 235–36
  metabolism of (see Cellular respiration)
  transformation of, into biomass, 422f
  use of fermentation to produce, 99
Food chains, 419f
  biological amplification of DDT in, 449f
Food vacuoles, 63
Food webs, 419, 420f, 421
Forams, 313f
Forensics and DNA fingerprinting, 18f, 224–29
  anthrax bioterrorism and, 231
  DNA fingerprinting, overview, 224f
  DNA fingerprinting, techniques, 226–29
  first case of, 225
  paternity questions and, 225–26
Forest
  biomes, 430f, 432f, 433f
  case study of chemical cycling in, 447, 448f
  conservation of coniferous, 321
  deforestation of (see Deforestation)
  fragmentation of, 457f
  old growth, 113f
  recovery of, after fire, 386f
  research in canopy of rain, 382f
Fossil(s), 245
  of extinct bird, 280f, 366–67
  gallery of, 282f
  of prokaryotes, 303f
  radiometric dating of, 284f
  whale, 250, 251f
Fossil fuels, 327, 451
Fossil record, 8f, 9, 250
  geologic time scale and, 282–86
  theorized asteroid impact and, 271, 286f
Founder effect, 262
Fox, Michael J., 217f
Franklin, Rosalind, 173
Freshwater biomes, 434–35
Frog, metamorphosis in, 364f
Fructose, 39f
Fruit, 331, 332f
Fruit fly, 162f
  homeotic genes in, 212f, 213f
  sex determination in, 163
  sex-linked genes and eye color in, 163, 164f
Fuel, hydrocarbons as, 37f, 327, 451
Functional groups, 38
Fungi, 334–39
  characteristics of, 336–37
  commercial uses of, 338–39

as decomposers, 337, 419f
ecological impact of, 337–39
examples of diversity in, 335f
lichen as symbiotic relationship between algae and, 340f
reproduction in, 337
symbiotic relationships with plant roots, 322f
Fungi, Kingdom, 6, 7f. See also Fungi

G

G₁ and G₂ phases, cell cycle, 123f
Galápagos finches
  beak adaptations of, 254f
  evolutionary tree for, 10f
Galápagos Islands, C. Darwin's voyage to, 246f, 247, 251, 271, 276
Gametangia, 322, 325
Gametes, 129, 322
Gametic isolation as reproductive barrier, 274
Gametophyte, 326, 328f, 329
Garrod, Archibald, 177
Gastropods, 354f
Gastrovascular cavity, 350
Gause, G. F., 409
Gel electrophoresis, 227–29
Gene(s), 60, 121
  alkaptonuria, 177
  alternative forms of (see Allele(s))
  cell division and transmittal of, 120
  DNA technology involving recombinant (see DNA technology)
  genetic disorders caused by single, 151–53
  genetic message in, translated into protein, 60f
  homeotic, 212–13
  inactivation of, 205
  independent assortment of, 147–48
  inherited characteristics linked to several (polygenic inheritance), 158
  linkage of, 160–62
  loci, 147
  mapping, 162f
  molecular biology of (see DNA (deoxyribonucleic acid); RNA (ribonucleic acid))
  multiple inherited characteristics linked to single (pleiotropy), 157–58
  mutations in, 186–88
  nucleic acid probes for finding and tagging, 222, 223f
  oncogenes, 209
  operon, 203
  regulation of (see Gene regulation)
  sex-linked, 163–65
  synthesis of, from eukaryotic mRNA, 223f
  transcription of, 180, 181f
  translation of, 177, 178–79, 182–86
  tumor-suppressor, 209–10
  viruses as packaged, 188 (see also Viruses)
Gene cloning, 221
Gene expression
  patterns of, in differentiated cells, 198f

regulation of (see Gene regulation)
visualizing, in DNA microarrays, 198, 199f, 200
Gene flow, human evolution and, 263
Gene pool
  analysis of, 258–59
  microevolution as changes in, 260–65
Gene regulation, 196–215
  in bacteria, 203–4
  cancer and faulty, 208–12
  cell diversity and reasons for, 197–202
  cell signaling and, 207–8
  in cytoplasm, 206–7
  homeotic genes, 212–13
  in nucleus of eukaryotic cells, 204–6
Gene therapy, 234
Genetically modified (GM) organisms
  defined, 218
  farm animals, 220
  plant foods, 219–20
  safety issues of, 235–36
Genetic code, 179–80
  dictionary of, listed by RNA codons, 179t
Genetic disorders
  autosomal, 151t
  carriers of, 151
  caused by single gene, 151–53
  deafness, 151f, 152
  dominant, 153
  errors in meiosis as origin of, 134–37
  gene therapy for, 234
  genetic drift and, 262–63
  hypercholesterolemia, 155
  phenylketonuria (PKU), 259
  pleiotropy and, 157–58
  population genetics and, 259–60
  recessive, 152–53
  sex-linked, 165
  sickle cell disease, 45f, 157–58, 186f
  Tay-Sachs, 63
Genetic drift, 261–63
  bottleneck effect as, 261–62
  founder effect as, 262
  hereditary disorders in humans and, 262–63
Genetic marker, 226–27
Genetic recombination
  crossing over and, 134f, 161f
  crossing over and linkage maps, 162
  technological (see Recombinant DNA technology)
Genetics
  cell reproduction and, 119–21
  chromosomes (see Chromosome(s))
  DNA (see DNA (deoxyribonucleic acid); DNA technology)
  evolution of new species from cell-division errors, 137–38 (see also Speciation)
  genes (see Gene(s); Gene regulation)
  meiosis, sexual reproduction, and, 128–37
  mitosis, cell cycle, and, 121–27
  patterns of inheritance (see Inheritance)
  population, 257
Genetic testing, 142

Genetic variation, origins of, 133–34
  crossing over, 134*f*
  inbreeding and reduced, 152
  independent assortment of chromosomes, 133–34
  random fertilization, 134
Genome, 5, 121
  mapping of, 232–33
  of select organisms, 230, 231*t*
Genomic library, 222
Genomics, 229–33
  Human Genome Project, 217, 230, 237
  some completed genomes, 231*t*
  techniques of mapping genomes, 232–33
  tracking anthrax using, 231–32
Genotype
  defined, 156, 177
  flow of genetic information to phenotype from, 177–78, 182–86, 198
  inheritance patterns and (*see* Inheritance)
  using testcross to determine unknown, 149
Genus, 287
Geographic isolation, allopatric speciation and, 275, 276*f*
Geologic time scale, fossil record and, 282, 283*t*, 284
Geology, impact of new theories in, on Charles Darwin, 247–48
Germination, 329
Giardia (protozoan), 312
Global warming, 114, 450*f*, 451
Glucose, 30, 40, 41
  oxidation of, in cellular respiration, 91–92
  as product of photosynthesis, 90*f*, 105, 111*f*
  ring structure of, 39*f*
Glyceraldehyde 3-phosphate (G3P), Calvin cycle and production of, 111*f*
Glycogen, 40
Glycolysis, 93, 94*f*
  as ancient metabolic pathway, 99
  fermentation and, 98*f*
  link between Krebs cycle and, 95*f*
Goiter, iodized salt for prevention of, 23*f*
Golgi apparatus, 62*f*
Gould, Stephen Jay, 11
Gradualist model of speciation, 279*f*
Grana, 65, 105
Grand Canyon
  allopatric speciation of squirrels in, 276*f*
  sedimentary rock strata at, 250*f*
Grasshopper, 358*f*
Grassland biome, temperate, 432*f*
Green algae
  origin of plants from, 323, 325
  seaweed, 316*f*
  unicellular, 315
Greenhouse effect, 113–14, 449–51
  factors influencing, 450*f*
  global warming and, 450*f*, 451
  photosynthesis and moderation of, 113
Growth factors, 209
Guanine (G), 47, 173

Guppie, study of life history traits and evolution in, 401–2, 403*f*
Gymnosperms, 325

H

Habitat
  conserving species and needs of, 458–59
  description of good, 458
  destruction or fragmentation of, as cause of biodiversity crisis, 454, 457–58
  of red cockaded woodpecker, 458*f*
  source, and sink, 458
Habitat isolation as reproductive barrier, 274
Haemophilus influenzae bacteria, 307*f*
Halophiles, 304*f*
Hantavirus, 192*f*, 193
Haploid cells, 129
  in animals, 345*f*
  formation of, 130, 131*f*
  in plants, 326*f*, 328*f*, 329
Hardy-Weinberg equilibrium, 260, 263
Hardy-Weinberg formula, 259
Health. *See also* Disease, human; Genetic disorders
  population genetics and, 259–60, 266–67
Heat, defined, 29, 74
Helium, two models of, 23*f*
Hemoglobin, 44*f*
  sickle cell disease and structure of, 45*f*, 157, 186*f*
Hemophilia, 165*f*
Hepatitis B, vaccine against, 219*f*
Herbivores
  plant defenses against, 411
  as primary consumers, 419
Heredity. *See* Inheritance
Herpes virus, 190
Heterotrophs, 89
Heterozygous organisms, 147
High-fructose corn syrup (HFCS), 40
Histones, 122
HIV (human immunodeficiency virus), 191
  attacking DNA of, with AZT drug therapy, 171
  behavior and structure of, 191*f*
Homeoboxes, 213
Homeotic genes, 212–13
Hominids, 371
Homo erectus, 373
Homo habilis, 372–73
Homologous chromosomes, 128–29
  halving chromosome number in, during meiosis, 129*f*, 130*f*, 131*f*
  relationship of alleles to, 146*f*, 147
Homology, 252
  analogy versus, 288–89
Homo sapiens, 373–74
  two hypotheses for origins of, 374*f*
Homozygous organisms, 147
Honey, simple sugars in, 39*f*
Hooke, Robert, 55
Host, symbiotic relationships and role of, 414
Hot spots, diversity, 456*f*, 457

Hubbard Brook Experimental Forest, New Hampshire, case study of chemical cycling and deforestation of, 447, 448*f*
Human(s)
  diseases of (*see* Diseases, human)
  evolution (*see* Human evolution)
  examples of inherited traits in, 150*f*
  genetic disorders of (*see* Genetic disorders)
  genetic testing of fetal, 143
  impact of, in ecosystems (*see* Human impact on environment)
  infertility and egg donation, 119
  lactose intolerance in, 36
  life cycle of, 129*f*
  population (*see* Human population)
  sex determination in, 163
  skull and brain size, compared to chimpanzees, 281*f*
  sugar consumption by, in U.S., 40*f*
  threat of biosphere alterations to, 455
Human body
  blood groups, 156
  chemical composition of, by weight, 22*f*
  chromosome number in, 60, 121, 129
  four types of cells in, 197*f*
  sex chromosomes in, 129, 136*t*, 137*f*, 163
  skin pigmentation in, 158
Human evolution, 370–75
  Australopithecus and bipedalism, 371–72
  common misconceptions about, 370–71
  cultural evolution in, 375
  Homo erectus and dispersion of humans, 373
  Homo habilis and brain evolution, 372–73
  Homo sapiens, origin of, 373–74
  microevolution in select populations, 261–63
  paedomorphosis and, 281–82
  primate phylogeny and, 370*f*
  timeline of, 371*f*
Human gene therapy, 234
Human Genome Project, 217, 230, 237
Human growth hormone (HGH), 219, 236*f*
Human impact on environment, 4, 442–64
  atmosphere, climate and, 449–51
  biodiversity crisis and, 452–55
  biological communities and, 443–46
  biophilia and environmental ethic as response to, 462
  chemical cycles in ecosystems and, 446–48
  conservation biology as response to, 456–61
  coral reefs and, 407
  disturbance of biological communities and, 444
  ecosystems and, 446–51
  on freshwater biomes and, 434, 435*f*
  introduction of invasive species and, 344, 443, 444–46, 454
Human nutrition, food energy available at different trophic levels, 423*f*
Human population
  centers of, 257*f*
  dispersion patterns, 391*f*
Human population growth, 381, 395–98
  age structure and, 396–97

history of global, 395
    sociology, economics, and politics of, 397–98
    in Sri Lanka, 395f
Humulin, production of, 218–19, 223
Huntingtone disease, 153
Hybrid(s), 144
Hybrid inviability as reproductive barrier, 275
Hybridization (genetic cross), 144
    dihybrid, 147–48
    monohybrid, 144–46
    polyploid formation by, 277f
Hybrid sterility as reproductive barrier, 275f
Hydrocarbons, 37
Hydrogen atom, 25f
Hydrogen bond, 28
Hydrolysis, 38
Hydrophilic molecules, 41
Hydrothermal vent communities, 301, 304, 437f
Hypercholesterolemia, 155
Hypertonic solutions, 81
Hyphae, fungal, 336
Hypothesis-driven science, 14–15
Hypotonic solutions, 81

I

Ice floats, 30
Immune system as defense against parasitism, 414
Inbreeding, 152–53
Incomplete dominance, 155
Independent assortment, principle of, 147–48, 161
Independent assortment of chromosomes, 133
Induced fit, enzymes, 78, 79f
Inductive reasoning, 14
Infants
    banking umbilical cord blood of, 197f
    testing of, for phenylketonuria, 259
Infertility, 66–67, 119
Influenza virus, 190f, 193
Ingestion, 344–45
Inheritance, 141–69
    chromosomal basis of, 159–62
    family pedigree and, 150–51
    genetic testing before birth and, 142
    human disorders controlled by single genes, 151–53
    G. Mendel's analysis of patterns of, 143–44
    beyond G. Mendel's principles of, 154–59
    principle of independent assortment and, 147–48
    principle of segregation and, 144–47
    rules of probability and, 149–50
    sex chromosomes, sex-linked genes, and, 163–65
    use of testcross to determine genotype, 149
    Y chromosomes, and evolutionary history, 166
Innocence Project, 225
Insecticides, 79, 243, 255, 384, 449f
Insects, 358
    anatomy, 358f
    coevolution between plants and, 438f
    color variation in Asian lady beetles, 255f

    diversity of, 358f
    flight in, 358
    metamorphosis in, 358, 359f
    pesticide resistant, 243, 255, 384
Insulin, 5
    formation of active molecule of, 207f
    production of, by genetically modified bacteria, 218–19
Intermediate disturbance hypothesis, 417
Interphase
    cell cycle, mitosis, and, 122, 123f, 124f
    meiosis and, 130f
Interspecific interactions in communities, 409–15
    competition between species, 409–10
    complexity of, 415
    predation, 410–14
    symbiotic relationships, 414–15
Intertidal zone, ocean, 436
Introduced species, 344, 443f, 444–46
    as cause of biodiversity crisis, 454
Introns, 181, 230
Invertebrates, 349–60
    annelids, 354–55
    arthropods, 356–59
    chordate, 361f
    cnidarians, 350–51
    echinoderms, 359–60
    flatworms, 352
    mollusks, 353–54
    roundworms, 352–53
    sponges, 350
In vitro fertilization, 119
Iodine as trace element, 23f
Ionic bonds, 25
Ions, 25
Ireland, 388
Isomers, 39
Isotonic solutions, 81
Isotopes, 24
Italy, age structure and population growth in, 396f

J

Jacob, Françoise, 203
Jeans, stonewashing with cellulase, 73
Jefferson, Thomas, 166, 225

K

Karyotype, 128f
Kenya, age structure and population growth in, 396f
Keystone predator, 413f, 414
Kinetic energy, 74
Kingdoms, classification into, 288, 290, 291f, 292
Klinefelter syndrome, 136, 137f
Krebs cycle, 93, 95f
    link between glycolysis and, 95f
Kudzu, 444, 445f

L

lac operon, 23, 204f
Lactase, 36
Lactic acid, fermentation and
    food production using, 99
    in muscle cells and accumulation of, 98
Lactose intolerance in humans, 36
Lactose metabolism in E. coli, gene regulation and, 203, 204f
Lakes and ponds
    as biome, 434f
    experimental study on ecology of, 382f
Lamarck, Jean Baptiste, 245
Lancelets, 361f
Land, plant colonization of, 299, 321–23
Landscape ecology, 459–61
Larva, 345
Lateral line system, 363
Leaf
    photosynthesis in, 104f, 105, 106–11
    veins in, 322f
Lederberg, Joshua, 217
Leeches, 354, 355f
Lemba people, South Africa, 166
Leopold, Aldo, 455
Leucine, 44f
Leukemia, 127
Lichens, 340f, 416
Liem, Karel, 281
Life
    diversity of, 6–7 (see also Biological diversity)
    domains and kingdoms of, 6–7 (see also Systematics; Taxonomy)
    evolution of (see Evolution)
    four-stage hypothesis for origins of, 300–302
    levels of, 2, 3, 4–5
    major episodes in history of, 297, 298f, 299
    regional distribution of, in biosphere, 384, 385f
    unity in, 7–8
Life cycle of organisms, 129
    angiosperm, 331f
    human, 129f
    sea star, 345f
    enveloped virus, 190f
Life histories and evolution, 398–401
    life history traits as evolutionary adaptations, 400–401
    life tables and survivorship curves, 399–400
    testing hypothesis on life history traits in guppies, 401–3
Lifestyle, human, linked to cancer, 211t, 212
Life table, 399t
Light microscope (LM), 54
Light reactions of photosynthesis, 105, 106–10
    chloroplasts pigments and, 107
    generation of ATP and NADPH in, 105, 109–10
    nature of sunlight and, 106–7
    photosystems and harvesting of light energy, 107–9
Lignin, 322
Limbs, homologous structures in, 252f
Linkage map, 162

Linked genes, 160–62
  crossing over and, 161*f*
Linnaeus, Carolus (1707-1778), 287, 291
Lipids, 42–43
  abiotic production of, 302*f*
  fats, 42–43
  steroids, 43
Liver
  smooth ER and effects of drug addiction in, 62
Lizards
  number of species of, in regions of U.S., 387*f*
  resource partitioning among populations of,
    411*f*
Lobe-finned fishes, 364
Lobster, 356*f*
Loci (locus), gene, 147
Locomotion
  amoeboid, 66*f*
  role of cilia in, 8*f*
Logistic population growth model, 391–92
  exponential growth compared to, 392*f*
Low-density lipoproteins (LDLs), 83*f*
Lucy, *Australopithecus afarensis*, 372*f*
Lungfishes, 364
Lyell, Charles, 247
Lyme disease, 308*f*
Lymphocytes, microscopic views of, 54*f*
Lymphomas, 127
Lynx, population cycles of snowshoe hare and,
    394*f*
Lysogenic cycle, 188, 189*f*
Lysosomal storage diseases, 63
Lysosomes, 62–63
  functions of, 63*f*
Lysozyme, 45*f*
Lytic cycle, 188, 189*f*

M

Macroevolution, 271–72, 282–86
  continental drift and, 285
  defined, 272
  geologic time, fossil record, and, 282–84
  mass extinctions, and explosive diversifications
    of life, 285–86
  speciation and, 271 (*see also* Speciation; Species)
Macromolecules, 38
  categories of biological, 39–48
  recycling of, in cells, 206
Magnification of microscopes, 54
Malaria, 157–58, 313*f*, 359
  relationship of sickle-cell disease to, 157–58,
    266–67
Malathion, 79
Males
  infertility in, 66–67
  Y chromosome and, 166 (*see also* Y
    chromosome)
Malignant tumor, 126
Mammals (Mammalia), 367–68
  three major groups of, 367*f*

Mantle, mollusk, 353
Margulis, Lynn, 312
Marine biomes, 435–37
Mark-recapture method of estimating wildlife
    populations, 389*f*
Marsupials, 367*f*
Mass as measure of matter, 23
Mass extinctions, 284, 285–86, 376, 381, 452–53
  causes of current, 454
Mass number, 23
Matter, 22–23
Mayr, Ernst, 272*f*, 273
Meat eating, ecosystem energetics and human,
    423–24
Mechanical isolation as reproductive barrier, 274
Mechanical work, 77*f*
Medicines, plants as sources of, 333*t*
Medusas, 350, 351*f*
Meiosis, 121, 130–32
  errors in, 134–37
  genetic variation and, 133–34
  halving number of chromosomes during, 129*f*
  mitosis compared to, 132*f*, 133
  stages of, 130–31*f*
Melanoma, 211
Membrane. *See* Endomembrane system; Plasma
    membrane
Membrane transport, 80–83
  active, 82–83
  passive, 80–82
  role of, in cell signaling, 83
Mendel, Gregor, patterns of inheritance
    determined by, 143–53, 160*f*
Mesophyll, 104
Mesozoic era, dinosaurs in, 365*f*
Messenger RNA (mRNA), 60, 181–82
  breakdown of, 206*f*
  gene synthesis from eukaryotic, 223*f*
  production of, 182*f*
  production of two different, 206*f*
  translation and, 182, 183*f*
Metabolic pathway of cellular respiration, 93–97
Metabolism, 93
  cellular respiration and, 93–97
  enzymes and, 78–79
  genetic errors in, 177
Metamorphosis, 345
  in insects, 358, 359*f*
Metaphase
  meiosis I and II, 130*f*, 131*f*
  mitosis, 124, 125*f*
Metastasis, 126
Methane, 26, 37*f*
Methanogens, 304
Microevolution
  as change in gene pool, 260
  mechanisms of, 260–65
Microscopy, 53–55
Microtubules, 66*f*. *See also* Cilia; Flagella
Migrations, bird, 388
Miller, Stanley, 300*f*

Millipedes, 358*f*
Mimicry, 15, 16*f*
Mining, damage to biological communities caused
    by, 444*f*
Missense mutations, 186
Mitochondria, 65
  electron transport chains in inner membranes
    of, 95–96
  endosymbiosis and origins of, 311*f*, 312
Mitosis, 123, 124–26
  eukaryotic cell cycle and, 123*f*
  meiosis compared to, 132*f*, 133
  phases of, 124–25*f*
Mitotic phase (M), cell cycle, 123*f*
Mitotic spindle, 124, 125*f*
Mobbing as behavior defense against predators,
    412*f*
Modern synthesis, Darwinism and genetics,
    256–60, 272*f*
Mold, 335*f*, 337
  penicillium, 339*f*
Molecular biology
  DNA structure and replication, 171–76
  as evidence for evolution, 253
  flow of genetic information from DNA to RNA
    to proteins, 176–87
  as tool in systematics, 289
  of viruses, 188–93
Molecule(s), 35–51
  abiotic synthesis of organic, 300–301
  abiotic synthesis of polymers, 301
  alternative ways to represent, 26*f*
  carbohydrates as biological, 39–41
  defined, 26
  DNA and proteins as tape measures of
    evolution, 49
  lipids as biological, 42–43
  nucleic acids as, 47–48
  organic, 36–38
  origin of cooperation between, 302*f*
  origin of self-replicating, 301*f*, 302
  proteins as biological, 44–47
Mollusks (Mollusca), 353–54
  general body plan, 353*f*
  major classes of, 354*f*
Monarch butterfly, 359*f*
Monod, Jacques, 203
Monohybrid crosses, 144–46
Monomers, 38
  abiotic synthesis of organic, 300–301
Monotremes, 367
Morgan, T. H., 163–65
Mosquitoes, pesticide resistant, 243
Mosses, 325–26, 417
  two forms of, 326*f*
Mouse, homeotic genes in, 213*f*
Movement corridor, 459–60
mRNA. *See* Messenger RNA (mRNA)
Müllerian mimicry, 412, 413*f*
Multicellularity, evolution of, 299, 317*f*
Multiple alleles, ABO blood groups, 156

Multiregional hypothesis for origins of *Homo sapiens*, 374*f*
Muscle cells, 197*f*
  fermentation and lactic acid accumulation in, 98
Mutagens, 187
Mutation, 186
  cancer and cellular, 210*f*
  causes of, 187
  evolution and, 263
  in hemoglobin, and sickle-cell disease, 186*f*
  types of, 186–87
Mutualism, 340, 414–15
Mycelium, fungal, 336*f*, 337, 340
Mycorrhizae, 322*f*

N

NAD+, role in cellular respiration, 92, 93, 98*f*
NADH, role in cellular respiration, 92, 95, 96*f*, 98*f*
NADPH, photosynthetic light reactions, 105–6, 109*f*, 110
Natural selection, 254–56
  antibiotic resistance as example of, 12–13, 255
  chemical evolution and, 302
  contemporary examples of, 255
  Charles Darwin on, 10–13, 244, 254–55
  microevolution and, 263–65
  three general outcomes of, 264–65
Neanderthals, 373
Nematodes, 352–53
Nerve cell, 197*f*
Nerve chord, dorsal, hollow, 361
Neutron, 23
Nirenberg, Marshall, 179
Nitrogen atom, 25*f*
Nitrogen cycle, 426*f*
  human impact on, 447
Nitrogenous base(s), 47, 173
  of DNA, 47*f*, 48*f*, 173–75
  insertions and deletions of, as mutation, 186*f*
  pairing rules, 175
  substitutions of, as mutation, 186*f*
Nondisjunction in meiosis, 135*f*, 136*f*
Non-native species, human-introduced, 443*f*, 444–46, 454
Nonsense mutations, 186
Notochord, 361
Novelty, evolution of biological, 280–82
Nuclear accidents, Chernobyl, 24
Nuclear envelope, 60
Nuclear transplantation, 200, 201*f*
Nucleic acid probes, 222, 223*f*
Nucleic acids, 47–48. *See also* DNA (deoxyribonucleic acid); RNA (ribonucleic acid)
Nucleolus, 60
Nucleosome, 122
Nucleotides, 47, 48*f*, 172–73
  amino acid sequences linked to sequences of DNA, 289
  DNA and RNA as polymers of, 172–73 (*see also*

DNA (deoxyribonucleic acid); RNA (ribonucleic acid))
  homeobox sequence of, 213
Nucleus (atom), 23
Nucleus (cell), 4*f*, 56, 57, 59–60
  DNA in, and cellular control, 60
  gene regulation in, 204–6
  ribosomes of, 60
  structure and function of, 60
  transplantation of, 200, 201*f*
Nutrient cycling in ecosystems, 3, 88–90
Nutrition
  animal, 344, 345*f*
  classification of organisms by mode of, 307*t*
  fungal, 336
  of prokaryotes, 307

O

Obligate aerobes, 99
Obligate anaerobes, 99
Ocean biomes, 436–37
  hydrothermal vent communities, 437*f*
  zones of, 436*f*, 437
Offspring, overproduction of, 254*f*
Oil spill, bacterial treatment of, 309*f*
Omnivores, 421
Oncogenes, 209
  alternative ways for making, from proto-oncogene, 209*f*
One gene-one protein hypothesis, 177–78
On the Origin of Species by Means of Natural Selection (Darwin), 9, 244, 248
Operator sequences, 203
Operculum, 363
Operon, 203
  lac, 203, 204*f*
Opportunistic life history, 400*f*, 401*t*
Oppossum, 367*f*
Orders, classification into, 288
Organ(s), defined, 22
Organelles, 55, 56
Organic chemistry, 36
Organic molecules, 36–38
  abiotic synthesis of, 300–301
Organism(s)
  asexual reproduction in, 120–21
  defined, 22
  diploid, and haploid, 129
  evolutionary adaptations of, 387–88
  genomes of select, 230, 231*t*
  life cycle of, 129, 331*f*, 345*f*
  nutritional classification of, 307*t*
  sexual reproduction in, 121, 128–37
Organismal ecology, 383, 387–88
Organ system, defined, 22
Osmoregulation, 81
Osmosis, 81*f*–82
Osteichthyes, 363
Out-of-Africa (replacement) hypothesis for origins of *Homo sapiens*, 374*f*

Ovaries, human, 163
Ovary, plant, 330*f*, 331
Overexploitation as cause of biodiversity crisis, 454
Ovule, plant, 329
Owl, northern spotted, 457*f*
Oxidation, 91
Oxygen
  atom, 25*f*
  evolution of atmospheric, 114, 298
  as product of photosynthesis, 90*f*, 105
  role of, in cellular respiration and harvest of food energy, 91–92
Ozone layer, depletion of atmospheric, 451*f*

P

Paedomorphosis, 281*f*
Paine, Robert, 413
Paleoanthropology, 370
Paramecium sp., 313*f*
  cilia of, 8*f*
  competitive exclusion principle applied to two populations of, 409*f*–10
Parasites, 414
  flatworms, 352
  fungi, 338
Parasitism, 414
Parkinson's disease, 217
Passive transport, 80*f*
  osmosis as, 81*f*–82
Paternity, determination of, 225
Pathogens, bacterial, 307–8
PCR (polymerase chain reaction), 226*f*
Pea plants, Gregor Mendel's analysis of inheritance patterns in, 143–53
  seven characteristics studied, 145*f*
  technique for cross fertilization, 144*f*
Peat moss bog, 325*f*
Pedigree, inheritance patterns and, 150–51
Pelagic zone, oceans, 436
Penicillin, 53, 79, 339*f*
Periodic table of elements, 22*f*
Permian extinctions, 286
Pesticide-resistant insects, 243, 255, 384
Pesticides, biological magnification of, 449*f*
Petals, flower, 330*f*
PET scans, 24*f*
Pfenning, David and Karin, 15–17
P generation, 144
Phages (bacteriophages), 188–89
  infection of bacteria by, 188*f*
  reproductive cycles, 189*f*
Phagocytosis, 82*f*
Pharmaceutical drugs
  antibiotics, 53 (*see also* Antibiotics)
  anti-cancer, 127
  AZT, 171
  DNA technology and production of, 218–19
  fungi as source of, 339
  plants as source of, 333*t*, 455*f*
Pharyngeal slits, 361

Phenotype
  blood type as, 156
  defined, 146, 177
  environment and human, 158–59
  family pedigrees and, 150–51
  flow of genetic information from genotype to, 177–78, 182–86, 198
  human disorders and (see Genetic disorders)
  skin pigmentation, 158
Phenylalanine, 259
Phenylketonuria (PKU), 259
Phloem, 322
Phospholipid bilayer of membrane, 58f
Phospholipids, 58
Phosphorus cycle, 427f
  human impact on, 447
Photic zone
  lakes and ponds, 434
  ocean, 436–37
Photoautotrophs, 307
Photoheterotrophs, 307
Photon, 107, 108f
Photosynthesis, 89, 102–16
  Calvin cycle in, 111–12
  chloroplasts as sites of, 64f–65, 103–5
  environmental impact of, 113–14
  light reactions in, 105, 106–10
  overall equation for, 105
  oxygen byproduct of, and Earth's atmosphere, 114
  road map for, 105f–6
Photosystem, 107, 108f
  light reactions of, 109f
pH scale, 31f
Phyla, classification into, 288
Phylogenetic tree, 288
  animals, 347f
  order Carnivora, 288f
  plants, 324f
  primates, 370f
  vertebrates, 362f
Phylogeny, 288–90
  cladistic analysis and, 290f
  relationship of classification and, 288f
Physiological adaptations of organisms to ecological conditions, 387–88
Phytoplankton, 434
Pigments, chloroplasts and photosynthetic, 107
Pinocytosis, 82f
Placenta, 367
Planarian, anatomy of, 352f
Plant(s)
  alternation of generations in, 326f, 328f
  anatomy of, 321f, 322
  angiosperms, 330–32
  bryophytes, 325–26
  C₃, C₄, and CAM, compared, 112
  cloning, 200f
  coevolution between insects and, 438f
  cohesion and water transport in, 29f
  contrasting environments of algae and, 321f

as energy source (biomass), 103
evolutionary highlights of, 324f, 325
gymnosperms, 328–29
ferns (seedless vascular plants), 327
idealized cell of, 57f
introduced, invasive species of, 444, 445f
leaves of (see Leaf)
origins of, from green algae, 323
photosynthesis in (see Photosynthesis)
polyploid, 137–38, 277f, 278f
prevalent, in communities, 408
reproductive adaptations of, for life on land, 322, 323f
seed, 328–29
storage of polysaccharides as starch in, 41f
structural adaptations of, for life on land, 321–22
sympatric speciation in, 277f, 278f
terrestrial adaptations of, 321–23, 328–29
toxins as defenses against predation in, 411
viruses of, 189–90
wheat, evolution of, 277, 278f
Plantae, Kingdom, 6, 7f
Plant cell(s)
  central vacuole in, 63f
  chloroplasts in, 64f–65, 103–5
  cytokinesis in, 126f
  prior to division, 121f
  water balance in, 81f, 82
Plant cell walls, 68
  cellulose in, 41f
  plasmodesmata in, 68f
Plant disease
  parasitic fungi, 338f
  tobacco mosaic virus, 189f
Plant turgor, 81f
Plasma membrane, 57, 58–59
  cell signaling and role of, 83
  fluid mosaic and selective permeability of, 58
  origin of, 69
  primary functions of proteins in, 59f
  structure, 58f
  transport across, 80–83 (see also Membrane transport)
Plasmids, bacterial, 220, 221f
Plasmodesmata, 68f
Plasmodial slime mold, 314f
Plasmodium, 313f, 414. See also Malaria
Plasmolysis, 82
Platyhelminthes, 352
Poisons, disruption of electron transport in mitochondria by, 96
Polar molecule, 28
Pollen, 322, 329
Polychaetes, 354, 355f
Polygenic inheritance, 157
Polymerase chain reaction (PCR), 226f
Polymers, 38
  abiotic synthesis of, 301
  synthesis and digestion of, 38f
Polynucleotide, structure of DNA, 172f
Polyp, 350, 351f

Polypeptides, abiotic production of, 302f
Polyploid species, 137–38
  sympatric speciation and, 277f, 278f
Polysaccharides, 40–41
Population(s)
  conservation biology at level of, 457–59
  defined, 22, 256, 389
  ecology and (see Population ecology)
  genetics of (see Population genetics)
  habitat fragmentation and subdivided, 457–58
  human centers of, 257f
  polymorphic, 258f
Population cycles, 394f–95
Population density, 389–90
Population ecology, 383, 388–98
  defined, 389
  dispersion patterns and, 390
  human population growth, 395–98
  models of population growth, 390–92
  population density and, 389–90
  regulation of population growth, 392–95
Population fragmentation, 457–59
Population genetics, 257–60
  gene pools and, 258–59
  genetic variation and, 257–58
  health science and, 259–60
  microevolution and, 260–65
  sickle-cell allele, 266–67
Population-limiting factors, 391, 392–95
  effect of, on growth of animal population, 392f
Populations, evolution of, 242–69
  Darwinian theory on descent with modification in, 244–48, 249f
  evidence of, 249–53
  genetics of sickle-cell allele in populations, 266–67
  mechanisms of microevolution, 260–65
  modern synthesis of Darwinism and genetics in study of, 256–60, 272f
  natural selection and adaptive evolution applied to, 254–56
Porifera (sponges), 350
Post-anal tail, 361
Post-zygotic reproductive barriers, 275
Potential energy, 74
Prairie dogs, indirect census of population of, 389f
Pre-cells, formation of, and origins of life, 302
Predation, 387, 410–14
  animal defenses against, 411–13
  defined, 410
  plant defenses against, 411
  predator adaptations and, 411
  species diversity in communities and, 413–14
Predator, 410
  adaptations of, 411
  defenses against, 411–13
  keystone, 413f, 414
Prey, 410
  defenses against, 411–13
Pre-zygotic reproductive barriers, 274
Primary electron acceptor, photosystems and, 109

Primary productivity of ecosystems, 421
Primary structure of proteins, 45*f*
Primary succession, 416
Primates, 368
　anthropoids, 369*f*
　emergence of humankind, 370–75
　evolution of, 368–69
　phylogeny, 370*f*
　prosiman, 368, 369*f*
Principles of Geology (Lyell), 247
Probability, inheritance patterns and rules of, 149–50
Producers in ecosystems, 89, 419
　photosynthetic autotrophs as, 104*f*
Productivity of ecosystems, 421, 422*f*
Products, chemical, 27
Prokaryotes, 55, 303–10. *See also* Bacteria
　Bacteria and Archaea as two main forms of, 6, 7*f*, 304
　ecological impact of, 307–10
　evolution of, 297, 298, 303*f*
　nutritional diversity of, 307
　structure, function, and reproduction of, 304–7
Prokaryotic cells, 4, 55. *See also* Bacteria
　contrasting size and complexity of eukaryotic and, 56*f*
　eukaryotic cells compared to, 56*f*
　flagella of, 306*f*
　idealized, 56*f*
　walls of, 304–5
Promoter (nucleotide sequence), 180, 203, 230
Prophage, 188
Prophase
　meiosis I and II, 130*f*, 131*f*
　mitosis, 124*f*
Prosimians, 368, 369*f*
Proteins, 44–47. *See also* Amino acids
　denaturation of, 47
　gene regulation by alteration or breakdown of, 207
　genetic message translated into, 60*f*
　four levels of structure in, 46*f*
　as monomers, 44
　as polymers, 44, 45*f*
　primary structure of, 45f
　some functions of, 44*f*
　shape of, 45–47
　translation of genetic information into synthesis of, 177, 178–79, 182–86
Protist(s), 311–16
　cilia on, 67*f*
　contractile vacuoles in, 63*f*
　evolution of, 299, 311–12
　hypothesized origin of animal from colonial, 346*f*
　protozoans as, 312–13
　seaweeds, 315–16
　slime molds, 314
　unicellular algae, 314–15
Protista, Kingdom, 6, 7*f. See also* Protists
Proton, 23
Proto-oncogene, 209

alternative ways to make oncogenes from, 209*f*
Protozoans, 312–13
Provirus, 191
Pseudocoelom, 348*f*, 349
Pseudopodia, 312, 313*f*
Punctuated equilibrium model of speciation, 279*f*
Punnett, Reginald, 160
Punnett square, 145*f*, 146, 147*f*
Pyruvic acid, conversion of, to acetyl-CoA in cellular respiration, 95*f*

Q

Quaternary consumers, 419
Quaternary structure of proteins, 46*f*
Quoll, 344*f*

R

Rabbits, myxoma virus and invasive populations of, in Australia, 414*f*, 444
Race, genetic disorders linked to, 152, 157, 266
Radial symmetry, 348*f*
Radiation as cause of mutations, 187
Radiation therapy, 127
Radioactive isotopes, 24
Radiometric dating, 284*f*
Radula, 353
Random population dispersion pattern, 390
ras gene, 209–10
Ray-finned fishes, 364
Reactants, chemical, 27
Reaction center of photosystems, 109
Reading frame, 186
Receptor-mediated endocytosis, 82*f*, 83*f*
Recessive allele, 146
　genetic disorders linked to, 152–53
Recessive traits, human, 150*f*
Recombinant DNA, 218
Recombinant DNA technology, 217–23
　defined, 218
　production of food products using, 219–20
　production of pharmaceuticals using, 218–19
Recombination frequency, 162
Red algae, 316*f*
Red blood cells, sickle cell anemia and, 45*f*, 157–58
Red-green color blindness, 165
Redox reactions, 91–92
Red tide, 315
Reduction, 91
Reefs, coral and artificial, 407
Regeneration, 200
Repetitive DNA, 230
Replacement hypothesis for origins of *Homo sapiens*, 374*f*
Reproductive adaptations for life on land, 322–23
Reproductive capability, biological species concept and, 273*f*
Reproductive cloning, 201
Reptiles (Reptilia), 365–66. *See also* Lizards; Snakes

Resolving power of microscopes, 54
Resource partitioning, 410
　in lizard group, 411*f*
Respiratory tract, cilia in, 66, 67*f*
Restriction enzymes, cutting and pasting DNA strands with, 222*f*
Restriction fragment length polymorphism (RFLP) analysis, 226, 227*f*
Restriction sites, 222
Retinitis pigmentosa, 263
Retrovirus, 191
Reverse transcriptase, 191
RFLP analysis, 226–27
Ribosomal RNA (rRNA), 183
Ribosome(s), 60, 183*f*
Ribozymes, 302
Rice, genetically modified, 220*f*
Ringworm, 338
Rivera Gonzales, Carlos, 462*f*
Rivers and streams biome, 434, 435*f*
RNA (ribonucleic acid), 47
　abiotic replication of, hypothetical, 301*f*, 302
　elongation of, in transcription, 180, 181*f*
　genetic code listed by codons of, 179*t*
　messenger, 60, 181, 182*f*, 183*f*
　processing of eukaryotic, 181–82
　ribosomal, 183
　transfer, 182*f*, 183
RNA nucleotide, 48*f*
RNA polymerase, 180, 203
RNA splicing, 182
　alternative, 206
RNA viruses, 192–93
Rocks and soil as abiotic components of biosphere, 386
Rodriguez, Eloy, 13*f*
Roots, 321*f*, 322
　symbiotic relationships with fungi, 322*f*
Rosy periwinkle, 455*f*
Rough ER, 61
　manufacture and packaging of secretory proteins in, 61*f*
Roundworms, 352, 353*f*
Rule of multiplication, inheritance patterns and, 149–50

S

Safety issues with genetically modified organisms, 235–36
Salt, *23*, 25, 31*f*
Sanitation, control of bacterial disease and, 308
Sarcomas, 126
SARS virus, 192
Saturated fats, 42
Savanna
　biome, 430*f*
　hierarchy of life on, 21*f*
Scanning electron microscope (SEM), 54*f*, 55
Science, 13–18
　case study in process of, 15–17

culture of, 17
discovery, 13–14
ecology as, 382
hypothesis-driven, 14–15
technology and, in society, 18
theories in, 17–18
Scientific method, 14f, 15f
Sea stars, 120f, 359f
as keystone predators, 413f, 414
life cycle, 345f
Seaweeds, 315, 316f
Secondary consumers, 419
Secondary structure of protein, 46f
Secondary succession, 417
Secretory proteins
rough ER manufacturing and packaging of, 61f
Seed(s), 325, 329, 331
dispersal of, 331f
Seedless vascular plants, 327f
Seed plants
reproduction in, 329f
terrestrial adaptations of, 328–29
Segregation, principle of, 144, 145f, 146–47
Selective permeability of plasma membranes, 58, 69
Self-fertilization, 144
Sepals, 330f
Serengo, Paul, 8f
Serine, 44f
Severe combined immunodeficiency (SCID), 234
Sewage treatment, use of prokaryotes in, 309f
Sex chromosomes, 129, 163
abnormalities in number of, in humans, 136t, 137f
sex determination and, 163
Sex determination, 163
Sex-linked genes, 163–65
human disorders caused by, 165
Sexual reproduction, 121, 128–37
errors in meiosis, 134–37
evolution of genetic variation due to, 258
gametes and life cycle of, 129
homologous chromosomes and, 128–29
origins of genetic variation in, 133–34
process of meiosis and, 130–32, 133
Sheep
cloned, 201f, 202
genetically modified, 220f
Shoots, plant, 321f, 322
Sickle cell disease, 157–58
allele mutation as cause of, 186f
population genetics of, 266–67
shape of red blood cells and amino acid order in hemoglobin, 45f
Signaling (transmitting) cell, 37f, 207
Signal transduction pathway, 83f, 208f
Silencer sequences, 206
Silent Spring (Carson), 384
Simpson, O.J., 226
Sink habitat, 458
Sister chromatids, 122

Sleeping sickness, 312, 313f, 359
Slime molds, 314
Smoking, 66, 211–12
Smooth ER, 61–62
Snakes
backbone, 361f
nutrition by ingestion, 345f
polymorphic, 258
warning coloration and mimicry in, 15, 16f, 17
Snowshoe hare, population cycles of lynx and, 394f
Sodium chloride, 25, 31f
Soil, 425
as abiotic component of biosphere, 386
Solute, 30
Solution, 30
hypertonic, hypotonic, and isotonic, 81
Solvent, 30
Somatic cell, 128
Sonoran Desert
biome, 431f
food web in, 420f
Source habitat, 458
Speciation, 272–80
biological species concept and, 272–73
cell division errors and, 137–38
defined, 272
earth history, macroevolution, and, 282–86
evolution of biological novelty and, 280–82
mechanisms of (allopatric, sympatric), 275–78
reproductive barriers and, 274–75
tempo of, 278–80
two patterns of, 272f
Species, 6
binomial name including, 287
competition between, 409–10, 411f
concept of biological, 272–73
conservation biology at level of, 457–59
diversity of, 408, 413–14 (See also Biological diversity)
endangered and threatened, 457–59
endemic, 456–57
extinct (see Extinct species)
human-introduced non-native, 344, 443, 444–46, 454
idea of fixed, before C. Darwin, 245
mass extinctions of (see Mass extinctions)
naming, 287
origin of, 272–80 (see also Speciation)
polyploid, 137–38, 277f, 278f
reproductive barriers between closely related, 274f, 275
reproductive capability of, 273f
taxonomy and classification of, 287–92
Sperm cell, 197f
fertilization of egg by, 129
movement of human, 66, 67f
plant, 322
S phase, cell cycle, 123f
Spirochetes (spirals), bacteria, 305f
Sponge (Porifera), anatomy of, 350f

Spontaneous generation, 300
Spores, 326
fungal, 337
overproduction of, 265f
Sporophyte, 326, 328f, 329
Squirrels, allopatric speciation of Grand Canyon, 276f
Sri Lanka, human population growth in, 395f
Stabilizing selection, 265
Stability of biological communities, 408–9
Stamens, flower, 330f
Starch, plant storage of polysaccharides as, 41f
Stem cells, 197, 202
Steroids, 43
"Sticky ends," DNA, 222
Stigma, flower, 330f, 331
Stomata, 104, 322
Stonewashed jeans, 73
Stop codon, 184
Storage proteins, 44f
Streptomycin, 53
Stroma, 64–65, 105
Structural adaptations for life on land, 321–22
Structural proteins, 44f
Style, flower, 330f
Substrate, enzyme, 78, 79f
Sucrose, 40, 41
Sugar-phosphate-backbone, DNA, 48, 172f, 173
Sugars, 39–41
Sunlight
as abiotic component in biosphere, 384–85
photosynthesis and nature of, 106–7
Survivorship curve, 399–400
Sustainable development, 461
Swim bladder, 363
Symbiont, 414
Symbiotic relationships, 414–15
endosymbiosis and development of eukaryotic cell, 311–12
lichen as, 340f
mutualism, 340, 414–15
mycorrhizae as, 322f
parasitism, 414
Sympatric speciation, 275f, 277–78
Systematics, 287–92
basics of taxonomy, 287–88
classification and phylogeny, 288–90
defined, 287
schemes of kingdoms and domains, 290–92

T

Tail nucleotide sequences, 181, 206
Tapeworms, 414
anatomy of, 352f
Target (receiving) cells, 37f, 207
Tatum, Edward, 177–78, 217
Taxonomy, 287–88
hierarchical classification in, 287, 288f
kingdom and domains as schemes in, 290–92
naming species, 287

phylogeny and, 288–90
Tay-Sachs disease, 63, 143
Technology, relationship to science, 18
Telemeres, 230
Telophase
    meiosis, 131f
    mitosis, 124, 125f
Temperate deciduous forest biome, 432f
Temperate grassland biome, 432f
Temperature
    as abiotic component of biosphere, 385
    defined, 29
    physiological adaptations of organisms to, 365, 366, 387
    water and moderation of, 29–30
Temporal isolation as reproductive barrier, 274
Terminator nucleotide sequence, 181
Terrarium ecosystem, 418f
Terrestrial adaptations of plants, 321–23, 328–29
Terrestrial biomes, 428–33
    chaparral, 431f
    coniferous forest, 433f
    desert, 431f
    savanna, 430f
    temperate deciduous forest, 432f
    temperate grassland, 432f
    tropical forest, 430f
    tundra, 433f
    world's major, 428f
Terrorism
    anthrax attack, 231–32, 297
    DNA fingerprinting of Sept. 11, 2001, victims of, 225
Tertiary consumers, 419
Tertiary structure of protein, 46f
Testcross, determining unknown genotype using, 149f
Testes, 163
Testosterone, 43f
Tetracycline, 53
Tetrad, 130
Tetrapod, origin of, 364f, 365
Theory, scientific, 17–18
    meaning of, 292–93
Therapeutic cloning, 201–2
Thermophiles, 304, 385f
Threatened species, 457–59
Thylakoids, 105
    conversation of light energy into NADPH and ATP by, 110f
Thymine (T), 47, 173
    AZT drug therapy for HIV and, 171f
Ticks as disease vector, 308f
Tight junctions, 68f
Tissues, defined, 22
Toads, cane, 344
Tobacco, cancer and, 211–12
Tobacco mosaic virus, 189f
Tobacco plant, expression of firefly gene in, 180f
Tooth decay, fluoride and, 21
Tortoises, giant, Galápagos Island, 271

Toxic chemicals, release of, into ecosystems, 448, 449f
Toxic sites, bacterial cleanup of, 309–10
Toxins in plants, defensive, 411
Trace element, 23
Transcription, 177, 180, 181f
    activators of, 205, 206
    of codons, 178f
    defined, 178
    gene regulation during, 205–6
    initiation of, 180, 203, 205–6
    repression of, 204
    RNA elongation in, 180
    summary of, 185f
    termination of, with terminator sequence, 180–81
Transcription factors, 206
Trans fats, 43
Transfer RNA (tRNA), 182–83
    structure of, 182f
Transgenic organism, defined, 218
Translation, 177, 178–79, 182–86
    amino acid elongation, 184f
    of codons, 178f
    defined, 178
    gene regulation and, 206–7
    initiation of, 183, 184f
    ribosomes as site of, 183
    role of mRNA in, 182
    role of tRNA in, 182–83
    summary of, 185f
    termination of, 184–86
Transmission electron microscope (TEM), 54f, 55
Transport across membranes, 80–83
Transport proteins, 44f, 58, 59f, 80
Transport vesicles, 62
Transport work, 77f
Triglyceride, synthesis and structure, 42f
Triplet code, 179
    reading frame, 186–87
Trisomy 21, 134, 135f
Tristan da Cunha, 262, 263f
Trophic levels, ecosystem, 409, 418
    food chains and, 419f
    human nutrition and, 423f
Tropical forest biome, 430f
True-breeding varieties, 144
Trypanosomes, 312, 313f
Tuberculosis, antibiotic resistant, 12f
Tumor, 126
Tumor-suppressor genes, 209f
Tundra biome, 433f
Tunicates, 361f
Turner syndrome, 136, 137f

U

Ultraviolet light as mutagen, 176f, 187
Unicellular algae, 314, 315f
Uniform population dispersion pattern, 390f

United States
    age structure and population growth in, 396f
    life table for population of, 399t
Unsaturated fats, 42
Uracil (U), 48f, 173

V

Vaccine, DNA technology and production of, 219
Vacuoles, 63–64
    two types of, 63f
Varmus, Harold, 209
Vascular plants, 325
Vectors, plasmids as, 220
Vegetation, prevalent form of, in communities, 408
Vertebrates, 360–68
    amphibians, 364–65
    birds, 366–67
    as chordates, 361–63
    evolutionary relationships of amino acids in six, 49f
    fishes, 363–64
    genealogy of, 362f
    mammals, 367–68
    reptiles, 365–66
Viruses, 188–93
    animal, 190–91
    bacteriophages, 188–89
    cancer caused by, 209
    emerging, 192–93
    enveloped, 190f
    HIV, 191
    plant, 189–90
Volcano, gaseous exhaust of, 32f
Volvox, 315f, 317

W

Wallace, Alfred, 248
Warning coloration, 15–16, 412
Wasps, in community networks, 415f
Water
    as abiotic component in biosphere, 385
    biological significance of ice floating, 30
    cohesion of, 29
    fluoride in drinking, 21
    moderation of temperature by, 29–30
    as solvent of life, 31–32
    structure of, 28
    as waste product of cellular respiration, 90f
Water balance in cells, 81–82
Water cycle, 427f
    human impact on, 447
Water transport in plants, 29f
Water vascular system, 360
Watson, James, 173–74
Wavelength of light, 106
Weapon, disease as, 297

Weather, density-independent regulation of
    population growth and, 394*f*
West Nile virus, 449
Whales, fossilized, 250, 251*f*
Wheat, evolution of, 277, 278*f*
White blood cells, microscopic views of, 54*f*
Whittaker, Robert H., 291
Whole Genome Shotgun Method, 233
Wilkins, Maurice, 173
Wilson, Edward O., 453, 462
Wind as abiotic component of biosphere, 385–86
Wings, 366*f*
Woodpecker, red-cockaded, 458*f*

X

X chromosomes, 129, 163
    abnormal number of, 136–37
    inactivation of, 205*f*
X-rays, 187
Xylem, 322

Y

Y chromosomes, 129, 163
    abnormal number of, 136–37
    evolutionary history revealed in, 166
Yeast, 230
    food production using, 99

Z

Zebra, 367*f*
Zebra mussels, as introduced, invasive species of,
    445*f*
Zoned reserves, 460*f*, 461
Zooplankton, 436
Zygote, 129
    animal, 345*f*
    plant, 331

STUDENT WEBSITE AND CD-ROM ACTIVITIES

Activities

1A   The Levels of Life Card Game
1B   Energy Flow and Chemical Cycling
1C   DNA Molecules: Blueprints of Life
1D   Classification Schemes
1E   Darwin and the Galápagos Islands
1F   Case Studies of Antibiotic Resistance
1G   Sea Horse Camouflage Video
1H   Science and Technology: DDT
2A   The Levels of Life Card Game
2B   The Structure of Atoms
2C   Electron Arrangement
2D   Build an Atom
2E   Ionic Bonds
2F   Covalent Bonds
2G   The Structure of Water
2H   The Cohesion of Water in Trees
2I   Acids, Bases, and pH
3A   Diversity of Carbon-Based Molecules
3B   Functional Groups
3C   Making and Breaking Polymers
3D   Models of Glucose
3E   Carbohydrates
3F   Lipids
3G   Protein Functions
3H   Protein Structure
3I   Nucleic Acid Functions
3J   Nucleic Acid Structure
4A   Metric System Review
4B   Prokaryotic Cell Structure and Function
4C   Comparing Cells
4D   Build an Animal Cell and a Plant Cell
4E   Membrane Structure
4F   Selective Permeability of Membranes
4G   Overview of Protein Synthesis
4H   The Endomembrane System
4I   Build a Chloroplast and a Mitochondrion
4J   Cilia and Flagella
4K   Cell Junctions
4L   Review: Animal Cell Structure and Function
4M   Review: Plant Cell Structure and Function
5A   Energy Concepts
5B   The Structure of ATP
5C   How Enzymes Work
5D   Membrane Structure
5E   Selective Permeability of Membranes
5F   Diffusion
5G   Facilitated Diffusion
5H   Osmosis and Water Balance in Cells
5I   Active Transport
5J   Exocytosis and Endocytosis
5K   Cell Signaling
6A   Build a Chemical Cycling System

6B   Overview of Cellular Respiration
6C   Glycolysis
6D   The Krebs Cycle
6E   Electron Transport
6F   Fermentation
7A   Plants in Our Lives
7B   The Sites of Photosynthesis
7C   Overview of Photosynthesis
7D   Light Energy and Pigments
7E   The Light Reactions
7F   The Calvin Cycle
7G   Photosynthesis in Dry Climates
7H   The Greenhouse Effect
8A   Asexual and Sexual Reproduction
8B   The Cell Cycle
8C   Mitosis and Cytokinesis Animation
8D   Mitosis and Cytokinesis Video
8F   Human Life Cycle
8G   Meiosis Animation
8H   Origins of Genetic Variation
8I   Polyploid Plants
9A   Monohybrid Cross
9B   Dihybrid Cross
9C   Gregor's Garden
9D   Incomplete Dominance
9E   Linked Genes and Crossing Over
9F   Sex-Linked Genes
10A   The Hershey-Chase Experiment
10B   DNA and RNA Structure
10C   DNA Double Helix
10D   DNA Replication
10E   Overview of Protein Synthesis
10F   Transcription and RNA Processing
10G   Translation
10H   Causes of Cancer
10I   Simplified Reproductive Cycle of a Virus
10J   Phage Lytic Cycle
10K   Phage Lysogenic and Lytic Cycles
10L   HIV Reproductive Cycle
11A   The lac Operon in E. coli
11B   Gene Regulation in Eukaryotes
11C   Review: Gene Regulation in Eukaryotes
11D   Signal Transduction Pathway
11E   Causes of Cancer
12A   Applications of DNA Technology
12B   DNA Technology and Golden Rice
12C   Cloning a Gene in Bacteria
12D   Restriction Enzymes
12E   DNA Fingerprinting
12F   Gel Electrophoresis of DNA
12G   Analyzing DNA Fragments Using Gel Electrophoresis
12H   The Human Genome Project: Human Chromosome 17
13A   The Voyage of the Beagle: Darwin's Trip Around the World
13B   Darwin and the Galápagos Islands

13C   Reconstructing Forelimbs
13D   Genetic Variation from Sexual Recombination
13E   Causes of Microevolution
14A   Mechanisms of Macroevolution
14B   Exploring Speciation on Islands
14C   Polyploid Plants
14D   Paedomorphosis: Morphing Chimps and Humans
14E   The Geologic Time Scale
14F   Classification Schemes
15A   The History of Life
15B   Prokaryotic Cell Structure and Function
15C   Diversity of Prokaryotes
16A   Terrestrial Adaptations of Plants
16B   Highlights of Plant Evolution
16C   Moss Life Cycle
16D   Fern Life Cycle
16E   Pine Life Cycle
16F   Angiosperm Life Cycle
16G   Madagascar and the Biodiversity Crisis
16H   Fungal Reproduction and Nutrition
17A   Overview of Animal Phylogeny
17B   Characteristics of Invertebrates
17C   Characteristics of Chordates
17D   Primate Diversity
17E   Human Evolution
18A   DDT and the Environment
18B   Evolutionary Adaptations
18C   Techniques for Estimating Population Density and Size
18D   Human Population Growth
18E   Analyzing Age-Structure Diagrams
18F   Investigating Life Histories
19A   Interspecific Interactions
19B   Primary Succession
19C   Energy Flow and Chemical Cycling
19D   Food Webs
19E   Energy Pyramids
19F   The Carbon Cycle
19G   The Nitrogen Cycle
19H   Terrestrial Biomes
19I   Aquatic Biomes
20A   Fire Ants: An Introduced Species
20B   Water Pollution from Nitrates
20C   DDT and the Environment
20D   The Greenhouse Effect
20E   Madagascar and the Biodiversity Crisis
20F   Conservation Biology Review

Case Studies in the Process of Science

1   How Do Environmental Changes Affect a Population of Leafhoppers?
2   How Are Space Rocks Analyzed for Signs of Life?
2   How Does Acid Rain Affect Trees?
3   What Factors Determine the Effectiveness of Drugs?
4   What Is the Size and Scale of Our World?
5   How Is the Rate of Enzyme Catalysis Measured?
5   How Does Osmosis Affect Cells?
5   How Do Cells Communicate with Each Other?
6   How Is the Rate of Cellular Respiration Measured?
7   How Does Paper Chromatography Separate Plant Pigments?
7   How Is the Rate of Photosynthesis Measured?
8   How Much Time Do Cells Spend in Each Phase of Mitosis?
8   How Can the Frequency of Crossing Over Be Estimated?
9   What Can Fruit Flies Reveal about Inheritance?
10   What Is the Correct Model for DNA Replication?
10   How Is a Metabolic Pathway Analyzed?
10   How Do You Diagnose a Genetic Disorder?
10   What Causes Infections in AIDS Patients?
10   Why Do AIDS Rates Differ Across the United States?
11   How Do You Design a Gene Expression System?
12   How Are Plasmids Introduced into Bacterial Cells?
12   How Can Gel Electrophoresis Be Used to Analyze DNA?
13   What Are the Patterns of Antibiotic Resistance?
13   How Do Environmental Changes Affect a Population of Leafhoppers?
13   How Can Frequency of Alleles Be Calculated?
14   How Do New Species Arise by Genetic Isolation?
14   How Is Phylogeny Determined Using Protein Comparisons?
15   How Did Life Begin on Early Earth?
15   What Are the Modes of Nutrition in Prokaryotes?
15   What Kinds of Protists Are Found in Various Habitats?
16   What Are the Different Stages of a Fern Life Cycle?
16   How Are Trees Identified by Their Leaves?
16   How Does the Fungus Pilobolus Succeed as a Decomposer?
17   How Are Insect Species Identified?
17   How Does Bone Structure Shed Light on the Origin of Birds?
18   Do Pillbugs Prefer Wet or Dry Environments?
18   How Do Abiotic Factors Affect the Distribution of Organisms?
19   How Are Impacts on Community Diversity Measured?
19   How Does Light Affect Primary Productivity?
20   How Are Potential Prairie Restoration Sites Analyzed?

Biology and Society on the Web

1   Find out how "wonky holes" affect coral reefs
2   Learn what the American Dental Association has to say about the importance of fluoride
3   Learn more about lactose intolerance
4   Learn about the downside of antibiotic use
5   Learn more about the commercial application of enzymes
6   Learn about the proper ways to exercise
7   Learn more about the potential of biomass energy
8   Learn more about controversial reproductive technologies
9   Read about socioeconomic factors related to fetal testing
10   Learn about other drugs that combat HIV
11   Learn about the use of umbilical cord blood in the treatment of disease
12   Learn more about the Human Genome Project
13   Learn about the evolution of bacteria that are resistant to antibiotics
14   Learn about the evidence for an asteroid impact 250 million years ago that may have contributed to mass extinctions at that time
15   Learn more about the history of biowarfare
16   Learn more about the impact of deforestation
17   Learn about how invasive species arrive in the United States
18   This U.S. census website will provide you with some impressive numbers
19   Find out how old subway cars can be made into reefs
20   Learn what the U.S. government has to say about introduced species

Evolution Connection on the Web

1 Examine issues surrounding the teaching of evolution

2 Learn more about what Earth was like before life appeared

3 Learn why mice can be used as models for medical research

4 Learn what pre-life membranes may have been like

5 Enzymes are very important to life. How did life and enzymes get together?

6 Learn about the problems life faced when oxygen first began to accumulate in the atmosphere

7 Learn more about the organisms that first released large amounts of oxygen into Earth's atmosphere

8 Learn more about the role of polyploidy in the evolution of plants

9 Learn more about how the Y chromosome is used to study ancestry

10 Discover why you can get the flu year after year after year . . .

11 You may be more like a fruit fly than you think

12 View animations of the steps of gel electrophoresis

13 Learn more about sickle-cell disease

14 Explore a website that explains common misconceptions about evolution

15 Learn about what may be the oldest fossils of multicellular organisms.

16 Discover more about lichens

17 Learn about what is being done to combat the effects of habitat destruction

18 Learn why the dominant male may not be the best choice when it comes to mating success

19 Investigate the relationship between endosymbiosis and coevolution

20 Read more about biophilia